Fundamentals of Radar Signal Processing

About the Author
Mark A. Richards, Ph.D., is an educator and consultant with over 35 years of experience in radar signal processing practice and education at the Georgia Institute of Technology, DARPA, and Lockheed Martin. He is also the co-editor of *Principles of Modern Radar: Basic Principles* and an IEEE fellow cited for "contributions in radar signal processing education."

Fundamentals of Radar Signal Processing

Mark A. Richards, Ph.D.

Third Edition

New York Chicago San Francisco
Athens London Madrid
Mexico City Milan New Delhi
Singapore Sydney Toronto

Library of Congress Control Number: 2022931805

McGraw Hill books are available at special quantity discounts to use as premiums and sales promotions or for use in corporate training programs. To contact a representative, please visit the Contact Us page at www.mhprofessional.com.

Fundamentals of Radar Signal Processing, Third Edition

Copyright © 2022, 2014, 2005 by McGraw Hill. All rights reserved. Printed in the United States of America. Except as permitted under the United States Copyright Act of 1976, no part of this publication may be reproduced or distributed in any form or by any means, or stored in a data base or retrieval system, without the prior written permission of the publisher.

2 3 4 5 6 7 8 9   LCR   27 26 25 24 23

ISBN 978-1-260-46871-7
MHID 1-260-46871-2

This book is printed on acid-free paper.

Sponsoring Editor
Lara Zoble

Editing Supervisor
Stephen M. Smith

Production Supervisor
Lynn M. Messina

Acquisitions Coordinator
Elizabeth M. Houde

Project Manager
Rishabh Gupta, MPS Limited

Copy Editor
Surendra Shivam, MPS Limited

Proofreader
Megha Dabral, MPS Limited

Indexer
Dianna Haught

Art Director, Cover
Jeff Weeks

Composition
MPS Limited

Information contained in this work has been obtained by McGraw Hill from sources believed to be reliable. However, neither McGraw Hill nor its authors guarantee the accuracy or completeness of any information published herein, and neither McGraw Hill nor its authors shall be responsible for any errors, omissions, or damages arising out of use of this information. This work is published with the understanding that McGraw Hill and its authors are supplying information but are not attempting to render engineering or other professional services. If such services are required, the assistance of an appropriate professional should be sought.

To Theresa,
whose support made possible these books, and so much more;
and to Jessica and Benjamin,
and Amelia and Emmeline,
of whom we are so proud.

Contents

Preface		xv
Acknowledgments		xix
Selected Symbols		xxi
Selected Acronyms		xxix

1 Introduction to Radar Systems and Signal Processing ... 1
- 1.1 History and Applications of Radar ... 1
- 1.2 Basic Radar Functions ... 2
- 1.3 Elements of a Radar ... 5
    - 1.3.1 Radar Frequencies ... 7
    - 1.3.2 Radar Waveforms and Transmitters ... 8
    - 1.3.3 Antennas ... 10
    - 1.3.4 Virtual Elements and Virtual Arrays ... 16
    - 1.3.5 Receivers ... 17
- 1.4 Common Threads in Radar Signal Processing ... 22
    - 1.4.1 Signal-to-Interference Ratio ... 22
    - 1.4.2 Resolution and Region of Support ... 23
    - 1.4.3 Integration and Phase History Modeling ... 28
- 1.5 A Preview of Basic Radar Signal Processing ... 30
    - 1.5.1 Radar Time Scales ... 30
    - 1.5.2 Phenomenology ... 32
    - 1.5.3 Signal Conditioning and Interference Suppression ... 32
    - 1.5.4 Detection ... 35
    - 1.5.5 Measurements and Track Filtering ... 37
    - 1.5.6 Imaging ... 38
- 1.6 Radar Literature ... 40
    - 1.6.1 Introductions to Radar Systems and Applications ... 40
    - 1.6.2 Basic Radar Signal Processing ... 40
    - 1.6.3 Advanced Radar Signal Processing ... 41
    - 1.6.4 Radar Applications ... 41
    - 1.6.5 Current Radar Research ... 42
- References ... 42
- Problems ... 43

2 Signal Models ... 47
- 2.1 Components of a Radar Signal ... 47
- 2.2 Modeling Amplitude ... 48
    - 2.2.1 Simple Point Target Radar Range Equation ... 48
    - 2.2.2 Distributed Target Forms of the Range Equation ... 51
    - 2.2.3 Radar Cross Section ... 57
    - 2.2.4 Radar Cross Section for Meteorological Targets ... 58

		2.2.5	Statistical Description of Radar Cross Section	60
		2.2.6	Target Fluctuation Models	76
		2.2.7	Swerling Models	79
		2.2.8	Effect of Target Fluctuations on Doppler Spectrum	80
	2.3	Modeling Clutter		81
		2.3.1	Behavior of σ^0	82
		2.3.2	Signal-to-Clutter Ratio	84
		2.3.3	Temporal and Spatial Correlation of Clutter	85
		2.3.4	Compound Models of Radar Cross Section	87
	2.4	Noise Model and Signal-to-Noise Ratio		89
	2.5	Jamming		92
	2.6	Electromagnetic Interference		92
	2.7	Frequency Models: The Doppler Shift		93
		2.7.1	Doppler Shift	93
		2.7.2	The Stop-and-Hop Approximation and Phase History	95
		2.7.3	Measuring Doppler Shift: Spatial Doppler	97
	2.8	Spatial Models		99
		2.8.1	Coherent Scattering	99
		2.8.2	Variation with Angle	101
		2.8.3	Variation with Range	104
		2.8.4	Noncoherent Scattering	105
		2.8.5	Projections	106
		2.8.6	Multipath	106
	2.9	Spectral Model		108
	2.10	Summary		109
	References			110
	Problems			112
3	**Radar Data Acquisition and Organization**			**117**
	3.1	A Signal Processor's Radar Architecture Model		118
	3.2	Measuring a Range Profile		119
		3.2.1	Pulsed Radar Range Profile: One Pulse in Fast Time	119
		3.2.2	FMCW Radar Range Profile: One Sweep in Fast Time	125
	3.3	Multiple Range Profiles: Slow Time and the CPI		131
	3.4	Multiple Channels: The Datacube		138
	3.5	Dwells		140
	3.6	Sampling the Doppler Spectrum		141
		3.6.1	The Nyquist Rate in Doppler	141
		3.6.2	Straddle Loss	143
	3.7	Sampling in the Spatial and Angle Dimensions		148
		3.7.1	Spatial Array Sampling	148
		3.7.2	Sampling in Angle	149
	3.8	I/Q Imbalance and Digital I/Q		151
		3.8.1	I/Q Imbalance and Offset	152
		3.8.2	Correcting I/Q Errors	154
		3.8.3	Digital I/Q	157

3.9	Summary	161
	References	162
	Problems	163

4 Radar Waveforms ... 167

- 4.1 Introduction ... 167
- 4.2 The Waveform Matched Filter ... 169
  - 4.2.1 The Matched Filter ... 169
  - 4.2.2 Matched Filter for the Simple Pulse ... 171
  - 4.2.3 All-Range Coherent Matched Filtering ... 173
  - 4.2.4 Straddle Loss ... 174
  - 4.2.5 Range Resolution of the Matched Filter ... 174
- 4.3 Matched Filtering of Moving Targets ... 175
- 4.4 The Ambiguity Function ... 177
  - 4.4.1 Definition and Properties of the Ambiguity Function ... 177
  - 4.4.2 Ambiguity Function of the Simple Pulse ... 181
- 4.5 The Pulse Burst Waveform ... 184
  - 4.5.1 Matched Filter for the Pulse Burst Waveform ... 184
  - 4.5.2 Pulse-by-Pulse Processing ... 186
  - 4.5.3 Range Ambiguity ... 187
  - 4.5.4 Doppler Response of the Pulse Burst Waveform ... 190
  - 4.5.5 Ambiguity Function for the Pulse Burst Waveform ... 191
  - 4.5.6 The Slow-Time Spectrum and the Periodic Ambiguity Function ... 195
- 4.6 Frequency-Modulated Pulse Compression Waveforms ... 196
  - 4.6.1 Linear Frequency Modulation ... 197
  - 4.6.2 The Principle of Stationary Phase ... 200
  - 4.6.3 Ambiguity Function of the LFM Waveform ... 202
  - 4.6.4 Range-Doppler Coupling ... 204
  - 4.6.5 Stretch Processing ... 205
- 4.7 Range Sidelobe Control for FM Waveforms ... 210
  - 4.7.1 Matched Filter Frequency Response Shaping ... 210
  - 4.7.2 Matched Filter Impulse Response Shaping ... 213
  - 4.7.3 Waveform Spectrum Shaping ... 213
- 4.8 Frequency-Coded Waveforms ... 216
  - 4.8.1 The Stepped Frequency Waveform ... 216
  - 4.8.2 The Stepped Chirp Waveform ... 220
  - 4.8.3 Costas Frequency Codes ... 221
- 4.9 Phase-Modulated Pulse Compression Waveforms ... 223
  - 4.9.1 Biphase Codes ... 225
  - 4.9.2 Polyphase Codes ... 230
  - 4.9.3 Mismatched Phase Code Filters ... 234
- 4.10 Continuous Wave Radar ... 235
  - 4.10.1 Single-Frequency CW ... 236
  - 4.10.2 Periodically Modulated CW ... 237
  - 4.10.3 Linear Frequency-Modulated CW ... 238
  - 4.10.4 "Fast Chirp" Linear Frequency-Modulated CW ... 240

Contents

4.10.5	Sidelobe Control in Linear FMCW	242
4.10.6	Other CW Waveforms	242
4.11	Frequency-Modulated versus Phase-Modulated Waveforms	242
4.12	Summary	243
References		244
Problems		245

5 Doppler Processing — 251

5.1	Introduction	251
5.2	Moving Platform Effects on the Doppler Spectrum	253
5.3	Moving Target Indication	256
	5.3.1 Pulse Cancellers	259
	5.3.2 Vector Formulation of the Matched Filter	262
	5.3.3 Matched Filters for Clutter Suppression	263
	5.3.4 Blind Speeds and Staggered PRFs	266
	5.3.5 MTI Figures of Merit	273
	5.3.6 Limitations to MTI Performance	278
5.4	Pulse Doppler Processing	280
	5.4.1 The Discrete-Time Fourier Transform of a Moving Target	281
	5.4.2 Sampling the DTFT: The Discrete Fourier Transform	283
	5.4.3 The DFT of Noise	285
	5.4.4 Pulse Doppler Processing Gain	286
	5.4.5 Matched Filter and Filterbank Interpretations of Pulse Doppler Processing with the DFT	286
	5.4.6 Fine Doppler Estimation	288
	5.4.7 Modern Spectral Estimation in Pulse Doppler Processing	294
	5.4.8 CPI-to-CPI Stagger and Blind Zone Maps	295
5.5	Pulse Pair Processing	300
5.6	Additional Doppler Processing Issues	305
	5.6.1 Range Migration and the Keystone Transform	305
	5.6.2 Combined MTI and Pulse Doppler Processing	309
	5.6.3 Transient Effects	310
	5.6.4 PRF Regimes	311
	5.6.5 PRF Selection	314
	5.6.6 Ambiguity Resolution	317
5.7	Clutter Mapping	321
5.8	The Moving Target Detector	322
5.9	MTI for Moving Platforms: Ground Moving Target Indication	323
	5.9.1 Simplified GMTI Clutter and Target Models	324
	5.9.2 DPCA and ATI	326
	5.9.3 Clutter Suppression Interferometry	330
	5.9.4 Analysis of Adaptive DPCA	331
5.10	Summary	334
References		335
Problems		337

6 Detection Fundamentals ... 343
6.1 Introduction ... 343
6.2 Radar Detection as Hypothesis Testing ... 344
6.2.1 The Neyman-Pearson Detection Rule ... 345
6.2.2 The Likelihood Ratio Test ... 345
6.3 Threshold Detection in Coherent Systems ... 354
6.3.1 The Gaussian Case for Coherent Receivers ... 355
6.3.2 Unknown Parameters and Threshold Detection ... 358
6.3.3 Linear and Square Law Detectors ... 364
6.3.4 Other Unknown Parameters ... 365
6.4 Threshold Detection of Radar Signals ... 366
6.4.1 Coherent, Noncoherent, and Binary Integration ... 366
6.4.2 Nonfluctuating Targets ... 368
6.4.3 Albersheim's Equation ... 373
6.4.4 Fluctuating Targets ... 375
6.4.5 Simplified Equations for P_D for Some Swerling Cases ... 378
6.4.6 Shnidman's Equation ... 381
6.4.7 Detection in Clutter ... 383
6.4.8 Binary Integration ... 385
6.4.9 Integration Summary ... 389
6.5 Constant False Alarm Rate Detection ... 390
6.5.1 The Effect of Unknown Interference Power on False Alarm Probability ... 390
6.5.2 Cell-Averaging CFAR ... 392
6.5.3 Analysis of Cell-Averaging CFAR ... 394
6.5.4 CA CFAR Limitations ... 398
6.5.5 Extensions to Cell-Averaging CFAR ... 404
6.5.6 Order Statistic CFAR ... 408
6.5.7 Adaptive CFAR ... 412
6.5.8 CFAR for Two-Parameter PDFs ... 413
6.5.9 Temporal CFAR ... 414
6.5.10 Distribution-Free CFAR ... 417
6.6 System-Level Control of False Alarms ... 418
6.7 Summary ... 419
References ... 420
Problems ... 422

7 Measurements and Introduction to Tracking ... 425
7.1 Estimators ... 426
7.1.1 Estimator Properties ... 426
7.1.2 The Cramèr-Rao Lower Bound ... 429
7.1.3 The CRLB and Signal-to-Noise Ratio ... 431
7.1.4 Maximum Likelihood Estimators ... 432
7.2 Range, Doppler, and Angle Estimators ... 434
7.2.1 Range Estimators ... 434
7.2.2 Doppler Signal Estimators ... 446
7.2.3 Angle Estimators ... 455

Contents

	7.3	Introduction to Tracking	469
		7.3.1 Optimal Combination of Two Noisy Measurements	470
		7.3.2 Sequential Least Squares Estimation	471
		7.3.3 The α-β Filter	475
		7.3.4 The Kalman Filter	480
		7.3.5 The Tracking Cycle	487
	7.4	Summary	492
	References		493
	Problems		495
8	**Introduction to Synthetic Aperture Imaging**		**499**
	8.1	Fundamental SAR Concepts and Relations	503
		8.1.1 Cross-Range Resolution in Radar	503
		8.1.2 The Synthetic Aperture Viewpoint	505
		8.1.3 Doppler Viewpoint	512
		8.1.4 SAR Coverage and Sampling	514
	8.2	Stripmap SAR Data Characteristics	518
		8.2.1 Stripmap SAR Geometry	518
		8.2.2 Stripmap SAR Data Set	521
	8.3	Stripmap SAR Image Formation Algorithms	524
		8.3.1 Doppler Beam Sharpening	525
		8.3.2 Quadratic Phase Error Effects	528
		8.3.3 Range-Doppler Algorithms	533
		8.3.4 Depth of Focus	539
		8.3.5 Range Migration Algorithm	540
	8.4	Spotlight SAR Data Characteristics	544
	8.5	The Polar Format Image Formation Algorithm for Spotlight SAR	550
	8.6	Backprojection	552
	8.7	Interferometric SAR	556
		8.7.1 The Effect of Height on a SAR Image	556
		8.7.2 IFSAR Processing Steps	559
	8.8	Other Considerations	564
		8.8.1 Motion Compensation	564
		8.8.2 Autofocus	567
		8.8.3 Speckle Reduction	574
		8.8.4 Moving Targets	575
	8.9	Summary	580
	References		581
	Problems		583
9	**Introduction to Array Processing**		**587**
	9.1	Virtual Arrays	587
	9.2	Beamforming and Beam Steering	591
		9.2.1 Time Delay Steering	592
		9.2.2 Phase Steering	593
		9.2.3 Narrowband Phase Beamforming	595

		9.2.4	Adaptive Beamforming	598
		9.2.5	Adaptive Beamforming with Preprocessing	603
	9.3	Space-Time Signal Environment		606
	9.4	Space-Time Signal Modeling		609
	9.5	Processing the Space-Time Signal		613
		9.5.1	Optimum Matched Filtering	613
		9.5.2	STAP Metrics	613
		9.5.3	Relation to Displaced Phase Center Antenna Processing	617
		9.5.4	Adaptive Matched Filtering	619
	9.6	Reduced-Dimension STAP		622
	9.7	Advanced STAP Algorithms and Analysis		623
	9.8	Limitations to STAP		625
	9.9	Summary		626
	References			626
	Problems			628
A	**Selected Topics in Probability and Random Processes**			**631**
	A.1	Probability Density Functions and Likelihood Functions		631
	A.2	Common Probability Distributions in Radar		632
		A.2.1	Power Distributions	633
		A.2.2	Amplitude Distributions	641
		A.2.3	The Unfortunate Tendency in Radar to Call Power Distributions by the Name of the Amplitude Distribution	644
		A.2.4	Phase Distributions	644
	A.3	Estimators and the Cramèr-Rao Lower Bound		645
		A.3.1	The Cramèr-Rao Lower Bound on Estimator Variance	646
		A.3.2	The CRLB for Transformed Parameters	648
		A.3.3	Signals in Additive White Gaussian Noise	648
		A.3.4	Signals with Multiple Parameters in AWGN	649
		A.3.5	Complex Signals and Parameters in AWGN	650
		A.3.6	Finding Minimum Variance Estimators	651
	A.4	Random Signals in Linear Systems		652
		A.4.1	Correlation Functions	652
		A.4.2	Correlation and Linear Estimation	653
		A.4.3	Power Spectrum	654
		A.4.4	White Noise	655
		A.4.5	The Effect of LSI Systems on Random Signals	655
	References			658
B	**Selected Topics in Digital Signal Processing**			**659**
	B.1	Fourier Transforms		659
	B.2	Windowing		664
	B.3	Sampling, Quantization, and A/D Converters		668
		B.3.1	Sampling	668
		B.3.2	Quantization	672
		B.3.3	A/D Conversion Technology	675

B.4	Spatial Frequency	676
B.5	Correlation	678
B.6	Vector-Matrix Representations and Eigenanalysis	680
	B.6.1 Basic Definitions and Operations	680
	B.6.2 Basic Eigenanalysis	682
	B.6.3 Eigenstructure of Sinusoids in White Noise	683
B.7	Instantaneous Frequency	685
B.8	Decibels	685
References		687
Problems		687

Index **689**

Preface

This third edition of *Fundamentals of Radar Signal Processing* (FRSP) shares with the first two editions the goal of providing an in-depth tutorial in the fundamental techniques of radar signal processing. The full spectrum of foundational methods on which virtually all modern radar systems rely is covered, including topics such as target and interference models, matched filtering, waveform design, Doppler processing, threshold detection, and measurement accuracy. Chapters or sections on tracking, adaptive array processing, and synthetic aperture imaging introduce those more advanced techniques and provide a bridge to dedicated texts.

The book is written from a digital signal processor's viewpoint. The techniques and interpretations of linear systems, filtering, sampling, Fourier analysis, and random processes are used throughout to provide a consistent and unified tutorial approach. Students should have a firm foundation in these areas to obtain the most benefit. The mathematical level is appropriate for college seniors and first-year graduate students and is leavened by extensive interpretation. Because this text concentrates on the signal processing, it does not address many other aspects of radar technology such as transmitter and receiver hardware technology or electromagnetic wave propagation. Familiarity with basic radar systems, perhaps from studying one of the books mentioned below, will help prepare the reader to get the most out of this text.

This book first came about in 2005 because I could not identify an appropriate textbook for Georgia Tech's ECE 6272, Fundamentals of Radar Signal Processing, a semester-length first-year graduate course I taught. There existed at that time a number of books on radar systems in general (e.g., Skolnik, Edde) that provided good qualitative and descriptive introductions to radar systems as a whole and could be enthusiastically recommended as first texts for anyone interested in the topic. Indeed, having worked on speech enhancement in graduate school, I read the first edition of Skolnik's classic *Introduction to Radar Systems* when I first accepted a job in radar, hoping to avoid appearing completely ignorant on my first day at the new job. (It didn't work, through no fault of Skolnik.) Some of these texts provided greater quantitative depth on basic radar systems and some signal processing topics. At the same time, a number of good texts were available on advanced topics in radar signal processing, principally synthetic aperture imaging and space-time adaptive processing. The problem, in my view, was the existence of a substantial gap between the systems books and the advanced signal processing books. Specifically, I believed the radar community lacked a current text providing a unified, modern treatment of the basic radar signal processing techniques mentioned above. The closest was probably Levanon's *Radar Principles*, which I used for early offerings of ECE 6272, but it was not comprehensive enough.

It was my hope that this book would fill that gap, and I believe it has largely been successful in doing so. However, it has now been over 16 years since the first edition was published. While new books continue to appear, particularly the excellent *Principles of*

Modern Radar series, to my surprise none has emerged that covers basic radar signal processing techniques with similar depth and breadth. In the meantime, radar technology and applications have continued to evolve, and rather rapidly. For instance, the last 10 years have seen a tremendous increase in the number of short-range continuous wave (CW) radars fielded, especially in the automotive industry. At the same time, complex new methods such as multi-input, multi-output (MIMO) processing, compressed sensing, artificial intelligence, and "deep learning" have crossed over from other application realms into advanced radar.

To continue meeting its goals, this text must also evolve. The second edition (2014) made one major addition, adding the chapter on measurement accuracy and the introduction to tracking, in addition to many more minor updates. This edition likewise has one major change. The first two editions assumed pulsed radar throughout, even though a number of the topics are also applicable to CW systems. In this edition, CW radars are now included explicitly, with an emphasis on "fast-chirp" linear frequency-modulated CW (FMCW) radars, the most common variety in current usage. Although data acquisition for pulsed and FMCW radars is very different, much of the basic processing that follows is essentially the same, a commonality that I have tried to emphasize in this edition.

Each chapter has one or two other significant updates and many small ones. Chapter 1 now includes a discussion of virtual antenna elements to set the stage for virtual arrays in Chap. 9. The discussion of Doppler shift in Chap. 2 has been simplified from that in previous editions. Also in Chap. 2, the K distribution has been added to the discussion of PDFs for describing target and clutter fluctuations.

The first portion of Chap. 3 has been significantly restructured and expanded to introduce FMCW radar, describing how range profiles are acquired and some of the range ambiguity and blind zone considerations in comparison to the pulsed case. The coherent processing interval is emphasized as a common data structure for both pulsed and FMCW, and so a common starting point for understanding subsequent processing steps.

The impact of incorporating FMCW continues in Chap. 4 with a brief discussion of CW waveforms in general before homing in on the fast-chirp linear FMCW variant of most interest. This chapter also now includes more information on mismatched filters for phase-coded waveforms and closes with a new comparison of frequency-modulated and phase-modulated waveforms.

Chapter 5 adds an introduction to the keystone transform for combatting range migration and introduces along-track interferometry (ATI) as a complement to DPCA for detecting ground movers in clutter. Chapter 6 has a new example of the effect of spiky interference on detection performance. Also new is the use of binary integration gain as an alternative way to quantify the impact of M-of-N processing.

Chapter 7 adds a brief section on the optimum combination of two noisy measurements to improve the motivation and understanding of the prediction-correction structure of most track filters. In Chap. 8, the sequence of SAR image formation algorithms has been extended to include the range migration algorithm, a workhorse in current practice, and a very basic introduction to the important emerging class of backprojection algorithms.

Chapter 9 now introduces the idea of virtual arrays (VAs), essential to the understanding of MIMO array systems. While MIMO processing itself is beyond the scope of this text, the discussion of VAs provides a base for its study in more specialized references. Also, both phase and time delay steering of arrays are now discussed and compared explicitly. The two appendices are largely unchanged except for the addition of the K distribution to the PDFs, discussed in App. A. Finally, some additional homework problems have been added to most chapters to improve the book's usefulness as an academic text.

Throughout the text, I once more attempt to do a better job of identifying and bringing out common themes that arise again and again in radar signal processing, if sometimes in disguise. These include phase history, coherent integration, matched filtering, integration and processing gain, and maximum likelihood estimation.

Several ancillary FRSP materials are available from the publisher. An errata list and a collection of MATLAB® demonstrations of various fundamental radar signal processing operations are available to all readers at www.mhprofessional.com/Richards3e. For instructors of classes using this book as a text, there is a solution manual for the end-of-chapter problems and a collection of MATLAB® mini-projects with sample solutions. While not directly related to FRSP, a series of technical memos on additional radar signal processing topics can be found by the reader at the author's website www.radarsp.com.

A one-semester course in radar signal processing can cover Chaps. 1 through 7, perhaps skipping some of the later sections of Chaps. 2 and 3 for time savings. Such a course provides a solid foundation for more advanced work in detection theory, adaptive array processing, synthetic aperture imaging, and more advanced radar concepts such as passive and bistatic systems. A quarter-length course could cover Chaps. 1 through 5 and the non-CFAR portion of Chap. 6 reasonably thoroughly. In either case, a firm background in basic continuous and discrete signal processing and an introductory exposure to random variables and processes are advisable.

I have tried in this edition to eliminate all known errors in the second edition, but because there is significant new material, there are likely new errors. I invite readers to help me keep the errata sheet up to date by sending any and all errors they find to me at mrichards@ieee.org.

Mark A. Richards, Ph.D.
January 2022

Acknowledgments

I remain indebted to many colleagues and students who have helped me over the years to learn this material and how to write and teach it. A few on whom I have relied the most, and who have in particular helped me with topics in this edition, merit special mention. Dr. Byron Keel continued to share his expertise in waveforms and CFAR, especially in comparing frequency- and phase-modulated waveforms. Dr. Greg Coxson and Jon Russo helped with the current state of the art in optimum phase codes. Samuel Piper helped me both directly and through his own publications with the new CW radar material. Dr. James Sangston helped me develop the mathematics and an example of detection in spiky clutter. Dr. Gregory Showman helped me to understand the SAR range migration and backprojection algorithms, providing the base for the RMA derivation and several simulation examples that appear in Chap. 8. Prof. Nadav Levanon of Tel Aviv University has exchanged many materials and ideas with me for some years now, especially regarding waveforms, Doppler processing, and radar education. I am grateful to each of them for their knowledge, assistance, and friendship both in preparing this edition and throughout my career.

Selected Symbols

The following definitions and relations between symbols are used throughout this text except as otherwise specifically noted. Some symbols, for example θ, have more than one usage; their meaning is generally clear from the context.

$*$	Convolution operator
$\otimes$	Kronecker product operator
$\odot$	Hadamard product operator
(x)	Continuous variable x
$[x]$	Discrete variable x
$((\cdot))_x$	Modulo x
$\mathbf{0}_N, \mathbf{1}_N$	N-element vector of zeroes, ones
$\sim$	"Is distributed as"
$\mathbf{x}$	Vector variable
$\mathbf{X}$	Matrix variable
x^*	Complex conjugate of x
$\mathbf{x}^H, \mathbf{X}^H$	Hermitian transpose of vector or matrix $\mathbf{x}, \mathbf{X}$
$\mathbf{x}^T, \mathbf{X}^T$	Transpose of vector or matrix $\mathbf{x}, \mathbf{X}$
α_q	Clutter temporal fluctuation vector
α	Threshold multiplier, cell-averaging CFAR
α_{GO}	Threshold multiplier, "greatest-of" CFAR
$\alpha_{\log}$	Threshold multiplier, log CFAR
α_{OS}	Threshold multiplier, order statistic CFAR
α_{SO}	Threshold multiplier, "smallest-of" CFAR
β	Bandwidth
β_3	3-dB bandwidth
β_D	Doppler bandwidth
β_{MLC}	Mainlobe clutter bandwidth
β_n	Noise-equivalent receiver bandwidth
β_{nn}	Null-to-null bandwidth
β_r	Rayleigh bandwidth
$\beta_{\rm rms}$	Root-mean-square bandwidth
$\beta_x, \beta_y, \beta_z$	Spatial frequency bandwidth in x, y, and z dimensions
γ	Interferogram coherence; Q channel DC offset
γ_c	Clutter ridge slope
Γ	Tracking index; gamma function
δ	Grazing angle

Selected Symbols

$\delta[\cdot]$	Discrete-variable impulse function
$\delta\theta$	Target angle relative to boresight
$\delta_D(\cdot)$	Continuous-variable Dirac impulse ("delta") function
δR	Range error
δR_s	Range bin spacing
δt	Differential delay
$\Delta\theta$	Angular resolution; lobing antenna squint
$\Delta\psi$	Change in squint angle
ΔCR	Cross-range resolution
ΔF	Frequency step size
ΔF_D	Doppler frequency resolution
Δh	Height displacement
ΔR	Range resolution
ΔR_b	Range relative to central reference point
ΔR_c	Range curvature
ΔR_w	Range walk
Δt	Time resolution
Δt_b	Time relative to central reference point delay
ε	I/Q amplitude mismatch; mismatch error
$\varepsilon_{\Delta/\Sigma}$	Error in lobing antenna ratio voltage $v_{\Sigma/\Delta}$
ζ	Baseband reflectivity amplitude ($\zeta \geq 0$)
$\bar{\zeta}$	Non-baseband reflectivity amplitude ($\bar{\zeta} \geq 0$)
η	Volume reflectivity
θ	Azimuth angle; phase; baseband transmitted signal phase
$\theta(t)$	Phase modulation of waveform
θ_3	3-dB beamwidth
θ_{az}	Azimuth beamwidth
θ_{el}	Elevation beamwidth
θ_{nn}	Null-to-null azimuth beamwidth
θ_R	Rayleigh beamwidth
θ_{SAR}	Effective beamwidth of synthetic aperture radar
Θ	Parameter to be estimated
$\mathbf{\Theta}$	Parameter vector to be estimated
$\hat{\Theta}$	Estimate of Θ
$\hat{\mathbf{\Theta}}$	Estimate of $\mathbf{\Theta}$
κ	I channel DC offset; Doppler spectrum oversampling factor
$\ell(a\|b)$	Likelihood function for parameter a given data b
λ	Wavelength; eigenvalue
Λ	Likelihood ratio
λ_t	Transmitted signal wavelength
$\rho = \zeta\exp(j\psi)$ $= \rho_I + j\rho_Q$	Complex baseband reflectivity
ρ_I	Baseband reflectivity in-phase (I) component
ρ_Q	Baseband reflectivity quadrature-phase (Q) component
ρ_f, ρ_{fg}	Normalized autocorrelation of function f, or cross-correlation of functions f and g
ρ'	Effective baseband complex reflectivity
$\tilde{\rho}$	Cross-range averaged effective baseband complex reflectivity
$\tilde{P}$	Range spatial spectrum (Fourier transform of $\tilde{\rho}$)

Selected Symbols

$\sigma = \|\rho\|^2 = \zeta^2$	Radar cross section (RCS)
σ^0	Area reflectivity
σ_h	Surface roughness
σ_x^2	Variance of random variable x
$\sigma_{\hat{\Theta}}$	Precision of estimate of Θ
ϕ	Elevation angle; phase
ϕ_3	3-dB elevation beamwidth
ϕ_{fg}	Interferometric phase difference
ϕ_n	Subpulse (chip) phase in phase-coded waveform
ϕ_{nn}	Null-to-null elevation beamwidth
τ	Pulse length
τ_c	Subpulse length in phase-coded waveform
φ	Baseband phase
χ	Signal-to-noise ratio
χ_1	Single sample signal-to-noise ratio
χ_N	N-sample signal-to-noise ratio
χ_{out}	Output signal-to-noise ratio
χ_∞	Signal-to-noise ratio with perfect noise level estimate
χ_Σ	Lobing antenna sum channel signal-to-noise ratio
ψ	Baseband reflectivity phase; squint angle; cone angle
ω	Normalized frequency (radians per sample)
ω_D	Normalized Doppler frequency shift (radians per sample)
ω_s	Sampling interval in normalized frequency ω (samples per radian)
Ω	Frequency (radians per second); solid angle
Ω_θ	Azimuth rotation rate (radians per second)
Ω_D	Doppler frequency shift (radians per second)
Ω_{diff}	Doppler frequency mismatch (radians per second)
Ω_i	Matched Doppler frequency shift (radians per second)
Ω_t	Transmitted or carrier frequency (radians per second)
Υ	Sufficient statistic
a	Baseband transmitted signal amplitude
$\mathbf{a}_s(\theta)$	Spatial steering vector
$\mathbf{a}_t(\theta)$	Temporal steering vector
$A(t, F_D), \hat{A}(t, F_D)$	Ambiguity function, complex ambiguity function
$A, \hat{A}, \overline{A}$	Signal amplitude
A_e	Effective antenna aperture size
$\mathbf{A}_q$	Covariance matrix of clutter temporal fluctuations
A_n	Complex amplitude of subpulse in phase-coded waveform
$AF(\theta), AF(\theta, \phi)$	Phased array antenna array factor
B	Number of bits; interferometric baseline
B_N	Length of Barker phase code
c	Speed of electromagnetic wave propagation
$\mathbf{c}_q$	Clutter space-time steering vector for patch q
CA	Clutter attenuation
$CN(a, b)$	Complex normal (Gaussian) distribution with mean a and variance b
CRLB	Cramèr-Rao lower bound
$C_x(\cdot)$	Characteristic function of random variable x; centroid of signal x
d	Phased array element spacing
$d_g(\cdot)$	Group delay function

Selected Symbols

d_M	Mahalanobis distance
d_{pc}	Phase center spacing
D	Antenna aperture size
D_{az}	Antenna size, azimuth dimension
D_{el}	Antenna size, elevation dimension
D_{SAR}	Synthetic aperture size
DOF	Degrees of freedom
DR	Dynamic range
D_x, D_y, D_z	Antenna aperture size in x, y, or z dimension
$\mathbf{e}$	Eigenvector
E, E_x	Energy; energy in signal x
$\mathbf{E}\{\cdot\}$	Expected value
$E(\theta, \phi)$	Electric field amplitude
$E_{el}(\theta), E_{el}(\theta, \phi)$	Phased array antenna element pattern
f	Normalized frequency (cycles per sample)
$\tilde{f}$	Quantized version of a function f
f_D	Normalized Doppler frequency shift (cycles per sample)
f_{Dt}	Target normalized Doppler frequency shift (cycles per sample)
f_θ	Normalized spatial frequency (cycles per sample)
F	Frequency (hertz); Fourier transform of a function f
$\mathbf{F}$	Fourier transform operator; track filter state transition matrix
F_θ	Spatial frequency (cycles per meter)
F_b	Beat frequency; blind Doppler frequency (hertz)
F_{bmax}	Maximum beat frequency (hertz)
F_{bmin}	Minimum beat frequency (hertz)
F_{bs}	Blind Doppler frequency using staggered PRIs
F_c	Corner frequency (hertz)
F_D	Doppler frequency shift (hertz)
F_{Da}	Apparent Doppler frequency (hertz)
F_{diff}	Doppler frequency mismatch (hertz)
F_{Dua}	Unambiguous Doppler frequency interval (hertz)
F_g	Greatest common divisor of a set of staggered PRFs
F_i	Instantaneous frequency (hertz)
F_n	Noise figure
F_r	Received frequency (hertz)
F_s	Sampling frequency (samples per second)
F_t	Transmitted or carrier frequency (hertz)
F_{us}	Unstaggered blind Doppler frequency
$\mathbf{g}, \mathbf{G}$	Tracking process noise gain
G	Antenna power gain
G_{nc}	Noncoherent integration gain
G_s	Maximum receiver gain
G_{sp}	Signal processing gain
h	Height
$\mathbf{h}$	Filter weight vector; beamformer weight vector
$\mathbf{h}, \mathbf{H}$	Tracking observation matrix
$h(t)$	Impulse response (continuous time)
$h[n]$	Impulse response (discrete time)
$h_p(t)$	Matched filter impulse response for individual pulse
H_0	Null hypothesis (interference only)
H_1	Non-null hypothesis (target plus interference)

Selected Symbols

$H(f), H(F), H(\omega), H(\Omega)$	Frequency response in various units
$H(z)$	Discrete-time system function
$H_N(z)$	System function of N-pulse canceller
$H_{N,P}(F)$	Frequency response of N-pulse canceller with P staggered PRFs
I; I	In-phase channel; in-phase channel signal; interference power; improvement factor
I_{opt}	Improvement factor for matched filter
$I_N(\cdot)$	Modified Bessel function of the first kind and order N
$\mathbf{I}_N$	Nth-order identity matrix
$I(\cdot, \cdot)$	Incomplete gamma function
$\mathbf{I}(\cdot)$	Fisher information matrix
$J_n(t)$	Jammer signal
$\mathbf{J}$	Jammer signal sample vector
k_p	Stagger ratio
$k_{\Delta/\Sigma}$	Lobing antenna Δ/Σ error slope
$k_\theta = -\tilde{k}_\theta$	Normalized spatial frequency (cycles per sample)
$K, \mathbf{k}, \mathbf{K}$	Tracking filter gain; Kalman filter gain (symbol varies with dimensionality)
K	Spatial frequency (radians per meter); DFT, IDFT, FFT, or IFFT size; normalized quantizer step size
K_R	Range spatial frequency (radians per meter)
K_u	Cross-range spatial frequency (radians per meter)
K_x, K_y, K_z	Spatial frequency in x, y, or z dimension (radians per meter)
K_θ	Spatial frequency corresponding to AOA θ (radians per meter)
k_θ	Normalized spatial frequency corresponding to AOA θ (radians per sample)
L	Number of fast-time samples per pulse
L_a	Atmospheric loss factor
L_d	Target depth as viewed from the radar
L_s	System loss factors; synthetic aperture radar swath length
L_{SIR}	Signal-to-interference ratio loss
L_w	Target width as viewed from the radar
LPG	Loss in processing gain
M	Number of slow-time samples per coherent processing interval
$\mathbf{M}$	Tracking mean-square error estimate matrix
M_{DD}	Minimum detectable Doppler shift
M_{DD+}, M_{DD-}	Minimum detectable positive, negative Doppler shift
M_{opt}	Optimum value of M in "M of N" detection rule
M_s	DPCA time slip
n_P	Matched filter output noise power
N	Noise power; number of phase code chips; number of phase centers
$N(a, b)$	Normal (Gaussian) distribution with mean a and variance b
N_γ	Number of spotlight SAR radial slices
N_R	Number of spotlight SAR range samples
N_{spot}	Number of spotlight SAR images per unit time
$p_x(\cdot)$	Probability density function for a random variable x
$P_x(\cdot)$	Cumulative density function for a random variable x
P	Power; degrees of freedom in space-time snapshot
$P(\theta, \phi)$	Antenna one-way power pattern
$P_\theta(\theta)$	Azimuth one-way antenna power pattern
$P_\phi(\phi)$	Elevation one-way antenna power pattern

Selected Symbols

P_b	Backscattered power
P_{BD}	Binary integrated probability of detection
P_{BFA}	Binary integrated probability of false alarm
P_{CD}	Cumulative probability of detection
P_{CFA}	Cumulative probability of false alarm
P_D	Probability of detection
P_{FA}	Probability of false alarm
P_M	Probability of miss
P_o	Output power
P_r	Received power; relative power of I/Q mismatch image
$\Pr\{\cdot\}$	Probability of argument occurring
P_t	Transmitted power
PL	Processing loss
PRF	Pulse repetition frequency (pulses per second)
q	Quantizer step size
Q, Q	Quadrature channel
Q	Power density; quadrature channel signal
Q_b	Backscattered power density
Q_k	Quadrature component, sample k
Q_M	Marcum Q function
Q_t	Transmitted power density
R, R_0	Range
R_a	Apparent range
R_b	Blind range
RI	Repetition interval
R_{max}	Maximum range
R_{min}	Minimum range
RR	Repetition rate
R_{ua}	Unambiguous range
R_{uas}	Unambiguous range using staggered PRIs
R_w	Range window; range swath
s_A	Autocorrelation of phase code complex amplitude sequence
s_f	Autocorrelation of a function or random signal f
s_{fg}	Cross-correlation of functions or random signals f and g
$s_p(t)$	Output of filter matched to single pulse in pulse train waveform
$\mathbf{S}$	Polarization scattering matrix
$S_f(\omega)$	Power spectrum of a function or random signal f
$S_{fg}(\omega)$	Cross-power spectrum of functions or random signals f and g
$\mathbf{S}_x$	Covariance matrix for a random vector $\mathbf{x}$
$\tilde{\mathbf{S}}_x$	Transformed covariance matrix
$\hat{\mathbf{S}}_x$	Estimated covariance matrix
SIR	Signal-to-interference ratio
$SQNR$	Signal-to-quantization noise ratio
t, t_0	Time
$\mathbf{t}$	Target model vector
$\tilde{\mathbf{t}}$	Transformed target model vector
T	Pulse or sweep repetition interval; detection threshold; track measurement update interval
$\mathbf{T}$	Transformation matrix
T'	Equivalent receiver temperature; detection threshold

T_θ	Sampling interval in θ
T_a	Aperture time
T_{avg}	Average PRI of a set of staggered PRFs
T_M	Time of matched filter output peak
T_p	pth PRI in set of staggered PRFs
T_s	Fast-time sampling interval; sampling interval in $s = \sin\theta$
T_{tot}	Sum of PRIs corresponding to staggered set of PRFs
T_w	Time corresponding to swath width
u	Along-track coordinate of synthetic aperture radar platform
$u, \mathbf{u}$	Tracking process noise
$UDSF$	Usable Doppler space fraction
$\text{var}(x)$	Variance of random variable x
v	Platform velocity
v_Σ, v_Δ	Sum and difference lobing voltages
$v_{\Sigma/\Delta}$	Lobing antenna ratio voltage
v_a	Apparent velocity
v_b	Blind speed
v_{bs}	Blind speed using staggered PRIs
v_L, v_R	Left and right lobing antenna voltages
v_{ua}	Unambiguous velocity interval
$w, \mathbf{w}$	Tracking measurement noise
$\mathbf{w}_f$	Temporal weight vector
$\mathbf{w}_\theta$	Spatial weight vector
(x, y, z)	Position in Cartesian coordinates
$(\dot{x}, \dot{y}, \dot{z})$	Velocity in Cartesian coordinates
$\bar{x}$	Mean of random variable x
$\hat{x}$	Estimated value of random variable x
$x = x_I + j\,x_Q$	Transmitted signal, baseband
$\overline{x} = \overline{x}_I + j\overline{x}_Q$	Transmitted signal, non-baseband
x_P	Along-track coordinate of synthetic aperture radar scatterer
$x_p(t)$	Subpulse of phase-coded waveform; single pulse of pulse train waveform
$y = y_I + jy_Q$	Received signal, baseband
$\mathbf{y}$	Baseband received signal sample vector
$\tilde{\mathbf{y}}$	Transformed baseband received signal sample vector
$\overline{y} = \overline{y}_I + j\overline{y}_Q$	Received signal, non-baseband
$y[l, m, n]$	Datacube for one coherent processing interval
$y[l, m]$	Fast time/slow time data matrix for one CPI
$y_s[m]$	Slow-time sequence for one CPI
z	Detected output
$\tilde{z}$	Transformed detected output
Z	Meteorological reflectivity; altitude

Selected Acronyms

The following acronyms are used in this text.

1D, 2D, 3D	One-, Two-, Three-Dimensional
ACF	Autocorrelation Function
A/D	Analog-to-Digital
ADC	Analog-to-Digital Converter
AF	Ambiguity Function
AGC	Automatic Gain Control
AL	Altitude Line
AM	Amplitude Modulation
AOA	Angle of Arrival
ASR	Airport Surveillance Radar
ATI	Along-Track Interferometry
BSR	Beam Sharpening Ratio
BT	Time-Bandwidth Product
CA	Clutter Attenuation
CA-CFAR	Cell-Averaging Constant False Alarm Rate
CCD	Coherent Change Detection
CDF	Cumulative Distribution Function
CF	Characteristic Function
CFAR	Constant False Alarm Rate
CMT	Covariance Matrix Taper
CNR	Clutter-to-Noise Ratio
CPI	Coherent Processing Interval
CRLB	Cramèr-Rao Lower Bound
CRP	Central Reference Point
CRT	Chinese Remainder Theorem
CSI	Clutter Suppression Interferometry
CUT	Cell under Test
CW	Continuous Wave
dB	Decibel
DBS	Doppler Beam Sharpening
dBsm	Decibels relative to 1 square meter
DFT	Discrete Fourier Transform
DOF	Degrees of Freedom
DPCA	Displaced Phase Center Antenna
DSP	Digital Signal Processing
DTED	Digital Terrain Elevation Data

xxx Selected Acronyms

DTFT	Discrete-Time Fourier Transform
EA	Electronic Attack
ECM	Electronic Countermeasures
EKF	Extended Kalman Filter
EM	Electromagnetic
EMI	Electromagnetic Interference
ENOB	Effective Number of Bits
EW	Electronic Warfare
FFT	Fast Fourier Transform
FIR	Finite Impulse Response
FM	Frequency Modulation
FMCW	Frequency-Modulated Continuous Wave
FSK	Frequency Shift Keying
FT	Fourier Transform
GMTI	Ground Moving Target Indication
GOCA CFAR	Greatest-of Cell-Averaging Constant False Alarm Rate
GPS	Global Positioning System
HF	High Frequency
HPRF	High Pulse Repetition Frequency
I	In-Phase
ICM	Internal Clutter Motion; Intrinsic Clutter Motion
IDFT	Inverse Discrete Fourier Transform
IF	Intermediate Frequency
IFFT	Inverse Fast Fourier Transform
IFSAR	Interferometric Synthetic Aperture Radar
i.i.d.	Independent Identically Distributed
IIR	Infinite Impulse Response
IMU	Inertial Measurement Unit
INS	Inertial Navigation System
InSAR	Interferometric Synthetic Aperture Radar
IPD	Interferometric Phase Difference
IPP	Interpulse Period
ISL	Integrated Sidelobe Level; Interference Subspace Leakage
JNR	Jammer-to-Noise Ratio
JSR	Jammer-to-Signal Ratio
KF	Kalman Filter
KT	Keystone Transform/Transformation
LCM	Least Common Multiple
LEO	Low Earth Orbit
LFM	Linear Frequency Modulation
LNA	Low Noise Amplifier
LO	Local Oscillator
LOS	Line of Sight
LPF	Lowpass Filter
LPG	Loss in Processing Gain
LPRF	Low Pulse Repetition Frequency
LRT	Likelihood Ratio Test
LSI	Linear Shift Invariant
MDD	Minimum Detectable Doppler
MDV	Minimum Detectable Velocity

Selected Acronyms

MIMO	Multiple Input, Multiple Output
MLC	Mainlobe Clutter
MLE	Minimum Likelihood Estimate/Estimator/Estimation
MMSE	Minimum Mean-Squared Error/Estimate
MMW	Millimeter Wave
MPRF	Medium Pulse Repetition Frequency
MTD	Moving Target Detector
MTI	Moving Target Indication
MVU	Minimum Variance Unbiased
NEXRAD	Next Generation Radar
NLFM	Nonlinear Frequency Modulation
OS CFAR	Order Statistic Constant False Alarm Rate
PAF	Periodic Ambiguity Function
PD	Pulse Doppler
PDF	Probability Density Function
PF, PFA	Polar Format, Polar Format Algorithm
PGA	Phase Gradient Algorithm
PL	Processing Loss
PM	Phase Modulation
PPP	Pulse Pair Processing
PPS	Pulses per Second
PRF	Pulse Repetition Frequency
PRI	Pulse Repetition Interval
PSD	Power Spectrum/Spectral Density
PSL	Peak Sidelobe Level
PSP	Principle of Stationary Phase
PSR	Point Spread Response
Q	Quadrature
RCS	Radar Cross Section
RD	Range-Doppler
RF	Radar Frequency, Radio Frequency
RFA	Rectangular Format Algorithm
RFI	Radio Frequency Interference
RI	Repetition Interval
RM, RMA	Range Migration, Range Migration Algorithm
RMB	Reed-Mallet-Brennan
RMS	Root Mean Square
ROC	Receiver Operating Characteristic
ROI	Region of Interest
ROS	Region of Support
RP	Range Profile
RR	Repetition Rate
RRE	Radar Range Equation
RV	Random Variable
RVP	Residual Video Phase
SAR	Synthetic Aperture Radar
SB	Sampling Bound
SCR	Signal-to-Clutter Ratio
SIR	Signal-to-Interference Ratio; Shuttle Imaging Radar
SMI	Sample Matrix Inverse

SLC	Sidelobe Clutter
SMTI	Surface Moving Target Indication
SNR	Signal-to-Noise Ratio
SOCA CFAR	Smallest-of Cell-Averaging Constant False Alarm Rate
SQNR	Signal-to-Quantization Noise Ratio
STAP	Space-Time Adaptive Processing
STC	Sensitivity Time Control
T/R	Transmit/Receive
UDSF	Usable Doppler Space Fraction
UHF	Ultra-High Frequency
ULA	Uniform Linear Array
UWB	Ultra Wideband
VA	Virtual Array
VE	Virtual Element
VHF	Very High Frequency
WGN	White Gaussian Noise
ZZB	Ziv-Zakai Bound

Fundamentals of Radar Signal Processing

CHAPTER 1
Introduction to Radar Systems and Signal Processing

1.1 History and Applications of Radar

The word "radar" was originally an acronym, RADAR, for "*r*adio *d*etection *a*nd *r*anging." Today, the technology is so common that the word has become a standard English noun. Many people have direct personal experience with radar in such applications as measuring fastball speeds or, often to their regret, traffic control.

The history of radar extends to the early days of modern electromagnetic theory (Swords, 1986; Skolnik, 2001). In 1886, Hertz demonstrated reflection of radio waves, and in 1900 Tesla described a concept for electromagnetic detection and velocity measurement in an interview. In 1903 and 1904, the German engineer Hülsmeyer experimented with ship detection by radio wave reflection, an idea advocated again by Marconi in 1922. In that same year, Taylor and Young of the U.S. Naval Research Laboratory (NRL) demonstrated ship detection by radar and in 1930 Hyland, also of NRL, first detected aircraft by radar (albeit accidentally), setting off a more substantial investigation that led to a U.S. patent for what would now be called a *continuous wave* (CW) radar in 1934.

The development of radar accelerated and spread in the middle and late 1930s with largely independent developments in the United States, Britain, France, Germany, Italy, Japan, and Russia. In the United States, R. M. Page of NRL began an effort to develop pulsed radar in 1934, with the first successful demonstrations in 1936. The year 1936 also saw the U.S. Army Signal Corps begin active radar work, leading in 1938 to its first operational system, the SCR-268 antiaircraft fire control system, and in 1939 to the SCR-270 early warning system, the detections of which were tragically ignored at Pearl Harbor. British development, spurred by the threat of war, began in earnest with work by Watson-Watt in 1935. The British demonstrated pulsed radar that year and by 1938 established the famous Chain Home surveillance radar network that remained active until the end of World War II. They also built the first airborne interceptor radar in 1939. In 1940, the United States and Britain began to exchange information on radar development. Up to this time, most radar work was conducted at *high frequency* (HF) and *very high frequency* (VHF) wavelengths; but with the British disclosure of the critical cavity magnetron microwave power tube and the United States' formation of the Radiation Laboratory at the Massachusetts Institute of Technology, the groundwork was laid for the successful development of radar at the microwave frequencies that have predominated ever since.

Each of the other countries mentioned also carried out CW radar experiments, and each fielded operational radars at some time during World War II. Efforts in France and Russia were interrupted by German occupation. On the other hand, Japanese efforts were aided by the capture of U.S. radars in the Philippines and by the disclosure of German technology. The Germans themselves deployed a variety of ground-based, shipboard, and airborne systems. By the end of the war, the value of radar and the advantages of microwave frequencies and pulsed waveforms were widely recognized.

Early radar development was driven by military necessity, and the military is still a major user and developer of radar technology. Military applications include surveillance, navigation, and weapons guidance for ground, sea, air, and space vehicles. Military radars span the range from huge ballistic missile defense systems to fist-sized tactical missile seekers.

Radar now enjoys an increasing range of applications. One of the most common is the police traffic radar used for enforcing speed limits (and measuring the speed of baseballs and tennis serves). Another is the "color weather radar" familiar to every viewer of local television news or numerous online sources. More sophisticated meteorological radar systems are used for large-scale weather monitoring and prediction and atmospheric research. Another radar application that affects many people is found in the air traffic control systems used to guide commercial aircraft both en route and in the vicinity of airports. Aviation also uses radar as one means for determining altitude and avoiding severe weather, and may soon use it to assist in imaging runway approaches in poor weather. Radar is commonly used by the shipping, heavy equipment, and automotive industries for collision avoidance, obstacle detection, and related safety functions. Indeed, one of the most important current drivers in radar technology is the automotive industry, which now places millions of small radars on the road each year in driver assistance systems. Radar is also an essential sensor for emerging autonomous driving systems. Finally, spaceborne and airborne radar is an important tool in mapping earth topology and environmental characteristics such as water and ice conditions, forestry conditions, land usage, and pollution. While this sketch of radar applications is far from exhaustive, it does indicate the breadth of applications of this remarkable technology.

This text tries to present a thorough, straightforward, and consistent description of the signal processing aspects of radar technology, focusing on fundamental range, Doppler, and angle processing techniques common to most radar systems. Previous editions emphasized pulsed over continuous wave radars. However, recent years have seen extensive proliferation of continuous wave radars in the automotive and other industries. Particularly common are linear frequency-modulated CW (linear FMCW) radars, which use many of the same signal processing and data organization techniques as pulsed systems. Consequently, this edition addresses both pulsed and linear FMCW systems, using the common data acquisition and organization construct of a *datacube* to unify their descriptions.

Similarly, because most radars are *monostatic*, meaning the transmitter and receiver antennas are collocated (and in fact are usually the same antenna), they are emphasized over *bistatic* radars where the antennas are significantly separated, though again many of the results apply to both. Finally, the subject is approached from a digital signal processing (DSP) viewpoint as much as practicable, both because most new radar designs rely heavily on digital processing and because this approach can unify concepts and results often treated separately.

1.2 Basic Radar Functions

Most uses of radar can be classified as *detection, tracking,* or *imaging*. Higher-level capabilities are built on top of these basic functions. This text addresses all three and the techniques of signal acquisition and interference reduction necessary to perform them.

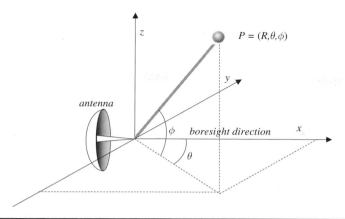

FIGURE 1.1 Spherical coordinate system for radar measurements.

The most fundamental problem in radar is detection of an object or physical phenomenon. This requires determining whether the receiver output at a given time includes the echo from a reflecting object or only noise or other interference. Detection decisions are usually made by comparing the amplitude $a(t)$ of the receiver output (where t represents time) to a threshold $T(t)$, which may be set a priori in the radar design or may be computed adaptively from the radar data. The time required for a signal to propagate a distance R and return, thus traveling a total distance $2R$, is $2R/c$, where c is the speed of electromagnetic (EM) wave propagation ("speed of light").[1] Therefore, if $a(t) > T(t)$ at some time delay t_0 after a signal is transmitted, it is assumed that a target is present at range

$$R = \frac{ct_0}{2} \text{ m} \qquad (1.1)$$

Once an object has been detected, it may be desirable to track its location and velocity. A monostatic radar naturally measures position in a spherical coordinate system with its origin at the radar antenna's phase center (defined in Sec. 1.3.4), as shown in Fig. 1.1. In this coordinate system, the antenna look direction, sometimes called the *boresight* direction, is along the $+x$ axis. The angle θ is called the *azimuth* angle, while ϕ is called the *elevation* angle.[2]

The range R to the object is obtained directly from the elapsed time from transmission to detection as just described. Elevation and azimuth angle ϕ and θ are determined from the antenna orientation, since the target must normally be in the antenna field of view to be detected. Velocity is estimated by measuring the Doppler shift of the target echoes. Doppler shift provides only the radial velocity component, but a series of measurements of position and radial velocity can be used to infer target dynamics in all three dimensions.

Because most people are familiar with the idea of following the movement of a "blip" on the radar screen, detection and tracking are the functions most commonly

---

[1] $c = 2.99792458 \times 10^8$ m/s in a vacuum. A value of $c = 3 \times 10^8$ m/s, normally used except where very high accuracy is required, is used exclusively in this text.

[2] In mathematics, the spherical coordinate system is often defined in terms of range, azimuth angle, and a *polar angle* ϕ_p (also called the *zenith* or *inclination* angle) measured from the z axis. The polar and elevation angles are related as $\phi = \pi/2 - \phi_p$ radians.

associated with radar. However, radars are increasingly used to generate two- and three-dimensional images of an area. Such images can be analyzed for intelligence and surveillance purposes, for topology mapping, or for "earth resources" applications such as analysis of land use, ice cover, deforestation, pollution spills, and so forth. They can also be used for "terrain following" navigation by correlating measured imagery with stored maps. While radar images have not achieved the resolution of optical images, the very low attenuation of electromagnetic waves at microwave frequencies gives radar the important advantage of "seeing" through clouds, fog, and precipitation very well. In addition, radar imagery works day or night because the transmitter provides the "illumination." Consequently, imaging radars generate useful imagery when optical instruments cannot be used at all.

The quality of a radar system is quantified with a variety of figures of merit, depending on the function being considered. In analyzing detection performance, the fundamental parameters are the *probability of detection* P_D and the *probability of false alarm* P_{FA}. If other system parameters are fixed, increasing P_D always requires accepting a higher P_{FA} as well. The achievable combinations are determined by the signal and interference statistics, especially the *signal-to-interference ratio* (SIR). When multiple targets are present in the radar field of view, additional considerations of resolution and sidelobes arise in evaluating detection performance. For example, if two targets cannot be resolved by a radar they will be registered as a single object. If sidelobes are high, the echo from one strongly reflecting target may mask the echo from a nearby but weaker target so that again only one target is registered when two are present. Resolution and sidelobes in range are determined by the radar waveform, while those in angle are determined by the antenna pattern.

In radar tracking, the basic figures of merit are *accuracy* (bias) and *precision* (standard deviation) of the range, angle, and velocity estimates. With appropriate signal processing the achievable accuracy is typically limited by a combination of resolution and SIR. For example, if noise is the primary interference source, the limiting precision often is proportional to $\sqrt{\Delta/SNR}$, where Δ is the resolution in the coordinate of interest and SNR is the value of the *signal-to-noise ratio* (SNR).

In imaging, the principal figures of merit are spatial resolution and dynamic range. Spatial resolution determines what size objects can be distinguished in the final image and therefore to what uses the image can be put. For example, a radar map with 1 km by 1 km resolution would be useful for large-scale land use studies but useless for detailed military surveillance of airfields or missile sites. Dynamic range determines image contrast, which also contributes to the amount of information that can be extracted from an image.

The purpose of signal processing in radar is to extract end products such as detections or images from the raw radar data and to maximize the quality of those products by maximizing resolvability, SIR, and other relevant figures of merit. SIR can be improved by integration of multiple measurements. Resolution and SIR can be jointly improved by matched filters and other waveform design and processing techniques such as frequency agility. Accuracy benefits from increased SIR and interpolation methods. Sidelobe behavior can be improved with the same windowing techniques used in virtually every application of signal processing. Each of these topics is explored in the chapters that follow.

Radar signal processing draws on many of the same techniques and concepts used in other signal processing areas, from such closely related fields as communications and sonar to very different applications such as speech and image processing. Linear filtering and statistical detection theory are central to radar's most fundamental task of target detection. Fourier transforms, implemented using *fast Fourier transform* (FFT) techniques, are ubiquitous, being used for everything from fast convolution implementations of matched filters, to Doppler spectrum estimation, to image formation. Modern model-based spectral estimation and adaptive filtering techniques are used for beamforming and jammer cancellation.

Pattern recognition and, more recently, machine learning techniques are used for target/clutter[3] discrimination and target identification.

At the same time, radar signal processing has several unique qualities that differentiate it from many other signal processing fields. Most modern radars are *coherent*, resulting in a received signal that, once demodulated to baseband, is complex-valued rather than real-valued. Radar signals have very high dynamic ranges of several tens of decibels, in some extreme cases approaching 100 dB. Thus, gain control schemes are common and sidelobe control is often critical to avoid having weak signals masked by stronger ones. In addition, received signals are many decibels weaker than transmitted signals due to propagation and other losses, so that SIR ratios at the receiver are often relatively low. For example, successful detection typically requires an SIR at the point of detection of 10 to 20 dB, but the SIR of the received signal at the antenna will be much less than 0 dB. Large signal processing gains are needed to overcome this deficit.

Another very important distinguishing feature of radar signal processing is the large signal bandwidths compared to most other DSP applications. Instantaneous bandwidths for an individual pulse or CW transmission are frequently on the order of a few megahertz. In some fine resolution[4] radars, they may reach several hundred megahertz and even low gigahertz levels. The difficulty of designing good converters at multi-megahertz or gigahertz sample rates has historically slowed the introduction of digital techniques into radar signal processing for two primary reasons. First, very fast *analog-to-digital* (A/D) converters are required to digitize the high-bandwidth data. Even now that digital techniques are standard in new designs, A/D converter effective word lengths in high-bandwidth systems (high tens of megahertz to ones of gigahertz) are usually a relatively short 8 to 12 bits, rather than the 16 bits common in many other areas.

Second, the high data rates require high-speed computing capability to implement the algorithms. Historically, this meant that it was often necessary to design custom hardware for the digital processor in order to obtain adequate throughput, that is, to "keep up with" the torrent of data. It also meant that radar signal processing algorithms had to be relatively simple compared to lower-bandwidth applications such as sonar in order to minimize the processing load. Only in the late 1990s and later have improved analog semiconductor technology and Moore's law[5] provided enough computing power to host radar algorithms for a wide range of high-performance systems on commercial hardware. This technological progress has enabled rapid development and introduction of new, more complex algorithms to radar signal processing, enabling major improvements in detection, tracking, and imaging capability.

1.3 Elements of a Radar

Figure 1.2 is one possible block diagram of a basic monostatic radar. The waveform generator output is the desired pulse or CW waveform to be transmitted. The transmitter comprises a series of *mixers* and *local oscillators* (LOs) to modulate this waveform to a desired *intermediate*

---

[3]"Clutter" is an interference signal consisting of unwanted echoes of the radar's transmitted signal from objects not of interest.

[4]Systems exhibiting good or poor resolution are commonly referred to as high- or low-resolution systems, respectively. Since better resolution means a *smaller* numerical value, in this text the terms "fine" and "coarse" are used instead to reduce confusion.

[5]Gordon Moore's famous 1965 prediction was that the number of transistors on an integrated circuit would double every 18 to 24 months. That prediction held remarkably true for at least 40 years, enabling the computing and networking revolutions that began in earnest in the 1980s. Whether it is being maintained, or can be, in the 2020s and later is a perennial subject of debate.

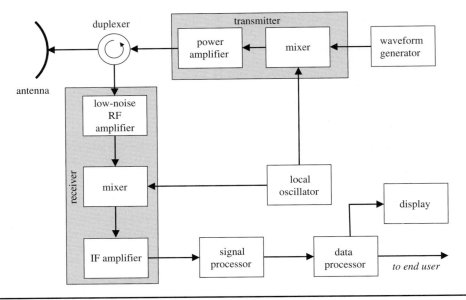

FIGURE 1.2 Block diagram of a monostatic radar.

frequency (IF) and then to the desired *radio frequency* or *radar frequency* (RF), followed by amplifiers to boost the signal power to a useful level. The transmitter output is routed to the antenna through a *duplexer*, *circulator*, or *T/R switch* (for transmit/receive).

The antenna focuses and concentrates the transmitted radiation into a narrow region of space in a particular direction, providing a power gain to the transmitted signal in that direction. Similarly, it selectively provides greater sensitivity on receive to echoes from that same direction. In doing so, the antenna both amplifies weak target echo signals and localizes them in azimuth and elevation.

The returning echoes are routed by the duplexer into the radar receiver. The receiver is usually a superheterodyne design (Bruder, 2010), and often the first stage is a low-noise RF amplifier. This is followed by one or more stages of demodulation of the received signal to successively lower IFs and ultimately to *baseband*, where the signal is not modulated onto any carrier frequency.

The baseband signal is next sent to the *signal processor*, which performs some or all of a variety of functions such as matched filtering, Doppler filtering, integration, and motion compensation. The output of the signal processor typically becomes the input to a *data processor*. Typical data processor functions might include target classification, target tracking, or image processing operations, depending on the radar purpose. The data processor output is sent to the system display, passed to other systems, or both as appropriate. The distinction between the signal processor and the data processor is somewhat arbitrary. Generally, the signal processor is associated with "lower" level, higher speed, streaming operations, while the data processor is associated with "higher" level operations that tend to be more data-dependent but require less computation.

The configuration of Fig. 1.2 is not unique. For example, many systems perform some of the signal processing functions at IF rather than baseband; matched filtering and some forms of Doppler filtering are common examples. Also, radars differ in which portions of the signal flow are analog and which are digital. Older systems are all analog, and many currently operational systems do not digitize the signal until it is converted to baseband. Thus, any signal processing performed at IF must be done with analog techniques. Increasingly, new designs digitize the received signal at an IF or even RF stage, moving the A/D converter closer to the radar front end and performing more of the processing digitally. Waveform generation is also done digitally in many modern systems.

The next few subsections provide some additional detail on these major radar subsystems and also discuss some additional radar systems issues.

1.3.1 Radar Frequencies

Radar systems have been operated at frequencies as low as 2 MHz and as high as 220 GHz (Skolnik, 2001); laser radars operate at frequencies on the order of 10^{12} to 10^{15} Hz, corresponding to wavelengths of 0.3 to 30 μm (Jelalian, 1992). However, most radars operate in the microwave frequency region of about 200 MHz to about 30 GHz, with corresponding wavelengths of 0.67 m to 1 cm. There are also numerous systems in the *millimeter wave* (MMW) region of 30 to 300 GHz (1 cm to 1 mm), especially in the 35 and 95 GHz regions. Table 1.1 summarizes the letter nomenclature used for the common nominal radar bands (IEEE, 2019).

Band	Frequencies	Wavelengths	Common Uses
HF	3–30 MHz	100–10 m	Over-the-horizon surveillance
VHF	30–300 MHz	10–1 m	Long range surveillance, foliage penetration, ground penetrating radar, counter-stealth
UHF	300 MHz–1 GHz	1–30 cm	Long range surveillance, foliage penetration
L	1–2 GHz	30–15 cm	Long range surveillance, long range air traffic control
S	2–4 GHz	15–7.5 cm	Moderate range surveillance, terminal air traffic control, airborne early warning, long range weather observation
C	4–8 GHz	7.5–3.75 cm	Long range tracking, weather observation, weapon location
X	8–12 GHz	3.75–2.5 cm	Short range tracking, missile guidance, marine radar, ground imaging, airborne intercept, weapon location
K_u	12–18 GHz	2.5–1.67 cm	High resolution mapping, satellite altimetry, UAV radar
K	18–27 GHz	1.67–1.11 cm	Police radar, automotive radar
K_a	27–40 GHz	1.11 cm–7.5 mm	Short-range fine resolution imaging, airport surveillance
V	40–75 GHz	7.5–4 mm	Scientific remote sensing
W	75–110 GHz	4–2.73 mm	Automotive radar, missile seekers, very fine resolution imaging
Millimeter wave (includes V and W)	30–300 GHz	1 cm–1 mm	Experimental

TABLE 1.1 Letter Nomenclature and Common Uses for Nominal Radar Frequency Bands

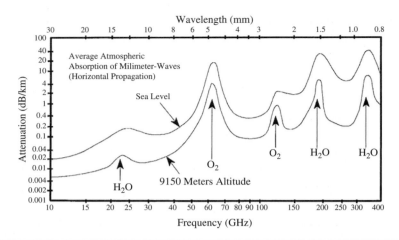

FIGURE 1.3 One-way atmospheric attenuation of electromagnetic waves. (*Source: EW and Radar Systems Engineering Handbook*, Naval Air Warfare Center, Weapons Division, http://ewhdbks.mugu.navy.mil/.)

Not all frequencies in these bands are suitable for or available to radar operation. Within the HF to K_a bands, specific frequencies are allocated by international agreement to radar operation. At frequencies above X band, atmospheric attenuation of electromagnetic waves becomes significant. Consequently, radars in these bands usually operate at one of several "atmospheric window" frequencies where attenuation is relatively low. Figure 1.3 illustrates the atmospheric attenuation for one-way propagation over the most common radar frequency ranges under approximately "clear air" atmospheric conditions. Most K_a band radars operate near 35 GHz and most W band systems operate near 95 GHz because of the relatively low atmospheric attenuation at these wavelengths.

Lower radar frequencies tend to be preferred for longer range surveillance applications because of the low atmospheric attenuation and high power available in transmitters at these frequencies. Higher frequencies tend to be preferred for finer resolution, shorter range applications due to the smaller achievable antenna beamwidths for a given antenna size, higher attenuation, and lower available transmitter powers.

Weather conditions can also have a significant effect on radar signal propagation. Figure 1.4 illustrates the additional one-way loss as a function of radar frequency for rain rates ranging from a drizzle to a tropical downpour. X-band frequencies (typically about 10 GHz) and below are affected significantly only by very severe rainfall, while MMW frequencies suffer severe losses for even light-to-medium rain rates.

1.3.2 Radar Waveforms and Transmitters

The radar *waveform* is the term for the signal that is modulated onto the RF carrier for transmission, echo reception, and demodulation. It plays a major role in determining the sensitivity and range resolution of the radar. There are many different waveforms in common use. They can all be classified as either pulsed or continuous wave. Figure 1.5 illustrates the difference between the two. The CW waveform of Fig. 1.5*a*, as its name suggests, simply radiates a sinusoidal signal at the desired RF continuously. The pulsed waveform in Fig. 1.5*b* transmits a series of finite length pulses. The series is described by the pulse length τ and either the *pulse repetition interval* (PRI) T between pulses or its inverse, the *pulse repetition frequency* (PRF).

Introduction to Radar Systems and Signal Processing   9

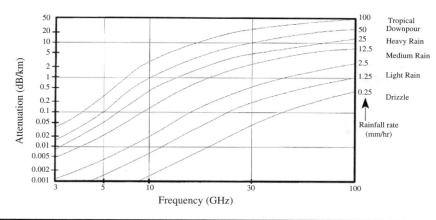

FIGURE 1.4 Effect of different rates of precipitation on one-way atmospheric attenuation of electromagnetic waves. (*Source: EW and Radar Systems Engineering Handbook*, Naval Air Warfare Center, Weapons Division, http://ewhdbks.mugu.navy.mil/.)

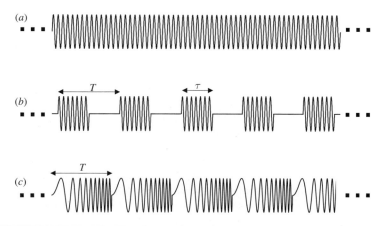

FIGURE 1.5 Three of the major classes of radar waveforms. (*a*) Continuous wave (CW). (*b*) Pulsed. (*c*) Frequency-modulated continuous wave (FMCW).

The pulsed waveform enables easy measurement of range by observing the time delay between transmission of a pulse and reception of an echo; see Eq. (1.1). Because the CW waveform lacks any distinguishing timing mark such as a pulse edge, another means is needed to measure range. One solution is to modulate the CW waveform with a repeating pattern. Currently the most common example is FMCW. Figure 1.5c illustrates an example wherein the frequency of the CW waveform is swept over some bandwidth repeatedly. The duration of one sweep is analogous to the PRI for a pulsed waveform. In the example shown, the sweep rate is constant so that the frequency increases linearly during the sweep interval.

Pulsed waveforms can also exhibit intra-pulse modulation. Both phase- and frequency-modulated pulses are common. Because of their widespread use, in this text the focus will be on pulsed waveforms with phase, frequency, or no modulation, and on linear FMCW waveforms. The details of each are the subject of Chap. 4.

The radar transmitter modulates the waveform to the RF and amplifies it to a useful level. Radar transmitters operate at peak powers ranging from milliwatts to in excess of 10 MW. A wide variety of technologies are used, from solid state sources at lower powers to various vacuum tube devices such as magnetrons and traveling wave tubes at high powers. An excellent survey of transmitter technologies and issues is given in Wallace et al. (2010). High peak power systems are invariably pulsed; CW systems have much lower peak powers. One of the more powerful existing pulsed transmitters is found in the AN/FPS-108 COBRA DANE radar, which has a peak power of 15.4 MW (Brookner, 1988). In pulsed radars the PRF varies widely but is typically between several hundred and several tens of thousands of pulses per second (PPS); in some modern integrated "radar-on-a-chip" systems, the PRF can be several hundred thousand PPS. The duty cycle of pulsed systems is usually relatively low and often well below 1 percent, so that average powers rarely exceed 10 to 20 kW. COBRA DANE again offers an extreme example with its high average power of 0.92 MW. Pulse lengths are most often between about 100 ns and 100 μs, though some systems use pulses as short as a few nanoseconds while others have extremely long pulses, on the order of 1 ms or more.

It will be seen in Chap. 6 that the detection performance achievable by a radar improves with the amount of energy in the transmitted waveform. Maximizing energy suggests that a radar waveform should be as long as feasible and be transmitted at maximum power. To satisfy the second condition, radars generally do not use amplitude modulation of the transmitted waveform since that implies that at least a part of the waveform is transmitted at less than full power.

Waveform length is a more complicated issue. It will be seen in Chap. 4 that the nominal range resolution ΔR is determined by the waveform bandwidth β in hertz:

$$\Delta R = \frac{c}{2\beta} \text{ m} \tag{1.2}$$

For an unmodulated pulse, the bandwidth is inversely proportional to its duration. Fine resolution therefore implies shorter pulses, in conflict with the need for longer pulses to maximize energy. To increase waveform bandwidth for a given pulse length without sacrificing energy, many radars routinely use phase or frequency modulation of the pulse. Similar issues apply to FMCW waveforms. Desirable values of range resolution vary from a few kilometers in long-range surveillance systems, which tend to operate at lower RFs, to a meter or less in very fine-resolution imaging systems, which tend to operate at high RFs. Corresponding waveform bandwidths are on the order of 100 kHz to 1 GHz. These bandwidths are typically 1 percent or less of the RF. Few radars achieve 10 percent bandwidth, though some achieve bandwidths of 25 percent of the RF or greater, qualifying them as *ultrawideband* (UWB) radars (IEEE, 2017). Nonetheless, most radar waveforms can be considered narrowband, bandpass functions.

1.3.3 Antennas

The antenna plays a major role in determining the sensitivity and angular resolution of the radar. A very wide variety of antenna types are used in radar systems. Some of the more common types are parabolic reflector antennas, scanning feed antennas, lens antennas, and phased array antennas.

From a signal processing perspective, the most important properties of an antenna are its gain, beamwidth, and sidelobe levels. Each of these follows from consideration of the antenna *power pattern*. The one-way power pattern $P(\theta, \phi)$ describes the radiation intensity during transmission in the direction (θ, ϕ) relative to the antenna boresight. Aside from scale factors, which are unimportant for normalized patterns, it is related to

Introduction to Radar Systems and Signal Processing   11

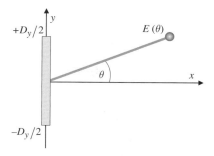

FIGURE 1.6 Geometry for one-dimensional electric field calculation on a linear aperture.

the radiated electric field intensity $E(\theta, \phi)$, known as the antenna one-way *voltage pattern*, according to

$$P(\theta, \phi) = |E(\theta, \phi)|^2 \qquad (1.3)$$

For a rectangular aperture with an illumination function that is separable in the two aperture dimensions, $P(\theta, \phi)$ can be factored as the product of separate one-dimensional patterns (Stutzman and Thiele, 2012):

$$P(\theta, \phi) = P_\theta(\theta) P_\phi(\phi) \qquad (1.4)$$

For most radar scenarios, only the *far field* (also called *Fraunhofer*) power pattern is of interest. The far-field is conventionally defined to begin at a range of D^2/λ_t or $2D^2/\lambda_t$ for an antenna of aperture size D. Consider the azimuth (θ) pattern of the one-dimensional linear aperture geometry shown in Fig. 1.6. From a signal processing viewpoint, an important property of aperture antennas such as flat plate arrays and parabolic reflectors is that the electric field intensity as a function of azimuth $E(\theta)$ in the far field is the inverse Fourier transform[6] of the distribution $A(y)$ of current across the aperture in the azimuth plane (Bracewell, 1999; Skolnik, 2001),

$$E(\theta) = \int_{-D_y/2}^{D_y/2} A(y) \exp[j(2\pi y/\lambda_t) \sin\theta] \, dy \qquad (1.5)$$

where the "frequency" variable is the *spatial frequency* or *wavenumber* $(2\pi/\lambda_t) \sin\theta$ and is in units of radians per meter. The idea of spatial frequency is discussed in App. B.

To be more explicit about this point, define $s = \sin\theta$ and $\zeta = y/\lambda_t$. Substituting these definitions in Eq. (1.5) gives

$$\frac{1}{\lambda_t} \int_{-D_y/2\lambda_t}^{D_y/2\lambda_t} A(\lambda_t \zeta) \exp(j2\pi \zeta s) \, d\zeta = E(s) \qquad (1.6)$$

which is clearly of the form of an inverse Fourier transform. (The finite integral limits are due to the finite support of the aperture.) Because of the definitions of ζ and s, this transform

---

[6]Whether it is the forward or inverse Fourier transform (FT) depends on the FT definition one uses. This text uses the electrical engineering convention, in which the sign of the argument of the exponential in the FT kernel is negative for the forward transform and positive for the inverse.

relates the current distribution as a function of aperture position normalized by the wavelength to a spatial frequency variable that is related to the azimuth angle through a nonlinear mapping. It of course follows that

$$A(\lambda_t \zeta) = \int_{-\infty}^{+\infty} E(s) \exp(-j2\pi\zeta s)\, ds \tag{1.7}$$

The infinite limits in Eq. (1.7) are misleading, since the variable of integration $s = \sin\theta$ can only range from -1 to $+1$. Because of this, $E(s)$ is taken to be zero outside of this range on s.

Equation (1.7) is a somewhat simplified expression that neglects a range-dependent overall phase factor and a slight amplitude dependence on range (Balanis, 2016). This Fourier transform property of antenna patterns will allow the use of linear system concepts in Chap. 2 to understand the effects of the antenna on cross-range resolution and the angular sampling densities needed to avoid spatial aliasing.

An important special case of Eq. (1.5) occurs when the aperture current illumination is a constant, $A(y) = A_0$. The far-field one-way voltage pattern, normalized to its peak, is then the familiar sinc function

$$E(\theta) = \frac{\sin[\pi(D_y/\lambda_t)\sin\theta]}{\pi(D_y/\lambda_t)\sin\theta} \tag{1.8}$$

The magnitude of $E(\theta)$ is illustrated in Fig. 1.7 for the case $D_y = 6\lambda_t$, along with the definitions for three important figures of merit of an antenna pattern. The pattern exhibits the typical structure of a high-gain mainlobe surrounded by low-gain sidelobes. The angular resolution of the antenna is determined primarily by the width of its mainlobe. The mainlobe width is conventionally quantified by the *3-dB beamwidth*, which is the width of the mainlobe

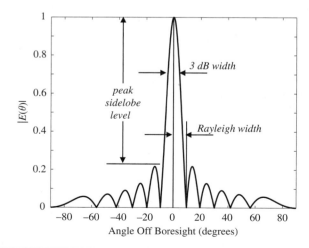

Figure 1.7 One-way radiation pattern of a uniformly illuminated aperture with $D_y = 6\lambda_t$. The 3-dB beamwidth, Rayleigh beamwidth, and peak sidelobe definitions are illustrated.

between the points where $|E(\theta)|^2$ is reduced by 3 dB (approximately one-half) from its peak at $\theta = 0$. This can be determined by setting $E(\theta) = 1/\sqrt{2} \approx 0.707$ and solving for the argument $\alpha = \pi(D_y/\lambda_t)\sin\theta$. The answer is found numerically to be $\alpha = 1.4$, which occurs at the angle $\theta_0 = \sin^{-1}(1.4\lambda_t/\pi D_y)$. The 3-dB beamwidth extends from $-\theta_0$ to $+\theta_0$ and is therefore

$$\text{3-dB beamwidth} = \theta_3 = 2\sin^{-1}\left(\frac{1.4\lambda_t}{\pi D_y}\right) \approx 0.89\frac{\lambda_t}{D_y} \text{ radians} \tag{1.9}$$

The small-angle approximation used in the last step holds for most radar antenna beamwidths.

Although 3-dB beamwidths are traditional, for analysis the Rayleigh beamwidth is often simpler to compute. This is the one-sided width of the mainlobe from its peak to its first null, and is given by (see Prob. 7)

$$\theta_R = \sin^{-1}\left(\frac{\lambda_t}{D_y}\right) \approx \frac{\lambda_t}{D_y} \text{ radians} \tag{1.10}$$

For the constant aperture illumination case, the Rayleigh beamwidth also happens to equal the 4-dB beamwidth. This is not true for antennas in general. Finally, the null-to-null beamwidth θ_{nn} is simply twice the Rayleigh beamwidth, encompassing the entire mainlobe.

Whichever metric is used, note that a smaller beamwidth requires a larger aperture or a shorter wavelength. Typical beamwidths range from as little as a few tenths of a degree to several degrees for a *pencil beam antenna,* where the beam is made as narrow as possible in both azimuth and elevation. Some antennas are deliberately designed to have broad vertical beamwidths of several tens of degrees for convenience in large volume search; these designs are called *fan beam antennas.*

The *peak sidelobe* of the pattern affects how echoes from one object affect the detection of neighboring objects. For the uniform illumination pattern, the peak sidelobe is 13.2 dB below the mainlobe peak. This is often considered insufficient in radar systems because strong unwanted signals entering the antenna from the sidelobe directions may not be attenuated enough to enable detection of weaker target echoes in the mainlobe direction. Antenna sidelobes can be reduced by use of a nonuniform aperture distribution (Skolnik, 2001), sometimes referred to as *tapering, shading, or apodization of* the antenna. In fact, this is no different from the window or weighting functions used for sidelobe control in other areas of signal processing such as spectrum analysis and digital filter design, and peak sidelobes can easily be reduced to around 25 to 40 dB at the expense of an increase in mainlobe width (see App. B). Lower sidelobes are possible but may be increasingly difficult to achieve due to manufacturing imperfections and inherent design limitations.

The antenna *power gain* G is the ratio of peak radiation intensity from the antenna to the radiation that would be observed from a lossless, isotropic (omnidirectional) antenna if both have the same input power. Power gain is determined by both the antenna pattern and by losses in the antenna. A useful rule of thumb for a typical antenna is (Stutzman, 1998)

$$\begin{aligned} G &\approx \frac{26{,}000}{\theta_3\phi_3} \quad (\theta_3, \phi_3 \text{ in degrees}) \\ &\approx \frac{7.9}{\theta_3\phi_3} \quad (\theta_3, \phi_3 \text{ in radians}) \end{aligned} \tag{1.11}$$

Though both higher and lower values are possible, typical radar antennas have gains from about 10 dB for a broad fan-beam search antenna to approximately 40 dB for a pencil beam that might be used for both search and track.

Effective aperture A_e is an important characteristic in describing the behavior of an antenna being used for reception. Suppose a wave with power density W W/m² incident on the antenna results in a power P delivered to the antenna load. The effective aperture is defined as the ratio (Balanis, 2016)

$$A_e = \frac{P}{W} \text{ m}^2 \tag{1.12}$$

Note that A_e is not the actual physical area of the antenna. It is the fictional area such that, if all of the power incident on that area was collected and delivered to the load with no loss, it would account for all of the observed power output of the actual antenna. Effective aperture is directly related to antenna *directivity*, which in turn is related to antenna gain and efficiency. For most antennas, the efficiency is near unity and the effective aperture and gain are related by (Balanis, 2016)

$$G = \frac{4\pi}{\lambda_t^2} A_e \tag{1.13}$$

Another important type of antenna is the *array antenna*. An array antenna is one composed of a collection of individual antennas called array *elements*. The elements are typically identical dipoles or other simple antennas with very broad patterns. Usually, the elements are evenly spaced to form a *uniform linear array* (ULA) as shown in one dimension in Fig. 1.8. Figure 1.9 illustrates examples of real array and aperture antennas. Many more examples are given in Chaps. 1 and 9 in Richards et al. (2010).

The voltage pattern for the linear array is most easily arrived at by considering the antenna in its receive mode. Suppose the rightmost element is taken as a reference point, there are N elements in the array, and the elements are isotropic (constant gain for all θ). The signal in branch n is weighted with the complex weight a_n. The incoming electric field voltage $E_0 \exp(j\Omega_t t)$ at the reference element will also appear at each of the other elements, but delayed in time by an additional $d \sin\theta/c$ from each element to the next due to the increasing propagation distance to each. The total output voltage will be the sum of the time-shifted and weighted outputs of all of the array elements. The amplitude of the output voltage is (Skolnik, 2001; Stutzman and Thiele, 2012; or see Sec. 9.2)

$$E(\theta) = E_0 \sum_{n=0}^{N-1} a_n \exp[j(2\pi/\lambda_t)nd \sin\theta] \tag{1.14}$$

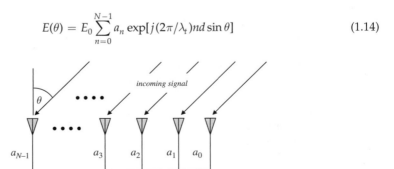

Figure 1.8 Geometry of the uniform linear array antenna.

Introduction to Radar Systems and Signal Processing   15

(a)

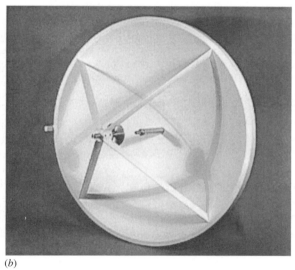

(b)

Figure 1.9 Examples of typical array and aperture antennas. (a) Slotted phased array in the nose of an F/A-18 aircraft. This antenna is part of the AN/APG-73 radar system. (b) A Cassegrain reflector antenna. [Image (a) courtesy of Raytheon Technologies. Image (b) courtesy of Quinstar Corp. Used with permission.]

This is similar in form to the inverse *discrete Fourier transform* (DFT) of the weight sequence $\{a_n\}$. Like the aperture antenna, the antenna pattern of the linear array thus involves a Fourier transform, this time of the weight sequence. For the case where all the $a_n = 1$, the pattern is the familiar "aliased sinc" (asinc) function,[7] whose magnitude is

---

[7]Also called the digital sinc (dsinc) or Dirichlet function. It is the discrete-variable equivalent of the usual continuous-variable sinc function.

$$|E(\theta)| = E_0 \left| \frac{\sin[N(\pi d/\lambda_t)\sin\theta]}{\sin[(\pi d/\lambda_t)\sin\theta]} \right| \qquad (1.15)$$

This function is very similar to that of Eq. (1.8) and Fig. 1.7. If the number of elements N is reasonably large (nine or more) and the product Nd is considered to be the total aperture size D, the 3-dB beamwidth is $0.89\lambda_t/D$, and the first sidelobe is 13.2 dB below the mainlobe peak; both numbers are the same as those of the uniformly illuminated aperture antenna. By varying the amplitudes of the weights a_n, it is possible to reduce the sidelobes at the expense of a broader mainlobe and a reduction in output SNR.

Actual array elements are not isotropic radiators. A simple model often used as a first-order approximation to a typical element pattern $E_{el}(\theta)$ is

$$E_{el}(\theta) \approx \cos\theta \qquad (1.16)$$

The right-hand side of Eq. (1.15) is then called the *array factor* $AF(\theta)$, and the composite radiation pattern becomes

$$E(\theta) = AF(\theta)E_{el}(\theta) \qquad (1.17)$$

Because the cosine function is slowly varying in θ, the beamwidth and first sidelobe level are not greatly changed by including the element pattern for signals arriving at angles near broadside (near $\theta = 0$). The element pattern does reduce distant sidelobes, thereby reducing sensitivity to waves impinging on the array from well off broadside.

The discussion so far has been phrased in terms of the transmit antenna pattern (for aperture antennas) or the receive pattern (for arrays), but not both. The patterns described have been *one-way antenna patterns*. The reciprocity theorem guarantees that the receive antenna pattern is identical to the transmit antenna pattern (Balanis, 2016). Consequently, for a monostatic radar, the *two-way antenna pattern* (power or voltage) is just the square of the corresponding one-way pattern. It also follows that the antenna phase center location is the same in both transmit and receive modes.

1.3.4 Virtual Elements and Virtual Arrays

Two more useful antenna concepts are the antenna *phase front* (or *wave front*) and *phase center* (Sherman, 2011; IEEE, 2014; Balanis, 2016; Richards, 2018). A phase front of a radiating antenna is any surface on which the phase of the field is a constant. In the far-field, the phase fronts are usually approximately spherical, at least over localized regions. The phase center of the antenna is the center of curvature of the phase fronts. Put another way, the phase center is the point at which an isotropic radiator should be located so that the resulting phase fronts best match those of the actual antenna. The phase center concept is useful because it defines an effective location of the antenna, which can in turn be used for analyzing effective path lengths, Doppler shifts, and so forth. For symmetrically illuminated aperture antennas, the phase center will be centered in the aperture plane but may be displaced forward or backward from the actual aperture. Referring to Fig. 1.6, the phase center would occur at $y = 0$ but possibly $x \neq 0$, depending on the detailed antenna shape.

The phase center idea is especially useful in developing the *virtual array* (VA) concept for analyzing situations where the transmit and receive antennas are not collocated due either to actual physical separation or to platform motion between transmission and reception times. Consider the situation in Fig. 1.10, which shows separate transmit and receive antenna elements at coordinates x_T and x_R and a point scatterer **P** in the far-field. It is straightforward to show that the phase shift of the signal transmitted from x_T and received at x_R (total path length equal to $R_R + R_T$) is, to a good approximation, the same as that of a signal transmitted from and received at the *virtual element* (VE) location **VE** halfway between the two (total path

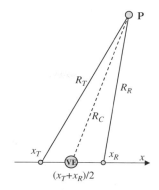

Figure 1.10 Virtual element corresponding to a transmit and receive element pair.

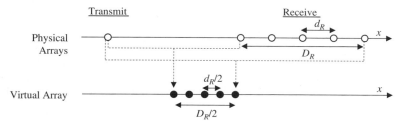

Figure 1.11 Virtual array formed by paring one transmit element with five independent receive elements. (The virtual array has been offset vertically for clarity.)

length $2R_C$ (Richards, 2018). Consequently, for analysis purposes the transmit and receive elements can be replaced by the single VE. This substitution will be useful in the discussion of ground moving target indication in Chap. 5.

Now consider the configuration of Fig. 1.11. The transmit antenna is the single isotropic white element on the left. The receive antenna is the five-element uniform linear array of white elements with element spacing d_R and total aperture size D_R on the right. The signal received by the leftmost element of the receive array will be essentially identical to one transmitted and received by a VE located halfway between it and the transmit element. This location is the leftmost of the five black elements forming the VA, as shown by the dotted line. (The VA will actually be located on the same x-axis line as the transmit and receive elements; it is offset vertically in the figure for clarity.) Each of the five transmit-receive element pairings generates a VE. These collectively form the five-element VA shown. Note that the element spacing and overall aperture size are half the physical receive array spacing and size.

The VE and VA concepts can be used to provide a common analysis framework for a variety of problems in antenna design, synthetic aperture analysis (Chap. 8), and emerging techniques such as multi-input, multi-output (MIMO) radar. Much more detail on these concepts is available in Richards (2018) and Richards (2019).

1.3.5 Receivers

It was shown in Sec. 1.3.2 that radar signals are usually narrowband, bandpass (because they are on a carrier frequency), phase- or frequency-modulated functions.

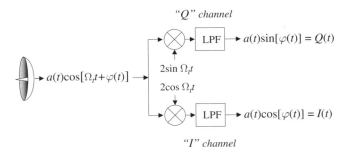

FIGURE 1.12 Quadrature or "I/Q" channel receiver model.

This means that the echo waveform $\bar{y}(t)$ received from a single scatterer can be modeled in the form

$$\bar{y}(t) = a(t)\cos[\Omega_t t + \varphi(t)] \qquad (1.18)$$

Here Ω_t is the nominal RF in radians per second; $\varphi(t)$ is the phase modulation (PM) function, which also includes frequency modulation as a special case; and $a(t)$ is the amplitude modulation (AM). In a CW system, $a(t)$ would be a constant. In the simplest pulsed systems, $a(t)$ would be just a rectangular pulse envelope. In a constant frequency pulsed or CW system (no FM or PM), $\varphi(t)$ would be a constant. The major function of the receiver processing is demodulation of the information-bearing part of the radar signal to baseband with the goal of estimating $a(t)$ and $\varphi(t)$.

Figure 1.12 illustrates the signal processor's simplified view of the receiver structure used in most classical radars. The lower channel mixes the received signal with a local oscillator at the radar frequency. The output of the mixer is the product of its two input signals. Applying a trigonometric identity, the mixer is seen to generate both sum and difference frequency components at its output:

$$2\cos(\Omega_t t) \cdot a(t)\cos[\Omega_t t + \varphi(t)] = a(t)\cos(\Omega_t t) + a(t)\cos[2\Omega_t t + \varphi(t)] \qquad (1.19)$$

The high-frequency sum frequency term is then removed by the lowpass filter (LPF), leaving only the difference frequency, also called the *beat frequency*. Note that the difference frequency is taken to be the received echo frequency minus the reference oscillator frequency. In this case, the beat frequency is zero so the output is just the modulation term $a(t)\cos[\varphi(t)]$. The upper channel mixes the signal with a *quadrature* oscillator having the same frequency but a 90° phase shift as compared to the lower channel oscillator. The upper channel mixer output is

$$2\sin(\Omega_t t) \cdot a(t)\cos[\Omega_t t + \varphi(t)] = a(t)\sin(\Omega_t t) + a(t)\sin[2\Omega_t t + \varphi(t)] \qquad (1.20)$$

which, after filtering, leaves the modulation term $a(t)\sin[\varphi(t)]$.

If the input $\bar{x}(t)$ is written as $a(t)\sin[\Omega_t t + \varphi(t)]$ instead, the lower and upper channel outputs are interchanged. Whichever form is used for the input, the receiver channel having an output proportional to the cosine of the input phase is called the *in-phase* or "I" channel; the other is called the *quadrature* phase or "Q" channel.

Both the I and Q channels are needed to unambiguously determine the amplitude and phase of the echo. Suppose only the I channel is implemented in the receiver, giving the single measured value $a(t)\cos[\varphi(t)]$. There are an infinite number of combinations of a and

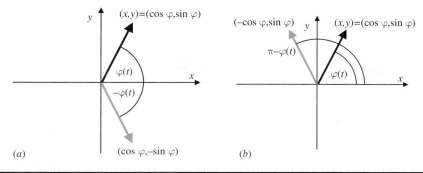

FIGURE 1.13 (a) The I channel of the receiver in Fig. 1.12 measures only the cosine of the phasor $\theta(t)$. (b) The Q channel measures only the sine of the phasor.

φ that produce the same product; the I measurement alone is not sufficient to specify both. However, if the Q channel measurement $a(t)\sin[\varphi(t)]$ is also available, the amplitude can be found as $a(t) = \sqrt{I^2(t) + Q^2(t)}$.

Knowing $a(t)$ is still not sufficient to determine $\varphi(t)$ using only one of the I or Q measurements. Figure 1.13 illustrates the problem. In Fig. 1.13a the signal phase $\varphi(t)$ is represented as a unit-magnitude black phasor in the complex plane; the amplitude $a(t)$ has been removed. If only the I channel is implemented in the receiver, only the cosine of $\varphi(t)$ will be measured. In this case, the true phasor will be indistinguishable from the gray phasor $-\varphi(t)$. Similarly, if only the Q channel is implemented so that only the sine of $\varphi(t)$ is measured, then the true phasor will be indistinguishable from the gray phasor of Fig. 1.13b, which corresponds to $\pi - \varphi(t)$. When both the I and Q channels are implemented, the phasor quadrant is determined unambiguously.[8]

In modern coherent radars, the signal processor will normally assign the I signal to be the real part and the Q signal to be the imaginary part of a new complex signal

$$x(t) = I(t) + jQ(t) = a(t)\exp[j\varphi(t)] \tag{1.21}$$

Equation (1.21) implies a more convenient way of representing the effect of an ideal coherent receiver on a transmitted signal. Instead of representing the transmitted signal by a cosine function, an equivalent complex exponential function is used instead.[9] The received echo signal of Eq. (1.18) is thus replaced by

$$\bar{x}(t) = a(t)\exp[\Omega_t t + \varphi(t)] \tag{1.22}$$

The receiver structure of Fig. 1.12 can then be replaced with the simplified model of Fig. 1.14, where the echo is demodulated by multiplication with a complex reference oscillator $\exp(-j\Omega_t t)$.

---

[8] This is analogous to the use of the two-argument `atan2()` function instead of the single-argument `atan()` function in many programming languages.

[9] Although these formalizations are not needed for the discussions in this text and are therefore avoided for simplicity, it is worthwhile to note that the complex signal in Eq. (1.22) is the *analytic signal* associated with the real-valued signal of Eq. (1.18). The imaginary part of Eq. (1.22) is the *Hilbert transform* of the real part (Papoulis, 1987; Bracewell, 1999).

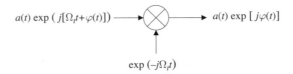

Figure 1.14 Simplified transmission and receiver model using complex exponential signals.

This technique of modeling the transmitted and received real-valued signals by equivalent complex signals using a corresponding complex demodulator produces exactly the same output result obtained in Eq. (1.21) by explicitly modeling the real-valued signals and the I and Q channels, but is much more compact and easier to manipulate. This complex exponential analysis approach is used throughout the remainder of the book. It is important to remember that this is an analysis technique; actual analog hardware must still operate with real-valued signals only. However, once signals are digitized, they may be treated explicitly as complex signals in the digital processor.

Figure 1.12 implies several requirements on a high-quality receiver design. For example, the local oscillator frequencies in the transmitter modulator and receiver demodulator must be identical. This is usually ensured by having a single *stable local oscillator* (STALO) in the radar system that provides a frequency reference for both. Furthermore, many types of radar processing require *coherent* operation. The IEEE *Standard Radar Definitions* defines "coherent signal processing" as "echo integration, filtering, or detection using amplitude *and phase* of the signal referred to a coherent oscillator" (emphasis added) (IEEE, 2017). Coherency is a stronger requirement than frequency stability. In practice, it means that the transmitted carrier signal must have a fixed phase reference for several, perhaps many, consecutive pulses or CW sweeps. Consider a pulse transmitted at time zero of the form $a(t)\cos(\Omega_t t + \varphi)$, where $a(t)$ is the pulse shape. In a coherent system, a pulse transmitted T seconds later will be of the form $a(t - T)\cos(\Omega_t t + \varphi)$. Note that both pulses have the same argument for their cosine term. Only the envelope term is delayed, shifting the pulse location on the time axis while keeping the same underlying sinusoid. An example of a noncoherently related second pulse would be $a(t - T)\cos[\Omega_t(t - T) + \varphi] = a(t - T)\cos[\Omega_t t + (\varphi - \Omega_t T)]$, which is nonzero over the same time interval as the coherent second pulse $a(t - T)\cos(\Omega_t t + \varphi)$ and has the same frequency, but has a different phase at any instant in time.

Figure 1.15 illustrates the difference. In the coherent case, the two pulses appear as if they were excised from the same underlying continuous, stable sinusoid; in the noncoherent case, the second pulse is not in phase with the extension of the first pulse. Another type of noncoherency arises when the starting phases φ of successive pulses are random, which occurs with some types of transmitters such as magnetrons (see Wallace et al., 2010). Because of the phase ambiguity discussed earlier, coherency also implies a system having both I and Q channels.

Another receiver requirement is that the I and Q channels have perfectly matched transfer functions over the signal bandwidth. Thus, the gain through each of the two signal paths must be identical, as must be the phase delay (electrical length of the two channels). Finally, a related requirement is that the oscillators used to demodulate the I and Q channels must be exactly in quadrature, that is, 90° out of phase with one another, not 89.9°. Of course, real receivers do not have perfectly matched channels. The effect of gain and phase imbalances will be considered in Chap. 3.

In the receiver structure shown in Fig. 1.12, the information-bearing portion of the signal is demodulated from the carrier frequency to baseband in a single mixing operation. This simple model is adequate to capture the receiver characteristics most important to radar signal processing. While convenient for analysis, radar receivers are virtually never

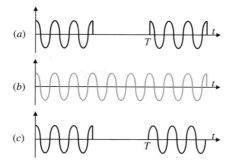

FIGURE 1.15 Illustration of the concept of a fixed phase reference in coherent signals. (*a*) Coherent pulse pair generated from the reference sinusoid. (*b*) Reference sinusoid. (*c*) Noncoherent pulse pair.

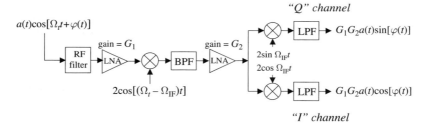

FIGURE 1.16 Structure of a superheterodyne radar receiver.

implemented this way in practice. One reason is that active electronic devices introduce various types of noise into their output signal, such as *shot noise* and *thermal noise* (Scheer, 1993). One noise component, known as *flicker noise* or $1/F$ noise, has a power spectrum that behaves approximately as F^{-1} and is therefore strongest near zero frequency. Since received radar signals are very weak, they can be corrupted by $1/F$ noise if they are translated to baseband before being amplified.

Figure 1.16 shows a more representative *superheterodyne* receiver structure. The key feature of the superheterodyne receiver is that the demodulation to baseband occurs in two or more stages. The received signal is amplified immediately upon reception using a *low-noise amplifier* (LNA). The LNA, more than any other component, determines the *noise figure* of the overall receiver. It will be seen in Sec. 2.4 that this is an important factor in determining the radar's *signal-to-noise* ratio (SNR), so good design of the LNA is important. The signal is next demodulated to an IF. The bandpass filter following the first mixer eliminates the sum frequency term, passing only the difference frequency term at the IF frequency Ω_{IF}. The RF filter eliminates any input signals at the *image frequency* $\Omega_t - 2\Omega_{IF}$; if present, these signals would generate undesired difference frequencies at the IF frequency. The signal is then amplified further. Amplification at IF is easier than at RF because of the greater percentage bandwidth of the signal and the lower cost of IF components compared to microwave components. In addition, modulation to IF rather than to baseband incurs a lower *conversion loss* (power loss in the mixer), improving the receiver sensitivity. The extra IF amplification also reduces the effect of flicker noise. Finally, the amplified signal is demodulated to baseband. Some receivers may use more than two demodulation stages so that there are two or more IF frequencies, but two stages is the most common choice. One final advantage of the superheterodyne configuration is its adaptability. The same IF stages can be used with variable RFs simply by tuning the local oscillator to track changes in the transmitted frequency. More information on radar receivers is in Bruder (2010).

1.4 Common Threads in Radar Signal Processing

A radar system's success or failure in detecting, tracking, and imaging objects or features of interest in the environment is affected by various characteristics of those objects, the environment, and the radar itself, and how they are reflected in the received signals available for processing. Two of the most basic and important signal quality metrics are the signal-to-interference ratio and the resolution. Improving SIR, resolution, or both is the major goal of many of the basic radar signal processing discussed in this text.

While subsequent chapters discuss a wide variety of signal processing techniques, there are a few basic ideas that underlie most of them. These include *coherent* and *noncoherent integration*, target *phase history modeling*, *region of support* (ROS) *expansion*, and *maximum likelihood estimation*. The remainder of this section defines SIR and resolution, and then gives simple examples of integration, phase history modeling, and ROS expansion and how they affect SIR and resolution. Maximum likelihood estimation is deferred to Chap. 9 and App. A.

1.4.1 Signal-to-Interference Ratio

Consider a discrete-time signal $x[n]$ that is the sum of a desired signal $s[n]$ and an interfering signal $w[n]$:

$$x[n] = s[n] + w[n] \tag{1.23}$$

The discussion is identical for continuous time signals. The SIR χ of this signal is the ratio of the power of the desired signal to that of the interference. If $s[n]$ is deterministic, the signal power is usually taken as the square of the peak signal amplitude and may therefore occur at a specific time t_0. In some deterministic cases, the average signal power may be used instead. The interference is almost invariably modeled as a random process so that its power is the mean-square $E\{|w[n]|^2\}$, where $E\{\cdot\}$ represents the expected value of a random variable or process. If the interference is zero mean, as is very often the case, then the power also equals the variance σ_w^2 of the interference. If the desired signal is also modeled as a random process, its power is also taken to be its mean square or variance.

As an example, let $s[n]$ be a complex sinusoid $A\exp(j\omega n)$ and let $w[n]$ be complex zero mean white Gaussian noise of variance σ_w^2. The SIR of their sum $x[n]$ is

$$\chi = \frac{A^2}{\sigma_w^2} \tag{1.24}$$

In this case, the peak and average signal power are the same. If $s[n]$ is a real-valued sinusoid $A\cos[\omega n]$ and $w[n]$ is real-valued zero mean white Gaussian noise of variance σ_w^2, the peak SIR would be the same, but the average SIR would be $A^2/2\sigma_w^2$ because the average power of a real cosine or sine function of amplitude A is $A^2/2$.

A variation is the "energy SIR," defined as the ratio of the total energy $E_s = \sum |s[n]|^2$ in the signal $s[n]$ to the average noise power:

$$\chi = \frac{E_s}{\sigma_w^2} \tag{1.25}$$

The proportionality between E_s and A depends on the signal shape. For a rectangular pulse or a complex exponential of amplitude A and duration N samples, $E_s = N \cdot A^2$. It will be seen in Chap. 6 that when matched filters are used, the peak SIR at the filter output equals the energy SIR of the original signal, so higher energy signals result in higher SIRs.

SIR affects detection, tracking, and imaging performance in different ways. In general, detection performance improves with SIR in the sense that P_D increases for a given P_{FA} as SIR increases. For instance, it will be seen in Chap. 6 that for one model of the target echo characteristics and radar detection algorithm, P_D is related to P_{FA} according to

$$(P_D) = (P_{FA})^{1/(1+\chi)} \tag{1.26}$$

which shows that $P_D \to 1$ as $\chi \to \infty$ for fixed P_{FA}. As another example, the limit on precision (standard deviation of measurement error) due to additive noise of typical estimators of range, angle, frequency, or phase tends to decrease as $1/\sqrt{\chi}$; this behavior will be demonstrated in Chap. 7. In radar imaging (Chap. 8), SIR directly affects the contrast or dynamic range (ratio of reflectivity of brightest to dimmest visible features) of the image.

These considerations make it essential to maximize the SIR of radar data, and many radar signal processing operations discussed in this text have as their primary goal increasing the SIR. The ways in which this is done will be discussed along with each technique.

1.4.2 Resolution and Region of Support

The closely related concepts of *resolution* and a *resolution cell* will arise frequently. Two equal-strength scatterers are considered to be *resolved* if they produce two separately identifiable signals at the system output, as opposed to combining into a single undifferentiated output.[10] The idea of resolution is applied in range, cross-range, Doppler shift or velocity, and angle of arrival. Two scatterers can simultaneously be resolved in one dimension, for example range, and be unresolved in another, perhaps velocity.

Figure 1.17 illustrates the concept of resolution, in this case in frequency. The Fourier transform of a T-second cosinusoidal pulse $x(t) = A\cos(\Omega_0 t), -T/2 \leq t \leq +T/2$ is

$$X(\Omega) = \frac{A}{2}\left\{\frac{\sin[(\Omega - \Omega_0)T/2]}{(\Omega - \Omega_0)T/2} + \frac{\sin[(\Omega + \Omega_0)T/2]}{(\Omega + \Omega_0)T/2}\right\} \tag{1.27}$$

This function has peaks at positive and negative Ω_0 rad/s as expected for a real-valued signal. Considering just the positive frequency peak, the Rayleigh width (peak to first null) is easily seen to be $2\pi/T$ rad/s or $1/T$ Hz. Figure 1.17a shows a portion of the positive frequency spectrum of the sum of two unit-amplitude cosine functions with a frequency separation of $\delta F = 500$ Hz, one at 1000 Hz and one at 1500 Hz, with both having zero initial phase. The observation time T is 10 ms so that the Rayleigh width of each mainlobe is $\Delta F = 1/T = 100$ Hz. The two vertical dotted lines mark the two cosine frequencies. This signal could represent the Doppler spectrum of two moving targets with the same echo strength but different radial velocities.

These two signal components are considered well-resolved: there are two distinct, well-separated peaks in the spectrum. The frequency of each peak is perturbed slightly from the actual frequency by the sidelobes of the other sinusoid. Figure 1.17b to d repeats this measurement with the frequency spacing reduced to 100, 75, and 50 Hz. At $\delta F = 100$ Hz the two

---

[10]The effects of unequal signal strength and noise on resolution are considered in Mir and Wilkinson (2008).

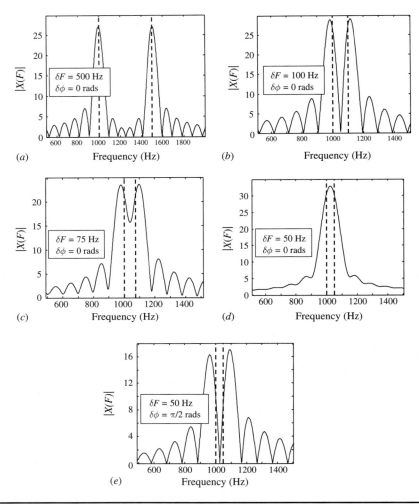

FIGURE 1.17 Resolution of two sinusoids in frequency, each having a Rayleigh width of 100 Hz and a 0° initial phase. (*a*) Well resolved at 500 Hz spacing. (*b*) Well resolved at 100 Hz spacing. (*c*) Marginally resolved at 75 Hz spacing. (*d*) Unresolved at 50 Hz spacing. (*e*) Well resolved at 50 Hz spacing and a phase difference of 90°.

spectral peaks are still well resolved, though with more perturbation of the apparent frequencies, but as the separation drops below the Rayleigh width ($\delta F < \Delta F$) to 75 and then to 50 Hz, they start to blur into a single spectral peak. At $\delta F = 50$ Hz they are no longer resolved; the spectrum measurement does not show two separate signals. At $\delta F = 75$ Hz they are marginally resolved, although a little noise added to the data would make that a precarious claim. It appears that a separation of about the Rayleigh width or greater is needed for clear resolution of the two frequencies. Although resolution is defined in terms of pairs of scatterers, this demonstration suggests that the mainlobe width of the signature of a single isolated target is the major determinant of the system's resolution.

Introduction to Radar Systems and Signal Processing

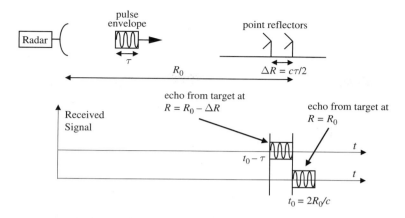

FIGURE 1.18 Geometry for describing conventional pulse range resolution. See text for explanation.

Figure 1.17*e* shows that phase also has an effect. In part *e* the sinusoid frequencies are the same as in part *d* but the initial phases differ by $\pi/2$ radians, whereas in part (*d*) the initial phases were the same. Although their locations are perturbed significantly and their amplitudes are imbalanced, nonetheless there are two clearly separated peaks even though the frequencies are separated by less than the Rayleigh width. As will be seen, echo phases are extremely sensitive to range and therefore effectively random in many cases. Consistent resolution of two signals independent of their relative phasing requires that they be separated by at least the Rayleigh width of the single-signal signature.

The resolution of a radar in turn determines the size of a resolution cell. A resolution cell in range, velocity, or angle is the interval in that dimension that contributes to the echo received by the radar at any one instant. Figure 1.18 illustrates resolution and the resolution interval in the range dimension for a simple constant-frequency pulse. If a pulse whose leading edge is transmitted at time $t = 0$ has duration τ seconds, then at time t_0 the echo of the leading edge of the pulse will be received from a scatterer at range $ct_0/2$. At the same time, echoes of the trailing edge of the pulse from a scatterer at range $c(t_0 - \tau)/2$ are also received. Any scatterers at intermediate ranges would also contribute to the measured voltage at time t_0. Thus, scatterers distributed over an interval of $c\tau/2$ in range contribute simultaneously to the received voltage. In order to resolve the contributions from two scatterers into different time samples, they must be spaced by more than $c\tau/2$ meters so that their individual echoes do not overlap in time. The quantity $c\tau/2$ is called the *range resolution* ΔR. Similarly, two- and three-dimensional resolution cells can be defined by considering the simultaneous resolution in range, azimuth angle, and elevation angle.

This description of range resolution applies only to unmodulated, constant frequency pulses. As will be seen in Chap. 4, both pulsed and CW waveforms, when combined with matched filtering, achieve range resolution determined by their bandwidth as in Eq. (1.2), namely $\Delta R = c/2\beta$ meters.

Angular resolution in the azimuth and elevation dimensions is determined by the antenna beamwidths in those planes. Two scatterers at the same range but different azimuth or elevation angles will contribute simultaneously to the received signal if they are within the antenna mainlobe and thus are both illuminated at the same time. For the purpose of estimating angular resolution, the mainlobe width is typically taken to be the one-way 3-dB beamwidth

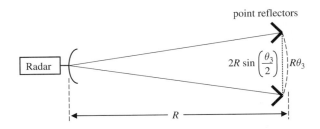

FIGURE 1.19 The angular and cross-range resolutions are determined by the 3-dB antenna beamwidth θ_3.

θ_3 of the antenna. Thus, the two point scatterers in Fig. 1.19 located at the 3-dB edges of the beam define the angular resolution of the radar. The figure illustrates the relation between the angular resolution in radians and the equivalent resolution in units of distance, which will be called the *cross-range resolution* ΔCR to denote resolution in a dimension orthogonal to range. The arc length between the two scatterers, indicated by the curved dashed line, is $R\theta_3$ meters. The cross-range resolution is the length of the straight dotted line in Fig. 1.19 and is given by

$$\Delta CR = 2R\sin\left(\frac{\theta_3}{2}\right) \approx R\theta_3 \text{ m} \tag{1.28}$$

The approximation error is less than 1 percent when the 3-dB beamwidth is less than about 14°, which is usually the case for pencil beam antennas.

Three details bear mentioning. First, the literature frequently fails to specify whether one- or two-way 3-dB beamwidth is required or given. The one-way pattern appears to be most commonly used for quantifying angular resolution in monostatic radar. Second, note that cross-range resolution increases (degrades) linearly with range, whereas range resolution was independent of range. Finally, as with range resolution, it will be seen later (Chap. 8) that signal processing techniques can be used to improve angular or cross-range resolution well beyond the conventional $R\theta_3$ limit and to make it independent of range as well.

The radar resolution cell volume ΔV is approximately the product of the total solid angle subtended by the 3-dB antenna mainlobe, converted to units of area, and the range resolution. For an antenna having an elliptical beam with azimuth and elevation beamwidths θ_3 and ϕ_3, this is

$$\Delta V = \pi\left(\frac{R\theta_3}{2}\right)\left(\frac{R\phi_3}{2}\right)\Delta R = \frac{\pi}{4}R^2\theta_3\phi_3\Delta R \approx R^2\theta_3\phi_3\Delta R \text{ m}^3 \tag{1.29}$$

The approximation in the second line of Eq. (1.29) is 27 percent larger than the expression in the first line but is widely used. Note that resolution cell volume increases with the square of range because of the two-dimensional angular spreading of the beam at longer ranges.

The region of support of a function is the portion of the independent variable axis (typically time, frequency, angle, or spatial coordinate in radar) over which the function is nonzero. Thus, a 100-μs pulse has a larger ROS in time than a 10-μs pulse. As seen above, antenna beamwidths are inversely proportional to the aperture length D and the mainlobe of the frequency spectrum of a sinusoid is inversely proportional to the length of the sinusoid in time. This reciprocal behavior of the signal domain and Fourier domain regions of support is illustrated in Fig. 1.20 and is sometime referred to as "reciprocal spreading." Part (*a*) shows a sinusoidal pulse with a frequency of 10 MHz and a duration of 1 μs and its Fourier transform, which is a

Introduction to Radar Systems and Signal Processing 27

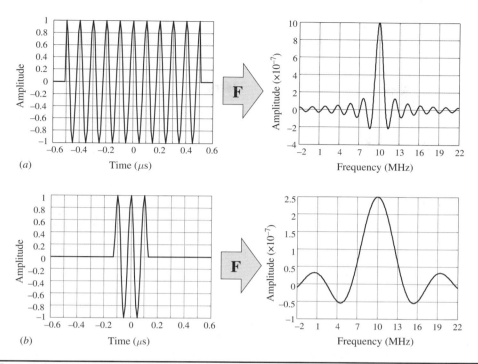

FIGURE 1.20 Illustration of reciprocal spreading property of Fourier transforms. (*a*) A sinusoidal pulse and the main portion of its Fourier transform. (*b*) A narrower pulse has a wider transform. See text for details.

sinc function centered on 10 MHz having a Rayleigh mainlobe width of 1 MHz, the reciprocal of the 1 μs pulse duration. In part (*b*) the pulse has the same frequency but only one-quarter the duration. Its spectrum is still a sinc centered at 10 MHz, but the Rayleigh width is now four times larger at 4 MHz. The spectrum amplitude is also reduced by a factor of 4.

Improving resolution in one domain requires increasing the size of the ROS in the complementary Fourier domain. For example, improving range resolution for simple pulses requires using shorter pulses, as was seen in Sec. 1.4.2; and Fig. 1.18 shows that a shorter pulse implies a larger ROS in frequency (wider spectrum), that is, more bandwidth. Conversely, improving resolution in the frequency domain requires a narrower spectrum mainlobe and thus a longer observation (larger ROS) in the time domain. This behavior holds for any two functions related by a Fourier transform: finer resolution in one domain requires a wider ROS in the complementary domain.

Radar designers have developed techniques for increasing the appropriate ROS to obtain improved resolution in various dimensions. For example, the increased waveform bandwidth required to improve range resolution has led to the use of wideband phase- and frequency-modulated waveforms in place of the simple pulse (Chap. 4). Improving cross-range resolution in radar imagery requires viewing a scene over a wide angular interval to increase cross-range spatial frequency bandwidth and leads to the *synthetic aperture* techniques of Chap. 8. Improving velocity (equivalently, Doppler frequency) resolution requires a longer time observation and is accomplished with multipulse or multisweep waveforms. Because the antenna far field pattern is the Fourier transform of the aperture current distribution, improved angular resolution requires larger apertures, that is, bigger antennas.

1.4.3 Integration and Phase History Modeling

A fundamental operation in radar signal processing is *integration* of samples to improve the SIR. Both *coherent integration* and *noncoherent integration* are of interest. The former refers to integration of complex (magnitude and phase) data, while the latter refers to integration based only on the magnitude (or possibly the squared or log magnitude) of the data.

Suppose a signal is transmitted, reflects off a target, and at the appropriate time the receiver output voltage signal is measured, yielding a complex echo amplitude $A\exp(j\varphi)$ corrupted by additive noise w. The noise is assumed to be a sample of a random process with power σ_w^2. The single-pulse SNR is

$$\chi_1 = \frac{\text{signal power}}{\text{noise power}} = \frac{A^2}{\sigma_w^2} \qquad (1.30)$$

Now suppose the measurement is repeated $N-1$ more times. One expects to measure the same deterministic echo response, but with an independent noise sample each time. Form a single measurement z by integrating (summing) the individual measurements. This complex sum operation, retaining the phase information, is a coherent integration:

$$\begin{aligned} z &= \sum_{n=0}^{N-1} \{A\exp(j\varphi) + w[n]\} \\ &= NA\exp(j\varphi) + \sum_{n=0}^{N-1} w[n] \end{aligned} \qquad (1.31)$$

The power in the integrated signal component is its amplitude squared, $(N\cdot A)^2$. Provided the noise samples $w[n]$ are independent of one another and zero mean, the power in the noise component is easily shown to be the sum of the power in the individual noise samples. Further assuming each has the same power σ_w^2, the total noise power is now $N\sigma_w^2$. The integrated SNR becomes

$$\chi_N = \frac{N^2 A^2}{N\sigma_w^2} = N\left(\frac{A^2}{\sigma_w^2}\right) = N\chi_1 \qquad (1.32)$$

Coherently integrating N measurements has improved the SNR by a factor of N compared to that of one individual measurement. This increase is called the *integration gain*. Later chapters show that, as one would expect, increasing the SNR improves detection and parameter estimation performance. The cost is the extra time, energy, and computation required to collect and combine the N pulses of data.

In coherent integration, the complex-valued (amplitude and phase) data samples are added. The hope is that the signal components will add in phase, that is, constructively. This is often described as adding on a *voltage* basis, since the amplitude (voltage) of the integrated signal component increased by a factor of N, with the result that signal power increased by N^2. The noise samples, whose phases varied randomly, will add on a *power* basis. It is the alignment of the signal component phases that allowed the signal power to grow faster than the noise power and create an integration gain.

Sometimes the data must be preprocessed to ensure that the signal component phases align so that the maximum coherent integration gain can be achieved. If the target had been moving in the previous example, the signal component of the measurements would have exhibited a Doppler shift and Eq. (1.31) would instead become

$$z = \sum_{n=0}^{N-1} A\exp[j(2\pi f_D n + \varphi)] + w[n] \qquad (1.33)$$

for some value of normalized Doppler frequency f_D. The signal power in this case will depend on the particular Doppler shift, but except in very fortunate cases will be less than N^2A^2. However, if the Doppler shift is known or can be estimated, the phase progression of the signal component can be compensated before summing by pre-multiplying the data by a countervailing phase progression:

$$\begin{aligned} z' = \exp(-j2\pi f_D n) \cdot z &= \sum_{n=0}^{N-1} \exp(-j2\pi f_D n)\{A\exp[j(2\pi f_D n + \varphi)] + w[n]\} \\ &= NA\exp(j\varphi) + \sum_{n=0}^{N-1} \exp(-j2\pi f_D n)w[n] \end{aligned} \quad (1.34)$$

The phase compensation aligns the signal component phases so that they add constructively. The noise phases remain random with respect to one another. The integrated signal power is again N^2A^2, the integrated noise power is again $N\sigma_w^2$, and therefore an integration gain of N is again achieved. Compensation for the phase progression so that the compensated target samples add in phase is an example of using *phase history modeling*: if the sample-to-sample pattern of target echo phases can be predicted or estimated (at least to within a constant overall phase), the data can be modified with a countervailing phase so that the full coherent integration gain is achieved. Phase history modeling is central to many radar signal processing functions and is essential for achieving adequate gains in SNR.

Coherent integration carries a risk. If the target echo phases are not constant or are not compensated successfully, they will not add in phase and the integration gain will be reduced. Worse, they may even combine destructively, actually reducing the integrated SNR relative to the single-measurement SNR. This risk can be avoided by considering noncoherent integration, in which the phases are discarded and the magnitudes, magnitudes-squared, or log magnitudes of the measured data samples are added. If the magnitude-squared is chosen, then z is formed as

$$\begin{aligned} z &= \sum_{n=0}^{N-1} |A\exp(j\varphi) + w[n]|^2 = \sum_{n=0}^{N-1} \{A\exp(j\varphi) + w[n]\}\{A\exp(j\varphi) + w[n]\}^* \\ &= \sum_{n=0}^{N-1} |A\exp(j\varphi)|^2 + \sum_{n=0}^{N-1} |w[n]|^2 + \sum_{n=0}^{N-1} 2\operatorname{Re}\{A\exp(j\varphi)w^*[n]\} \\ &= NA^2 + \sum_{n=0}^{N-1} |w[n]|^2 + \sum_{n=0}^{N-1} 2\operatorname{Re}\{A\exp(j\varphi)w^*[n]\} \end{aligned} \quad (1.35)$$

The important fact is that phase information in the received signal samples is discarded so the result is not sensitive to phase alignment.

The first line of Eq. (1.35) defines noncoherent square-law integration. The next two lines show that, because of the nonlinear magnitude-squared operation, z cannot be expressed as the sum of a signal-only part and a noise-only part due to the presence of the third term involving cross products between signal and noise components. A similar situation exists if the magnitude or log magnitude is chosen for the noncoherent integration. Consequently, a noncoherent integration gain cannot be defined as simply as it was for the coherent case.

However, it is possible to define a noncoherent gain indirectly. For example, in Chap. 6 it will be seen that detection of a constant-amplitude target signal in complex Gaussian noise with a probability of detection of 0.9 and a probability of false alarm of 10^{-8} requires a single-sample SNR of 14.2 dB (about 26.3 on a linear scale). The same probabilities can be obtained by integrating the magnitude of 10 samples each having an individual SNR of only 5.8 dB (3.8 on a linear scale). The reduction of 8.4 dB (a factor of $26.3/3.8 = 6.9$) in the required single-sample SNR when 10 samples are noncoherently integrated is the implied noncoherent integration gain. The coherent integration gain for this example would be a factor of $N = 10$.

Computing the noncoherent integration gain typically requires derivation of the probability density functions of the noise-only and signal-plus-noise cases. Chapter 6 will show that in many useful cases the noncoherent integration gain is approximately N^α, where α ranges from about 0.7 or 0.8 for small N to about 0.5 ($\sqrt{N}$) for large N, rather than in direct proportion to N ($\alpha = 1$). Thus, noncoherent integration is less efficient than coherent integration. This should not be surprising, since not all the signal information is used.

1.5 A Preview of Basic Radar Signal Processing

There are a number of instances where the design of a component early in the radar signal processing chain is driven by properties of some later component. For example, in Chap. 4 it will be seen that the matched filter maximizes SNR; but it is not until the performance curves for the detectors that follow the matched filter are derived that it will be verified that maximizing SNR also optimizes detection performance. Until the detector is considered, it is hard to see precisely how performance depends on SNR. Having seen the major components of a typical coherent radar system, the most common signal processing operations in the radar signal processing chain are now described heuristically. Sketching out this preview of the "big picture" from beginning to end may make it easier to understand the motivation for and interrelation of many of the processing operations to be described in later chapters.

Figure 1.21 illustrates one possible sequence of operations in a generic radar signal processor. Each major category of processing has a few more specific representative operations listed adjacent. The left branch is for search and track radars intended mainly to detect and locate discrete targets and track them over time. The right branch is for imaging radars. Many other types of radars, for example, weather radars, follow flows with many similarities to the search and track column. The sequences shown and their ordering are not unique or exhaustive. In addition, the point in the chain at which the signal is digitized varies in different systems; it might occur as late as the output of the clutter filtering step, though the trend in modern systems is to digitize closer and closer to the antenna. Radar signal *phenomenology* must also be considered.

1.5.1 Radar Time Scales

Radar signal processing operations take place on time scales ranging from less than a nanosecond to tens of seconds or longer, a range of 10 orders of magnitude or more. Different classes or levels of operations tend to operate on significantly different time scales. Figure 1.22 illustrates one possible association of operations and time scale.

Operations that are applied to data from a single pulse or FMCW sweep occur on the shortest time scale, often referred to as *fast time* because the PRI or FMCW sweep rate, which limits the time available for processing before data from the next pulse or sweep arrives, is relatively short. The PRI or sweep rate is typically on the order of tens of microseconds to tens of milliseconds. Furthermore, the sample rate of the fast time data, determined by the instantaneous pulse bandwidth or FMCW sweep bandwidth (see Chap. 2), is on the order of hundreds of kilohertz to as much as a few gigahertz in some cases. Corresponding sampling intervals range from a few microseconds down to a fraction of a nanosecond. Typical fast time operations are digital I/Q signal formation, beamforming, matched filtering, sensitivity time control, and some forms of jammer suppression.

The next level up in signal processing operations acts on data from multiple pulses or sweeps. The available processing time is simply the number of pulses or sweeps M times the interval between them and is much longer than the single-interval time scale of the

Introduction to Radar Systems and Signal Processing

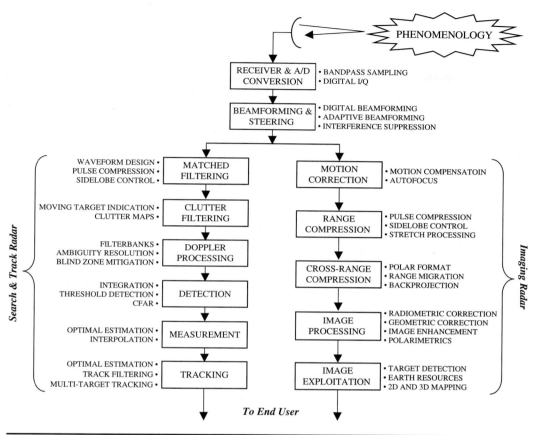

FIGURE 1.21 One example of a generic radar signal processor flow of operations.

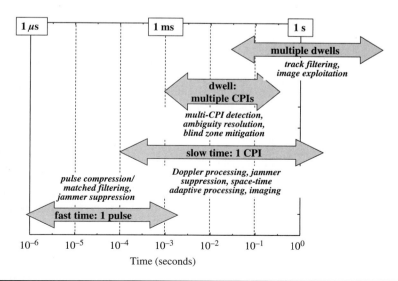

FIGURE 1.22 The range of time scales over which radar signal processing is performed.

fast time operations. Typical values of M range from single digits to a few tens of pulses or sweeps. Combined with the typical interval values mentioned above, the time scale for most operations at this level is hundreds of microseconds to tens or hundreds of milliseconds.

These multipulse or multisweep operations are said to act in *slow time*. Note that the repetition interval becomes the sampling rate of the slow time data and is much lower than the fast-time sampling rate. Typical slow-time operations include coherent and noncoherent integration, Doppler processing of all types, and space-time adaptive processing. Synthetic aperture imaging is also a slow-time operation; however, the number of pulses involved in an operation may be hundreds to many thousands, so that the total time interval can be hundreds of milliseconds to many seconds. The idea of slow and fast time will be revisited in the discussion of the data organizational concept of the datacube in Chap. 3.

A group of pulses or sweeps that are to be somehow combined coherently, for example, via Doppler processing or synthetic aperture radar (SAR) imaging, are said to form a *coherent processing interval* (CPI). A still higher level of radar processing acts on data from multiple CPIs and therefore operates on an even longer time scale often called a *dwell* and typically lasting milliseconds to ones or tens of seconds. Operations on this scale include multiple-CPI detection, range and Doppler ambiguity resolution, blind zone mitigation, and multilook SAR imaging.

So long as a target stays within a radar's detection range, the radar track can be continued. Some radars may track detected targets for many seconds or minutes using data from multiple dwells. Track filtering operates in this open-ended regime. Some imaging radars, especially in earth resources and strategic intelligence applications, may monitor an area over days, months, or even years.

1.5.2 Phenomenology

To design a successful signal processor, one must understand how the target or scene information of interest is encoded in those signals. *Phenomenology* refers to the characteristics of the signals received by the radar. Relevant characteristics include signal power, frequency, phase, polarization, or angle of arrival; variation in time and spatial location; and randomness. The received signal phenomenology is determined by both intrinsic features of the physical object(s) giving rise to the radar echo, such as their physical size or their orientation and velocity relative to the radar, and the characteristics of the radar itself, such as its transmitted waveform, polarization, or antenna gain. For example, because of the details of its shape, a target may reflect more power to the radar when it is illuminated from the front than from the side.

Models of the behavior of typical measured signals that are relevant to the design of signal processors will be developed in Chap. 2. The radar range equation will give a means of predicting received signal power. The Doppler phenomenon will predict received frequency. It will be seen that the complexity of the real world gives rise to very complex variations in the amplitude of radar echoes; this will lead to the use of random processes to model the signals and to particular probability density functions that match measured behavior well. A (very) brief overview of the behavior of the variation of ground and sea echo with sensing geometry and radar characteristics will be given. It will also be shown that measured signals can be represented as the convolution of an idealized arbitrarily fine resolution signal with the radar waveform (in the range dimension) or its antenna pattern (in the azimuth or elevation dimension, both also called the cross-range dimension). Thus, a combination of random process and linear systems theory will be used to describe radar signals and to design and analyze radar signal processors.

1.5.3 Signal Conditioning and Interference Suppression

The first several blocks after the antenna in the search and track radar column of Fig. 1.21 can be considered as signal conditioning operations whose purpose is to improve the SIR of the data prior to detection, parameter measurement, or tracking operations. That is, the intent of these blocks is to "clean up" the radar data as much as possible. This is done in general with a combination of fixed and adaptive *beamforming, matched filtering, integration, clutter filtering,* and *Doppler processing.*

Beamforming as a signal processing operation is applicable when the radar antenna is an array, that is, when there are multiple phase center signals, or *channels*, available to the signal processor. Fixed beamforming is the process of combining the outputs of the various available phase centers to form a directive gain pattern similar to that shown in Fig. 1.7. The high gain mainlobe and low sidelobes selectively enhance the echo strength from scatterers in the antenna look direction while suppressing the echoes from scatterers in other directions, typically clutter. The sidelobes also provide a measure of suppression of jamming signals so long as the angle of arrival of the jammer is not within the mainlobe of the antenna. By proper choice of the sample weights used to combine the channels, the mainlobe of the beam can be steered to various look directions and the tradeoff between the sidelobe level and the mainlobe width (angular resolution) can be varied.

Adaptive beamforming takes this idea a step further. By examining the correlation properties of the received data across channels, it is possible to recognize the presence of jamming and clutter entering the antenna pattern sidelobes. One can then design a set of weights for combining the channels such that the antenna not only has a high-gain steerable mainlobe and generally low sidelobes but also has a null in the antenna pattern at the angle of arrival of the jammer. Much greater jammer suppression can be obtained in this way. Similarly, it is also possible to increase clutter suppression by this technique. *Space-time adaptive processing* (STAP) combines adaptive beamforming in both angle and Doppler for simultaneous suppression of clutter and jammer interference. Figure 1.23 illustrates interference suppression using STAP, allowing a previously invisible target signal to be seen and perhaps detected. The figure shows the distribution of received energy in an airborne sidelooking radar in angle of arrival and Doppler frequency coordinates at a particular range. The two vertical bands in Fig. 1.23*a* represent jammer energy, which comes from a fixed angle of arrival but is usually in the form of relatively wideband noise; thus it is spread across all Doppler frequencies observed by the radar. The diagonal band in Fig. 1.23*a* is due to ground clutter, for

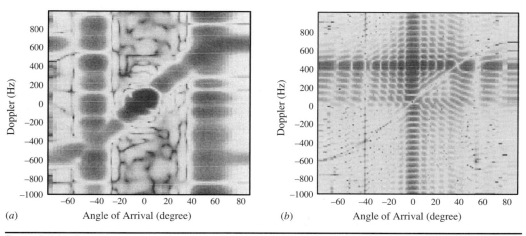

FIGURE 1.23 Example of effect of adaptive beamforming. (*a*) Map of received signal power as a function of angle of arrival and Doppler shift. (*b*) Angle-Doppler map after adaptive processing. A target is now visible at 400 Hz and 0°; the horizontal and vertical bars are the sidelobes of the target response. (Images courtesy of Dr. W. L. Melvin. Used with permission.)

which the Doppler shift depends on the angle from the radar to the ground patch contributing energy. Figure 1.23*b* shows that the adaptive filtering has created nulls along the loci of the jammer and clutter energy, allowing the target peak at 0° angle of arrival and 400 Hz Doppler shift to be seen. (The vertical and horizontal bars are sidelobes of the target signature.) Adaptive interference suppression will be introduced in Chap. 9.

Many radar system designs strive for both high sensitivity in detecting targets and fine range resolution (the ability to distinguish closely spaced targets). Upcoming chapters show that target detectability improves as the transmitted energy increases, and that range resolution improves as the transmitted waveform's instantaneous bandwidth increases. If a pulsed radar employs a simple, constant-frequency rectangular envelope pulse as its transmitted waveform, the pulse must be lengthened to increase the transmitted energy for a given power level. However, lengthening the pulse also decreases its instantaneous bandwidth, degrading the range resolution. Sensitivity and fine range resolution appear to be conflicting goals.

Pulse compression provides a way out of this dilemma by decoupling the waveform bandwidth from its duration, allowing both to be independently specified. This is done by abandoning the constant-frequency pulse and instead designing a modulated waveform. A very common choice is the linear frequency modulated (linear FM, LFM, or "chirp") waveform. The instantaneous frequency of an LFM pulse is swept linearly over the desired bandwidth during the pulse duration. The frequency may be swept either up or down, but the rate of frequency change is constant. Figure 1.24*a* shows the real part of a complex LFM chirp that sweeps over 20 MHz in 1 μs (black curve); the imaginary part is similar except for a phase shift (gray curve).

The matched filter is by definition a filter in the radar receiver designed to maximize the SNR at its output. Chapter 4 will show that the impulse response of the filter having this property turns out to be a replica of the transmitted waveform's modulation function that has been reversed in time and conjugated; thus the impulse response is "matched" to the particular transmitted waveform modulation. In pulsed radar, pulse compression is the process of designing a modulated waveform and its corresponding matched filter so that the matched filter output in response to the echo from a single point scatterer concentrates most of its energy in a very short duration, thus providing good range resolution while still allowing the high transmitted energy of a long pulse. Figure 1.24*b* shows the output of the matched filter corresponding to the LFM pulse of Fig. 1.24*a*. The mainlobe of the response, measured by its Rayleigh width, is only 1/20th the duration of the original pulse. The concepts of matched filtering, pulse compression, and waveform design, as well as the properties of linear FM and other common waveforms, are described in Chap. 4. There it is seen that the 3-dB width of the mainlobe in time is approximately $1/\beta$ seconds, where β is the instantaneous bandwidth of the waveform used. This width determines the ability of the waveform to resolve targets in range. Converted to equivalent range units, the range resolution is given by $\Delta R = c/2\beta$, as noted earlier in Eq. (1.2).

Clutter filtering and Doppler processing are closely related. Both are techniques for improving the detectability of moving targets by suppressing interference from clutter echoes, usually from the terrain in the antenna field of view, based on differences in the Doppler shift of the echoes from the clutter and from the targets of interest. The techniques differ primarily in whether they are implemented in the time or frequency domain and in historical usage of the terminology.

Clutter filtering usually takes the form of *moving target indication* (MTI), which is simply pulse-to-pulse or FMCW sweep-to-sweep highpass filtering of the radar echoes at a given range to suppress constant components, which are assumed to be due to non-moving clutter. Extremely simple, very low-order digital filters are applied in the time domain to samples taken at a fixed range but on successive transmitted pulses.

The term "Doppler processing" generally implies the use of the Fourier transform, or occasionally some other spectral estimation technique, to explicitly compute the spectrum of

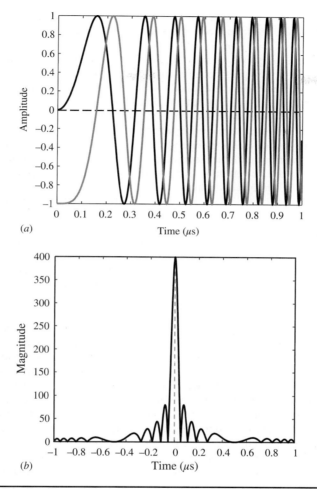

Figure 1.24 (a) Complex linear FM waveform modulation function. The instantaneous frequency sweeps from 0 Hz to 20 MHz at a constant rate over a 1-μs interval. The black curve is the real part, the gray curve is the imaginary part. (b) Magnitude of the output of the matched filter for the LFM waveform of (a).

the echo data for a fixed range across multiple pulses or FMCW sweeps. Due to their different Doppler shifts, energy from moving targets concentrates in different parts of the spectrum from the clutter energy, allowing separation and detection of the targets. Doppler processing obtains more information from the radar signals than does MTI filtering, such as the number and approximate velocity of moving targets. The cost is more required radar pulses or sweeps, consuming more energy and timeline, and greater processing complexity. Many systems use both techniques in series. Clutter filtering and Doppler processing are the subjects of Chap. 5.

1.5.4 Detection

The most basic function of a radar signal processor is detection of the presence of one or more targets of interest. Information about the presence of targets is contained in the echoes of the radar pulses. These echoes compete with receiver noise, undesired echoes from clutter signals, and possibly intentional jamming or unintentional electromagnetic interference

(EMI), for example from other radars or cellular phone services. The signal processor must somehow analyze the total received signal and determine whether it contains a target echo and, if so, at what range, angle, and velocity.

Because the complexity of radar signals will lead to the use of statistical models, detection of target echoes in the presence of competing interference signals is a problem in statistical decision theory. The theory as applied to radar detection will be developed in Chap. 6. There it will be seen that in most cases optimal performance can be obtained using *threshold detection*. In this method, the magnitude of each complex sample of the radar echo signal, after any signal conditioning and interference suppression, is compared to a precomputed threshold. If the signal amplitude is below the threshold it is assumed to be due to interference signals only. If it is above the threshold it is assumed due to the presence of a target echo in addition to the interference, and a detection or "hit" is declared. In essence, the detector decides whether the energy in each received signal sample is too large to likely have resulted from interference alone; if so, it is assumed a target echo contributed to that sample. Figure 1.25 illustrates the concept. The "clutter + target" signal might represent the variation in received signal strength versus range (fast time) for a single transmitted pulse or CW sweep. It crosses the threshold at three different times, suggesting the presence of three targets at different ranges.

Because they are the result of a statistical process, threshold detection decisions have a non-zero probability of being wrong. For example, a noise spike could cross the threshold, leading to a false target declaration, commonly called a *false alarm*. These errors are minimized if the target spikes stand out strongly from the background interference, that is, if the SIR is as large as possible. If this is the case the threshold can be set relatively high, resulting in few false alarms while still detecting most targets. This fact also accounts for the importance of matched filtering in radar systems. It will be seen in Chap. 4 that the matched filter maximizes the SIR, thus providing the best threshold detection performance. Furthermore, the achievable SIR increases monotonically with the transmitted signal energy E, thus encouraging the use of longer pulses or FMCW sweeps to get more energy on the target. Since longer pulses or sweeps degrade range resolution, the frequency modulation will also be important so that fine resolution can be obtained while maintaining good detection performance.

The concept of threshold detection can be applied to many different radar signal processing systems. Figure 1.25 illustrates its application to a fast-time (range) signal trace, but it can be equally well applied to a signal composed of measurements at different Doppler frequencies for a fixed range, or in a two-dimensional form to combined range-Doppler data or to SAR imagery.

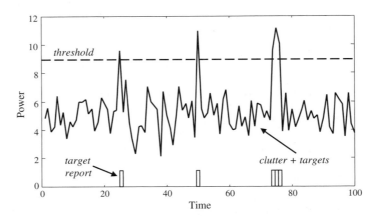

FIGURE 1.25 Illustration of threshold detection.

There are numerous significant details in implementing threshold detection. Various detector designs work on the magnitude, squared magnitude, or even log magnitude of the complex signal samples. The threshold is computed from knowledge of the interference statistics so as to limit false alarms to a selected acceptable rate. However, in many real systems the interference statistics are rarely known accurately enough to allow for precomputing a fixed threshold. Instead, the required threshold is set using interference statistics estimated from the data itself, a process called *constant false alarm rate* (CFAR) detection. Detection processing is described in detail in Chap. 6.

1.5.5 Measurements and Track Filtering

Radar systems employ a wide variety of processing operations after the point of target detection. One of the most common post-detection processing steps is *tracking* of targets, an essential component of many radar systems. Tracking comprises multiple *measurements* of the position of detected targets followed by *track filtering*.

The radar signal processor detects the presence of targets using signal conditioning and threshold detection methods. The range, angle, and Doppler resolution cell in which a target is detected provide a coarse estimate of its location in those coordinates. Once detected, the radar will seek to refine the estimated position by using signal processing methods to more precisely estimate the time delay after signal transmission at which the threshold crossing occurred, the angle of the target relative to the antenna mainbeam direction, and its radial velocity. Individual measurements will always still have some error due to interference, and so provide a noisy snapshot of the target location and motion at one instant in time.

The term "track filtering" describes a higher-level, longer time scale process of combining a series of such measurements to estimate a complete trajectory of the target over time. It is often categorized as data processing rather than signal processing. Because there may be multiple targets with crossing or closely spaced trajectories, track filtering must deal with the problems of determining which new measurements to associate with which targets already being tracked, and with correctly resolving nearby and crossing trajectories. A variety of optimal estimation techniques have been developed to perform track filtering. An excellent reference in this area is Bar-Shalom (1988).

Figure 1.26 illustrates a series of noisy measurements in one dimension of the position of two targets and the filtering of that noisy trajectory using an extremely simple alpha-beta

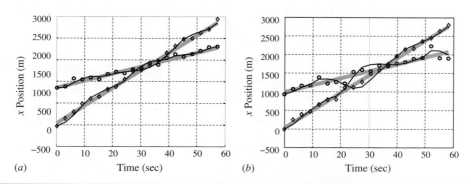

FIGURE 1.26 Track filtering of noisy measurements for two targets in one dimension using an alpha-beta filter. Markers show individual measurements. Gray lines are the actual position, black lines are the filtered position estimates. (*a*) Tracks follow the current targets in low measurement noise even when the trajectories cross. (*b*) Tracks incorrectly switch targets in higher measurement noise.

filter, to be discussed in Chap. 7. The actual position in the *x* dimension versus time for each target (called the *ground truth*) is shown by the gray lines. The two targets are moving at different velocities along the *x* axis and one passes the other at around time 33 seconds. The circle and diamond markers indicate the noisy radar measurements of position for each. The solid black lines are the smoothed estimates of position produced by the alpha-beta filter. In part (*a*) of the figure, the filter correctly associates the measurements with each target when they cross, so that each smoothed estimate follows the same target over the observation time. In Fig. 1.26*b* the noise variance is higher, causing the filter to make multiple association errors in the 30 to 40 second time interval. From time 42 seconds onward the tracks are swapped, each following a different target from when they started. This represents an error in measurement-to-track *data association*. A variety of techniques attempt to address association problems; a few basic ones are discussed in Chap. 7.

1.5.6 Imaging

Most people are familiar with the idea of a radar producing "blips" on a screen to represent targets, and in fact systems designed to detect and track moving targets may do exactly that. However, radars can also be designed to compute fine-resolution images of a scene. Figure 1.27 compares the quality routinely obtainable in SAR imagery in the mid-1990s to that of an aerial color photograph of the same scene (here shown in black and white); close examination reveals many similarities and many significant differences in the appearance of the scene at radar and visible wavelengths. Not surprisingly, the photograph is easier for a human to interpret and analyze, since the imaging wavelengths (visible light) and phenomenology are the same as those observed by the human visual system. In contrast, the radar image, while remarkable, is monochromatic, offers less detail, and exhibits a "speckled" texture, some contrast reversals, and some missing features such as the runway stripes. Given these drawbacks, why is radar imaging of interest?

While radars do not obtain the resolution or image quality of photographic systems, they have two powerful advantages. First, they can image a scene through clouds and inclement weather due to the superior propagation of RF wavelengths. Second, they can

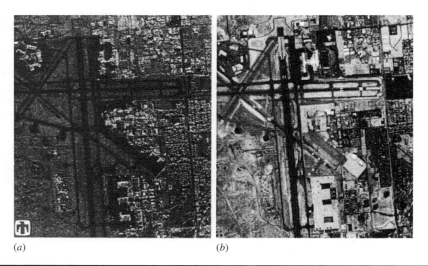

FIGURE 1.27 Comparison of optical and SAR images of the Albuquerque airport. (*a*) K_u band (15 GHz) SAR image, 3-m resolution. (*b*) Aerial photograph. (Images courtesy of Sandia National Laboratories. Used with permission.)

image equally well 24 hours a day since they do not rely on the sun or ground sources for illumination; they provide their own "light" via the transmitted signal. If the example of Fig. 1.27 were repeated in the middle of a rainy night, the SAR image would not be affected in any noticeable way, but the optical image would disappear entirely.

To obtain fine-resolution imagery, radars use a combination of high-bandwidth waveforms to obtain good resolution in the range dimension and the synthetic aperture radar technique to obtain good resolution in the cross-range dimension. The desired range resolution is obtained while maintaining adequate signal energy by using modulated waveforms, usually linear FM. A waveform that is swept over a large enough bandwidth β and processed using a matched filter can provide very good range resolution in accordance with Eq. (1.2). For example, range resolution of 1 m can be obtained with a waveform swept over 150 MHz. Depending on their applications, modern imaging radars usually have range resolution of 30 m or better; many systems have 10 m or better resolution, and some advanced systems have resolution under 1 m.

For a conventional non-imaging radar, referred to as a *real aperture* radar, the resolution in cross-range is determined by the width of the antenna beam at the range of interest and is given by $R\theta_3$, as was shown in Eq. (1.28). Realistic antenna beamwidths for narrow-beam antennas are typically on the order of 0.5° (8.7 mrad) for a low earth orbit satellite at a range of 700 km, or 3° (52 mrad) for an airborne system at 10 km range. The cross-range resolutions that result would be 520 m for the airborne system and about 6 km for the satellite, orders of magnitude worse than typical range resolutions and far too coarse to produce useful imagery. This poor cross-range resolution is overcome by using SAR techniques.

The synthetic aperture technique refers to the concept of synthesizing the effect of a very large antenna by having the actual physical radar antenna move in relation to the area being imaged. Thus, SAR is associated with moving airborne or space-based radars rather than with fixed ground-based radars. Figure 1.28 illustrates the concept. By transmitting pulses or

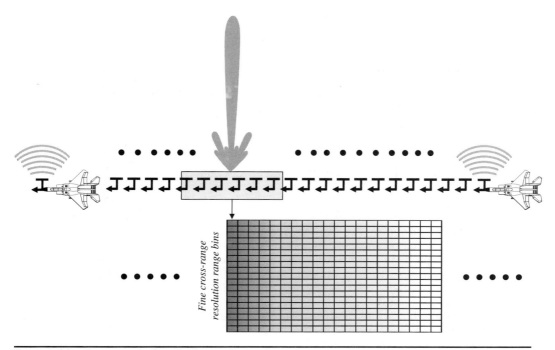

Figure 1.28 The concept of synthetic aperture radar.

FMCW sweeps at each indicated location, collecting the fast time (range) data, and properly processing it together, a SAR system creates the effect of a much larger phased array antenna being flown along the aircraft flight path. These "synthetic arrays" may be hundreds of meters to a few kilometers long. As suggested by Eq. (1.10) (though some details differ in the SAR case), a very large aperture size produces a very narrowly focused effective antenna beam, making possible very fine cross-range resolution. The SAR concept is explained more fully in Chap. 8.

1.6 Radar Literature

This text covers a middle ground in radar technology. It focuses on basic radar signal processing from a digital signal processing point of view. It does not address radar systems, components, or phenomenology in any great depth except where needed to explain the signal processing aspects; nor does it provide in-depth coverage of advanced radar signal processing specialties. Fortunately, there are many excellent radar reference books that address both needs. Good books appear every year; those listed in the paragraphs that follow are current as of the year 2021.

1.6.1 Introductions to Radar Systems and Applications

One of the newest and possibly best "radar 101" introductions is volume 1 of the three-volume *Principles of Modern Radar* (POMR) series, which addresses a full range of basic radar principles, phenomenology, components, and signal processing (Richards et al., 2010). The second volume by Melvin and Scheer (2012) addresses advanced radar techniques, providing more depth to the basic signal processing methods of volume 1 as well as surveying numerous advanced modern processing techniques. An up-to-date survey of a broad range of traditional and modern radar applications is given in the third volume, also by Melvin and Scheer (2013). It shows how many of the techniques discussed in these texts and those below are brought together into complete systems. Collectively, the series provides a comprehensive textbook and reference for studying modern radar systems. At this writing a significantly updated second edition of all three volumes is in preparation, with publication expected in 2022 and 2023.

There are several other good introductory books on radar systems. A classic introduction to radar systems, now in its third edition, is by Skolnik (2001). The comprehensive text by Peebles (1998) is more academic in its approach. Mahafza (2013) provides a number of useful MATLAB® files to aid in simulation and experimentation. Unique in the field, *Stimson's Introduction to Airborne Radar* (Stimson et al., 2014) is a very accessible introduction to radar with a strong emphasis on airborne pulsed Doppler systems and featuring extensive illustrations and examples. Alabaster (2012) provides another excellent introduction to airborne pulsed Doppler systems and signal processing. Morris and Harkness (1996) is also a good introduction to these systems.

1.6.2 Basic Radar Signal Processing

It is this author's opinion that there are a number of excellent books about radar systems in general, including coverage of components and system designs, and several on advanced radar signal processing topics, especially in the areas of synthetic aperture imaging and adaptive processing. There have been few books that address the middle ground of basic radar signal processing, such as matched filtering, Doppler filtering, and CFAR detection. Such books are needed to provide greater quantitative depth than is available in the radar system books while at the same time covering a range of topics found in most radar systems, instead of being restricted to in-depth coverage of a single advanced application area. This text aims to fill that gap.

There are a few texts that fit somewhat into this middle area. Nathanson (1991) purports to cover radar systems in general but in fact concentrates on signal processing issues, especially RCS and clutter modeling, waveforms, MTI, and detection. Probably the closest text in intent to this one is by Levanon (1988), which provides excellent and very concise analyses of many basic signal processing functions. The text by Levanon and Mozeson (2004) addresses the widening variety of radar waveforms in detail. Sullivan (2000) is interesting especially for its introductory coverage of both SAR and space-time adaptive processing (STAP), thus providing a bridge between basic signal processing and more advanced texts specializing in SAR and STAP.

1.6.3 Advanced Radar Signal Processing

Two very active areas of advanced radar signal processing research are SAR imaging and STAP. SAR research extends back to 1951, but only in the 1990s did open literature textbooks begin to appear in the market. There are now many good textbooks on SAR. The first comprehensive text was by Curlander and McDonough (1991). Based on experience gained at the NASA Jet Propulsion Laboratory, it emphasizes space-based SAR and includes a strong component of scattering theory as well. Cumming and Wong (2005) is a newer text based on experience with the Canadian RADARSAT satellite SARs. The spotlight SAR mode received considerable development in the 1990s, and two major groups published competing texts in the mid-1990s. Carrara et al. (1995) represented the work of the group at the Environmental Research Institute of Michigan (ERIM, ultimately becoming a part of MacDonald Dettwiler and Associates); Jakowatz, Jr., et al. (1996) represented the work of a group at Sandia National Laboratories, a unit of the U.S. Department of Energy. Franceschetti and Lanari (1999) provide a compact, unified treatment of both major modes of SAR imaging, namely, stripmap and spotlight. The book by Soumekh (1999) is the most complete academic reference on synthetic aperture imaging and includes a number of MATLAB simulation resources. The newest SAR book is by Jansing (2021) and features many more modern examples and algorithms.

STAP, one of the most active radar signal processing research areas, began in earnest in 1973. Klemm (1998) wrote the first significant open literature text on the subject. The book by Guerci (2014) is the newest primer on this subject at this writing, while Van Trees (2002) prepared a detailed text that continues his classic series on detection and estimation. Additionally, there are other texts on more limited forms of adaptive interference rejection. A good example is the text by Nitzberg (1999), which discusses several forms of sidelobe cancellers.

New signal processing methods continue to find applications in radar. Volume 2 by Melvin and Scheer (2012) of the POMR series mentioned above surveys a wide range of these and other advanced radar signal processing techniques, including such new topics as MIMO radar and *compressive sensing* (CS). For example, MIMO radar may (or may not) prove to have advantages in many areas of radar search, tracking, imaging, and interference rejection. A good primer is Bergin and Guerci (2018), while more in-depth coverage is available in Li and Stoica (2008). The first book on CS in radar is by De Maio et al. (2019). As in many other fields of technology, applications of artificial intelligence-based ideas such as "deep learning" and "machine learning" to radar are active areas of current research, but no book-level treatments of this area have appeared as of this writing (2021).

1.6.4 Radar Applications

The preceding sections have cited several books addressing general radar applications, such as imaging or pulse Doppler. There are a number of books in the literature devoted to more specific application areas. Volume 3 of the POMR series by Melvin and Scheer (2013) provides an excellent survey of and introduction to a wide range of applications in a single text.

1.6.5 Current Radar Research

Current radar research appears in a number of scientific and technical journals. The most important in the United States are the Institute of Electrical and Electronics Engineers (IEEE) *Transactions on Aerospace and Electronic Systems, Transactions on Geoscience and Remote Sensing, Transactions on Signal Processing,* and *Transactions on Image Processing*. Radar-related material in the latter is generally limited to papers related to SAR processing, especially interferometric three-dimensional SAR. In the United Kingdom, radar technology papers are often published in the Institution of Engineering and Technology (IET) [formerly the Institution of Electrical Engineers (IEE)] journal *IET Radar, Sonar, and Navigation*. Another good source of imaging radar research is the various publications of SPIE (formerly the Society of Photo-Optical Instrumentation Engineers, now just SPIE).

References

Alabaster, C. M., *Pulse Doppler Radar: Principles, Technology, Applications*. SciTech Publishing, Raleigh, NC, 2012.
Balanis, C. A., *Antenna Theory*, 4th ed. Harper & Row, New York, 2016.
Bar-Shalom, Y., and T. E. Fortmann, *Tracking and Data Association*. Academic Press, Boston, MA, 1988.
Bergin, J., and J. R. Guerci, *MIMO Radar: Theory and Application*. Artech House, Boston, MA, 2018.
Bracewell, R. N., *The Fourier Transform and Its Applications*, 3d ed. McGraw-Hill, New York, 1999.
Brookner, E. (ed.), *Aspects of Modern Radar*. Artech House, Boston, MA, 1988.
Bruder, J. A., "Radar Receivers," Chap. 11 in M. A. Richards, J. A. Scheer, and W. A. Holm (eds.), *Principles of Modern Radar: Basic Principles*. SciTech Publishing, Raleigh, NC, 2010.
Carrara, W. G., R. S. Goodman, and R. M. Majewski, *Spotlight Synthetic Aperture Radar*. Artech House, Norwood, MA, 1995.
Cumming, I. G., and F. N. Wong, *Digital Processing of Synthetic Aperture Radar Data*. Artech House, Norwood, MA, 2005.
Curlander, J. C., and R. N. McDonough, *Synthetic Aperture Radar*. Wiley, New York, 1991.
De Maio, A., Y. C. Eldar, and A. M. Haimovich, *Compressed Sensing in Radar Signal Processing*. Cambridge University Press, Cambridge, UK, 2019.
EW and Radar Systems Engineering Handbook, 4th ed., Oct. 2013. Naval Air Warfare Center, Weapons Division. Available at https://apps.dtic.mil/docs/citations/ADA617071.
Franceschetti, G., and R. Lanari, *Synthetic Aperture Radar Processing*. CRC Press, New York, 1999.
Guerci, J. R., *Space-Time Adaptive Processing for Radar*, 2d ed. Artech House, Norwood, MA, 2014.
Institute of Electrical and Electronics Engineers (IEEE), "IEEE Standard Definitions of Terms for Antennas," Standard 145-2013, Mar. 6, 2014.
Institute of Electrical and Electronics Engineers (IEEE), "IEEE Standard Radar Definitions," Standard 686-2017, Mar. 23, 2017.
Institute of Electrical and Electronics Engineers (IEEE), "IEEE Standard Letter Designations for Radar-Frequency Bands," Standard 521-2019, Nov. 7, 2019.
Jakowatz, C. V., Jr., et al., *Spotlight-Mode Synthetic Aperture Radar: A Signal Processing Approach*. Kluwer, Boston, MA, 1996.
Jansing, E. D., Introduction to Synthetic Aperture Radar: Concepts and Practice. McGraw-Hill, New York, 2021.
Jelalian, A. V., *Laser Radar Systems*. Artech House, Boston, MA, 1992.
Klemm, R., *Space-Time Adaptive Processing: Principles and Applications*. INSPEC/IEEE, London, 1998.
Levanon, N., *Radar Principles*. Wiley, New York, 1988.

Levanon, N., and E. Mozeson, *Radar Signals*. Wiley, New York, 2004.

Li, J., and P. Stoica, *MIMO Radar Signal Processing*. Wiley-IEEE Press, Hoboken, NJ, 2008.

Mahafza, B. R., *Radar Systems Analysis and Design Using MATLAB®*, 3d ed. Chapman & Hall/CRC, New York, 2013.

Melvin, W. L., and J. A. Scheer (eds.), *Principles of Modern Radar: Advanced Techniques*. SciTech Publishing, Raleigh, NC, 2012.

Melvin, W. L., and J. A. Scheer (eds.), *Principles of Modern Radar: Radar Applications*. SciTech Publishing, Raleigh, NC, 2013.

Mir, H. S., and J. D. Wilkinson, "Radar Target Resolution Probability in a Noise-Limited Environment," *IEEE Transactions on Aerospace and Electronic Systems*, vol. 44, no. 3, pp. 1234–1239, Jul. 2008.

Morris, G. V., and L. Harkness (eds.), *Airborne Pulsed Doppler Radar*, 2d ed. Artech House, Boston, MA, 1996.

Nathanson, F. E. (with J. P. Reilly and M. N. Cohen), *Radar Design Principles*, 2d ed. McGraw-Hill, New York, 1991.

Nitzberg, R., *Radar Signal Processing and Adaptive Systems*, 2d ed. Artech House, Boston, MA, 1999.

Papoulis, A., *The Fourier Integral and Its Applications*, 2d ed. McGraw-Hill, New York, 1987.

Peebles, Jr., P. Z., *Radar Principles*. Wiley, New York, 1998.

Richards, M. A., "Virtual Arrays, Part 1: Phase Centers and Virtual Elements," technical memorandum, Mar. 2018. Available at http://radarsp.com.

Richards, M. A., "Virtual Arrays, Part 2: Virtual Arrays and Coarrays," technical memorandum, Feb. 2019. Available at http://radarsp.com.

Richards, M. A., J. A. Scheer, and W. A. Holm (eds.), *Principles of Modern Radar: Basic Principles*. SciTech Publishing, 2010.

Scheer, J. A. (ed.), *Coherent Radar Performance Estimation*. Artech House, Boston, MA, 1993.

Sherman, S. M., *Monopulse Principles and Techniques*, 2d ed. Artech House, Boston, MA, 2011.

Skolnik, M. I., *Introduction to Radar Systems*, 3d ed. McGraw-Hill, New York, 2001.

Soumekh, M., *Synthetic Aperture Radar Signal Processing with MATLAB Algorithms*. Wiley, New York, 1999.

Stimson, G. W. et al., *Stimson's Introduction to Airborne Radar*, 3rd ed. SciTech Publishing, Raleigh, NC, 2014.

Stutzman, W. L., "Estimating Gain and Directivity of Antennas," IEEE *Transactions on Antennas and Propagation*, vol. 40, no. 4, pp. 7–11, Aug. 1998.

Stutzman, W. L., and G. A. Thiele, *Antenna Theory and Design*, 3d ed. Wiley, New York, 2012.

Sullivan, R. J., *Microwave Radar: Imaging and Advanced Concepts*. Artech House, Boston, MA, 2000.

Swords, S. S., *Technical History of the Beginnings of RADAR*. Peter Peregrinus Ltd., London, 1986.

Van Trees, H. L., *Optimum Array Processing: Part IV of Detection, Estimation, and Modulation Theory*. Wiley, New York, 2002.

Wallace, T. V, R. J. Jost, and P. E. Schmid, "Radar Transmitters," Chap. 10 in M. A. Richards, J. A. Scheer, and W. A. Holm (eds.), *Principles of Modern Radar: Basic Principles*. SciTech Publishing, Raleigh, NC, 2010.

Problems

1. Compute the range R corresponding to echo delays t_0 of 1 ns, 1 μs, 1 ms, and 1 second.

2. Compute the time delays for two-way propagation to targets at distances of 100 km, 100 statute miles, and 100 ft.

3. Radar is routinely used as one means of measuring the distance to objects in space. For example, it has been used to calculate the orbital parameters and rate of rotation

of the planet Jupiter. The distance from Earth to Jupiter varies from 588.5×10^6 to 968.1×10^6 km. What are the minimum and maximum time delays *in minutes* from the time a pulse is transmitted in the direction of Jupiter until the time the echo is received? If pulses are transmitted at a rate of 100 pulses per second, how many pulses are in flight, either on their way to Jupiter or back again, at any given instant?

4. Table 1.1 defines the millimeter wave (MMW) band to extend from 30 to 300 GHz. Only certain frequencies in this band are widely used for radar; others are deliberately avoided. This is partly due to frequency allocation rules (which frequencies are allotted to which services), but also due to atmospheric propagation. Based on Fig. 1.3 and assuming longer range capability is preferable, identify a frequency in the MMW band that might *not* be a good choice for a MMW radar operating frequency. Explain.

5. Consider two equal-power transmitters, one at 10 GHz and the other at 30 GHz. The power received from each is measured at a distance of 10 km and compared. Based on Figs. 1.3 and 1.4, estimate how many decibels weaker than the 10 GHz signal will the 30 GHz signal be in sea level clear air conditions. Repeat for a medium rainfall of 2.5 mm/h and a tropical downpour of 50 mm/h.

6. Compute the bandwidth β needed to achieve range resolutions of 1 m, 1 km, and 100 km. What is the length of a constant-frequency rectangular pulse having this Rayleigh bandwidth for each value of resolution?

7. Derive Eq. (1.10) from Eq. (1.8).

8. Using the result from Prob. 7, find the value of D_y/λ_t for which the error in the small-angle approximation of the Rayleigh beamwidth is 10 percent. What are the exact and approximate Rayleigh beamwidths for that value of D_y/λ_t?

9. How large must a uniformly illuminated aperture antenna be (value of D_y) in terms of wavelengths so that its one-way 3-dB beamwidth is 1°? What is the estimated gain in decibels of an antenna having azimuth and elevation beamwidths $\theta_3 = \phi_3 = 1°$, based on the approximation in Eq. (1.11)?

10. The adjoining figure shows a physical one-dimensional uniform linear array comprising two transmit elements, two receive elements, and one element that both transmits and receives. This array is "sparse" in that not all elements needed to form a conventional "filled" ULA are present; the two dashed circles represent missing elements. Construct the virtual array and show that it is filled. What is the spacing and overall size of the VA elements compared to those of the physical array?

physical array element positions:

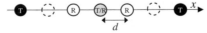

11. Suppose a police "speed gun" radar has a rectangular antenna. It is desired to have a cross-range resolution ΔCR of 10 ft at a distance of one-quarter mile. What is the required antenna width in inches if the radar frequency is 9.4 GHz? Repeat for 34.4 GHz.

12. Continuing Prob. 11, what is the actual cross-range resolution in feet at each RF if the antenna width is 6 inches?

13. Starting from Eq. (1.14) and setting $a_n = 1$, derive Eq. (1.15). *Hint:* You will need the finite geometric sum formula.

14. What is the maximum 3-dB beamwidth θ_3 in degrees such that the approximation for the cross-range resolution, $R\theta_3$, in the last step of Eq. (1.28) has an error of no more than 1 percent?

15. Determine the cross-range resolution ΔCR in meters at ranges of 10, 100, and 1000 km for a 3-dB beamwidth $\theta_3 = 3°$.

16. Determine the approximate size of a volume resolution cell in cubic meters, ΔV, for $R = 20$ km, $\Delta R = 100$ m, and $\theta_3 = \phi_3 = 3°$.

17. Equation (1.2) is a general result applicable to any waveform provided a matched filter is used. Specialize this to a common formula for the range resolution of a simple pulse waveform in terms of the pulse length τ.

18. Equation (1.35) showed that the magnitude-squared of signal-plus-noise data cannot be expressed as the sum of only a signal portion and a noise portion; cross products of the signal and noise samples were also present, complicating the definition of post-integration SNR and integration gain. Show that this is also true for the magnitude (not magnitude-squared) of the data. It is sufficient to use $N = 1$, that is, a single sample. *Hint*: Apply a Taylor series to the square root of the magnitude-squared result.

19. Repeat Prob. 18 for a log-magnitude detector.

20. How many signal-plus-noise samples must be coherently integrated to achieve an integration gain of at least 3 dB? 10 dB? 20 dB? The result N must be an integer.

21. Repeat Prob. 20 for the case of noncoherent integration assuming an integration gain of $N^{0.7}$. Summarize the main advantage and disadvantage of noncoherent integration relative to coherent integration.

22. In a threshold detection scheme like that pictured in Fig. 1.25 and with a fixed SNR, would the probability of false alarm P_{FA} be expected to increase, decrease, or be unchanged if the threshold level was raised? What about the probability of detection, P_D?

23. Suppose that a system for which Eq. (1.26) applies requires $P_D = 0.9$ and $P_{FA} = 10^{-6}$. What is the required SNR in linear units and in decibels?

CHAPTER 2
Signal Models

2.1 Components of a Radar Signal

While a radar transmits a controlled, well-defined signal, the signal measured at the receiver output in response is the superposition of several major components, none of them entirely under the control of the designer. The major components are the target, clutter, noise, and, in some cases, jamming and electromagnetic interference (EMI). These signals are sometimes subdivided further. For instance, clutter can be separated into ground clutter and weather clutter (such as rain), while jamming can be separated into active jamming (hostile transmitters) and passive jamming (such as chaff clouds). Signal processing is applied to this composite signal; the goal is to extract useful information regarding the presence of targets and their characteristics, or to form a radar image. Noise, jamming, and EMI are interference signals; they degrade the ability to detect targets and accurately measure their position and velocity. Clutter may be interference in some cases, such as when detecting aircraft, or may be the desired signal itself, as with a ground imaging radar. The effectiveness of the signal processing is measured by the improvement it provides in various figures of merit such as detection probability, signal-to-interference ratio, or angle accuracy.

It was shown in Chap. 1 that conventional pulsed radars transmit narrowband bandpass signals. Transmitted energy is maximized by restricting amplitude modulation, if any, to on-off pulsing; phase modulation is used to expand the instantaneous bandwidth when needed to improve resolution. A transmitted radar signal can be written as

$$\bar{x}(t) = a(t)\cos[2\pi F_t t + \varphi(t)] \tag{2.1}$$

where F_t is the radar carrier frequency and $\varphi(t)$ may be a constant or may represent phase or frequency modulation of the pulse. The overbar on $\bar{x}$ indicates a signal that is not at baseband, that is, is on a carrier. It will usually be assumed that $a(t)$ is an ideal square pulse envelope of amplitude A and duration τ seconds for a pulsed radar. For a continuous wave radar $a(t)$ is simply a constant amplitude A for all time. The average power of this signal while it is on is $P_s = A^2/2$. The signal at the receiver output will be a combination of echoes of $\bar{x}(t)$ from targets and clutter, possibly jamming and EMI, and additive noise.

Because the target and clutter components are delayed echoes of the transmitted signal, they are also narrowband signals, although their amplitude and phase modulation will in general be altered, for example, by propagation loss and Doppler shift. Receiver noise appears as an additive random signal. Assuming a coherent receiver, the received signal echoed from a single scatterer at range $R_0 = ct_0/2$ with only noise as interference can be modeled as

$$\begin{aligned}\bar{y}(t) &= k \cdot a(t - t_0)\exp(j[2\pi F_t(t - t_0) + \varphi(t - t_0) + \psi(t)]) + n(t) \\ &= k \cdot \{a(t - t_0)\exp[j\varphi(t - t_0)]\}\exp(-j2\pi F_t t_0)\exp[j\psi(t)]\exp(j2\pi F_t t) + n(t)\end{aligned} \tag{2.2}$$

where $n(t)$ = receiver noise
k = echo amplitude factor due to propagation losses and target reflectivity
$\psi(t)$ = echo phase modulation due to target interaction

The important parameters of $\bar{y}(t)$ are the delay time t_0, the echo component amplitude $k \cdot |a(t)|$ and its power relative to the noise component, and the echo phase or frequency modulation function $\varphi(t - t_0) + \psi(t)$. These characteristics are used to estimate target range scattering strength and radial velocity, suppress interference, form images, and so forth.

The second line of Eq. (2.2) separates the various phase factors in the echo signal. The first term in curly brackets is the delayed waveform echo, while the third term is any phase shift or modulation due to the target reflection, such as Doppler shift or just a phase reversal. The fourth term is the carrier frequency, which will be removed in the receiver. The $\exp(-j2\pi F_t t_0)$ term is a phase shift due to the delay to the target. Using $t_0 = 2R_0/c$ and $c = \lambda_t F_t$, this term is more often expressed as $\exp(-j4\pi R_0/\lambda_t)$ or $\exp(-j2\Omega_t R_0/c)$. As will be seen, this phase is a very sensitive indicator of radar-target range. The variation in this phase term over time will be critical data used for many radar signal processing operations.

The amplitude and phase or frequency modulation functions also determine the range resolution ΔR of a measurement. For example, $\Delta R = c\tau/2$ if $\varphi(t)$ is a constant and $\bar{x}(t)$ is a simple constant-frequency pulse of length τ seconds. Resolution in angle or cross range is determined by the width of the antenna pattern in a non-imaging radar.

In order to design good signal processing algorithms, good models of the signals to be processed are needed. In this chapter, an understanding of common radar signal characteristics is developed by presenting models of the effect of the scattering process on the amplitude, phase, and frequency properties of radar measurements. While deterministic models suffice for simple scatterers, it will be seen that complicated real targets require statistical descriptions of the scattering process.

2.2 Modeling Amplitude

2.2.1 Simple Point Target Radar Range Equation

The *radar range equation* (RRE) is a deterministic model that relates received echo power to transmitted power in terms of a variety of system design parameters (Skolnik, 2001; Richards et al., 2010). It is a fundamental relation used for basic system design and analysis. The received signal power P_r estimated by the range equation can be directly related to the received signal amplitude.

To derive the range equation, assume that an isotropic radiating element transmits a waveform of power P_t watts into a lossless medium. Because the transmission is isotropic and no power is lost in the medium, the power density at a range R is the total power P_t divided by the surface area of a sphere of radius R:

$$\text{Isotropic transmitted power density} = \frac{P_t}{4\pi R^2} \text{ W/m}^2 \qquad (2.3)$$

Instead of isotropic radiators, real radars use directive antennas to focus the outgoing energy. As described in Chap. 1, the antenna gain G is the ratio of maximum power density to isotropic density. Thus, in the direction of maximum radiation intensity the power density at range R becomes

$$\text{Peak transmitted power density} = Q_t = \frac{P_t G}{4\pi R^2} \text{ W/m}^2 \qquad (2.4)$$

This is the power density incident upon the target if it is aligned with the antenna's axis of maximum gain.

When the electromagnetic wave with power density given by Eq. (2.4) is incident upon a single discrete scattering object (*point target*) at range R, the incident energy is scattered in various directions; some of it may also be absorbed by the scatterer itself. In particular, some of the incident power is reradiated toward the radar (*backscattered*). Imagine that the target collects all of the energy incident upon a collector of area σ square meters and reradiates it isotropically. The backscattered power is

$$\text{Backscattered power} = P_b = \frac{P_t G \sigma}{4\pi R^2} \text{ W} \tag{2.5}$$

The quantity σ is called the *radar cross section* (RCS) of the target. It is *not* equal to the physical cross-sectional area of the target; it is an equivalent area used to relate incident power density at the target to the reflected power density that results at the receiver. RCS will be discussed further in Sec. 2.2.3.

RCS is defined under the assumption that the backscattered power is reradiated isotropically. The density of the backscattered power at range R is found by dividing the power of Eq. (2.5) by the surface area of a sphere of radius R as was done in Eq. (2.3), giving the backscattered power density at the radar receiver as

$$\text{Backscattered power density} = Q_b = \frac{P_t G \sigma}{(4\pi)^2 R^4} \text{ W/m}^2 \tag{2.6}$$

If the effective aperture size of the radar antenna is A_e square meters the total backscattered power collected by the receiving antenna will be

$$\text{Received power} = P_r = A_e Q_b = \frac{P_t G A_e \sigma}{(4\pi)^2 R^4} \text{ W} \tag{2.7}$$

Recall from Chap. 1 that the effective aperture of an antenna is related to its gain and operating wavelength according to $A_e = \lambda_t^2 G / 4\pi$ m^2. Then

$$P_r = \frac{P_t G^2 \lambda_t^2 \sigma}{(4\pi)^3 R^4} \text{ W} \tag{2.8}$$

Equation (2.8) describes the power that would be received if an ideal radar operated in free space and used no signal processing techniques to improve sensitivity. Various loss and gain factors are customarily added to the formula to account for a variety of additional considerations. For example, losses incurred in various components such as the duplexers, power dividers, waveguide, and *radome* (a protective covering over the antenna) can be lumped into a *system loss factor* L_s that reduces the received power. System losses are typically in the range of 3 to 10 dB but can vary widely.

One of the most important loss factors, particularly at X band and higher frequencies, is atmospheric attenuation $L_a(R)$. Unlike system losses, atmospheric losses are a function of range. If the one-way loss in decibels per kilometer of Fig. 1.3 is denoted by α, the loss in decibels for a target at range R meters (not kilometers) is

$$L_a(R)(\text{dB}) = 2\alpha(R/1000) = \alpha R / 500 \text{ dB} \tag{2.9}$$

In linear units this is

$$L_a(R) = 10^{\alpha R / 5000} \tag{2.10}$$

Atmospheric loss can be inconsequential at 10 GHz and moderate ranges, or tens of decibels at 60 GHz and a range of a few kilometers. (This is why 60 GHz is not a popular radar frequency.) This example also shows that, like system losses, atmospheric loss is a strong function of radar frequency. Losses due to propagation effects not found in free space propagation may also be lumped into L_a (Saffold, 2010).

Incorporating atmospheric and system losses in Eq. (2.8) finally gives

$$P_r = \frac{P_t G^2 \lambda_t^2 \sigma}{(4\pi)^3 R^4 L_s L_a(R)} \text{ W} \qquad (2.11)$$

Equation (2.11) is one simple form of the RRE. It relates received echo power to fundamental radar system and target parameters such as transmitted power, operating frequency, and antenna gain; radar cross section; and range. Because the power of the radar signal is proportional to the square of the electric field amplitude, the range equation also serves as a model of the amplitude of the target and clutter components of the signal. Note that all variables in Eq. (2.11) are in linear units, not decibels, even though several of the parameters are often specified in decibels; frequent examples include the system and atmospheric losses, antenna gain, and RCS. Also note that P_r is instantaneous, not average, received power. Finally, realize that for a scatterer at range R, the backscattered EM wave will be received with a time delay of $2R/c$ seconds after transmission.

As an example, consider an X-band (10 GHz) radar with a peak transmitted power of 1 kW and a pencil beam antenna with a 1° beamwidth in both azimuth and elevation. Suppose an echo is received from a jumbo jet aircraft with an RCS of 100 m² at a range of 10 km. The received power can be determined using Eq. (2.11). The antenna gain can be estimated from Eq. (1.11) to be $G = 26{,}000/(1)(1) = 26{,}000 = 44$ dB. The wavelength is $\lambda_t = c/F_t = 3 \times 10^8 / 10 \times 10^9 = 3 \times 10^{-2}$ m $= 3$ cm. Assuming atmospheric and system losses are negligible, the received power is

$$P_r = \frac{(1000)(26{,}000)^2(0.03)^2(100)}{(4\pi)^3(10{,}000)^4} = 3.07 \times 10^{-9} \text{ W} \qquad (2.12)$$

Even though this example is a large target at short range, the received power is only 3.07 nW, nearly 12 orders of magnitude less than the transmitted power! Nonetheless, this signal level is adequate for reliable detection if the receiver is sufficiently sensitive and the signal is stronger than the interference, or can be made so by the signal processing. This example illustrates the huge dynamic ranges observed in radar between transmitted and received signal powers.

An important consequence of Eq. (2.11) is that for a point target the received power decreases as the fourth power of range from the radar to the target. Thus, the ability to detect a target of a given radar cross section decreases rapidly with range. Detection range can be increased by increasing transmitted power, but because of the R^4 dependence the power must be raised by a factor of 16 (12 dB) just to double the detection range. Alternatively, the antenna gain can be increased by a factor of 4 (6 dB), implying an antenna area larger by a factor of 4. On the other hand, designers of "stealth" aircraft and other target vehicles must reduce the RCS σ by a factor of 16 in order to halve the range at which they can be detected by a given radar system.

The range equation is a fundamental radar system design and analysis tool. More elaborate or specialized versions of the equation can be formulated to show the effect of other variables, such as pulse length, intermediate frequency bandwidth, or signal processing gains. Several such variations are given in Scheer (2010). The range equation also provides the basis for calibrating a radar system. If the system power, gain, and losses are carefully characterized, the expected received power of echoes from test targets of known RCS can be computed. Calibration tables equating receiver voltage observed to incident power density can then be constructed.

Signal processing techniques can increase the effective received power and therefore increase the obtainable range. The effect of each technique on received power is discussed as they are introduced in later chapters.

2.2.2 Distributed Target Forms of the Range Equation

Not all scattering phenomena can be modeled as a reflection from a single point scatterer. Ground clutter, for example, is best modeled as distributed scattering from a surface, while meteorological phenomena such as rain or hail are modeled as distributed scattering from a three-dimensional volume. The radar range equation can be rederived in a generalized way that accommodates all three cases.

Equation (2.3) is still applicable as a starting point. To consider distributed scatterers and account for the variation of antenna gain with azimuth angle θ and elevation angle ϕ, Eq. (2.4) must be replaced with a version that accounts for the effect of the one-way antenna power pattern $P(\theta, \phi)$ on the power density radiated in a particular direction (θ, ϕ):

$$Q_t(\theta, \phi) = \frac{P_t P(\theta, \phi)}{4\pi R^2} \text{ W/m}^2 \qquad (2.13)$$

Assume that the antenna boresight corresponds to $\theta = \phi = 0$. The antenna boresight is normally the axis of maximum gain so that $P(0, 0) = G$.

Now consider the scattering from an incremental volume dV located at range and angle coordinates (R, θ, ϕ). Suppose the incremental RCS of the volume element at those coordinates is $d\sigma(R, \theta, \phi)$ square meters. The incremental backscattered power from dV is

$$dP_b(\theta, \phi) = \frac{P_t P(\theta, \phi) d\sigma(R, \theta, \phi)}{4\pi R^2} \qquad (2.14)$$

As before, $d\sigma$ is defined such that it is assumed this power is reradiated isotropically and then collected by the antenna effective aperture. After substituting for effective aperture and accounting for losses, the incremental received power is

$$dP_r = \frac{P_t \, P^2(\theta, \phi) \lambda_t^2 d\sigma(R, \theta, \phi)}{(4\pi)^3 R^4 L_s L_a(R)} \text{ W} \qquad (2.15)$$

The total received power is obtained by integrating over all space to obtain a generalized RRE

$$P_r = \frac{P_t \lambda_t^2}{(4\pi)^3 L_s} \int_V \frac{P^2(\theta, \phi)}{R^4 L_a(R)} d\sigma(R, \theta, \phi) \text{ W} \qquad (2.16)$$

In Eq. (2.16), the volume of integration V is all of three-dimensional space. However, the backscattered energy from all ranges does not arrive simultaneously at the radar. As discussed in Sec. 1.4.2, only scatterers within a single range resolution cell of extent ΔR contribute significantly to the radar receiver output at any given instant. Thus, a more appropriate form of the generalized radar range equation gives the received power at a particular range or time,

$$P_r\left(t_0 = \frac{2R_0}{c}\right) = \frac{P_t\lambda_t^2}{(4\pi)^3 L_s} \int_{\Delta R, \Omega} \frac{P^2(\theta,\phi)}{R^4 L_a(R)} d\sigma(R,\theta,\phi) \quad \text{W} \tag{2.17}$$

where ΔR is the range interval of the resolution cell centered at range R_0, and Ω represents integration over the angular coordinates.

By integrating power, it is being assumed that the backscatter from each volume element adds noncoherently rather than coherently. This means that the power of the composite electromagnetic wave formed from the backscatter of two or more scattering centers is the sum of the individual powers, as opposed to the voltage (electric field amplitude) being the sum of the individual voltages, in which case the power would be the square of the voltage sum. Noncoherent addition occurs when there are many scattering centers having random individual phases that are uncorrelated with one another, as opposed to the coherent case when the individual echoes add in phase. This issue will be revisited in Sec. 2.8.

The general result of Eq. (2.17) is more useful if evaluated for the special cases of point, volume, and area scatterers. Beginning with the point scatterer, the differential RCS in the resolution cell volume is represented by a Dirac impulse function $\delta_D(\cdot)$ of weight σ,

$$d\sigma(R,\theta,\phi) = \sigma \cdot \delta_D(R - R_0, \theta - \theta_0, \phi - \phi_0) dV \quad \text{(point scatterer)} \tag{2.18}$$

Using Eq. (2.18) in Eq. (2.17) gives the range equation for a point target at (R_0, θ_0, ϕ_0),

$$P_r(t_0) = \frac{P_t P^2(\theta_0,\phi_0)\lambda_t^2 \sigma}{(4\pi)^3 R_0^4 L_s L_a(R_0)} \quad \text{W (point scatterer)} \tag{2.19}$$

If the point scatterer is located on the antenna boresight $\theta_0 = \phi_0 = 0$, $P(\theta_0, \phi_0) = G$ and Eq. (2.19) is identical to Eq. (2.11) evaluated at range R_0.

Notice that $d\sigma$ is modeled as the product of a term dV that depends only on the radar spatial resolution characteristics and a term η that depends only on the nature of the external scatterers.[1] This decomposition of the RCS into radar and external terms will also be seen in the modeling of both surface and volume clutter signal power.

Next consider the *volume scattering* case where the RCS seen by the radar is presumed to be due to a distribution of scatterers distributed throughout the volume, rather than associated with a single point. In this case, σ is expressed in terms of RCS per cubic meter, called *volume reflectivity* and denoted as η. The units of volume reflectivity are $m^2/m^3 = m^{-1}$. The RCS of a differential volume element dV is then

$$d\sigma = \eta \cdot dV = \eta \cdot R^2 dR\, d\Omega \quad \text{(volume scatterer)} \tag{2.20}$$

where $d\Omega$ is a differential solid angle element.

The range equation now becomes

$$P_r(t_0) = \frac{P_t \lambda_t^2 \eta}{(4\pi)^3 L_s} \int_{\Delta R, \Omega} \frac{P^2(\theta,\phi)}{R^2 L_a(R)} dR\, d\Omega \quad \text{W} \tag{2.21}$$

---

[1]This overstates the case a bit, since the reflectivity of a given scattering object such as rain, ground or sea surface, or man-made objects depends on radar frequency and polarization. Once the RF and polarization are fixed, η depends only on the scatterer characteristics such as material, size, shape, and orientation.

If it is assumed that atmospheric loss is slowly varying over the extent of a range resolution cell, then $L_a(R)$ can be replaced by $L_a(R_0)$, where R_0 is the center of the range resolution cell. $L_a(R_0)$ can then be removed from the integral. The integral over range that remains is

$$\int_{R_0-\Delta R/2}^{R_0+\Delta R/2} \left(\frac{dR}{R^2}\right) = \frac{\Delta R}{R_0^2 - (\Delta R/2)^2} \approx \frac{\Delta R}{R_0^2} \tag{2.22}$$

provided the range resolution is small compared to the absolute range, which is usually the case. Using Eq. (2.22) in Eq. (2.21) gives

$$P_r(t_0) = \frac{P_t \lambda_t^2 \eta \Delta R}{(4\pi)^3 R_0^2 L_s L_a(R_0)} \int_\Omega P^2(\theta,\phi)\, d\Omega \quad \text{W} \tag{2.23}$$

Integration over the angular coordinates requires knowledge of the antenna pattern. One common approximate model of the mainlobe of many antennas is a Gaussian function (Sauvageot, 1992). It can be shown that a good approximation to the integral in Eq. (2.23) over the angular variables for the Gaussian case is (Probert-Jones, 1962)

$$\iint_\Omega P^2(\theta,\phi) \sin\phi\, d\theta\, d\phi \approx \frac{\pi \theta_3 \phi_3}{8\ln(2)} G^2 = 0.57\, \theta_3\, \phi_3 G^2 \tag{2.24}$$

where θ_3 and ϕ_3 are the 3-dB beamwidths in azimuth and elevation, and G is the peak gain. For first-order calculations the much simpler assumption is frequently made that the antenna power pattern $P(\theta,\phi)$ is a constant equal to the gain G over the 3-dB beamwidths and zero elsewhere so that the integral reduces to $G^2 \theta_3 \phi_3$, a value 2.5 dB higher than that of Eq. (2.24). Using this approximation, Eq. (2.23) reduces to the range equation for volume scatterers:

$$P_r(t_0) = \frac{P_t G^2 \lambda_t^2 \eta \Delta R\, \theta_3 \phi_3}{(4\pi)^3 R_0^2 L_s L_a(R_0)} \quad \text{W (volume scatterers)} \tag{2.25}$$

Comparing to Eq. (2.11) shows again that the RCS σ has been replaced by the product of a radar resolution cell volume term $\Delta R \cdot R_0 \theta_3 \cdot R_0 \phi_3$ and a term η representing the reflectivity per unit volume.

Unlike the point scatterer case described by Eq. (2.11) or (2.19), the received power in the volume scattering case of Eq. (2.25) decreases only as R^2 instead of R^4. The reason is that the volume of the radar resolution cell, which determines the extent of the scatterers contributing to the received power at any one instant, increases as R^2 due to the spreading of the antenna beam in angle at longer ranges.

Finally, the *area scattering* case will be considered. This model is used for the RCS of backscatter from the ground, forest canopy, ocean, and other primarily two-dimensional surfaces. The area scattering case must further be divided into two subcases depending on whether the range extent of the scatterers contributing to the echo is limited by the antenna elevation beamwidth or by the range resolution.

First assume that the scattering surface is represented by a flat plane[2] and consider the intersection of the antenna mainlobe with the surface, called the mainlobe *footprint*. Assuming a typical pencil beam (conical) mainlobe, the footprint is an ellipse with a cross-range extent of

---

[2] This ignores earth curvature effects that may be significant in very long range or spaceborne radars. See Nathanson (1991) or Skolnik (2001) for additional details.

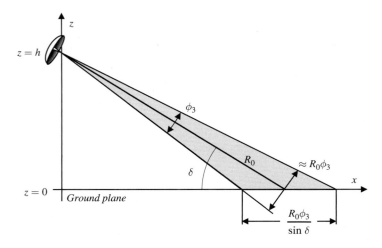

FIGURE 2.1 Projection of elevation beamwidth onto a horizontal plane at a slant range R_0 and grazing angle δ.

$R_0\theta_3$ meters at its widest point, where R_0 is the range to the center of the footprint. To estimate the down-range extent consider Fig. 2.1, which shows the boresight vector intersecting the scattering plane at a *grazing angle* of δ radians. The extent of the beam footprint in the down-range dimension is well-approximated as $R_0\phi_3/\sin\delta$ meters. Now suppose a waveform of range resolution ΔR is transmitted and consider the geometry in Fig. 2.2. The processed waveform range resolution ΔR along the direction of propagation is called the *slant range* resolution. However, the relevant range resolution is along the surface. This value is obtained by projecting the interval ΔR into the ground plane, degrading (increasing) the resolution to a value $\Delta R/\cos\delta$ meters. This is the range extent of scatterers within the resolution cell, and therefore contributing backscattering energy to the receiver output at any instant.

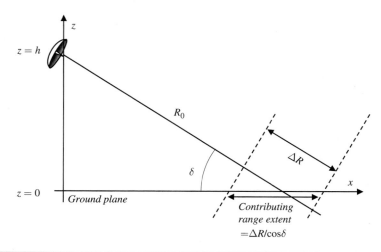

FIGURE 2.2 Projection of slant range resolution onto a horizontal plane at a slant range R_0 and grazing angle δ.

Scatterers will not contribute significantly to the received signal at a particular instant t_0 unless two conditions hold. First, they must be within the ground range interval generating an echo at the receiver at t_0. Second, they must be within the antenna mainlobe footprint so they are illuminated with significant energy and also enjoy significant receive gain. Consequently, the effective down-range extent of the surface that contributes significantly to the echo at a given time is the lesser of the slant range resolution and the elevation beamwidth as each is projected onto the scattering surface. Depending on the relative values of range, slant range resolution, and grazing angle, either could be the limiting factor. If the range resolution limits the effective extent, the resolution cell is said to be *resolution limited*; if the mainlobe footprint extent is the limiting factor, it is said to be *beam limited*. These two cases are shown in Fig. 2.3. The boundary between the two cases is obtained by equating the ground range extent in the two cases to see which is shorter. The result is

$$\text{Beam limited:} \quad \frac{\Delta R}{R_0} \tan \delta > \phi_3$$
$$\text{Resolution limited:} \quad \frac{\Delta R}{R_0} \tan \delta < \phi_3 \qquad (2.26)$$

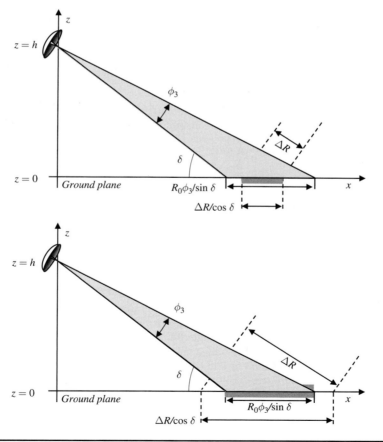

FIGURE 2.3 Relative range extent of antenna mainlobe and ground range resolution footprints. (*a*) Beam-limited case. (*b*) Resolution-limited case.

In area scattering the differential RCS is proportional to the differential area of the scattering surface and can be represented as

$$d\sigma = \sigma^0(\delta) \cdot \delta_D(R - R_0) dA \tag{2.27}$$

where σ^0 (called "sigma nought") is the *area reflectivity* in m^2/m^2 and is therefore dimensionless. The area reflectivity of many surface types is a significant function of the grazing angle δ. The generalized range equation [Eq. (2.17)] becomes

$$P_r(t_0) = \frac{P_t \lambda_t^2 \sigma^0}{(4\pi)^3 R_0^4 L_s L_a(R_0)} \int_{\Delta A} P^2(\theta, \phi) dA \quad W \tag{2.28}$$

where ΔA is the illuminated area at range R_0.

If the scattering area is beam limited, the geometry of Fig. 2.3a shows that the area contributing to the backscatter at one instant is approximately $R^2 \theta_3 \phi_3 / \sin\delta$.[3] Thus, a differential area contributing to the received power is of the form

$$dA = \frac{R_0^2}{\sin\delta} d\theta \, d\phi \quad m^2 \quad \text{(beam-limited case)} \tag{2.29}$$

Applying this to Eq. (2.28) and again using the constant-gain approximation to the antenna 3-dB beamwidth gives the beam-limited range equation for area scatterers:

$$P_r(t_0) = \frac{P_t G^2 \lambda_t^2 \phi_3 \theta_3 \sigma^0}{(4\pi)^3 R_0^2 L_s L_a(R_0) \sin\delta} \quad W \text{ (area scatterers, beam-limited case)} \tag{2.30}$$

If the illuminated area is resolution limited, the geometry of Fig. 2.3b shows that the area contributing to the backscatter at one instant is approximately $R\theta_3 \Delta R / \cos\delta$. The differential contribution is thus

$$dA = \frac{R_0 \Delta R}{\cos\delta} d\theta \quad m^2 \quad \text{(resolution-limited case)} \tag{2.31}$$

Using the first-order approximation of constant gain over the mainlobe once more, Eq. (2.28) becomes

$$P_r(t_0) = \frac{P_t G^2 \lambda_t^2 \sigma^0 \Delta R \theta_3}{(4\pi)^3 R_0^3 L_s L_a(R_0) \cos\delta} \quad W \text{ (area scatterers, resolution-limited case)} \tag{2.32}$$

Note that power varies as R^{-2} in the beam-limited case because, as with the volume scattering, the resolution cell size grows in both cross-range and down-range extent with increasing range. In the resolution-limited case, power varies as R^{-3} because the resolution cell extent increases in only the cross-range dimension with increasing range.

If the range of interest varies by a large amount, there will be a significant variation in the grazing angle δ and therefore in the down-range extent of both the antenna beam footprint and the ground range resolution cell. For instance, for a radar at a constant altitude h and a slant range R to the ground, $\sin\delta = h/R$. As R increases, the beam-limited antenna footprint area will then increase as R^3 instead of R^2 so that the clutter power would be

---

[3]The area of an ellipse with axes of length $R\theta_3$ and $R\phi_3/\sin\delta$ is $(\pi/4)R^2\theta_3\phi_3/\sin\delta$. The rectangular approximation $R^2\theta_3\phi_3/\sin\delta$ is greater by a moderate factor of 27% (1.05 dB).

expected to fall only as R^{-1}. However, σ^0 may also vary significantly with grazing angle (see Sec. 2.3.1). Additional complications occur when R increases so much that a radar that was beam limited at a relatively short range and steep grazing angle becomes resolution limited at a longer range and shallower grazing angle, or the grazing angle falls below the "critical angle" at which the area reflectivity tends to decrease rapidly with decreasing δ. Consequently, the received surface clutter power may fall off at various rates from R^{-1} to R^{-3} or even more rapidly at very shallow angles (Long, 2001; Currie, 2010).

2.2.3 Radar Cross Section

Section 2.2.1 introduced the radar cross section to account for the amount of power reradiated by a scatterer back toward the radar transmitter. To restate the concept, denote the incident power density at the scatterer as Q_t and the backscattered power density at the transmitter as Q_b. If that backscattered power density resulted from isotropic radiation from the target, it would have to satisfy

$$Q_b = \frac{P_b}{4\pi R^2} \quad \text{W/m}^2 \tag{2.33}$$

for some total backscattered power P_b. RCS is the *fictional* area over which the transmitted power density Q_t must be intercepted to collect that total power P_b that would account for the received power density. In other words, σ must satisfy

$$P_b = \sigma Q_t \quad \text{W} \tag{2.34}$$

Combining Eqs. (2.33) and (2.34) gives

$$\sigma = 4\pi R^2 \frac{Q_b}{Q_t} \quad \text{m}^2 \tag{2.35}$$

This definition is usually written in terms of electric field amplitude. Also, in order to make the definition dependent only on the scatterer characteristics, range is eliminated by taking the limit as R tends to infinity. Thus, the formal definition of radar cross section becomes (Knott et al., 1993)

$$\sigma = 4\pi \lim_{R \to \infty} \left[R^2 \frac{|E^b|^2}{|E^t|^2} \right] \quad \text{m}^2 \tag{2.36}$$

where E^b and E^t are the backscattered and transmitted electric field complex amplitudes, respectively.

The RCS just defined is a single real scalar number. Implicit in the definition is the use of a single polarization of the transmitted wave and a single receiver polarization, usually the same as the transmitted polarization. However, the polarization state of a transverse electromagnetic plane wave is a two-dimensional vector, so two orthogonal polarization basis vectors are required to fully describe the wave. The most common basis choices are linear (horizontal and vertical polarizations) and circular (left and right rotating polarizations). Furthermore, a general scatterer will modify the polarization of an incident wave so that the energy backscattered from, say, the vertical component of the incident wave may have both vertical and horizontal components. To account fully for polarization effects, RCS must be generalized to the *polarization scattering matrix* (PSM) **S**, which relates the complex amplitudes of the incident and backscattered fields for all four combinations of transmit and receive polarization. For a radar using a linear polarization basis, this relation is (Mott, 1986; Holm, 1987; Knott et al., 1993)

$$\begin{bmatrix} E_H^b \\ E_V^b \end{bmatrix} = \begin{bmatrix} S_{HH} & S_{HV} \\ S_{VH} & S_{VV} \end{bmatrix} \begin{bmatrix} E_H^t \\ E_V^t \end{bmatrix} = \mathbf{S} \begin{bmatrix} E_H^t \\ E_V^t \end{bmatrix} \quad (2.37)$$

Instead of a single real number, the target backscattering characteristics are now described by four complex numbers. If the radar transmitted and received, say, only the vertical component, then the observed RCS σ would be related to $\mathbf{S}$ by

$$\sigma = |S_{VV}|^2 \ \ \mathrm{m}^2 \quad (2.38)$$

Radars can be designed to measure the full complex PSM. Other designs measure the magnitudes but not the phases of the elements of the PSM, or the magnitudes of two of the PSM elements. A radar that measured $|S_{VV}|$ and $|S_{HV}|$ (vertical transmitted polarization, both vertical and horizontal received polarization) would be said to measure the vertical *co-polarized* (*co-pol*) and *cross-polarized* (*cross-pol*) components. These *polarimetric* measurements can be used for a variety of target analysis purposes. A discussion of polarimetric techniques is beyond the scope of this book. Henceforth, it will be assumed that a single fixed polarization is transmitted and a single fixed polarization received so that RCS is described by a scalar function rather than a matrix. The reader is referred to Holm (1987) and Mott (1986) for discussions of polarimetric radars and polarimetric signal processing.

Typical values of RCS for targets of interest range from 0.01 m² (−20 dB with respect to 1 m², or −20 dBsm) to hundreds of square meters (≥+20 dBsm). Both larger and smaller values are sometimes observed. Table 2.1 lists representative RCS values for various types of targets.

2.2.4 Radar Cross Section for Meteorological Targets

Weather targets are an example of volume clutter. The actual observed echo is the composite backscatter of many raindrops, suspended water particles, hailstones, or snowflakes in the radar's resolution cell. The field of radar meteorology expresses the reflectivity of weather targets such as rain or snow in terms of a normalized factor called the *meteorological reflectivity factor* (not to be confused with the *volume reflectivity* η used here) and usually represented with the symbol Z (Sauvageot, 1992; Doviak and Zrnic, 1993).

Suppose the RCS of the *i*th individual scatterer is σ_i and assume noncoherent addition. Then the total RCS of a volume ΔV containing N such scatterers is $\sum \sigma_i$ and the volume reflectivity is

$$\eta = \frac{1}{\Delta V} \sum_{i=1}^{N} \sigma_i \ \ \mathrm{m}^{-1} \quad (2.39)$$

Water droplets are often modeled as small conducting spheres. When the ratio of the sphere radius a to the radar wavelength λ_t is small, specifically $2\pi a/\lambda_t \ll 1$, the radar cross section associated with the *i*th scatterer can be expressed as

$$\sigma_i = \frac{\pi^5 |K|^2 d_i^6}{\lambda_t^4} \ \ \mathrm{mm}^6/\mathrm{m}^2 \ \mathrm{or}\ \mathrm{m}^2 \quad (2.40)$$

Signal Models

	Target	RCS, m²	RCS, dBsm
Aircraft	Conventional unmanned winged missile	0.5	−3
	Small single-engine aircraft	1	0
	Small fighter aircraft or 4-passenger jet	2	3
	Large fighter aircraft	6	8
	Medium bomber or jet airliner	20	13
	Large bomber or jet airliner	40	16
	Jumbo jet	100	20
Ships	Small open boat	0.02	−17
	Small pleasure boat	2	3
	Cabin cruiser	10	10
	Large ship at zero grazing angle	10,000+	40+
Wheeled	Bicycle	2	3
	Automobile	100	20
	Pickup truck	200	23
Biological	Insect	0.00001	−50
	Bird	0.01	−20
	Man	1	0

Source: After Skolnik (2001).

TABLE 2.1 Typical RCS Values at Microwave Frequencies

where d_i is the drop diameter. While the obvious units of σ_i are square meters, the meteorological community normally expresses droplet sizes in millimeters, so that σ_i is often given in the mixed units of mm⁶/m⁴. Both types of units are given in Eq. (2.40). The constant K is determined from the complex index of refraction m according to

$$K = \frac{m^2 - 1}{m^2 + 2} \tag{2.41}$$

The index of refraction is a function of both the temperature and wavelength. However, for wavelengths between 3 cm (10 GHz, X band) and 10 cm (3 GHz, C band) and temperatures between 0 and 20°C, the value of $|K|^2$ is approximately a relatively constant 0.93 for scatterers composed of water and 0.197 for ice.

Substituting Eq. (2.40) in Eq. (2.39) gives

$$\eta = \frac{1}{\Delta V} \sum_{i=1}^{N} \frac{\pi^5 |K|^2 d_i^6}{\lambda_t^4} = \frac{\pi^5 |K|^2}{\lambda_t^4} \frac{1}{\Delta V} \sum_{i=1}^{N} d_i^6 \quad \text{mm}^6/\text{m}^7 \text{ or m}^{-1} \tag{2.42}$$

The quantity

$$Z \equiv \frac{1}{\Delta V} \sum_{i=1}^{N} d_i^6 \quad \text{mm}^6/\text{m}^3 \text{ or m}^3 \tag{2.43}$$

Level	Rain Fall Rate, mm/h	Reflectivity, dBZ	Category
1	0.49 to 2.7	18 to <30	Light mist
2	2.7 to 13.3	30 to <41	Moderate
3	13.3 to 27.3	41 to <46	Heavy
4	27.3 to 48.6	46 to <50	Very heavy
5	48.6 to 133.2	50 to <57	Intense
6	133.2 and greater	57 and above	Extreme

TABLE 2.2 Correspondence between dBZ Meteorological Reflectivity Factor and Volume Reflectivity

is called the *meteorological reflectivity factor* and is usually expressed in mixed units of mm^6/m^3. The value of Z is determined primarily by the *drop size distribution*, that is, the relative frequency of larger- and smaller-diameter drops. Because of its dependence on the sixth power of diameter, Z rises rapidly as drop size increases. Due to the large range of values observed for Z, it is commonly expressed on a decibel scale and denoted as dBZ. Using this definition in Eq. (2.42) gives the following relation between η and Z:

$$\eta = \frac{\pi^5 |K|^2}{\lambda_t^4} Z \text{ m}^{-1} \tag{2.44}$$

Given a measured echo power, the radar range equation for volume scatterers [Eq. (2.25)] can be used to estimate η, and then Eq. (2.44) can be used to convert η to Z.

Because it is related only to the volume density and size of scatterers, meteorologists prefer to express radar echo strength in terms of the meteorological reflectivity factor Z rather than the volume reflectivity η. The value of Z can then be related to the amount of water in the air or the precipitation rate. A number of empirical models are used to relate observed values of Z to rain rates. These models depend on the type of precipitation, for example, snow versus thunderstorm rain versus orographic[4] rain. A common model is that of Table 2.2, which shows the six-level equivalence between observed Z values (in dBZ) and rainfall rates used in the U.S. NEXRAD national weather radar system. Very similar scales are used in the commercial "Doppler weather radar" systems familiar to every viewer of television and online weather reports.

It is important to note that the dBZ values in Table 2.2 are 10 times the base 10 logarithm of Z in the mixed units mm^6/m^3. When Z is given in m^6/m^3 = m^3, it must be multiplied by 10^{18} to convert it to units of mm^6/m^3 before converting to a decibel scale and using Table 2.2.

2.2.5 Statistical Description of Radar Cross Section

The radar cross section of real targets cannot be effectively modeled as a simple constant. In general, RCS is a complex function of aspect angle, frequency, and polarization, even for relatively simple scatterers. For example, the conducting trihedral corner reflector of Fig. 2.4

---

[4]A form of rain that occurs when moist air is lifted over an obstacle such as a mountain range, cooling as it rises and condensing into rainfall.

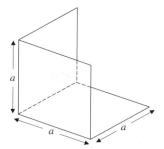

FIGURE 2.4 Square trihedral corner reflector.

is often used as a calibration target in field measurements. Its RCS when viewed along its axis of symmetry (looking "into the corner") can be determined analytically; it is (Knott et al., 1993)

$$\sigma = \frac{12\pi a^4}{\lambda_t^2} \text{ m}^2 \tag{2.45}$$

The RCS increases with increasing reflector size but also as the square of frequency. On the other hand, at least one frequency- and aspect-independent scatterer exists: a conducting sphere having a radius r that is large relative to the radar wavelength, $r \gg \lambda_t$. The RCS of a conducting sphere is independent of aspect angle because of the spherical symmetry. Far from being constant, its RCS depends strongly on frequency for wavelengths long relative to r. However, the RCS does converge to a constant πr^2 when the sphere is large relative to wavelength.

A simple example of frequency and aspect dependence is the two-scatterer "dumbbell" target of Fig. 2.5. If the nominal range R is much greater than the separation D, the range to the two scatterers is approximately

$$R_{1,2}(\theta) \approx R \pm \frac{D}{2}\sin\theta \text{ m} \tag{2.46}$$

If the signal $A\exp(2\pi F_t t)$ is transmitted, the echo from each scatterer will be proportional to $A\exp[j2\pi F_t(t - 2R_{1,2}(\theta)/c)]$. The voltage $\bar{y}(t)$ of the composite echo is therefore proportional to

$$\begin{aligned}
\bar{y}(t) &\propto A\exp[j2\pi F_t(t - 2R_1(\theta)/c)] + A\exp[j2\pi F_t(t - 2R_2(\theta)/c)] \\
&= A\exp[j2\pi F_t(t - 2R/c)] \cdot [\exp(-j2\pi F_t D\sin\theta/c) + \exp(+j2\pi F_t D\sin\theta/c)] \\
&= 2A\exp[j2\pi F_t(t - 2R/c)]\cos(2\pi F_t D\sin\theta/c) \\
&= 2A\exp(-j4\pi R/\lambda_t)\exp(j2\pi F_t t)\cos(2\pi F_t D\sin\theta/c)
\end{aligned} \tag{2.47}$$

RCS is proportional to the power of the composite echo. Taking the squared magnitude of Eq. (2.47) and simplifying leads to the result

$$\sigma \propto 4A^2 \left|\cos(2\pi F_t D\sin\theta/c)\right|^2 = 4A^2\left|\cos(2\pi D\sin\theta/\lambda_t)\right|^2 \text{ m}^2 \tag{2.48}$$

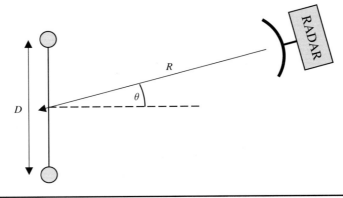

FIGURE 2.5 Geometry for determining relative RCS of a "dumbbell" target.

Equation (2.48) shows that the RCS is a periodic function of both radar frequency and aspect angle. The larger the scatterer separation in terms of wavelengths, the more rapidly the RCS varies with angle or frequency. The variation in RCS of the dumbbell target is plotted in Fig. 2.6 for $D = 5\lambda_t$. The plot has been normalized so that the maximum value corresponds to 0 dB. Notice the multilobed structure as the varying path lengths traversed by the echoes from the two scatterers cause their echoes to shift between constructive (in phase with another) and destructive (out of phase) interference. Also note that the maxima at aspect angles of 90° and 270° (the two "end fire" cases) are the broadest, while the maxima at the two "broadside" cases of 0° and 180° are the narrowest. Figure 2.7 plots the same data in a more traditional polar format.

The relative RCS of a target comprising multiple scatterers can be computed as a function of θ and λ_t using a generalization of Eq. (2.47). Imagine a target composed of N scatterers

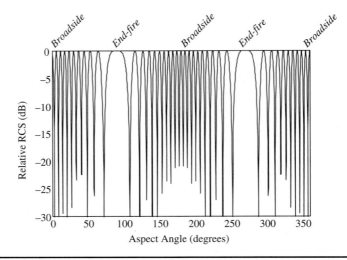

FIGURE 2.6 Relative radar cross section of the "dumbbell" target of Fig. 2.5 when $D = 5\lambda_t$.

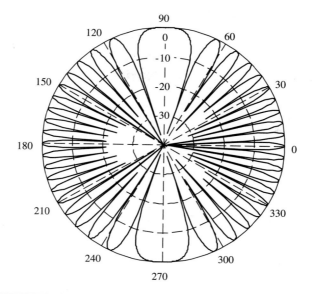

FIGURE 2.7 Polar plot of the data of Fig. 2.6.

clustered in a compact region. This might represent a man-made object, for example, a vehicle, with the scatterers being individual sources of reflection from the vehicle, or it might represent a resolution cell of clutter, with the scatterers being multiple blades of grass, ocean ripples, or rain drops. The complex reflectivity of the ith scatterer is $\rho_i = \sqrt{\sigma_i}\exp(j\psi_i)$ corresponding to an RCS of $\sigma_i = |\rho_i|^2$. Let the radar be located at a nominal range R and arbitrary aspect angle θ from the center of the cluster. While the nominal range R remains constant as the aspect angle varies, the ranges R_i to the individual scatterers within the target vary with aspect angle θ. The complex voltage of the echo will be, to within a proportionality constant,

$$\begin{aligned}\bar{y}(t) &\propto \sum_{i=1}^{N} \sqrt{\sigma_i}\exp(j\psi_i)\exp[j2\pi F_t(t - 2R_i(\theta)/c)] \\ &= \exp(j2\pi F_t t)\sum_{i=1}^{N}\sqrt{\sigma_i}\exp(j\psi_i)\exp[-j4\pi F_t R_i(\theta)/c] \\ &= \exp(j2\pi F_t t)\sum_{i=1}^{N}\sqrt{\sigma_i}\exp(j\psi_i)\exp[-j4\pi R_i(\theta)/\lambda_t]\end{aligned} \qquad (2.49)$$

This expression assumes the target size (variation in range among the scatterers) is small compared to the nominal range so that the R^{-4} term is approximately the same for all scatterers in the summation and may be factored out into the proportionality constant.

Each term in Eq. (2.49) represents the echo from a single scatterer. The echoes from each of the scatterers add coherently to form the resultant voltage $\bar{y}(t)$. The individual echoes exhibit various degrees of constructive and destructive addition, depending on the details of their individual phases. The RCS σ is proportional to $|\bar{y}|^2$. Define the real-valued reflectivity amplitude (within a proportionality constant) as

$$\zeta \equiv |\bar{y}| \propto \left| \sum_{i=1}^{N} \sqrt{\sigma_i} \exp(j\psi_i) \exp[-j4\pi R_i(\theta)/\lambda_t] \right| \qquad (2.50)$$

Then the RCS is proportional to

$$\sigma = \zeta^2 \propto \left| \sum_{i=1}^{N} \sqrt{\sigma_i} \exp(j\psi_i) \exp[-j4\pi R_i(\theta)/\lambda_t] \right|^2 \quad \text{m}^2 \qquad (2.51)$$

For signal processing purposes, the proportionality constant can usually be ignored so that Eqs. (2.50) and (2.51) can be taken as equalities.

Equation (2.51) exposes an extremely important characteristic of the scatterer echoes. In addition to being scaled by the scatterer reflectivity, each term is additionally phase-shifted relative to the carrier by the amount $-4\pi R_i(\theta)/\lambda_t = -4\pi F_t R_i(\theta)/c$ radians. The phase shift is a very sensitive indicator of range. For example, a change in range of only one-quarter wavelength shifts the echo phase by 180°, changing whether a given scatterer adds constructively or destructively to the total response. A quarter wavelength is only 7.5 cm at an RF of 1 GHz, 7.5 mm at 10 GHz, and 0.79 mm at 95 GHz. This suggests that phase measurements may provide a way to estimate very fine radar-target range changes and, indeed, phase measurements provide the basis for most coherent radar signal processing operations such as Doppler processing, imaging, and adaptive beamforming. By the same token, however, echo phase is also a highly *ambiguous* indicator of range changes since every $\lambda_t/2$ change in range produces a 2π change in phase. Consequently, phase measurements will be more useful for estimating range *changes* from one measurement to the next rather than total range, which will likely be thousands or millions of wavelengths.

RCS variations like those of Fig. 2.6 become very complicated for complex targets having many scatterers of varying individual RCS. Figure 2.8 shows a "target" consisting of 50 point scatterers randomly distributed within a rectangle 5 m wide and 10 m long. The RCS of each individual point scatterer is a constant, $\sigma_i = 1.0$. Figure 2.9 shows the relative RCS computed at 0.2° increments using Eq. (2.51) that results when this target is viewed at varying aspect angles 10 km from its center at a frequency of 10 GHz. The dynamic range

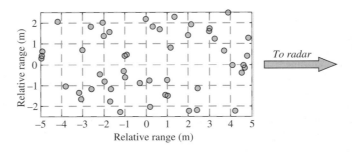

FIGURE 2.8 Random distribution of 50 scatterers used to obtain Fig. 2.9. See text for details.

Signal Models   65

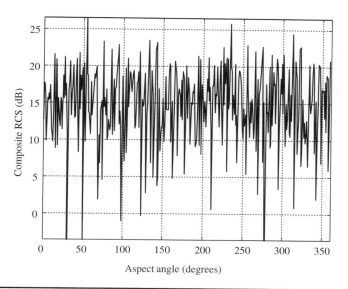

FIGURE 2.9 Relative RCS versus aspect angle of the complex target of Fig. 2.8 at a range of 10 km and radar frequency of 10 GHz.

observed for RCS is similar to that of the simple dumbbell target, but the lobing structure is much more complicated.[5]

This variation in composite RCS is called target RCS *fluctuations*. Fluctuations occur because, when the radar views the target from a different aspect angle, the ranges to each of the individual scatterers change, changing their echo phases. This in turn changes how the scatterers combine constructively and destructively at the receiver, causing the variations in observed RCS. In real-world targets, changing aspect angles will also change the number and location of the scattering centers on a target that are visible to the radar and so contributing to the RCS.

Changing the RF frequency also changes the scatterer phases and induces RCS fluctuations even when there is no change in aspect angle. For Fig. 2.10, a 20-scatterer, 5 m by 10 m-random target was observed from an unchanging aspect angle. If a fixed RF was used the RCS would be exactly the same on each measurement. However, in this case the RF frequency was increased by 18.5 MHz from one measurement to the next, starting at 10.0 GHz. The resulting relative RCS values fluctuate by 38 dB, a factor of about 6300.

The complicated variation of RCS with radar frequency and target aspect angle observed for even moderately complex targets leads to the use of a statistical description for radar cross section (Levanon, 1988; Nathanson, 1991; Skolnik, 2001). This means that the RCS σ of the scatterers within a single resolution cell is considered to be a random variable described by a specified *probability density function* (PDF). The mean or median RCS is typically used for radar range equation calculations but the full PDF is needed for detection probability calculations, as will be seen in Chap. 6.

It is important to understand that RCS is not considered to actually be a random quantity in any physical sense; it is an entirely deterministic target characteristic that could be predicted quite accurately, including its fluctuations with aspect angle and RF, given a

---

[5]RCS also varies similarly with angle over the full 3D sphere, not just in a 2D plane.

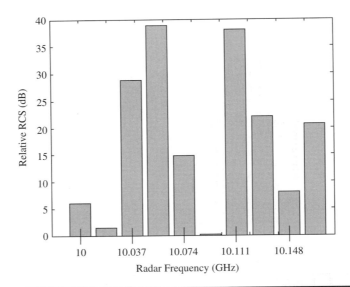

FIGURE 2.10 Variation in RCS due to frequency agility for a constant viewing angle. See text for details.

sufficiently detailed description of the target and sufficiently sophisticated electromagnetic scattering simulation codes. Statistical models are used because they offer a much simpler way to capture the complex behavior of RCS in a manner amenable to calculating radar detection performance and other system characteristics, both analytically and in simulations.

A variety of PDFs are used to describe the statistical model of the RCS for different targets. Consider first a target similar to that of Fig. 2.8 consisting of a large number of individual scatterers randomly distributed within the target boundaries, each with its own individual but fixed RCS. Because of its high sensitivity to small range changes, the phase of the echoes from the various scatterers is assumed to be a random variable distributed uniformly on $(0, 2\pi]$. Under these circumstances, the central limit theorem guarantees that the real and imaginary parts of the composite echo can each be assumed to be independent, zero mean Gaussian random variables with the same variance, say α^2 (Beckmann and Spizzichino, 1963; Papoulis and Pillai, 2001). In this case, the squared-magnitude σ necessarily exhibits an exponential PDF:

$$p_\sigma(\sigma) = \begin{cases} \dfrac{1}{\bar{\sigma}} \exp\left[\dfrac{-\sigma}{\bar{\sigma}}\right], & \sigma \geq 0 \\ 0, & \sigma < 0 \end{cases} \qquad (2.52)$$

where $\bar{\sigma} = 2\alpha^2$ is the mean value of the RCS σ. The corresponding amplitude ζ has a Rayleigh PDF,

$$p_\zeta(\zeta) = \begin{cases} \dfrac{2\zeta}{\bar{\sigma}} \exp\left(\dfrac{-\zeta^2}{\bar{\sigma}}\right), & \zeta \geq 0 \\ 0, & \zeta < 0 \end{cases} \qquad (2.53)$$

The phase of the complex echo will be uniformly distributed over $(0, 2\pi]$.

Signal Models

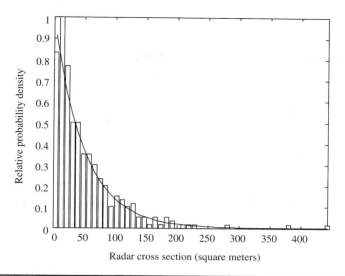

FIGURE 2.11 Histogram of RCS data of Fig. 2.9 and exponential PDF with the same mean.

Figure 2.11 compares a histogram of the RCS values from Fig. 2.9 to an exponential curve of the form of Eq. (2.52) having the same mean $\bar{\sigma}$. The fit of the total RCS histogram to the exponential PDF is quite good, suggesting that it is a good predictor of the relative frequency of high and low RCS values as the radar-target aspect angle changes. It works equally well for fluctuations due to RF changes. Similar results are observed when the randomly distributed scatterers also have random individual cross sections drawn from the same Gaussian distribution, a somewhat more general and plausible situation than the fixed-RCS case. While the exponential model for RCS is only strictly accurate in the limit of a very large number of scatterers, in practice it can be a good model for a target having as few as 6 to 10 significant scatterers.

Instead of an ensemble of equal-strength scatterers, some radar targets are better modeled as consisting of one or a few dominant scatterers contributing most of the RCS, modified by the contributions of a number of smaller scatterers. Many PDFs have been advocated as models for these targets. Table 2.3 summarizes several of the more common PDFs for modeling the variability of RCS with aspect angle and radar frequency for targets with and without dominant scatterers. The mean value $\bar{\sigma}$ of RCS is given for each case in which the PDF is not written explicitly in terms of $\bar{\sigma}$. The variance var(σ) is also given for each case. Several other PDFs have also been used to model RCS fluctuations but are less common and so are not discussed here. Additional information on these PDFs and their relationships, including more conventional forms and in some cases the characteristic functions, is given in App. A. Additional discussion of some of these PDFs in the context of clutter modeling is also given in Sec. 2.3.

One fundamental difference among the various RCS models of Table 2.3 is whether the PDF has one or two free parameters. The nonfluctuating, exponential, and all chi-square PDFs (once the order is stated) are *one-parameter distributions*. That parameter in the form given earlier is the mean RCS $\bar{\sigma}$. The non-central chi-square, Weibull, log-normal, and K power PDFs are *two-parameter distributions*, as is the chi-square with variable degree. In the forms given, the parameters are $\bar{\sigma}$ and a^2 for the noncentral chi-square, $\bar{\sigma}$ and m for the variable-order chi-square, B and C for the Weibull, σ_m and s for the log-normal, and a and c

Chapter Two

Model Name	PDF for RCS σ	Comment
One-Parameter PDFs		
Nonfluctuating, Marcum, Swerling 0, or Swerling 5	$p_\sigma(\sigma) = \delta_D(\sigma - \bar{\sigma})$ $\text{var}(\sigma) = 0$	Constant echo power, e.g., calibration sphere or perfectly stationary target with no radar or target motion.
Exponential, chi-square of degree 2	$p_\sigma(\sigma) = \dfrac{1}{\bar{\sigma}} \exp\left[\dfrac{-\sigma}{\bar{\sigma}}\right]$ $\text{var}(\sigma) = \bar{\sigma}^2$	Many scatterers, randomly distributed, none dominant. Used in Swerling case 1 and 2 models.
Chi-square of degree 4	$p_\sigma(\sigma) = \dfrac{4\sigma}{\bar{\sigma}^2} \exp\left[\dfrac{-2\sigma}{\bar{\sigma}}\right]$ $\text{var}(\sigma) = \bar{\sigma}^2/2$	Approximation to case of many small scatterers + one dominant. Used in Swerling case 3 and 4 models.
Two-Parameter PDFs		
Chi-square of degree $2m$, Weinstock	$p_\sigma(\sigma) = \dfrac{m}{\Gamma(m)\bar{\sigma}} \left[\dfrac{m\sigma}{\bar{\sigma}}\right]^{m-1} \exp\left[\dfrac{-m\sigma}{\bar{\sigma}}\right]$ $\text{var}(\sigma) = \bar{\sigma}^2/m$	Generalization of the two preceding cases. Weinstock cases correspond to $0.6 \leq 2m \leq 4$. Higher degrees correspond to presence of a more dominant single scatterer.
Noncentral chi-square of degree 2	$p_\sigma(\sigma) = \dfrac{1}{\bar{\sigma}}(1+a^2)\exp\left[-a^2 - \dfrac{\sigma}{\bar{\sigma}}(1+a^2)\right]$ $I_0\left[2a\sqrt{(1+a^2)(\sigma/\bar{\sigma})}\right]$ $\text{var}(\sigma) = \dfrac{(1+2a^2)}{(1+a^2)^2}\bar{\sigma}^2$	Exact solution for one dominant scatterer plus many small ones. Corresponds to Rice amplitude PDF. Ratio of dominant RCS to sum of small RCS is a^2. $I_0(\cdot)$ is the modified Bessel function of the first kind and order zero.
Weibull	$p_\sigma(\sigma) = CB\sigma^{C-1}\exp[-B\sigma^C]$ $\bar{\sigma} = B^{-1/C}\Gamma(1+1/C)$ $\text{var}(\sigma) = B^{-2/C}\left[\Gamma(1+2/C) - \Gamma^2(1+1/C)\right]$	Empirical fit to many measured target and clutter distributions. Can model long-tailed or "spiky" data. Not readily expressible in terms of $\bar{\sigma}$.
Log-normal	$p_\sigma(\sigma) = \dfrac{1}{\sqrt{2\pi}\,s\sigma}\exp\left[-\ln^2(\sigma/\sigma_m)/2s^2\right]$ $\bar{\sigma} = \sigma_m \exp(s^2/2)$ $\text{var}(\sigma) = \sigma_m^2 \exp(s^2)[\exp(s^2)-1]$ $= \bar{\sigma}^2[\exp(s^2)-1]$	Empirical fit to many measured target and clutter distributions. Can model long-tailed of "spiky" data. σ_m is the median value of σ. Not readily expressible in terms of $\bar{\sigma}$.
K power	$p_\sigma(\sigma) = \dfrac{c}{\sqrt{\sigma}\cdot\Gamma(a)}\left(\dfrac{c\sqrt{\sigma}}{2}\right)^a K_{a-1}(c\sqrt{\sigma})$ $\bar{\sigma} = \dfrac{4a}{c^2}, \quad \text{var}(\sigma) = \dfrac{16a(a+2)}{c^4}$	Compound PDF for "spiky" target and clutter distributions. $K_{a-1}(\cdot)$ is the modified Bessel function of the second kind and order $a-1$.

TABLE 2.3 Common Statistical Models for Radar Cross Section

Signal Models

for the K power. For a one-parameter distribution, specifying the mean is sufficient to characterize the complete PDF. For the two-parameter case estimates of two parameters, usually the variance and either the mean or median, must be specified to characterize the PDF. This distinction is important in the design of automatic detection algorithms in Chap. 6.

The shape of the PDF of RCS directly affects detection performance, as will be seen in Chap. 6. Figure 2.12*a* compares the exponential, fourth-degree chi-square, second-degree noncentral chi-square, Weibull, log-normal, and K power density functions when all have the same RCS mean of 1.0. The exponential distribution then necessarily has a variance of 1, while the fourth-degree chi-square necessarily has a variance of 0.5. The parameters of the remaining density functions have been chosen to give them a variance of 1.5.

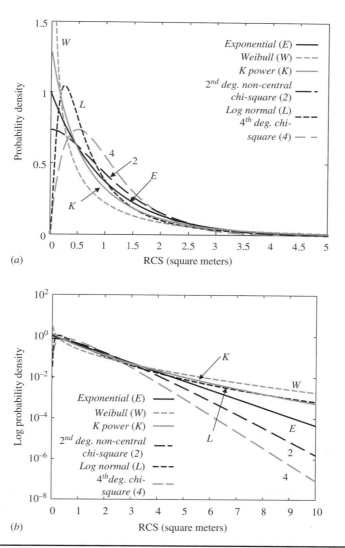

Figure 2.12 Comparison of six models for the probability density function of radar cross section. (*a*) Linear scale. (*b*) Log scale. See text for details.

For these parameter choices, the exponential, Weibull, and K power PDFs do not exhibit a peak at a particular RCS. Rather, they all decay monotonically as RCS increases. Target fluctuations described by these distributions would have a higher frequency of occurrence for lower RCS values. The log-normal and fourth-degree chi-square distributions show distinct peaks, so that RCS values close to the peak will occur most often, with both larger and smaller values being less frequent. In this example, the second-degree noncentral chi-square peaks at $\sigma = 0$, but shows a leveling off of the PDF as $\sigma \to 0$. These observations suggest that, for these parameters, the log-normal and fourth-degree chi-square PDFs might be better for modeling target echoes having a dominant scatterer, while the others might be candidates for targets lacking such a component. For some other values of the parameter a, the second-degree noncentral chi-square PDF exhibits a distinct peak. The same is true of the Weibull for some other choices of its parameters. Thus, these two PDFs could also be considered for modeling the RCS of targets having dominant scatterers in some cases.

Figure 2.12b repeats the same data on a semilogarithmic scale so that the behavior of the PDF "tails" is more evident. Notice that for these parameters, the probability density at $\sigma = 10$ m^2 of the Weibull PDF is a factor of about 1120 higher than that of the second-degree noncentral chi-squared PDF even though both have the same mean and variance. Consequently, modeling the target with the Weibull PDF will result in a much more frequent occurrence of relatively high RCS values. This behavior is often referred to as "spiky" RCS. For these parameters, the "longest" PDF tail, meaning the most slowly decaying and therefore the "spikiest" model, is the Weibull, while the shortest tail is the fourth-degree chi-squared. However, different choices of the parameters can change the relative tail lengths of the various PDFs.

Most radar analysis and measurement programs emphasize RCS measurements, which are proportional to received power. Sometimes the corresponding amplitude (magnitude of voltage) ζ is of interest, particularly for use in simulations where Eq. (2.50) is used explicitly to model the composite echo from a multiple scatterer target. The probability density function for the voltage is then required in order to properly model the probabilistic variations of the complex sum. The PDF of ζ is easily derived from the PDF of σ using basic results of random variables (Papoulis and Pillai, 2001). Because RCS is nonnegative, the transformation[6]

$$\zeta = \sqrt{\sigma} \tag{2.54}$$

from RCS to voltage has only one real solution for σ, namely $\sigma = \zeta^2$. It follows that the PDF of ζ is given by

$$p_\zeta(\zeta) = \frac{p_\sigma(\zeta^2)}{d\zeta/d\sigma} = 2\zeta p_\sigma(\zeta^2) \tag{2.55}$$

Equation (2.55) can be used to write the voltage PDFs by inspection from Table 2.3. The results, given in Table 2.4, are expressed in terms of the parameters of the corresponding RCS distribution from Table 2.3. Additional information is given in App. A.

As has been seen, the RCS of a complex target varies with both transmitted frequency and aspect angle. Another important characteristic of a target's signature is the *decorrelation interval* in time, frequency, and angle. This is the *change* in time, frequency, or angle

---

[6]Because the power P of a real sinusoid is related to its amplitude A according to $P = A^2/2$ instead of just $P = A^2$ as with a complex sinusoid, some authors present a slightly different form for the voltage distributions.

RCS Model Name	PDF for Voltage ζ	Comment
One-Parameter PDFs		
Nonfluctuating, Marcum, Swerling 0, or Swerling 5	$p_\zeta(\zeta) = \delta_D(\zeta - \bar{\zeta})$ $\bar{\zeta} = \sqrt{\bar{\sigma}}, \quad \text{var}(\zeta) = 0$	Also nonfluctuating model, one parameter changed: $\bar{\sigma} \to \sqrt{\bar{\sigma}}$.
Rayleigh, central chi of degree 2	$p_\zeta(\zeta) = \dfrac{2\zeta}{\bar{\sigma}} \exp\left[\dfrac{-\zeta^2}{\bar{\sigma}}\right]$ $\bar{\zeta} = \dfrac{1}{2}\sqrt{\pi\bar{\sigma}}, \quad \text{var}(\zeta) = \bar{\sigma}(1 - \pi/4)$	Corresponding RCS has exponential distribution. Voltage equivalent of Swerling 1 and 2 models.
Central chi of degree 4	$p_\zeta(\zeta) = \dfrac{8\zeta^3}{\bar{\sigma}^2} \exp\left[\dfrac{-2\zeta^2}{\bar{\sigma}}\right]$ $\bar{\zeta} = \dfrac{3}{4}\sqrt{\dfrac{\pi\bar{\sigma}}{2}}, \quad \text{var}(\zeta) = \left(1 - \dfrac{9}{32}\pi\right)\bar{\sigma}$	Corresponding RCS has fourth-degree chi-square distribution. Voltage equivalent of Swerling 3 and 4 models.
Two-Parameter PDFs		
Central chi of degree 2m	$p_\zeta(\zeta) = \dfrac{2\zeta m}{\Gamma(m)\bar{\sigma}}\left(\dfrac{m\zeta^2}{\bar{\sigma}}\right)^{m-1} \exp\left[-\dfrac{m\zeta^2}{\bar{\sigma}}\right]$ $\bar{\zeta} = \dfrac{\Gamma(m+0.5)}{\Gamma(m)}\sqrt{\dfrac{\pi\bar{\sigma}}{m}}, \quad \text{var}(\zeta)$ $= \bar{\sigma}\left\{1 - \dfrac{1}{m}[\Gamma(m+0.5)/\Gamma(m)]^2\right\}$	Voltage equivalent of Weinstock RCS models.
Rice or Rician, noncentral chi of degree 2	$p_\zeta(\zeta) = \dfrac{2\zeta(1+a^2)}{\bar{\sigma}} \exp\left[-a^2 - \dfrac{\zeta^2}{\bar{\sigma}}(1+a^2)\right] I_0\left(2a\zeta\sqrt{(1+a^2)/\bar{\sigma}}\right)$ $\bar{\zeta} = \dfrac{1}{2}\sqrt{\dfrac{\pi\bar{\sigma}}{1+a^2}} \exp(-a^2)\,_1F_1[1.5, 1; a^2]$ $\text{var}(\zeta) = \left(\dfrac{\bar{\sigma}}{1+a^2}\right)\exp(-a^2)\left[\,_1F_1(2,1;a^2) - \dfrac{\pi}{4}\exp(-a^2)\,_1F_1^2(1.5,1;a^2)\right]$	Voltage equivalent of noncentral chi-square of degree 2. $_1F_1(\alpha,\beta,\chi)$ is the confluent hypergeometric function, also called Kummer's function.
Weibull	$p_\zeta(\zeta) = 2CB\zeta^{2C-1}\exp[-B\zeta^{2C}]$ $\bar{\zeta} = B^{-1/2C}\,\Gamma(1+1/2C), \quad \text{var}(\zeta)$ $= B^{-1/C}[\Gamma(1+1/C) - \Gamma^2(1+1/2C)]$	Also Weibull, one parameter changed: $C \to 2C$.
Log-normal	$p_\zeta(\zeta) = \dfrac{2}{\sqrt{2\pi}\,s\zeta}\exp[-4\ln^2(\zeta/\sqrt{\sigma_m})^2/s^2]$ $\bar{\zeta} = \sqrt{\sigma_m}\exp(s^2/8)$ $\text{var}(\zeta) = \sigma_m\exp(s^2/4)[\exp(s^2/4) - 1] = \bar{\zeta}^2[\exp(s^2/4) - 1]$	Also log-normal, both parameters (s, σ_m) changed: $s \to s/2$, $\sigma_m \to \sqrt{\sigma_m}$.
K	$p_\sigma(\zeta) = \dfrac{2c}{\Gamma(a)}\left(\dfrac{c\sigma}{2}\right)^a K_{a-1}(c\sigma)$ $\bar{\sigma} = \dfrac{\sqrt{\pi}}{c}\cdot\dfrac{\Gamma(a+1/2)}{\Gamma(a)}, \quad \overline{\sigma^2} = \dfrac{4a}{c^2}, \quad \sigma_\sigma^2 = \overline{\sigma^2} - \bar{\sigma}^2$	Compound PDF for "spiky" target and clutter distributions. $K_{a-1}(\cdot)$ is the modified Bessel function of the second kind and order $a-1$.

TABLE 2.4 Voltage Distributions Corresponding to Common Statistical Models of Radar Cross Section

required to cause the echo amplitude to decorrelate to a specified degree. If a rigid target such as a building is illuminated with a series of identical radar signals and there is no motion between the radar and target, one expects the same received complex voltage y from each pulse (ignoring receiver noise). If motion between the two is allowed, however, the relative path length between the radar and the various scatterers comprising the target will change, causing the composite echo amplitude to fluctuate as in Fig. 2.9. Thus, for rigid targets, decorrelation of the RCS is induced by changes in range and aspect angle. If natural clutter such as the ocean surface or a stand of trees is illuminated, the signature may decorrelate even if the radar and target do not move relative to each other. In this case, the decorrelation is caused by "internal motion" (sometimes called "intrinsic motion") of the clutter, such as the wave motion on the sea surface or the blowing leaves and limbs of the trees. The rate of decorrelation due to internal motion is influenced by factors external to the radar such as wind speed. Range or aspect changes also induce decorrelation of clutter signatures.

Although the behavior of real targets can be quite complex, a useful estimate of the change in frequency or angle required to decorrelate a target or clutter patch can be obtained by the following simple argument. Consider a target consisting of a uniform line array of point scatterers tilted at an angle θ with respect to the antenna boresight and separated by Δx from one another, as shown in Fig. 2.13. Assume an odd number $2M+1$ of scatterers indexed from $-M$ to $+M$, as shown for $M = 5$ in the figure. The total target extent is considered to be $L = (2M+1)\Delta x$. If the nominal distance to the radar R_0 is much larger than the target extent, $R_0 \gg L$, then the incremental distance an EM plane wave must travel from one scatterer to the next is $\Delta x \cdot \sin\theta$. If the target is illuminated with the waveform $A\exp(j\Omega_i t)$, the received signal will be

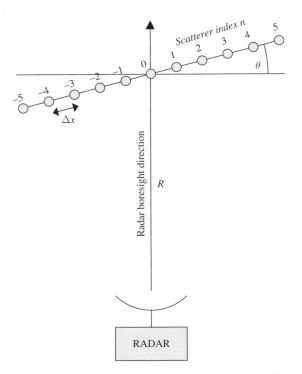

Figure 2.13 Geometry for calculation of RCS decorrelation interval in frequency and aspect angle.

$$\bar{y}(t) = \sum_{n=-M}^{M} A\exp\{j\Omega_t[t - 2(R_0 + n \cdot \Delta x \sin\theta)/c]\}$$
$$= A\exp[j\Omega_t(t - 2R_0/c)] \sum_{n=-M}^{M} \exp(-j4\pi n \cdot \Delta x \sin\theta \cdot F_t/c) \quad (2.56)$$

To simplify the notation, define

$$K_\theta = \frac{2\pi F_t}{c}\sin\theta = \frac{2\pi}{\lambda_t}\sin\theta \text{ rads/m}, \qquad \alpha = 2 \cdot \Delta x \quad (2.57)$$

K_θ is a spatial frequency (see App. B). Then $\bar{y}(t)$ can be considered as a function $\bar{y}(t; K_\theta)$ of both t and K_θ. The correlation in the variable K_θ, which includes both aspect angle and radar frequency, is of interest. Note that $\bar{y}(t; K_\theta)$ is periodic in K_θ with period $2\pi/\alpha$. The deterministic autocorrelation of $\bar{y}(t; K_\theta)$ as a function of the lag δK_θ is

$$s_{\bar{y}}(\delta K_\theta) = \int_{-\pi/\alpha}^{\pi/\alpha} \bar{y}(t; K_\theta)\bar{y}^*(t; K_\theta + \delta K_\theta) dK_\theta$$
$$= \int_{-\pi/\alpha}^{\pi/\alpha} \left\{ A\exp[j\Omega_t(t - 2R_0/c)] \sum_{n=-M}^{M} \exp(-j\alpha K_\theta n) \right\} \times \cdots \quad (2.58)$$
$$\cdots \times \left\{ A^*\exp[-j\Omega_t(t - 2R_0/c)] \sum_{l=-M}^{M} \exp[+j\alpha(K_\theta + \delta K_\theta)l] \right\} dK_\theta$$

The complex exponential terms outside the summations cancel. Interchanging integration and summation and collecting terms then gives

$$s_{\bar{y}}(\delta K_\theta) = |A|^2 \sum_{l=-M}^{M} \exp(+j\alpha \cdot \delta K_\theta l) \sum_{n=-M}^{M} \left[\int_{-\pi/\alpha}^{\pi/\alpha} \exp[-j\alpha(n-l)K_\theta] dK_\theta \right] \quad (2.59)$$

A change of variables $K'_\theta = \alpha K_\theta$ makes it clear that the integral has the form of the inverse discrete-time Fourier transform of a constant spectrum $S(K_\theta) = 2\pi/\alpha$. Therefore, the integral is just the discrete impulse function $(2\pi/\alpha)\delta[n-l]$. Using this fact reduces Eq. (2.59) to a single summation over l. Evaluating that sum and recalling that $L = (2M+1)\Delta x$ gives

$$s_{\bar{y}}(\delta K_\theta) = \frac{2\pi|A|^2}{\alpha} \frac{\sin[\alpha(2M+1)\delta K_\theta/2]}{\sin[\alpha \cdot \delta K_\theta/2]} = \frac{\pi|A|^2}{\Delta x}\frac{\sin(L \cdot \delta K_\theta)}{\sin(\Delta x \cdot \delta K_\theta)} \quad (2.60)$$

The decorrelation interval can now be determined by evaluating Eq. (2.60) to find the value of δK_θ which reduces s_y to a given level. This value of δK_θ can then be converted into equivalent changes in frequency or aspect angle.

One criterion is to choose the value of δK_θ corresponding to the first zero of the correlation function in Eq. (2.60), which occurs when the argument of the numerator equals π. The resulting value is the Rayleigh width of the autocorrelation function,

$$\pi = L \cdot \delta K_\theta \Rightarrow \delta K_\theta = \frac{\pi}{L} \text{ radians} \quad (2.61)$$

Recall that $K_\theta = (2\pi/c)F_t\sin\theta$. The total differential of z is then $dK_\theta = (2\pi/c)(\sin\theta \cdot dF_t + F_t\cos\theta \cdot d\theta)$. To determine the decorrelation interval in angle for a fixed radar frequency, set $dF_t = 0$ to give $dK_\theta = (2\pi/c)F_t\cos\theta \cdot d\theta$, so that $\delta K_\theta \approx (2\pi/c)F_t\cos\theta \cdot \delta\theta$. Similarly, the frequency step required to decorrelate the target is obtained by fixing the aspect angle θ so that $d\theta = 0$, leading to $\delta K_\theta \approx (2\pi/c)\sin\theta \cdot \delta F_t$. Combining these relations with Eq. (2.61) then gives the desired result for the change in angle or frequency required to decorrelate the echo amplitude:

$$\delta\theta = \frac{c}{2F_t L\cos\theta} \text{ radians}, \quad \delta F_t = \frac{c}{2L\sin\theta} \text{ Hz} \tag{2.62}$$

Note that $L\cos\theta$ is the projection of the target extent orthogonal to the radar boresight, while $L\sin\theta$ is the projection along the radar boresight. Thus, the decorrelation interval in aspect angle is driven by the width of the target as viewed from the radar, while the interval in frequency is driven by the depth. This observation suggests a more general pair of expressions that can be applied to more arbitrary many-scatterer targets:

$$\delta\theta = \frac{c}{2F_t \cdot L_w} \text{ radians}, \quad \delta F_t = \frac{c}{2L_d} \text{ Hz} \tag{2.63}$$

where L_w and L_d are the target width and depth, respectively, as viewed from the radar. These formulas show that the larger the target, the more rapid the decorrelation in frequency and angle.

As an example, consider a broadside view of an automobile having L_w about 5 m long. At L band (1 GHz), the target signature can be expected to decorrelate in $(3 \times 10^8)/(2 \times 5 \times 10^9) = 30$ mrad (about 1.7°) of aspect angle rotation while at W band (95 GHz) this is reduced to only 0.018°. The frequency step required for decorrelation from a head-on aspect (depth of $L_d = 5$ m) is 30 MHz. This result does not depend on the nominal transmitted frequency.

As another example, Fig. 2.14a shows the central portion of the autocorrelation function in angle for many-scatterer targets similar to that of Fig. 2.8, using only the data for aspect angles over a range ±3°. Each of the two autocorrelation functions shown is the average of the autocorrelations of 20 different random targets,[7] each having 20 randomly placed scatterers in a 5 m by 10-m box. The black curve is the autocorrelation of the data around a nominal boresight orthogonal to the 5 m side of the target, while the gray curve is the autocorrelation of the data viewed from the 10-m side. These aspect angles correspond to viewing the target nominally from the right and from the top in Fig. 2.8. At $F_t = 10$ GHz, the expected decorrelation interval in angle when viewed from the right is 0.17°; when viewed from the top it is 0.09°. These expected decorrelation intervals are marked by the vertical dashed lines in Fig. 2.14a. In both cases, the first minimum of the correlation function occurs at the predicted amount of change in the aspect angle. Figure 2.14b shows the average autocorrelation function in frequency over 30 similar random targets. The predicted decorrelation intervals closely approximate the first zero crossing.

It will be seen in Chap. 6 that in certain cases detection performance is improved when successive target measurements are uncorrelated. For this reason, some radars use a technique called *frequency agility* to force decorrelation of successive measurements (Ray, 1966). In this process, the radar frequency is increased by δF Hz or more between successive pulses, where δF is given by Eq. (2.63), ensuring that the target echo decorrelates from one pulse or

---

[7]The autocorrelation function for any one many-scatterer target can vary significantly from the results predicted by this analysis and the average shapes seen in Fig. 2.14. The analysis in this section is indicative of the expected value of the autocorrelation function and the decorrelation intervals.

Signal Models 75

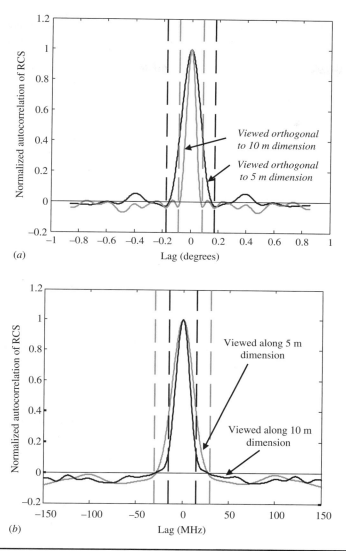

FIGURE 2.14 Average autocorrelation function for many-scatterer targets. (*a*) Angle autocorrelation functions. (*b*) Frequency autocorrelation functions. See text for details.

sweep to the next. Once the desired number of uncorrelated measurements is obtained, the cycle of increasing frequencies is repeated for the next set of measurements.

Equation (2.63) is based on a highly simplified target model and an assumption about what constitutes the decorrelation interval. A different definition, for example, defining the interval by the point at which the correlation function first drops to $1/2$ or $1/e$ of its peak, would result in a smaller estimate of the required change in angle or frequency to decorrelate the target. Also, many radars operate on the magnitude-squared of the echo amplitude, rather than the magnitude as has been assumed in this derivation. A square law detector produces a correlation function proportional to the square of Eq. (2.60) (Birkmeier and

Wallace, 1963). The first zero therefore occurs at the same value of δK_θ, and the previous conclusions still apply. However, if a different definition of the correlation interval (such as the 50 percent decorrelation point) is used, the required change in δK_θ is less for the square law than for the linear detector.

2.2.6 Target Fluctuation Models

Radar detection decisions are generally based on collecting one or more samples of target-plus-interference echo voltage ("measurements"), noncoherently integrating them to form a single resultant measurement, and comparing the magnitude, magnitude squared, or log magnitude of that result to a threshold to produce one detection decision. If the target, the radar, or both are moving during the time a set of N measurements is collected, the aspect angle will change. If the radar uses frequency agility, the RF will also change. A natural question is whether the target RCS during the measurement collection time should be considered constant or fluctuating. That is, assuming frequency agility is not used, does the radar-target aspect angle vary little enough that the RCS should be modeled as the same random variable from the appropriate PDF during the entire set of N measurements? Or is the aspect changing sufficiently rapidly that the RCS decorrelates from one measurement to the next, and so should be modeled as independent random variables from that PDF? In this text, the former case is called *slow decorrelation* while the latter is called *fast decorrelation*. Which situation applies has a significant impact on both the procedure and the results for computing detection probabilities, as will be seen in Chap. 6.

One scenario in which it would make sense to combine a set of N measurements into a single integrated measurement is that of a ground-based rotating surveillance radar. (This is possibly the original motivation for the fast and slow decorrelation models). Consider a pulsed radar transmitting at a pulse repetition frequency of PRF pulses per second (PPS) with an antenna having an azimuth beamwidth of θ radians and rotating at a constant angular velocity of Ω radians per second. Suppose that a target is present at a particular location. The geometry is shown in Fig. 2.15a. Assume that significant returns are received only when the target is in the antenna mainlobe. Every complete 360° scan of the antenna results in a new set of $N = (\theta/\Omega)PRF$ echoes during the time the target is in the mainbeam as

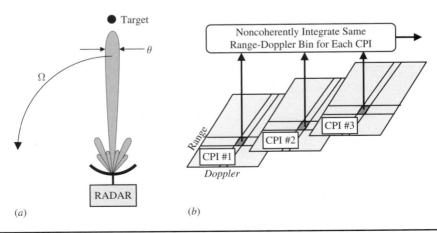

FIGURE 2.15 Sample scenarios for collection of multiple noncoherently related measurements. (a) Rotating surveillance antenna with noncoherent radar. (b) Multiple CPIs with a coherent radar.

it scans by. Consequently, it would seem reasonable to integrate the amplitude or power measurements from a given range bin from each of those N pulses in an attempt to improve signal-to-noise ratio before performing a detection test. Early radars could only do this noncoherently. In this scenario, a "measurement" is the power or amplitude of the echo signal observed in a particular range bin on one pulse.

If the target moves slowly enough that the N pulses from one scan reflect approximately the same RCS value drawn from the appropriate PDF, but fast enough that by the time the radar scans through a full circle and returns to the target again, the next group of N pulses reflects a different RCS value, the data would be considered to exhibit slow decorrelation. If pulse-to-pulse frequency agility was used to force target RCS fluctuations, or radar-target motion during the collection time was such that aspect angle changes caused each pulse to exhibit a different RCS value, the data would be said to exhibit fast decorrelation.

Now consider a very different radar data collection protocol. As will be seen in Chap. 4, a burst of M pulses or FMCW sweeps is a common waveform well-suited to Doppler measurements and interference suppression. Many modern systems are designed to transmit a rapid burst of M pulses or sweeps at a constant repetition rate, often with the antenna staring in a fixed or nearly fixed direction, forming a *coherent processing interval* (CPI) of data. A common use of data collected during a CPI is to combine it coherently using a one-dimensional discrete Fourier transform (DFT) in the slow time (pulse or sweep) dimension to compute a Doppler spectrum for each range bin, forming a two-dimensional *range-Doppler map*. The radar may then repeat the signal burst, collecting a series of N CPIs in the same look direction. The total time to collect all N CPIs is called the *dwell time*. Successive CPIs may share the same radar parameters, or the radar may change the repetition rate, the waveform, or the RF from one burst to the next. Frequency agility within a CPI is *not* used. Combining data across CPIs must generally be done noncoherently. A typical example would be noncoherent integration of N measurements of the same range-Doppler bin in the N CPIs, as shown in Fig. 2.15b, followed by threshold testing that cell for the presence of a target. The length of a CPI is usually short enough that the target exhibits little decorrelation. Significant decorrelation may or may not occur from one CPI to the next. If it does occur, it would be considered fast decorrelation; if not, slow decorrelation. In this example, each "measurement" is the amplitude or power in a particular range-Doppler bin and is the result of coherent processing of multiple pulses or sweeps. This is very different from the scanning surveillance radar in which each measurement was obtained from a single pulse or sweep.

Modeling the expected correlation behavior requires consideration of the dynamics of the radar-target encounter in light of the decorrelation interval in angle given in Eq. (2.63). As an example, consider the crossing encounter of Fig. 2.16a. Aircraft #1 views aircraft #2 at broadside from a range of 5 km with an X-band (10 GHz) radar. Assume aircraft #2 is traveling at 100 m/s and has a length (width as viewed from the radar) of 10 m. Assume that aircraft #1 transmits a burst of $M = 10$ pulses at a 1-kHz PRF. In the resulting 10 ms CPI aircraft #2 will travel 1 m, resulting in an angular change with respect to aircraft #1 of approximately $1/5000 = 0.2$ mrad. From Eq. (2.63), the decorrelation interval in angle is expected to be $(3 \times 10^8)/(2 \cdot 10 \cdot 10 \times 10^9) = 1.5$ mrad. Because the actual angle change within a CPI is less than the angular decorrelation interval, one would expect all the pulses within a CPI to experience essentially the same target RCS. Now suppose that the radar transmits a series of such pulse bursts, each one starting 100 ms after the previous burst. The angular change between aircraft #1 and #2 from one CPI to the next is then 2 mrad, greater than the 1.5 mrad decorrelation interval. Consequently, it is expected that the aircraft RCS during a given CPI will be uncorrelated with that during other CPIs in the dwell.

Figure 2.16b illustrates these effects using another 10 × 5-m random complex target model with the radar and motion parameters just described. The RCS observed for the target on each pulse is plotted for a dwell of five CPIs. Notice that the RCS is nearly constant within

78   Chapter Two

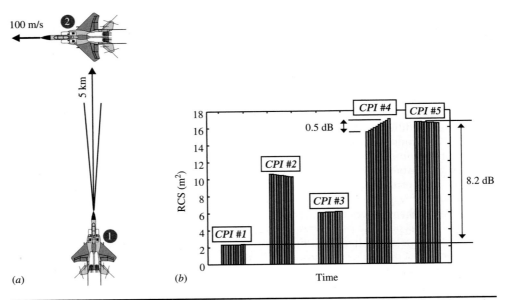

FIGURE 2.16 Crossing target scenario. (a) Encounter geometry. (b) Target RCS for five 10-sample CPIs. See text for motion and radar details.

each CPI. The greatest intra-CPI variation is only about 0.5 dB. Significantly greater variation is seen from one CPI to the next, with the total range in this example being approximately 8.2 dB. Consequently, the full set of five CPIs of data could reasonably be modeled by drawing one random value of RCS from an exponential PDF to represent each CPI. The same RCS value is used for all pulses within a CPI. This is an example of fast decorrelation.

A caution on some historical terminology is in order. In older radar literature slow decorrelation is often called *scan-to-scan decorrelation* and fast decorrelation is called *pulse-to-pulse decorrelation*. This terminology probably originates from scenarios like that of Fig. 2.15a. There, each measurement results from a single pulse. Fast decorrelation is then equivalent to pulse-to-pulse decorrelation. If the measurements do not decorrelate within the group of N pulses, it is assumed that radar-target motion will change the RCS significantly by the time the radar scans around again, so slow decorrelation is equivalent to scan-to-scan decorrelation.

This terminology does not relate well to the actual data collection and processing methods used in modern coherent radars. In the CPI-based data collection protocol a single "measurement" is not based on a single pulse, but is the signal strength in a particular range-Doppler bin. The noncoherent combination occurs from one CPI to the next in the dwell. Thus, a more appropriate terminology might be CPI-to-CPI and dwell-to-dwell decorrelation.

In interpreting older radar literature for modern radars, the reader is cautioned to consider carefully the correlation properties of the N measurements that will be noncoherently combined for a single detection decision. The critical point is whether those measurements are expected to be highly correlated, that is, all approximately the same target RCS value, or whether they are expected to be highly decorrelated (different values). If the measurements are highly correlated (slow decorrelation), published results on "scan-to-scan" mathematical models may be applicable. If they are uncorrelated (fast decorrelation), "pulse-to-pulse" models may be applicable. Newer literature is less likely to use the "scan-to-scan" and "pulse-to-pulse" terminology, obviating this problem over time. Finally, note that coherent combinations of data such as the intra-CPI DFT used to form a range-Doppler cell are generally considered to form a single measurement for threshold detection purposes.

A *target fluctuation model* is a combination of a PDF describing the target RCS variation with angle, RF, or other important parameters and a decorrelation model for measurements to be combined noncoherently. Any PDF that adequately models the RCS distribution for the targets and radar of interest could be used. Examples include any of the PDFs in Table 2.3. For man-made targets, the decorrelation model is usually taken as one of the extremes of either the fully correlated or fully decorrelated models. Analysis carried out using these two models produces bounding results for detection performance. In reality, the noncoherently combined measurements will often be partially correlated. Partial correlation models specified with a measurement-to-measurement correlation coefficient or an autocorrelation function are sometimes given, though this is more common in clutter modeling.

2.2.7 Swerling Models

The four *Swerling models* are examples of specific, widely used models of target RCS fluctuation and decorrelation (Swerling, 1960; Meyer and Mayer, 1973; Nathanson, 1991; Skolnik, 2001). An extensive body of radar detection theory has been built up using them. They are formed from the four combinations of two choices for the PDF and two for the correlation properties. The two density functions used are the exponential and the chi-square of degree 4. The exponential model describes the behavior of a complex target consisting of many scatterers. The fourth-degree chi-square models targets having many scatterers of similar strength with one dominant scatterer. Although the Rice distribution is the exact PDF for this case, the chi-square is an approximation based on matching the first two moments of the two PDFs (Meyer and Mayer, 1973). These moments match when the RCS of the dominant scatterer is $1 + \sqrt{2} = 2.414$ times that of the sum of the RCS of the small scatterers. More generally, a chi-square of degree $2m = 1 + [a^2/(1 + 2a)]$ is a good approximation to a Rice distribution with a ratio of a^2 of the dominant scatterer to the sum of the small scatterers. However, only the specific case of the fourth-degree chi-square is considered a Swerling model.

The Swerling models are denoted as "Swerling 1," "Swerling 2," and so forth. Table 2.5 defines the four cases. A nonfluctuating target is sometimes identified as the "Swerling 0" or "Swerling 5" model.

Figures 2.17 and 2.18 illustrate the difference in the behavior of two of the Swerling models. In both cases, the received power from a single point scatterer having a unit mean RCS is plotted, and in both it is assumed that 10 samples are obtained on each of three scans or CPIs of the radar. Figure 2.17 is a sample Swerling 1 series (exponential PDF, slow decorrelation). Within a group of 10 samples the RCS is constant, but it varies from one group to the next. In contrast, Fig. 2.18 illustrates a Swerling 4 case (fourth-degree chi-square PDF, fast decorrelation) in which each individual sample within each group is independent of the others.

Probability Density Function of RCS	Correlated (Slow Decorrelation)	Uncorrelated (Fast Decorrelation)
Rayleigh/exponential	Case 1	Case 2
Chi-square, degree 4	Case 3	Case 4

TABLE 2.5 Swerling Models

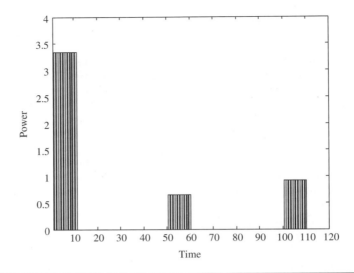

FIGURE 2.17 Three scans or CPIs, each having 10 samples of a unit mean Swerling 1 power sequence.

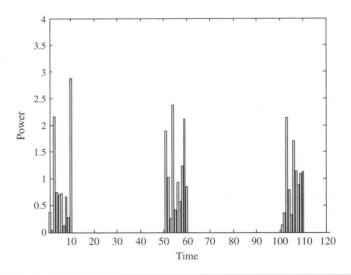

FIGURE 2.18 Three scans or CPIs, each having 10 samples of a unit mean Swerling 4 power sequence.

2.2.8 Effect of Target Fluctuations on Doppler Spectrum

A common operation in radar signal processing is computing the *discrete-time Fourier transform* (DTFT) of the data in a particular range bin for one CPI. The DTFT is a coherent combination of measurements, usually over a sufficiently short CPI that the target echo RCS and thus amplitude do not decorrelate significantly. As will be seen in Sec. 2.7.3, the series of samples within a CPI for a constant-velocity target will form a discrete-time sinusoid. Thus,

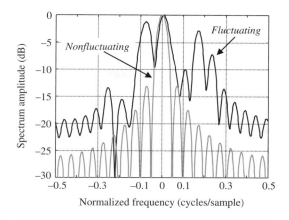

FIGURE 2.19 Effect of intra-CPI target fluctuations on Doppler spectrum.

the usual model for the DTFT of a target is an asinc function with its mainlobe centered at the appropriate frequency and sidelobes that peak 13.2 dB below the mainlobe peak and decay at frequencies further from the mainlobe.

In cases where there are significant RCS fluctuations within the CPI, the amplitude and phase of the target data will vary within the CPI so that the input to the DTFT is no longer a discrete sinusoid with a constant complex amplitude. Figure 2.19 illustrates one example of the resulting effect on the DTFT. The gray spectrum is that of an unwindowed zero-frequency sinusoid, modeling the returns from 20 pulses echoed from a stationary 10 × 5-m simulated many-scatterer target viewed at a constant aspect angle. The black line is the spectrum of data observed using the same target and waveform but with the aspect angle changing 0.7 mrad per pulse. The total angle change over 20 pulses is then 13.3 mrad, nearly nine times the decorrelation interval of 1.5 mrad. The target echo amplitude and phase then both fluctuate, raising the sidelobes and smearing the target energy over a wider frequency range, effectively whitening the spectrum significantly.

2.3 Modeling Clutter

In radar the term "clutter" refers to a component of the received signal due to echoes from volume or surface scatterers. Such scatterers include the earth's surface, both terrain and sea; weather echoes such as rain clouds or snow; and man-made distributed clutter, such as so-called *chaff* clouds of airborne scatterers, typically made out of lightweight strips of reflecting material and used for electronic warfare (EW) purposes. Clutter echoes are sometimes interference and sometimes the desired signal. For instance, synthetic aperture imaging radars are designed to image the earth surface, thus the terrain clutter is the target for a SAR. For an airborne or space-borne surveillance radar trying to detect moving vehicles on the ground, clutter echo from the surrounding terrain is an interference signal.

From a signal processing point of view, the major concern is how to model clutter echoes. As with man-made targets, clutter is a complex target with many scatterers per resolution cell so that the echoes are highly sensitive to radar parameters and encounter geometry. Like complex targets, clutter is therefore modeled as a random process. In addition to temporal correlation, clutter reflectivity can also exhibit spatial correlation: the reflectivity samples from adjacent resolution cells may be correlated. Two excellent general references on land

and sea clutter phenomenology are Ulaby and Dobson (2019) and Long (2001). A good brief introduction is Currie (2010).

Clutter echoes differ from target echoes in that they will typically exhibit different PDFs, temporal and spatial correlation properties, Doppler characteristics, and power levels. These differences can be exploited to separate target and clutter signals. Means to do so are the principal concern of Chaps. 5 and 9. Clutter differs from noise in two major ways: because it is correlated interference, its power spectrum is not white; and since it is an echo of the transmitted signal, the received clutter power is affected by such radar and scenario parameters as the antenna gain, transmitted power, and the range from the radar to the terrain. Because noise is not an echo of the transmitted signal, its power is affected by none of those factors, but is affected by the radar receiver noise figure and bandwidth.

2.3.1 Behavior of σ^0

Area clutter (land and sea surface) reflectivity is characterized by its mean or median value of area reflectivity σ^0 (dimensionless), the probability density function of the reflectivity variations, and their correlation in space and time. Many of the same PDFs described in Sec. 2.2.5 are applied to modeling σ^0 as well. Popular examples include the exponential, log-normal, Weibull, and K power distributions.

The values of σ^0 of terrain observed by the radar are a strong function of terrain type and condition (e.g., surface roughness and moisture), weather (wind speed and direction, precipitation), engagement geometry (especially grazing angle), and radar parameters (wavelength, polarization). Consequently, selection of a PDF is not sufficient to model clutter; it is also necessary to model the dependence of σ^0 on these parameters. Consider land clutter. Values of σ^0 commonly range from −60 to −10 dBsm. Extensive measurement programs over the years have collected statistics of land clutter under various conditions and resulted in many tabulations of σ^0 for various terrain types and conditions, as well as empirical models for the variation of σ^0. Figure 2.20 shows one set of representative data for the area reflectivity of desert terrain versus radar frequency and grazing angle. Note that σ^0 generally increases with radar frequency and decreases at shallower grazing angles. For a given frequency, the variation with grazing angle over the range shown is about 20 to 25 dBsm. For a given grazing

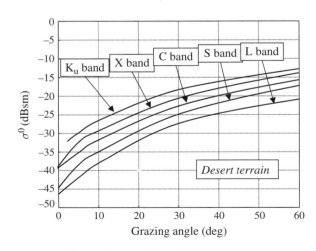

Figure 2.20 Behavior of σ^0 of desert terrain versus radar frequency and grazing angle. (Data from Currie, 2010.)

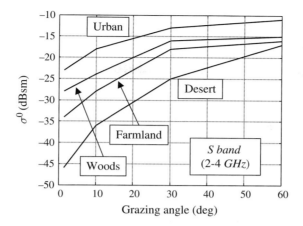

FIGURE 2.21 Behavior of σ^0 versus terrain type and grazing angle at S band. (Data from Currie, 2010.)

angle, the variation across frequency in this example is about 8 to 12 dBsm. Figure 2.21 is one example of the variation in σ^0 versus grazing angle for different terrain types at a fixed frequency, in this case in S band (2 to 4 GHz). Generally, reflectivity increases with terrain roughness, from the presumably smoother desert terrain to the complex, rough urban terrain.

As seen in those figures, σ^0 varies significantly with grazing angle. Generally, it decreases rapidly at very low grazing angles and increases rapidly at very high grazing angles (radar look direction approaching normal to the clutter surface), with a milder variation in a middle "plateau region." Figure 2.22 is a notional diagram of this behavior. A common model for the behavior of σ^0 over the plateau region is the "constant gamma" model (Long, 2001),

$$\sigma^0 = \gamma \sin \delta \qquad (2.64)$$

where γ is a characteristic of the particular clutter type at the radar frequency and polarization of interest. This model predicts that σ^0 is maximum at normal incidence and becomes

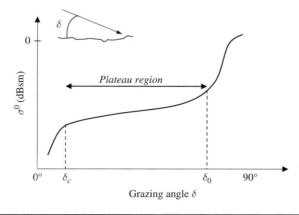

FIGURE 2.22 General behavior of σ^0 with grazing angle for land clutter. (After Long, 2001.)

	Clutter Type						
Parameter	Soil/Sand	Grass	Tall Grass, Crops	Trees	Urban	Wet Snow	Dry Snow
A	0.25	0.023	0.006	0.002	2.0	0.0246	0.195
B	0.83	1.5	1.5	0.64	1.8	1.7	1.7
C	0.0013	0.012	0.012	0.002	0.015	0.0016	0.0016
D	2.3	0	0	0	0	0	0

Source: Adapted from Currie (2010).

TABLE 2.6 GTRI Land Clutter Model Parameters for X Band

vanishingly small as the grazing angle tends to zero. However, it does not adequately reflect the behavior of σ^0 often observed at near-normal or near-zero incidence angles so additional models are often used at these two extremes.

Various predictive models for land clutter σ^0 as a function of important parameters have been presented in the literature. One well-known example is the Georgia Tech Research Institute (GTRI) model given by (Currie, 2010),

$$\sigma^0 = A \cdot (\delta + C)^B \exp\left[\frac{-D}{1 + (\sigma_h/10\lambda_t)}\right] \quad (2.65)$$

where σ_h is the RMS surface roughness and the parameters A, B, C, and D depend on the clutter type and radar frequency. Sample values for X band are given in Table 2.6.

Models of a similar spirit exist for sea clutter σ^0. Important parameters for sea clutter reflectivity in addition to frequency, grazing angle, and polarization include wind speed and direction relative to the radar boresight, wave height, and multipath (Sec. 2.8.6). Details of one representative model developed at GTRI are given in Currie (2010).

2.3.2 Signal-to-Clutter Ratio

In many scenarios, the dominant interference is not noise, but clutter. Consequently, the *signal-to-clutter ratio* (SCR) is often of more importance than the *signal-to-noise ratio* (SNR). The SCR is easily obtained as the ratio of the received target power given by Eq. (2.11) to the received clutter power, given by Eq. (2.25), (2.30), or (2.32) for the volume clutter, beam-limited area clutter, and resolution-limited area clutter cases, respectively. The resulting equations are

$$\begin{aligned}
SCR &= \frac{\sigma}{R^2 \eta \cdot \Delta R \cdot \theta_3 \phi_3} \quad \text{(volume clutter case)} \\
&= \frac{\sigma \sin \delta}{R^2\, \theta_3 \phi_3\, \sigma^0} \quad \text{(beam-limited area clutter case)} \\
&= \frac{\sigma \cos \delta}{R \sigma^0 \cdot \Delta R \cdot \phi_3} \quad \text{(resolution-limited area clutter case)}
\end{aligned} \quad (2.66)$$

In each case, such system parameters as the transmitted power and the antenna gain cancel out. This occurs because both the clutter and target signals are echoes of the radar

	Radar Frequency, GHz		
Target	10	35	95
Rain, 5 mm/h	35	80	140
Rain, 100 mm/h	70	120	500
Trees, 6–15 mph wind	9	21	35

Source: Currie, N. C. "Clutter Characteristics and Effects," Chap. 10 in J. L. Eaves and K. E. Reedy (eds.), *Principles of Modern Radar*. Van Nostrand Reinhold, New York, 1987.

TABLE 2.7 Cubic Power Spectrum Corner Frequencies (Hz) for Rain and Tree Clutter

signal; increasing power or antenna gain increases the strength of both types of echoes equally. Thus, the SCR just becomes the ratio of the target RCS to the total RCS of the contributing clutter.

2.3.3 Temporal and Spatial Correlation of Clutter

Clutter decorrelation in time is induced by internal motion of the clutter such as tree leaves moving in the wind or waves on the sea surface, and by changes in radar-clutter patch geometry. Various investigators have experimentally characterized the decorrelation characteristics of clutter echoes due to internal motion by modeling their autocorrelation function, or equivalently their power spectrum.[8] For example, one model suggested to estimate the power spectrum of the RCS of foliated trees or rain uses a cubic spectrum (Currie, 2010),

$$S_\sigma(F) = \frac{A}{1 + (F/F_c)^3} \qquad (2.67)$$

The *corner frequency* F_c is the cutoff frequency of this lowpass spectrum. It is a function of the wavelength and either wind speed (for trees) or rain rate (for rain). Some sample measured values are given in Table 2.7. A higher corner frequency (wider power spectrum) implies a shorter decorrelation interval (narrower autocorrelation function). Shorter decorrelation times render the clutter signals more like white noise and degrade the effectiveness of some of the clutter suppression techniques of Chap. 5. Notice that for a given weather condition, the clutter decorrelates more rapidly at higher radar frequencies. Figure 2.23 plots additional windblown tree clutter data that also shows the decrease in decorrelation time for both increased clutter motion and increased radar frequency.

Another model frequently used to model generic power spectra is the Gaussian given by

$$S_\sigma(F) = A \exp\left[-\alpha\left(\frac{F}{F_0}\right)^2\right] \qquad (2.68)$$

The Gaussian model is very commonly used in weather radar and is the basis of the pulse pair Doppler velocity estimation technique to be discussed in Chap. 5.

---

[8]See App. A for a discussion of the properties of the autocorrelation and power spectrum of random signals.

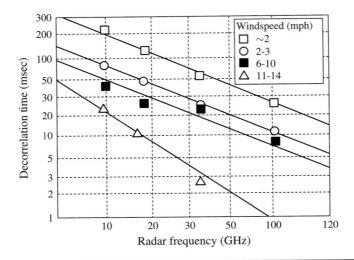

Figure 2.23 Decorrelation time of windblown tree clutter versus frequency and wind speed. (Data from Currie, 2010.)

Both the cubic and Gaussian power spectral models can be well matched by a low-order autoregressive (AR, or all-pole) spectrum model of the form (Haykin et al., 1982)

$$S_\sigma(F) = \frac{A}{1 + \sum_{k=1}^{N} \alpha_k F^{2k}} \quad (2.69)$$

Real clutter measured from ground-based radars appears to be well matched using an order N of only two to four. Other studies of clutter measured by airborne radars in a landing scenario indicate that orders up to 10 may be required (Baxa, 1991). The AR clutter spectrum model has the advantage that its parameters can be efficiently computed directly from measured data and adapted in real time using the Levinson-Durbin or similar algorithms (Kay, 1988). Furthermore, the AR parameters can be used to construct optimal adaptive clutter suppression filters, as is seen in Chap. 5. The disadvantage is that the calculations rapidly become computationally intensive as the model order increases.

The Billingsley model, more recently developed and popular in studies of detection of ground targets from moving platforms, represents the correlation properties of windblown tree clutter and other vegetative cover, said to be the "most pervasive" ground clutter (Billingsley, 2001). It assumes that the clutter temporal power spectrum is the sum of a two-sided decaying exponential function and an impulse at the origin in Doppler frequency space:

$$S_\sigma(F) = \sigma_c^2 \left[\underbrace{\frac{\alpha}{\alpha+1} \delta_D(F)}_{\text{"DC term"}} + \underbrace{\frac{1}{\alpha+1}\left(\frac{\beta\lambda_t}{4}\right)\exp\left(-\frac{\beta\lambda_t}{2}|F|\right)}_{\text{"AC term"}} \right] \quad (2.70)$$

The parameter α establishes the ratio of the DC to AC components and is a function of both wind and radar frequency, while β determines the width of the AC power spectral component and is dependent primarily on wind conditions. The corresponding autocorrelation function is

$$s_\sigma(\tau) = \sigma_c^2 \left(\frac{\alpha}{\alpha+1} + \frac{1}{\alpha+1} \frac{(\beta\lambda_t)^2}{(\beta\lambda_t)^2 + (4\pi\tau)^2} \right) \quad (2.71)$$

where τ is the autocorrelation lag.

Based on extensive measurements, Billingsley proposed empirical formulas for α and β:

$$\alpha = 489.8 \cdot w^{-1.55} F_t^{-1.21} \quad (2.72)$$

$$\beta^{-1} = 0.1048(\log_{10} w + 0.4147) \quad (2.73)$$

where w is the wind speed in statute miles per hour and F_t is the radar carrier frequency in GHz. Caution is needed in applying Eq. (2.73) due to mixed units. Specifically, w is in statute miles per hour but β is in meters per second.

The "DC term" in Eqs. (2.70) and (2.71) represents a constant, nonrandom component of the clutter echo that is sometimes called a "persistent component" of the received signal. For such a component to exist, both the amplitude and phase of the reflectivity of the clutter scatterers involved must be constant. Thus, the DC component is attributable to backscatter from elements such as bare ground, rocks, and tree trunks. The AC term accounts for backscatter from moving elements such as leaves, branches, and blades of grass. Simple autoregressive filters can be used to implement the model in simulations (Mountcastle, 2004).

2.3.4 Compound Models of Radar Cross Section

Radar detection performance predictions depend strongly on the details of target and clutter RCS models. Because RCS statistics vary significantly with a host of factors such as geometry, resolution, wavelength, and polarization, the development of good statistical RCS models is a very active area of empirical and analytical research. Following are three brief examples of an extension to the basic modeling approach described earlier, all motivated by the complexities of modeling clutter. Because the literature regarding these models is developed primarily in terms of the echo amplitude (voltage) ζ instead of RCS σ (power), the remainder of this section also concentrates on amplitude PDFs.

Some amplitude PDFs are physically motivated, especially the Rayleigh (exponential RCS), which follows from the "many equal-strength scatterer" model and the central limit theorem argument, and the Rice or Rician model, which corresponds to a Rayleigh model with an additional persistent scattering source. Others such as the log-normal and Weibull have been developed empirically by fitting distributions to measured data. One attempt to provide a physical justification for a non-Rayleigh model abandons the single-PDF approach, instead assuming that the random variable representing echo amplitude can be written as the product of two independent random variables, $\zeta = x \cdot y$. The PDF of ζ can then be represented in a Bayesian formulation as

$$p_\zeta(\zeta) = p_x(x) p_{\zeta|x}(\zeta|x) \quad (2.74)$$

This model has been used to describe sea clutter (Jakeman and Pusey, 1976; Ward, 1981). The random variable x is identified with a slowly decorrelating component having a voltage distribution following a central chi-square of degree $2m$ with $m \geq 2.5$. This component is introduced to account for "bunching" of scatterers due to ocean swell structure and radar-clutter geometry and represents variation in the mean of the amplitude over time.

The distribution $p_{\zeta|x}(\zeta|x)$ is assumed to represent the composite of a large number of independent scatterers. Its amplitude distribution is therefore Rayleigh. The resulting composite PDF $p_\zeta(\zeta)$ can be shown to be the *K distribution*, which is given by

$$p_\zeta(\zeta) = \begin{cases} \dfrac{2k}{\Gamma(a)}\left(\dfrac{k\zeta}{2}\right)^a K_{a-1}(k\zeta), & \zeta \geq 0 \\ 0, & \zeta < 0 \end{cases} \quad (2.75)$$

where $K_{a-1}(\cdot)$ is the modified Bessel function of the second kind and order $a-1$ and k is a constant. Thus, the product formulation suggests that modulation of a standard Rayleigh variable by a central chi-distributed geometric term can account for observed sea clutter distributions. Additional information on the K distribution and the corresponding "K power" PDF for power is given in App. A.

The compound PDF model is one approach to bridging the gap between the physics of scattering and empirical clutter PDFs. Another approach starts with the model of Eq. (2.50) but lets the number of scatterers N be a random variable instead of a fixed constant. This representation is referred to as a *number fluctuations* model. Depending on the choice of the statistics of the number N of scatterers contributing to the return at any given time, this modified version of Eq. (2.50) can result in K, Weibull, gamma, Nakagami-*m*, or any of a number of other distributions in the class of so-called Rayleigh mixtures (Sangston and Gerlach, 1994). The number fluctuations model is intuitively appealing in this case because it can be related to the physical behavior of waves. Specifically, scattering theory suggests that the principal scatterers on the ocean surface are the small capillary waves, as opposed to the large swells. These small scattering centers tend to cluster near the crest of the swells, with fewer of them in between. In other words, they are nonuniformly distributed over the sea surface. Consequently, a radar illuminating the sea will receive echoes from a variable number N of scatterers as the crests of the swells move into and out of a given resolution cell. By summing echoes from a variable number of scatterers the number fluctuation model predicts the Weibull and K distributions and provides a link between a phenomenological model of sea scatter and these empirically observed statistics.

All of the statistical models described in Sec. 2.2.5 apply to the scattering observed from a single resolution cell. That is, they represent the variations in RCS observed by measuring the same region of physical space multiple times, for example, by transmitting multiple pulses or sweeps in the same direction and measuring the received power in the same range bin on each transmission. Another use of the product model of Eq. (2.74) is to describe the spatial variation of clutter reflectivity. If the scene being viewed by the radar is nonhomogeneous, then the characteristics of the RCS observed in one resolution cell might vary significantly from those of another. For example, the dominant clutter observed by a scanning radar at a coastal site might be an urban area in one look direction and the sea in another. Another example occurs when scattered rain cells occupy only part of the scanned region so that some resolution cells contain rain while others are clear.

These situations can be modeled by letting the slowly decorrelating term x in the product model represent spatial variations in the local mean of the received voltage. If the PDF of x is log-normal with a large variance and the PDF of ζ conditioned on x is gamma distributed (which includes Rayleigh as a special case), then the overall PDF of the product $\zeta \cdot x$ has a log-normal distribution (Lewinski, 1983). Consequently, the product model implies that log-normal variations of the local mean from one resolution cell to another could account for the log-normal variation often used to model ground clutter returns. A similar argument can be used to justify the log-normal model for fine-resolution target RCS by modeling the variation of RCS with aspect angle as a log-normal process.

2.4 Noise Model and Signal-to-Noise Ratio

The echo signal received from a target or clutter inevitably competes with noise. There are two sources of noise: that received through the antenna from external sources, and that generated in the radar receiver itself.

External noise is a strong function of the direction in which the radar antenna is pointed. The primary contributor is the sun. If the antenna is directed toward the night sky and there are no interfering microwave sources, the primary source is *galactic* (also called *cosmic*) *noise*. Internal noise sources include *thermal noise* (also called *Johnson* or *Nyquist noise*) due to ohmic losses, *shot noise* and *partition noise* due to the quantum nature of electric current, and *flicker noise* due to surface leakage effects in conducting and semiconducting devices (Carlson and Crilly, 2009).

Of these various sources, thermal noise is normally dominant. The theories of statistical and quantum mechanics dictate that the thermal noise voltage in an electronic circuit is a zero-mean Gaussian random process (Curlander and McDonough, 1991). The mean energy of the random process is $kT/2$ joules, where T is the temperature of the noise source in kelvins (absolute temperature) and $k = 1.38 \times 10^{-23}$ J/K is *Boltzmann's constant*. The power spectrum $S_n(F)$ of the thermal noise delivered to an impedance-matched load is

$$S_n(F) = \frac{hF}{\exp(hF/kT) - 1} \quad \text{W/Hz} \tag{2.76}$$

where $h = 6.6254 \times 10^{-34}$ J/s is *Planck's constant*. If $hF/kT \ll 1$, a series approximation gives $\exp(hF/kT) \approx 1 + hF/kT$ so that Eq. (2.76) reduces to the white noise power spectrum

$$S_n(F) = kT \quad \text{W/Hz} \tag{2.77}$$

Equation (2.77), when integrated over frequency, implies infinite power in the white noise process. In reality, however, the noise is not white [Eq. (2.76)] and in any event it is observed in any real system only over a finite bandwidth. For frequencies below 100 GHz, the approximation of Eq. (2.77) requires the equivalent noise temperature T' (to be defined below) to be larger than about 50 K, which is almost always the case. Consequently, thermal noise is well-modeled by a white power spectrum. For many practical systems it is reasonable to choose the temperature of the system to be the "standard" temperature $T_0 = 290$ K = 62.3°F; then $kT_0 \approx 4 \times 10^{-21}$ W/Hz.

In a coherent radar receiver, the noise present at the front end of the system contributes to both the I and Q channels after the quadrature demodulation. The I and Q channel noises are both zero-mean Gaussian random processes with equal power. Since the total noise spectral density is kT W/Hz, the noise density in each channel individually is a zero-mean Gaussian random process with spectral density $kT/2$ W/Hz. Furthermore, if the power spectrum of the input noise is white, then the I and Q noise processes can be shown to have white power spectra and to be uncorrelated with one another. Since the I and Q noise processes are Gaussian and uncorrelated, it follows that they are also independent (Papoulis and Pillai, 2001). Finally, since the I and Q signals are independent zero-mean Gaussian processes, it also follows from standard transformations in random variables that the magnitude of the complex signal $I + jQ$ is Rayleigh distributed, the magnitude-squared is exponentially distributed, and the phase angle $\tan^{-1}(Q/I)$ is uniformly distributed over $(0, 2\pi]$.

The bandwidths of the various components of a receiver vary, but the narrowest bandwidth is generally approximately equal to the bandwidth of the transmitted signal. If the receiver contains any component of narrower bandwidth than the signal, echo energy will be lost, reducing sensitivity. If the most narrowband component has a bandwidth

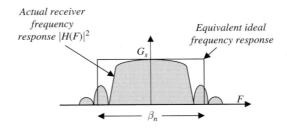

FIGURE 2.24 The concept of noise equivalent bandwidth of a filter.

appreciably wider than the signal bandwidth, the signal will have to compete against more noise power than necessary, again reducing sensitivity. Thus for the purpose of noise power calculation, the frequency response of the receiver can be approximated as a bandpass filter centered at the transmit frequency with a bandwidth equal to the signal bandwidth.

Real filters do not have perfectly rectangular passbands. For analyzing receiver output noise power the *noise-equivalent bandwidth* β_n of the receiver transfer function $H(F)$ is used. Figure 2.24 illustrates the concept. Imagine an ideal rectangular spectrum with gain equal to the peak gain G_s of the actual frequency response $H(F)$. The noise equivalent bandwidth β_n is the bandwidth the rectangular spectrum must have so that the areas under the two squared frequency responses are equal. This condition guarantees that given a white noise input, both filters exhibit the same output noise power.[9] Thus

$$\beta_n = \frac{\int_{-\infty}^{\infty} |H(F)|^2 \, dF}{\max\left(|H(F)|^2\right)} = \frac{1}{G_s} \int_{-\infty}^{\infty} |H(F)|^2 \, dF \quad \text{Hz} \tag{2.78}$$

The area of the squared frequency response will be $\beta_n G_s$. The total noise power N present at the output of the filter $H(F)$ is then given by

$$N = \int_{-\infty}^{\infty} |H(F)|^2 S_n(F) \, dF = kT \int_{-\infty}^{\infty} |H(F)|^2 \, dF \tag{2.79}$$
$$= kT \beta_n G_s \quad \text{W}$$

White noise passed through a frequency response $H(F)$ is no longer white but instead has a power spectrum proportional to $|H(F)|^2$. If $|H(F)|^2$ is approximated as a rectangular filter of two-sided bandwidth β_n Hz, the autocorrelation function of the noise at the filter output is approximately a sinc function with its first zero at lag $1/\beta_n$ seconds. It will be seen in Chap. 3 that the receiver output is normally sampled at intervals of approximately $1/\beta_n$ seconds. Consequently, the noise components of successive receiver output samples are still uncorrelated with one another.

The power spectral density of white noise at the output of any source or circuit can be described as the product of Boltzmann's constant and some equivalent temperature T', mimicking the simple formulation of Eq. (2.77). Source noise power is usually referenced to the input of a system so that the power gain G_s (or loss if $G_s < 1$) of the system must also be

---

[9] See App. A, Sec. A.4.5.

taken into account. That is, if the observed receiver output power spectral density (still assumed white over the receiver bandwidth) at the output of a receiver is some value S_n, then an equivalent temperature T' of the noise source at its input is defined to be

$$T' \equiv \frac{S_n}{kG_s} \text{ K} \qquad (2.80)$$

so that $S_n = kT'G_s$ and the total noise power is

$$N = kT'\beta_n G_s \text{ W} \qquad (2.81)$$

The total output noise power at the receiver output is the primary quantity of interest. In a radar system, the contributors to this noise include the external noise, the intrinsic $kT_0\beta_n$ thermal noise, and additional thermal noise due to losses in the antenna structure and nonideal receivers. Detailed noise analyses assign individual equivalent noise temperatures to each stage in the system; a good introductory description is given in Curlander and McDonough (1991). When considering the system as a whole, it is common to express the total output noise power as the sum of the power that would be observed due to the minimum noise density kT_0 at the input and a second term that accounts for the additional noise due to the nonideal system,

$$N = kT_0\beta_n G_s + kT_e\beta_n G_s \text{ W} \qquad (2.82)$$

In this equation, G_s is now the power gain of the complete receiver system, including antenna loss effects. The equivalent temperature T_e used to account for noise above the theoretical minimum is called the *effective temperature* of the system.

The noise temperature description of noise power is most useful for low-noise receivers. An alternative description common in radar is based on the idea of *noise figure* F_n, which is the ratio of the actual noise power at the output of a system to the minimum power $kT_0\beta_n G_s$ (Skolnik, 2001). As with noise temperatures, various noise figures can be defined to include the effects of just the receiver, or of the entire antenna and receiver system, and so forth. Here, the term *noise figure* used without qualification will mean the noise figure of the complete receiver system so that

$$F_n = \frac{N}{kT_0\beta_n G_s} \qquad (2.83)$$

Equation (2.83) shows that knowledge of the noise equivalent bandwidth, gain, and noise figure of the receiver system is sufficient to calculate the output noise power using $N = kT_0\beta_n F_n G_s$. It also follows from using Eq. (2.82) in Eq. (2.83) that $T_e = (F_n - 1)T_0$. Typical noise figures for radars can be as low as 2 or 3 dB and as high as 10 dB or more. Corresponding effective temperatures range from about 170 K to over 2600 K.

In Sec. 2.2, the term "radar range equation" was applied to Eqs. (2.11), (2.25), (2.30), and (2.32). These expressions described the echo power received by the radar given various system and propagation conditions. As will be seen in Chap. 6, the detection performance of a radar depends not on the received power per se but on the SNR at the point of detection. Equation (2.83) can be used to convert the power range equations to SNR range equations.

To illustrate, consider the point target range equation [Eq. (2.11)], which expresses the power P_r of the signal available at the input to the receiver. The signal power at the output will be $P_o = G_s P_r$ provided the signal bandwidth is entirely contained within the receiver bandwidth β_n. From Eq. (2.83), the output noise power is $N = kT_0\beta_n F_n G_s$. The SNR is therefore

$$\chi = \frac{P_o}{N} = \frac{G_s P_t G^2 \lambda_t^2 \sigma}{(4\pi)^3 R^4 L_s L_a(R)} \cdot \frac{1}{kT_0 \beta_n F_n G_s}$$
$$= \frac{P_t G^2 \lambda_t^2 \sigma}{(4\pi)^3 R^4 kT_0 \beta_n F_n L_s L_a(R)} \quad (2.84)$$

The last expression in Eq. (2.84) gives the SNR in terms of transmitter and receiver characteristics, target RCS, range, and loss factors. Modifications of Eqs. (2.25), (2.30), and (2.32) for volume and area scatterers to express them in terms of signal to noise ratio are obtained in the same manner by simply including the quantity $kT_0 \beta_n F_n$ in their denominators. In these cases, the SNR becomes the clutter-to-noise ratio (CNR).

Equation (2.84) represents the SNR at the receiver output prior to any signal processing. A primary goal of many of the techniques discussed in this text is to increase the SNR above that value through signal processing means so as to obtain better detection, measurement, and imaging results. The impact of signal processing on the SNR can be modeled by simply adding a signal processing gain term G_{sp} to the range equation:

$$\chi = \frac{P_t G^2 \lambda_t^2 \sigma G_{sp}}{(4\pi)^3 R^4 kT_0 \beta_n F_n L_s L_a(R)} \quad (2.85)$$

In ensuing chapters, G_{sp} will be expressed in terms of the parameters of specific techniques such as matched filtering and Doppler processing.

Like Eq. (2.11), Eq. (2.85) is also often called the radar range equation. In the remainder of this text, the term "range equation" or "radar range equation" usually refers to the SNR form of Eq. (2.85) and its analogues for volume and area scatterers.

2.5 Jamming

Jamming refers to intentional interference directed at the radar system from a hostile emitter. Jamming is an example of *electronic countermeasures* (ECM) or *electronic attack* (EA), which are in turn elements of electronic warfare. While the purpose of many radar signal processing is to improve the SIR, the purpose of many jamming techniques is just the opposite: to reduce the SIR so that the radar performance is degraded.

The most basic form of jamming is simple noise jamming. A hostile emitter directs an amplified noise waveform of appropriate RF and bandwidth at the victim radar, essentially increasing the noise level out of the receiver. If the noise power spectrum fills the entire radar receiver bandwidth, then the noise out of the receiver will appear like any other white noise process and is modeled in the same way. More advanced forms of noise jamming use various amplitude and frequency modulations. Instead of noise, other jamming techniques use waveforms designed to mimic target echoes and fool the radar into detecting and tracking nonexistent targets, wasting resources and drawing attention away from legitimate targets.

Even a limited discussion of ECM is outside the scope of this text, due both to the breadth of the topic and the limited amount of material publishable in the open literature. The reader is referred to Adamy (2001) and Lothes et al. (1990) for good general references on jamming signals in radar.

2.6 Electromagnetic Interference

Like jamming, electromagnetic interference is an externally generated signal. Unlike jamming, the intent of an EMI emitter is not overtly hostile. Rather, the interference is just the consequence of the radar sharing a crowded spectrum with other radars and communication

Signal Models

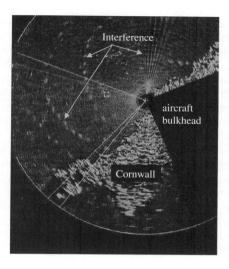

FIGURE 2.25 Effect of low- and high-duty cycle EMI on a radar ground image. See text for details. (Image courtesy of Dr. Simon Watts. Additional information available in Griffiths, 2015.)

services in the vicinity. EMI is a concern both for other radars and communications systems interfering with one's own radar, and for one's own radar interfering with other systems.

As an example, many mobile phone frequency bands in the United States fall in UHF to Ka band ranges. Wi-Fi signals are found in the L, S, C, and MMW bands. Specific frequencies are allocated to specific uses by international regulation so as to avoid direct conflicts, but interference between services does occur. In areas where multiple radars are operating in close proximity, for example, where there are multiple aircraft using the same bands, interference between like systems can be a problem. For instance, as the number of automotive radars on the road rapidly becomes very large, avoidance of EMI between vehicles will become increasingly difficult.

Figure 2.25 shows an example of the effect of an interfering radar on a coarse resolution radar ground image. The spiral groups of "blips" in the left-central portion of the image are short-duration pulses from another radar at a PRF similar to but different from that of the imaging radar. (If the PRFs were the same, the EMI would be expected to show up in the same range bin at every azimuth, instead of "walking" through different range bins.) The linear interference "spokes" extending toward the top of the image are from an interferer that is transmitting most of the time, and so is likely either a high duty cycle pulsed system or a CW system.

While there are numerous techniques for recognizing and mitigating EMI, they are beyond the scope of this text. The reader is referred to Griffiths et al. (2015) as an entry point into the literature.

2.7 Frequency Models: The Doppler Shift

2.7.1 Doppler Shift

Consider a radar that radiates a sinusoidal waveform

$$\bar{x}(t) = a(t)\exp[j(2\pi F_i t + \varphi_0)] \qquad (2.86)$$

The envelope a(t) can be either a constant (CW waveform) or a finite-length pulse. If the radar and scatterer are not at rest with respect to each other, the frequency F_r of the received echo will differ from the transmitted frequency F_t due to the Doppler effect. The change in frequency, $F_r - F_t$, is called the *Doppler shift* F_D. Doppler frequency shifts can be used to advantage to detect echoes from moving targets in the presence of much stronger echoes from clutter or to drastically improve cross-range resolution. Uncompensated Doppler shifts can also have harmful effects, particularly measurement errors and a loss of sensitivity for some types of waveforms. Thus, characterization and measurement of Doppler shifts is an important topic in radar.

To quantify the Doppler shift, suppose the waveform is reflected from a target at an arbitrarily time-varying range $R(t)$. For instance, a constant-range target would have $R(t)$ equal to a fixed R_0 meters, while a constant-velocity target would have $R(t) = R_0 - vt$ meters.[10] It makes no difference whether the radar, the target, or both are moving such that the range between the two is $R(t)$, so it can be assumed without loss of generality that the radar is stationary and the target is moving. Also note that it is virtually always the case in radar that the relative velocity v is very small compared to the speed of light c. For example, a car traveling at 60 mph (26.82 m/s) has a ratio $|v/c|$ of 8.94×10^{-8}; an aircraft at Mach 1 (about 340.3 m/s at sea level) has $|v/c| = 1.13 \times 10^{-6}$; and even a low-earth orbit (LEO) satellite with a velocity of 7800 m/s has $|v/c| = 2.6 \times 10^{-5}$.

A detailed analysis of the received waveform $\bar{y}(t)$ under these conditions is summarized in Richards (2020). The key result is that the received signal is well-approximated as

$$\bar{y}(t) \approx k \cdot \bar{x}\left[t - \frac{2R(t)}{c}\right] \tag{2.87}$$

where k absorbs all amplitude factors, which are not of concern here. More specifically, if the transmitted waveform is the standard RF signal of Eq. (2.86), the received echo waveform will be

$$\begin{aligned}\bar{y}(t) &= k \cdot a\left(t - \frac{2R(t)}{c}\right)\exp\left[j\left(2\pi F_t\left(t - \frac{2R(t)}{c}\right) + \varphi_0\right)\right] \\ &= k \cdot a\left(t - \frac{2R(t)}{c}\right)\exp\left(-\frac{j4\pi R(t)}{\lambda_t}\right)\exp[j(2\pi F_t t + \varphi_0)]\end{aligned} \tag{2.88}$$

The case of a constant-velocity target is of special interest. Setting $R(t) = R_0 - vt$ in Eq. (2.88) gives, after some manipulation,

$$\begin{aligned}\bar{y}(t) &= k \cdot a\left[\left(1 + \frac{2v}{c}\right)t - \frac{2R_0}{c}\right]\exp\left[j\left(2\pi F_t\left[\left(1 + \frac{2v}{c}\right)t - \frac{2R_0}{c}\right] + \varphi_0\right)\right] \\ &= k \cdot a\left[\left(1 + \frac{2v}{c}\right)t - \frac{2R_0}{c}\right]\exp\left(-\frac{j4\pi R_0}{\lambda_t}\right)\exp\left[j\left(2\pi F_t\left(1 + \frac{2v}{c}\right)t + \varphi_0\right)\right]\end{aligned} \tag{2.89}$$

Equation (2.89) shows that constant-velocity radar-target motion affects the waveform envelope, frequency, and phase. Because $|v/c|$ is very small the term $(1 + 2v)/c$ in the envelope argument represents a very slight expansion ($v < 0$) or contraction ($v > 0$) of the

---

[10]It is probably more common to define a constant-velocity target so that positive v corresponds to increasing range, but the preference here is to define v so that a positive v gives a positive Doppler shift.

Signal Models

Band	Frequency, GHz	Doppler Shift (Hz) for v = 1 m/s
L	1	6.67
C	6	40.0
X	10	66.7
K_a	35	233
W	95	633

TABLE 2.8 Doppler Shift Resulting from a Velocity of 1 m/s

envelope time scale, and therefore of the pulse or sweep length; this effect is usually negligible. The phase of the sinusoidal signal is reduced by $2R_0 F_t/c = 4\pi R_0/\lambda_t$ radians as usual.

The effect of the $(1 + 2v)/c$ term in the argument of the exponential is *not* negligible. Equation (2.89) becomes

$$\bar{y}(t) = k \cdot a\left[\left(1 + \frac{2v}{c}\right)t - \frac{2R_0}{c}\right]\exp\left(-j\frac{4\pi R_0}{\lambda_t}\right)\exp[j(2\pi(F_t + F_D)t + \varphi_0)] \quad (2.90)$$

where F_D is the Doppler shift, which can be expressed in terms of either the transmitted frequency or wavelength:

$$F_D = \frac{2vF_t}{c} = \frac{2v}{\lambda_t} \text{ Hz} \quad (2.91)$$

The Doppler shift is positive for approaching targets ($v > 0$) and negative for receding targets as expected.

Numerical values of Doppler shift are small compared to the RF frequencies because velocities are small compared to the speed of light. Table 2.8 gives the magnitude of the Doppler shift for each 1 m/s of velocity at various radar frequencies. The Mach 1 aircraft observed with the L-band radar would cause a Doppler shift of only 2.27 kHz in the 1 GHz carrier frequency.

For a monostatic radar and a constant-velocity target, the observed Doppler shift is proportional to the *radial velocity*, which is the component of velocity along the *line of sight* (LOS) connecting the radar and target. If the angle between the velocity vector of a target traveling at v meters per second and the vector from the radar position to the target position is ψ (sometimes called the *cone angle*), the radial velocity is $v \cdot \cos\psi$ meters per second. The geometry is illustrated in two dimensions in Fig. 2.26. The magnitude of the Doppler shift is maximum when the target is traveling directly toward or away from the radar. Regardless of the target velocity, the Doppler shift is zero when the target is crossing tangentially to the radar boresight.

2.7.2 The Stop-and-Hop Approximation and Phase History

The envelope contraction or expansion due to the $(1 + 2v/c)$ term in its argument in Eq. (2.90) is not significant, even though that same factor is important in the exponential term. Neglecting it gives the additional simplification

$$\bar{y}(t) = k \cdot a\left(t - \frac{2R_0}{c}\right)\exp\left(-j\frac{4\pi R_0}{\lambda_t}\right)\exp(j2\pi F_D t)\exp[j(2\pi F_t t + \varphi_0)] \quad (2.92)$$

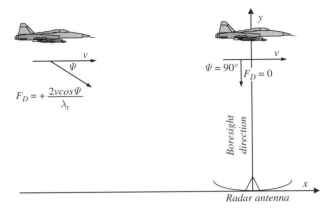

Figure 2.26 Doppler shift is determined by the radial component of relative velocity between the target and radar.

Equation (2.92) is a simplified but very useful model of reflection of a radar pulse from a target moving relative to the radar. The envelope term $a(\cdot)$ states that the echo is received with a time delay corresponding to the range at the beginning of the pulse transmission, effectively ignoring the effect on the echo timing of the motion between the radar and target while the transmitted signal is in flight. This is the "stop" part of the *stop-and-hop* assumption common in radar analysis: the envelope of the echo appears as if the target motion effectively stopped while the signal was in transit. The "hop" portion will be discussed shortly.

For a more interesting example of the use of Eq. (2.88), consider Fig. 2.25 again. Let the radar be located at (x, y) coordinates $(x_r = 0, y_r = 0)$ with its antenna aimed in the $+y$ direction, and let the coordinates of the target aircraft be $(x_t = vt, y_t = R_0)$. This means that the target aircraft is on the radar boresight at a range R_0 at time $t = 0$ and is crossing orthogonal to the radar line of sight at a velocity v meters per second. The range between radar and aircraft is

$$R(t) = \sqrt{R_0^2 + (vt)^2} = R_0\sqrt{1 + \left(\frac{vt}{R_0}\right)^2} \quad \text{m} \tag{2.93}$$

While it is possible to work with Eq. (2.93) directly, it is common to expand the square root in a power series:

$$R(t) = R_0\left[1 + \frac{1}{2}\left(\frac{vt}{R_0}\right)^2 - \frac{3}{8}\left(\frac{vt}{R_0}\right)^4 - \cdots\right] \quad \text{m} \tag{2.94}$$

In evaluating this expression, the range of t that must be considered may be limited by any of several factors, such as the time the target is within the radar mainbeam or the coherent processing interval duration over which pulses will be collected for subsequent processing.

Assume that the distance traveled by the target within this time of interest is much less than the nominal range R_0 so that higher-order terms in (vt/R_0) can be neglected. Then

$$R(t) \approx R_0 + \left(\frac{v^2}{2R_0}\right)t^2 \quad \text{m} \tag{2.95}$$

Equation (2.95) shows that the range is approximately a quadratic function of time for the crossing target scenario of Fig. 2.26. Using this truncated series in Eq. (2.88) gives

$$\tilde{y}(t) \approx ka\left(t - \frac{2R_0}{c}\right)\exp\left(-j\frac{4\pi}{\lambda_t}R_0\right)\exp\left[-j2\pi\left(\frac{v^2}{R_0\lambda_t}\right)t^2\right]\exp[j(2\pi F_t t + \varphi_0)] \quad (2.96)$$

All of the terms are the same as in the constant-velocity case of Eq. (2.92) except for the middle exponential. In App. B it is shown that the *instantaneous frequency* of a real or complex sinusoid is proportional to the time derivative of its phase. The quadratic phase function therefore represents a Doppler frequency shift that varies linearly with time due to the changing radar-target geometry:

$$F_D(t) = \frac{1}{2\pi}\frac{d}{dt}\left[-2\pi\left(\frac{v^2}{R_0\lambda_t}\right)t^2\right] = -\frac{2v^2}{R_0\lambda_t}t \quad \text{Hz} \quad (2.97)$$

As the target aircraft approaches from the left in Fig. 2.26 ($t < 0$) the instantaneous Doppler shift is positive. When the aircraft is abreast of the radar ($t = 0$), the Doppler shift is zero because the radial component of velocity is zero. Finally, as the aircraft passes by the radar ($t > 0$) the Doppler shift becomes negative, as would be expected for a receding target. This quadratic range case is important in synthetic aperture radar and will be revisited in Chap. 8.

The exponential term $\exp[-j4\pi R(t)/\lambda_t]$ in Eq. (2.88), or just its argument $-4\pi R(t)/\lambda_t$, is called the *phase history* of the received signal. The phase history encodes the variation of the range between the target and radar during the data collection time. For the constant-velocity example [Eq. (2.92)], the phase history is a linear function of time corresponding to a constant frequency sinusoid, that is, a constant Doppler shift. For the crossing target example of Eq. (2.96), it is approximately a quadratic function of time, producing a Doppler shift sinusoid having a frequency that varies linearly with time. Other radar-target motions will produce other functional forms for the phase history.

More generally, the term phase history can refer to the variation of phase (or the corresponding complex exponential) in any dimension of the radar data. Two other common uses are to describe the phase function of a frequency- or phase-modulated waveform or the spatial phase variation across the face of an array antenna at a fixed time. As will be seen, the phase history is central to radar signal processing. The design of many important radar signal processing operations depends critically on accurately estimating and modeling the phase history of the collected data. Examples include pulse compression, adaptive interference cancellation, and imaging.

2.7.3 Measuring Doppler Shift: Spatial Doppler

The Doppler shifts observed in radar are small fractions of the transmit frequency. In pulsed radars, they are too small to be measured from a single pulse echo in most cases. In Chap. 7 it will be seen that a lower bound on the standard deviation σ_F of the error in a single measurement (no integration) of the frequency of a complex sinusoid of SNR χ but unknown amplitude, frequency, and phase using an observation of length T seconds is $\sigma_F \geq \sqrt{6/[(2\pi)^2 \chi \cdot T^2]}$ Hz. This value should be much less than the Doppler shift, $\sigma_F \ll F_D$, if that shift is to be measured with reasonable precision, leading to a requirement that $T \gg \sqrt{6/[(2\pi)^2 \chi \cdot F_D^2]}$. Since the intent is to estimate the Doppler shift on a single-pulse basis, the SNR will likely be fairly low. A relatively favorable case having a Doppler shift of

10 kHz and an SNR of 0 dB ($\chi = 1$) gives $T \gg 39$ μs so that the pulse length, which sets the single-measurement observation time limit, must be at least 400 μs and preferable a few milliseconds. A less favorable case of −3 dB SNR and a 1-kHz Doppler shift requires $T \gg 552$ μs, requiring a pulse length in the ones to tens of milliseconds, impractically long in most pulsed systems. For this reason, most pulsed radars do not measure Doppler shift on an intrapulse basis, although a few designed for very high speed targets (satellites and missiles) and using very long pulses can do so.

CW radars do not share this limitation. Their continuous observation allows them to directly observe the frequency difference between transmitted and received sinusoids. However, frequency-modulated CW (FMCW) radars make measurements based on a single, usually brief, frequency sweep and therefore have the same limitations in measuring Doppler as pulsed radars.

The long observation time needed for pulsed or FMCW radars can be obtained by combining multiple pulse or sweeps. Suppose a pulsed radar transmits a series of M distinct pulses of duration τ beginning at times $t_m = mT$, where T is the pulse repetition interval (PRI). The mth transmitted pulse and received echo are

$$\tilde{x}_m(t) = a(t - mT)\exp[j(2\pi F_t t + \varphi_0)] \tag{2.98}$$

$$\tilde{y}_m(t) \approx k' \cdot a\left(t - mT - \frac{2R(mT)}{c}\right)\exp\left[j\left(2\pi F_t\left(t - \frac{2R(t)}{c}\right) + \varphi_0\right)\right] \tag{2.99}$$

After demodulation, the baseband received signal is

$$y_m(t) \approx k' \cdot a\left(t - mT - \frac{2R(mT)}{c}\right)\exp\left(-j\frac{4\pi}{\lambda_t}R(t)\right) \tag{2.100}$$

where k' includes the $\exp(-j\varphi_0)$ term. Assume each baseband pulse echo is sampled $2R_s/c$ seconds after transmission, corresponding to a range R_s. Also assume a target is present within the range bin corresponding to that sample time for the entire data collection time of mT seconds, meaning that $R(t)$ remains in the range interval $[R_s - c\tau/2, R_s]$.[11] The mth sample in this range bin is then

$$\begin{aligned}y_m\left(mT + \frac{2R_s}{c}\right) &= k' \cdot a\left(\frac{2}{c}[R_s - R(mT)]\right)\exp\left[-j\frac{4\pi}{\lambda_t}R\left(mT + \frac{2R_s}{c}\right)\right] \\ &= \hat{k} \cdot \exp\left[-j\frac{4\pi}{\lambda_t}R\left(mT + \frac{2R_s}{c}\right)\right] \\ &\equiv y[m]\end{aligned} \tag{2.101}$$

The constant $\hat{k}$ combines k' and the amplitude of the sampled pulse envelope $a(\cdot)$. The echo samples $y[m]$ form a "slow time" series of samples for that range bin, as will be described in Chap. 3.

The "stop" assumption applied in Eq. (2.92), when used across a series of pulses as in Eq. (2.101), is called the *stop-and-hop* approximation. The target is modeled as if it "stops" at the time of each pulse transmission at the corresponding range $R(mT)$, remains stationary

---

[11]Movement of a target across multiple range bins during the series of pulses or sweeps due to high rates of radar-target motion is known as *range migration*. It is discussed in Chap. 4 and, because it is common in imaging radar due to their much longer observation times, in Chap. 8.

until the echo is received from the longest range of interest, and then "hops" to the range at the next pulse transmission time, rather than moving continuously.

Consider again a constant velocity target, $R(t) = R_0 - vt$. The slow-time data series becomes

$$\begin{aligned} y[m] &= \hat{k} \cdot \exp\left[-j\frac{4\pi}{\lambda_t}\left[R_0 - v \cdot \left(mT + \frac{2R_s}{c}\right)\right]\right] \\ &= \hat{k} \cdot \exp\left[-j\frac{4\pi}{\lambda_t}\left(R_0 - \frac{2R_s v}{c}\right)\right]\exp\left[+j2\pi\left(\frac{2v}{\lambda_t}\right)mT\right] \end{aligned} \qquad (2.102)$$

The first exponential in Eq. (2.102) is a constant phase shift for all of the slow-time samples $y[m]$ and is of little consequence. The second exponential is a discrete complex sinusoid with normalized frequency $2vT/\lambda_t$ cycles/sample, corresponding to the expected Doppler frequency of $2v/\lambda_t$ Hz. Thus, the phase history obtained from a moving target using a series of pulses provides a way to measure the Doppler shift with good precision by observing the signal over an observation time much longer than that of a single pulse. A similar result is obtained for multiple sweeps in an FMCW radar.

The manifestation of the target Doppler shift in the slow-time phase history is sometimes referred to as *spatial Doppler*. This terminology emphasizes the fact that the Doppler shift is measured not from intrapulse or intrasweep frequency changes, but rather from the change of phase of the echoes at a given range bin over a series of transmissions. Because the change in phase from one pulse or sweep to the next is proportional to the change in range, the procedure effectively measures the change in range over time, that is, velocity. Because of the impracticality of measuring intrapulse or intrasweep Doppler frequency shifts in most systems, the term "Doppler processing" in radar usually refers to sensing and processing this spatial Doppler information.

2.8 Spatial Models

Previous sections have dealt with models of Doppler shift and the received power (both mean value and statistical fluctuations) of radar echoes from a single resolution cell. In this section, the variation in received complex voltage or power as a function of the spatial dimensions of range and angle will be considered. It will be seen that the observed complex voltage can be viewed as the output of a linear filter with the weighted variation in reflectivity over range or angle as its input. A similar result holds for power when the reflectivity field has a random phase variation. These relationships will lay the groundwork for an analysis of data sampling requirements and range and angle resolution in subsequent chapters.

2.8.1 Coherent Scattering

Consider a stationary radar. At time zero it transmits the complex signal

$$\bar{x}(t) = \sqrt{P_t} \cdot a(t) \cdot \exp[j(2\pi F_t t + \varphi_0)] \qquad (2.103)$$

Assume that $a(t)$ has unit amplitude so that the transmitted signal amplitude is represented by the term $\sqrt{P_t}$. This signal echoes off a differential scatterer of cross section $d\sigma(R,\theta,\phi)$ at coordinates (R,θ,ϕ). The *baseband complex reflectivity* or just *reflectivity* of the differential scatterer is, analogous to Eqs. (2.50) and (2.51), $d\rho = d\zeta(R,\theta,\phi)\exp[j\psi(R,\theta,\phi)]$ so that

$d\sigma = |d\rho|^2 = (d\zeta)^2$. The antenna is assumed to be mechanically scanning[12] in either or both angle coordinates with one-way voltage pattern $E(\theta, \phi)$. At the time of transmission it is steered in the direction (θ_0, ϕ_0). The differential received voltage is[13]

$$d\bar{y}\left(t = \frac{2R}{c}, \theta, \phi\right) = \sqrt{\frac{P_t \lambda_t^2}{(4\pi)^3 R^4 L_s L_a(R)}} E^2(\theta - \theta_0, \phi - \phi_0) \cdot d\rho(R, \theta, \phi) \times \cdots \\ \cdots \times a\left(t - \frac{2R}{c}\right) \exp\left\{j\left[2\pi F_t\left(t - \frac{2R}{c}\right) + \varphi_0\right]\right\} \quad (2.104)$$

Equation (2.104) can be simplified by defining $A_r = \exp(j\varphi_0)\sqrt{P_t \lambda_t^2/(4\pi)^3 L_s}$ and removing the carrier term with coherent demodulation, leaving only the baseband complex received voltage dy for the single differential scattering element:

$$dy(t, \theta, \phi) = A_r \cdot d\rho(R, \theta, \phi) \left[\frac{E^2(\theta - \theta_0, \phi - \phi_0)}{R^2 \sqrt{L_a(R)}} \cdot a\left(t - \frac{2R}{c}\right) \exp\left(-j\frac{4\pi}{\lambda_t} R\right)\right] \quad (2.105)$$

Equation (2.105) gives the contribution to the received echo voltage of a differential scatterer element at coordinates (R, θ, ϕ). The total received voltage is obtained by integrating these differential contributions over all space,

$$y(t, \theta, \phi) = \int_{\phi=-\pi/2}^{\pi/2} \int_{\theta=-\pi}^{\pi} \int_{R=0}^{\infty} dy(t, \theta, \phi) \quad (2.106)$$

Equation (2.106) is a *coherent* scattering model: the differential scatterers are assumed to add as complex voltages. This is most appropriate for reflectivity fields characterized by relatively static configurations of scatterers, for example, man-made vehicles and urban areas. The case where the scatterers are not static is considered in Sec. 2.8.4.

Now write $d\rho(R, \theta, \phi) = \rho(R, \theta, \phi) \cdot dV = \rho(R, \theta, \phi) R^2 \cos\phi \cdot dR\, d\theta\, d\phi$ to obtain

$$y(t, \theta, \phi) = A_r \int_{\phi=-\pi/2}^{\pi/2} \int_{\theta=-\pi}^{\pi} \int_{R=0}^{\infty} \frac{\exp[-j(4\pi/\lambda_t)R]}{\sqrt{L_a(R)}} \rho(R, \theta, \phi) \times \cdots \\ \cdots \times E^2(\theta - \theta_0, \phi - \phi_0) a\left(t - \frac{2R}{c}\right) \cos\phi\, dR\, d\theta\, d\phi \quad (2.107)$$

Define the *effective reflectivity* ρ' to include the attenuation due to atmospheric loss, the phase rotation due to two-way propagation range, and the $\cos\phi$ term of the differential volume element:

$$\rho'(R, \theta, \phi) \equiv \frac{\exp[-j(4\pi/\lambda_t)R]}{\sqrt{L_a(R)}} \rho(R, \theta, \phi) \cos\phi \quad (2.108)$$

---

[12]The results in this section must be modified for an electronically scanned antenna, for which the antenna pattern is a function of the scan angles.

[13]In order to avoiding writing $2R/c$ and $ct/2$ many times, equations will be written somewhat casually in the remainder of this section in terms of both t and R, though they are linked according to $t = 2R/c$.

Signal Models

Applying Eq. (2.108) to Eq. (2.107), the received signal is seen to be similar to a *three-dimensional convolution* of this effective reflectivity with a convolution kernel comprising the antenna two-way voltage pattern in the angle coordinates and the signal waveform function in the range coordinate. Specifically,

$$y(t) \approx A_r \rho'\left(\frac{ct}{2}, \theta_0, \phi_0\right) *_t *_\theta *_\phi \left[E^2(-\theta_0, -\phi_0) a(t)\right] \qquad (2.109)$$

where the symbols $*_t$, $*_\theta$, and $*_\phi$ denote convolution over the indicated coordinate. Now assume the antenna pattern is symmetric in the two angular coordinates, as is often the case; rescale the time variable to units of range; and replace θ_0 and ϕ_0 with general angular variables θ and ϕ. These substitutions finally give

$$y\left(t = \frac{2R}{c}, \theta, \phi\right) \approx A_r \rho'(R, \theta, \phi) *_R *_\theta *_\phi \left[E^2(\theta, \phi) a\left(\frac{2R}{c}\right)\right] \qquad (2.110)$$

Equation (2.110) is stated as an approximate convolution because of the finite integration limits in the angular variables, which arise due to the periodicity in angle of the antenna pattern and scene reflectivity. Nonetheless, like a linear convolution, Eq. (2.110) computes the output at a given point in space as a local average of the reflectivity distribution weighted by the antenna pattern and waveform. For most antennas and waveforms these patterns concentrate most of their energy in a relatively small finite region defined by the mainlobe for the antenna pattern and the range resolution cell extent of the waveform. Consequently, the output signal can be expected to behave like a true linear convolution. A full discussion of a spherical convolution-like equation similar to Eq. (2.110), including development of the Fourier transform relations, is given in Baddour (2010).

The convolutional model of Eq. (2.110) is an important result. Its significance is that it allows interpretation of the measured data as the result of a linear filtering process, so that Fourier transform relations between $y(t)$, $\rho'(R,\theta,\phi)$, $E^2(\theta,\phi)$, and $a(t)$ can be established and applied to model signal properties, determine sampling rates, and so forth. For example, the range resolution of the measured reflectivity function is seen to be limited by the waveform $a(t)$. (The introduction of matched filtering in Chap. 4 will significantly change this statement.) Similarly, for a conventional scanning radar, the angular resolution will be determined by the antenna beamwidth. (The introduction of synthetic aperture techniques in Chap. 8 significantly changes this statement as well.) It also follows from the filtering action of $a(t)$ and $E^2(\theta, \phi)$ that the bandwidth of the measured reflectivity function in range and angle is limited by the bandwidth of the waveform modulation function and antenna power pattern. This observation will be used in Chap. 3 to determine the range and angle sampling requirements.

2.8.2 Variation with Angle

Now consider more specifically the variation in reflectivity with angle for a fixed range, say R_0. Define the range-averaged effective reflectivity

$$\hat{\rho}(\theta, \phi; R_0) = \int_R a\left(\frac{2}{c}(R_0 - R)\right) \rho'(R, \theta, \phi) dR = \left[\rho'(R, \theta, \phi) *_R a\left(\frac{2R}{c}\right)\right]_{R=R_0} \qquad (2.111)$$

This is the reflectivity variation in angle, taking into account the range averaging at each angle due to the finite waveform length. Note that in the limit of very fine range resolution, that is, if the waveform modulation $a(2R/c) \to \delta_D(R - R_0)$, then $\hat{\rho}(\theta, \phi; R_0) \to \rho'(R_0, \theta, \phi)$.

That is, the "range-averaged" reflectivity would exactly equal the effective reflectivity evaluated at the range of interest R_0.

Applying Eq. (2.111) to Eq. (2.110) gives

$$y\left(\theta,\phi;t_0 = \frac{2R_0}{c}\right) = A_r \int_{\phi'=-\pi/2}^{\pi/2} \int_{\theta'=-\pi}^{\pi} E^2(\theta-\theta',\phi-\phi')\hat{\rho}(\theta',\phi';R_0)\,d\theta'\,d\phi' \qquad (2.112)$$
$$\approx \hat{\rho}(\theta,\phi;R_0) *_\theta *_\phi E^2(\theta,\phi)$$

where again symmetry of the antenna pattern has been assumed in the second line. Equation (2.112) is a special case of Eq. (2.110) showing that the complex voltage at the output of a coherent receiver for a fixed range and a scanning antenna is approximately the convolution in the angle dimensions of the range-averaged effective reflectivity function evaluated at the range R_0, $\hat{\rho}(\theta,\phi;R_0)$, with the antenna two-way voltage pattern $E^2(\theta,\phi)$.

As mentioned earlier, the interpretation of Eq. (2.112) as a linear convolution is an approximation. Suppose that the elevation angle ϕ is fixed, and consider only the variation in azimuth angle θ. Because the integration is over a full 2π radians and the integrand is periodic in θ with period 2π, the integration over azimuth is a circular convolution of periodic functions. This would not appear to be the case if instead θ is fixed and ϕ varies because the integrand is over a range of only π radians. However, one could equally well replace Eq. (2.112) with

$$y(\theta,\phi;t_0) = A_r \int_{\phi'=-\pi}^{\pi} \int_{\theta'=-\pi/2}^{\pi/2} E^2(\theta-\theta',\phi-\phi')\hat{\rho}(\theta',\phi';R_0)\,d\theta'\,d\phi' \qquad (2.113)$$
$$\approx \hat{\rho}(\theta,\phi;R_0) *_\theta *_\phi E^2(\theta,\phi)$$

For fixed azimuth, this is now a circular convolution of periodic functions in elevation. Taken together, the two integrals over the angular variables implement a two-dimensional weighting and averaging over the (θ, ϕ) space. So long as the antenna beamwidths are small, this circular convolution will closely approximate a linear convolution in the vicinity of (θ, ϕ).

Figure 2.27 illustrates intuitively in one angle dimension the process described by Eq. (2.112) or (2.113). Assume that the elevation angle is fixed at $\phi = 0°$ and consider only the azimuth variation. An array of three ideal point scatterers is illuminated by a radar that scans in azimuth across the target field. The response to any one scatterer is maximum when the radar boresight is aimed at that scatterer; as the radar boresight moves away, the strength of the echo declines because less energy is directed to the scatterer on transmission and the antenna is also less sensitive to echoes from directions other than the boresight on reception. For an isolated scatterer, the amplitude of the coherent baseband received signal $y(\theta, 0; R_0)$ at the receiver output will be proportional to $E^2(\theta, 0)$. Thus, a graph of the received signal mimics the antenna two-way azimuth voltage pattern.

Assuming a linear receiver so that superposition applies, the response to two closely spaced point scatterers is proportional to two replicas of the antenna pattern, overlapped and added to get a composite response. If the two scatterers are close enough together the individual responses are not resolved, instead blurring together into a single peak, as illustrated in Fig. 2.27. The details of the combined response depend on the relative phase of the two individual responses; they may combine in or out of phase, yielding significantly different composites as was seen in Fig. 1.27. However, the separation at which scatterers are consistently resolved regardless of relative phase clearly depends on the antenna pattern $E^2(\theta, 0)$, and in particular on the mainlobe beamwidth.

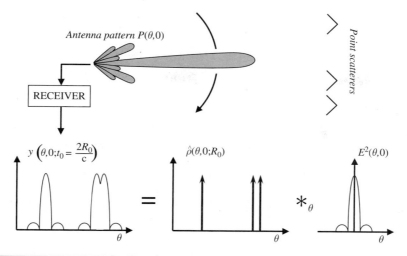

FIGURE 2.27 When scanning past an array of point scatterers, the receiver output is the convolution of the scatterer location map and the antenna pattern.

Because of the approximately linear convolution relation of Eq. (2.112), the spatial Fourier transform of the observed signal is approximately the input spatial Fourier transform multiplied by the Fourier transform of the antenna pattern. Practical antenna patterns have lowpass spectra. The ideal two-way azimuth voltage patterns for circular and rectangular apertures of width D are (Balanis, 2016)

$$E^2(\theta, 0) = \left[\frac{J_1(\pi D \sin\theta/\lambda_t)}{\pi D \sin\theta/\lambda_t} \right]^2 \quad \text{(circular aperture)}$$

$$E^2(\theta, 0) = \left[\frac{\sin(\pi D \sin\theta/\lambda_t)}{\pi D \sin\theta/\lambda_t} \right]^2 \quad \text{(rectangular aperture)}$$

(2.114)

Figure 2.28 plots these patterns on a decibel scale for the case $D = 40\lambda_t$.

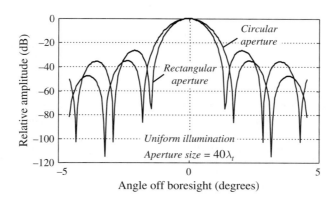

FIGURE 2.28 Two-way antenna voltage patterns for ideal, uniformly illuminated circular and rectangular apertures.

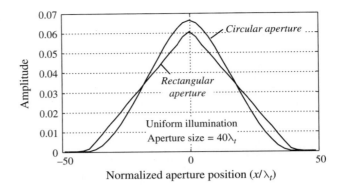

FIGURE 2.29 Spatial spectra corresponding to the antenna patterns of Fig. 2.28.

The corresponding spatial spectra are shown in Fig. 2.29. For the rectangular aperture, it is a triangle function with a support of twice the aperture width. The reason is easy to see: the one-way voltage pattern is just the inverse Fourier transform of the aperture function, which for uniform illumination is a sinc pattern. The two-way pattern is therefore a sinc-squared. Its Fourier transform is then the self-convolution of the rectangular aperture function, giving a triangle of twice the aperture width. The spectrum for the circular aperture has a similar but smoother shape.

The spatial spectra of these idealized but typical antenna patterns are lowpass functions. Thus, the upper frequencies in the spatial spectrum of the observed data will be strongly attenuated and in fact effectively removed. Since resolution is inversely proportional to bandwidth, the antenna pattern's lowpass spatial spectrum degrades resolution.

2.8.3 Variation with Range

A development similar to that in Sec. 2.8.2 can be carried out to specialize Eq. (2.110) for the variation of received voltage in the range dimension along the boresight look direction (θ_0, ϕ_0).[14] The resulting function is called the *range profile* (RP) in the look direction. First, interchange the order of integration in Eq. (2.106) so that the outer integral is over range. Next, define the new quantity

$$\tilde{\rho}(R;\theta_0,\phi_0) = \iint_{\theta,\phi} E^2(\theta - \theta_0, \phi - \phi_0) \rho'(R,\theta,\phi) \, d\theta \, d\phi$$
$$= \rho'(R,\theta,\phi) *_\theta *_\phi E^2(\theta,\phi)\big|_{\theta=\theta_0, \phi=\phi_0} \qquad (2.115)$$

This is the reflectivity variation in range, taking into account the azimuth and elevation averaging at each range due to the antenna power pattern. Note that in the limit as the antenna power pattern tends to the ideal, $E^2(\theta,\phi) \to G\delta_D(\theta,\phi)$, then $\tilde{\rho}(R;\theta_0,\phi_0) \to \rho'(R,\theta_0,\phi_0)$. That is, the "angle-averaged" reflectivity exactly equals the effective reflectivity along the antenna look direction, as expected.

---

[14]The analysis can be carried out equally easily for an off-boresight look direction. The only difference is to substitute an antenna gain value other than the peak gain G.

Applying Eq. (2.115) to Eq. (2.110) leaves (Munson and Visentin, 1989)

$$y\left(t = \frac{2R}{c}; \theta_0, \phi_0\right) = A_r \tilde{\rho}(R; \theta_0, \phi_0) *_R \left[a\left(\frac{2R}{c}\right)\right]$$

$$= A_r \int_{R'=0}^{\infty} a\left(\frac{2}{c}(R - R')\right) \tilde{\rho}(R'; \theta_0, \phi_0) \, dR' \quad (2.116)$$

or an equivalent equation, using time units instead of range units,

$$y(t; \theta_0, \phi_0) = A_r \tilde{\rho}\left(\frac{ct}{2}; \theta_0, \phi_0\right) *_t a(t) = A_r \int_{t'=0}^{\infty} a(t - t') \tilde{\rho}\left(\frac{ct'}{2}; \theta_0, \phi_0\right) dt' \quad (2.117)$$

Equation (2.116) or (2.117) shows that the complex voltage at the output of a coherent receiver versus time for a given antenna look direction is the convolution in the range dimension of the angle-averaged effective reflectivity function in that look direction, $\tilde{\rho}(R; \theta_0, \phi_0)$, with the waveform modulation function $a(t)$. Consequently, the bandwidth of the range profile is set by the bandwidth of the radar waveform.

2.8.4 Noncoherent Scattering

Equation (2.107) and its approximate form Eq. (2.110) assume coherent addition of individual differential scatterer echoes; that is, the complex amplitude (magnitude and phase) of the total response is the complex sum of the differential complex echoes. For distributed area or volume clutter comprising very large numbers of scatterers with essentially random phases such as rain or natural ground clutter (grass, trees, water, etc.), it is more useful to model the scatterer reflectivity as having a random phase with either a random or nonrandom magnitude. The total received signal is then also a random variable, and the expected value of the received power becomes of interest.

From Eqs. (2.107) and (2.108), the power of the integrated received power can be written

$$|y(t, \theta, \phi)|^2 = y(t, \theta, \phi) \cdot y^*(t, \theta, \phi)$$

$$= \left\{A_r \int_V \rho'(R_1, \theta_1, \phi_1) E^2(\theta_1 - \theta_0, \phi_1 - \phi_0) a\left(t - \frac{2R_1}{c}\right) dR_1 \, d\theta_1 \, d\phi_1\right\} \times \cdots \quad (2.118)$$

$$\cdots \times \left\{A_r^* \int_V [\rho'(R_2, \theta_2, \phi_2)]^* [E^2(\theta_2 - \theta_0, \phi_2 - \phi_0)]^* a^*\left(t - \frac{2R_2}{c}\right) dR_2 \, d\theta_2 \, d\phi_2\right\}$$

The subscripts "1" and "2" distinguish the spatial variables in the two integrals. Taking the expected value of $|y(t)|^2$, the mean received power becomes

$$\mathbf{E}\left\{|y(t, \theta, \phi)|^2\right\} =$$

$$|A_r|^2 \iint_{V_1, V_2} s_\rho(R_1 - R_2, \theta_1 - \theta_2, \phi_1 - \phi_2) E^2(\theta_1 - \theta_0, \phi_1 - \phi_0) a\left(t - \frac{2R_1}{c}\right) \times \cdots \quad (2.119)$$

$$\cdots \times [E^2(\theta_2 - \theta_0, \phi_2 - \phi_0)]^* a^*\left(t - \frac{2R_2}{c}\right) dR_2 \, d\theta_2 \, d\phi_2 \, dR_1 \, d\theta_1 \, d\phi$$

where $s_\rho(\cdot)$ is the autocorrelation of $\rho'(\cdot)$.

Again express the effective reflectivity distribution $\rho'(R, \theta, \varphi)$ as the product of its phase $\exp[j\psi'(R, \theta, \phi)]$ and amplitude $\zeta'(R,\theta,\phi)$. Model ψ' as uniformly distributed over $(0, 2\pi]$ and white in all three spatial variables, while ζ' may be random or deterministic; if random, it is statistically independent of ψ'. The autocorrelation function of ρ then takes the form

$$\begin{aligned} s_\rho(R_1 - R_2, \theta_1 - \theta_2, \phi_1 - \phi_2) &= \mathrm{E}\{\rho'(R_1,\theta_1,\phi_1)[\rho'(R_2,\theta_2,\phi_2)]^*\} \\ &= \mathrm{E}\{\zeta'(R_1,\theta_1,\phi_1)\zeta'(R_2,\theta_2,\phi_2)\}\mathrm{E}\{\exp[j(\psi'(R_1,\theta_1,\phi_1) - \psi'(R_2,\theta_2,\phi_2))]\} \\ &= \mathrm{E}\{\zeta'(R_1,\theta_1,\phi_1)\zeta'(R_2,\theta_2,\phi_2)\}\,\delta_D(R_1 - R_2) \cdot \delta_D(\theta_1 - \theta_2) \cdot \delta_D(\phi_1 - \phi_2) \end{aligned} \quad (2.120)$$

(See Prob. 32.) Again assuming a symmetric antenna pattern and integrating over the impulse functions to eliminate one volume integral in Eq. (2.119), the mean received power becomes

$$\begin{aligned} \mathrm{E}\{|y(t,\theta,\phi)|^2\} &= |A_r|^2 \int_{V_1} \zeta^2(R,\theta,\phi)|E^2(\theta_1 - \theta_0, \phi_1 - \phi_0)|^2 \left|a\!\left(t - \frac{2R_1}{c}\right)\right|^2 dR_1\,d\theta_1\,d\phi \\ &\approx |A_r|^2\,\zeta^2(R,\theta,\phi) *_R *_\theta *_\phi \left|E^2(\theta_1 - \theta_0, \phi_1 - \phi_0)\right|^2 \left|a\!\left(\frac{2R}{c}\right)\right|^2 \\ &= |A_r|^2 |\rho'(R,\theta,\phi)|^2 *_R *_\theta *_\phi \left|E^2(\theta_1 - \theta_0, \phi_1 - \phi_0)\right|^2 \left|a\!\left(\frac{2R}{c}\right)\right|^2 \end{aligned} \quad (2.121)$$

where the last line recognizes that $\zeta^2 = |\rho|^2$.

Equation (2.121) is the noncoherent equivalent of Eq. (2.110). It shows the expected value of the received power in the case of noncoherent scattering to be the 3D convolution of the squared effective reflectivity with the two-way antenna power pattern and the waveform power envelope. Thus, the received power still obeys a convolutional model. Similar results are developed for weather clutter in Doviak and Zrnic (1993, Sec. 4.4). An example of the use of a noncoherent convolutional model of radar clutter data can be found in Richards et al. (1986).

2.8.5 Projections

The range-averaged reflectivity $\hat{\rho}(\theta,\phi;R_0)$ of Eq. (2.111) and the angle-averaged reflectivity $\tilde{\rho}(R;\theta_0,\phi_0)$ of Eq. (2.115) are examples of *projections*. In each case, the three-dimensional reflectivity is reduced in dimension by integrating over one or more dimensions. The range-averaged reflectivity was reduced to a two-dimensional function by integrating over range, while the angle-averaged reflectivity was reduced to a one-dimensional function by integrating over both angle coordinates.

The idea of projections, particularly the angle-averaged projection $\tilde{\rho}(R;\theta_0,\phi_0)$, will be important in deriving the polar format spotlight SAR algorithm in Chap. 8. The projections that will be needed are integrals over straight lines or planar surfaces. The averaging in Eq. (2.115) is over the surface of a sphere. However, only a region of θ_3 radians in azimuth and ϕ_3 radians in elevation contributes significantly to the integral, and for small beamwidths at long ranges this limited region is nearly planar.

2.8.6 Multipath

The convolutional model of the measured range profile is based on the assumption of superposition of backscattered fields and a one-to-one mapping of echo arrival time to range, $t \rightarrow R = ct/2$. The superposition of electromagnetic fields is a valid assumption, but the mapping of time to range may not be. To illustrate, consider Fig. 2.30, which diagrams

Signal Models

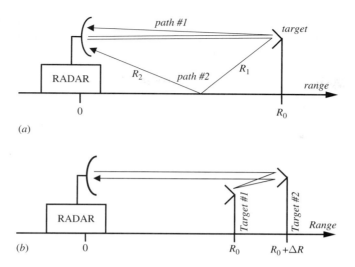

FIGURE 2.30 Illustration of two scattering phenomena which violate the one-to-one mapping of time to range. (a) Multipath. (b) Multiple bounce.

two phenomena that violate this assumption. Figure 2.30a illustrates the problem of *multipath*, in which echoes from the same target arrive at the radar receiver via two different paths. The first is the direct path of total length $2R_0$. The second is the "multipath" or "ground bounce" path with length $R_0 + R_1 + R_2 > 2R_0$. Though not shown, it is also possible for a portion of the transmitted wave to arrive at the target via the ground bounce and be scattered back along both paths, meaning that there may also be an echo with a time delay corresponding to a two-way path length of $2(R_0 + R_1 + R_2)$. Consequently, one scatterer may produce echoes at three different apparent ranges if multipath is present. Whether these appear as distinct echoes depends on the relationship between the path length difference and the range resolution. The ground bounce echoes are often, but not always, significantly attenuated with respect to the direct path echo. The degree of attenuation depends on the bistatic scattering characteristics of the surface, the antenna pattern characteristics (because the multipath bounce is not on the peak of the mainlobe and may be in the sidelobes) and the problem geometry. As the range between target and radar varies, the path length difference also varies, so that the direct and multipath bounces may alternately add in and out of phase, provided the path length difference is such that the two received echoes overlap. Multipath is generally most significant for targets located at low altitude over a good reflecting surface such as a relatively smooth terrain or calm ocean and at long range, so that the grazing angles involved are small. More information on multipath is available in Saffold (2010).

Figure 2.30b illustrates the effect of *multiple bounce* echoes in a situation involving two scatterers. A portion of the energy reflected from the more distant scatterer bounces off the nearer scatterer, then reflects a second time off the distant scatterer and returns to the radar. Obviously additional multiple bounces are also possible. For the situation sketched, three apparent echoes will again result, with the third appearing to be due to a phantom scatterer at range $R_0 + 2\Delta R$. As with multipath, the amplitude of multiple bounce echoes often falls off rapidly, and the same considerations of constructive and destructive superposition apply.

These possible differences in the measured and actual reflectivity distributions do not mean that range profile measurements are not useful. They do mean that in situations where significant multipath or multiple bounce phenomena are possible, the range profiles must be interpreted with care.

2.9 Spectral Model

There is one more interpretation of the received radar signal that proves useful in subsequent chapters. The preceding two sections have emphasized linear filtering models of the spatial reflectivity distribution as observed through the received complex baseband signals. However, it was pointed out previously that radar cross section is a function of, among many other things, the radar frequency. Thus, it is useful to investigate the significance of the radar transmitted frequency F_t on the reflectivity measurements.

To understand the role of transmitted frequency, it is necessary to deal with the radar signal while it is still at the radar frequency F_t. If the development from Eq. (2.103) to Eq. (2.117) is repeated without demodulating the signals to baseband and the range variation is considered, the RF version of Eq. (2.117) can be obtained as

$$\bar{y}(t) = A_r \int_{t'=0}^{\infty} a(t-t') \tilde{\rho}\left(\frac{ct'}{2}; \theta_0, \phi_0\right) \exp(j2\pi F_t t') \, dt' \qquad (2.122)$$
$$= A_r \left[\tilde{\rho}\left(\frac{ct'}{2}; \theta_0, \phi_0\right) \exp\left[j(2\pi F_t t + \varphi)\right] \right] *_t a(t)$$

Now consider the Fourier transform of $\bar{y}(t; \theta_0, \phi_0)$ with respect to the time (range) variable t. Using simple properties of Fourier transforms gives

$$\bar{Y}(F; \theta_0, \phi_0) = \frac{2A_r}{c} \exp(j\phi) \cdot A(F) \cdot \tilde{P}\left[\frac{2(F-F_t)}{c}\right] \qquad (2.123)$$

Figure 2.31 provides a pictorial interpretation of this equation under the assumption that the transmitted waveform $a(t)$ is a narrowband waveform. In this case,

$$\bar{Y}(F; \theta_0, \phi_0) \approx \frac{2A_r}{c} \exp(j\phi) \cdot \tilde{P}\left(\frac{-2F_t}{c}\right) \cdot A(F) \qquad (2.124)$$

so that the amplitude of the spectrum of the received signal, and therefore of the time domain signal itself, is proportional to the amplitude of the spectrum of the angle averaged range profile, evaluated at the transmitted frequency. Since it is the complex spectrum that appears

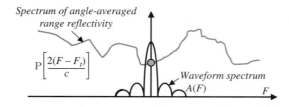

FIGURE 2.31 Pictorial interpretation of Eq. (2.124), illustrating the spectral sampling effect of a narrowband radar pulse.

Signal Models

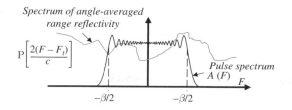

FIGURE 2.32 Pictorial interpretation of Eq. (2.125) illustrating the spectral windowing effect of a wideband radar pulse.

in Eq. (2.124), both the amplitude and phase of the returned waveform are affected by the amplitude and phase of the reflectivity spectral sample. Equation (2.124) shows that a narrowband radar pulse can be interpreted as measuring a frequency sample of the spectrum of the angle-averaged reflectivity range variation at the RF F_t. The stepped frequency waveform to be discussed in Sec. 4.8.1 will take advantage of this property to develop fine-resolution range profiles of targets and clutter.

Another case of interest occurs when $a(t)$ is a wideband signal of bandwidth β Hz. For some waveforms such as the linear frequency-modulated pulse, the magnitude of the spectrum $A(F)$ is approximately a rectangle, as shown in Fig. 2.32. The spectrum of the receiver waveform $\bar{y}(t; \theta_0, \phi_0)$ is then approximately that of the angle-averaged range profile over the bandwidth of the waveform, modified by the phase of the signal spectrum:

$$\bar{Y}(F; \theta_0, \phi_0) \approx \frac{2A_r}{c} \tilde{P}\left[\frac{2(F-F_t)}{c}\right] \exp\{j \cdot \arg[A(F)]\}, \quad -\frac{\beta}{2} \leq F \leq +\frac{\beta}{2} \quad (2.125)$$

In other words, the pulse spectrum acts as a window selecting a portion of the spectrum of the angle-averaged reflectivity. This result will be useful in understanding the use of linear FM and other modulated waveforms to achieve fine range resolution in Chap. 4, and in understanding the spotlight synthetic aperture radar data model in Sec. 8.4.

2.10 Summary

An understanding of the nature of the signals of interest is prerequisite to the design of successful signal processing systems. This chapter reviews the most common signal models used in designing and analyzing radar signal processors. It has been seen that multiple views of the radar echo are used: its variation in amplitude, space, time, and frequency, and deterministic and statistical interpretations of these variations.

Radar signal modeling traditionally focuses most strongly on amplitude models, that is, on radar cross section. RCS is viewed as a deterministic quantity, predictable in principle through the use of Maxwell's equations if the scattering is modeled accurately enough. The radar range equation in its many forms (only a very small subset of which has been introduced here) is the radar engineer's most fundamental tool for estimating received signal amplitude or SNR or, conversely, determining required system characteristics such as transmitted power or antenna gain. However, equally important in modern systems is measurement and modeling of target phase history.

The radar system is a measuring instrument, used to observe the variation of RCS in space. Its waveform (modulation and carrier term) and antenna power pattern determine its measurement characteristics, which in turn determine the achievable resolution and required sampling rates. The effect of the radar measurement system on the spatial variation of observed RCS is well modeled by the convolution of the combined waveform-and-antenna pattern measurement kernel with the three-dimensional reflectivity function. This important observation means that the tools of linear system analysis can be brought to bear to help analyze and understand the performance of radar systems. The carrier frequency in combination with any Doppler shifts determines what portion of the reflectivity frequency spectrum is sampled by the pulse. This observation reinforces the need for frequency domain analyses of radar measurements. Linear systems and frequency domain viewpoints are relied on heavily throughout the remainder of the book.

Even though RCS is a deterministic quantity, its sensitivity to radar frequency, aspect angle, and range coupled with the complexity of typical targets results in very complex behavior of observed amplitude measurements. Statistical models are used to describe this complexity. A variety of statistical models, comprising both probability density functions and correlation properties, have gained acceptance for various scenarios and form the basis for much analysis, particularly in calculations of probabilities of detection and false alarm, two of the most important radar performance measures.

References

Adamy, D., *EW 101: A First Course in Electronic Warfare*. Artech House, Boston, MA, 2001.

Baddour, N., "Operational and Convolution Properties of Three-Dimensional Fourier Transforms in Spherical Polar Coordinates," *Journal of the Optical Society of America*, vol. 27, no. 3, pp. 2144–2155, Oct. 2010.

Balanis, C. A., *Antenna Theory*, 4th ed. Wiley, New York, 2016.

Baxa, E. G., Jr., "Airborne Pulsed Doppler Radar Detection of Low-Altitude Windshear—A Signal Processing Problem," *Digital Signal Processing*, vol. 1, no. 4, pp. 186–197, Oct. 1991.

Beckmann, P., and A. Spizzichino, *The Scattering of Electromagnetic Waves from Rough Surfaces*. MacMillan, New York, 1963.

Billingsley, J. B., *Radar Clutter*. Artech House, Norwood, MA, 2001.

Birkmeier, W. P., and N. D. Wallace, *AIEE Transactions on Communication Electronics*, vol. 81, pp. 571–575, Jan. 1963.

Carlson, A. B., and P. B. Crilly, *Communication Systems*, 5th ed. McGraw-Hill, New York, 2009.

Curlander, J. C., and R. N. McDonough, *Synthetic Aperture Radar*. Wiley, New York, 1991.

Currie, N. C., "Characteristics of Clutter," Chap. 5 in M. A. Richards, J. A. Scheer, and W. A. Holm (eds.), *Principles of Modern Radar: Basic Principles*. SciTech Publishing, Raleigh, NC, 2010.

Doviak, D. S., and R. J. Zrnic, *Doppler Radar and Weather Observations*, 2d ed. Academic Press, San Diego, CA, 1993.

Griffiths, H. et al., "Radar Spectrum Management: Technical and Regulatory Issues," *Proceedings of the IEEE*, vol. 103, no. 1, pp. 85–102, Jan. 2015.

Haykin, S., B. W. Currie, and S. B. Kesler, "Maximum Entropy Spectral Analysis of Radar Clutter," *Proceedings of the IEEE*, vol. 70, no. 9, pp. 953–962, Sep. 1982.

Holm, W. A., "MMW Radar Signal Processing Techniques," Chap. 6 in N. C. Currie and C. E. Brown (eds.), *Principles and Applications of Millimeter-Wave Radar*. Artech House, Boston, MA, 1987.

Jakeman, E., and P. N. Pusey, "A Model for Non-Rayleigh Sea Echo," *IEEE Transactions on Antennas and Propagation*, vol. 24, no. 6, pp. 806–814, Nov. 1976.

Kay, S. M., *Modern Spectral Estimation*. Prentice Hall, Englewood Cliffs, NJ, 1988.

Knott, E. F., J. F. Shaeffer, and M. T. Tuley, *Radar Cross Section*, 2d ed. Artech House, Boston, MA, 1993.

Levanon, N., *Radar Principles*. Wiley, New York, 1988.

Lewinski, D. J., "Nonstationary Probabilistic Target and Clutter Scattering Models," *IEEE Transactions on Antennas and Propagation*, vol. AP-31, no. 3, pp. 490–498, May 1983.

Long, M. W., *Radar Reflectivity of Land and Sea*, 3d ed. Artech House, Boston, MA, 2001.

Lothes, R. N., M. B. Szymanski, and R. G. Wiley, *Radar Vulnerability to Jamming*. Artech House, Boston, MA, 1990.

Meyer, D. P., and H. A. Mayer, *Radar Target Detection*. Academic Press, New York, 1973.

Mott, H., *Polarization in Antennas and Radar*. Wiley, New York, 1986.

Mountcastle, P. D., "New Implementation of the Billingsley Clutter Model for GMTI Data Cube Generation," *Proceedings of the IEEE 2004 Radar Conference*, pp. 398–401, Apr. 2004.

Munson, D. C., and R. L. Visentin, "A Signal Processing View of Strip-Mapping Synthetic Aperture Radar," *IEEE Transactions on Acoustics, Speech, and Signal Processing*, vol. 27, no. 12, pp. 2131–2147, 1989.

Nathanson, F. E., *Radar Design Principles*, 2d ed. McGraw-Hill, New York, 1991.

Papoulis, A., and S. U. Pillai, *Probability, Random Variables and Stochastic Processes*, 4th ed. McGraw-Hill, New York, 2001.

Probert-Jones, J. R., "The Radar Equation in Meteorology," *Quarterly Journal of the Royal Meteorological Society*, vol. 88, pp. 485–495, 1962.

Ray, H., "Improving Radar Range and Angle Detection with Frequency Agility," *Microwave Journal*, vol. 8, p. 64ff, May 1966.

Richards, M. A., "Doppler Shift in Radar," technical memorandum, Apr. 2020. Available at http://radarsp.com.

Richards, M., C. Morris, and M. Hayes, "Iterative Enhancement of Noncoherent Radar Data," *Proceedings of the IEEE International Conference on Acoustics, Speech, and Signal Processing* (ICASSP '86), pp. 1929–1932, Tokyo, Japan, 1986.

Richards, M. A., J. A. Scheer, and W. A. Holm (eds.), *Principles of Modern Radar: Basic Principles*. SciTech Publishing, Raleigh, NC, 2010.

Saffold, J. A., "Propagation Effects and Mechanisms," Chap. 4 in M. A. Richards, J. A. Scheer, and W. A. Holm (eds.), *Principles of Modern Radar: Basic Principles*. SciTech Publishing, Raleigh, NC, 2010.

Sangston, K. J., and K. R. Gerlach, "Coherent Detection of Radar Targets in a Non-Gaussian Background," *IEEE Transactions on Aerospace and Electronic Systems*, vol. AES-30, no. 2, pp. 330–340, Apr. 1994.

Sauvageot, H., *Radar Meteorology*, Artech House, Boston, MA, 1992.

Scheer, J. A., "The Radar Range Equation," Chap. 2 in M. A. Richards, J. A. Scheer, and W. A. Holm (eds.), *Principles of Modern Radar: Basic Principles*. SciTech Publishing, Raleigh, NC, 2010.

Skolnik, M. I., *Introduction to Radar Systems*, 3d. ed. McGraw-Hill, New York, 2001.

Swerling, P., "Probability of Detection for Fluctuating Targets," *IRE Transactions on Information Theory*, vol. IT-6, pp. 269–308, Apr. 1960.

Ulaby, F. T., and M. C. Dobson, *Handbook of Radar Scattering Statistics for Terrain*. Artech House, Norwood, MA, 2019.

Ward, K. D., "Compound Representation of High Resolution Sea Clutter," *Electronics Letters*, vol. 17, no. 16, pp. 561–563, Aug. 6, 1981.

Problems

1. Find the received power P_r expected from a radar and target having the following parameters: RF = 95 GHz (W band), transmitted power = 100 W, antenna beamwidth = 2° in azimuth and 5° in elevation, system losses = 5 dB, target range = 3 km, target RCS = 20 m². Use Fig. 1.3 and assume operation near sea level to estimate atmospheric losses. What is the ratio in decibels of the received power to the transmitted power?

2. By how many dB will the received power be reduced (not the absolute power received, but the *reduction* in power) in Prob. 1 if the weather changes from clear to a heavy rain of 25 mm/h? Estimate the needed parameters from Fig. 1.4.

3. Suppose the parameters of a radar are such that the power of the echo from a particular target is just detectable at a range of 50 miles. If the target RCS is reduced by 10 dB, what will be the new detection range? By how many dB must the RCS be reduced to reduce the detection range to 5 miles?

4. According to the Smithsonian Air and Space Museum, the RCS of a B-52 bomber is about 1000 m², while that of a B-2 stealth bomber is 10^{-6} m². If a given radar system could detect the B-52 at a range of 100 km, at what range could the same radar system detect the B-2 stealth bomber? Assume that atmospheric losses are negligible. If the B-2 flies at 550 mph, how much warning time would the radar give?

5. The example around Eq. (2.12) in the text calculated a received power $P_r = 3.07 \times 10^{-9}$ W at a range of 10 km. If the radar noise figure is 10 dB and the receiver noise equivalent bandwidth is 10 MHz, what is the expected signal-to-noise ratio in dB for the same target and range? Assume the receiver is at the standard temperature $T_0 = 290$ K.

6. Consider a millimeter wave (MMW) seeker, which is a small radar typically used on a small missile. The RF is 95 GHz and the 3 dB beamwidth is 1° in both azimuth and elevation. The range resolution ΔR is 5 m. The grazing angle of the antenna beam with respect to the ground is 20°, and the slant range to the ground, that is, along the line of sight of the antenna, is 5 km. The terrain has a reflectivity of $\sigma^0 = -10$ dB. Is the clutter cell range extent of this system beam limited or resolution limited? What is the approximate area of the clutter cell on the ground? What is the total RCS σ of the clutter cell?

7. A radar is attempting to detect a point target in the presence of ground clutter. The parameters of the radar and its environment are such that the SNR at a range of $R = 10$ km is 30 dB, while the clutter-to-noise ratio (CNR) at the same range is 20 dB. The detection performance at this range is "clutter limited" because the clutter is the dominant interference. Assuming resolution-limited clutter interference, at what range will the SNR and CNR be equal? (Equivalently, at what range will the SCR equal 1 (0 dB)?)

8. Consider the WSR-88D "NEXRAD" weather radar used by the U.S. National Weather Service. It is a pulsed S-band (3 GHz) system with elevation and azimuth beamwidths of 0.88° and pulse length in its "short pulse" mode of $\tau = 1.57$ μs. Suppose the radar measures the RCS in a region of a rain storm 50 km away as $\sigma = 20$ m². Assume the simplified model of the resolution cell volume used to get Eq. (2.25), namely $\Delta V = (\Delta R)(R\theta_3)(R\phi_3) = \Delta R \cdot R^2 \theta_3 \phi_3$ What are the range, azimuth, and elevation resolutions and the resolution cell volume ΔV at the 50 km range? What is the volume reflectivity η? Using Eq. (2.44), compute the value of the

meteorological reflectivity factor Z in mm^6/m^3. Using Table 2.2, how hard is it raining? *Note*: The value of Z that results from Eq. (2.44) will be in m^6/m^3 = m^3 if wavelength is in meters. Convert this to units of mm^6/m^3 by multiplying by 10^{18} before converting to a decibel scale to get dBZ.

9. In terms of wavelengths, by how much must the range between the radar and a scatterer change in order for the received echo phase to change by 180°? How far is this at RF frequencies of 1 GHz (L band), 10 GHz (X band), and 95 GHz (MMW band)?

10. The fourth-degree chi-square PDF used to model the case of one dominant scatterer with many small scatterers is an approximation to the exact model for this case, which is the noncentral chi-square PDF of degree 2. Both PDFs are listed in Table 2.3 along with the formulas for their variances. Show that when they both have the same mean $\bar{\sigma}$, their variances will also be the same if the noncentral chi-square parameter $a^2 = 1 + \sqrt{2}$.

11. Suppose a target was modeled as consisting of one large scatterer and many small ones, but that the ratio a^2 of the large scatterer RCS to the sum of the small scatterer RCS values is 1 (instead of $1 + \sqrt{2}$ as assumed by the fourth-degree chi-square model). Assuming the means of the two distributions are the same, what degree $2m$ should be chosen for the chi-square so as to match the variance of the noncentral chi-square of degree 2 (see Table 2.3). Repeat for $a^2 = 10$. Note: m does not have to be an integer.

12. Part of the significance of choosing the probability density function used to model target RCS (or clutter or other interference) is that the differences in the "tails" of the PDF can have a significant impact on the probability of observing relatively large signal values, sometimes called signal "spikes." The probability that a random variable x described by a PDF $p_x(x)$ exceeds some value T is given by

$$P\{x > T\} = \int_T^{+\infty} p_x(x)dx$$

Consider a set of RCS data with a mean value (linear scale) of 1.0. Compute the probability that the RCS σ is greater than 2 when an exponential PDF is a good model for the RCS statistics, and again when a fourth-degree chi-square is a good model for the statistics. What is the ratio of the exponential PDF value to the chi-square PDF value? Repeat for $\sigma > 10$.

13. A ground-based airport surveillance radar has an antenna that rotates at 10 rpm (revolutions per minute). The 3-dB azimuth bandwidth of the antenna is 3°. Assume the PRF is chosen such that the maximum target range for which the echo will arrive prior to transmission of the next pulse is 150 km. (This is called the *unambiguous range*.) How many pulses will be transmitted during the time a given target is within the antenna mainbeam during a single rotation? (This is the number of "hits" the radar gets on the target on each rotation.)

14. Consider a complex target with dimensions of 2 m by 4 m. What is the maximum change in aspect angle needed to decorrelate the target RCS, assuming the radar frequency remains fixed at $F_t = 3$ GHz? Repeat for $F_t = 35$ GHz.

15. For the same target considered in the previous problem, what is the maximum frequency step needed to decorrelate the target RCS, regardless of radar-target aspect angle?

16. A stationary, fixed-frequency radar collects several CPIs of data, each consisting of several pulses, from a stationary complex target. What is an appropriate correlation model for this data within a CPI? What is the appropriate model from one CPI to the next? Justify your answers.

17. Consider a radar at a fixed height above the ground of h m, with the boresight intercepting the ground at a slant range of R m. Show that in this scenario, the constant-gamma model for σ^0 given by Eq. (2.64) implies that the received clutter power will be proportional to R^{-2} if the clutter echo is beam-limited. What will be the proportionality if σ^0 is independent of grazing angle?

18. Use the GTRI clutter model of Eq. (2.65) with $\lambda_t = 3$ cm and the X-band data in Table 2.6 to compute and plot or sketch on a single graph the variation in σ^0 in dBsm for grass, trees, and urban clutter as the grazing angle δ varies from 10° to 70°. Assume $\sigma_h = 10\lambda_t$.

19. Show that the GTRI model predicts that σ^0 becomes independent of surface roughness when $\sigma_h \ll 10\lambda_t$.

20. Consider the four radar system parameters of antenna azimuth plane size, antenna elevation plane size, transmitter power, and pulse length. Which of these affect the signal-to-clutter ratio (SCR) in a beam-limited area clutter scenario? Repeat for a resolution-limited area clutter scenario. Justify the answers.

21. Use the Billingsley clutter model to estimate the time lag required for the AC term only of the clutter to decorrelate to 10 percent of its power (maximum correlation value) for wind speeds of 5, 15, and 25 mph. Assume $F_t = 1$ GHz. Repeat for 10 GHz.

22. At the standard temperature of $T_0 = 290$ K, at what frequency in hertz does $S_n(F)$ of Eq. (2.76) fall to 3 dB below its value at $F = 0$ Hz? It may be necessary to find the answer numerically, but a small number of terms in the series approximation to the exponential can be used to develop a good initial estimate.

23. What is the effective noise temperature in kelvins for a system with a noise figure of 3 dB? Repeat for 6 dB and 10 dB.

24. Suppose two aircraft are flying straight and level at the same altitude. At a particular instant one is traveling due north at 100 m/s, while the other is flying directly at the

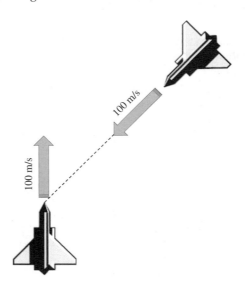

first but in the southwesterly direction, also at 100 m/s. The scenario is shown in the adjoining figure. What is the radial velocity between the two aircraft at that instant? What is the Doppler shift in hertz, including the sign, at that instant? A moment after the instant shown, will the Doppler shift be increased, decreased, or the same?

25. A stationary radar with a rotating antenna (typical of an airport approach radar, for instance) observes an aircraft moving through its airspace in a straight line at a speed of 200 mph. The aircraft approaches from the east, flies directly in front of the radar, and continues to the west. Sketch the general behavior of the radial velocity of the target relative to the radar as it flies from east to west through the airspace. Label significant values.

26. Derive an expression for the maximum radial velocity of a constant-radial velocity target such that the total range migration in a pulsed radar CPI of M pulses collected with a PRI of T seconds is less than the range resolution $c\tau/2$. The answer will be in terms of M, T, and τ. What is this velocity for a relatively long (for Doppler processing purposes) CPI of 30 pulses at $T = 1$ ms and a pulse length of 10 μs?

27. In Eqs. (2.93) to (2.97) the time-varying range between a moving radar and a stationary target was computed. A series expansion was used to approximate the range as a quadratic function (specifically, a parabola), and then the instantaneous Doppler shift was shown to be a linear function of time. What kind of conic section curve describes the range variation if the approximation to the square root is *not* made? Derive the formula for the instantaneous Doppler shift in this case.

28. Consider again the series approximation to range referred to in the previous problem. Find the maximum absolute value of t such that the magnitude of the dropped fourth-order term in t in the series approximation [see Eq. (2.94)] is less than 10 percent of the magnitude of the retained second-order term. (This condition is a limit on the amount of data that can be collected while still using the approximation to the range.) Give the numerical value of the maximum allowable t when $v = 100$ m/s and $R_0 = 10$ km.

29. Suppose the reflectivity distribution $\rho(R,\theta,\phi)$ consists of a single isolated point scatterer at coordinates (R_t,θ_t,ϕ_t), that is, $\rho(R,\theta,\phi) = \rho_t \delta_D(\theta - \theta_t)\delta_D(\phi - \phi_t)\delta_D(R - R_t)$. Determine $y(t)$ of Eq. (2.107). What determines the shape of this function in the azimuth (θ) dimension for fixed ϕ and t? Repeat for the elevation and fast time dimensions with the other two variables fixed.

30. The first zero of the function $J_1(x)$ occurs at $x \approx 3.8317$. What is the ratio of the Rayleigh azimuth beamwidth of a circular aperture of diameter D with uniform illumination to the azimuth beamwidth of a rectangular antenna of the same width D, also with uniform illumination? Figure 2.28 can be used as a crude approximate check on the result.

31. Show that $\mathbf{E}\{\exp(j\psi)\} = 0$ when ψ is a random variable uniformly distributed on the interval $[0, 2\pi)$. Use this to show that $\mathbf{E}\{\exp[j(\psi'(R_1,\theta_1,\phi_1) - \psi'(R_2,\theta_2,\phi_2))]\} = \delta_D(R_1 - R_2) \cdot \delta_D(\theta_1 - \theta_2) \cdot \delta_D(\phi_1 - \phi_2)$ in Eq. (2.120).

32. Based on Eq. (2.113), what is the approximate relationship between the Fourier transforms of $y(\theta, \phi, t_0)$, $\hat{\rho}(\theta,\phi,R_0)$, and $E^2(\theta,\phi)$? Use the result to explain why antenna beamwidth determines angular resolution.

33. The Fourier transform relationship from the previous problem suggests an approach to improving resolution by signal processing means. If $\tilde{P}(\theta,\phi;R_0)$ is the spectrum of the ideal high-resolution reflectivity distribution in angle, and $Y(\theta,\phi;R_0)$ is the spectrum of the observed data actually available to the radar, suggest a simple idea for recovering $\tilde{P}(\theta,\phi;R_0)$ from $Y(\theta,\phi;R_0)$. Discuss whether this procedure, called

"deconvolution," could potentially improve the angular resolution and how. Identify some potential practical limitations to this idea.

34. Consider a scatterer at elevation h above the ground plane, and an incoming EM plane wave at a grazing angle of δ radians. What is the difference in path lengths between the "single bounce" direct reflection (path #1 in the adjoining figure) and the "double bounce" multipath reflection (path #2), as a function of δ and h? If $h = 50$ m and the radar slant range resolution is $\Delta R = 20$ m, will the double-bounce echo appear in the same range bin as the single-bounce echo?

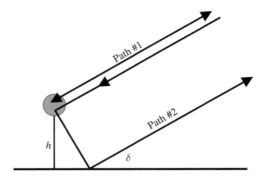

35. Continuing with the previous problem, assume the range bin spacing is greater than the path length difference so that both echoes combine in a single range bin. Develop a formula for the phase difference between the two echoes in terms of δ, h, and the wavelength λ_t. If the amplitudes of the direct and double-bounce reflections are the same (very unlikely), at what values of h will there be a maximum total signal amplitude? At what values of h will there be a null in the signal amplitude?

36. δR in the figure below is the maximum path length difference between a constant-range path of length R over an integration angle θ and a straight-line path of equal range R at the center of the integration angle. Find the maximum integration angle θ such that $\delta R \leq \lambda_t/8$. The answer will depend on R. A path length difference of this amount would lead to an echo phase change of $4\pi\delta R/\lambda_t = \pi/2$ radians, often used as an upper bound on the tolerable phase error in various calculations.

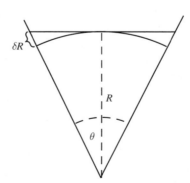

CHAPTER 3
Radar Data Acquisition and Organization

Many radars fundamentally seek to locate a scatterer in a three-dimensional spherical coordinate system of range, azimuth angle, and elevation angle, and to measure its reflectivity and radial velocity. The orientation of the directional antenna beam mainlobe when detection occurs provides the coarse location of the scatterer in azimuth and elevation angle. Range is measured by observing time delay of the echo relative to the transmitted signal. Radial velocity is inferred from measuring Doppler shift.

The means by which pulsed and continuous wave (CW) radars make these measurements are partly common to both, partly unique to each. Specifically, the two radar classes diverge in their methods of measuring range, as will be seen in Sec. 3.2. Pulsed radars emit a series of individual pulses and record the received voltage as a function of time, equivalent to range. In order to provide a timing reference, CW radars use some sort of periodic modulation. Here we consider only the most common form of frequency-modulated CW (FMCW), which transmits a repeating series of linear frequency sweeps and deduces range by measuring the frequency difference between the transmitted and reflected waveforms. Although very different from the simple pulsed radar range measurement process, each sweep of the FMCW system produces essentially the same range measurement as each pulse in the pulsed system.

The approach to measuring Doppler shift or, more generally, phase history is common to both pulsed and most FMCW systems, combining a series of range measurements from multiple pulses or frequency sweeps. Section 3.3 will develop the two-dimensional coherent processing interval (CPI) data structure that is the basis of most range-Doppler and imaging radar processing. Some radars further organize the data from multiple CPIs into a three-dimensional *datacube* format, typically to enable digital beamforming in angle. The datacube is the subject of Sec. 3.4.

Whether pulsed or CW, modern radars use coherent receivers so that the measured voltage is complex-valued (I and Q channels). They also record and process the data digitally. As with any digital data acquisition system, the selection of sampling rates and quantization strategies are crucial design decisions affecting signal fidelity, resolution, aliasing, and noise properties, as well as processor memory and computational requirements. Sampling rate issues in building the datacube are discussed as they arise, along with measurement complications such as ambiguities and blind zones. Quantization effects are relegated to App. B. The less common application of the Nyquist criterion to sampling in Doppler and spatial angle is the subject of Secs. 3.6 and 3.7. Finally, Sec. 3.8 will address some error issues in digitizing radar I/Q data and introduce digital generation of the I and Q signals as an alternative to traditional analog circuits.

3.1 A Signal Processor's Radar Architecture Model

Most practical radar systems have sophisticated transmitter and receiver architectures featuring multiple stages of modulation and demodulation, filtering, and amplification. These designs provide high signal quality through good noise minimization, rejection of unwanted signals from adjoining frequency bands, and minimization of imperfections such as unbalanced signal channels. However, for discussing fundamental radar signal processing operations it is usually sufficient to consider simplified models of the transmitter and receiver like those shown in Fig. 3.1. The transmitter (Fig. 3.1a) is modeled as a baseband waveform generator, a mixer and reference oscillator to modulate the baseband waveform onto the carrier frequency of Ω_t rad/s, and an amplifier to boost the signal before radiating it from the transmit antenna. The receiver (Fig. 3.1b) accepts the weak echo signal from the receive antenna, amplifies it with a low noise amplifier (LNA), and demodulates it back down to baseband for subsequent signal processing.

In this text, the emphasis is on monostatic radars and coherent processing. Figure 3.2 is a specialization of the architecture of Fig. 3.1 to implement these two features while retaining the simplified signal processor's view. The antenna is now shared by the transmitter and receiver using a circulator or similar device and the receiver now uses the coherent quadrature (I/Q) structure first introduced in Sec. 1.3.5. The various $\cos(\Omega_t t)$ and $\sin(\Omega_t t)$

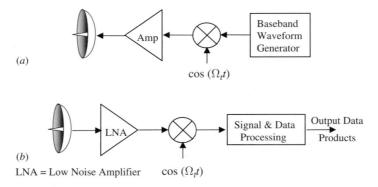

Figure 3.1 A signal processor's model of a basic pulsed or continuous wave radar.

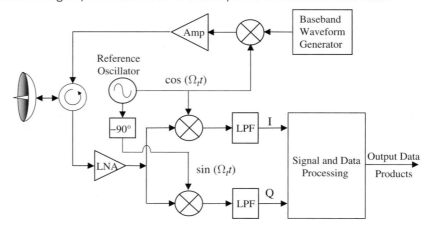

Figure 3.2 A more sophisticated radar model having a monostatic (shared) antenna and a quadrature receiver.

reference signals are all generated from the same stable oscillator. Applicable to both pulsed and CW waveforms, this architecture will be assumed unless otherwise stated.

3.2 Measuring a Range Profile

The first task of a radar is measurement of the range profile (angle-averaged distribution of reflectivity vs. range) $\tilde{\rho}(R;\theta_0,\phi_0)$ in the direction illuminated by the antenna mainbeam. This is done in very different ways by pulsed and CW radars. Pulsed radar range profiles are introduced first along with various details and complications in their measurement. CW measurement of a range profile requires a more complicated waveform and system architecture and so is discussed after the simpler pulsed case.

3.2.1 Pulsed Radar Range Profile: One Pulse in Fast Time

In Fig. 3.3a, a radar transmits a single pulse of length τ seconds. The leading edge of the pulse is emitted from the antenna at time $t = 0$. Suppose the environment in the look direction

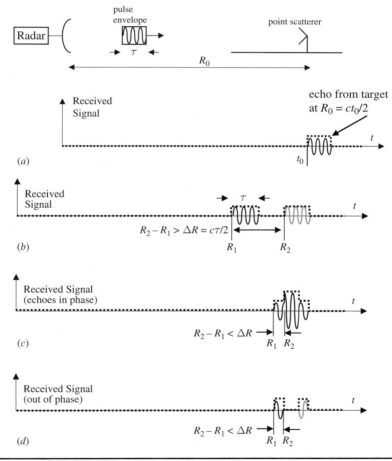

FIGURE 3.3 Received echo signal for one transmitted pulse. (a) Received signal for a scatterer at range $R_0 = c \cdot t_0/2$. (b) Echoes from two scatterers separated by more than the range resolution are well-resolved. (c) Echoes from two scatterers separated by less than the range resolution and constructively interfering, or (d) destructively interfering.

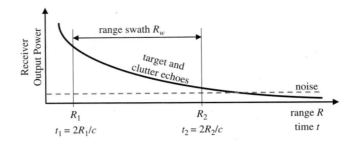

FIGURE 3.4 Signal level versus range, and range swath or window R_w.

of the antenna contains a single scatterer at range R_0. Ignoring amplitude factors and noise, the signal observed at the antenna will be a replica of the transmitted pulse beginning at time $t_0 = 2R_0/c$ as shown. The dotted line represents the envelope of the echo signal. The range to the scatterer can be estimated as $R_0 = c \cdot t_0/2$ by detecting the arrival time t_0.

Recall from Chap. 1 that the range resolution of a simple pulse is $\Delta R = c\tau/2$, and that a measurement of the echo signal at a given time t_0 is affected by all scatterers within an interval of length ΔR ending at $R = c \cdot t_0/2$. If there are multiple scatterers in the scene the signal at the antenna will be a superposition of echoes. If they are separated by more than τ seconds ($c\tau/2$ meters) the echoes will be nonoverlapping; that is, the scatterers will be *resolved* in range. This case is shown for two equal-strength echoes in Fig. 3.3*b*. Echoes from scatterers separated by less than τ seconds will overlap and combine either constructively (Fig. 3.3*c*) or destructively (Fig. 3.3*d*), depending on their respective phases. In the latter two cases, the scatterers are considered *unresolved* and a voltage measurement taken during the time their individual responses overlap will be the sum of both complex amplitudes.

More generally, the received echo signal will be the convolution of the pulse waveform with the range profile, as discussed in the previous chapter. The echo power at the receiver will decay with range or time, while the noise power generated within the receiver will be constant. Figure 3.4 is a notional illustration of this behavior. Note that near-in clutter and target echoes can be quite strong due to their inverse dependence on powers of R. Depending on the goals of its particular mode of operation, the radar will measure the received power over some interval of interest in range, say from R_1 to R_2. This interval is called the *range swath* or the *range window*; its length is $R_w = R_2 - R_1$. The range R_1 to the beginning of the range swath may be influenced by a number of factors. For example, for an airborne ground imaging radar it would likely be determined by the range to the nearest edge of the antenna mainlobe's footprint on the ground, since no significant ground echoes would occur at a shorter range. Similarly, the end of the range swath, R_2, might be set in different radars by the far edge of the mainlobe footprint or by the maximum expected detection range for targets of interest. In automotive radar it is common to have multiple radars with different ranges, for example, a "long range" (perhaps 250 m) forward-looking system, a 70-m-range wide-angle sidelooking system, and a rear-looking system, perhaps also on the order of 70 m maximum range. The near edge of the range swath might be as little as 1 or 2 m. Another constraint on R_1 is the radar minimum range, discussed next, while R_2 is also constrained by the unambiguous range, to be addressed in Sec. 3.3.

In a monostatic pulsed radar, the radar receiver must be isolated from the antenna during the τ-second duration of pulse transmission so as to avoid damaging the sensitive receiver circuits with leakage of the high-power transmitted signal. It follows that the *closest*

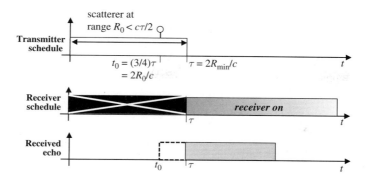

FIGURE 3.5 Partially eclipsed pulse echo resulting from a target at less than the minimum range R_{min}.

range from which a full echo can be received, called the *minimum range*, is $R_{min} = c\tau/2$ m.[1] In practice, R_{min} will be somewhat larger to allow for the finite switching time needed to reconnect the receiver to the antenna after pulse transmission, and in some environments to protect against particularly strong *near-in clutter* echoes. Any clutter or target scatterer closer to the radar than R_{min} will produce an echo that arrives in part during the initial τ seconds after transmission begins, and that portion of the pulse echo will not be seen at the receiver. A pulse echo that is not received in whole or part because it arrives during the time the receiver is isolated is said to be wholly or partially *eclipsed*. Figure 3.5 illustrates the idea of a partially eclipsed pulse. In this case, the target is at 3/4 of the blind range and 1/4 of the echo will be eclipsed.

The received signal is demodulated to baseband by the coherent receiver, as described in Chap. 1. In modern digital radar systems the resulting complex-valued baseband signal is sampled at a high rate, typically in the range of hundreds of kilohertz to a few tens of megahertz and sometimes higher. The desired range swath is implemented by beginning sampling at time $t_1 = 2R_1/c$ after pulse transmission and ending at time $t_2 + \tau = 2R_2/c + \tau$. The additional τ seconds at the end of the sampling period are needed to capture the end of the pulse echo from the far edge of the swath. The resulting samples are stored in digital memory as shown in Fig. 3.6, where each small cube represents a single complex-valued baseband signal sample. A set of L samples from a single transmitted pulse are referred to as *range bins, range gates, range cells,* or *fast-time samples*.[2] These complex samples form the sampled range profile (RP) in the look direction. Their phases constitute the fast-time phase history of the pulse echo data.

The term "fast time" arises because the time interval corresponding to the range window for a single RP is usually quite short, often on the order of tens to hundreds of microseconds, though it can be shorter or longer. This is in contrast to the *slow time* axis to be introduced in Sec. 3.3 with the use of multiple RPs.

---

[1]Some authors call this the "blind range," but there are multiple blind ranges with periodic modulations such as pulse trains and linear FMCW. Here the term "minimum range" is used to refer to only the first blind range due to pulse transmission.

[2]The terms "resolution bins" and "resolution cells" are also sometimes used synonymously with range bins, and in many cases are synonymous; but the sampling interval in range does not always equal the range resolution, so caution should be used in interpreting these terms.

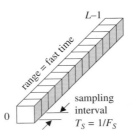

FIGURE 3.6 Range profile formed as a vector of fast-time samples for one pulse or FMCW sweep.

How rapidly should one sample the echo from a single received pulse, that is, what should the spacing of the range bins be? Assuming a quadrature receiver, the Nyquist theorem states that the sampling rate F_s should equal or exceed the total bandwidth (*Nyquist bandwidth*) β of the received complex-valued signal (see App. B). In Chap. 2 it was shown that the received signal in the range dimension can be modeled as the convolution of the effective range reflectivity function and the modulation function $a(t)$ of the transmitted waveform. The spectrum of the received signal is then the product of the spectra of the range reflectivity function and the modulation function so that the bandwidth of the received fast-time signal will be limited by the bandwidth β of the transmitted signal. Therefore, the Nyquist rate in fast time is simply β samples per second and the range sampling interval becomes

$$\delta R_s = \frac{cT_s}{2} = \frac{c}{2\beta}\,\text{m} \tag{3.1}$$

Equation (3.1) is a general result,[3] but different pulse waveforms require different interpretations of what exactly "the bandwidth" β means. At one extreme, consider the simple rectangular constant-frequency pulse. Recall that the spectrum of the complex exponential pulse of frequency F_0 Hz is a sinc function in the frequency domain centered at F_0 (see App. B). This spectrum is most definitely *not* strictly bandlimited, so a Nyquist bandwidth cannot be unambiguously defined. One very common approximate bandwidth definition in radar is the 3-dB bandwidth, which is the total width of the function at a level 3 dB below the spectrum maximum. For the rectangular pulse this is $0.89/\tau$ Hz, which encompasses only 72 percent of the total spectrum energy. More conservative definitions are the 4-dB bandwidth of $1/\tau$ Hz for the rectangular pulse (78 percent of energy) and the null-to-null bandwidth $\beta_{nn} = 2/\tau$ Hz (91 percent). These bandwidth metrics are shown in Fig. 3.7 with $F_0 = 0$ for convenience. Another common metric is the Rayleigh (peak-to-first-null) bandwidth, also shown in Fig. 3.7. This also equals $1/\tau$ Hz for the rectangular pulse. Notice that the 3-dB, 4-dB, and null-to-null metrics are all two-sided metrics, while the Rayleigh is a one-sided metric. Nonetheless, since it is the same as the 4-dB bandwidth, the Rayleigh bandwidth is sometimes used as a reasonable measure of two-sided spectrum width when discussing ideal pulses.

Another approach is to define the bandwidth as the two-sided frequency interval beyond which the spectrum amplitude is "insignificant," but this approach is not very

---

[3]Notice that this is the same as the range resolution ΔR of Eq. (1.2). Again, sampling interval and resolution in various domains (range, angle, Doppler) are closely related but not necessarily identical.

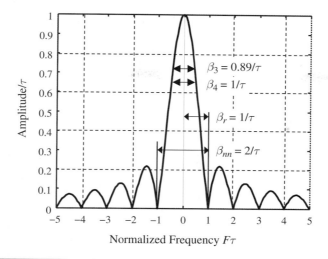

Figure 3.7 Four common definitions of bandwidth for the spectrum of a simple baseband rectangular pulse of duration τ.

useful with the rectangular pulse because of the slow decay of the sinc sidelobes. For example, a criterion that defines the cutoff as the minimum frequency such that the spectrum is reduced 40 dB or more from its peak gives an approximate Nyquist bandwidth of about 66 times the 3-dB bandwidth, far too high to be useful. This problem is somewhat mitigated by the fact that real pulses are not perfectly rectangular and their somewhat rounded edges result in sidelobes that decay faster than the sinc function. In practice 3-dB and Rayleigh bandwidths are most commonly used in radar to characterize the rectangular pulse.

The Rayleigh bandwidth $\beta_r = 1/\tau$ is used here to estimate appropriate range sampling rates for the rectangular pulse. Equation (3.1) becomes

$$\delta R_s = \frac{cT_s}{2} = \frac{c}{2\beta_r} = \frac{c\tau}{2} \text{ m (rectangular pulse)} \tag{3.2}$$

This is the same result obtained graphically in Sec. 1.4.2. In practice, the fast-time signal is often sampled at some margin above the Nyquist rate. This compensates both for the transition band of receiver anti-aliasing filters and for some of the non-bandlimited nature of common pulse waveforms. Sampling rate margins of 20 to 50 percent are common.

It will be seen in Chap. 4 that pulses are often phase modulated in order to increase their bandwidth. The pulse spectrum is then no longer a sinc function. In fact, many phase modulated pulses are designed to have a spectrum that is approximately constant magnitude (but with complicated phase characteristics) over some desired bandwidth β that is much larger than the simple pulse bandwidth of approximately $1/\tau$. At the other extreme from the sinc spectrum of the rectangular pulse, an idealized model of the spectrum of an ideal received phase-modulated radar pulse after demodulation to baseband is

$$Y(F) = \begin{cases} A\exp[j\Phi(F)], & |F| < \beta/2 \\ 0, & |F| > \beta/2 \end{cases} \tag{3.3}$$

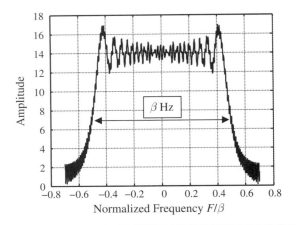

Figure 3.8 Magnitude of the Fourier transform of a linear FM "chirp" waveform having a time-bandwidth product of 100.

where $\Phi(F)$ is some frequency domain phase function. Unlike the rectangular pulse spectrum, this idealized spectrum has an unambiguous Nyquist bandwidth of β hertz, making application of the Nyquist criterion to range sampling straightforward: simply choose $F_s = \beta$. In fact, β is also the 3-dB, 4-dB, and null-to-null bandwidth of this spectrum. However, it is not the Rayleigh bandwidth; that is $\beta/2$. Basing the fast-time sampling rate on the Rayleigh bandwidth for this signal would result in severe undersampling in range.

Figure 3.8 shows an example that approaches the ideal phase-modulated case, the magnitude spectrum of a linear frequency-modulated (LFM) or "chirp" waveform with a time-bandwidth product $\beta\tau = 100$; this waveform will be studied in Chap. 4. On the normalized frequency scale shown, the spectrum is approximately rectangular with support $f \in (-0.5, +0.5)$, corresponding to $\pm\beta/2$ Hz. Choosing $F_s = \beta$ would be appropriate for this waveform. As with the rectangular pulse, in practice that rate might be increased by perhaps 20 percent to allow for the imperfect bandlimiting of the pulse.

Figure 3.4 pointed out that the power of echoes from near-in clutter or targets can be quite large, while distant target and clutter echoes may be quite weak. For example, the power of resolution-limited ground clutter echoes is expected to vary as R^{-3} while discrete targets vary as R^{-4}. The difference in the maximum echo power received from a single pulse and the noise power floor, which sets the minimum useful signal strength, is called the *dynamic range* of the echo signal. It can be many tens of decibels, making it difficult to design a receiver sensitive enough to detect long-range echoes near the noise floor while still avoiding saturation and possible damage from strong near-in echoes.

Sensitivity time control (STC) is one response to this problem. STC applies a time-varying attenuation to the receiver input after each pulse. The initial attenuation for near-in targets may be on the order of 20 to 40 dB. The attenuation is then reduced over the range window extent, limiting the dynamic range at the receiver input. The attenuation might be proportional to R^{-3} or R^{-4} m, depending on the expected scenarios, or a compromise value such as $R^{-3.5}$ might be implemented.

STC should not be confused with *automatic gain control* (AGC), which varies the receiver gain over a longer time, often by varying an intermediate frequency (IF) amplifier gain. AGC typically operates over a time interval of multiple pulses, CPIs, or dwells with the goal of maintaining near-constant target amplitude to enhance tracking or display performance. More information on both STC and AGC is available in Bruder (2010).

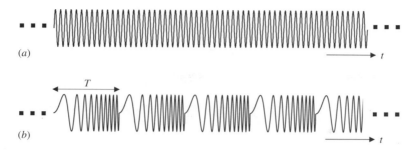

FIGURE 3.9 (a) Unmodulated continuous wave (CW) waveform. (b) Periodically repeating linear frequency modulated CW (FMCW) waveform.

3.2.2 FMCW Radar Range Profile: One Sweep in Fast Time

In Sec. 1.3.2 it was noted that a basic constant-frequency continuous wave signal like that in Fig. 3.9a provides no way to measure time delay of the echo because the waveform lacks a timing mark such as a pulse edge. It was also noted that this could be remedied by modulating the CW signal with a periodically repeating pattern. A common example is a linear frequency sweep, resulting in the signal in Fig. 3.9b. An expression for the complex form of one sweep of that linear FMCW waveform is

$$\bar{x}(t) = \exp\left[j\left(2\pi F_t t + \pi \frac{\beta}{T} t^2\right)\right] = \exp[j\Phi(t)], \quad 0 \leq t \leq T \tag{3.4}$$

This is repeated every T seconds to form the continuous wave transmit signal. Applying the definition of instantaneous frequency $F_i(t)$ from App. B gives

$$F_i(t) \equiv \frac{1}{2\pi} \frac{d\Phi(t)}{dt} = F_t + \frac{\beta}{T} t \text{ Hz}, \quad 0 \leq t \leq T \tag{3.5}$$

which sweeps upward in frequency linearly from F_t to $F_t\beta$ Hz in T seconds. T is called the FMCW *sweep interval*; its inverse $1/T$ is the *sweep rate*.

A useful graphical representation of the linear FMCW waveform is Fig. 3.10a, which plots its instantaneous frequency versus time at baseband ($F_t = 0$). Based on this graph, the

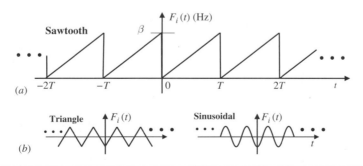

FIGURE 3.10 Instantaneous frequency patterns. (a) Linear FMCW ("sawtooth") waveform similar to that of Fig. 3.9b. (b) Triangle FMCW waveform and a sinusoidal FMCW waveform.

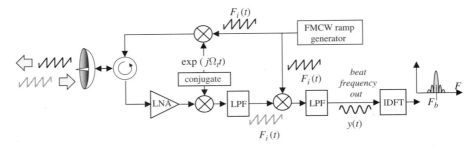

Figure 3.11 Simplified transmitter-receiver architecture for measuring a range profile with a sawtooth FMCW radar.

linear FMCW waveform is often referred to as a "ramp" or "sawtooth" FMCW signal. A signal that sweeps down in frequency rather than up can be defined by simply changing the sign of the quadratic phase term in Eq. (3.4). Both upsweeps and downsweeps are used in practice.

Other frequency modulation rules are also used with CW waveforms. Figure 3.10b illustrates the instantaneous frequency patterns, relative to the nominal carrier frequency F_t, of two: a "triangular" FMCW and a "sinusoidal" FMCW. There are also modulated CW waveforms that do not use continuous frequency sweeps. One example is frequency shift keying, which alternates between two or more constant frequencies in a processing interval (Rohling and Möller, 2008). This text concentrates on the widely used linear sawtooth FMCW waveform. The reader is referred to Baker and Piper (2013) for a thorough introduction to CW radar systems and applications.

To see how linear FMCW measures a range profile, consider the FMCW transceiver architecture sketched in Fig. 3.11. The complex signal model is assumed in this figure; in practice a two-channel quadrature receiver would be implemented.[4] While time waveforms $x(t)$ and $y(t)$ are shown at the beginning and end of the transmit-receive chain, the intermediate stages are annotated with the instantaneous frequency pattern of the signal at various points in the chain.

An important feature of this architecture is that after the received signal is demodulated to baseband by the first mixer and lowpass filter (LPF) it is then mixed again with a replica of the baseband sawtooth FMCW transmit waveform, followed again by an LPF to remove the sum frequency at the mixer output. This second demodulation is sometimes called a *dechirp* or *deramp* operation. The output signal will be a sinusoid at the instantaneous difference frequency between the transmitted and received sawtooth waveforms. Since both the echo and reference signal instantaneous frequencies are chirping at the same rate, the difference between them is a constant frequency. Furthermore, the value of that frequency increases as the time delay of the echo increases. That is, the frequency difference encodes range to the scatterer.

To quantify this observation, Fig. 3.12 compares the instantaneous frequency versus time for two sweeps of the reference and received sawtooth signals, assuming the received signal is an echo from a target at range $R_0 = c \cdot t_0/2$. The beat frequency at any instant is defined here as the received echo frequency minus the transmitted frequency at that instant.

---

[4]The lowpass filters are not actually necessary in this complex signal version, but are included here as a reminder that they are required in the practical quadrature receiver.

Radar Data Acquisition and Organization

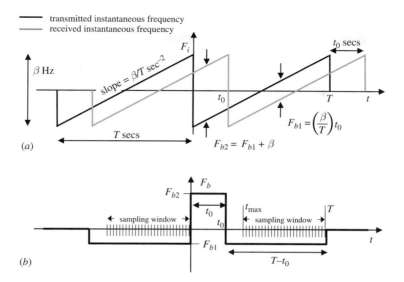

FIGURE 3.12 (a) Transmitted and received instantaneous frequency patterns for a sawtooth FMCW radar and a scatterer at delay t_0. (b) Beat frequency pattern and sampling window resulting from the signals in (a). The beat frequency is the received frequency minus the transmitted frequency.

Two different beat frequencies are observed. The longer duration beat frequency F_{b1} is easily seen by consideration of the slope of the sawtooth to be

$$F_{b1} = -\left(\frac{\beta}{T}\right)t_0 = -\left(\frac{t_0}{T}\right)\beta = -\left(\frac{2R_0}{cT}\right)\beta \text{ Hz}, \quad t_0 \leq t \leq T \tag{3.6}$$

The shorter duration beat frequency is

$$F_{b2} = \beta + F_{b1} = \left(1 - \frac{t_0}{T}\right)\beta = \left(1 - \frac{2R_0}{cT}\right)\beta \text{ Hz}, \quad 0 \leq t \leq t_0 \tag{3.7}$$

In most FMCW systems of interest in this text $(t_0/T) \ll 1$ and only F_{b1} is used for measuring the range profile.[5]

Time delay or range is always nonnegative. Equation (3.6) then shows that F_{b1} is proportional to the negative of range and so is always a negative frequency. Consequently, an inverse discrete Fourier transform (IDFT) is used instead of a discrete Fourier transform (DFT) at the receiver output because the IDFT will map negative beat frequency signals into peaks at positive IDFT indices to form the range profile (see Prob. 5). Furthermore, because range is nonnegative instead of taking on both positive and negative values, all K indices of a K-point IDFT can be interpreted as positive ranges.

The FMCW radar does not isolate the receiver during a portion of the sweep period T, so in principle there is no minimum range as there is with the pulsed system. However, in practice some system imperfections such as phase noise may limit the ability to measure

---

[5]If the target is moving with radial velocity v relative to the radar both beat frequencies will be increased by the Doppler shift $2v/\lambda_t$ Hz. This complication is ignored here but will be confronted in Chap. 4.

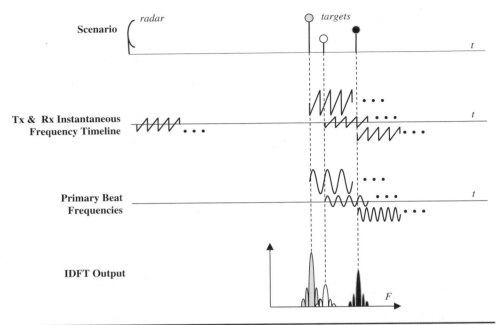

Figure 3.13 Sequence by which scatterers at different ranges and amplitudes map into the IDFT frequency axis using the linear FMCW architecture of Fig. 3.11.

very low beat frequencies corresponding to very short range targets, creating a de facto minimum range.

If there are multiple scatterers present, there will be multiple beat frequencies at the deramp mixer/lowpass filter output and multiple peaks in the IDFT output spectrum, as illustrated in Fig. 3.13. The amplitudes and frequencies of the IDFT spectral peaks are proportional to the amplitudes and ranges of the scatterers. Thus, the system of Fig. 3.11 outputs a range profile for each sweep of the transmitted FMCW signal. A single sweep plays the same role in measuring a range profile as does a single pulse in a pulsed radar.

Equation (3.4) gave the expression for one period of the transmitted RF linear FMCW signal. The echo received from a scatterer at range $R_0 = ct_0/2$ will be

$$\bar{y}(t) = \bar{x}(t - t_0) = \exp\left[j\left(2\pi F_t(t - t_0) + \pi \frac{\beta}{T}(t - t_0)^2\right)\right], \quad t_0 \leq t \leq t_0 + T \tag{3.8}$$

Mixing (multiplying) with the RF reference $\exp(-j2\pi F_t t)$ gives the intermediate baseband FMCW signal during the time in which they overlap,

$$\exp\left[-j\frac{4\pi R_0}{\lambda_t}\right] \cdot \exp\left[j\pi \frac{\beta}{T}(t - t_0)^2\right], \quad t_0 \leq t \leq T \tag{3.9}$$

The signal is next mixed with a complex baseband reference chirp $\exp[-j\pi(\beta/T)t^2]$ to give the fully demodulated baseband signal

$$y(t) = \exp\left[-j\frac{4\pi R_0}{\lambda_t}\right] \cdot \exp\left[-j2\pi\left(\frac{2R_0}{cT}\right)\beta t\right] \cdot \exp\left[j\pi \frac{\beta}{T}t_0^2\right], \quad t_0 \leq t \leq T \tag{3.10}$$

Note that this signal is present only over the primary beat frequency interval.

The first complex exponential in Eq. (3.10) is the usual fixed phase shift due to the range to the scatterer. The phase of the second term is a linear function of time, representing a sinusoid at a constant frequency of $-2R_0\beta/cT$ Hz. This is the beat frequency of Eq. (3.6). The last term is a phase constant sometimes termed *residual video phase* (RVP), especially in the context of radar imaging. While of no particular concern here, it can be a nuisance in imaging algorithms, as will be seen in Chap. 8.

To understand the spacing of the range bins in the FMCW system, it is necessary to consider the required time sampling rate at the output of the deramp mixer and the spacing of the IDFT output samples. The range profile will be measured from the F_{b1} beat frequency signal. Equation (3.6) shows that F_{b1} can vary from zero (for very close targets, $t_0 \to 0$) to $-\beta(t_0 \to T)$. However, it is also assumed that $t_0 \ll T$. Suppose the range window of interest has a maximum range $R_{\max} = c \cdot t_{\max}/2$. Then $t_{\max}$ is the maximum value of interest for t_0 and the largest magnitude (most negative) beat frequency of interest is

$$F_{b\min} = -\left(\frac{t_{\max}}{T}\right)\beta = -\left(\frac{2R_{\max}}{cT}\right)\beta \text{ Hz} \qquad (3.11)$$

This minimum beat frequency, and thus the maximum range $R_{\max}$, can be enforced by a lowpass filter with cutoff frequency $F_{b\min}$ at the deramp mixer output.

The range of beat frequencies is $|F_{b\min}|$ Hz. This is the Nyquist sampling rate F_N required for the deramp mixer output. The actual sampling rate must satisfy $F_s \geq F_N$ samples per second. Sampling of the beat frequency signal should not begin before the echo from the longest range of interest so that only the primary beat frequency F_{b1} is captured. The maximum sampling window extent is therefore $(t_{\max}, T)$, as shown in Fig. 3.12, and the duration of the sampled F_{b1} beat signal is limited to $T - t_{\max}$ seconds. The number of time samples of the beat frequency is $N_b = F_s(T - t_{\max})$. This will upper bound the length of the time signal at the input to the IDFT.

It follows from the above that the IDFT output spectrum spans F_s Hz and that the IDFT size must be $K \geq N_b$. Since the beat frequency is $(-2\beta/cT)$ times range [Eq. 3.6], the IDFT frequency axis can be relabeled in terms of range. The IDFT sample spacing is $\delta F_s = F_s/K$ Hz or $\delta R_s = (cT/2\beta)(F_s/K)$ m. If F_s is the Nyquist value $|F_{b\min}| = F_N = (2R_{\max}\beta/cT)$ then $\delta F_s = (2R_{\max}\beta/KcT)$ and $\delta R_b = R_{\max}/K$.

The IDFT samples are interpreted as representing positive ranges from zero to $(K-1)\delta R_b$, corresponding to a range window of zero to $[(K-1)/K]R_{\max}$ meters. A minimum swath range R_1 greater than zero can be implemented by simply ignoring IDFT samples corresponding to shorter ranges. Finally, because the duration of the primary beat frequency is $T - t_{\max}$, the IDFT frequency resolution will be $\Delta F_b = 1/(T - t_{\max})$ Hz. Again converting the IDFT frequency axis to range, the range resolution is

$$\Delta R = \left(\frac{T}{T - t_{\max}}\right)\left(\frac{c}{2\beta}\right) \text{ m} \qquad (3.12)$$

Equation (3.12) shows that unlike the pulsed case, in the FMCW case the range resolution is slightly greater than $c/2\beta$. That is, linear FMCW systems degrade (increase) the achievable range resolution. Larger values of $t_{\max}$ suffer greater degradation. Pulsed radars do not share this limitation. However, remembering that the intent is $t_{\max} \ll T$, the degradation is usually small.

Table 3.1 summarizes these relationships. A numerical example will help illustrate them. Suppose the FMCW sweep covers $\beta = 100$ MHz at a repetition interval of $T = 10$ μs. Choose $R_{\max} = 150$ m, corresponding to a maximum echo time delay of $t_{\max} = 1$ μs. From Eq. (3.12), the range resolution will be $(10 \text{ }\mu\text{s}/9 \text{ }\mu\text{s})(3 \times 10^8/200 \times 10^6) = 1.67$ m. From Eq. (3.11), the

Relationship or Quantity	Formulas	Comments
General relation between range and beat frequency	$R = -\left(\dfrac{cT}{2\beta}\right) F_b$ m, $F_b = -\left(\dfrac{2\beta}{cT}\right) R$ Hz	T = sweep repetition interval β = swept bandwidth
Maximum beat frequency duration	$T - t_{max} = T - \dfrac{2R_{max}}{c}$ seconds	t_{max}, R_{max} = maximum target time delay or range of interest
Nyquist sampling rate at deramp mixer output (maximum beat frequency magnitude)	$F_N = \left(\dfrac{t_{max}}{T}\right)\beta = \left(\dfrac{2R_{max}}{cT}\right)\beta$ Hz	Actual sampling rate F_s should satisfy $F_s \geq F_N$
Number of range samples	$N_b = (T - t_{max}) F_s = \left(T - \dfrac{2R_{max}}{c}\right) F_s$	
Range sample spacing	$\delta R_s = \left(\dfrac{cT}{2\beta}\right)\left(\dfrac{F_s}{K}\right)$ m $\delta R_s = \left(\dfrac{R_{max}}{K}\right)$ m (if $F_s = F_N$)	K = IDFT size
Range resolution	$\Delta R = \left(\dfrac{c}{2\beta}\right)\left(\dfrac{T}{T - t_{max}}\right)$ m	
Theoretical minimum range	$R_{min} = 0$ m	Implementation limitations such as phase noise will result in nonzero R_{min} in practice

TABLE 3.1 Sawtooth Linear FMCW Single-Sweep Relationships

magnitude of the minimum (most negative) beat frequency will be $|F_{bmin}| = |-(1\ \mu s/10\ \mu s)(100\ \text{MHz})| = 10$ MHz. Choose the sampling rate at the output of the second demodulation stage to be the 10 megasamples per second Nyquist rate. The number of beat frequency time samples collected on each sweep will then be $N_b = |F_b| \cdot (T - t_{max}) = (10\ \text{MHz})(9\ \mu s) = 90$. Conventionally, the DFT size K would be chosen as the least power of two greater than or equal to 90, giving $K = 128$.[6] The spacing of the range bins will then be $\delta R_s = (150\ \text{m})/128 = 1.17$ m. Notice that this range profile is somewhat oversampled in range, since there are $1.67/1.17 \approx 1.5$ range samples per range resolution cell. If the IDFT size were chosen as the minimum 90, the range sample spacing would equal the resolution of 1.67 m.

---

[6]This choice is made to enable use of the highly efficient power-of-2 fast Fourier transform (FFT) algorithms but is overly restrictive in two ways. First, good FFT algorithms are available to compute IDFTs at sizes that are not powers of two. Second, the technique called *data turning* (see Sec. 3.6.1) can be used to compute IDFTs for K less than the number of input points if desired.

3.3 Multiple Range Profiles: Slow Time and the CPI

Both pulsed and FMCW radars transmit a periodic series of signals. If pulsed, the time between pulses is called the *pulse repetition interval (PRI)* or *inter-pulse period* (IPP) and denoted as T. Its inverse is the *pulse repetition frequency* (PRF). The PRF may range from a few hundred pulses per second (also called, casually, hertz) to tens and sometimes a few hundreds of kilohertz. The FMCW sweep interval is also denoted T. Its inverse is called the *sweep rate* and is analogous to the pulsed PRF. The FMCW sweep rate varies at least from hundreds of sweeps per second ("hertz") to perhaps 100,000 Hz. The term "repetition rate" (RR) is used in this text as a common term to refer to either the pulsed PRF or FMCW sweep rate, while *repetition interval* (RI) provides a common term for the PRI and sweep interval.[7]

In some cases (e.g., a rotating weather or surveillance radar) the series may be continuously ongoing, but in many cases it is organized into groups of M range profiles to be processed together. The vectors of L fast-time samples collected for each of the M RPs are typically organized into a two-dimensional matrix $y[l, m]$, as shown in Fig. 3.14. The time required to collect this data is MT seconds. Because the data samples along the pulse or sweep number axis for a given range bin represent a time interval more than M times that of the fast-time axis, that dimension is called the *slow time* axis in comparison.

If a coherent series of RPs was used, the slow-time duration of MT seconds is called the *coherent processing interval* (CPI). In practice, the term "CPI" refers both to the matrix of data and the time required to collect it. While there are exceptions, a CPI of data is usually collected using a constant repetition interval, constant radar frequency, and the same pulse or sweep waveform for all repetitions within the CPI. On the other hand, it is common to change some or all of these from one CPI to the next.

Although the data for a single CPI is collected by columns (RPs), once it is stored in memory it may be accessed in any fashion. In Fig. 3.14, the fourth range bin for each RP is shaded gray. This row of samples in the data matrix is the *slow-time signal* for that range bin. Assuming the antenna boresight is not moving significantly during the CPI duration, these samples represent the reflectivity from the same range and angle, that is, the same region in three-dimensional space, sampled at the signal repetition interval T.

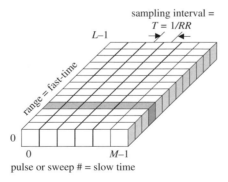

FIGURE 3.14 Coherent processing interval (CPI) data matrix formed as a concatenation of multiple range profiles.

---

[7]In this text, the abbreviations PRF, PRI, RR, and RI are used both as acronyms and as mathematical variables. When used as acronyms, they are not italicized (PRF, PRI, RR, RI); when used as mathematical variables, they are italicized (*PRF, PRI, RR, RI*).

How should the RR be chosen? It affects, and is affected by, many aspects of the radar and environment. As was seen in the discussion of spatial Doppler in Chap. 2, the slow-time phase history reflects the Doppler components in the received signal. One criterion for choosing the RR is to avoid aliasing of the spectrum replicas so as to preserve the information in the Doppler spectrum for subsequent processing such as target detection or synthetic aperture imaging. Thus, the Nyquist requirement in slow time is that the RR be at least as large as the slow-time signal bandwidth.

A nonzero Doppler bandwidth results from two sources: intrinsic motion of the scatterers in the area being measured, and motion of the radar platform relative to those scatterers. If the scatterer of interest in the area being measured is a target in the conventional sense of a man-made vehicle or object, its intrinsic motion is simply the motion of the vehicle relative to the radar.[8] If it is clutter, then intrinsic motion can be due to wind blowing the leaves of trees or blades of grass, waves on the ocean, falling and swirling rain, air-conditioning fans on tops of buildings, and so forth. For instance, the Doppler power spectrum corner frequencies in Table 2.7 imply an intrinsic velocity spread on the order of 0.5 to 1.0 m/s for rain at X band. The intrinsic velocity spread of moving man-made objects can be much larger. Consider an urban clutter scene where a stationary radar observes automobile traffic with a maximum speed of 60 mph both toward and away from the radar. The radar therefore sees targets with a velocity spread of 120 mph, or about 53.6 m/s. For a more extreme example, consider a moving radar installed on one of two subsonic (200 m/s) jet aircraft flying toward one another. As they approach, the closing rate is 400 m/s; once they pass, they separate at 400 m/s. The change in velocities observed by the radar on one of the aircraft over time is 800 m/s.

Suppose a 24-GHz ($\lambda_t = 12.5$ mm) automotive radar operates in the urban scene described above. The Doppler spread corresponding to the 53.6 m/s velocity spread is 8.6 kHz. At 77 GHz ($\lambda_t = 3.9$ cm), another common automotive radar frequency, this rises to 27.5 kHz. At 10 GHz, a common airborne radar frequency, the spread is 3.6 kHz. The repetition rate must equal or exceed these bandwidths to avoid Doppler ambiguities. These values are unremarkable for pulsed radars. For linear FMCW radars, RRs reaching into the tens of thousands of sweeps per second are considered fairly high. Such systems are often referred to as *fast chirp* FMCW. In Chap. 4, another requirement will arise that also contributes to a fast chirp requirement for FMCW.

Radar motion also induces a spread in the Doppler bandwidth of the echo from stationary objects distributed across the beam. This is most relevant in air-to-ground and space-borne radars. Figure 3.15 illustrates in two dimensions an approach to estimating the Doppler bandwidth of a patch of terrain induced by radar platform motion. The 3-dB radar beamwidth is θ_3 radians. Recall from Chap. 2 that the Doppler shift for a radar moving at velocity v with its boresight squinted ψ radians off the velocity vector is

$$F_D = \frac{2v}{\lambda_t} \cos\psi \text{ Hz} \tag{3.13}$$

Now consider three point scatterers **P1**, **P2**, and **P3**, each at the same range from the radar. **P1** and **P3** are at the 3-dB edges of the antenna beam, while **P2** is on boresight. The received echo at a delay corresponding to their range is the superposition of the echoes from all three

---

[8]The velocity of portions of a vehicle may differ from the nominal velocity of the vehicle as a whole. The Doppler frequency corresponding to the nominal velocity is frequently termed the *skin return*. However, the lug nuts on the wheels on a wheeled vehicle have velocities ranging from zero to twice the nominal velocity of the vehicle. Some laser radars even attempt to measure Doppler shifts due to vehicle vibration. The spectrum of Doppler shifts from different portions of a vehicle is sometimes called a *micro-Doppler spectrum*.

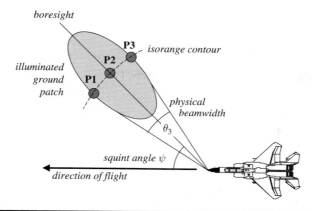

FIGURE 3.15 Geometry for estimating the Doppler bandwidth component due to radar platform motion.

scatterers. However, each is at a slightly different angle with respect to the aircraft velocity vector. **P2** is on the boresight at the squint angle of ψ, but **P1** and **P3** are at $\psi \pm \theta_3/2$ radians. The *difference* in the Doppler shift of the echoes from **P1** and **P3** is

$$\beta_D = \frac{2v}{\lambda_t}[\cos(\psi - \theta_3/2) - \cos(\psi + \theta_3/2)] = \frac{4v}{\lambda_t}\sin\left(\frac{\theta_3}{2}\right)\sin\psi \quad \text{Hz} \quad (3.14)$$

Many radar antenna beamwidths are small, typically less than 5°. Applying a small angle approximation to the $\sin(\theta_3/2)$ term in Eq. (3.14) gives a simple expression for Doppler bandwidth due to platform motion

$$\beta_D \approx \frac{2v\theta_3}{\lambda_t}\sin\psi \quad \text{Hz} \quad (3.15)$$

Equations (3.14) and (3.15) assume the radar is squinted sufficiently that the main beam does not include the velocity vector, that is, $|\psi| > \theta_3/2$. If the radar is forward looking or nearly so, then the $\cos(\psi - \theta_3/2)$ term in Eq. (3.14), which represents the largest Doppler shift in the mainbeam, is replaced by 1. A more complete expression for the platform motion-induced Doppler bandwidth is

$$\beta_D \approx \begin{cases} \dfrac{2v\theta_3}{\lambda_t}\sin\psi \text{ Hz,} & |\psi| \geq \dfrac{\theta_3}{2} \text{ (squinted)} \\ \dfrac{2v}{\lambda_t}[1 - \cos(\psi + \theta_3/2)] \text{ Hz,} & |\psi| < \dfrac{\theta_3}{2} \text{ (forward looking)} \end{cases} \quad (3.16)$$

For example, an L-band (1 GHz) sidelooking ($\psi = 90°$) radar with a beamwidth of 3° traveling at 100 m/s will induce $\beta_D \approx 35$ Hz, while an X-band (10 GHz) sidelooking radar with a 1° beam flying at 200 m/s will induce $\beta_D \approx 233$ Hz. The same two radars in a forward-looking configuration induce negligible Doppler bandwidths of only 0.9 Hz and 0.5 Hz, respectively. Thus, while absolute Doppler shift due to platform motion is highest for a forward-looking system, the Doppler bandwidth spread is highest for a sidelooking system.

In the previous example, the radar was viewing a static patch of ground and the Doppler bandwidth observed by a stationary radar would be zero hertz. The nonzero Doppler bandwidth β_D is entirely due to the motion of the observing radar, not to the characteristics of the target scene itself. The total Doppler bandwidth observed is approximately the sum of the bandwidth induced by platform motion and the intrinsic bandwidth of the scene being measured. The RR should be chosen equal to or greater than this value if possible to meet the Nyquist sampling criterion for the slow-time signal.

Although the Doppler spectrum of the illuminated terrain is both shifted according to Eq. (3.13) and broadened according to Eq. (3.16) by relative motion between the terrain and platform, the shift in center frequency is not relevant to selection of the RR; only the bandwidth determines the Nyquist rate. Also, antenna patterns are not strictly limited in angular extent and therefore the motion-induced Doppler spectrum is not strictly bandlimited to the value based on the 3-dB bandwidth given in Eq. (3.16). Nonetheless, that equation provides a good basis for estimating motion-induced bandwidth.

For any sampled time signal, the sampling frequency determines the aliasing interval of the discrete-time Fourier transform (DTFT) of that signal. The DTFT of the slow-time signal in a given range bin is the Doppler spectrum. The RR is the slow-time sampling rate and therefore determines the aliasing interval of the Doppler spectrum. That *unambiguous Doppler interval* or, equivalently, *unambiguous velocity interval* is

$$F_{Dua} = RR = \frac{1}{T} \text{ Hz, or}$$
$$v_{ua} = \frac{\lambda_t}{2} RR = \frac{\lambda_t}{2T} \text{ m/s} \quad (3.17)$$

For reasons to be seen in Chap. 5, this interval is also called the *blind Doppler shift* or *blind velocity*, denoted as F_{Db} or v_b.

Because the Doppler or radial velocity spectrum is usually plotted on the interval $(-F_{Dua}/2, +F_{Dua}/2]$ or $(-v_{ua}/2, +v_{ua}/2]$, the quantities $F_{Dua}/2$ and $v_{ua}/2$ are sometimes designated the unambiguous Doppler or velocity. Care must be taken to determine if a stated value refers to the full unambiguous interval F_{Dua} or v_{ua}, or to the one-sided frequency (velocity) cutoff point $F_{Dua/2}$ or $v_{ua/2}$. In this text those terms refer to the full two-sided intervals of Eq. (3.17).

Because the Doppler spectrum is periodic in frequency, targets having a Doppler shift F_D or radial velocity v outside of the corresponding unambiguous interval will be aliased into that interval. The apparent (aliased) Doppler shift F_{Da} or radial velocity v_a will satisfy

$$F_D = F_{Da} + n \cdot F_{Dua}, \text{ or}$$
$$v = v_a + n \cdot v_{ua} \quad (3.18)$$

where the integer n is chosen such that F_{Da} and v_a fall in the unambiguous interval.

As a numerical example, consider a 10-GHz radar viewing a target with radial velocity $v = +100$ m/s. The Doppler shift will be $F_D = 6.67$ kHz. If the radar collects a CPI of data using an RR of 3 kHz and obtains the Doppler spectrum via a DTFT of the slow time signal in the appropriate range bin, the unambiguous Doppler and velocity intervals are $F_{Dua} = 3$ kHz and $v_{ua} = 45$ m/s, respectively. The DTFT will usually be interpreted as covering $(-1.5, +1.5]$ kHz and $(-22.5, +22.5]$ m/s. The target's Doppler shift and velocity are outside of these intervals and will therefore be aliased into them. With $n = 2$ in Eq. (3.18), it is seen that the apparent Doppler and velocity are $F_{Da} = 0.67$ kHz and $v_a = +10$ m/s. Techniques to determine the true Doppler or velocity from these aliased measurements are discussed in Chap. 5.

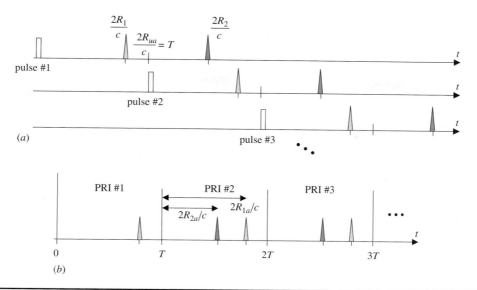

FIGURE 3.16 Illustration of range ambiguous target returns. (a) Pattern of received data for three pulses. (b) Total received signal.

The use of multiple RPs in a CPI raises additional issues not encountered with a single RP. First, the repetition rate sets a limit on the range swath for a pulsed radar. In particular, the maximum range from which the leading edge of the echo of one pulse can be received before the next pulse is transmitted is the *unambiguous range* R_{ua}. It must satisfy $2R_{ua}/c < T$. Specializing RR to the pulsed term PRF,

$$R_{ua} = \frac{cT}{2} = \frac{c}{2PRF} \text{ m (pulsed radar)} \qquad (3.19)$$

The far edge of the sampled range swath (R_2 in Fig. 3.3) must be limited to R_{ua} m or less so the pulse echo from a scatterer at R_2 is received before the next pulse is transmitted. The interpretation of echoes from targets at ranges greater than R_{ua} will be discussed shortly.

To understand another impact of multiple RPs on range measurements, consider the idealized signals of Fig. 3.16a for a pulsed radar. In the first line, a pulse is transmitted at time $t = 0$. It is assumed that there are two targets present at ranges R_1 and R_2, and that the PRF is such that the unambiguous range falls between these two. The target echoes occur $2R_1/c$ and $2R_2/c$ seconds after transmission as shown. Now suppose a second pulse is transmitted at a PRI $T = 2R_{ua}/c$. The same target echo profile is repeated, simply delayed by T seconds as shown on the second line. This pattern continues for the third and any subsequent pulses.

Figure 3.16b shows the resulting total return observed by the radar. In the first PRI ($0 < t < T$), only one target echo is observed because target #2 is past the unambiguous range. In the second PRI ($T < t < 2T$), two target echoes are observed: target #1 from pulse #2 and target #2 from pulse #1. This pattern repeats in the third and subsequent PRIs until the pulse train ends. If the radar receives detectable echoes from ranges up to N times the unambiguous range, the pattern of returns observed after each pulse will reach steady state in the Nth PRI. In Fig. 3.16, it reaches steady state in the second PRI.

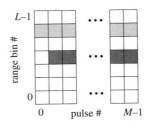

Figure 3.17 Steady-state range ambiguous return for the scenario of Fig. 3.16.

Once steady state is achieved, each pulse appears to result in two detections at the apparent ranges R_{1a} and R_{2a}, as shown in Fig. 3.16b. For target #1, the apparent range is the actual range. However, target #2 was beyond R_{ua} and so aliases or "folds over" to the apparent range $R_{2a} = R_2 - R_{ua}$. Anytime the radar is sensitive enough to detect targets beyond the unambiguous range for a given PRF the apparent ranges are potentially ambiguous. In particular, given a detection at an apparent range R_a, the target's actual range could be any value R_0 that satisfies

$$R_0 = R_a + n \cdot R_{ua} \quad \text{m} \tag{3.20}$$

for some integer n and is within the plausible maximum detection range of the radar. The target is said to have folded over n times. In the example of Fig. 3.16 $n = 0$ for target #1 and $n = 1$ for target #2. Furthermore, a detected target may have several plausible ranges if the maximum detection range is several times R_{ua}. Techniques to determine true ranges from range-ambiguous measurements are discussed in Chap. 5.

Figure 3.17 illustrates the structure of the CPI of data that would result from this example. Suppose the unambiguous range corresponds to seven range bins.[9] Assume the ranges to targets #1 and #2 correspond to the sixth and eleventh range bins. During PRI #1, only the first target is detected, so the first fast-time column of data has only one detection in range bin #6. Target #2 will alias to range bin $11 - 7 = 4$, so the second and subsequent pulses will show detections in range bins #4 and #6.

This example illustrates the existence of a start-up transient in processing when the returns are range ambiguous. If the scenario and PRF are such that targets can be detected at ranges of at least $(N - 1)R_{ua}$ but no further than $N \cdot R_{ua}$ ($N = 2$ in the example), the received signal will not achieve steady state until the Nth PRI (second PRI in this example).

Range aliasing also affects clutter returns. If the clutter echo power is above the receiver noise levels at ranges exceeding R_{ua}, the clutter return will also fold over so that the clutter competing with targets in the steady state may actually be the combined clutter of several range ambiguity intervals. Like targets, the clutter level versus range in a given RI will not reach steady state until multiple pulses have been transmitted if the clutter is range ambiguous. If the clutter reaches steady state in the Nth RI, the first $N - 1$ pulses are often called *clutter fill* pulses. Processing of this nonstationary data generally gives degraded results; it is often better to discard the data from the clutter fill pulses and use only the data received after steady state is achieved. Conversely, if it is desired to have a certain number of steady state RIs for subsequent processing, the number of pulses is often augmented by the necessary number of clutter fill pulses. The same considerations apply to FMCW sweeps.

---

[9]Seven range bins is unrealistically few in most situations but is used here for ease of illustration.

It is tempting to conclude that range ambiguities could be resolved by observing whether or not a target detection appears in all of the pulses. A target detection missing from the first n pulses in a CPI suggests that the actual range is the apparent range plus n times the unambiguous range. For example, in Fig. 3.17 the detection in range bin #4 does not occur in the first RP but does occur in all later RPs ($n = 1$), suggesting that the actual range of that target is the apparent range plus one multiple of R_{ua}. This idea will work if detection algorithms are applied to the fast-time data for each RP separately and the signal-to-noise ratio (SNR) is high enough that the probability of missed detections is small. However, it is rare to use a CPI of data in this manner. More commonly, the SNR of the single-pulse data is not adequate for reliable detection so that it is not known whether the target is absent in the first n pulses. Instead, the slow time data will be coherently or noncoherently integrated in order to obtain an adequate SNR.

In addition to creating the possibility of range ambiguities, the use of multiple pulses also aggravates the eclipsing phenomenon in pulsed radars. This is illustrated by Fig. 3.18, which shows on the upper axis three pulses at a PRI of T seconds. The lower axis shows echoes of the first pulse only from three targets at different ranges. When the first pulse is transmitted, the complete echo from a target at any range less than $R_{ua} - \tau/2$ will arrive before any subsequent transmissions and will not be eclipsed; the first echo on the lower axis illustrates this case. A target at an integer multiple of the unambiguous range $R_{ua} = cT/2$ will produce an echo that arrives as a later pulse is being transmitted so long as the pulse train continues. During this interval, the receiver will again be isolated and the target echo will be fully eclipsed. The second target echo shows this case. Targets at other time delays within the interval $(nT - \tau/2, nT + \tau/2)$ for some integer n will be partially eclipsed, as shown by the third target echo. Thus, the pulse burst creates a series of *blind zones* in range or time delay centered on multiples of the unambiguous range or delay. Targets in these blind zones will be difficult or impossible to detect even when they have adequate SNR because their echoes will be received only partially or not at all. Techniques to overcome this limitation are also discussed in Chap. 5.

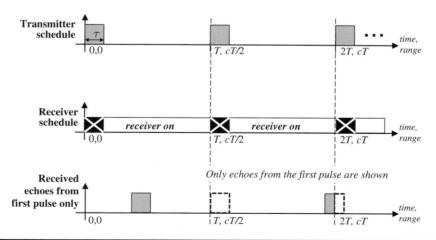

Figure 3.18 Blind zones with a pulsed radar. (a) Sequence of pulse transmissions at a PRI T. (b) The receiver is isolated during each pulse transmission, creating blind zones in which target detectability is degraded. (c) Echoes of the first pulse only from three targets. The first echo is not eclipsed, the second is fully eclipsed, and the third is partially eclipsed.

FMCW radars also exhibit a range ambiguity due to the periodic repetition of the transmitted waveform. Sweep echoes from targets at ranges longer than $cT/2$ will overlap with a later sweep, producing the same beat frequency as a much closer target. Equation (3.19) applies to these systems as well.

Some sawtooth linear FMCW systems suffer a second source of range ambiguity. Range is measured by rescaling the frequency of the beat signal at the deramp mixer output. The sampling rate F_s determines the maximum representable range (far edge of the range swath) according to

$$R_2 = \frac{cT}{2} \cdot \frac{F_s}{\beta} \text{ m} \quad \text{(sawtooth linear FMCW radar)} \tag{3.21}$$

Since F_s is typically much less than β, the range of Eq. (3.21) is much shorter than the unambiguous range of Eq. (3.19).

What happens to echoes of targets at ranges from R_2 to R_{ua} depends on the system design. Targets at ranges longer than R_2 but less than $(cT/2)$ will produce beat frequencies greater than F_s. These will alias to lower frequencies corresponding to shorter ranges; this range interval therefore produces ambiguous range measurements. In this case, R_2 becomes a second unambiguous range,

$$\begin{aligned} R_{ua1} &= \frac{cT}{2} \cdot \frac{F_s}{\beta} \text{ m} \\ R_{ua2} &= \frac{cT}{2} \text{ m} \end{aligned} \quad \text{(sawtooth linear FMCW radar)} \tag{3.22}$$

However, many implementations will implement an analog lowpass filter at the output of the deramp mixer, prior to the sampling and IDFT, with a cutoff frequency equal to or slightly greater than the Nyquist rate F_N. In these systems targets at ranges greater than $(cTF_s/2\beta)$ will have their beat frequencies filtered out by the lowpass filter and will not be observed. The ranges from $(cTF_s/2\beta)$ to $cT/2$ will then be blind ranges.

Notice that for either pulsed or FMCW radars, a lower RR (larger RI) allows a longer unambiguous range [Eq. (3.19) or (3.22)], while a higher RR allows unaliased measurement of a wider range of Doppler shifts [Eq. (3.17)]. The product of R_{ua} with F_{Dua} or v_{ua} defines a range-Doppler coverage space whose size is independent of the RR or RI:

$$\begin{aligned} R_{ua}F_{Dua} &= \frac{c}{2} \text{ m/s}, \quad R_{ua}v_{ua} = \frac{\lambda_t c}{4} \text{ m}^2/\text{s} \quad \text{(pulsed radar)} \\ R_{ua1}F_{Dua} &= \frac{c}{2} \cdot \frac{F_s}{\beta} \text{ m/s}, \quad R_{ua1}v_{ua} = \frac{\lambda_t c}{4} \cdot \frac{F_s}{\beta} \text{ m}^2/\text{s} \quad \text{(sawtooth linear FMCW radar)} \end{aligned} \tag{3.23}$$

The coverage space for the FMCW case is that of the pulsed radar scaled by (F_s/β). This scale factor is normally much less than one; for instance, it equals 0.1 in the example following Eq. (3.12). Consequently, pulsed radars are often better suited to long range and/or fast target applications.

3.4 Multiple Channels: The Datacube

Some radars, but by no means all, have antennas that provide multiple simultaneous outputs. The most obvious example is a system using a phased array antenna with multiple subarrays, each having its own receiver, or even with one receiver per array element in some cases. Each receiver will generate a matrix of data like that of Fig. 3.14 for every pulse or

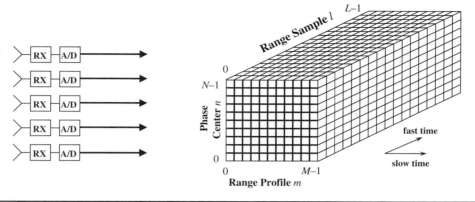

Figure 3.19 Datacube illustrating one CPI of data from a multichannel radar.

sweep burst. The complete set of data $y[l, m, n]$ from all N channels is called a *datacube* and is illustrated in Fig. 3.19. The third dimension is often referred to as the *receiver channel* or *phase center* dimension. Another type of system that generates a datacube uses a monopulse antenna, common in some types of tracking radars. A monopulse tracking antenna has three output channels and so generates a datacube having $N = 3$ layers. Radar data is often explicitly organized in the processor memory in a datacube format, that is, as a three-dimensional structure of complex-valued data.

How large is a datacube? The number of samples in each dimension is determined by the characteristics of the desired radar measurements. The number L of range samples is simply the length of the range swath divided by the range bin spacing, $L = R_w/\delta R_s$. The swath length is determined by mission requirements, while the range bin spacing is determined primarily by the waveform bandwidth as seen in Eq. (3.1). Both may vary significantly for the same radar as it switches between various operating modes with different search ranges and range resolutions.

One important determinant of the number of range profiles M in a CPI is the desired Doppler resolution (equivalently, velocity resolution). The Doppler spectrum is the DTFT of the slow-time data. The duration of the slow-time signal is the CPI length of MT seconds. The Doppler resolution will therefore be on the order of $\Delta F_D = 1/(MT)$,[10] giving the required number of RPs as $M = 1/(\Delta F_D T) = RR/\Delta F_D$. Thus, M depends on the repetition rate as well as the desired Doppler resolution and can vary widely. In pulse Doppler processing for basic target detection and tracking, M is frequently a small number of tens of RPs. However, in fine-resolution imaging it can be hundreds or even thousands of RPs.

For a multichannel receiver the number N of channels is more difficult to characterize. A phased array antenna with a receiver per element may have hundreds or thousands of phase centers, each constituting a receiver channel. A subarrayed architecture may have many fewer, perhaps ranging from as little as three or four to a few tens. A monopulse antenna has three phase centers. The antenna type, size, and architecture all significantly influence N.

The datacube view of a CPI of data from a multichannel radar provides a good conceptual model for understanding most digital radar signal processing operations. Many of the

---

[10]The precise value depends on the definition of mainlobe width that is used and whether or not weighting for sidelobe control is employed.

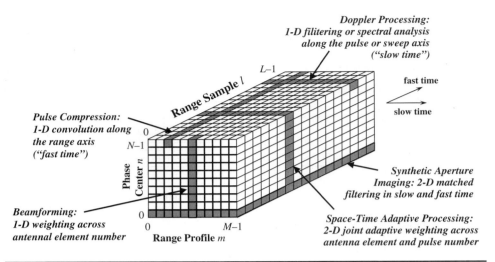

Figure 3.20 Correspondence between key radar signal processing functions and operations on the radar datacube.

basic radar signal processing operations considered in the remainder of this text correspond to processing one-dimensional subvectors or two-dimensional submatrices of the datacube in various dimensions. Figure 3.20 illustrates these relationships. The particular operations depicted are discussed in upcoming chapters. For example, pulse compression (Chap. 4) is implemented as a one-dimensional convolution on a single vector in the fast-time (range) dimension. It can be performed independently on each such range vector for every pulse and receiver channel. Computation of the Doppler spectrum requires one-dimensional spectrum estimation (usually a DFT) on a single vector in the slow time dimension, and can be repeated for every range bin and receiver channel. Other examples are shown in Fig. 3.20.

3.5 Dwells

The *dwell time* or just *dwell*, sometimes also called the *time on target* or *look time*, is another term for a radar data collection interval. It is defined in the IEEE standard for radar definitions as "a data acquisition interval during which the data is usually processed together for detection or measurement" (IEEE, 2017). Like the term "CPI," "dwell" is often used to refer to both the time interval and the data collected within that time.

Consider a rotating radar like that shown in Chap. 2, Fig. 2.15a, and suppose the beamwidth is 3° and the antenna scans at a rate of 60° per second. A point target would be in the beam for 50 ms during each scan. If the radar is pulsed with a PRF of 2000 pulses per second, a target would be illuminated with 100 pulses during the traversal of the mainbeam. Since it is known that the echo of a target, if present, would be present in 100 successive pulses, it might be sensible to integrate 100 pulses at a time for SNR improvement before performing a detection test. In this scenario, the dwell time would be the full 50 ms that the target was in the antenna mainlobe. If only 50 pulses at a time were integrated, the dwell time would be considered to be 25 ms. The same ideas and calculations apply to sweeps if an FMCW radar is used.

The idea of a dwell is not limited to mechanically scanning radars. Consider a pulsed Doppler radar with an electronically scanned antenna pointed in some particular direction.

The radar emits a burst of 20 pulses at a 2-kHz PRF. The CPI is then 10 ms. Suppose the radar collects three such CPIs while illuminating the same region, with 50 ms from the start of one CPI to the start of the next. The total data collection time of 110 ms from the beginning of the first CPI to the end of the third would be the dwell time for the radar in that look direction.

In practice the term "dwell" is somewhat vague. It is sometimes used synonymously with "CPI," but the preceding example illustrates that they are not the same. For a coherent radar that organizes its data into CPIs, a dwell can correspond to multiple CPIs provided the target remains illuminated. For rotating and similar mechanically scanned radars, a dwell is usually the time that the antenna illuminates a target as it scans, because that time upper bounds the maximum time over which one would reasonably want to combine data for a single detection. The reader is cautioned to be alert to the intended meaning.

3.6 Sampling the Doppler Spectrum

Selecting a value for the pulse or sweep repetition rate determines the sampling rate for the slow-time signal. The frequency spectrum of the slow-time signal is traditionally called the Doppler spectrum because the nonzero frequency components are due to the spatial Doppler effect arising from the relative motion between the radar and the target scene. Doppler processing, which is the analysis or modification of the information about the target scene contained in the Doppler spectrum, will be the subject of Chap. 5. Doppler processing will sometimes be performed directly in the slow-time domain, that is, directly on the time signal represented by a row of $y[l, m]$; but frequently the spectrum of each row will be explicitly calculated. In a digital processor, this must be done with a discrete Fourier transform or other discrete spectral analysis technique. In this section, it is assumed that the spectrum is computed using conventional DFT techniques; no nonlinear spectral estimation methods or other alternatives are considered.

3.6.1 The Nyquist Rate in Doppler

The question immediately arises as to how closely successive samples of the computed Doppler spectrum should be spaced, that is, what should be the Doppler sampling interval? The Nyquist criterion can be applied to sampling in frequency as well as the more usual application to sampling in time. The result will be a frequency sampling rate that is dependent on a "bandwidth" in the time domain.

The Nyquist sampling rate in the frequency domain can be determined by reviewing the relation between the sampled Doppler spectrum and the slow-time signal. Consider a possibly infinite-length slow-time signal $y_s[m]$. Its DTFT is (Oppenheim and Schafer, 2010)

$$Y_s(\omega) \equiv \sum_{m=-\infty}^{\infty} y_s[m]\exp[-j\omega m], \quad \omega \in (-\pi, \pi] \tag{3.24}$$

$Y_s(\omega)$ is a function of a continuous frequency variable despite the fact that the signal $y_s[m]$ is discrete. Furthermore, it is periodic in ω with period 2π radians per sample.

Consider the K-point discrete spectrum $Y_s[k]$ formed by sampling $Y_s(\omega)$ at K evenly spaced points along the interval $[0, 2\pi)$,

$$Y_s[k] = Y_s\left(\frac{2\pi k}{K}\right), \quad k \in [0, K-1] \tag{3.25}$$

Interpret $Y_s[k]$ as a K-point DFT. To find the relation between it and the original signal $y_s[m]$, compute its inverse DFT:

$$\begin{aligned}\hat{y}_s[m] &= \frac{1}{K}\sum_{k=0}^{K-1}Y_s\left(\frac{2\pi k}{K}\right)\exp(j2\pi mk/K), \quad m \in [0, K-1] \\ &= \frac{1}{K}\sum_{k=0}^{K-1}\left\{\sum_{p=-\infty}^{\infty}y_s[p]\exp(-j2\pi pk/K)\right\}\exp(j2\pi mk/K) \\ &= \sum_{p=-\infty}^{\infty}y_s[p]\left\{\frac{1}{K}\sum_{k=0}^{K-1}\exp[j2\pi(m-p)k/K]\right\}\end{aligned} \quad (3.26)$$

The inner sum can be evaluated as

$$\frac{1}{K}\sum_{k=0}^{K-1}\exp[-j2\pi(m-p)k/K] = \sum_{q=-\infty}^{\infty}\delta[m-p-qK] \quad (3.27)$$

where $\delta[\cdot]$ is the discrete unit impulse function.[11] Substituting Eq. (3.27) in Eq. (3.26) gives

$$\hat{y}_s[m] = \sum_{q=-\infty}^{\infty}y_s[m-qK], \quad m \in [0, K-1] \quad (3.28)$$

Equation (3.28) shows that, in the dual of time domain sampling behavior, sampling the frequency spectrum replicates the signal in the time domain with a period proportional to the frequency sampling rate. Specifically, if the slow-time signal spectrum is computed at K frequency points, the time domain signal obtained by an inverse DFT of those frequency samples is the original slow-time signal replicated at intervals of K samples. If $y_s[m]$ in fact is finite and its length M is less than or equal to K, the replications of $y_s[m]$ will not overlap; that is, no time domain aliasing occurs. Then $y_s[m-qK] = 0$ in the interval of interest $[0, K-1]$ for all $q \neq 0$, so that $\hat{y}_s[m] = y_s[m]$ in that interval. This means that the IDFT output will be the original signal $y_s[m]$ as desired. This is the usual case where the DFT size is at least as long as the data sequence size.

To summarize, the Nyquist rate for sampling in the frequency dimension is

$$K \geq M \text{ samples per Doppler spectrum period}^{12} \quad (3.29)$$

Note that the width of the signal's region of support (i.e., its length or "bandwidth") in the time domain of M samples plays the same role for sampling in frequency as does the width of a signal's region of support in frequency (its actual bandwidth) for sampling in time. The corresponding normalized radian frequency Nyquist sampling interval is

$$\omega_s \leq \frac{2\pi}{M} \text{ rad/sample} \quad (3.30)$$

---

[11] Not to be confused with the "Dirac delta" impulse function $\delta_D(\cdot)$ used in continuous analysis.

[12] Using a DFT size K that is strictly greater than the number of time samples M is often referred to as *zero padding*. This term is a relic of early software for computing the fast Fourier transform, which often returned the vector of frequency samples in the same variable used to input the time samples. When $K > M$, the input vector was "padded" to the required output length of K by adding $K - M$ zeros at the end. While no longer important in software, the term still implies sampling the spectrum at greater than a Nyquist rate.

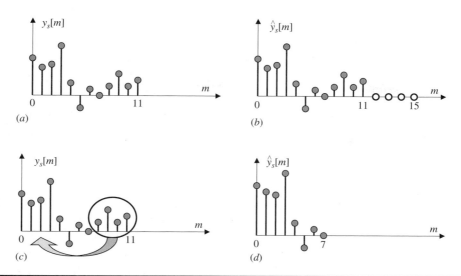

FIGURE 3.21 Illustration of the "zero padding" and "data turning" operations. (*a*) Original 12-point data sequence. (*b*) Zero-padded to 16 points for use in a 16-point DFT. (*c*) Data turning to create an aliased 8-point sequence shown in (*d*) for use in an 8-point DFT.

In some systems the number of Doppler samples computed is less than the number of data samples available, that is, $K < M$. This can occur if only a limited number of spectrum samples are required by the system design. In early (1960s to 1980s) digital radar processors, it was more likely motivated by the difficulty of implementing a larger DFT at radar data rates, a problem mostly obviated by computing advances enabled by Moore's law. One way to compute a K-point DFT from an M-point sequence when $K < M$ is to simply retain only K data samples and compute their K-point DFT. This is not desirable when there are $M > K$ samples available for two reasons. First, the DTFT of the K samples used is not the same as that of the full M-point sequence, so the DFT will compute samples of a modified, degraded-resolution DTFT. Second, the SNR of the calculated spectrum is reduced because only K samples instead of all M available samples are coherently integrated by the DFT. It is rarely a good idea to discard measured data if the highest possible measurement quality is desired.

If the Doppler spectrum samples are still to be equal to samples of the DTFT of $y_s[m]$ in this case, Eqs. (3.26) to (3.28) imply that it is necessary to form a new, reduced-length K-point sequence $\hat{y}_s[m]$ from the slow-time data sequence $y_s[m]$ by aliasing it according to Eq. (3.28). This operation, depicted pictorially in Fig. 3.21, is sometimes called *data turning*. It maximizes the SNR of the Doppler spectrum samples by using all of the available samples and is in fact used in some older operational radars.

3.6.2 Straddle Loss

The previous section established the Nyquist sampling rate in Doppler frequency. When actually computing the sampled spectrum, whether by the DFT or other means, one would like to be confident that the sampled spectrum captures all of the important features of the underlying DTFT. For example, if the DTFT exhibits significant peaks, it is hoped that one of the spectral samples will fall on or very near those peaks so that the sampled spectrum reflects those peaks.

An appropriate signal model to consider this issue is a pure complex sinusoid, corresponding for example to a target moving at constant velocity relative to the radar over the

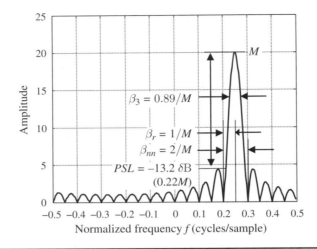

FIGURE 3.22 The magnitude of the DTFT of a sampled pure complex sinusoid of 20 samples length, normalized frequency 0.25 cycles per sample, and amplitude 1. Common mainlobe width and peak sidelobe metrics are shown.

observation interval and therefore exhibiting a constant Doppler shift. The slow-time signal $y_s[m]$ is modeled as

$$y_s[m] = A\exp(j\omega_D m), \quad m \in [0, M-1] \quad (3.31)$$

where ω_D is the Doppler frequency shift in normalized radian frequency units. Its DTFT is

$$Y_s(\omega) = A\frac{\sin[(\omega-\omega_D)M/2]}{\sin[(\omega-\omega_D)/2]}\exp\left[-j\left(\tfrac{M-1}{2}\right)(\omega-\omega_D)\right], \quad \omega \in (-\pi, \pi] \quad (3.32)$$

That is, $Y_s(\omega)$ is an asinc function, shifted in the frequency domain so that its peak occurs at $\omega = \omega_D$. An example is shown in Fig. 3.22 for the case $\omega_D = \pi/2$ (corresponding to $f_D = \omega_D/2\pi = 0.25$) and $M = 20$. Significant features of this DTFT include the amplitude and frequency of its peak, the mainlobe bandwidth, and the sidelobe structure. Equation (3.32) shows that the M-point DTFT of a pure complex sinusoid of amplitude A has a peak value of MA with the peak sidelobe 13.2 dB below the peak (22% of peak value). The 3-dB width of the mainlobe in normalized frequency units is $\beta_3 = 0.89/M$ cycles per sample, the Rayleigh width is $\beta_r = 1/M$ cycles per sample, and the null-to-null mainlobe width is $\beta_{nn} = 2/M$ cycles per sample. These metrics are illustrated in Fig. 3.22.

The K-point DFT computes samples of this spectrum at normalized frequencies $2\pi k/K$ rad/sample. Figure 3.23 shows the result when $K = M$ and the sinusoid frequency exactly equals one of the DFT frequencies, that is, $\omega_D = 2\pi k_0/K$ for some k_0. In this example, $k_0 = 5$ and $K = 20$, corresponding to $\omega_D = \pi/2$ rad/sample. Also, an "fftshift" operation has been performed to place zero frequency in the center of the graph, and the frequency axis is in units of cycles per sample. One DFT sample falls on the peak of the asinc function, while all of the others fall on its zeroes, so that the DFT becomes a discrete impulse function. This could be viewed as an ideal measurement since the discrete spectrum indicates a single sinusoid at the correct frequency and nothing else; but it is misleading in that it does not reveal the mainlobe width or sidelobe structure of the underlying DTFT.

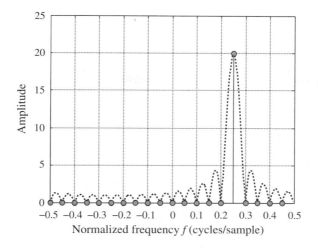

Figure 3.23 The 20-point DFT of a sampled pure complex sinusoid of 20 samples length, normalized frequency 0.25 cycles per sample, and amplitude 1. The dotted line shows the underlying DTFT of the same data from Fig. 3.22.

More importantly, the good result of Fig. 3.22 depends critically on the actual sinusoid frequency exactly matching one of the DFT sample frequencies. If this is not the case, the DFT samples will fall somewhere on the asinc function other than the peak and zeros. Figure 3.24 shows the result when the example of Figs. 3.22 and 3.23 is modified by changing the normalized frequency from 0.25 to 0.275 (equivalently, changing ω_D to 0.55π), exactly halfway between two DFT sample frequencies. Now a pair of DFT samples straddle the actual underlying peak of the asinc function, while the other samples fall near the sidelobe peaks. Even though the underlying asinc function is identical in shape in both cases, differing only by a half-bin shift on the frequency axis, the effect on the apparent spectrum measured

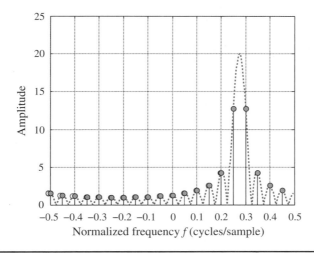

Figure 3.24 Same as Fig. 3.23 except for a frequency shift of the sinusoid by one-half DFT bin to a normalized frequency of 0.275 cycles per sample.

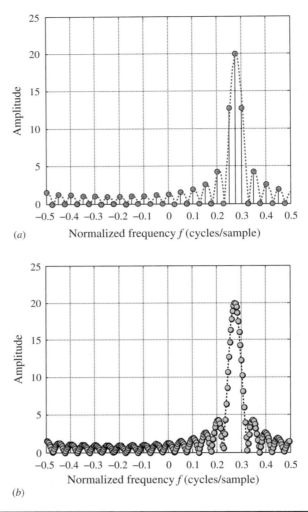

FIGURE 3.25 Continuation of the example of Fig. 3.23. (a) 40-point DFT of the 20-point sinusoid of normalized frequency 0.275. (b) 256-point DFT of the same sequence.

by the DFT is dramatic: a broadened and attenuated mainlobe and the appearance of significant sidelobes where before there apparently were none.

Because the DFT sample frequencies straddle the true peak of the underlying DTFT, the apparent peak amplitude of the spectrum in Fig. 3.24 is about 13, whereas the peak amplitude of the underlying DTFT $Y_s(\omega)$ and thus of the DFT in Fig. 3.23 is 20. This reduction in measured peak signal amplitude is called a *straddle loss* (because the samples straddle the true peak location).[13]

One obvious way to reduce straddle loss is to sample the Doppler frequency axis more densely, that is, to choose the number of spectrum samples $K > M$. The resulting samples are more closely spaced so that the maximum amount by which a sample frequency can miss the peak frequency of the DTFT is reduced, thus reducing the straddle loss. Figure 3.25

---

[13]Straddle loss is also called *scallop loss* by some authors, for example, Harris (1978).

continues the example of Fig. 3.23, but with the sampling density doubled to 2M samples per Doppler spectrum period (40 samples in this case), and then to 12.8M samples per spectrum period (256 samples). Increasing the sample density causes the apparent spectrum measured by the DFT to begin to resemble the underlying asinc of the DTFT even at as little as 2M samples per period. At 12.8M samples per spectrum period the DFT gives an excellent representation of the details of the underlying DTFT.

The straddle loss for a sinusoidal signal can be limited to a specified value by appropriate choice of the spectrum sampling rate K. For example, the loss can be limited to 3 dB or less by choosing K such that the interval $2\pi/K$ between samples does not exceed the 3-dB width of the asinc function. The 3-dB width can be found by considering just the magnitude of Eq. (3.32) with $\omega_D = 0$ for convenience. The peak value of the asinc function is MA. Therefore, the −3-dB frequency is the value ω_3 of ω such that the asinc function has the value $MA/\sqrt{2}$. This is best found numerically. The answer is a strong function of M for small M but rapidly approaches an asymptotic value of $\omega_3 = 2.79/M$ for $M \geq 10$. It follows that the two-sided 3-dB width of the asinc function is $\Delta\omega = 5.58/M$ rad/sample.

The sampling interval for a rate of K samples per period is $2\pi/K$ rad/sample. Equating this to the 3-dB width and solving gives the sampling rate required to limit off-peak sampling attenuation to 3 dB in the Doppler spectrum in terms of the Nyquist rate of M samples per Doppler spectrum period,

$$K \geq \frac{2\pi}{5.58} M = 1.13M \text{ samples per Doppler spectrum period} \qquad (3.33)$$

which is 13 percent higher than the Nyquist sampling rate in Doppler. If the off-peak sampling loss is to be kept significantly less than 3 dB, the Doppler spectrum must be oversampled still more.

The analysis leading to Eq. (3.33) can be repeated for any specified level of tolerable straddle loss. Figure 3.26 shows the worst-case straddle loss as a function of the oversampling factor κ (i.e., $K = \kappa M$) for the case $M = 100$. Both undersampled ($\kappa < 1$) and oversampled ($\kappa > 1$) cases are shown. The loss is somewhat less for very short duration sequences ($M < 10$) but varies little for larger M.

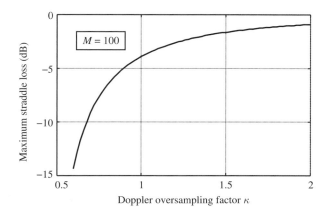

FIGURE 3.26 Maximum off-peak Doppler spectrum sampling loss for a sinusoidal slow-time signal sampled at κM samples per Doppler spectrum period.

3.7 Sampling in the Spatial and Angle Dimensions

Two distinct types of spatial sampling are of concern in a radar system. The first concerns the design of phased array antennas. A phased array samples the incoming wavefront at the individual array element locations. The spacing of these elements must be chosen to adequately sample the wavefront for any incidence angle. The second type concerns beam steering. Mechanically or electronically steered antennas can change the pointing direction of their antenna beam. As the beam is scanned to search or map a region in space, a choice must be made as to how far it is permissible to scan before another pulse or sweep (or entire CPI) must be transmitted by the radar so that the external environment is adequately sampled. The next two subsections address these questions.

3.7.1 Spatial Array Sampling

Chapter 1 introduced the concept of spatial frequency and wavenumber. Consider a uniform linear array with element spacing d, as shown in Fig. 3.27. The wavenumber (spatial radian frequency) of an RF signal with wavelength λ_t impinging on the array antenna from a direction of arrival θ radians off the normal to the array as shown in the figure is

$$K_x = \frac{2\pi}{\lambda_t} \sin\theta \quad \text{rad/m} \tag{3.34}$$

The equivalent spatial frequency in cyclical units is

$$F_x = \frac{1}{\lambda_t} \sin\theta \quad \text{cycles/m} \tag{3.35}$$

The angle of arrival θ can vary between $-90°$ and $+90°$, so the spatial frequency bandwidth becomes

$$\beta_x = \frac{1}{\lambda_t}\sin\left(\frac{\pi}{2}\right) - \frac{1}{\lambda_t}\sin\left(-\frac{\pi}{2}\right) = \frac{2}{\lambda_t} \quad \text{cycles/m} \tag{3.36}$$

It follows immediately by the Nyquist criterion that the required spatial sampling interval is

$$d \leq \frac{1}{\beta_x} = \frac{\lambda_t}{2} \quad \text{m} \tag{3.37}$$

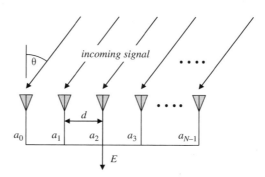

FIGURE 3.27 Geometry of a uniform linear array antenna.

Thus, the elements of the array should be spaced no more than $\lambda_t/2$ meters apart to avoid aliasing of spatial frequencies.[14]

Many practical arrays, particularly large wideband systems, employ more complicated architectures effectively having two spatial sampling intervals. The complete antenna array is broken into a relatively small number of subarrays, each of which is populated with elements obeying the Nyquist spacing of Eq. (3.37). Since the multielement subarrays are necessarily separated by multiples of d, they contribute a term to the complete antenna pattern that does exhibit spatial aliasing. As a result, the overall antenna pattern can exhibit aliasing in some circumstances, depending on whether phase or time delay steering is used for individual elements and across subarrays, the steering direction of the array, and the bandwidth of the radar waveform. An introduction to these issues is given in Bailey (2010).

3.7.2 Sampling in Angle

Consider a steerable or scanning antenna, whether mechanically steered (typically a parabolic dish or slotted flat-plate array, and others) or electronically steered (phased array), with a 3-dB beamwidth θ_3 radians. Each transmitted pulse or sweep samples the reflectivity of the environment in the direction in which the antenna is pointed. If a region in angular (elevation and azimuth) space is to be searched, the question arises: how densely in angle must the space be sampled? That is, how much can the antenna be steered before another pulse or sweep should be transmitted? Smaller angular sampling intervals provide a better representation of the search volume, but also require more transmissions and therefore more time and energy to search a given volume. Since the antenna voltage pattern suppresses returns more than about $\pm\theta_3/2$ radians from the antenna boresight, one intuitively expects that to adequately sample the reflectivity of the scene scanned by the antenna it will be necessary to make a new measurement every time it scans by some angle on the order of θ_3. The Nyquist criterion can be applied to this spatial sampling problem to quantify this expectation.

It was seen in Chap. 2 [Eq. (2.112)] that the observed reflectivity in angle at a constant range is the convolution of the range-averaged reflectivity with the two-way antenna voltage pattern. An equivalent expression in just one angle dimension for simplicity, say azimuth, is

$$y(\theta; R_0) = \hat{\rho}(\theta; R_0) *_\theta E^2(\theta)$$
$$= A_r \int_{-\pi}^{\pi} E^2(\zeta - \theta)\hat{\rho}(\zeta; R_0) d\zeta \qquad (3.38)$$

where $y(\theta; R_0)$ is the complex coherent receiver output as a function of azimuth angle θ at range R_0, $\hat{\rho}(\theta; R_0)$ is the range-averaged reflectivity evaluated at range R_0, and $E^2(\theta)$ is the two-way voltage pattern in the angular dimension θ. It follows that the Fourier transform in the angle dimension of y is the product of the Fourier transforms of the antenna pattern and the range-averaged reflectivity.

Taking the pattern of the ideal rectangular aperture as representative, the normalized two-way antenna voltage pattern is the square of the one-way pattern from Eq. (1.8):

$$E^2(\theta) = \left[\frac{\sin(\pi D \sin\theta/\lambda_t)}{\pi D \sin\theta/\lambda_t}\right]^2 \qquad (3.39)$$

---

[14]This same result is often derived in antenna literature by requiring that the antenna pattern not contain *grating lobes*, which are aliased replicas of the antenna pattern caused by sampling of the aperture of a phased array antenna by the elements.

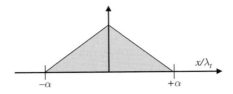

FIGURE 3.28 Fourier transform of the two-way antenna voltage pattern for an ideal rectangular antenna aperture with uniform illumination.

Defining $s = \sin\theta$ and $\alpha = D/\lambda_t$, Eq. (3.39) can be rewritten as

$$E^2(s) = \left[\frac{\sin(\pi\alpha s)}{\pi\alpha s}\right]^2 \tag{3.40}$$

which is a sinc-squared function. It follows immediately that its Fourier transform is a triangle function of width 2α in the normalized variable (x/λ_t), where x is the spatial dimension of the antenna aperture (Bracewell, 1999). This function, illustrated in Fig. 3.28, is the autocorrelation of the rectangular antenna illumination pattern that gives the one-way sinc pattern.

Because the Fourier transform of the antenna pattern has a width of 2α, the Nyquist sampling interval in s must be

$$T_s \leq \frac{1}{2\alpha} = \frac{\lambda_t}{2D} \quad \text{(dimensionless)} \tag{3.41}$$

Recall that $s = \sin\theta$. To convert T_s into a sampling interval in θ, consider the differential $ds = \cos\theta \cdot d\theta$ so that $d\theta = ds/\cos\theta$. Thus, a small interval T_s in s corresponds approximately to an interval $T_\theta = T_s/\cos\theta$ in θ. The minimum value for T_θ occurs when $\theta = 0$ and is $T_\theta = T_s$. The sampling interval in angle should therefore obey

$$T_\theta \leq \frac{\lambda_t}{2D} \text{ rad} \tag{3.42}$$

This is the Nyquist sampling interval in angle for a rectangular aperture of size D with uniform illumination.

This result can also be expressed in terms of 3-dB beamwidths. The 3-dB beamwidth of an aperture antenna is of the form (Balanis, 2015)

$$\theta_3 = k\frac{\lambda_t}{D} = \frac{k}{\alpha} \text{ rad} \tag{3.43}$$

Combining Eqs. (3.42) and (3.43) gives

$$T_\theta \leq \frac{\theta_3}{2k} \text{ rad} \tag{3.44}$$

In the uniformly illuminated case $k = 0.89$, giving a Nyquist sampling rate of 0.56 times the 3-dB beamwidth or 1.8 samples per 3-dB beamwidth. In practice, many systems sample in angle at approximately one sample per 3-dB beamwidth. The search space is then undersampled in angle, at least according to the Nyquist criterion.

While derived for the uniformly illuminated case, these results apply to all aperture antennas. For a finite aperture of size D, different antenna patterns (for instance, with lower sidelobes at the expense of a wider mainlobe) are obtained by changing the aperture illumination function, typically by tapering it in a manner similar to windowing operations in signal processing. Perhaps surprisingly, so long as the aperture width remains the same, so does the Nyquist rate regardless of the aperture weighting and resulting one-way pattern. This occurs because even though the detailed shape of the autocorrelation of the illumination function (the Fourier transform of the two-way pattern) will change, it will still be limited to a width of 2α in s, and it is that width that determines the angular Nyquist sampling rate.

What will change for different illumination functions and antenna patters is the factor k. Lower sidelobe antennas typically have values of k in the range of approximately 1.4 to 2.0, giving corresponding Nyquist sampling rates on the order of 2.8 to 4 samples per 3-dB beamwidth for low sidelobe antennas.

For a rotating radar, the angular sampling rate of Eq. (3.44) implies a lower bound on the repetition interval. Suppose the rate of rotation is Ω_0 radians per second. In order that successive pulses or sweeps be transmitted in directions differing by no more than the T_θ of Eq. (3.44), the RI and RR must satisfy

$$RI \leq \frac{(\theta_3/2k)}{\Omega_0} = \frac{\theta_3}{2k\Omega_0} \text{ seconds} \quad \rightarrow \quad RR \geq \frac{2k\Omega_0}{\theta_3} \text{ s}^{-1} \qquad (3.45)$$

Equations (3.45) and (3.19) illustrate a conflict between volume coverage and search rate in a rotating search radar. For a given antenna design, θ_3 and k are fixed. Increasing the sweep rate Ω_0 will increase the volume search rate, requiring an increased RR; but a higher RR (smaller T) reduces the unambiguous range, reducing the volume that can be searched without ambiguities.

3.8 I/Q Imbalance and Digital I/Q

In Chap. 1, it was shown that the output of a quadrature receiver given a real-valued bandpass signal as input is the same as would be obtained by using the equivalent analytic (one-sided spectrum) complex signal with complex demodulation by the reference oscillator $\exp(-j\Omega_t t)$. In other words, the quadrature receiver acts to select the upper band of the bandpass signal and shift it to baseband. Any system that accomplishes this same result can be used to derive the in-phase and quadrature signals needed for further signal processing.

The quadrature receiver could, in principle, be implemented entirely digitally. The input signal would be converted to a digital signal after the low-noise amplifier. The mixing operations would be replaced by multiplications and the analog lowpass filters by digital filters. Historically, this has not done in practice because a straightforward implementation would require the A/D converter to operate at about twice the carrier frequency rather than twice the information bandwidth of the signal (specifically, $2F_t + \beta$ rather than just β samples per second), until recently a technologically unreasonable requirement.[15] On the other hand, the conventional analog quadrature receiver also has technological limitations. Correct operation assumes that the two channels are perfectly matched in delay and gain across the frequency band of interest, there are no DC biases in either channel, and the two reference oscillators are exactly 90° out of phase. In this section, the effect of I/Q imbalances is investigated. Two digital I/Q receiver structures that combat imbalance errors are then described.

---

[15]Sampling at RF and implementing the receiver entirely digitally is now done in some systems, especially at lower frequency bands.

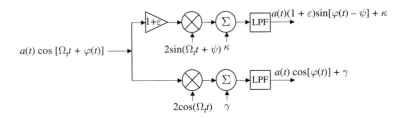

FIGURE 3.29 Conventional coherent receiver of Fig. 1.12 with amplitude and phase mismatch errors and DC offsets.

3.8.1 I/Q Imbalance and Offset

Figure 1.12 describing the conventional quadrature receiver is repeated below as Fig. 3.29, but with the addition of an amplitude mismatch factor $(1+\varepsilon)$, a phase mismatch ψ, and DC offsets γ and κ in the in-phase (I) and quadrature (Q) channels, respectively. The I channel can be taken as the gain and phase reference without loss of generality, so the gain and phase errors are placed entirely in the Q (upper) channel. As shown in Fig. 3.29, the introduction of these errors is reflected as an undesired gain and phase shift in the Q channel output, along with the DC offset in each channel. For processing, the I and Q channel outputs are combined as usual into a single complex signal, $x(t) = I(t) + jQ(t)$. In the absence of mismatch errors, $x(t) = a(t)\exp[j\varphi(t)]$.

How are the mismatch errors manifested in $x(t)$? Inspection of Fig. 3.29 gives

$$\begin{aligned} x(t) &= a(t)\cos\varphi(t) + \gamma + j\{a(t)(1+\varepsilon)\sin[\varphi(t)-\psi] + \kappa\} \\ &= a(t)\{[1 - j(1+\varepsilon)\sin\psi]\cos\varphi(t) + j(1+\varepsilon)\cos\psi\sin\varphi(t)\} + (\gamma + j\kappa) \\ &\equiv a(t)[\alpha\cos\varphi(t) + j\beta\sin\varphi(t)] + (\gamma + j\kappa) \end{aligned} \quad (3.46)$$

Note that the constant α is complex but β is not, and that $\alpha = \beta = 1$ in the absence of gain and phase errors, that is, if $\varepsilon = \psi = 0$. Using the identities

$$\begin{aligned} \alpha &= \frac{\alpha+\beta}{2} + \frac{\alpha-\beta}{2} \\ \beta &= \frac{\alpha+\beta}{2} - \frac{\alpha-\beta}{2} \end{aligned} \quad (3.47)$$

in Eq. (3.46) and collecting terms of equal amplitude gives

$$\begin{aligned} x(t) &= a(t)\left\{\frac{\alpha+\beta}{2}[\cos\varphi(t) + j\sin\varphi(t)] + \frac{\alpha-\beta}{2}[\cos\varphi(t) - j\sin\varphi(t)]\right\} + (\gamma + j\kappa) \\ &= a(t)\left\{\frac{\alpha+\beta}{2}\exp[+j\varphi(t)] + \frac{\alpha-\beta}{2}\exp[-j\varphi(t)]\right\} + (\gamma + j\kappa) \end{aligned} \quad (3.48)$$

Equation (3.48) shows that in the presence of amplitude or phase errors, the complex signal $x(t)$ will not only contain the desired signal component $a(t)\exp[j\varphi(t)]$ (with a slightly modified amplitude) but also an *image* component with a different amplitude and a conjugated phase function, as well as a complex DC term. The image component is an error resulting from the amplitude and phase mismatches; the DC component is the direct result of the individual channel DC offsets.

Recall that the phase function $\exp[j\varphi(t)]$ can represent phase modulation of the radar waveform, the effect of the environment on the waveform (such as a phase shift due to spatial Doppler), or both. In the case of a spatial Doppler phase shift, $\varphi(t)$ on the mth pulse will be of the form $\omega_D m$ for some normalized Doppler radian frequency ω_D. The image component will then have a phase shift of the form $-\omega_D m$. Thus, over a series of M pulses the mismatches will give rise to a false signal at the negative of each actual Doppler frequency component in addition to the desired signal. Furthermore, the DC component is equivalent to a false signal at a Doppler shift of zero, that is, clutter or a stationary target.

As another example, suppose $\varphi(t)$ represents the intentional quadratic phase modulation used to construct a linear FM chirp signal, $\varphi(t) = \alpha t^2$. Then the image component will have a phase modulation of $-\alpha t^2$, which represents a linear frequency modulation (FM) signal with a slope opposite to the transmitted pulse. This signal will not be properly compressed by the matched filter, instead causing an apparent increase in the noise floor (Sinsky and Wang, 1974).

To judge the significance of the gain and phase mismatch errors, consider the ratio P_r of the power in the image component relative to that in the desired component. From Eq. (3.48), this is

$$P_r = \frac{|(\alpha - \beta)/2|^2}{|(\alpha + \beta)/2|^2}$$
$$= \frac{[1 - (1+\varepsilon)\cos\psi]^2 + [(1+\varepsilon)\sin\psi]^2}{[1 + (1+\varepsilon)\cos\psi]^2 + [(1+\varepsilon)\sin\psi]^2} \quad (3.49)$$

Figure 3.30 illustrates the value of P_r as a function of the phase and amplitude imbalance.

It is also useful to consider simplifications of Eq. (3.49) for the cases of small amplitude mismatch only and small phase mismatch only. First consider the case of small amplitude

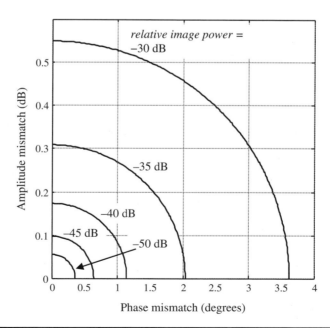

Figure 3.30 Relative power of I/Q mismatch-induced signal image as a function of amount of phase and amplitude mismatch.

mismatch only, so that $\psi = 0$ but $\varepsilon \ll 1$. Then

$$P_r = \frac{[1-(1+\varepsilon)]^2}{[1+(1+\varepsilon)]^2} = \frac{\varepsilon^2}{(2+\varepsilon)^2} \qquad (3.50)$$
$$\approx \frac{\varepsilon^2}{4}$$

Amplitude mismatch is often specified in decibels, as is the relative power of the image signal component. A mismatch of k dB implies that $20 \cdot \log_{10}(1+\varepsilon) = k$. Substituting this relation in Eq. (3.50) and expressing the result in decibels gives

$$P_r(\text{dB}) = 20\log_{10}(10^{k/20} - 1) - 6.02 \quad (k \text{ in dB}) \qquad (3.51)$$

For example, an amplitude mismatch of 0.1 dB gives rise to an image component 44.7 dB below the desired component of $x(t)$.

A result similar to Eq. (3.50) holds for the case of small phase mismatch only, that is, $\varepsilon = 0$ and $\psi \ll 1$. In this case, Eq. (3.49) reduces to

$$P_r = \frac{[1-\cos\psi]^2 + \sin^2\psi}{[1+\cos\psi]^2 + \sin^2\psi} = \frac{1-\cos\psi}{1+\cos\psi} \qquad (3.52)$$
$$\approx \frac{\frac{1}{2}\psi^2}{2 + \frac{1}{2}\psi^2} \approx \frac{\psi^2}{4}$$

where the second line is obtained using the small angle approximation $\cos\phi \approx 1 - \phi^2/2$. Note that ϕ is in radians. The relative power of the image component in decibels is then

$$P_r(\text{dB}) = 20\log_{10}(\psi) - 6.02 \quad (\psi \text{ in radians}) \qquad (3.53)$$

As an example, a phase mismatch of 1° (17.5 mrad) gives an image component approximately 41.2 dB below the desired response.

3.8.2 Correcting I/Q Errors

As shown in Fig. 3.29, the I and Q signals in the presence of mismatch can be modeled as

$$I(t) = a(t)\cos\varphi(t) + \gamma, \qquad Q(t) = a(t)(1+\varepsilon)\sin(\varphi(t) - \psi) + \kappa \qquad (3.54)$$

The desired in-phase signal $I(t)$ is $a(t)\cos\varphi(t)$, and in the quadrature channel $Q(t)$ is $a(t)\sin\varphi(t)$. Is it possible to recover the desired outputs from the available measurements of Eq. (3.54)?

Consider forming new signals I' and Q' as a linear combination of the measured I and Q. Specifically, require that $I' = a(t)\cos\varphi(t)$ and $Q' = a(t)\sin\varphi(t)$. Although it is straightforward to solve the general problem, it is obvious that the DC offsets should simply be subtracted. Suppressing the t dependence of $a(t)$ and $\varphi(t)$ for brevity, form a linear combination of the zero-offset data:

$$\begin{bmatrix} I' \\ Q' \end{bmatrix} = \begin{bmatrix} a\cos\varphi \\ a\sin\varphi \end{bmatrix} = \begin{bmatrix} a_{11} & a_{12} \\ a_{21} & a_{22} \end{bmatrix} \left[\begin{bmatrix} I \\ Q \end{bmatrix} - \begin{bmatrix} \gamma \\ \kappa \end{bmatrix} \right] \qquad (3.55)$$
$$= \begin{bmatrix} a_{11} & a_{12} \\ a_{21} & a_{22} \end{bmatrix} \begin{bmatrix} a\cos\varphi \\ a(1+\varepsilon)\sin(\varphi - \phi) \end{bmatrix}$$

By inspection, $a_{11} = 1$ and $a_{12} = 0$. The remaining equation is

$$Q' = a\sin\varphi = a_{21}I + a_{22}Q$$
$$= a_{21}a\cos\varphi + a_{22}a(1+\varepsilon)\sin(\varphi - \psi) \qquad (3.56)$$

Applying a trigonometric identity for $\sin[\varphi - \psi]$ and equating terms in $\sin\varphi$ and $\cos\varphi$ on both sides of Eq. (3.56) leads to the following solution for a_{21} and a_{22}:

$$a_{21} = \tan\psi, \quad a_{22} = \frac{1}{(1+\varepsilon)\cos\psi} \qquad (3.57)$$

Using Eq. (3.57) in Eq. (3.55) gives the final transformation required

$$\begin{bmatrix} I' \\ Q' \end{bmatrix} = \begin{bmatrix} 1 & 0 \\ \tan\psi & 1/(1+\varepsilon)\cos\psi \end{bmatrix} \left[\begin{bmatrix} I \\ Q \end{bmatrix} - \begin{bmatrix} \gamma \\ \kappa \end{bmatrix} \right] \qquad (3.58)$$

Once the I/Q errors ε, ψ, γ, and κ are determined, Eq. (3.58) can be used to compute a new value Q' for the quadrature channel sample for each measured I-Q sample pair. The difficulty, of course, is in actually determining the errors; the correction is then easy. The errors are generally estimated by injecting a known *pilot signal*, usually a pure sinusoid, into the receiver and observing the outputs. Details for one specific technique to estimate gain and phase errors are given in Churchill et al. (1981); that paper also derives limits to mismatch correction (and thus to image suppression) caused by noise, which introduces errors into the estimates of ε and ψ.

A second method for eliminating I/Q error is based on the idea of transmitting multiple pulses or sweeps, stepping the starting phase of each pulse through a series of evenly spaced values, and then integrating the measured range profiles. To see how this technique works, suppose the input signal in Fig. 3.29 is changed to $a(t)\sin[\Omega t + \varphi(t) + k(2\pi/N)]$ for some fixed integer N and variable integer k; that is, the pulse is one of a series of N pulses, where the initial phase is increased by $2\pi/N$ radians on each successive pulse. The extra phase shift propagates to the output signals (still suppressing the t dependence of φ)

$$\begin{aligned} I_k &= a\cos\left[\varphi + k\frac{2\pi}{N}\right] + \gamma \\ Q_k &= a(1+\varepsilon)\sin\left[\varphi + k\frac{2\pi}{N} - \psi\right] + \kappa \end{aligned} \qquad (3.59)$$

for $k = 0, 1, \ldots, N-1$. The development leading to Eq. (3.48) can be repeated to obtain the complex signal for this case, which is

$$x_k(t) = a\left\{\frac{\alpha+\beta}{2}\exp\left(j\left[\varphi + \frac{2\pi k}{N}\right]\right) + \cdots \right. \\ \left. \cdots + \frac{\alpha-\beta}{2}\exp\left(-j\left[\varphi + \frac{2\pi k}{N}\right]\right)\right\} + (\gamma + j\kappa) \qquad (3.60)$$

Now coherently integrate the N RPs $x_k(t)$ to form a single composite measurement, applying a counter phase rotation to each to realign their phases:

$$\begin{aligned} x(t) &= \frac{1}{N}\sum_{k=0}^{N-1} x_k(t)\exp(-jk2\pi/N) \\ &= \frac{a}{N}\frac{\alpha+\beta}{2}\sum_{k=0}^{N-1}\exp(j\varphi) + \frac{a}{N}\frac{\alpha-\beta}{2}\sum_{k=0}^{N-1}\exp(-j\varphi)\exp\left(-j\frac{4\pi k}{N}\right) + \cdots \\ &\quad + \frac{(\gamma+j\kappa)}{N}\sum_{k=0}^{N-1}\exp\left(-j\frac{2\pi k}{N}\right) \\ &= a\frac{\alpha+\beta}{2}\exp(j\varphi) + a\frac{\alpha-\beta}{2}\exp(-j\varphi)\sum_{k=0}^{N-1}\exp\left(-j\frac{4\pi k}{N}\right) + \cdots \\ &\quad + \frac{(\gamma+j\kappa)}{N}\sum_{k=0}^{N-1}\exp\left(-j\frac{2\pi k}{N}\right) \end{aligned} \qquad (3.61)$$

The summations in the middle and last terms of the last form of Eq. (3.61) can be evaluated in closed form to give

$$\sum_{k=0}^{N-1} \exp\left(-j\frac{4\pi k}{N}\right) = \begin{cases} N, & N=1,2 \\ 0, & N \geq 3 \end{cases} \quad (3.62)$$

and

$$\sum_{k=0}^{N-1} \exp\left(-j\frac{2\pi k}{N}\right) = \begin{cases} 1, & N=1 \\ 0, & N \geq 2 \end{cases} \quad (3.63)$$

so that (restoring the t dependency)

$$x(t) = a(t)\frac{\alpha+\beta}{2}\exp[j\varphi(t)] \quad (N \geq 3) \quad (3.64)$$

Thus, as long as at least three pulses or sweeps are used, the process of rotating the transmitted phase, compensating the received RPs, and integrating will suppress both the undesired image component and the DC component.

The algebraic correction technique of Eq. (3.58) is applied to individual I/Q sample pairs, requiring two real multiplies and three real additions per time sample (assuming the correction coefficients have been precomputed). The major advantage of this technique is that it can be applied individually to each sample of data. Its major disadvantage is that it requires that transmitter/receiver control and analog hardware be augmented to allow pilot signal insertion for determining the correction coefficients. The pilot signal operation is performed relatively infrequently on the assumption that ε and ψ vary only slowly.

The phase rotation and integration technique of Eqs. (3.59) to (3.61), in contrast, requires integration of at least three pulses or sweeps with the transmitted phase adjusted for each. Thus, the technique requires both high-speed transmitter phase control and more time to complete a measurement since multiple RPs must be collected. The increase in measurement time requires that the scene not vary during multiple repetition intervals; decorrelation of the scene degrades the effectiveness of the technique. This method also places a heavier load on the signal processor since the integration requires N complex multiplies and $N-1$ complex additions per fast-time sample, or a total of $4N$ real multiplies and $4N-2$ real additions, with $N \geq 3$. However, the integration method has one very important advantage: it does not require knowledge of any of the errors ε, ϕ, γ, and κ. It also has the side benefit that the integration of multiple RPs increases the signal-to-noise ratio of the final result $x(t)$. Given these considerations, it is often used in instrumentation systems at fixed site installations such as turntable RCS measurement facilities. In these systems, N is often on the order of 16 to 64 and may even be as high as 65,536 (64K) in some cases.

Note also that Eqs. (3.59) to (3.61) implicitly assume that the phase modulation $\varphi(t)$ is the same for each pulse or sweep $x_k(t)$. If $\varphi(t)$ represents waveform modulation (e.g., a linear FM chirp), this will be true; but if $\varphi(t)$ contains a term representing environmental phase modulation, for example due to Doppler shift, then the technique assumes that the appropriate component of $\varphi(t)$ is the same on each of the pulses integrated. This is the case for stationary targets (assuming the radar is also stationary). For constant Doppler targets, the frequency implied by $\varphi(t)$ will be the same from one RP to the next, but the absolute phase will change in general, so that the target response does not integrate properly. For accelerating targets, the assumption will fail entirely. The phase rotation and integration technique is therefore most appropriate for stationary or nearly stationary (over N PRIs) targets. The algebraic technique does not have this limitation, since it operates on individual samples only.

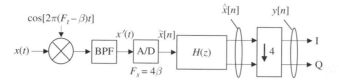

FIGURE 3.31 Architecture of Rader's system for digital generation of in-phase and quadrature signals. (After Rader, 1984.)

3.8.3 Digital I/Q

Digital I/Q or *digital IF* is the name given to a collection of techniques that form the I and Q signals digitally in order to overcome the channel matching limitations of analog receivers. Many variations have been described in the literature. In general, they all share two characteristics. First, they use analog mixing and filtering to shift the single real-valued input signal to a low intermediate frequency prior to A/D conversion, greatly relaxing the A/D speed requirements compared to RF sampling. Furthermore, the intermediate frequency (IF) is chosen so that required complex multiplications by functions of the form $\exp(j\omega_0 n)$ reduce to particularly simple forms, saving computation. Second, they use a combination of digital filtering and down sampling to obtain a final output consisting only of the desired sideband of the original spectrum, sampled at or near the appropriate Nyquist rate of β complex samples per second. Two approaches are briefly described here.

The first method, which is particularly elegant, is described in Rader (1984). The RF signal is assumed to have a bandpass spectrum with a total information bandwidth of β Hz. Figure 3.31 is a block diagram of the system, and Fig. 3.32 sketches the signal spectrum at various stages. The first step is an analog frequency shifting operation that translates this spectrum to a low IF of β Hz. The bandpass filter rejects the double frequency terms created by the mixer. The spectrum is therefore bandlimited to $\pm 3\beta/2$ Hz, so the Nyquist rate is 3β samples per second. However, for reasons that will become clear shortly a higher sampling rate of 4β samples per second is used, giving a discrete-time signal with the spectrum shown in Fig. 3.32c.

Recall that the goal of quadrature demodulation is to select one sideband of the bandpass signal and translate it to baseband. Assume that the upper sideband is to be retained. The next step is therefore to filter the real signal $\tilde{x}[n]$ to eliminate the lower sideband. Since the resulting spectrum will not be Hermitian, the output signal must be complex; this is shown in Fig. 3.31 as a one-input, two-output filter. The required frequency response is clear from the spectrum diagrams in Fig. 3.32; it is

$$H(\omega) = \begin{cases} 1, & \frac{\pi}{4} < \omega < \frac{3\pi}{4}, \\ 0, & -\frac{3\pi}{4} < \omega < -\frac{\pi}{4}, \\ \text{don't care}, & \text{otherwise} \end{cases} \quad (3.65)$$

This asymmetric filter frequency response corresponds to a complex-valued impulse response, giving rise to the complex output from the single real input.

While Eq. (3.65) states that the value of $H(\omega)$ around DC is unconstrained, in fact it should be close to zero. The filter will then also suppress any DC component in the

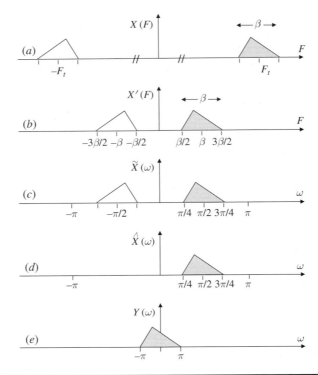

Figure 3.32 Spectra corresponding to successive signals in the digital I/Q system of Fig. 3.30. (a) Spectrum of bandpass input signal with information bandwidth β Hz. (b) Result of translation to an IF frequency also equal to β. (c) One period of spectrum on normalized frequency scale after A/D conversion. (d) Only the upper sideband remains after filtering. (e) A replica of the upper sideband is centered at DC after decimation. (After Rader, 1984.)

signal (not sketched in Fig. 3.32) that may have been introduced by nonideal mixing in the first analog frequency translation. Thus, this digital I/Q architecture also makes it easier to suppress mixer bias terms. This would not be possible if the spectrum had been translated to the lowest possible IF frequency, namely β Hz, since there would then be no region of the spectrum around DC that did not contain signal components of interest.

A particularly efficient design for realizing the filter $H(z)$ as a pair of low-order recursive filters is based on the mathematics of phase-splitting networks; details are given in Rader (1984). However, the particular design of the filters is not central to the architecture of the approach.

The final step is to translate the remaining spectral sideband centered at $\omega_0 = \pi/2$ to baseband and to reduce the sampling rate from 4β to the final Nyquist rate β. This can be accomplished by multiplying the complex filter output $\hat{x}[n]$ by the sequence $\exp[-j(\pi/2)n] = (-j)^n$ and then simply discarding three of every four samples. Because of the special form of the multipliers, the complex multiplications could be implemented simply with sign changes and interchanges of real and imaginary parts, rather than with actual complex multiplications. This is a consequence of having selected the original sampling rate to be 4β instead of 3β.

However, this multiplication is not shown in Fig. 3.31 because, in fact, it is not necessary at all. The spectrum of the decimator output y[n] is related to the spectrum of $\hat{x}[n]$ according to (Oppenheim and Schafer, 2010):

$$Y(\omega) = \frac{1}{4}\sum_{k=0}^{3}\hat{X}\left(\frac{\omega - 2\pi k}{4}\right) = \frac{1}{4}\sum_{k=0}^{3}\hat{X}\left(\frac{\omega}{4} - k\frac{\pi}{2}\right) \qquad (3.66)$$

Equation (3.66) states that the decimation process causes the spectrum to replicate at intervals of $\pi/2$ radians. Since the nonzero portion of the spectrum is bandlimited to $\pi/2$ radians, these replications abut but do not alias; furthermore, since the spectrum prior to decimation is centered at $\omega = \pi/2$, one of the replications ($k = 3$, specifically) is centered at $\omega = 2\pi$ radians. The periodicity of the spectrum of a discrete-time signal therefore guarantees that there is a replica centered at $\omega = 0$ as well; this replica is the final desired spectrum. Thus, the real and imaginary outputs of the decimator are the desired I and Q signals. Another digital I/Q system that uses the spectrum replicating properties of decimation to advantage is described in Rice and Wu (1982).

The success of the decimation operation in eliminating the need for a final complex frequency translation depended on the proper relationship between the bandwidth and center frequency of the signal, and the decimation factor. This is the major reason for choosing the IF to be β instead of $\beta/2$ (or some other permissible value), and the sampling frequency as 4β instead of 3β (or some other value).

Rader's digital I/Q architecture has reduced the number of analog signal channels from two to one, making the issues of oscillator quadrature and gain and phase matching completely moot, while also providing a natural opportunity to filter out DC biases introduced by the remaining analog mixer. Furthermore, the two A/D converters required at the output of the conventional quadrature receiver to enable subsequent digital processing have been reduced to one. There are two major costs to these improvements. The first is an increase by a factor of four in the A/D converter speed requirement, from β samples per second for conventional baseband sampling to 4β samples per second for Rader's system; this may be difficult at higher radar signal bandwidths but becomes less so as A/D converter technology advances. The second is the introduction of the need for computationally expensive high-rate digital filtering (although Rader's efficient filter design lessens this cost).

Figures 3.33 and 3.34 sketch the processor conceptual architecture and the relevant signals of the second digital I/Q architecture (Shaw and Pohlig, 1995). In this case, analog frequency translation is used to shift the signal spectrum to a lower IF than used by Rader, namely 0.625β. The signal is then A/D converted at a rate of 2.5β samples per second, resulting in the signal $\hat{x}[n]$ having a spectrum centered at $\omega = \pi/2$, as shown in Fig. 3.34. An explicit complex modulation by $\exp(+j\pi n/2) = j^n$ then shifts one of the sidebands, in this case the lower one, to baseband, resulting in the spectrum shown in Fig. 3.34d. Clearly $\bar{x}[n]$ is complex as a result of this complex modulation.

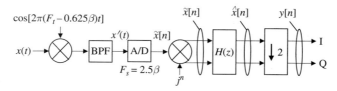

FIGURE 3.33 Conceptual architecture of Lincoln Laboratory system for digital generation of in-phase and quadrature signals. (After Shaw and Pohlig, 1995.)

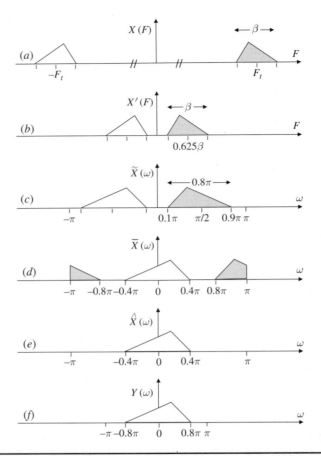

FIGURE 3.34 Spectra corresponding to successive signals in the digital I/Q system of Fig. 3.32. (a) Spectrum of bandpass input signal with information bandwidth of β Hz. (b) Result of translation to an IF frequency of 0.625β. (c) One period of the spectrum on the normalized frequency scale after A/D conversion. (d) Digital complex modulation centers the lower sideband at DC. (e) Only the lower sideband remains after lowpass filtering. (f) Decimation by two reduces the sampling rate to 1.25β. (After Shaw and Pohlig, 1995.)

The next step is to lowpass filter $\tilde{x}[n]$ to remove the upper sideband, leaving only the baseband portion of the spectrum. A 16-point finite impulse response (FIR) digital filter is used for this task in Shaw and Pohlig (1995). Once the lowpass filtering is completed, the spectrum is nonzero only for $\omega \in (-0.4\pi, +0.4\pi)$. The sampling rate is then reduced by a factor of two by discarding every other output sample. The final result is the desired digital I and Q signals, sampled at a rate of 1.25β samples per second.

As with Rader's system, the computational complexity is actually reduced by taking advantage of the properties of decimation and FIR filters. The decimation is performed immediately after the A/D conversion by splitting the data into even- and odd-numbered sample streams. The complex modulation by j^n, which implies both sign changes and real/imaginary interchanges, then reduces only to sign changes on every other sample in each channel, and the 16-point FIR filters are replaced with 8-point FIR filters in each channel without any reduction in filtering quality.

A significant advantage of this system over Rader's is that the A/D converter must operate at only 2.5 times the signal information bandwidth, rather than four times the bandwidth. This is an important savings at higher radar bandwidths. There are three disadvantages. The first is that the lower IF and sampling rate require sharper transitions in the digital filter, increasing the filter order necessary to achieve a given stopband suppression and thus the computational complexity of the filter. The second is the requirement for an explicit multiplication by j^n. Although this reduces to switching and sign changes, it nonetheless represents extra processing. Third, the final sampling rate exceeds the signal Nyquist rate by 25 percent, whereas in Rader's system it equaled the Nyquist rate. This increases the computational load by 25 percent over the minimum necessary throughout the remainder of the digital processing. This may not be a problem in practice. Sampling rates are usually set somewhat above Nyquist rates anyway to provide a margin of safety, since real signals are never perfectly bandlimited.

Two other details merit mention. It may appear that modulating the sideband to baseband before filtering eliminates the possibility of using the digital filter to suppress DC bias errors from the analog mixer. However, that same modulation will move any DC term contributed by the mixer to $\omega = \pi/2$, where it can still be removed by the lowpass filter. Finally, in the Rader system the I and Q signals were derived from the upper sideband of the original bandpass signal, while in the Shaw and Pohlig system the lower sideband was used. Because the original signal was real valued, its spectrum was Hermitian, and consequently the spectra of the complex outputs of the two systems, say $Y_1(\omega)$ and $Y_2(\omega)$, are related according to $Y_2(\omega) = Y_1^*(-\omega)$ so that $y_2[n] = y_1^*[n]$. Clearly, either system could be modified to use the opposite sideband.

3.9 Summary

Many modern radars, both pulsed and FMCW, organize collected data into coherent processing intervals or datacubes. This text concentrates on such systems. Development of a datacube begins with acquisition of a range profile using a single pulse (pulsed radar) or sweep (linear FMCW radar). A range profile comprises a vector of samples of the complex (I and Q) voltage at the receiver output. Multiple RPs are assembled to form a CPI. If the radar antenna has multiple output channels, typically the case for a phased array antenna, a CPI is formed for each channel. These are then assembled into the radar datacube representing one CPI or dwell of data.

The method for acquiring an RP is very different for pulsed and FMCW radars. Pulsed systems are very simple: a pulse is transmitted and the echo is sampled at regular intervals to form the RP. FMCW systems are more complex, requiring a deramp operation followed by spectrum analysis of the beat frequency signal to form the range profile. Once the RP is formed, however, construction of a CPI and a datacube is the same for both types of radar.

The Nyquist sampling theorem can be applied in all three datacube dimensions: range (fast time), pulse or sweep number (slow time), and antenna channel. The fast-time sampling rate is determined by the pulse bandwidth for a pulsed system, and by the swept bandwidth and the maximum range of interest in FMCW systems. The pulse or sweep repetition rate is the slow time sampling interval and is determined by the Doppler bandwidth of the data, among other considerations. The spacing of antenna channel phase centers is determined by the range of signal angles of arrival of interest. The angular sampling density for a scanning antenna is determined by the aperture size in wavelengths.

Parameter	Pulsed	Linear Sawtooth FMCW	Comments
Minimum range R_{min}	$\dfrac{c\tau}{2}$ m	0 m	Implementation limitations such as phase noise will result in non-zero R_{min} in practical FMCW
Unambiguous range	$R_{ua} = \dfrac{cT}{2}$ m	$R_{ua1} = \dfrac{cT}{2}$ m, $R_{ua2} = \left(\dfrac{cT}{2}\right)\left(\dfrac{F_s}{\beta}\right)$ m	FMCW has two sources of range ambiguity: the sweep repetition interval and the fast-time Nyquist sampling
Doppler resolution	$\Delta F_D = \dfrac{1}{MT}$ Hz $\Delta v = \dfrac{\lambda_t}{2MT}$ m/s	$\Delta F_D = \dfrac{1}{MT}$ Hz $\Delta v = \dfrac{\lambda_t}{2MT}$ m/s	
Unambiguous Doppler interval	$F_{Dua} = RR = \dfrac{1}{T}$ Hz $v_{ua} = \dfrac{\lambda_t}{2} RR = \dfrac{\lambda_t}{2T}$ m/s	$F_{Dua} = RR = \dfrac{1}{T}$ Hz $v_{ua} = \dfrac{\lambda_t}{2} RR = \dfrac{\lambda_t}{2T}$ m/s	

TABLE 3.2 Comparison of Multipulse or Multisweep Parameters

The periodicity of the waveforms in a typical CPI or dwell creates a number of secondary problems. These include a minimum range for pulsed radars, range and velocity ambiguities for both pulsed and FMCW, and blind zones in both range and Doppler for both types of systems. Table 3.2 collects and compares the formulas for these metrics. Means of mitigating some of these limitations will be introduced in later chapters.

Practical receivers suffer imperfections in collecting the raw data and developing the I and Q signal due to the difficulty of matching the two channels. These channel mismatches can result in the appearance of false targets in the radar data. Techniques for compensating for these errors exist; two were described here. Another approach to avoiding mismatch errors is to digitize the data at a high RF or IF and construct the I and Q signals digitally, making moot the issue of matching channels since there is then only one. Two efficient methods for this "digital I/Q" process make clever choices of IF frequencies to simplify decimation and modulation; one also makes clever use of the spectrum replication due to sampling to avoid demodulation.

References

Bailey, C. D., "Radar Antennas," Chap. 9 in M. A. Richards, J. A. Scheer, and W. A. Holm (eds.), *Principles of Modern Radar: Basic Principles*. SciTech Publishing, Raleigh, NC, 2010.

Baker, C., and S. O. Piper, "Continuous Wave Radar," Chap. 2 in W. L. Melvin and J. A. Scheer (eds.), *Principles of Modern Radar: Radar Applications*. SciTech Publishing, Raleigh, NC, 2013.

Balanis, C. A., *Antenna Theory*, 4th ed. Harper & Row, New York, 2015.

Bracewell, R. N., *The Fourier Transform and Its Applications*, 3d ed. McGraw-Hill, New York, 1999.

Bruder, J. A., "Radar Receivers," Chap. 10 in M. A. Richards, J. A. Scheer, and W. A. Holm (eds.), *Principles of Modern Radar: Basic Principles*. SciTech Publishing, Raleigh, NC, 2010.

Churchill, F. E., G. W. Ogar, and B. J. Thompson, "The Correction of I and Q Errors in a Coherent Processor," *IEEE Transactions on Aerospace and Electronic Systems*, vol. 17, no. 1, pp. 131–137, Jan. 1981. See also "Corrections to 'The Correction of I and Q Errors in a Coherent Processor,'" *IEEE Transactions on Aerospace and Electronic Systems*, vol. 17, no. 2, p. 312, Mar. 1981.

Harris, F. J., "On the Use of Windows for Harmonic Analysis with the Discrete Fourier Transform," *Proceedings of the IEEE*, vol. 66, no. 1, pp. 51-83, Jan. 1978.

IEEE Standard Radar Definitions. IEEE Standard 686-2017, Sep. 13, 2017, New York.

Oppenheim, A. V., and R. W. Schafer, *Discrete-Time Signal Processing*, 3d ed. Prentice Hall, Englewood Cliffs, NJ, 2010.

Rader, C. M., "A Simple Method for Sampling In-Phase and Quadrature Components," *IEEE Transactions on Aerospace and Electronic Systems*, vol. 20, no. 6, pp. 821–824, Nov. 1984.

Rice, D. W., and K. H. Wu, "Quadrature Sampling with High Dynamic Range," *IEEE Transactions on Aerospace and Electronic Systems*, vol. 18, no. 4, pp. 736–739, Nov. 1982.

Rohling, H., and C. Möller, "Radar Waveform for Automotive Radar Systems and Applications," *Proceedings 2008 IEEE Radar Conference*, Rome, 2008, pp. 1–4.

Shaw, G. A., and S. C. Pohlig, "I/Q Baseband Demodulation in the RASSP SAR Benchmark," Project Report RASSP-4, Massachusetts Institute of Technology Lincoln Laboratory, Aug. 24, 1995.

Sinsky, A. I., and P. C. P. Wang, "Error Analysis of a Quadrature Coherent Detector Processor," *IEEE Transactions on Aerospace and Electronic Systems*, vol. 10, no. 6, pp. 880–883, Nov. 1974.

Problems

1. Compute the minimum range for a pulsed radar using pulse lengths of 1 ns, 1 μs, and 1 ms.

2. A radar transmits a series of 10 μs long pulses at a PRI of 100 μs. Determine the maximum and minimum target range such that at least a portion of the echo from one pulse will arrive back at the receiver during the transmission of the next pulse. Targets in this range interval will be completely or partially eclipsed. What target range produces a completely eclipsed echo?

3. Suppose a radar has a pulse length of 100 ns. What is the Rayleigh bandwidth of the pulse spectrum, in Hz? What is the 3-dB bandwidth in Hz?

4. Consider two signals: a square pulse $x_1(t)$ of length τ seconds, and a triangle $x_2(t)$ of length 2τ seconds obtained by convolving the square pulse with itself, $x_2(t) = x_1(t) * x_1(t)$. What is the relationship between the two spectra $X_1(F)$ and $X_2(F)$? Determine the relationship between the 3-dB bandwidths of $X_1(F)$ and $X_2(F)$. How does this compare to the relationship between the Rayleigh bandwidths of $X_1(F)$ and $X_2(F)$?

5. Consider a sample complex sinusoidal signal $y[m] = A\exp(j\omega_0 m)$, $m = 0,\ldots, M-1$. Show that increasingly negative values of w_0 will cause the peak at the output of the IDFT to occur at increasingly positive indices k.

6. Equations (3.6) and (3.7) gave the beat frequencies observed in one cycle of the sawtooth linear FMCW waveform for a stationary target. How will these beat

frequencies be changed if the target has a constant velocity of v_r m/s toward the radar? Will it still be possible to measure target range using only the beat frequency?

7. Continuing the previous problem, suppose the primary purpose of the system is to measure range of both stationary and moving targets; velocity is not a concern. For a given bandwidth, which will result in more accurate range measurements, a faster sweep (smaller T) or a slower sweep (larger T)? Explain.

8. The example FMCW system at the end of Sec. 3.2 had a bandwidth of $\beta = 100$ MHz, an RI of $T = 10$ μs, a maximum target delay of $t_{max} = 1$ μs, and a Nyquist sampling rate of $F_N = 10$ MHz. Assume that the system does not filter out beat frequencies above F_N. Sketch the apparent target range that will be observed as the actual target range increases from zero to 2 km.

9. Consider two RF pulses at frequencies of 5.0 GHz and 5.01 GHz. Assume two pulses are resolvable in frequency if their center frequencies are separated by at least the Rayleigh resolution of the individual pulses. What is the minimum pulse length required so that the two pulses could be resolved in frequency?

10. A finite pulse train waveform is composed of 20 pulses, each of 10 μs length and separated by a PRI of 1 ms. What is the coherent processing interval for this waveform?

11. Consider an X-band (10 GHz) radar on an aircraft traveling at 100 m/s. Assume the 3-dB azimuth beamwidth of the antenna is 3°. Compute the Doppler shift F_D of a scatterer on the antenna boresight and the Doppler bandwidth β_D across the beam for squint angles of $\psi = 0°, 30°, 60°$, and $90°$.

12. Consider a radar with a PRF of 5 kHz. What is the maximum unambiguous range R_{ua} of this radar in kilometers? If a target is located at a range of 50 miles, how many pulses will the radar have transmitted before the first echo from the target arrives? What will be the apparent range of the target in kilometers?

13. A radar has a repetition interval of 100 μs. The sensitivity of the radar is such that it could plausibly detect a target up to a maximum range of 50 km. A target is detected at an apparent range $R_a = 2$ km. What are the possible actual ranges of the target?

14. Continuing the previous problem, suppose the measurement is repeated at an RI of 30 μs. The target is now detected at an apparent range $R_a = 3.5$ km. What is the actual range of the target? (This problem foreshadows a method for resolving range ambiguities to be discussed in Chap. 5.)

15. A 77-GHz sawtooth FMCW automotive radar observes a target approaching. The velocity of the car and target are such that the closing speed is 35 m/s (about 78 mph). The radar's repetition rate is 10,000 sweeps per second. The Doppler spectrum observed by the radar is interpreted as occupying the interval $(-F_{Dua}/2, +F_{Dua}/2]$ Hz. What will be the apparent Doppler shift in hertz of the target? Will the target appear to be approaching or receding from the car?

16. Assuming a sidelooking radar ($\psi = 90°$) and Nyquist sampling in slow time, determine the relationship between the maximum unambiguous range R_{ua} and the antenna beamwidth θ_{az}. Ignore eclipsing. Using $\theta_{az} \approx \lambda/D_{az}$, determine the relationship between R_{ua} and the antenna azimuth dimension D_{az}.

17. Suppose a target is at a range of 10 km from a C-band (5-GHz) radar and has a radial velocity of +50 m/s with respect to the radar. Determine whether the target is ambiguous in range and velocity, and the apparent range and velocity R_a and v_a for

PRFs of 1, 10, and 100 kHz. The unambiguous velocity interval is considered to be $[-v_{ua}/2, +v_{ua}/2]$.

18. Suppose a C-band (5-GHz) radar has a fast-time sampling rate of 2M samples per second and a PRF of 5000 pulses per second. A CPI of data is to be collected that will cover a range interval of 30 km and support velocity resolution of 10 m/s. What are the dimensions of the data matrix for one CPI of data?

19. There is sometimes a concern as to whether a target will stay in the same range bin while during the time it takes to collect a CPI of data. Sometimes this is an issue, but often it is not. Using the PRF and number of pulses in the CPI from the previous problem, what is the total duration of the CPI in seconds? Consider a target moving at 100 m/s (about 224 mph). How far does the target move during one CPI, and how does that compare to the range bin size? What is the minimum target velocity in meters per second such that the target would move more than one range bin in range during the CPI?

20. Suppose N samples of time-domain data are collected at a sampling rate of F_s samples per second. The K-point DFT of the data is computed. Depending on the relative values of K and N, zero padding or data turning is used as required. Develop a formula for the spacing of the DFT bins in hertz.

21. Consider a sequence of 20 slow-time data samples collected at a PRF of 2 kHz. If a 1000-point DFT of this sequence is computed, what is the spacing between DFT frequency samples in hertz?

22. Derive a condition on the DFT size K similar to that of Eq. (3.33) for a maximum straddle loss of 1 dB. The result will depend on the value of M. Instead of solving the appropriate equations numerically, use the first two terms of the Taylor series for $\sin(x)$ to get a closed-form result. Figure 3.26 can be used as an approximate check on the result for the case $M = 100$.

23. Consider a search radar at 1 GHz (L band) with a rotating $D = 10$ m dish antenna. Suppose the beamwidth is $\theta_3 = 2°$. What is the antenna parameter k in Eq. (3.43)? What is the Nyquist sampling rate in degrees for this antenna? If the antenna rotates at a rate of one revolution every 6 seconds, what is the PRF required to achieve this angular sampling rate?

24. Compute the relative power ratio P_r in the image component of the output of an I/Q receiver when there is a simultaneous mismatch of 0.1 dB in gain and 1° in phase. Express the answer in dB. Use Fig. 3.30 to check the answer.

25. Consider a digital I/Q architecture similar to Rader's. Starting with the original signal spectrum of Fig. 3.32a, assume that the signal is demodulated from the original center frequency F_t to an IF of β Hz. What will be the minimum required sampling rate F_s of the real-valued data? Assuming this value for F_s and also that the sampling rate is reduced to the minimum possible without aliasing in the last step, sketch the complete set of spectra from the original analog spectrum $X(F)$ to the final discrete time spectrum $Y(\omega)$, similar to Fig. 3.32. Also show the required frequency response $H(\omega)$ of the digital filter, assuming the upper sideband is the one that is retained. Discuss the spectrum recentering step in going from the equivalent of Fig. 3.32d to the equivalent of Fig. 3.32e: can demodulation by decimation be used and, if not, why not? If multiplication by complex exponentials is used, do the multipliers assume a simplified form? Summarize the pros and cons of this system versus Rader's system.

CHAPTER 4
Radar Waveforms

4.1 Introduction

A pulsed radar transmits a waveform typically modeled as

$$\bar{x}(t) = a(t)\cos[\Omega_t t + \varphi(t)] \qquad (4.1)$$

The term "Ω" in the argument of the sine function is the radar frequency (RF) in radians per second. The term $a(t)$ represents amplitude modulation (AM) of the RF carrier. In a pulsed radar this is typically just a rectangular function that pulses the waveform on and off; more general pulse AM is rare in radar. In continuous wave radar, $a(t)$ is just a constant A. The term $\varphi(t)$ models any phase or frequency modulation of the carrier. It can be zero, a nonzero constant, or a nontrivial function. The overbar on $\bar{x}(t)$ denotes that the signal is on a carrier, that is, it has not yet been demodulated to baseband. As discussed in Chap. 1, the real-valued waveform of Eq. (4.1) is more conveniently modeled by its complex equivalent

$$\bar{x}(t) = a(t)\exp(j[\Omega_t t + \varphi(t)]) \qquad (4.2)$$

The portion of $\bar{x}(t)$ other than the carrier term, or equivalently the complex baseband signal after demodulation, is called the *complex envelope* of the waveform

$$x(t) = a(t)\exp[j\varphi(t)] \qquad (4.3)$$

It is this function that describes the amplitude and phase or frequency modulation applied to the RF carrier and is considered to be "the waveform" in this chapter.

A taxonomy of radar waveforms can be developed in several ways. Perhaps first is whether the waveform is continuous wave (CW) or pulsed; sometimes variations such as "interrupted CW" are defined as well (Baker and Piper, 2014). Pulsed waveforms can be defined based on a single pulse, or "the waveform" can be considered to be a multipulse burst. Both pulsed and CW waveforms can be further categorized based on the presence or absence of frequency or phase modulation (FM or PM). If present, pulse modulation may be intrapulse (applied to individual pulses), interpulse (applied across the pulses of a multipulse waveform), or both. Phase modulation can be biphase (two possible states) or polyphase (more than two phase states); frequency modulation can be linear or nonlinear. Intrapulse amplitude modulation may be used but usually is not.

Figure 4.1 illustrates three example waveform types common in pulsed radar. The "simple pulse" is a constant-amplitude burst at the RF frequency. The frequency of the *linear frequency modulated* (LFM) pulse increases at a constant rate during the time the pulse is on. LFM pulses can also have decreasing frequency during the pulse. The third example is a

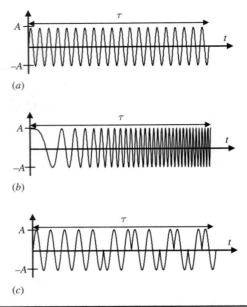

FIGURE 4.1 Examples of common pulsed radar waveforms. (a) Simple pulse. (b) L linear frequency modulated (LFM) pulse. (c) Binary phase-coded pulse.

binary phase-coded (biphase) pulse. In this waveform, the frequency is constant but the phase $\varphi(t)$ of the waveform changes between zero and π radians at several specific times within the pulse. Other types of modulation are also used. Examples include nonlinear frequency modulation (NLFM) and more complex phase sequences.

Similar options exist for continuous wave radars. A constant-frequency CW is equivalent to a simple pulse of infinite length. Of more interest in this text are periodically modulated CW radars, which continuously repeat a finite-duration modulated segment to form a continuous modulated waveform.[1] Like the pulsed case, the segment modulation could be LFM, a biphase code, or a more complex modulation.

The choice of waveform directly determines or is a major contributor to several fundamental radar system performance metrics. These include the signal-to-noise ratio (SNR) χ, the range resolution ΔR, the Doppler or velocity resolution ΔF_D or Δv, ambiguity intervals in range and Doppler, range and Doppler sidelobe magnitudes, and range-Doppler coupling. These metrics are determined by such waveform attributes as pulse duration or sweep length, bandwidth, amplitude, and phase or frequency modulation.

Two other important metrics of the radar receiver's response to the waveform are the *peak sidelobe level* (PSL) and *integrated sidelobe level* (ISL) in range and Doppler. The PSL is the ratio of the largest sidelobe of the receiver output to the peak of the mainlobe. Lower PSLs reduce target masking, the phenomenon where sidelobes of a strong target may prevent detection of a nearby weaker target. ISL is the ratio of the total energy in all sidelobes to that in the mainlobe. A low ISL indicates a waveform that admits less clutter interference power through the sidelobes.

While all of these metrics are discussed, the primary emphasis is on SNR, range resolution, and Doppler resolution because these are the most fundamental drivers in choosing the

---

[1]Periodically modulated CW waveforms are sometimes described as pulse trains where the pulse length τ equals the repetition interval T so that the successive pulses are contiguous, forming a continuous waveform.

waveform. As an example, the simple pulse of Fig. 4.1a has a duration of τ seconds and an amplitude of A volts. The SNR achieved will prove to be proportional to the waveform energy, which is the product $A^2\tau$ of its power and duration. The range resolution of $c\tau/2$ is proportional to the pulse length. It will be shown shortly that both the waveform bandwidth and the Doppler resolution of the simple pulse are inversely proportional to the pulse length.

Two classic references on radar waveforms are Cook and Bernfeld (1993) and Rihaczek (1996). Most radar system books cover the fundamentals of radar waveforms (e.g., Nathanson, 1991; Peebles, 1998). Good chapter-level surveys of basic and advanced waveforms are in Keel (2010) and Keel and Baden (2012), respectively. A thorough modern reference on radar waveforms is Levanon and Mozeson (2004). In addition to covering the mainstream waveforms such as pulse bursts and LFM, that text covers many developments in phase codes in recent decades. Another recent text that focuses more on advanced waveforms and emerging applications is Gini et al. (2012).

4.2 The Waveform Matched Filter

4.2.1 The Matched Filter

So far, it has been implicitly assumed that the overall frequency response of the radar receiver is a bandpass characteristic with a bandwidth equal to or greater than that of the transmitted signal. Equivalently, once the carrier is demodulated out, the effective frequency response is a low-pass filter with a bandwidth equal to that of the complex envelope. It will be shown in Chap. 6 that, unsurprisingly, detection performance improves with increasing SNR. Thus, it is reasonable to ask what overall receiver frequency response $H(\Omega)$ will maximize the SNR at its output.

The signal component of the spectrum of the receiver output $y(t)$ will be $Y(\Omega) = H(\Omega)X(\Omega)$, where $X(\Omega)$ is the spectrum of the transmitted waveform and therefore, except for a phase shift due to the overall delay which can be ignored here, of a received target echo. Consider maximizing the output SNR at a specific time T_M. The power of the signal component of the output at that instant is

$$|y(T_M)|^2 = \left| \frac{1}{2\pi} \int_{-\infty}^{\infty} X(\Omega) H(\Omega) \exp(j\Omega T_M) \, d\Omega \right|^2 \qquad (4.4)$$

To determine the output noise power, consider the case where the interference is white noise with power spectral density σ_w^2 W/Hz. The noise power spectral density (PSD) at the output of the receiver will be $\sigma_w^2 |H(\Omega)|^2$ W/Hz. The total output noise power is obtained by integrating the PSD over all frequency

$$n_p = \frac{\sigma_w^2}{2\pi} \int_{-\infty}^{\infty} |H(\Omega)|^2 \, d\Omega \qquad (4.5)$$

The SNR measured at time T_M is

$$\chi = \frac{|y(T_M)|^2}{n_p} = \frac{\left| \frac{1}{2\pi} \int_{-\infty}^{\infty} X(\Omega) H(\Omega) \exp(j\Omega T_M) \, d\Omega \right|^2}{\frac{\sigma_w^2}{2\pi} \int_{-\infty}^{\infty} |H(\Omega)|^2 \, d\Omega} \qquad (4.6)$$

Clearly χ depends on the receiver frequency response. The choice of $H(\Omega)$ that will maximize χ can be determined via the Schwarz inequality. One of its many forms is

$$\left|\int A(\Omega)B(\Omega)\,d\Omega\right|^2 \leq \left\{\int |A(\Omega)|^2\,d\Omega\right\}\left\{\int |B(\Omega)|^2\,d\Omega\right\} \qquad (4.7)$$

with equality if and only if $B(\Omega) = \alpha A^*(\Omega)$ for any arbitrary constant α. Identifying $A(\Omega) = X(\Omega)\exp(j\Omega T_M)$ and $B(\Omega) = H(\Omega)$ and applying Eq. (4.7) to the numerator of Eq. (4.6) gives the upper bound on SNR as

$$\chi \leq \frac{(1/2\pi)^2 \int_{-\infty}^{\infty} |X(\Omega)\exp(j\Omega T_M)|^2\,d\Omega \int_{-\infty}^{\infty} |H(\Omega)|^2\,d\Omega}{(\sigma_w^2/2\pi)\int_{-\infty}^{\infty} |H(\Omega)|^2\,d\Omega} \qquad (4.8)$$

The SNR is maximized when

$$\begin{aligned} H(\Omega) &= \alpha X^*(\Omega)\exp(-j\Omega T_M) \quad \text{or} \\ h(t) &= \alpha x^*(T_M - t) \end{aligned} \qquad (4.9)$$

This particular choice of the receiver filter frequency or impulse response is called the *matched filter*, because it is "matched" to the particular signal waveform. Thus, the waveform and the receiver filter needed to maximize the output SNR are a matched pair. If the radar changes waveforms, it must also change the receiver filter response in order to stay in a matched condition. The impulse response of the matched filter is obtained by time-reversing and conjugating the complex waveform. The gain constant α is usually set equal to unity; it has no impact on the achievable SNR, as seen shortly. The time T_M at which the SNR is maximized is arbitrary. However, $T_M \geq \tau$ is required for $h(t)$ to be causal.

Given a baseband input signal $x'(t)$ consisting of both target and noise components, the output of the matched filter is given by the convolution

$$\begin{aligned} y(t) &= \int_{-\infty}^{\infty} x'(s)h(t-s)\,ds \\ &= \alpha \int_{-\infty}^{\infty} x'(s)x^*(s + T_M - t)\,ds \end{aligned} \qquad (4.10)$$

The second line of Eq. (4.10) is recognized as the cross-correlation of the target-plus-noise signal $x'(t)$ with the transmitted waveform $x(t)$, evaluated at lag $T_M - t$. Thus, the matched filter implements a *correlator* with the transmitted waveform as the reference signal.

Two details of the matched filter output signal bear notice. First, the energy E in the waveform $x(t)$ is

$$E = \int_{-\infty}^{\infty} |x(t)|^2\,dt = \frac{1}{2\pi}\int_{-\infty}^{\infty} |X(\Omega)|^2\,d\Omega \qquad (4.11)$$

where the second version follows from Parseval's relation. It follows that the signal component of the filter output at the peak time T_M is the signal energy scaled by the filter constant α,

$$y(T_M) = \int x(s)\alpha x^*(s)\,ds = \alpha \int |x(s)|^2\,ds = \alpha E \qquad (4.12)$$

Second, the duration of the signal component of the matched filter output is exactly 2τ seconds, since it is the convolution of the τ-second pulse with the τ-second matched filter impulse response.

What is the maximum value of SNR achieved by the matched filter? Using $H(\Omega) = \alpha X^*(\Omega)\exp(-j\Omega T_M)$ in Eq. (4.6) gives

$$\chi = \frac{\left|(1/2\pi)\int_{-\infty}^{\infty} X(\Omega)[\alpha X^*(\Omega)\exp(-j\Omega T_M)]\exp(j\Omega T_M)\,d\Omega\right|^2}{(\sigma_w^2/2\pi)\int_{-\infty}^{\infty}|\alpha X^*(\Omega)\exp(-j\Omega T_M)|^2\,d\Omega}$$

$$= \frac{\left|(1/2\pi)\alpha\int_{-\infty}^{\infty}|X(\Omega)|^2\,d\Omega\right|^2}{|\alpha|^2(\sigma_w^2/2\pi)\int_{-\infty}^{\infty}|X(\Omega)|^2\,d\Omega} \qquad (4.13)$$

$$= \frac{1}{2\pi\sigma_w^2}\int_{-\infty}^{\infty}|X(\Omega)|^2\,d\Omega$$

Comparing Eq. (4.13) to Eq. (4.11) gives

$$\chi = \frac{E}{\sigma_w^2} \qquad (4.14)$$

Equation (4.14) states the remarkable result that the maximum achievable SNR depends only on the energy of the waveform and not on other details such as its modulation. Two waveforms having the same energy will produce the same maximum SNR, provided each is processed through its respective matched filter. Although it is the ratio of the peak signal component power to the noise power [Eq. (4.6)], the SNR of Eq. (4.14) is called the *energy SNR* because it is proportional to the energy of the transmitted signal.

The previous results can be generalized to develop a filter that maximizes output *signal-to-interference ratio* (SIR) when the interference power spectrum is not white. In radar, this is useful for example in cases where the dominant interference is clutter, which generally has a colored power spectrum. The result can be expressed as a two-stage filtering operation. The first stage is a *whitening filter* that converts the interference power spectrum to a flat spectrum (and also modifies the signal spectrum in the process); the second stage is then a conventional matched filter as described earlier, but designed for the now-modified signal spectrum. Details are given by Kay (1998). An example of whitening in a vector-matrix framework common in array processing is given in Sec. 9.2.4.

4.2.2 Matched Filter for the Simple Pulse

To illustrate the previous ideas, consider a simple pulse of duration τ,

$$x(t) = \begin{cases} 1, & 0 \leq t \leq \tau \\ 0, & \text{otherwise} \end{cases} \qquad (4.15)$$

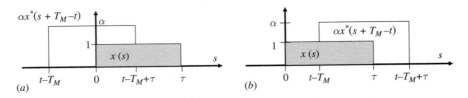

FIGURE 4.2 Convolution of simple pulse and its matched filter. (a) $T_M - \tau \leq t \leq T_M$. (b) $T_M \leq t \leq T_M + \tau$.

The corresponding matched filter impulse response is

$$h(t) = \alpha x^*(T_M - t)$$
$$= \begin{cases} \alpha, & T_M - \tau \leq t \leq T_M \\ 0, & \text{otherwise} \end{cases} \tag{4.16}$$

where $T_M > \tau$ for causality. Because $x(t)$ is a much simpler function than its Fourier transform (a sinc function), it is easier to work with the correlation interpretation of Eq. (4.10) to compute the output. Figure 4.2 illustrates the two terms in the integrand, helping to establish the regions of integration. Figure 4.2a of the figure shows that

$$y(t) = \begin{cases} 0, & t < T_M - \tau \\ \int_0^{t-T_M+\tau} (1)(\alpha)\, ds, & T_M - \tau \leq t \leq T_M \end{cases} \tag{4.17}$$

while Fig. 4.2b is useful in identifying the next two regions,

$$y(t) = \begin{cases} \int_{\tau-T_M}^{\tau} (1)(\alpha)\, ds, & T_M \leq t \leq T_M + \tau \\ 0, & t > T_M + \tau \end{cases} \tag{4.18}$$

The result, illustrated in Fig. 4.3, is

$$y(t) = \begin{cases} \alpha[t - (T_M - \tau)], & T_M - \tau \leq t \leq T_M \\ \alpha[(T_M + \tau) - t], & T_M \leq t \leq T_M + \tau \\ 0, & \text{otherwise} \end{cases} \tag{4.19}$$

The matched filter output is a triangle function of duration 2τ seconds with its peak at $t = T_M$ as expected. The peak value is $\alpha\tau$; since the energy of the unit amplitude pulse is just τ, the peak value equals αE as predicted.

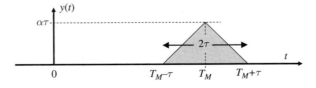

FIGURE 4.3 Matched filter output for a simple pulse.

The noise power at the output of the matched filter is

$$n_p = \frac{\sigma_w^2}{2\pi} \int_{-\infty}^{\infty} |H(\Omega)|^2 \, d\Omega$$

$$= \sigma_w^2 \int_{-\infty}^{\infty} |h(t)|^2 \, dt \quad \text{(Parseval's relation)} \quad (4.20)$$

$$= \sigma_w^2 |\alpha|^2 \tau$$

The SNR is therefore

$$\chi = \frac{|\alpha \tau|^2}{\sigma_w^2 |\alpha|^2 \tau} = \frac{\tau}{\sigma_w^2} = \frac{E}{\sigma_w^2} \quad (4.21)$$

consistent with Eq. (4.14). Note that the filter gain α has no effect on the SNR since it affects both noise and signal components equally.

4.2.3 All-Range Coherent Matched Filtering

The matched filter was designed to maximize the output SNR at a particular time instant T_M. This raises several questions. How should T_M be chosen, and how can the range of a target be related to the resulting output? What happens if the received signal contains echoes from multiple targets at different ranges?

Start by choosing $T_M = \tau$, the minimum value that results in a causal matched filter. Now suppose the input to the matched filter is the echo from a target at an unknown range R_0 corresponding to a time delay $t_0 = 2R_0/c$. That input signal after coherent demodulation will be $\exp(-j4\pi R_0/\lambda_t)x(t - t_0)$. Choose $\alpha = 1$. The signal component of the output of the matched filter will be

$$y(t) = \int_{-\infty}^{\infty} [\exp(-j4\pi R_0/\lambda_t)x(s - t_0)]x^*(s + \tau - t) \, ds$$

$$= \exp(-j4\pi R_0/\lambda_t) \int_{-\infty}^{\infty} x(s - t_0)x^*(s + \tau - t) \, ds \quad (4.22)$$

This is just the correlation of the received, delayed echo and the matched filter impulse response. The output waveform will again be a triangle with its peak at correlation lag zero. That peak occurs when $s - t_0 = s + \tau - t$, or $t = t_0 + \tau$. The matched filter output will appear as in Fig. 4.4. The time of the peak, t_{peak}, corresponds to the actual delay to the target plus the delay of the causal matched filter. The target range can be easily determined from observation of the matched filter output as $R_0 = c(t_{peak} - \tau)/2$.

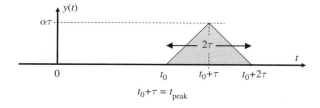

FIGURE 4.4 Output of the matched filter for a target at range $R_0 = ct_0/2$ with $T_M = \tau$.

This discussion shows that the matched filter parameter T_M can be chosen arbitrarily (typically as $T_M = \tau$). Once T_M is known, the range of a target can be determined by detecting the time at which a peak occurs at the matched filter output, subtracting T_M to get the delay to the target and back, and converting to units of range. Thus, a single choice of T_M allows detection of targets at all ranges. One simply samples the matched filter output at a series of fast-time sample instants t_k; if a peak occurs at time t_k, it corresponds to a target at range $c(t_k - T_M)/2$. If the received signal contains echoes from multiple targets at different ranges, by superposition, the matched filter output will contain multiple copies of the single-pulse triangle response, one centered at the time delay (plus filter delay) of each of the various targets.

4.2.4 Straddle Loss

In modern practice, matched filtering is carried out digitally so that $y(t)$ is sampled at some fast-time sampling rate $F_s = 1/T_s$. Typically, F_s equals or is slightly greater than the waveform bandwidth β. The range sample spacing is then $cT_s/2$ meters. Unfortunately, targets do not arrange themselves precisely at ranges corresponding to the range samples. The receiver then may not sample the matched filter output precisely at its peak. The result is a reduction in the measured signal amplitude and therefore an SNR loss.

This is exactly the issue of straddle loss that was discussed in Chap. 3 with regard to the discrete Fourier transform (DFT) or inverse discrete Fourier transform (IDFT), which compute a sampled spectrum. The finite sampling rate allows the processor to "miss" the peak response, whether it is the matched filter output in fast time or the spectrum of a slow-time signal. Straddle loss also arises in angular sampling with scanning antennas. In any of these cases it can be reduced with higher sampling rates or various interpolation methods. Consideration of these methods is deferred to the discussion of pulse Doppler analysis in Chap. 5 and the analyses of time delay, frequency, and angle estimation in Chap. 7. All of the methods there can be applied to the fast-time straddle loss for the various waveforms in this chapter.

4.2.5 Range Resolution of the Matched Filter

By determining the range separation that would result in nonoverlapping echoes, it was shown in Chap. 1 that the range resolution achieved by a simple pulse of duration τ seconds is $c\tau/2$ meters. When a matched filter is used, the output due to each scatterer is now 2τ seconds long, but is also triangular rather than rectangular in shape. Does the longer matched filter output result in a larger value of range resolution?

Before considering this question, it is useful to recall that the demodulated echo from a scatterer at range R_0 meters has not only a delay of $t_0 = 2R_0/c$ seconds, but also an overall phase shift of $\exp[j(-4\pi/\lambda_t)R_0]$ radians.[2] A change of only $\lambda_t/4$ in range will cause a change of 180° in the received echo phase. Two overlapping target responses may therefore add either constructively or destructively in phase, and small changes in their spacing can result in large changes in the composite response. Consider two targets at ranges $ct_0/2$ and $ct_0/2 + c\tau/2$ and assume τ is such that the two matched filter responses add in phase. Then the composite response at the matched filter output is a flat-topped trapezoid as shown in Fig. 4.5a. Clearly, if the separation between the two scatterers increases, a dip will begin to develop in the composite response, even when the separation is such that they remain in phase. If the separation decreases, the in-phase response will still be a trapezoid, but with a higher peak and a shorter flat region as the responses overlap more. Because any increase in separation will result in a dip between the two responses, the separation of $c\tau/2$ meters is still considered to be the range resolution of the matched filter output. Thus, using a matched filter does

---

[2]This phase shift term was absorbed into the effective reflectivity ρ' in Chap. 2.

Radar Waveforms   175

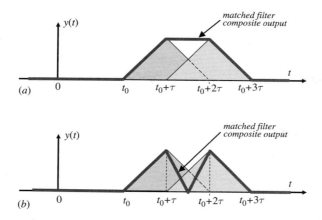

FIGURE 4.5 Composite matched filter response due to two scatterers separated by $c\tau/2$ meters. (a) T target responses in phase. (b) Target responses 180° out of phase.

not degrade the range resolution. To reinforce this further, recall that the definition of the Rayleigh resolution is the peak-to-first null distance. Inspection of Fig. 4.3 shows that $c\tau/2$ is also the Rayleigh resolution of the simple pulse matched filter output.

Scatterers that are closer together than the Rayleigh resolution may still be resolved if the spacing is such that the individual responses add out of phase. Figure 4.5b illustrates the case where the two responses differ in phase by 180°. Destructive interference in the region of overlap causes a deep null in the composite response. However, this null is very sensitive to the fine spacing of the scatterers and cannot be relied on to resolve two targets.

4.3 Matched Filtering of Moving Targets

Suppose a simple pulse is transmitted, $x(t) = 1$, $0 \leq t \leq \tau$, and it echoes from a target moving toward the radar with a radial velocity of v meters per second. After demodulation, the received waveform (ignoring the overall time delay) will be $x'(t) = x(t)\exp(j\Omega_D t)$ with $\Omega_D = 4\pi v/\lambda_t$. Because the echo is different from $x(t)$, a filter matched to $x(t)$ will *not* be matched to $x'(t)$. If the target velocity is known, the matched filter for $x'(t)$ can be constructed:

$$h(t) = \alpha x'^*(-t) = \alpha x^*(-t)\exp(+j\Omega_D t) \quad (4.23)$$

The frequency response of this matched filter is

$$H(\Omega) = \alpha \int_{-\infty}^{\infty} x^*(-t)\exp(+j\Omega_D t)\exp(-j\Omega t)\, dt \; ; \text{ define } t' = -t; \text{ then}$$

$$= \alpha \left[\int_{-\infty}^{\infty} x(t')\exp[-j(\Omega - \Omega_D)t']\, dt' \right]^* \quad (4.24)$$

$$= \alpha X^*(\Omega - \Omega_D).$$

Thus, the matched filter for $x'(t)$ can be obtained by simply shifting the center frequency of the matched filter for $x(t)$ to the expected Doppler shift.

A more interesting situation occurs when the velocity is not known in advance so that the receiver filter cannot be matched to the target Doppler shift. More generally, suppose the filter is matched to some Doppler shift Ω_i radians per second but the actual Doppler shift of the echo is Ω_D. Choosing $T_M = 0$ for simplicity, the matched filter output will be zero for $|t| > \tau$. For $0 \le t \le \tau$ the response is

$$y(t) = \alpha \int_t^\tau \exp(j\Omega_D s)\exp[-j\Omega_i(s-t)]\,ds \qquad (4.25)$$

If the filter is in fact matched to the actual Doppler shift, $\Omega_i = \Omega_D$, the output becomes

$$\begin{aligned} y(t) &= \alpha\exp(j\Omega_D t)\int_t^\tau (1)\,ds \\ &= \alpha\exp(j\Omega_D t)(\tau - t), \quad 0 \le t \le \tau \end{aligned} \qquad (4.26)$$

The analysis is similar for negative t, $-\tau \le t \le 0$. The complete result is

$$y(t) = \begin{cases} \alpha\exp(j\Omega_D t)(\tau - |t|), & -\tau \le t \le \tau, \\ 0, & \text{otherwise} \end{cases} \qquad (4.27)$$

Thus, $|y(t)|$ is the usual triangular function, peaking as expected at $t = 0$.

If there is a Doppler mismatch, $\Omega_i \ne \Omega_D$, the response at the expected peak time $t = 0$ is

$$\begin{aligned} y(t)\big|_{t=0} &= \alpha \int_0^\tau \exp(j\Omega_D s)\exp(-j\Omega_i s)\,ds \\ &= \alpha \int_0^\tau \exp[+j(\Omega_D - \Omega_i)s]\,ds \\ &= \frac{\alpha}{j(\Omega_D - \Omega_i)}\exp[j(\Omega_D - \Omega_i)s]\Big|_0^\tau \end{aligned} \qquad (4.28)$$

Defining $\Omega_{\text{diff}} \equiv \Omega_D - \Omega_i$,

$$|y(0)| = \left|\frac{2\alpha\sin(\Omega_{\text{diff}}\tau/2)}{\Omega_{\text{diff}}}\right| \qquad (4.29)$$

Equation (4.29) is plotted in Fig. 4.6. The first zero of this sinc function occurs at $F_{\text{diff}} = 1/\tau$ Hz.[3] Relatively small Doppler mismatches ($F_{\text{diff}} \ll 1/\tau$) will cause only slight reductions in the matched filter output peak amplitude. Large mismatches, however, can cause very substantial reductions.

The effect of Doppler mismatch can be either good or bad. If targets are moving and the velocities are unknown, mismatch will cause reductions in observed peaks and, if severe enough, may prevent detection. The signal processor must either estimate the target Doppler so that the matched filter can be adjusted or construct matched filters for a number of different possible Doppler frequencies and observe the output of each to search for targets. On the other hand, if the goal is to be selective in responding only to targets of a particular Doppler shift, it is desirable to have a matched filter that suppresses targets at other Doppler shifts.

---

[3] Note that a frequency component of $1/\tau$ hertz goes through exactly one full cycle during a pulse of duration τ seconds.

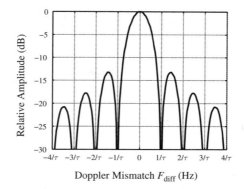

FIGURE 4.6 Effect of Doppler mismatch on matched filter response at expected peak time.

From Fig. 4.6, it is clear that the Rayleigh resolution of the Doppler mismatch response is also $1/\tau$ Hz. The resolution in velocity is therefore $\lambda/2\tau$ meters per second. Again, for typical pulse lengths these are fairly large values. For example, a 10-μs pulse would exhibit a Rayleigh resolution in Doppler of 100 kHz or in velocity at 10 GHz of 1500 m/s. This example shows that targets at the same range usually cannot be resolved in Doppler on a single pulse. If finer Doppler resolution is desired, a very long pulse may be needed. For example, velocity resolution of 1 m/s at 10 GHz using a single pulse requires $\tau = 15$ ms. The range resolution is then a very poor 2250 km. This conflict between good range resolution and good Doppler resolution can be resolved using a pulse burst waveform (Sec. 4.5) or a linear frequency modulated waveform (Sec. 4.6.1).

4.4 The Ambiguity Function

4.4.1 Definition and Properties of the Ambiguity Function

In the preceding sections, the matched filter response for the simple pulse waveform has been analyzed to show its behavior both in time and in response to Doppler mismatches. The *ambiguity function* (AF) is an analytical tool for waveform design and analysis that succinctly characterizes the behavior of a waveform paired with its matched filter. The AF is useful for examining resolution, sidelobe behavior, and ambiguities in both range and Doppler for a given waveform, as well as phenomena such as range-Doppler coupling (to be introduced in Sec. 4.6.4).

Consider the output of a matched filter for a waveform $x(t)$ when the input is a Doppler-shifted response $x(t)\exp(j2\pi F_D t)$. Also assume that the filter has unit gain ($\alpha = 1$) and is designed to peak at $T_M = 0$; this merely means that the time axis at the filter output is relative to the expected peak output time for the range of the target. The filter output will be

$$y(t; F_D) = \int_{-\infty}^{\infty} x(s)\exp(j2\pi F_D s)x^*(s-t)\,ds \qquad (4.30)$$
$$\equiv \hat{A}(t, F_D)$$

which is defined as the *complex ambiguity function* $\hat{A}(t, F_D)$. An equivalent definition can be given in terms of the signal spectrum by applying basic Fourier transform properties to get

$$\hat{A}(t, F_D) = \int_{-\infty}^{\infty} X^*(F)X(F-F_D)\exp(j2\pi Ft)\,dF \qquad (4.31)$$

The *ambiguity function*[4] is defined as the magnitude of $\hat{A}(t, F_D)$,

$$A(t, F_D) \equiv \left| \hat{A}(t, F_D) \right| \quad (4.32)$$

It is a function of two variables: the time delay relative to the expected matched filter peak output, and the Doppler mismatch between the transmitted waveform and the received echo. For example, the AF evaluated at $t = 0$ and $F_D = 0$ corresponds to the output of the actual matched filter at time $t = 2R_0/c + \tau$ for a target at range R_0 that is not in motion relative to the radar (no Doppler shift). The particular form of the AF is determined entirely by the complex waveform $x(t)$.

Three properties of the ambiguity function are of immediate interest. The first states that if the waveform has energy E, then

$$|A(t, F_D)| \leq |A(0,0)| = E \quad (4.33)$$

Thus, when there is no Doppler mismatch and the echo is sampled at a delay corresponding to the target range, the matched filter output will be maximum. If there is a Doppler mismatch or the filter output is sampled at a different delay, then the response will be less than or equal to (usually less than) the maximum. The second property states that total area under any ambiguity function is constant and is given by

$$\int_{-\infty}^{\infty} \int_{-\infty}^{\infty} |A(t, F_D)|^2 \, dt \, dF_D = E^2 \quad (4.34)$$

This "conservation of mass" statement implies that in the design of waveforms one cannot remove energy from one portion of the ambiguity surface without placing it somewhere else; it can only be moved around on the ambiguity surface. The third property is a symmetry relation:

$$A(t, F_D) = A(-t, -F_D) \quad (4.35)$$

In order to prove the first property, start with the squared magnitude of Eqs. (4.32) and (4.30):

$$|A(t, F_D)|^2 = \left| \int_{-\infty}^{\infty} x(s) x^*(s-t) \exp(j 2\pi F_D s) \, ds \right|^2 \quad (4.36)$$

Applying the Schwartz inequality to Eq. (4.36) yields

$$|A(t, F_D)|^2 \leq \int_{-\infty}^{\infty} |x(s)|^2 \, ds \int_{-\infty}^{\infty} |x^*(s-t) \exp(j 2\pi F_D s)|^2 \, ds$$

$$= \int_{-\infty}^{\infty} |x(s)|^2 \, ds \int_{-\infty}^{\infty} |x^*(s-t)|^2 \, ds \quad (4.37)$$

---

[4]Some authors define the term "ambiguity function" as $\left| \hat{A}(t, F_D) \right|^2$ or as $\hat{A}(t, F_D)$ itself. Also, some authors define the ambiguity function as $\left| \int_{-\infty}^{\infty} x(s) \exp(j 2\pi F_D s) x^*(s + t) ds \right|$ instead of $\left| \int_{-\infty}^{\infty} x(s) \exp(j 2\pi F_D s) x^*(s - t) ds \right|$. The definition used here is consistent with that given in Rihaczek (1996).

Each integral is just the energy E in $x(t)$, so that

$$|A(t, F_D)|^2 \leq E^2 \tag{4.38}$$

The equality holds only if $x(s) = x(s - t)\exp(-j2\pi F_D s)$ for all s, which occurs if and only if $t = F_D = 0$. Making these substitutions in Eq. (4.38) gives the equality in Eq. (4.33).

The proof of the second property starts with the complex conjugate of the complex ambiguity function,

$$\hat{A}^*(t, F_D) = \int_{-\infty}^{\infty} x^*(s)x(s - t)\exp(-j2\pi F_D s)\,ds$$
$$= \int_{-\infty}^{\infty} X(F)X^*(F - F_D)\exp(-j2\pi F t)\,dF \tag{4.39}$$

The squared magnitude of the ambiguity function can then be written as

$$|A(t, F_D)|^2 = \hat{A}(t, F_D)\hat{A}^*(t, F_D)$$
$$= \int_{-\infty}^{\infty}\int_{-\infty}^{\infty} x(s)x^*(s - t)X(F)X^*(F - F_D)\exp[j2\pi(F_D s - Ft)]\,ds\,dF \tag{4.40}$$

The total energy in the ambiguity surface is

$$\int_{-\infty}^{\infty}\int_{-\infty}^{\infty}|A(t, F_D)|^2\,dt\,dF_D = \frac{1}{2\pi}\int_{-\infty}^{\infty}\int_{-\infty}^{\infty}\int_{-\infty}^{\infty}\int_{-\infty}^{\infty} x(s)x^*(s - t)\,X(F)X^*(F - F_D) \times \cdots$$
$$\cdots \times \exp[j2\pi(F_D s - Ft)]\,ds\,dF\,dt\,dF_D \tag{4.41}$$

Isolating those terms integrated over t and F_D, recognizing the results as Fourier integrals, and applying Fourier properties yields the following two relationships:

$$\int_{-\infty}^{\infty} x^*(s - t)\exp(-j2\pi Ft)\,dt = \exp(-j2\pi Fs)X^*(F) \quad \text{and} \tag{4.42}$$

$$\int_{-\infty}^{\infty} X^*(F - F_D)\exp(j2\pi F_D s)\,dF_D = \exp(j2\pi Fs)x^*(s) \tag{4.43}$$

Substituting these into Eq. (4.41) yields

$$\int_{-\infty}^{\infty}\int_{-\infty}^{\infty}|A(t, F_D)|^2\,dt\,dF_D = (1/2\pi)\int_{-\infty}^{\infty}\int_{-\infty}^{\infty} x(s)X^*(F)X(F)x^*(s)\,ds\,dF$$
$$= \left\{\int_{-\infty}^{\infty}|x(s)|^2\,ds\right\}\left\{(1/2\pi)\int_{-\infty}^{\infty}|X(F)|^2\,dF\right\} \tag{4.44}$$

The first integral on the right-hand side of Eq. (4.44) is just the energy E of the waveform measured in the time domain; the second, by Parseval's theorem, is also the energy. Thus

$$\int_{-\infty}^{\infty}\int_{-\infty}^{\infty} |A(t, F_D)|^2 \, dt \, dF_D = E^2 \tag{4.45}$$

The symmetry property can be proved by substituting $-t$ and $-F_D$ for t and F_D, respectively, in the definition in Eq. (4.30):

$$\hat{A}(-t, -F_D) = \int_{-\infty}^{\infty} x(s)\exp(-j2\pi F_D s)x^*(s+t)\, ds \tag{4.46}$$

Now make the change of variables $s' = s + t$ to get

$$\hat{A}(-t, -F_D) = \int_{-\infty}^{\infty} x(s'-t)\exp[-j2\pi F_D(s'-t)]x^*(s')\, ds'$$

$$= \exp(j2\pi F_D t) \int_{-\infty}^{\infty} x(s'-t)\exp(-j2\pi F_D s')x^*(s')\, ds' \tag{4.47}$$

$$= \exp(j2\pi F_D t)\hat{A}^*(t, F_D)$$

Since $A(t, FD) \equiv |\hat{A}(t, F_D)|$, Eq. (4.35) follows immediately.

What would be an ideal ambiguity function? The answer varies depending on the intent of the system design, but a commonly cited goal is the "thumbtack" ambiguity function of Fig. 4.7a, which features a single central peak at coordinates (0, 0) with the remaining energy spread uniformly throughout the delay-Doppler plane. The location of the peak implies that if there is no Doppler mismatch ($F_D = 0$) the peak will occur at the true target relative delay ($t = 0$). The narrow mainlobe of the central peak implies fine resolution in both range and Doppler. The lack of any secondary peak implies that there will be no range or Doppler ambiguities. The uniform plateau suggests low and uniform sidelobes, minimizing target masking and sidelobe clutter effects. All of these features are beneficial for a system designed to make fine-resolution measurements of targets in range and Doppler or to perform radar imaging. On the other hand, a waveform intended to be used for target search might be preferred to be more tolerant of Doppler mismatch so that the Doppler shift of targets whose velocity is not yet known does not prevent their detection due to a weak response at the matched filter output. A good ambiguity function for this case would be more like Fig. 4.7b. This AF retains the fine resolution and lack of ambiguities in range. However, the relatively wide ridge in Doppler means that small-to-moderate Doppler mismatches will not reduce the matched filter output peak

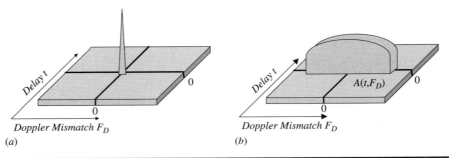

FIGURE 4.7 (a) "Thumbtack" ambiguity function. (b) Doppler-tolerant ridge ambiguity function.

4.4.2 Ambiguity Function of the Simple Pulse

As a first example of an ambiguity function, consider a simple pulse centered on the origin and normalized to have unit energy ($E = 1$) for convenience

$$x(t) = \frac{1}{\sqrt{\tau}}, \quad -\frac{\tau}{2} \leq t \leq +\frac{\tau}{2} \tag{4.48}$$

Applying Eq. (4.30) gives for $t > 0$

$$\begin{aligned}
\hat{A}(t, F_D) &= \int_{-\tau/2+t}^{\tau/2} \frac{1}{\tau} \exp(j2\pi F_D s) \, ds \\
&= \frac{\exp[j2\pi F_D \tau/2] - \exp[j2\pi F_D(-\tau/2 + t)]}{j2\pi F_D \tau} \\
&= \frac{1}{j2\pi F_D \tau} \exp(j2\pi F_D t/2) \left\{ \exp\left[j2\pi F_D\left(\frac{\tau}{2} - \frac{t}{2}\right)\right] - \exp\left[-j2\pi F_D\left(\frac{\tau}{2} - \frac{t}{2}\right)\right] \right\} \\
&= \exp(j2\pi F_D t/2) \frac{\sin[\pi F_D(\tau - t)]}{\pi F_D \tau}
\end{aligned} \tag{4.49}$$

The ambiguity function for $t > 0$ is the magnitude of Eq. (4.49), which evaluates to

$$A(t, F_D) = \left|\hat{A}(t, F_D)\right| = \left|\frac{\sin[\pi F_D(\tau - t)]}{\pi F_D \tau}\right|, \quad 0 \leq t \leq \tau \tag{4.50}$$

Repeating the derivation for $t < 0$ gives a similar result, but with the quantity $(\tau - t)$ replaced by $(\tau + t)$. The complete AF of the simple pulse is therefore

$$\begin{aligned}
A(t, F_D) &= \left|\frac{\sin[\pi F_D(\tau - |t|)]}{\pi F_D \tau}\right| \\
&= \left(1 - \frac{|\tau|}{t}\right) \left|\frac{\sin[\pi F_D \tau(1 - |\tau|/t)]}{\pi F_D \tau(1 - |\tau|/t)}\right|, \quad -\tau \leq t \leq \tau
\end{aligned} \tag{4.51}$$

Equation (4.51) is plotted in Fig. 4.8 in a three-dimensional surface plot and in Fig. 4.9 as a contour plot, which is often easier to interpret and is therefore used in most cases in the remainder of this chapter. The AF for a simple pulse is a triangular ridge oriented along the delay axis. Doppler mismatches on the order of $1/\tau$ Hz or more drastically reduce and spread the matched filter output peak, as was shown previously.

The zero-Doppler response $A(t, 0)$ gives the matched filter output when there is no Doppler mismatch. Setting $F_D = 0$ in Eq. (4.51) and using L'Hôpital's rule to resolve the indeterminate form gives

$$\begin{aligned}
A(t, 0) &= \left|\frac{\pi(\tau - |t|)\cos[\pi F_D(\tau - |t|)]}{\pi \tau}\right|_{F_D=0} \\
&= \frac{\tau - |t|}{\tau}, \quad -\tau \leq t \leq \tau
\end{aligned} \tag{4.52}$$

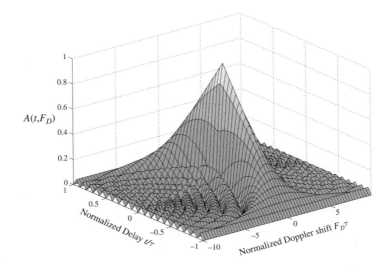

FIGURE 4.8 Ambiguity function of a unit-energy simple pulse of length τ.

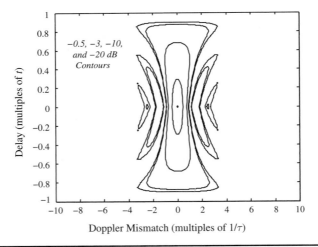

FIGURE 4.9 Contour plot of the simple pulse ambiguity function of Fig. 4.8.

Similarly, the zero-delay cut $A(0, F_D)$ gives the output of the matched filter at the expected peak time $t = 0$ as a function of Doppler mismatch. Using $t = 0$ in Eq. (4.51) immediately gives

$$A(0, F_D) = \left| \frac{\sin(\pi F_D \tau)}{\pi F_D \tau} \right| \quad (4.53)$$

Equations (4.52) and (4.53) are the expected triangle and sinc function responses derived previously. They are illustrated in Fig. 4.10.

A Doppler mismatch not only reduces the peak amplitude of the matched filter output but also alters the shape of the range response of the matched filter. Figure 4.11 shows the

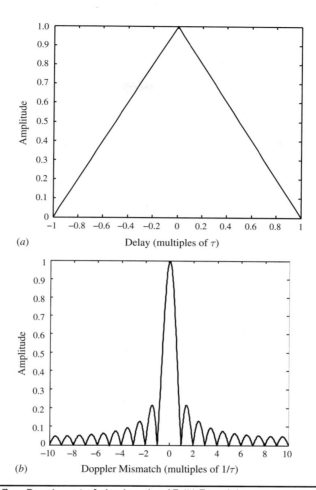

FIGURE 4.10 (a) Zero-Doppler cut of simple pulse AF. (b) Zero-delay cut.

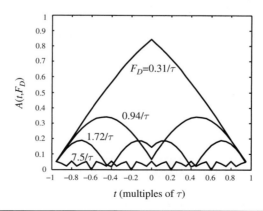

FIGURE 4.11 Effect of Doppler mismatch on the range response of the matched filter for the simple pulse.

effect of varying degrees of Doppler mismatch on the matched filter range response. These curves should be compared to Fig. 4.10a. A mismatch of $0.31/\tau$ Hz results in a reduction of about 16 percent in the peak amplitude, but the peak remains at the correct time delay. A larger shift, for example $0.94/\tau$, not only reduces the maximum output amplitude by 65 percent but also eliminates the central peak altogether. This curve indicates that there will instead be two reduced amplitude peaks at the matched filter output for a single target, neither one at the correct time delay. If they are large enough, they may result in two false target detections. By the time the mismatch is several times $1/\tau$, the response becomes completely unstructured. Note that a mismatch of n/τ Hz means that there will be n cycles of the Doppler frequency during the pulse duration τ. Also recall that for typical pulse lengths, $1/\tau$ is a large Doppler shift, so that the simple pulse still ranks as a relatively Doppler-tolerant waveform. For instance, if $\tau = 10\,\mu s$, a Doppler shift of $0.31/\tau$ is 31 kHz, corresponding at an RF of 10 GHz to a velocity of 465 m/s, or 1040 mph. Even with this very large Doppler mismatch, the simple pulse matched filter output retains its basic shape, correct peak location, and suffers only the 16 percent (1.5 dB) amplitude loss.

4.5 The Pulse Burst Waveform

The flip side of the Doppler tolerance of the simple pulse described in the preceding example is that its Doppler resolution is very poor. If the designer wants the radar system to respond to targets only at certain velocities and reject targets at nearby velocities, the simple pulse is not adequate as a waveform. Finer frequency resolution requires a longer observation time. The *pulse burst waveform* is one way to meet this requirement. It is defined as

$$x(t) = \sum_{m=0}^{M-1} x_p(t - mT) \qquad (4.54)$$

where $x_p(t)$ = single pulse of length τ
 M = number of pulses in the burst
 T = pulse repetition interval

While the constituent pulse $x_p(t)$ can be any single-pulse waveform, for the moment only the simple pulse will be considered. Figure 4.12 illustrates this waveform. The solid line forming the envelope of the sinusoidal pulses is the actual baseband waveform $x(t)$. The train of RF pulses that results when it is impressed upon a carrier is denoted as usual as $\tilde{x}(t)$. The total duration MT (which includes the dead time after the last pulse) is the coherent processing interval (CPI).

4.5.1 Matched Filter for the Pulse Burst Waveform

The matched filter for the pulse burst is (with $\alpha = 1$ and $T_M = 0$)

$$h(t) = x^*(-t) = \sum_{m=0}^{M-1} x_p^*(-t - mT) \qquad (4.55)$$

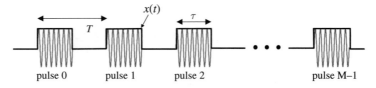

FIGURE 4.12 Pulse burst waveform and the resulting train of RF pulses.

and the matched filter output, given an echo from a range corresponding to a time delay t_0, is therefore

$$\begin{aligned} y(t) &= \int_{-\infty}^{\infty} \left[\sum_{m=0}^{M-1} x_p(s - t_0 - mT) \right] \left[\sum_{n=0}^{M-1} x_p^*(s - t - nT) \right] ds \\ &= \sum_{m=0}^{M-1} \sum_{n=0}^{M-1} \int_{-\infty}^{\infty} x_p(s - t_0 - mT) x_p^*(s - t - nT) \, ds \end{aligned} \tag{4.56}$$

The inner integral is the matched filter output for the constituent simple pulse. Let $t_0 = 0$ for simplicity; the results can be adjusted for any other delay t_0 by shift invariance. Renaming the simple pulse matched filter output from Eq. (4.19) as $s_p(t)$, Eq. (4.56) becomes

$$y(t) = \sum_{m=0}^{M-1} \sum_{n=0}^{M-1} s_p(-t - (m-n)T) = \sum_{m=0}^{M-1} \sum_{n=0}^{M-1} s_p^*(t - (n-m)T) \tag{4.57}$$

where the symmetry of $s_p(t)$ has been used in the last step. Equation (4.57) states that the matched filter output is a superposition of shifted copies of $s_p(t)$. The double summation can be simplified by noting that all terms that have the same value of $(n - m)$ are identical and can be combined. There are M combinations of m and n such that $m - n = 0$, namely, all those where $m = n$. There are $M - 1$ cases, where $m - n = +1$ and another $M - 1$ cases, where $m - n = -1$. Continuing in this vein gives

$$y(t) = \sum_{m=-(M-1)}^{M-1} (M - |m|) s_p^*(t - mT) \tag{4.58}$$

The matched filter output for the pulse burst waveform is simply a sum of scaled and shifted replicas of the output of the filter matched to a single constituent pulse.

Since the constituent pulse $x_p(t)$ is of duration τ, $s_p(t)$ is of duration 2τ. If $T > 2\tau$ as is usually the case, none of the replicas of $s_p(t)$ overlap one another. Figure 4.13 illustrates a pulse

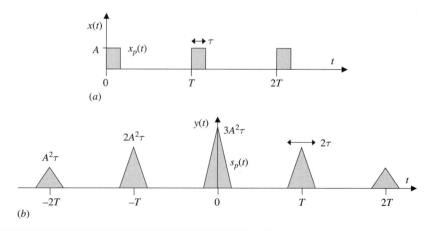

FIGURE 4.13 (a) Pulse burst waveform, $M = 3$. (b) Matched filter output.

burst waveform and the corresponding matched filter output for this case and $M = 3$. The peak output occurs at $t = T_M = 0$:

$$y(0) = \sum_{m=-(M-1)}^{M-1} (M - |m|)s_p^*(-mT) = \sum_{m=-(M-1)}^{M-1} (M - |m|)s_p(mT) \qquad (4.59)$$
$$= M \cdot s_p(0) = M \cdot E_p = E$$

where the last step uses $s_p(mT) = 0$ when $T > \tau$. In this equation E_p is the energy in the single pulse $x_p(t)$, while E is the energy in the entire M-pulse waveform. Note that the peak response is M times that achieved with a single pulse of the same amplitude. Recall the radar range equation signal processing gain factor G_{sp} of Eq. (2.85). The increase in the matched filter output peak for the pulse burst waveform represents a coherent signal processing gain $G_{sp} = M$ that will improve the SNR compared to a single-pulse waveform, aiding detection probability and measurement precision.

4.5.2 Pulse-by-Pulse Processing

The structure of Eq. (4.58) suggests that it is not necessary to construct an explicit matched filter for the entire pulse burst waveform $x(t)$, but rather that the matched filter can be implemented by filtering the data from each individual pulse with the single-pulse matched filter and then combining those outputs. This process, here called *pulse-by-pulse* processing, uses separable two-dimensional processing in fast time and slow time. It provides a much more convenient implementation and describes how pulse burst waveforms are processed in real systems.

Define the matched filter impulse response for the individual pulse in the burst, assuming $T_M = 0$, as

$$h_p(t) = x_p^*(-t) \qquad (4.60)$$

The output from this filter for the mth transmitted pulse, assuming a target at some delay t_l, is

$$y_m(t) = x_p(t - t_l - mT) * h_p(t)$$
$$= s_p^*(t - t_l - mT) = s_p(-t + t_l + mT), \quad 0 \le m \le M - 1 \qquad (4.61)$$

Assume that the echo from the individual pulse matched filter for the first pulse ($m = 0$) is sampled at $t = t_l$; that value will be $y_0(t_l) = s_p(0)$. Now sample the filter response to each succeeding pulse at the same delay after its transmission (i.e., sample the same range bin for each pulse). The filter output for pulse m is sampled at $t = t_l + mT$, giving $y_m(t_l + mT) = s_p(0)$ again.

If the sample taken at time t_l after pulse transmission is associated with range bin l, the M samples so obtained form a discrete constant-valued sequence $y[l, m] = s_p(0)$, $0 \le m \le M - 1$. The discrete-time causal matched filter in the slow-time (m) dimension for such a sequence is $h[m] = \alpha y^*[M - 1 - m]$; with $\alpha = 1/s_p(0)$, $h[m] = 1$ for $0 \le m \le M - 1$. The output of this discrete-time matched filter is

$$z[m] = \sum_{r=0}^{M-1} y[l, r]h[m - r]$$
$$= \begin{cases} \displaystyle\sum_{r=0}^{m} y[l, r](1), & 0 \le m \le M - 1, \\ \displaystyle\sum_{r=m-M+1}^{M-1} y[l, r](1), & M - 1 \le m \le 2(M - 1) \end{cases} \qquad (4.62)$$

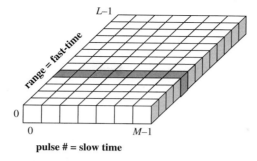

FIGURE 4.14 Slow-time sequence to be integrated for matched filtering of a pulse burst waveform.

The peak output will occur when the two functions in the summand completely overlap, which requires $m = M - 1$; then

$$z[M-1] = \sum_{r=0}^{M-1} y[l,r] = M \cdot s_p(0) = M \cdot E_p = E \qquad (4.63)$$

Equation (4.63) indicates that in pulse-by-pulse processing, matched filtering of the slow-time sequence from a given range bin reduces to coherently integrating the slow-time samples in each range bin. The resulting peak output is identical to that obtained with a whole-waveform continuous matched filter of Eq. (4.55). Figure 4.14 illustrates the row of slow-time samples that are integrated (after matched filtering of the single pulse in fast time) to complete the matched filtering process for the pulse burst. This operation is performed independently for each range bin.

4.5.3 Range Ambiguity

Evaluating the pulse burst matched filter output at $t = 0$ gave the peak output for a target at the time delay t_0 under consideration. Suppose $t_0 < T$. If a peak is observed it will be interpreted as implying the presence of a target at range $R_0 = ct_0/2$ m. However, suppose the data instead contain echoes from a target an additional T seconds of delay further away. The received waveform will be unchanged except for a delay of T seconds and a reduced amplitude according to the range equation. The amplitude reduction is not pertinent to the discussion and is ignored. By shift invariance, the matched filter output of Eq. (4.58) will also be delayed by T seconds

$$y(t) = \sum_{m=-(M-1)}^{M-1} (M - |m|) s_p^*(t - (m+1)T) \qquad (4.64)$$

Now when the matched filter output is evaluated at $t = 0$, the result is

$$y(0) = \sum_{m=-(M-1)}^{M-1} (M - |m|) s_p^*(-(m+1)T) \qquad (4.65)$$

In this expression (and continuing to assume $T > 2\tau$) only the $m = -1$ term is nonzero, so that

$$y(0) = (M-1) s_p(0) = (M-1) E_p \qquad (4.66)$$

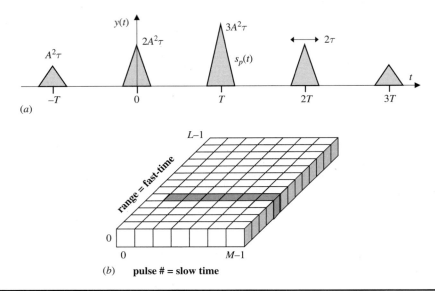

FIGURE 4.15 Pulse burst matched filtering when target is range ambiguous by one PRI. (a) Whole-waveform matched filter output. Compare to Fig. 4.13b. (b) Slow-time data in pulse-by-pulse processing viewpoint. Compare to Fig. 4.14.

This equation shows that the output at the sample time is reduced from $M \cdot E_p$ to $(M-1)E_p$. The situation is illustrated in Fig. 4.15 from both the whole-waveform matched filter and pulse-by-pulse viewpoints. From the former viewpoint, a local peak of the matched filter output is sampled, but the global peak is missed because the filter is "tuned" for the wrong delay. The result, while not zero, is a reduced-amplitude sample. From the latter viewpoint, the echo appears in only $M-1$ of the M slow-time samples integrated because it first returns within the sampling window following transmission of the second pulse rather than the first.

This behavior creates two problems. The reduced amplitude of the target component of the matched filter output reduces the SNR and thus the probability of detecting the target. Assuming the reduced-amplitude response does prove large enough to be detected, the processor will assume the target is at delay t_0 when in fact it is at $t_0 + T$. This phenomenon whereby there is more than one possible range that can be associated with a detection is called a *range ambiguity*. First discussed in Chap. 3, it is a characteristic of pulse burst waveforms. It is not readily apparent if a peak at the matched filter output is due to a target at the apparent range or at that range plus a multiple of the unambiguous range $R_{ua} = cT/2$ meters.

As will be seen in Chap. 5, it is common in some radars to operate at a PRF for which the unambiguous range is less than the maximum detection range, so methods are needed to counter these two problems. Range ambiguities can be resolved using multiple pulse burst waveforms at different PRFs, to be discussed in Chap. 5. The reduction in matched filter output amplitude and SNR for range-ambiguous targets is countered by noting that it occurs because the pulse burst echo is not fully overlapped with the matched filter reference pulse burst at the output sampling time when the target time delay $t_0 > T$. The solution to this problem is to extend the transmitted waveform. Suppose the radar can be expected to detect targets at ranges up to $P \cdot R_{ua}$ for some integer P. Extend the transmitted waveform from M to $M + P - 1$ pulses. The receiver matched filter remains the same M-pulse waveform. Still using $t_0 = 0$ and $T_M = 0$, the matched filter output will be the waveform shown in Fig. 4.16a.

Radar Waveforms 189

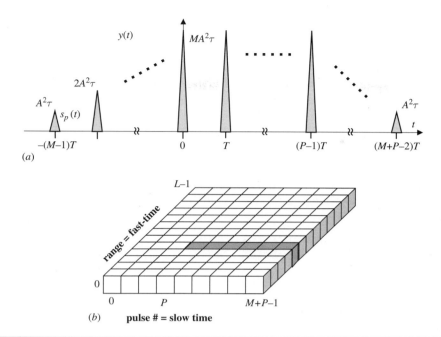

FIGURE 4.16 Effect of extending the transmitted waveform with additional pulses when P range ambiguities can occur. (a) Whole-waveform matched filter output. (b) Slow-time data in pulse-by-pulse processing viewpoint.

It indicates that the full integrated target power of $MA^2\tau = M \cdot E_p$ will be obtained for a target at the delay T_M (zero in the figure) for which the matched filter is tuned, but also for targets at the ambiguous delays $T_M + pT$, $p = 0, \ldots, P - 1$. Evaluating the matched filter output for delays T_M between zero and T allows full-SNR matched filtering of targets at delays up to $T_M + (P - 1)T$, corresponding to ranges up to $P \cdot R_{ua}$ as desired. The pulse-by-pulse viewpoint is shown in Fig. 4.16b. A target in the Pth delay interval, $(P - 1)T \leq t_0 < PT$, will produce a response in the appropriate range bin in the Pth and later slow time samples. By integrating only samples P through $M + P - 1$ in each range bin the design integration gain of M can be achieved for all targets up to ranges $P \cdot R_{ua}$.

Another issue arises with pulse burst waveforms and clutter echoes. Suppose the radar can be expected to receive significant clutter returns at ranges up to $P \cdot R_{ua}$ and consider the clutter component of the slow-time signal for a given range bin in the pulse-by-pulse processing viewpoint. When the range bin of interest is sampled at delay $t_0 < T$ after the first pulse is transmitted, only clutter echoes from the corresponding range $R_0 = ct_0/2$ will be sampled at the receiver. When the range bin is sampled again after the second pulse, the clutter component will include echoes from the second pulse and range R_0 as well as from the first pulse and range $c(t_0 + T)/2$. These two contributions represent echo from two physically different patches of clutter scatterers. The first slow-time sample, which includes echoes from only the nearer patch, may differ significantly in power and statistical behavior from the second slow-time sample, which includes echoes from both. The Pth and subsequent slow-time samples will contain contributions from all P contributing range intervals and therefore exhibit the consistent clutter power levels and statistical behavior needed for effective clutter filtering and target detection. Extending the transmitted waveform to $M + P - 1$ pulses as above therefore allows collection of M steady-state clutter measurements. In

Chap. 5 these additional pulses will be called *clutter fill* pulses. The first $P - 1$ slow-time samples will be discarded in each range bin and only the remaining M samples will be used in clutter filtering, coherent integration, and detection processing.

4.5.4 Doppler Response of the Pulse Burst Waveform

To evaluate the effect of a Doppler mismatch on the pulse burst waveform and its matched filter, consider a target moving toward the radar at velocity v meters per second so that its range is $R_0 - vt$ meters at time t. Assume that the "stop-and-hop" approximation is valid and that the target motion does not exceed one range bin over the CPI, that is, $MvT < c\tau/2$; this ensures that all echoes from a given target appear in the same range bin over the course of a CPI.[5] The demodulated echoes will have a phase shift of $-(4\pi/\lambda_t)R(t) = -(4\pi/\lambda_t)(R_0 - vt)$. Adopting the pulse-by-pulse processing viewpoint and absorbing the phase $\exp(-j4\pi R_0/\lambda_t)$ due to the nominal range R_0 into the overall gain, the individual matched filtered outputs for each pulse become

$$y_m(t) = x_p(t - mT) * h_p(t)$$
$$= \exp[j(4\pi v/\lambda_t)mT]s_p(-t + mT), \quad 0 \le m \le M - 1 \qquad (4.67)$$

The resulting slow-time sequence is

$$y[l, m] = y_m(mT) = \exp[j(4\pi vT/\lambda_t)mT]s_p(0)$$
$$= \exp(j\omega_D m)E_p, \quad 0 \le m \le M - 1 \quad (\omega_D = 4\pi vT/\lambda_t) \qquad (4.68)$$

Integrating the slow-time samples gives

$$\sum_{m=0}^{M-1} y[l, m] = E_p \sum_{m=0}^{M-1} \exp(j\omega_D m) \equiv Y[l, \omega_D]$$
$$= E_p \frac{\sin(\omega_D M/2)}{\sin(\omega_D/2)} \exp\left[-j\left(\frac{M-1}{2}\right)\omega_D\right] \qquad (4.69)$$

Equation (4.69) shows that the system response to the pulse burst waveform in an arbitrary range bin l as a function of the normalized Doppler mismatch ω_D is the familiar asinc function.

Figure 4.17 shows the central portion of the magnitude of this function. The zeros occur at intervals of $1/M$ cycles per sample in normalized frequency; thus, the Rayleigh resolution in Doppler is $1/M$ cycles per sample or $1/MT$ Hz. MT is the duration of the entire pulse burst waveform. The Doppler resolution is therefore determined by the duration of the entire waveform instead of the duration of a single pulse. In this manner, the pulse burst waveform achieves much finer Doppler resolution than a single pulse of the same duration while maintaining the same range resolution. The cost is the time and energy required to transmit and receive M pulses instead of one, and the (minor) computational load of integrating M samples in each range bin.

Integrating the slow-time samples of the pulse burst echo corresponds to implementing a matched filter in slow time for a signal with zero Doppler shift; in this case the expected slow-time signal is simply a constant. A matched filter for a Doppler-shifted pulse burst can

---

[5]When the target motion during the CPI exceeds one range bin it is said to exhibit *range migration*. A technique for compensating range migration in Doppler processing is briefly introduced in Chap. 5.

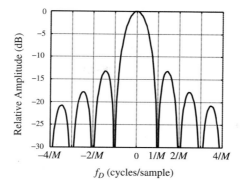

FIGURE 4.17 Central portion of the Doppler mismatch response of the slow-time signal using a pulse burst waveform.

be implemented by continuing to use the single-pulse matched filter in fast time and constructing the appropriate slow-time matched filter for a given Doppler shift.

Suppose the normalized Doppler shift of interest is ω_D radians per sample. The expected slow-time signal is then of the form $A\exp(j\omega_D m)$. After conjugation and time-reversal the slow-time matched filter coefficients will be $h[m] = \exp(+j\omega_D m)$. Consider the response of this filter when the actual Doppler shift of the signal is ω. The matched filter peak output occurs when the impulse response and data sequence are fully overlapped, giving

$$\begin{aligned} Y[l,\omega;\omega_D] &= \sum_{m=0}^{M-1} \exp(-j\omega_D m)\, y[l,m] \\ &= E_p \sum_{m=0}^{M-1} \exp(-j\omega_D m)\exp(j\omega m) = \sum_{m=0}^{M-1} \exp[-j(\omega-\omega_D)m] \\ &= E_p \frac{\sin[(\omega-\omega_D)M/2]}{\sin[(\omega-\omega_D)/2]} \exp\left[-j\left(\frac{M-1}{2}\right)(\omega-\omega_D)\right] \end{aligned} \qquad (4.70)$$

which is identical to Eq. (4.69) except that the peak of the asinc function has been shifted to $\omega = \omega_D$ radians per sample.

Note that the first line of Eq. (4.70) is simply the discrete time Fourier transform (DTFT) of the slow-time data sequence. Thus, a matched filter for a pulse burst waveform and a Doppler shift of ω_D radians can be implemented with a single-pulse matched filter in fast time and a DTFT in slow time, evaluated at ω_D. If ω_D is a discrete Fourier transform frequency, that is, it is of the form $2\pi k/K$ for some integers k and K, the slow-time matched filter can be implemented with a DFT calculation. It follows that a K-point DFT of the data $y[l, m]$ in the slow-time dimension simultaneously computes the output of K matched filters, one at each of the DFT frequencies. These frequencies correspond to Doppler shifts of $F_k = k/KT$ hertz or radial velocities $v_k = \lambda k/2KT$ meters per second, $k = 0, \ldots, K-1$. The fast Fourier transform (FFT) algorithm then allows very efficient search of the data for targets at various Doppler shifts by simply applying an FFT to each slow-time row of the data matrix.

4.5.5 Ambiguity Function for the Pulse Burst Waveform

Inserting the definition of the pulse burst waveform of Eq. (4.54) into the definition of the complex ambiguity function of Eq. (4.30) gives

$$\hat{A}(t, F_D) = \int_{-\infty}^{\infty} \left(\sum_{m=0}^{M-1} x_p(s - mT)\right) \left(\sum_{n=0}^{M-1} x_p^*(s - t - nT)\right) \exp(j2\pi F_D s)\, ds \qquad (4.71)$$

$$= \sum_{m=0}^{M-1} \sum_{n=0}^{M-1} \int_{-\infty}^{\infty} x_p(s - mT) x_p^*(s - t - nT) \exp(j2\pi F_D s)\, ds$$

Substituting $s' = s - mT$ gives

$$\hat{A}(t, F_D) = \sum_{m=0}^{M-1} \exp(j2\pi F_D mT) \sum_{n=0}^{M-1} \int_{-\infty}^{\infty} x_p(s') x_p^*(s' - t - nT + mT) \exp(j2\pi F_D s')\, ds' \qquad (4.72)$$

Denoting the complex ambiguity function of the single simple pulse $x_p(t)$ as $\hat{A}_p(t, F_D)$, the integral in Eq. (4.72) is $\hat{A}_p(t + (n - m)T, F_D)$. Thus

$$\hat{A}(t, F_D) = \sum_{m=0}^{M-1} \exp(j2\pi F_D mT) \sum_{n=0}^{M-1} \hat{A}_p(t - (m - n)T, F_D) \qquad (4.73)$$

The double sum in Eq. (4.73) is somewhat difficult to deal with. Obviously, all combinations of m and n having the same difference $m - n$ result in the same summand in the second sum, but the dependence of the exponential term on m only prevents straightforward combining of all such terms. Defining $n' = m - n$, it can be shown by simply enumerating all of the combinations that the double summation of some function $f[m, n]$ can be written[6] (Rihaczek, 1996)

$$\sum_{m=0}^{M-1} \sum_{n=0}^{M-1} f[m, n] = \sum_{n'=-(M-1)}^{0} \sum_{m=0}^{M-|n'|-1} f[m, m - n'] + \sum_{n'=1}^{M-1} \sum_{m=0}^{M-|n'|-1} f[m + n', m] \qquad (4.74)$$

Applying the decomposition of Eq. (4.74) to Eq. (4.73) gives

$$\hat{A}(t, F_D) = \sum_{n'=-(M-1)}^{0} \hat{A}_p(t - n'T, F_D) \sum_{m=0}^{M-|n'|-1} \exp(j2\pi F_D mT)$$
$$+ \sum_{n'=1}^{M-1} \exp(j2\pi F_D n'T) \hat{A}_p(t - n'T, F_D) \sum_{m=0}^{M-|n'|-1} \exp(j2\pi F_D mT) \qquad (4.75)$$

The geometric series that appears in both halves of the right-hand side of this equation sums to

$$\sum_{m=0}^{M-|n'|-1} \exp(j\pi F_D mT) = \exp[j\pi F_D(M - |n'| - 1)T] \frac{\sin(\pi F_D(M - |n'|)T)}{\sin(\pi F_D T)} \qquad (4.76)$$

Using this result in Eq. (4.75) and combining the two remaining sums over n' into one while renaming the index of summation as m gives

---

[6]This amounts to stepping through the various (m, n) index pairs along diagonals instead of along rows or columns.

$$\hat{A}(t, F_D) = \sum_{m=-(M-1)}^{M-1} \hat{A}_p(t - mT, -F_D) \exp[j\pi F_D(M - 1 + m)T] \frac{\sin[\pi F_D(M - |m|)T]}{\sin(\pi F_D T)} \quad (4.77)$$

Equation (4.77) expresses the complex ambiguity function of the coherent pulse train in terms of the complex ambiguity function of its constituent simple pulses and the pulse repetition interval (PRI).

Recall that the support in the delay axis of $\hat{A}_p(t, F_D)$ is $|t| \leq \tau$. If $T > 2\tau$, which is almost always the case, the replications of $\hat{A}_p$ in Eq. (4.77) will not overlap and the magnitude of the sum of the terms as m varies will be equal to the sum of the magnitude of the individual terms. The ambiguity function of the pulse burst waveform can then be written as

$$A(t, F_D) = \sum_{m=-(M-1)}^{M-1} A_p(t - mT, F_D) \left| \frac{\sin[\pi F_D(M - |m|)T]}{\sin(\pi F_D T)} \right| \quad (T > 2\tau) \quad (4.78)$$

To understand this ambiguity function, it is convenient to first look at the zero Doppler and zero delay responses. The zero Doppler response is obtained by setting $F_D = 0$ in Eq. (4.78) and recalling that $A_p(t, 0) = 1 - |t|/\tau$:

$$A(t, 0) = \begin{cases} \sum_{m=-(M-1)}^{M-1} (M - |m|) \left(1 - \frac{|t - mT|}{\tau}\right), & |t - mT| < \tau, \\ 0, & \text{elsewhere} \end{cases} \quad (4.79)$$

Equation (4.79) describes the triangular output of the single-pulse matched filter, repeated every T seconds and weighted by an overall triangular function $M - |m|$. Figure 4.18 illustrates this function for the case $M = 5$ and $T = 4\tau$. The ambiguity function has been normalized by the signal energy E so that it has a maximum value of 1. Note that, as with any waveform, the maximum of the AF occurs at $t = 0$ and the duration is twice the total waveform duration ($2MT$ in this case). The local peaks every T seconds represent the range

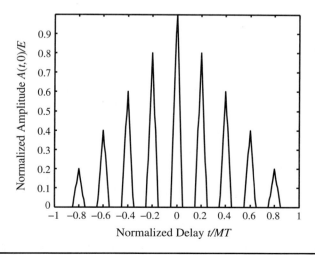

FIGURE 4.18 Zero-Doppler cut of the ambiguity function of a pulse burst. $M = 5$ pulses, $T = 4\tau$.

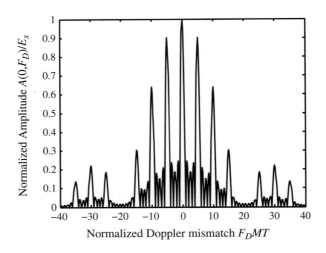

FIGURE 4.19 Zero-delay cut of the pulse burst ambiguity function with $M = 5$ and $T = 4\tau$.

ambiguities discussed previously in Sec. 4.5.3 and illustrated in Fig. 4.15a. If the transmitted waveform were extended by P pulses while the reference waveform remained M pulses long as discussed above, there would be P consecutive spikes with the full amplitude of 1, similar to Fig. 4.16a.

The zero delay cut is obtained by setting $t = 0$ in Eq. (4.78) and recalling that $A_p(0, F_D) = |\sin(\pi F_D \tau)/\pi F_D \tau|$ (assuming a unit energy simple pulse), giving

$$A(0, F_D) = \left| \frac{\sin(\pi F_D \tau)}{\pi F_D \tau} \right| \left| \frac{\sin(\pi F_D MT)}{\sin(\pi F_D T)} \right| \quad (4.80)$$

The response is an asinc function with a first zero at $F_D = 1/MT$ Hz, repeating with a period of $1/T$ Hz. This basic behavior is weighted by a more slowly varying true sinc function with its first zero at $1/\tau$ Hz. This structure is evident in Fig. 4.19, which shows a portion of the zero delay cut for the same case with $M = 5$ and $T = 4\tau$. The $1/T$ spacing of the principal peaks in the zero-delay response is the unambiguous Doppler interval first defined in Chap. 3.

Figure 4.20 is a contour plot of a portion of the complete ambiguity function for this waveform. Note the broadening of the response peaks in Doppler when sampling at the range-ambiguous delays such as 4τ and 8τ (corresponding to ± 0.2 and ± 0.4 on the normalized delay scale of the contour plot). This phenomenon is caused by the $(M - |m|)$ term in the asinc term of Eq. (4.78) and reflects the fact that at these range-ambiguous delays fewer than M pulses are contributing to the matched filter local output peak. The reduced observation time results in degraded Doppler resolution. Again, if the transmitted waveform were extended by P pulses, there would be P consecutive range peaks which retained the full Doppler resolution. This plot also illustrates the breakup of the well-defined peaks in delay when the Doppler mismatch reaches $1/\tau$ hertz (corresponding to 20 on the normalized Doppler scale of the plot).

Figure 4.21 is a diagram of the structure of the central peak of the pulse burst ambiguity function and the first repeated peaks in Doppler and range. This figure summarizes how the various waveform parameters determine the resolution in range and Doppler, the range ambiguity interval, and the blind Doppler interval. The individual pulse length τ is chosen to achieve

Radar Waveforms

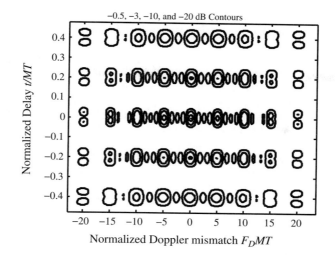

FIGURE 4.20 A portion of the ambiguity function for the pulse burst waveform with $M = 5$ and $T = 4\tau$.

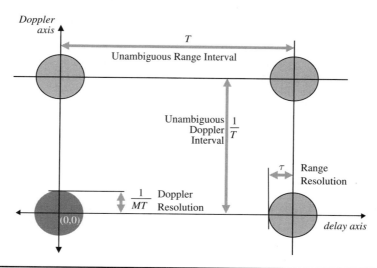

FIGURE 4.21 Relationship between pulse burst waveform parameters and range and Doppler resolution and ambiguities.

the desired range resolution ($c\tau/2$ meters). The pulse repetition interval T sets the ambiguity interval in range ($cT/2$ meters) and the blind interval in Doppler ($1/T$ Hz). Finally, once the PRI is chosen, the number of pulses in the burst determines the Doppler resolution ($1/MT$ Hz).

4.5.6 The Slow-Time Spectrum and the Periodic Ambiguity Function

For a fixed range bin index l, it would seem that the DTFT $Y[l, \omega_D]$ of the slow-time sequence $y[l, m]$ should be related to the variation of the complex ambiguity function $\hat{A}_p(t, F_D)$ in

Doppler. The slow-time sequence $y[l, m]$ is obtained by sampling the simple pulse matched filter output $s_p(t)$ at the same delay after transmission on each pulse. If the target motion across the CPI is small compared to the range resolution (i.e., the target moves only a small fraction of a range bin during the CPI), the amplitude of the sample taken on each pulse will be approximately the same. This amplitude will be the maximum value $s_p(0)$ if the sampling time exactly corresponds to the target range; if it differs by δ_t seconds, the measured amplitude on each pulse will be $s_p(\delta t)$. Thus, the slow-time sequence in a given range bin will have an approximately constant amplitude determined by the ambiguity function of the waveform and the alignment of the target range and the range bin sampling times.

If there is relative motion between the radar and target, there will be a sample-to-sample decrease in the phase of the slow-time samples of the form $-4\pi m v T/\lambda_t$. If the target is within the first unambiguous range interval the target echo will be present in all M slow-time samples for the appropriate range bin and the magnitude of the DTFT will have the $|\sin(\pi F_D MT)/\pi F_D T|$ form seen in Eq. (4.80). However, the $|\sin(\pi F_D \tau)/\pi F_D \tau|$ term of the AF Doppler response due to the individual pulse shape will not be observed in the DTFT; rather, this term will weight the overall amplitude of the DTFT. Finally, if the target range exceeds R_{ua}, the target echo will not be present in all of the slow-time samples and the Doppler resolution will degrade in $Y[l, \omega_D]$ in the same manner it did in $\hat{A}(t, F_D)$.

When the transmitted pulse burst waveform is extended by P pulses to provide full integration gain of a factor of M for targets extending over P range ambiguities, the matched filter output maintains its full maximum peak value of $M \cdot E_p$ over the P range ambiguities [delay interval 0 to $(P-1)T$] of interest, as shown in Fig. 4.16a. The same result over only that delay interval could be obtained by at least two equivalent calculations: correlation of an M-pulse transmitted waveform with an infinitely extended reference, evaluated over $[0, (P-1)T]$; or circular correlation of an M-pulse waveform with an M-pulse reference. The *periodic ambiguity function* (PAF) is a modification of the complex AF of Eq. (4.30) that, when applied to a pulse burst waveform, produces the full-gain AF over this delay interval. A typical definition is (Levanon and Mozeson, 2004; Levanon, 2010)

$$PA(t, F_D) \equiv \left| \int_0^{MT} x(s) \exp(j2\pi F_D s) x^*(s-t) \, ds \right|, \quad 0 \leq t < P \cdot T \quad (4.81)$$

A significant property of the PAF is its relation to the AF of the single constituent pulse in the pulse burst when $T > 2\tau$:[7]

$$PA(t, F_D) = A_p(t, F_D) \left| \frac{\sin(\pi F_D MT)}{\sin(\pi F_D T)} \right|, \quad 0 \leq t < P \cdot T, \quad T \geq 2\tau \quad (4.82)$$

That is, the PAF is the AF of the single pulse multiplied by the DTFT of a discrete M-sample pulse. This is exactly the DTFT $Y[l, \omega_D]$ that will result from the pulse-by-pulse processing approach as described above.

4.6 Frequency-Modulated Pulse Compression Waveforms

A simple pulse has only two parameters, its amplitude A and its duration τ. The range resolution $c\tau/2$ is directly proportional to τ; finer resolution requires a shorter pulse. Most modern radars operate with the transmitter in saturation. That is, any time the pulse is on, its

---

[7] A slightly more complicated version of this result holds when $T \leq 2\tau$, see Levanon and Mozeson (2004).

amplitude is kept at the maximum value of A; amplitude modulation other than on/off switching is not used. This mode of operation maximizes the pulse energy, which is then $A^2\tau$ and is directly proportional to τ. As will be seen in Chaps. 6 and 7, increasing pulse energy improves detection and estimation performance. Thus, improving resolution requires a shorter pulse while improving detection and estimation performance requires a longer pulse. The two metrics are coupled in this unfortunate way because there is effectively only one free parameter τ in the design of the simple pulse waveform.

Pulse compression waveforms decouple energy and resolution. Recall that a simple pulse has a Rayleigh bandwidth $\beta = 1/\tau$ Hz and a Rayleigh resolution in time at the matched filter output of τ seconds. Thus, the *time-bandwidth product* (BT product) $\beta\tau$ of the simple pulse is $\tau(1/\tau) = 1$. A pulse compression waveform, in contrast, has a bandwidth β that is much greater than $1/\tau$. Equivalently, it has a duration τ much greater than that of a simple pulse with the same bandwidth, $\tau \gg 1/\beta$. Either condition is equivalent to stating that a pulse compression waveform has a BT product much greater than 1.

Pulse compression waveforms are obtained by adding frequency or phase modulation to a simple pulse. There are a vast number of pulse compression waveforms in the literature. In this text, only the most commonly used types will be described. These include linear frequency modulation, biphase codes, and certain polyphase codes. Nonlinear FM will also be briefly introduced. Many other waveforms are described in Levanon and Mozeson (2004) and Keel and Baden (2012).

4.6.1 Linear Frequency Modulation

A linear frequency modulated waveform is defined by

$$x(t) = \cos\left(\pi \frac{\beta}{\tau} t^2\right), \quad 0 \leq t \leq \tau \qquad (4.83)$$

The complex equivalent is

$$x(t) = \exp\left(j\pi \frac{\beta}{\tau} t^2\right) = \exp[j\varphi(t)], \quad 0 \leq t \leq \tau \qquad (4.84)$$

The instantaneous frequency in hertz of this waveform is the time derivative of the phase function

$$F_i(t) = \frac{1}{2\pi} \frac{d\varphi(t)}{dt} = \frac{\beta}{\tau} t \quad \text{Hz} \qquad (4.85)$$

This function is shown in Fig. 4.22, assuming $\beta > 0$. $F_i(t)$ sweeps linearly across a total bandwidth of β Hz during the τ-second pulse duration. The waveform $x(t)$ [Eq. (4.83), or the real part

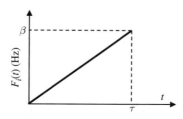

FIGURE 4.22 Instantaneous frequency of an LFM pulse.

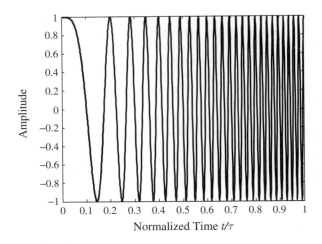

FIGURE 4.23 Real-valued LFM upchirp waveform, BT product $\beta\tau = 50$.

of Eq. (4.84)] is shown in Fig. 4.23 for $\beta\tau = 50$. The LFM waveform is often called a *chirp* waveform in analogy to the sound of an acoustic sinusoid with a linearly changing frequency. When β is positive the pulse is an *upchirp*; if β is negative, it is a *downchirp*. The BT product of the LFM pulse is simply $\beta\tau$; $\beta\tau \gg 1$ if the LFM pulse is to qualify as a pulse compression waveform.

Figure 4.24 shows the magnitude spectrum of the LFM waveform for a relatively low BT product case ($\beta\tau = 10$), and again for a higher BT product case ($\beta\tau = 100$). For low BT, the spectrum is relatively poorly defined. As the BT product increases, the spectrum takes on a more rectangular shape. This is intuitively reasonable: because the sweep is linear, the waveform spreads its energy uniformly across the spectrum.

Figure 4.25 shows the output of the matched filter for the same two chirp waveforms. The dotted line superimposed on the output waveform is the output of a matched filter for a simple pulse of the same duration. As always, the total duration of the matched filter output is 2τ seconds. In both cases, the LFM waveform results in a matched filter output with a

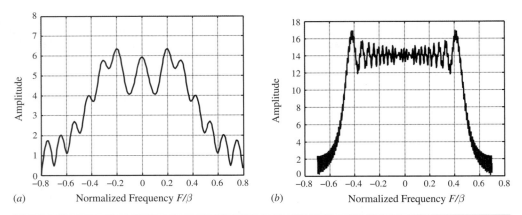

FIGURE 4.24 Magnitude spectrum of an LFM waveform. (a) $\beta\tau = 10$. (b) $\beta\tau = 100$.

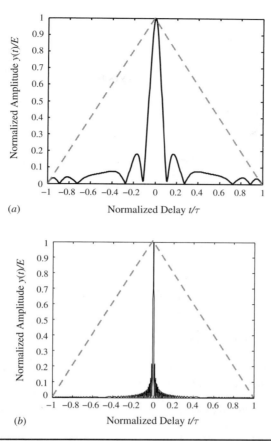

FIGURE 4.25 Output of matched filter for an LFM waveform. (a) $\beta\tau = 10$. (b) $\beta\tau = 100$. The dotted line is the output of a matched filter for a simple pulse of the same duration.

Rayleigh resolution much narrower than τ. In fact, the Rayleigh resolution is very nearly $1/\beta$ in each case (this will be confirmed shortly), an improvement over the simple pulse by a factor of the time-bandwidth product $\beta\tau$.

Simple and LFM pulses of the same amplitude and duration will have the same peak power at the matched filter output and achieve the same output SNR in accordance with Eq. (4.14). However, for an LFM pulse and a simple pulse of the same amplitude to have the same fast-time Rayleigh resolution at the output of their matched filters, the simple pulse must be shorter than the LFM pulse by the factor $\beta\tau$. The energy in the simple pulse and the SNR achieved are then also less by the factor of $\beta\tau$. In other words, the LFM waveform with proper matched filtering achieves a signal processing gain of $G_{sp} = \beta\tau$ compared to a simple pulse of the same Rayleigh resolution.

Unlike the simple pulse case, the matched filter output for the LFM pulse exhibits a sidelobe structure. Figure 4.26 expands the central portion of Fig. 4.25b, showing the distinctly sinc-like mainlobe and first few sidelobes. This should not be surprising: the waveform spectrum $X(F)$ (Fig. 4.24b) is approximately a rectangle of width β Hz.

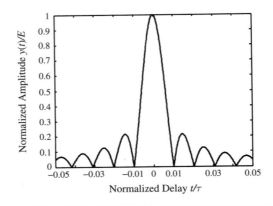

FIGURE 4.26 Expanded view of central portion of Fig. 4.25b.

Consequently, the spectrum of the matched filter output, $|X(F)|^2$, will also be approximately a rectangle of width β. The time-domain output of the matched filter is therefore expected to be approximately a sinc function with a Rayleigh resolution of $1/\beta$ seconds.

To summarize, the LFM waveform enables separate control of pulse energy (through its duration) and range resolution (through its swept bandwidth). The possibility of pulse compression is created by the use of matched filters. The output of the matched filter is not a replica of the transmitted waveform $x(t)$ but of its autocorrelation function $s_x(t)$. Therefore, if a waveform can be designed that has a long duration but a narrowly concentrated autocorrelation, both high energy and fine range resolution can be obtained simultaneously. This in turn is accomplished by modulating a long pulse to spread its bandwidth beyond the usual $1/\tau$. Since the spectrum of the autocorrelation function is just the squared magnitude of the waveform spectrum, a spectrum spread over β Hz will tend to produce a filter output with most of its energy concentrated in a mainlobe of about $1/\beta$ seconds duration. The linear FM pulse is the first example of such a waveform, but phase-coded waveforms will provide more examples of this approach.

4.6.2 The Principle of Stationary Phase

The Fourier transform of Eq. (4.84) is a relatively complicated calculation involving the sine integral $Si(F)$ (Rihaczek, 1996). A very useful and much simpler approximation can be derived using the *principle of stationary phase* (PSP), an advanced technique in Fourier analysis. The PSP is useful for approximate evaluation of integrals with highly oscillatory integrands; thus, it applies particularly well to Fourier transforms of sinusoidally varying signals. Write $x(t)$ in amplitude and phase form, $x(t) = a(t)\exp[j\varphi(t)]$, and consider its Fourier transform

$$X(\Omega) = \int_{-\infty}^{+\infty} \underbrace{a(t)\exp[j\varphi(t)]}_{x(t)} \exp(-j\Omega t)\, dt \qquad (4.86)$$

Define the phase $\phi(\tau, \Omega)$ of the entire Fourier integrand as the combination of the signal phase and the Fourier kernel phase, then

$$X(\Omega) = \int_{-\infty}^{+\infty} a(t)\exp[j(\varphi(t) - \Omega t)]\, dt \equiv \int_{-\infty}^{+\infty} a(t)\exp[j\phi(t,\Omega)]\, dt \qquad (4.87)$$

Of course, the exact Fourier transform is known for many signals having relatively simple phase functions $\varphi(t)$. The PSP is most useful when the signal phase function and thus the total integral phase $\phi(t, \Omega)$ is continuous but nonlinear or otherwise complicated.

Define a *stationary point* of the integrand as a value of $t = t_0$ such that the first time derivative of the integral phase $\phi'(t_0, \Omega) = 0$. Then the PSP approximation to the spectrum is (Born and Wolf, 1959; Raney, 1992)

$$X(\Omega) \approx \sqrt{\frac{-2\pi}{\phi''(t_0, \Omega)}} \exp(-j\pi/4) a(t_0) \exp[j\phi(t_0, \Omega)] \qquad (4.88)$$

where $\phi''(t_0, \Omega)$ is the second time derivative of $\phi(t, \Omega)$ evaluated at $t = t_0$. If there are multiple stationary points the spectrum is the sum of such terms for each stationary point. Equation (4.88) states that the magnitude of the spectrum at a given frequency Ω is proportional to the amplitude of the signal envelope at the time that the stationary point occurs and, more importantly, is inversely proportional to the square root of the rate of change of the frequency at that time, $\phi''(t_0, \Omega)$. The PSP also implies that only the stationary points significantly influence $X(\Omega)$.

The PSP can be applied to estimate the spectrum of the LFM waveform. The waveform is defined as

$$x(t) = a(t) \exp(j\alpha t^2), \quad a(t) = \begin{cases} 1, & -\tau/2 \le t \le +\tau/2 \\ 0, & \text{otherwise} \end{cases}, \quad \alpha \equiv \pi \frac{\beta}{\tau} \qquad (4.89)$$

The Fourier transform of $x(t)$ is

$$X(\Omega) = \int_{-\infty}^{+\infty} x(t) \exp(-j\Omega t)\, dt = \int_{-\infty}^{+\infty} a(t) \exp[j(\alpha t^2 - \Omega t)]\, dt \qquad (4.90)$$

The integrand phase and its derivatives are then

$$\phi(t, \Omega) = \alpha t^2 - \Omega t, \quad \phi'(t, \Omega) = 2\alpha t - \Omega, \quad \phi''(t, \Omega) = 2\alpha \qquad (4.91)$$

The stationary points are found by setting $\phi'(t, \Omega) = 0$ and solving for t. In this case, there is only one stationary point:

$$0 = \phi'(t_0, \Omega) \Rightarrow t_0 = \frac{\Omega}{2\alpha} \qquad (4.92)$$

Inserting Eq. (4.92) into Eq. (4.88) gives

$$\begin{aligned} X(\Omega) &\approx \sqrt{\frac{-2\pi}{\phi''(t_0, \Omega)}} \exp(-j\pi/4) a(t_0) \exp[j\phi(t_0, \Omega)] \\ &= \sqrt{\frac{-2\pi}{2\alpha}} \exp(-j\pi/4) a\left(\frac{\Omega}{2\alpha}\right) \exp(j[\alpha(\Omega/2\alpha)^2 - \Omega(\Omega/2\alpha)]) \\ &= j\sqrt{\frac{\pi}{\alpha}} \exp(-j\pi/4) a\left(\frac{\Omega}{2\alpha}\right) \exp(-j\Omega^2/4\alpha) \end{aligned} \qquad (4.93)$$

Recalling the finite support of the signal envelope $a(t)$, the term $a(\Omega/2\alpha)$ becomes (using $\alpha = \pi\beta/\tau$)

$$a\left(\frac{\Omega}{2\alpha}\right) = \begin{cases} 1, & -\tau/2 \le \frac{\Omega}{2\alpha} \le +\tau/2 \\ 0, & \text{otherwise} \end{cases} = \begin{cases} 1, & -2\pi\left(\frac{\beta}{2}\right) \le \Omega \le +2\pi\left(\frac{\beta}{2}\right) \\ 0, & \text{otherwise} \end{cases} \qquad (4.94)$$

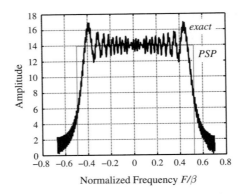

FIGURE 4.27 Comparison of actual magnitude spectrum and PSP approximation for an LFM pulse with $\beta\tau = 100$.

The final result is

$$X(\Omega) \approx j\sqrt{\frac{\pi}{\alpha}}\exp(-j\pi/4)\exp(-j\Omega^2/4\alpha), \quad -2\pi\left(\frac{\beta}{2}\right) \leq \Omega \leq +2\pi\left(\frac{\beta}{2}\right) \quad (4.95)$$

Figure 4.27 compares this approximation with the exact spectrum when $\beta\tau = 100$. Equation (4.95) estimates that $|X(\Omega)|$ is constant over the range $\pm\beta/2$ Hz and is zero outside of this range. This is both intuitively satisfying, since this is exactly the range over which the instantaneous frequency of the LFM pulse sweeps, and consistent with the increasingly rectangular shape of the exact spectrum as the BT product increases. The PSP result also gives an estimate of the phase of the spectrum which, like the temporal phase of the waveform $x(t)$, is seen to be quadratic.

4.6.3 Ambiguity Function of the LFM Waveform

The ambiguity function of an LFM pulse can be obtained by direct calculation, similar to the simple pulse, but with a good deal more tedium. An easier way is to introduce the "chirp property" of the ambiguity function and then apply it to the LFM case. Suppose that a waveform $x(t)$ has an ambiguity function $A(t, F_D)$. Create a modified waveform $x'(t)$ by modulating $x(t)$ with a linear FM complex chirp, $x'(t) \equiv x(t)\exp(j\pi\beta t^2/\tau)$, and compute its complex ambiguity function:

$$\begin{aligned}
\hat{A}'(t, F_D) &= \int_{-\infty}^{\infty} x'(s)x'^{*}(s-t)\exp(j2\pi F_D s)\,ds \\
&= \int_{-\infty}^{\infty} x(s)\exp(j\pi\beta s^2/\tau)x^{*}(s-t)\exp(-j\pi\beta(s-t)^2/\tau)\exp(j2\pi F_D s)\,ds \\
&= \exp(-j\pi\beta t^2/\tau)\int_{-\infty}^{\infty} x(s)x^{*}(s-t)\exp[j2\pi(F_D + \beta t/\tau)s]\,ds \\
&= \exp(-j\pi\beta t^2/\tau)\hat{A}\left(t, F_D + \frac{\beta}{\tau}t\right)
\end{aligned} \quad (4.96)$$

Taking the magnitude of $\hat{A}'(t, F_D)$ gives the ambiguity function of the chirp signal in terms of the ambiguity function of the original signal without the chirp:

$$A'(t, F_D) = A\left(t, F_D + \frac{\beta}{\tau}t\right) \qquad (4.97)$$

Equation (4.97) states that adding a chirp modulation to a signal skews its ambiguity function in the delay-Doppler plane. Applying this property to the simple pulse AF [Eq. (4.51)] gives the AF of the LFM waveform

$$\begin{aligned}A(t, F_D) &= \left(1 - \frac{|t|}{\tau}\right)\left|\frac{\sin[\pi(F_D + \beta t/\tau)\tau(1 - |t|/\tau)]}{\pi(F_D + \beta t/\tau)\tau(1 - |t|/\tau)}\right| \\ &= \left(1 - \frac{|t|}{\tau}\right)\left|\frac{\sin[\pi(F_D\tau + \beta t)(1 - |t|/\tau)]}{\pi(F_D\tau + \beta t)(1 - |t|/\tau)}\right|, \quad -\tau \leq t \leq \tau \end{aligned} \qquad (4.98)$$

Figure 4.28 is a contour plot of the AF of an LFM pulse of duration $\tau = 10$ μs and swept bandwidth $\beta = 1$ MHz; thus, the BT product is 10. The AF retains the triangular ridge of the simple pulse but is now skewed in the delay-Doppler plane as predicted by Eq. (4.97).

The zero-Doppler cut of the LFM ambiguity function is the matched filter output when there is no Doppler mismatch,

$$A(t, 0) = \left|\frac{\sin[\pi\beta t(1 - |t|/\tau)]}{\pi\beta t}\right|, \quad -\tau \leq t \leq \tau \qquad (4.99)$$

This function was illustrated for BT products of both 10 and 100 in Fig. 4.25. The Rayleigh resolution of the LFM pulse is obtained by examination of Eq. (4.99). The peak of $A(t, 0)$ occurs at $t = 0$. The first zero occurs when the argument of the numerator equals π, which occurs when $\beta t(1 - |t|/\tau) = 1$. For positive t, this becomes

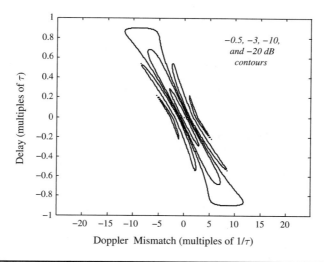

FIGURE 4.28 Contour plot of the ambiguity function of an LFM waveform with $\beta\tau = 10$.

$$\beta t - \frac{\beta t^2}{\tau} = 1 \quad \Rightarrow \quad t^2 - \tau \cdot t + \tau/\beta = 0 \tag{4.100}$$

The roots of this equation are $t = \left(\tau \pm \sqrt{\tau^2 - 4\tau/\beta}\right)/2 = \tau\left(1 \pm \sqrt{1 - 4/\beta\tau}\right)/2$. So long as $\beta\tau > 4$, the square root in the last expression is a positive real number between 0 and 1, so taking the negative sign gives the positive root closest to zero, which is the Rayleigh resolution in time. This result can be simplified with the following series expansion of the square root

$$\sqrt{1-x} = 1 - \frac{x}{2} - \frac{x^2}{8} - \cdots \approx 1 - \frac{x}{2} \quad (x \ll 4) \quad \overset{\Rightarrow}{x = 4/\beta\tau}$$

$$t \approx \frac{\tau}{2}\left[1 - \left(1 - \frac{2}{\beta\tau}\right)\right] \approx \frac{1}{\beta} \quad (\beta\tau \gg 1) \tag{4.101}$$

The Rayleigh resolution in time is therefore approximately $1/\beta$ seconds, corresponding to a Rayleigh range resolution ΔR of

$$\Delta R = \frac{c}{2\beta} \quad \text{m} \tag{4.102}$$

The zero-delay response is

$$A(t, F_D) = \left|\frac{\sin(\pi F_D \tau)}{\pi F_D \tau}\right| \tag{4.103}$$

which is simply a standard sinc function. The Doppler resolution of the LFM pulse is the same as that of a simple pulse, namely

$$\Delta F_D = \frac{1}{\tau} \quad \text{Hz} \tag{4.104}$$

Equation (4.103) shows that, like the simple pulse, the Doppler resolution of an LFM pulse is inversely proportional to the pulse length. Furthermore, the energy in the LFM pulse is still $A^2\tau$, directly proportional to the pulse length. Equation (4.102) shows that, *unlike* the simple pulse, the range resolution is not determined by the pulse width τ but instead is inversely proportional to the swept bandwidth. Thus, the LFM waveform's two parameters, bandwidth and duration, can be used to independently control pulse energy and range resolution. The pulse length is chosen (along with the pulse amplitude) to set the desired energy, while the swept bandwidth is chosen to obtain the desired range resolution.

The expression $c/2\beta$ for range resolution is quite general. For instance, the Rayleigh bandwidth of a simple pulse is $\beta = 1/\tau$ Hz; using this in $c/2\beta$ gives $\Delta R = c\tau/2$ as before. While bandwidth and pulse length are directly related in the simple pulse, modulation of the LFM waveform has decoupled them. If $\beta\tau > 1$ for the LFM pulse the range resolution will be better than that of a simple pulse of the same duration by the factor $\beta\tau$. Alternatively, the range resolution of a simple pulse of length τ can be matched by an LFM pulse that is longer (and thus higher energy, given the same transmitted power) by the factor $\beta\tau$.

4.6.4 Range-Doppler Coupling

The skew in the ambiguity function for the LFM pulse gives rise to an interesting phenomenon. Consider the AF of Eq. (4.98). The peak of this sinc-like function will occur when

$$F_D + \frac{\beta}{\tau} t = 0 \quad \text{Hz} \quad \Rightarrow \quad t = -\frac{\tau}{\beta} F_D \quad \text{seconds} \tag{4.105}$$

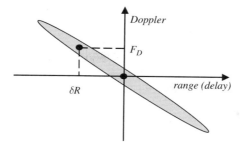

FIGURE 4.29 Illustration of the effect of range-Doppler coupling on apparent target range.

That is, when there is a Doppler mismatch, $F_D \neq 0$, the peak of the matched filter output will not occur at $t = 0$ as desired. Instead, it will be shifted by an amount proportional to the Doppler shift. Because the target range will be estimated based on the time of occurrence of this peak, a Doppler mismatch will induce an error in measuring range of

$$\delta R = -\frac{c\tau F_D}{2\beta} \text{ m} \qquad (4.106)$$

The amplitude of the peak will also be reduced by the factor $(1 - |t|/\tau) = (1 - F_D/\beta)$. Figure 4.29 illustrates the skewed ridge of the LFM ambiguity function and the relationship between Doppler shift and range measurement error.

While an incorrect range measurement is certainly undesirable, range-Doppler coupling is a useful phenomenon in some systems. A simple pulse with duration τ will have a Doppler Rayleigh resolution of $1/\tau$ Hz; targets with Doppler mismatches approaching this value or larger will produce a greatly attenuated output from the matched filter and will likely go undetected. An LFM pulse of the same duration will produce a significant output peak for a much broader range of Doppler shifts, even though the peak will be mislocated in range. Nonetheless, the target will be more likely to be detected. The LFM waveform is said to be more *Doppler tolerant* than the simple pulse. This makes it a good choice for surveillance applications because a relatively large range of Doppler shifts can be searched with an LFM pulse. The range error can be eliminated, at least for isolated targets, by repeating the measurements with an LFM pulse of the opposite slope, for example, an upchirp followed by a downchirp. In this case, the sign of the range error will be reversed. Averaging the two measurements will give the true range and also allow determination of the Doppler shift.

4.6.5 Stretch Processing

LFM waveforms are often the waveform of choice for exceptionally wideband radar systems where the swept bandwidth β may be hundreds of megahertz or even exceed 1 GHz. Digital processing can be difficult to implement in such systems because the high instantaneous bandwidth of the waveform requires equally high sampling rates in the A/D converter. It was difficult to obtain high-quality A/D converters at these rates with dynamic ranges corresponding to an *effective number of bits* (ENOB)[8] more than perhaps 9.5 bits with then-current technology in 2012; the ENOB at 1 GHz was expected to reach only about 11 bits by 2020

---

[8]ENOB is a better measure of A/D performance than number of bits. One can always clock *something* into the output register at a high rate; the issue is, how many of those bits are accurate? ENOB takes into account circuit noise and other distortions to give a measure of the number of reliable bits.

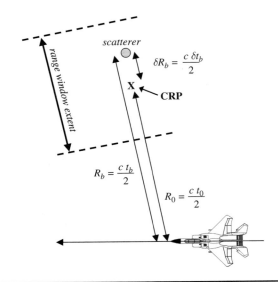

FIGURE 4.30 Scenario for stretch processing analysis.

(Jonsson, 2010; Jonsson, 2012). In addition, the sheer number of samples generated can be stressing for the signal processor.

Stretch processing is a specialized technique for matched filtering of wideband LFM waveforms. It is also called *deramp processing*, *deramp on receive*, *dechirp*, and *one-pass processing*. It is essentially the same as the processing used with linear frequency-modulated continuous wave (FMCW) radar, described in Chap. 3 and to be revisited in Sec. 4.10. Stretch processing is most appropriate for applications seeking very fine range resolution over relatively short-range windows.

Figure 4.30 shows the scenario for analyzing stretch processing. The *central reference point* (CRP) is at the middle of the range window of interest at a range of R_0 meters, corresponding to a time delay of t_0 seconds. Consider a scatterer at range R_b and time delay $t_b = t_0 + \delta t_b$. The problem will be analyzed in terms of differential range or delay relative to the CRP, denoted δR_b and δt_b. The transmitted waveform is the LFM pulse of Eq. (4.84). The echo from the scatterer, with the carrier frequency included, is

$$\bar{y}(t) = \rho \exp\left[j\pi \frac{\beta}{\tau}(t - t_b)^2\right] \exp[j\Omega_t(t - t_b)], \quad 0 \leq t - t_b \leq \tau \qquad (4.107)$$

where ρ is proportional to the scatterer reflectivity. This echo is processed with the modified coherent receiver in complex equivalent form, shown in Fig. 4.31. The unique aspects of this stretch receiver are the reference oscillator and the Fourier transform. The oscillator contains

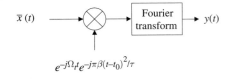

FIGURE 4.31 Complex equivalent receiver for stretch processor.

a conventional term $\exp(j\Omega_t t)$ to remove the carrier. However, it also contains a replica of the transmitted chirp, referenced to the time delay t_0 corresponding to the CRP. The Fourier transform performs a spectral analysis of the mixer output signal.

After some algebra and using $t_b = t_0 + \delta t_b$, the output $y(t)$ can be expressed as

$$y(t) = \rho \exp\left[-j\frac{4\pi R_b}{\lambda_t}\right]\exp\left[-j2\pi\frac{\beta}{\tau}\delta t_b(t-t_0)\right]\exp\left[j\pi\frac{\beta}{\tau}(\delta t_b)^2\right], \quad t_0 \leq t - \delta t_b \leq t_0 + \tau \quad (4.108)$$

The phase term that is quadratic in δt_b is a complex constant. In synthetic aperture imaging it is called the *residual video phase* (RVP). The middle complex exponential contains a term that is linear in t and therefore represents a constant-frequency complex sinusoid. By inspection, the sinusoid frequency is $F_b = -\beta \cdot \delta t_b/\tau$ Hz. F_b is proportional to δt_b and thus to the range of the scatterer relative to the CRP. The differential range can be obtained from the mixer output frequency as

$$\delta R_b = -\frac{cF_b\tau}{2\beta} \text{ m} \quad (4.109)$$

Heuristically, the scatterer produces a constant frequency tone at the output of the stretch receiver because the receiver not only removes the carrier from the LFM echo but also combines it in a mixer with a replica of the LFM with a delay corresponding to the CRP. In the conventional real-signal receiver of Fig. 1.12, the mixer produces sum and difference "beat" frequencies. The sum frequency is removed by a lowpass filter. (This lowpass filter is not needed in the complex representation and is therefore not shown in Fig. 4.31.) The difference frequency is the difference between the instantaneous frequency of the LFM echo and the LFM reference. Since both have the same sweep rate, this beat frequency is a constant. This process was described in more detail for the analogous linear FMCW waveform in Chap. 3.

If there are several scatterers distributed at ranges R_i and delays δt_i, the stretch receiver output is simply the superposition of several terms of the form of Eq. (4.108),

$$y(t) = \sum_i \rho_i \exp\left[-j\frac{4\pi R_i}{\lambda_t}\right]\exp\left[-j2\pi\frac{\beta}{\tau}\delta t_i(t-t_0)\right]\exp\left[j\pi\frac{\beta}{\tau}(\delta t_i)^2\right] \quad (4.110)$$

Thus the output of the stretch receiver contains a different beat frequency tone for each scatterer. Spectral analysis of $y(t)$ can identify the beat frequencies present in the mixer output and therefore the ranges and amplitudes of the scatterers present in the composite echo. Figure 4.32 illustrates the instantaneous frequencies and timing of the signals involved for three scatterers, one in the middle and one at each edge of the scene.

It is desirable that the reference LFM chirp completely overlap the echo from a scatterer anywhere within the range window. If the range window is $R_w = cT_w/2$ meters long, the leading edge of the echo from a scatterer at the nearest range, $R_0 - R_w/2$, will arrive $t_0 - T_w/2$ seconds after transmission, as shown in Fig. 4.32. The trailing edge of the echo from the scatterer at the far limit of the range window, $R_0 + R_w/2$, will arrive $t_0 + T_w/2 + \tau$ seconds after transmission. Thus, data from the range window have a total duration of $T_w + \tau$ seconds. To ensure complete overlap of the reference chirp with echoes from any part of the range window the reference chirp must be $T_w + \tau$ seconds long and so will sweep over $(1 + T_w/\tau)\beta$ Hz.

Another issue evident in Fig. 4.32 is *range skew*. This is the phenomenon whereby the beat frequencies for scatterers at different ranges, while all of the same duration (provided the reference chirp is lengthened), start and stop at different times. This complicates

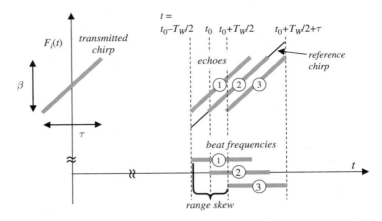

FIGURE 4.32 Instantaneous frequency versus time for an LFM transmitted pulse and echoes from three scatterers. See text for details.

weighting of the mixer output prior to spectral analysis for range sidelobe control. If the window is aligned with the beat frequency for the center scatterer response, it will be misaligned with earlier and later scatterer responses. If it is lengthened to cover the full mixer output duration of $T_w + \tau$ seconds, none of the beat frequencies will be weighted by the full window and each will have a different effective window function. In either case, sidelobe suppression will be poor.

This problem can be solved by placing an additional filter between the mixer output and the Fourier transform. Notice that the scatterer at delay δt_b relative to the patch center generates a beat frequency of $-\beta \cdot \delta t_b / \tau$ Hz. What is needed is a filter whose frequency response has unit magnitude for all frequencies so as not to distort the scatterer amplitudes, but also a *group delay*[9] of $+\delta t_b$ seconds at the frequency $-\beta \cdot \delta t_b / \tau$ so as to align the beat frequencies in time. It can be shown that the required frequency response in analog radian frequency units is $H(\Omega) = \exp(-j\Omega^2 \tau / 4\pi\beta)$, which has unit magnitude for all frequencies and a quadratic phase in the frequency domain (see Prob. 15). All of the beat frequencies will be aligned in time at the output of this filter. As an extra benefit, this filter also corrects RVP (Carrara et al., 1995).

The bandwidth of the stretch receiver output can be obtained by considering the difference in beat frequencies for scatterers at the near and far edges of the range window. This gives

$$F_{near} - F_{far} = \left[-\frac{\beta}{\tau}\left(-\frac{T_w}{2}\right)\right] - \left[-\frac{\beta}{\tau}\left(+\frac{T_w}{2}\right)\right]$$
$$= \frac{T_w}{\tau}\beta$$
(4.111)

If $T_w < \tau$, the bandwidth at the receiver output is less than the original signal bandwidth β. The mixer output can then be sampled with slower A/D converters and the number of range

---

[9]Group delay in seconds is the negative of the derivative with respect to Ω of the frequency-domain phase function $\Phi(\Omega)$ of $H(\Omega) = |H(\Omega)|\exp[j\Phi(\Omega)]$. It is a measure of the filter delay for inputs of a given frequency. See Oppenheim and Schafer (2010).

samples needed to represent the range window data is reduced. Thus, the stretch technique is most useful for systems performing fine range resolution analysis over limited range windows. Also note that while the digital processing rates have been reduced, the analog receiver hardware up through the LFM mixer must still be capable of handling the full instantaneous signal bandwidth.

As an example, consider a 100-μs pulse with a swept bandwidth of 750 MHz, giving a BT product of 75,000. Suppose the desired range window is $R_w = 1.5$ km, corresponding to a sampling window of $T_w = 10$ μs. In a conventional receiver the sampling rate will be 750 megasamples per second. Data from scatterers over the extent of the range window will extend over $T_w + \tau$ seconds, requiring (750 MHz)(10 μs + 100 μs) = 82,500 samples to represent the range profile. In contrast, the bandwidth at the output of the stretch receiver will be $(T_w/\tau)\beta = 75$ MHz. The sampled time interval remains the same, so only 8250 samples are required. Restricting the analysis to a delay window one-tenth the length of the pulse and using the stretch technique has resulted in a factor of 10 reduction in both the sampling rate required and the number of samples to be digitally processed.

Stretch processing of linear FM waveforms preserves both the resolution and the range-Doppler coupling properties of conventionally processed LFM. Consider the output of the stretch mixer for a scatterer at differential range δt_b from the central reference point. This signal will be a complex sinusoid at a frequency $F_b = -\beta \cdot \delta t_b/\tau$ Hz observed for a duration of τ seconds. In the absence of windowing, the Fourier transform of this signal will be a sinc function with its peak at F_b and a Rayleigh resolution of $1/\tau$ Hz. The processor will be able to resolve scatterers whose beat frequencies are at least $\Delta F_b = 1/\tau$ Hz apart. The time delay spacing that gives this frequency separation satisfies

$$\frac{1}{\tau} = \left|\frac{\beta}{\tau}\delta t_b\right| \Rightarrow \delta t_b = \frac{1}{\beta} \text{ seconds} \tag{4.112}$$

The corresponding range separation is then the usual result for range resolution,

$$\delta R_b = \frac{c}{2}\delta t_b = \frac{c}{2\beta} \text{ m} \tag{4.113}$$

If the reference oscillator sweep is not lengthened as discussed above to fully overlap the echo from scatterers at any location in the range window, the range resolution will be degraded. Specifically, the duration τ' of the beat frequency will be less than τ seconds for scatterers at any delay other than the center of the window due to the incomplete overlap. The Rayleigh resolution of the Fourier transform of that scatterer's beat frequency will increase to a value $1/\tau' > 1/\tau$, causing the range resolution of Eq. (4.113) to increase proportionately. The processing gain will similarly be reduced from the ideal factor of $\beta\tau$.

To consider the effect of Doppler shift on the stretch processor, replace $\bar{y}(t)$ in Eq. (4.107) with

$$\bar{y}(t) = \rho \exp\left[j\pi\frac{\beta}{\tau}(t - t_b)^2\right]\exp(j2\pi F_D t)\exp[j\Omega_t(t - t_b)], \quad 0 \leq t - t_b \leq \tau \tag{4.114}$$

Repeating the previous analysis, Eq. (4.108) becomes

$$y(t) = \rho \exp\left[-j\frac{4\pi R_b}{\lambda_t}\right]\exp\left[-j2\pi\left(\frac{\beta}{\tau}\delta t_b - F_D\right)t - 2\pi\frac{\beta}{\tau}\delta t_b \cdot t_0\right]\exp\left[j\pi\frac{\beta}{\tau}(\delta t_b)^2\right] \tag{4.115}$$

Equation (4.115) shows that the effect of a Doppler shift is to increase the beat frequency F_b by F_D Hz. Since beat frequency is mapped to differential range by the stretch processor according to $\delta R_b = -cF_b\tau/2\beta$, this implies a measured range shift of

$$\delta R = -\frac{c\tau}{2\beta}F_D \quad \text{m} \tag{4.116}$$

which is the same range-Doppler coupling relationship obtained previously [Eq. (4.106)].

Stretch processing and especially Eq. (4.110) will be revisited and extended further in Chap. 8, where the technique is central to the polar format algorithm for spotlight synthetic aperture radar imaging. Additional details of stretch processing are given in Keel and Baden (2012).

4.7 Range Sidelobe Control for FM Waveforms

It was seen in the previous section that the output of the LFM matched filter exhibits sidelobes in range. These are a consequence of the approximately rectangular LFM matched filter output spectrum, which produces a sinc-like range response. The first range sidelobe is approximately 13 dB below the output peak for moderate-to-high BT products, and about -15 dB for small BT products. Sidelobes this large are unacceptable in many systems that will encounter multiple targets in range due to *target masking*. This phenomenon is shown in Fig. 4.33a, where the smaller target is barely visible above the sidelobes of the stronger target despite being separated by approximately 16 times the Rayleigh resolution. The smaller target could not be reliably detected in this scenario. If the sidelobes could be reduced, this masking effect could be greatly reduced as shown in Fig. 4.33b.

For a simple pulse, the matched filter output is a triangle function, which exhibits no sidelobes. Thus, sidelobe reduction is not an issue for that waveform, and it will not be further discussed. For the LFM pulse, there are two basic approaches to range sidelobe reduction: shaping the receiver frequency response and shaping the waveform spectrum.

4.7.1 Matched Filter Frequency Response Shaping

Recall from finite impulse response (FIR) digital filter design that to reduce sidelobes of the frequency response of a digital filter, a window function is applied in the time domain to the

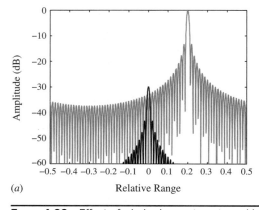
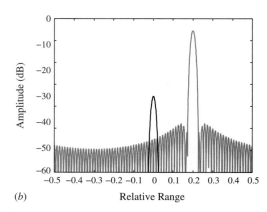

FIGURE 4.33 Effect of windowing on target masking. (a) No windowing. (b) Hamming window applied.

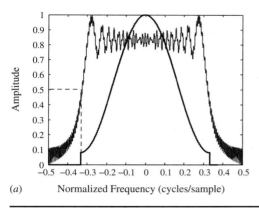 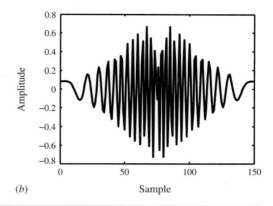

FIGURE 4.34 Hamming weighting of the LFM receiver frequency response. (a) Hamming window overlaid on matched filter frequency response. (b) Resulting filter impulse response $h'(t)$.

impulse response. The goal here is to reduce sidelobes in range, corresponding to the time domain, so the analogous approach is to window the receiver frequency response in the frequency domain. Basic properties of windowing are discussed in App. B.

The matched filter frequency response is $H(F) = X^*(F)$, and at least for larger BT products is approximately rectangular. A modified frequency response $H'(F)$ can be obtained by multiplying $H(F)$ by a window function $w(F)$[10]

$$H'(F) = w(F)H(F) = w(F)X^*(F) \qquad (4.117)$$

Figure 4.34a shows a Hamming window function overlaid on the matched filter frequency response for an LFM waveform with $\beta\tau = 100$. The window cutoff was placed at $\pm\beta/2$ Hz. $H'(F)$ is the product of these two functions. The resulting impulse response $h'(t)$ is shown in Fig. 4.34b. The responses of both the matched filter and the filter of Fig. 4.34b to the unwindowed LFM echo are overlaid in Fig. 4.35. The peak sidelobe has dropped 23.7 dB, from 13.5 dB below the mainlobe peak in the unwindowed case to 37.2 dB below the mainlobe peak for the windowed case. This comes at a cost of the mainlobe peak gain dropping 5.35 dB and the Rayleigh time (range) resolution increasing by 93 percent.

Since the matched frequency response does not have a perfectly sharp cutoff frequency, there is some uncertainty as to where in frequency to place the window cutoff. In the example, the support of the window equals the instantaneous frequency span of $\pm\beta/2$ Hz, which is ± 0.36 cycles per sample on the normalized frequency scale used for this particular sampled LFM waveform. However, this choice cuts off some of the waveform energy in the sidelobes, increasing the mismatched filtering losses. A case could be made for a narrower support so that the window is applied only over the relatively flat portion of the spectrum. This would provide range sidelobes more closely matching those expected for the chosen window but would reduce the effective bandwidth, further degrading the range resolution. A case could also be made for increasing the support to maximize the output energy, but this

---

[10] The lower case w is used for the frequency-domain window function $w(F)$ to emphasize that the multiplying function is the window function itself (e.g., a Hamming window) rather than its Fourier transform.

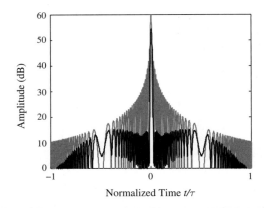

FIGURE 4.35 Comparison of the receiver filter output with (black curve) and without (gray curve) frequency-domain Hamming weighting of the matched filter. See text for details.

choice might increase the range sidelobes by "wasting" some of the window shape on the skirts of the LFM spectrum.

Since $H'(F) \neq X^*(F)$, the modified receiver is not matched to the transmitted LFM pulse and therefore the output peak and SNR will be reduced from their maximum values. This effect was evident in Fig. 4.33b, where the peak of the dominant target response is several dB lower than the unwindowed case in Fig. 4.33a. The losses in output peak amplitude and SNR can be estimated from the window function $w(F)$. In practice, a discrete window $w[k]$ will be applied to a discrete-frequency version of $H(F)$, $H[k]$. From App. B, the loss in the peak signal output from the matched filter, called the *loss in processing gain* (LPG), is

$$LPG = \frac{K^2}{\left|\sum_{k=0}^{K-1} w[k]\right|^2} \quad (4.118)$$

where K is the window length. The loss in SNR at the matched filter output is called the *processing loss* (PL) and is

$$PL = \frac{K \sum_{k=0}^{K-1} |w[k]|^2}{\left|\sum_{k=0}^{K-1} w[k]\right|^2} \quad (4.119)$$

With these definitions *LPG* and *PL* are both greater than 1 so that the losses in decibels are positive numbers. For a relatively long Hamming window, the LPG is approximately 5.4 dB while the PL is approximately 1.4 dB. Both are weak functions of K and are slightly larger for small K. These formulas are approximate when applied to windowing of the LFM spectrum due to the finite-width transition of the LFM spectrum and the designer's discretion in choosing the cutoff of the window in frequency. In the example of Fig. 4.34, the LPG is 5.35 dB. These formulas are derived in App. B. They will arise again in the context of Doppler processing and where in Chap. 5, where the results will be exact.

4.7.2 Matched Filter Impulse Response Shaping

The impulse response of the filter just obtained and illustrated in Fig. 4.34b suggests that similar results could have been obtained by windowing the LFM waveform in the time domain. Consider again the signal in Eq. (4.89) having an arbitrary amplitude function $a(t)$ and a quadratic phase function. The PSP approximation to its spectrum was given in Eq. (4.93). The magnitude of the spectrum is proportional to the *time*-domain amplitude function $a(\cdot)$:

$$|X(\Omega)| \propto \left|a\left(\frac{\Omega}{2\alpha}\right)\right| \qquad (4.120)$$

If $a(t)$ has finite support on $-\tau/2 \leq t \leq \tau/2$, it follows that $X(\Omega)$ will have finite support on $-\beta/2 \leq F \leq \beta/2$ and in that interval $|X(\Omega)|$ has the same shape as the window magnitude $|a(t)|$. Thus, a Hamming-shaped (for example) spectrum can be obtained by applying a Hamming window to the impulse response $h(t)$ instead of the frequency response $H(F)$. This result is specific to the use of a linear FM waveform.

The output of the resulting filter is overlaid on the matched filter response in Fig. 4.36. It has the same general character as the frequency-domain weighting result but with some differences in details of the sidelobe structure. The peak is reduced from 60 to 54.64 dB with weighting, a nearly identical LPG of 5.36 dB. The peak sidelobe of the weighted response is 40.7 dB below the corresponding mainlobe peak, 3.5 dB better than the frequency-domain case. The Rayleigh width has increased 97 versus 93 percent in the frequency domain weighted case. This is consistent with the better sidelobe performance of the time-domain case. Additional detail on frequency- and time-domain weighting of LFM waveforms is available in Richards (2006).

4.7.3 Waveform Spectrum Shaping

The principal limitation of the receiver weighting approach to range sidelobe control is that the resulting filter is not matched to the transmitted waveform, resulting in an SNR loss. An alternative approach is to design a modified pulse compression waveform whose matched filter output inherently has lower sidelobes than the standard LFM. The waveform should be designed to have a spectrum shaped like that of a window function with the desired

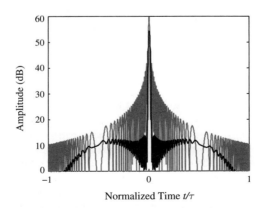

FIGURE 4.36 Comparison of the receiver filter output with (black curve) and without (gray curve) time-domain Hamming weighting of the matched filter. See text for details.

sidelobe behavior. Such a waveform would combine the maximized SNR of a truly matched filter with low sidelobes.

There are two common ways to shape the spectrum. Both start with the idea that the LFM spectrum's relatively square shape is the result of a linear sweep rate combined with a constant pulse amplitude, resulting in a fairly uniform distribution of the signal energy across the spectral bandwidth. Instead, the spectral energy could be reduced at the edges, giving a "window-shaped" spectrum, by either reducing the signal amplitude at the pulse edges while maintaining a constant sweep rate, using a faster sweep rate at the edges with a constant pulse amplitude so as to spend less time in each spectral interval near the band edges, or both. The technique using variable sweep rates is referred to as *nonlinear FM* (NLFM).

The amplitude modulation technique implies operating the power amplifier at less than full power over the pulse length. This requires more complicated transmitter control but, more importantly, results in a pulse with less than the maximum possible energy for the given pulse length. This technique is not discussed further in this book; see Levanon and Mozeson (2004) for more information.

Two methods that have been proposed for NLFM waveform design are the principle of stationary phase method and empirical techniques. The PSP technique is used to design a temporal phase function from a prototype spectral amplitude function; the instantaneous frequency function is then obtained from the temporal phase. The method is fairly involved. It is discussed in Keel and Baden (2012), where examples are given for deriving NLFM waveforms from common window functions such as Hamming or Taylor functions.

One empirically developed design gives the instantaneous frequency function as (Price, 1979)

$$F_i(t) = \frac{t}{\tau}\left(\beta_L + \beta_C \frac{1}{\sqrt{1 - 4t^2/\tau^2}}\right), \quad |t| \leq \frac{\tau}{2} \text{ Hz} \qquad (4.121)$$

The term $\beta_L t/\tau$ represents a linear FM component, while the term involving β_C is designed to achieve a result that approximates a Chebyshev-shaped spectrum, which produces a constant sidelobe level. Since $F_i(t) = (1/2\pi)(d\varphi(t)/dt)$, integrating and scaling this instantaneous frequency function gives the required phase modulation,

$$\varphi(t) = \frac{\pi \beta_L}{\tau} t^2 - \frac{\pi \beta_C \tau}{2}\sqrt{1 - 4t^2/\tau^2}, \quad |t| \leq \frac{\tau}{2} \text{ rads} \qquad (4.122)$$

Figure 4.37 illustrates the behavior of the resulting nonlinear FM waveform for $\beta_L \tau = 50$ and $\beta_C \tau = 20$. The waveform is sampled at 10 times the bandwidth of the linear term, $T_s = 1/10\beta_L$. The instantaneous frequency (Fig. 4.37a) is nearly linear in the center of the pulse but sweeps much more rapidly near the pulse edges. This reduces the spectral density at the pulse edge, resulting in the spectrum shown in Fig. 4.37c, which has a window-like tapered shape instead of the usual nearly square LFM spectrum. The resulting matched filter output, shown in Fig. 4.37d, has most of its sidelobes between −48 and −51 dB with the first sidelobe at −29 dB. In contrast, Fig. 4.38 illustrates the spectrum and matched filter output of the same waveform with β_C set to zero. This results in a linear FM waveform with the usual nearly square spectrum. The spectra in Figs. 4.37c and 4.38a are on the same normalized frequency scale. Comparing them illustrates how the nonlinear term has spread and tapered the LFM spectrum to lower the matched filter sidelobes. The LFM matched filter output has a peak sidelobe of −13.5 dB, decaying approximately as $1/F$ at higher frequencies. The Rayleigh resolution in time of the NLFM waveform is approximately $0.8/\beta_L$, less than the $1/\beta_L$ value observed for the LFM case but greater than $1/(\beta_L + \beta_C)$.

Radar Waveforms 215

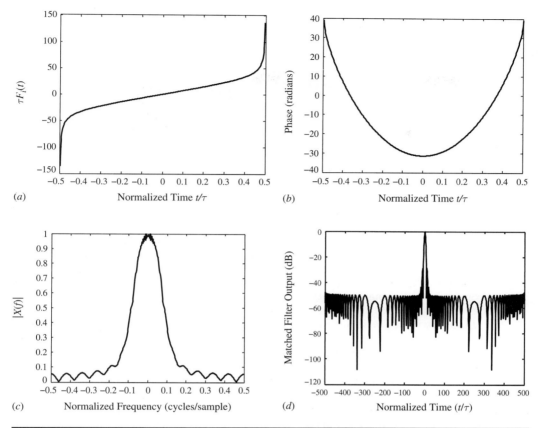

FIGURE 4.37 Nonlinear FM waveform. (a) Normalized instantaneous frequency $\tau \cdot F_i(t)$. (b) Resulting phase modulation function. (c) Magnitude of Fourier spectrum. (d) Magnitude of matched filter output.

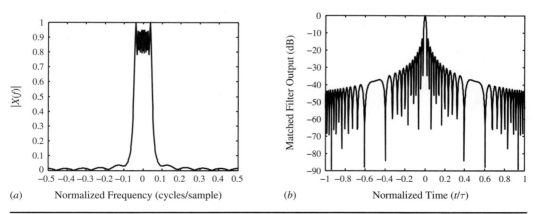

FIGURE 4.38 FM waveform having same linear component as that of Fig. 4.33, but no nonlinear component. (a) Magnitude of Fourier spectrum. (b) Magnitude of matched filter output.

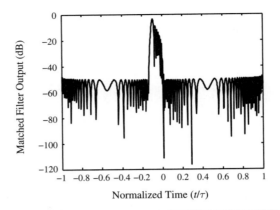

FIGURE 4.39 Output of NLFM matched filter when $F_D = 7/\tau$ Hz.

An example of a hybrid technique that combines a similar frequency modulation function with amplitude tapering of the matched filter impulse response is described in De Witte and Griffiths (2004). It is claimed there that the far sidelobes are controlled primarily by the maximum instantaneous frequency, while the near-in sidelobes are controlled by the amplitude weighting.

In addition to the more difficult phase control required, the major drawback of nonlinear FM pulses is their Doppler intolerance. Figure 4.39 shows the matched filter output for the waveform of Fig. 4.37 when a Doppler mismatch of $7/\tau$ Hz is present. While the general sidelobe level remains largely unchanged, the mainlobe is seriously degraded, exhibiting both range-Doppler coupling (a shift of the peak) and severe spreading and ambiguity caused by very high near-in sidelobes. A second disadvantage is lack of a stretch processing option for very wideband cases. The major advantage of NLFM over linear FM with receiver weighting is that the receiver filter for the NLFM waveform is a matched filter so that lower sidelobes are achieved with no reduction of the matched filter output peak.

4.8 Frequency-Coded Waveforms

4.8.1 The Stepped Frequency Waveform

The LFM waveform increases resolution well beyond that of a simple pulse by sweeping the instantaneous frequency over the desired range β within the pulse. This technique is very effective and very common but does have drawbacks in some systems, particularly those using very large bandwidths on the order of hundreds of megahertz or more. First, the transmitter hardware must be capable of generating the wideband LFM sweep. Second, all of the analog components must be able to support an instantaneous bandwidth of β Hz without introducing distortion. Even if stretch processing is used, the same is true of the receiver components up to and including the dechirp mixer and reference oscillator.

A second issue arises in systems using phase-steered array antennas. Recall from Chap. 1 that the antenna pattern of a phase-steered array antenna is determined primarily by the array factor

$$E(\theta) = E_0 \sum_{n=0}^{N-1} a_n \exp[j(2\pi/\lambda_t)nd\sin\theta] \qquad (4.123)$$

where d is the element spacing and the $\{a_n\}$ are the complex weights on each subarray output. The antenna is steered to a particular look direction θ_0 by setting the steering weights an according to[11]

$$a_n = |a_n| \exp[-j(2\pi/\lambda_t)nd \sin \theta_0] \quad (4.124)$$

The magnitudes of the weights are chosen to provide the desired sidelobe level. $E(\theta)$ will exhibit a peak at $\theta = \theta_0$; for example, if $|a_n| \equiv 1$, $E(\theta)$ will be an asinc function with its peak at θ_0. Note that the phases of the required weights $\{a_n\}$ are a function of the wavelength λ_t. If an LFM pulse is transmitted, the effective wavelength changes during the pulse sweep. If the system is wideband, this wavelength change will be significant and the value of θ at which $E(\theta)$ peaks will change as well. That is, the antenna look direction will actually change during the LFM sweep (see Prob. 18). This undesired frequency steering effect is an additional source of SNR loss.

Stepped frequency pulsed waveforms are an alternative technique for obtaining a large bandwidth and thus fine range resolution without requiring intrapulse frequency modulation. A stepped frequency waveform is a pulse burst waveform. Each pulse in the burst is a simple, constant-frequency pulse; however, the RF is changed from one pulse to the next. The waveform essentially combines the scene spectrum sampling behavior of a narrowband pulse discussed in Sec. 2.9 with a pulse-to-pulse frequency change to construct a sampled wideband spectrum and achieve fine range resolution.

The most common stepped frequency waveform employs a linear frequency stepping pattern, where the RF of each pulse is increased by ΔF Hz from the preceding pulse. Factoring out the starting RF gives the following baseband waveform

$$x(t) = \sum_{m=0}^{M-1} x_p(t - mT) \exp[j2\pi m \cdot \Delta F(t - mT)] \quad (4.125)$$

Figure 4.40 illustrates the linearly stepped frequency waveform.

Because only simple pulses are used for each constituent pulse, the instantaneous bandwidth capability of the transmitter and receiver need be only on the order of $1/\tau$ Hz. The total bandwidth of the waveform as a whole is $M \cdot \Delta F + 1/\tau \approx M \cdot \Delta F$. When used with a phase-steered array antenna, the time between pulses can be used to reset the phase shifters to update the $\{a_n\}$ sequence and maintain a nearly constant steering direction θ_0 as the effective wavelength changes from pulse to pulse. In addition, the M-fold coherent integration of the matched filter gives an additional processing gain. The major disadvantages of this waveform are that it requires a pulse-to-pulse tunable transmitter and receiver, and that M PRIs are required to collect data over the desired bandwidth instead of just one.

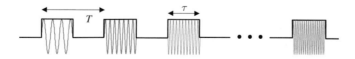

FIGURE 4.40 Linearly frequency-stepped waveform.

---

[11]Array antennas can also be steered using time delay units at each element or a combination of phase steering within a subarray and time delay steering across subarrays. Pure time-delay-steered arrays do not suffer antenna steering errors due to wideband waveforms. Time delay and phase steering are discussed in Chap. 9.

The pulse-by-pulse processing viewpoint applied to the constant-frequency pulse burst waveform can be applied again to analyze the matched filter response for the stepped frequency waveform. Suppose the radar is stationary, and a stationary target is located at a range corresponding to a delay $t_l + \delta t$, where δt represents an incremental delay relative to the nominal delay t_l corresponding to range bin l. Individual pulses are processed through the filter matched to that particular pulse, including the frequency modulation. This produces the output waveforms (assuming $T_M = 0$)

$$y_m(t) = s_p^*[t - (t_l + \delta t) - mT]\exp(j2\pi m \cdot \Delta F[t - (t_l + \delta t) - mT]) \quad (4.126)$$

This output is then sampled at $t = t_l + mT$ (i.e., t_l seconds after the current pulse was transmitted), corresponding to range $R_l = ct_l/2$. The resulting sample becomes the lth coarse range bin sample for the current pulse,

$$\begin{aligned} y[l,m] &= y_m(t_l + mT) \\ &= s_p^*(\delta t)\exp(j2\pi m \cdot \Delta F \cdot \delta t) \end{aligned} \quad (4.127)$$

Equation (4.127) shows that the slow-time sequence at a fixed coarse range bin l when using a linearly stepped frequency waveform is a discrete time sinusoid. The frequency is proportional to the displacement of the scatterer from the nominal range bin location of $R_l = ct_l/2$ meters. The amplitude of the sequence is weighted by the triangular simple pulse matched filter response evaluated at the incremental delay $s_p(\delta t)$.

Following the earlier discussion of pulse-by-pulse processing for the conventional pulse burst waveform, the slow-time matched filter impulse response for a target located at the nominal delay $t_l + \delta t$ is $h[m] = \exp(-j2\pi m \cdot \Delta F \cdot \delta t)$. Thus, the matched filter impulse response is different for every value of δt. Consider a DTFT of the slow-time data,

$$\begin{aligned} Y[l,\omega] &= \sum_{m=0}^{M-1} y[l,m]\exp(-j\omega m) = \sum_{m=0}^{M-1} s_p^*(\delta t)\exp(j2\pi m \cdot \Delta F \cdot \delta t)\exp(-j\omega m) \\ &= s_p^*(\delta t) \sum_{m=0}^{M-1} \exp[-j(\omega - 2\pi \cdot \Delta F \cdot \delta t)m] \end{aligned} \quad (4.128)$$

The summation will yield an asinc function having its peak at $\omega = 2\pi \cdot \Delta F \cdot \delta t$. Thus, the peak of the DTFT of the slow-time data in a fixed range bin with a linearly stepped frequency waveform provides a measure of the delay of the scatterer relative to the nominal delay t_l. Specifically, if the peak of the DTFT is at $\omega = \omega_p$, the scatterer is at an incremental delay

$$\delta t = \frac{\omega_p}{2\pi \cdot \Delta F} = \frac{f_p}{\Delta F} \text{ seconds} \quad (4.129)$$

Note also that the DTFT evaluated at ω_p is the matched filter for the slow-time sequence, so that the data samples are integrated in phase

$$\begin{aligned} Y[l,\omega_p] &= \sum_{m=0}^{M-1} s_p^*(\delta t)\exp(j2\pi m \cdot \Delta F \cdot \delta t)\exp(-j\omega_p m) \\ &= s_p^*(\delta t) \sum_{m=0}^{M-1} \exp(+j\omega_p m)\exp(-j\omega_p m) \\ &= s_p^*(\delta t) \sum_{m=0}^{M-1} (1) = M \cdot s_p^*(\delta t) \end{aligned} \quad (4.130)$$

The factor of M is the coherent integration gain from using M pulses. If $\delta t = 0$, meaning the matched filter output was sampled at its peak, then $Y[l, \omega_p] = M \cdot E_p = E$, the total waveform energy. If $\delta t \neq 0$ the ambiguity function of the individual pulses reduces the amplitude of the slow-time samples by $|s_p(\delta t)|$. This represents a straddle loss.

It follows that applying a K-point DFT to the slow-time sequence implements K filters, each matched to a different incremental delay δt. The DFT of the slow-time data within a single range bin for a stepped frequency waveform is then a map of echo amplitude versus incremental range within that coarse range bin.

The DTFT of an M-point sinusoid has a Rayleigh frequency resolution of $\Delta f = 1/M$ cycles per sample. Using the scaling between f and t from Eq. (4.129), the corresponding time resolution is $\Delta t = 1/(M \cdot \Delta F)$ seconds; the range resolution is therefore

$$\Delta R = \frac{c}{2M \cdot \Delta F} = \frac{c}{2\beta} \text{ m} \qquad (4.131)$$

where β is the total stepped bandwidth $M \cdot \Delta F$. Thus, the linearly stepped frequency waveform achieves the same range resolution as a single pulse of bandwidth β. If a K-point DFT is used to process the slow-time data, the DFT output will provide range measurements at intervals of

$$\delta R = \frac{c}{2K \cdot \Delta F} = \frac{M}{K} \Delta R \text{ m} \qquad (4.132)$$

Since $K \geq M$ normally, the DFT output provides echo amplitude samples at intervals equal to or less than the range resolution. This fine-resolution reflectivity map is often called a *high-resolution range* (HRR) *profile*.[12]

The total bandwidth β of the stepped frequency waveform is determined by the desired range resolution. It can be realized by various combinations of the number of frequency steps M and the step size ΔF. To determine how to choose these parameters, note that the DTFT of the slow-time data is periodic in ω with period 2π radians per sample. Because the DTFT peak is at $\omega_p = 2\pi \cdot \Delta F \cdot \delta t$, the range profile is periodic in δt with period $1/\Delta F$ seconds. This periodicity establishes the required coarse range bin spacing. Specifically, avoiding range ambiguities in the range profile requires $c/2\Delta F > L_t$, where L_t is the maximum target length of interest. Once ΔF is chosen, M is selected to span the bandwidth required to provide the desired fine range resolution. The DFT range profile then effectively breaks each relatively large coarse-resolution range bin ($c/2\Delta F$ meters) into M fine-resolution range bins ($c/2\beta$ meters) sampled at K points within the coarse-resolution range bin. If $K = M$, the range sample spacing equals the range resolution. If $K > M$, the range profile is oversampled compared to the resolution by the factor K/M. The pulse length τ is chosen to balance straddle losses and range ambiguities. Recall that the single-pulse matched filter output $s_p(t)$ is 2τ seconds long. Choosing $\tau < 1/(2\Delta F)$ means that $s_p(t)$ will be no more than $1/\Delta F$ seconds long so that a scatterer will only influence measurements in one coarse range bin, avoiding range ambiguities. On the other hand, the shorter τ is made, the greater the potential straddle loss for targets located between coarse range samples. A detailed consideration of these tradeoffs is in Keel and Baden (2012).

Details of the Doppler response and ambiguity function of the linearly stepped frequency waveform are available in Levanon and Mozeson (2004). A small central portion of the ambiguity function is shown in Fig. 4.41 for the case $M = 8$ pulses, PRI $T = 10\tau$, and a

---

[12]While "fine" is preferred in this text to "high" to describe small values of resolution, the term "high-range-resolution profile" is well-established in the literature.

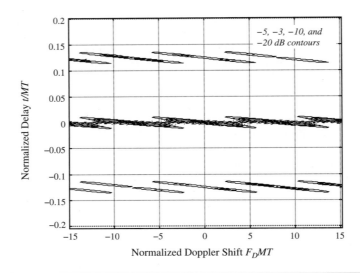

FIGURE 4.41 Contour plot of the central portion of the ambiguity function of a linear frequency-stepped waveform. $M = 8$, $T = 10\tau$, and $\Delta F = 0.8/\tau$.

frequency step size of $\Delta F = 0.8/\tau$. The resulting bandwidth is $\beta = M \cdot \Delta F = 6.4/\tau$ Hz. The AF displays both the skewed response typical of a linear FM modulation and the range and Doppler ambiguities typical of pulse burst waveforms. Ambiguities in delay (range) are evident at intervals of T seconds, corresponding to $1/8 = 0.125$ on the normalized scale of the figure. The first zero in Doppler of the main ridge occurs at $1/MT$ Hz, corresponding to 1 on the normalized Doppler scale.

Figure 4.42a further magnifies the delay coordinate of this AF. The delay coordinate now covers the interval $\tau \in [-0.0125, +0.0125]$ on this normalized scale. The zero-delay and zero-Doppler axes are highlighted by the heavier gray lines. The expected Rayleigh resolution in delay is $1/\Delta F = \tau/6.4$, which becomes 0.002 on this scale. The dotted heavy gray line marks the $+0.002$ delay coordinate. It can be seen that this intersects the first null on the zero-Doppler axis of the AF, confirming that the intended resolution is achieved.

Choosing $\Delta F > 1/\tau$ allows generation of a wide total bandwidth with fewer pulses and therefore a shorter data collection time, but the resulting undersampling creates aliasing that appears as extra range ambiguities (also called grating lobes). Figure 4.42b shows a similar view of another case with $\Delta F \cdot \tau = 2.5$ but the other parameters unchanged. The bandwidth is now $20/\tau$ Hz so the resolution in delay is correspondingly finer. However, there are now five peaks along the zero-Doppler axis within ± 1 pulse length, representing five range ambiguities.

4.8.2 The Stepped Chirp Waveform

The stepped chirp waveform is a stepped frequency pulsed waveform that substitutes an LFM constituent pulse for the constant-frequency pulse used above. It can achieve very wideband operation without resorting to stretch processing, thereby avoiding the restriction of short-range windows. In addition, it avoids the array frequency steering effects mentioned previously so long as the individual pulse bandwidth is not too large.

The stepped chirp waveform can allow a large frequency step $\Delta F > 1/\tau$ without suffering the aliasing seen in the conventional stepped frequency waveform. Careful design is needed to relate the LFM pulse bandwidth and length to the RF step size in order to achieve

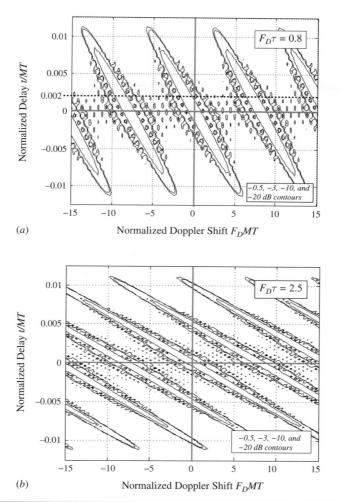

FIGURE 4.42 Contour plot of the central portion of the ambiguity function of a pulse burst waveform with $M = 8$ and $T = 10\tau$. (a) $\Delta F = 0.8/\tau$ (same waveform as Fig. 4.41), (b) $\Delta F = 2.5/\tau$.

effective suppression of the ambiguities. Details and sample parameter sets are given in Levanon and Mozeson (2004). Processing of the waveform requires individually demodulating and matched filtering each individual pulse, and then post-processing the ensemble to construct a new signal with the full bandwidth. This post-processing can be performed in either the time or frequency domain. Details are given in Keel and Baden (2012).

4.8.3 Costas Frequency Codes

Costas waveforms are a class of pulse compression waveforms having aspects of both phase-coded and stepped frequency pulse burst waveforms (Costas, 1984). A Costas waveform is similar to a polyphase waveform (to be discussed in Sec. 4.9.2) in that it is a single pulse waveform divided into N subpulses. It is similar to the linearly stepped frequency waveform in that, rather than maintaining a constant frequency and altering the phase of each subpulse, it alters the subpulse frequencies, stepping through a set of N frequencies that differ by ΔF Hz. Unlike the stepped frequency pulse burst, however, the Costas waveform does not step through the frequencies in linear order. The Costas pulse can be expressed as

$$x(t) = \sum_{n=0}^{N-1} x_n(t - \eta \tau_c),$$
$$x_n(t) = \begin{cases} \exp(jc[n] \cdot \Delta F \cdot t), & 0 \leq t \leq \tau_c \\ 0, & \text{elsewhere} \end{cases} \quad (4.133)$$

where the sequence $c[n]$ denotes the ordering of the stepped frequencies.

Figure 4.43 shows the frequency sequence for a typical low-order Costas waveform. With proper design of the frequency step sequence, the Costas waveform can be designed to have a more thumbtack-like ambiguity function than the linearly stepped waveform. Figure 4.44 illustrates the ambiguity function of a Costas waveform with $N = 15$; the

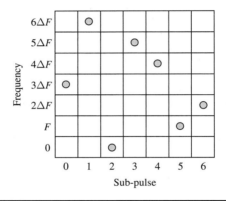

FIGURE 4.43 Frequency sequence for Costas waveform with $N = 7$.

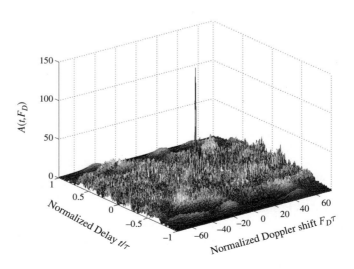

FIGURE 4.44 Ambiguity function for a Costas waveform with $N = 15$.

frequency step sequence was $c[n] = \{1,7,8,11,3,13,9,14,12,6,5,2,10,0,4\}$. Note the generally low and relatively uniform sidelobe structure throughout the delay-Doppler plane. The construction and properties of Costas waveform are discussed and more examples given in Levanon and Mozeson (2004).

4.9 Phase-Modulated Pulse Compression Waveforms

The second major class of pulse compression waveforms is *phase-coded* waveforms. A phase-coded pulse has a constant RF but an absolute phase that is switched between two or more fixed values at regular intervals within the pulse length. Such a pulse can be modeled as a collection of N contiguous subpulses $x_n(t)$ of duration τ_c, each with the same frequency but a (possibly) different phase:

$$x(t) = \sum_{n=0}^{N-1} x_n(t - n\tau_c),$$
$$x_n(t) = \begin{cases} \exp(j\varphi_n), & 0 \leq t \leq \tau_c \\ 0, & \text{elsewhere} \end{cases}$$
(4.134)

The total pulse length is $\tau = N\tau_c$. Individual subpulses are often referred to as *chips*. Phase-coded waveforms are divided into *biphase codes* and *polyphase codes*. A biphase code has only two possible choices for the phase state φ_n, typically 0 and π; a polyphase code has more than two phase states. There are several common subcategories of each. Figure 4.1c was an example of a biphase-coded waveform.

Recall that pulse compression waveforms have a bandwidth $\beta \gg 1/\tau$. Because phase-coded waveforms are constant frequency, it may not be obvious that their spectrum is similarly spread. However, the discontinuities caused by the phase transitions do spread the signal spectrum. As an example, Fig. 4.45 shows the effect of a single phase switch of 180° on the spectrum of a constant-frequency waveform. While the effect depends on the point in the pulse at which the switch occurs, clearly it significantly spreads the signal energy in frequency. Multiple phase transitions increase this effect: Fig. 4.46 compares the spectra of the 13-bit biphase Barker-coded waveform (to be defined shortly) with that of a simple pulse of

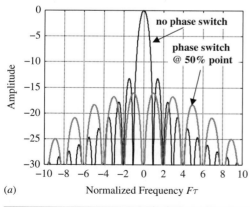
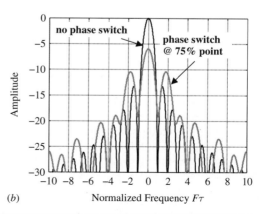

FIGURE 4.45 Effect of a single 180° phase switch on the spectrum of a constant-frequency pulse. (a) Phase switch occurs at $t = \tau/2$. (b) Phase switch occurs at $t = 3\tau/4$.

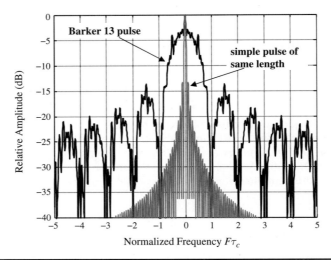

FIGURE 4.46 Spectra of a 13-bit Barker coded pulse and a simple pulse of the same length.

the same duration. The Rayleigh bandwidth of the Barker spectrum is $\beta = 1/\tau c$ Hz, 13 times as wide as that of the simple pulse in this specific case. In addition, the sidelobes of the Barker waveform spectrum decay much more slowly than those of the simple pulse.

Because of the increased bandwidth, the signal processing gain due to pulse compression with phase codes is the same factor $G_{sp} = \beta\tau$ that applied for the LFM waveform. This is again because a simple pulse must be shorter than a phase-coded pulse by the factor $\beta\tau$ to achieve the same Rayleigh resolution in time. Note that in the case of phase codes, $\beta\tau$ can also be expressed as τ/τ_c.

The matched filter output for a phase-coded pulse is derived in detail in Levanon and Mozeson (2004); the result is now summarized. Denote the sequence of complex amplitudes of the individual pulse chips $x_n(t)$ of Eq. (4.134) as $\{A_n\} = \{\exp(j\varphi n)\}$. Express the time variable t in terms of the chip duration τ_c, an offset $\eta \in [0, \tau_c)$, and an integer $k \in [0, N-1]$ as $t = k\tau_c + \eta$. The matched filter output is the autocorrelation of $x(t)$, which can be shown to be

$$y(t) = s_x(t) = s_x(k\tau_c + \eta) = \left(1 - \frac{\eta}{\tau_c}\right)s_A[k] + \frac{\eta}{\tau_c}s_A[k+1] \quad (4.135)$$

where $s_A[k]$ is the discrete autocorrelation of the complex amplitude sequence $\{A_n\}$. Equation (4.135) shows that $s_x(t)$ takes on the value $s_A[k]$ at $t = k\tau_c$ ($\eta = 0$) and is linearly interpolated in the complex plane between adjacent samples as η varies from 0 to τ_c. Thus, the matched filter output can be determined by computing the autocorrelation of the complex amplitude sequence and interpolating between those values. One consequence of this result and the fact that the $\{A_n\}$ have unit magnitude is that the peak value of the autocorrelation will always be $s_x(0) = N$. Another is that the PSL and ISL levels can be computed from the values of $s_x[k]$. Recall that for a length-N code the support of $s_x[k]$ is $[-N+1, N-1]$ and that $s_x[0] = N$ is the central peak. Then

$$\begin{aligned} PSL &= \max_{k \neq 0}\left(|s_x[k]|\right); & PSL \text{ (normalized, dB)} &= 20\log_{10}(PSL/N) \\ ISL &= \sum_{\substack{k=-N+1 \\ k \neq 0}}^{N-1} |s_x[k]|^2; & ISL \text{ (normalized, dB)} &= 10\log_{10}(ISL/N^2) \end{aligned} \quad (4.136)$$

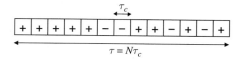

FIGURE 4.47 Binary sequence describing the Barker code of Fig. 4.48 in "+/−" notation.

4.9.1 Biphase Codes

There are 2^N different biphase codes of length N. Because there are only two phase states, the waveform is often represented by a diagram such as the one illustrated in Fig. 4.47 using either "+" and "−" symbols as shown, or +1 and −1 symbols. Note that biphase codes do not necessarily change phase state at every subpulse transition.

For a given length N, many of the 2^N codes are equivalent to one another in the sense that their matched filter outputs exhibit the same PSL and ISL. Specifically, given one biphase code in the ± 1 notation with the chips indexed by $n = 0$ to $N - 1$, a new code obtained by reversal of the original code, negation of all its elements, or element-wise multiplication of its elements by either $(-1)^n$ or $(-1)^{n+1}$ will have the same PSL and ISL. Some of these equivalent codes may be identical to one another. Consider the length-3 code $[1, -1, 1]$. Reversing it produces $[1, -1, 1]$ again, negating it produces $[-1, 1, -1]$, multiplying by $(-1)^n$ produces $[1, 1, 1]$, and multiplying by $(-1)^{n+1}$ produces $[-1, -1, -1]$. These four codes all produce a PSL of 2 and an ISL of 10. The other four length-3 codes also are equivalent to one another, producing a PSL of 1 and ISL of 2. Consequently, there are effectively only two unique codes of length 3.

The most common biphase codes in radar are the *Barker codes*. Barker codes are a specific set of biphase sequences that have a maximum sidelobe magnitude of 1 at the matched filter output and therefore attain an N:1 ratio of the peak to the highest sidelobe. A low-frequency (one cycle per chip length) Barker-coded waveform for $N = 13$ is shown in Fig. 4.48. The phase switches are visible at $t = 5\tau_c, 7\tau_c, 9\tau_c, 10\tau_c, 11\tau_c,$ and $12\tau_c$.

One of the major disadvantages of Barker codes is that there are not very many of them. Barker codes have been found only for N up to 13. Sample Barker codes of all known lengths are listed in Table 4.1; more than one code exists for some lengths, while none exist for

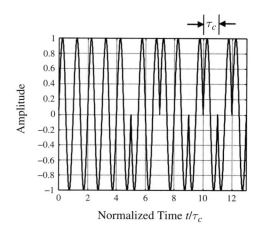

FIGURE 4.48 Barker coded waveform, $N = 13$.

	Code Sequence		
N	+/− Format	Octal	PSL, dB
2	+ +	2	−6.0
2	+ +	3	−6.0
3	+ + −	6	−9.5
4	+ + − +	15	−12.0
4	+ + + −	16	−12.0
5	+ + + − +	35	−14.0
7	+ + + − − + −	162	−16.9
11	+ + + − − − + − − + −	3422	−20.8
13	+ + + + + − − + + − + − +	17,465	−22.3

TABLE 4.1 Barker Codes

$N = 6, 8 - 10$, and 12. The table also lists the PSL relative to the mainlobe peak in decibels. Because of the modest code lengths, Barker codes are limited to modest PSLs and $\beta\tau$ products.

As an example, consider the Barker code with $N = 13$. Representing the code sequence of Table 4.1 as the sequence $\{A_n\} = \{1,1,1,1,1,-1,-1,1,1,-1,1,-1,1\}$ gives the autocorrelation sequence $\{1,0,1,0,1,0,1,0,1,0,1,0,13,0,1,0,1,0,1,0,1,0,1,0,1\}$. Figure 4.49 illustrates the resulting autocorrelation function obtained by interpolating between the discrete autocorrelation samples. In addition to a peak autocorrelation value of N and sidelobe peaks equal to 1, the discrete autocorrelation sequence sidelobes of a Barker code always follow an alternating pattern of zeros and ones. Consequently, the Rayleigh resolution is always τ_c seconds in time or $c\tau_c/2$ meters in range and is therefore controlled by the chip length instead of the complete pulse length.

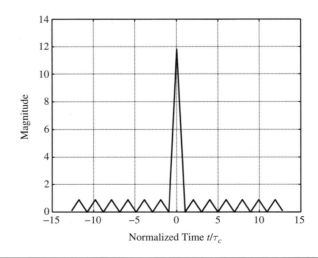

FIGURE 4.49 Matched filter output for a 13-bit Barker code.

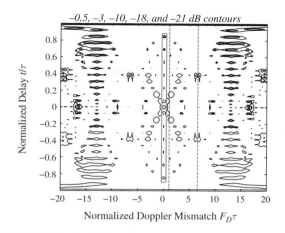

Figure 4.50 Contour plot of ambiguity function of a 13-bit Barker code.

Barker codes have two major disadvantages. The first is the limited attainable sidelobe suppression and processing gain discussed above. The second is that they are relatively Doppler intolerant, a fact illustrated by Figs. 4.50 and 4.51. The 13-bit Barker code ambiguity function is shown in Fig. 4.50. The zero Doppler cut (matched filter output in time) at $F_D\tau = 0$, enclosed by a gray box, shows the expected central peak and, just barely visible, the symmetric equal-amplitude time sidelobes to either side, consistent with Fig. 4.49. This AF also shows that Doppler mismatches badly degrade the filter output. For example, the delay cut at $F_D\tau \approx 6.5$, marked with a vertical gray dashed line, has two peaks surrounding $\tau = 0$ instead of one central peak at that delay. This corresponds to a Doppler mismatch frequency having 6.5 cycles across the duration of the pulse. A much smaller Doppler mismatch phase rotation of only 360° ("one cycle of Doppler") across the full pulse is also more than sufficient to completely break up the structure of the matched filter output. This case, shown by the other gray dashed line at $F_D\tau = 1$ in Figs. 4.50 and 4.51a, also exhibits bifurcation of the mainlobe.

It is common to restrict the use of Barker-coded waveforms to situations that limit the likely Doppler mismatch to one-quarter cycle or less across the pulse. This case is shown in Fig. 4.51b. The matched filter range response retains a correctly located central peak, though the range resolution is degraded by the widening of that peak. There is also a modest reduction in peak amplitude (LPG) of 0.9 dB. This constraint requires that the maximum expected Doppler shift and target velocity satisfy

$$F_{D_{\max}}\tau < \frac{1}{4} \quad \Rightarrow \quad v_{\max} < \frac{\lambda_t}{8\tau} \tag{4.137}$$

This can become a tight constraint for longer pulses and higher RFs. For instance, a radar using a 100-μs pulse must limit the Doppler mismatch to 2.5 kHz. If the radar is X band (10 GHz), the corresponding maximum velocity mismatch becomes 37.5 m/s (83.9 mph).

The limited number and length of Barker codes has led to various attempts to construct longer biphase codes with good sidelobe properties. *Combined* or *nested* Barker codes form a longer code as the Kronecker product of two shorter Barker codes. If an N-bit Barker code sequence is denoted as B_N, an MN-bit code can be constructed as $B_M \otimes B_N$. The Kronecker product is simply the B_N code repeated M times, with each repetition multiplied by the

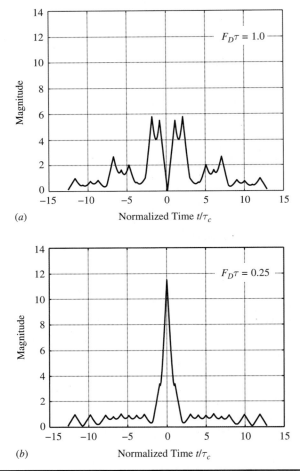

FIGURE 4.51 Effect of Doppler mismatch on matched filter output (zero-Doppler cut of AF) for a Barker 13 code. (a) One cycle of Doppler across the pulse. (b) One-quarter cycle.

corresponding element of the B_M code. For example, a 20-bit code can be constructed as the product $B_4 \otimes B_5$

$$B_4 \otimes B_5 = \{1, 1, 1, -1\} \otimes \{1, 1, 1, -1, 1\}$$
$$= (1)\{1, 1, 1, -1, 1\} + (1)\{1, 1, 1, -1, 1\} + (1)\{1, 1, 1, -1, 1\} + (-1)\{1, 1, 1, -1, 1\}$$
$$= \{1, 1, 1, -1, 1, 1, 1, 1, -1, 1, 1, 1, 1, -1, 1, 1, -1, -1, -1, 1, -1\}$$

(4.138)

These codes have a peak sidelobe higher than 1. The autocorrelation of the code of Eq. (4.138) is shown in Fig. 4.52. Notice that the magnitude of the peak sidelobes is 5, so that the PSL compared to the autocorrelation peak is only 1/4 instead of the 1/20 that would be obtained

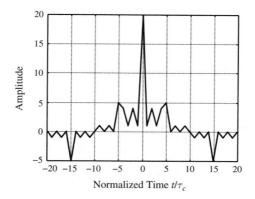

Figure 4.52 Autocorrelation of combined $B_4 \otimes B_5$ code.

if a 20-bit Barker code existed. A better code that obtains a PSL = 2 for $N = 20$ is discussed momentarily.

While the Barker codes are the only biphase codes with a peak sidelobe value of 1 that are known to exist, one can seek longer codes with *minimum peak sidelobe* (MPS) levels for the length of interest. These MPS codes are found by exhaustive search techniques, taking advantage of certain properties of biphase code autocorrelations such as the equivalences mentioned earlier to prune the search somewhat. As an example, the MPS code for $N = 20$ has a maximum sidelobe level of 2, giving a PSL ratio of $1/10$ instead of the $1/4$ obtained by the nested Barker code above. The state of the art is summarized and references given in Keel (2010). The peak sidelobe for MPS codes of lengths 2 through 5, 7, 11, and 13 (Barker codes) is 1; for $N = 6, 8 - 10, 12, 14 - 21, 25,$ and 28 is 2; for $N = 22 - 24, 26 - 27, 29 - 48,$ and 51 is 3; for $N = 49 - 50$ and $52 - 82$ is 4; and for $N = 83 - 105$ is 5. Table 4.2 lists one sample code for the longest code length in each of these sidelobe level regimes. It is evident that the sidelobe level in dB improves only very slowly as the code length increases.

MPS codes have not been determined at this writing (2021) for lengths greater than 105, although the search continues. For example, PSL = 6 codes have been found for $N = 106$ through 112. These have not been proven to be MPS but at least some likely are (Coxson et al., 2020).

For most lengths for which MPS codes have been found, there is more than one code of that length (along with its equivalents discussed earlier) that achieves the same minimum PSL. For instance, Keel (2010) states that there are eight length-8 unique MPS codes and 858

Code Length N	Sample Code (Hexadecimal)	Peak Sidelobe Level (PSL)	PSL, dB
13 (Barker)	1F35	1	−22.3
28	8F1112D	2	−22.9
51	0E3F88C89524B	3	−24.6
82	3CB25D380CE3B7765695F	4	−26.2
105	1C6387FF5DA4FA325C895958DC5	5	−26.4

Table 4.2 Sample Minimum Peak Sidelobe Biphase Codes

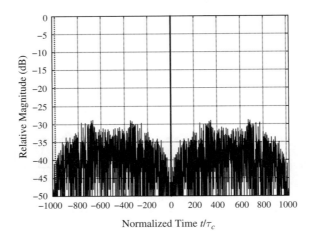

FIGURE 4.53 Matched filter output for a 1023-bit pseudorandom biphase code.

unique length-24 MPS codes. While MPS codes of a given length share the same PSL, they do not necessarily all have the same ISL. As an example, the codes $[-1,-1,-1,1,1,-1,1,-1]$ and $[1,1,1,1,-1,1,-1,-1]$ are not related by any of the earlier equivalences, yet both have the same unnormalized PSL value of 2 (-12 dB relative to the autocorrelation peak of 8). However, the unnormalized ISL of the first code is 16 (-6 dB relative to the squared peak of 64), while that of the second code is 32 (-3 dB). This difference in ISL could be an important design consideration if sidelobe clutter is expected to be a significant interference source.

Another technique uses pseudorandom noise sequences to generate much longer biphase codes. Pseudorandom sequences have length $N = 2^P - 1$ for some integer P and generally exhibit range sidelobes on the order of $-10 \log_{10}(N)$. For example, the matched filter output of Fig. 4.53 for a typical $N = 1023$ ($P = 10$) code has peak sidelobes just above -30 dB.

4.9.2 Polyphase Codes

Biphase codes, as noted previously, have poor Doppler tolerance. They also suffer from precompression bandlimiting effects. As was seen in Fig. 4.46, the spectrum of a typical biphase code not only exhibits the desired mainlobe spreading but also a very slow falloff of the far sidelobes. This is a direct consequence of the sharp phase discontinuities. Practical receivers will have a noise-limiting bandpass filter that will bandlimit the biphase waveform spectrum, smoothing the phase transitions. This has the effect of mismatching the received waveform relative to the correlator, reducing the peak gain and widening the mainlobe.

Polyphase codes allow arbitrary values for the chip phases φ_n. Compared to biphase codes, they can exhibit lower sidelobe levels and greater Doppler tolerance. A number of polyphase codes are in common use. These include "polyphase Barker codes," the Frank codes, and the P1, P2, P3, P4, and P(n, k) codes. All of these except the polyphase Barker codes are related to LFM or NLFM waveforms. Several are special cases of quadratic phase codes developed by Frank, Zadoff, and Chu. Numerous other polyphase codes have also been proposed. Many of these codes are described in Levanon and Mozeson (2004) along with some of their history and relationships.

Frank codes are codes whose length is a square, $N = M^2$, for some M. The phase sequence for a Frank code is given by

$$\varphi_n = \varphi(Mp + q) = \frac{2\pi}{M} pq, \quad p = 0, 1, 2, \ldots, M-1, \quad q = 0, 1, 2, \ldots, M-1 \quad (4.139)$$

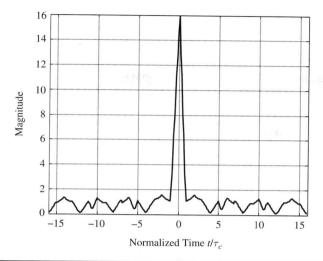

FIGURE 4.54 Matched filter output for a 16-bit Frank code.

As an example, if $M = 4$ so $N = 16$, the sequence of phases becomes

$$\varphi_n = \{\underbrace{0,0,0,0}_{p=0},\ \underbrace{0,\frac{\pi}{2},\pi,\frac{3\pi}{2}}_{p=1},\ \underbrace{0,\pi,0,\pi}_{p=2},\ \underbrace{0,\frac{3\pi}{2},\pi,\frac{\pi}{2}}_{p=3}\} \quad (4.140)$$

Figure 4.54 shows the magnitude of the matched filter output for the case $N = 16$.[13] Note that while the mainlobe has a local minimum at $t = \tau_c$, it does not go to zero at that point as the Barker code autocorrelations do. The largest sidelobe in this example is $\sqrt{2}$, larger than the Barker codes. The peak sidelobe power level is $\left(\sqrt{2}/16\right)^2$ or -21.1 dB. Figure 4.55 shows the ambiguity function in contour plot form. The main ridge is skewed in the delay-Doppler plane, similar to the range-Doppler coupling of an LFM ambiguity function.

The *P3* and *P4 codes* of length N are given respectively by (Levanon and Mozeson, 2004; Keel, 2010)

$$\begin{aligned} \text{P3}: \ & \varphi_n = \frac{\pi}{N} n^2, & n = 0,1,2,\ldots, N-1 \\ \text{P4}: \ & \varphi_n = \frac{\pi}{N} n^2 - \pi n, & n = 0,1,2,\ldots, N-1 \end{aligned} \quad (4.141)$$

Unlike the Frank code, these codes can be generated for any length N. Figure 4.56 shows the matched filter output for the $N = 20$ P3 code, while Fig. 4.57 shows the corresponding ambiguity function. Again, range-Doppler coupling is evident.

---

[13]This figure may appear to violate the earlier claim that the continuous autocorrelation function is a linear interpolation between the discrete autocorrelation values of the code sequence exp($j\varphi_n$). However, linear interpolation of the complex values does not result in linear interpolation of the magnitude (see Prob. 27).

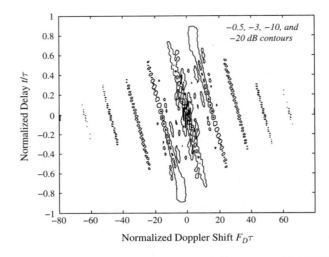

FIGURE 4.55 Contour plot of 16-bit Frank code ambiguity function.

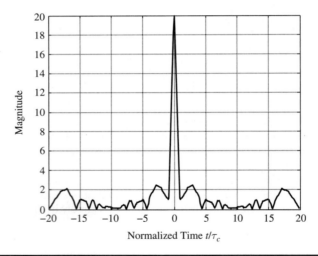

FIGURE 4.56 Matched filter output for a 20-bit P3 code.

The Frank, P3, and P4 codes all are based on quadratic phase progressions as is evident from Eqs. (4.139) and (4.141) and are therefore related to LFM waveforms. Figure 4.58 shows the (unwrapped) phase progression of these three codes for the case $N = 16$. The P3 and P4 codes are truly quadratic, the difference being whether the minimum phase slope occurs at the beginning (P3) or the middle (P4) of the waveform. The smallest phase increments, and therefore the minimum discontinuities in the actual RF waveform, occur where the phase slope is least. The Frank code uses a piecewise linear approximation to a quadratic phase progression. The phase increment is constant for M bits at a time and then increases for the next M bits. This can be viewed as a phase code approximation to a stepped frequency waveform having M steps and M bits per step (Lewis and Kretschmer, 1982). As a result, the Frank code is less Doppler tolerant than the P3 and P4 codes.

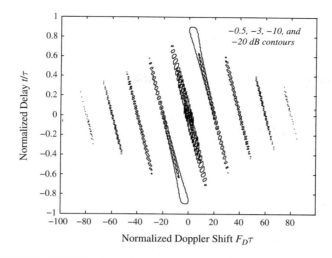

Figure 4.57 Contour plot of ambiguity function of 20-bit P3 code.

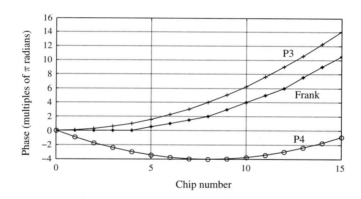

Figure 4.58 Unwrapped phase sequences of 16-bit Frank, P3, and P4 codes.

Bandlimiting of the phase-coded waveform prior to matched filtering results in an increase in mainlobe width but a decrease in PSL in codes that have the largest phase increments at the ends of the codes (Lewis and Kretschmer, 1982; Levanon and Mozeson, 2004). Codes with the largest phase increments in the middle of the code exhibit the opposite behavior. Of the three codes shown, the P4 will consequently show the greatest tolerance to precompression bandlimiting in the sense of maintaining or improving its sidelobe level at the matched filter output, though at the cost of degraded range resolution.

Just as phase codes can be designed based on linear frequency modulation waveforms, they can also be designed based on nonlinear frequency modulation waveforms. A class of codes based on NLFM waveforms designed using the PSP technique mentioned earlier is given in Felhauer (1994). No closed form expression is known for these $P(n, k)$ codes; they must be found numerically. Typical results are very similar to those for the empirical NLFM waveforms described earlier. The effect of Doppler mismatch is similar to that observed in Fig. 4.39. This is an improvement over conventional polyphase codes, which are prone to

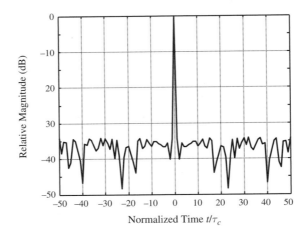

FIGURE 4.59 Autocorrelation of an $N = 51$ polyphase Barker code.

exhibiting significantly increased sidelobes near the ends of the code and, in many cases, large spurious peaks well above the general sidelobe level. $P(n, k)$ codes also exhibit better tolerance to precompression bandlimiting than do codes based on linear FM, since their spectra are already shaped by the basic NLFM design approach. Their chief disadvantage is the difficulty of their design.

Another approach to reducing spectral sidelobes and thus improving precompression bandlimiting tolerance is the use of quadriphase codes. These codes are obtained from biphase codes by mapping the binary phase progression to a four-phase code using a specified transformation, and also by replacing the rectangular subpulse chips with half-cosine chips of twice the width. Compared to the biphase code, the resulting codes have significantly lower spectral sidelobes, nearly the same autocorrelation sidelobes, but a significant loss of time (range) resolution. Details are given in Keel and Baden (2012).

Polyphase Barker codes are polyphase codes that exhibit a maximum sidelobe peak level of 1. The phases of a length-N code are either unrestricted or restricted to a Pth root of unity, $\varphi_n = 2\pi p_n / P$ for some integer P with $p_n \in [0, P-1]$ and $n \in [0, N-1]$. Figure 4.59 shows the discrete autocorrelation of a polyphase Barker code having $N = 51$, $P = 50$, and the following $\{p_n\}$ sequence:

0, 0, 4, 4, 18, 20, 27, 25, 25, 26, 24, 15, 15, 14, 9, 32, 36, 2, 21, 17, 9, 27, 46, 49, 19, 29, 9, 32, 7, 45, 21, 46, 22, 47, 18, 35, 0, 22, 9, 31, 44, 5, 29, 21, 4, 49, 33, 24, 9, 49, 29

The PSL is −34.2 dB, significantly better than the −24.6 dB for the 51-point MPS code in Table 4.2.

4.9.3 Mismatched Phase Code Filters

The sidelobe structure of phase-coded waveforms can be improved with the use of *mismatched filters*, just as is done with stepped frequency and linear FM waveforms to improve their sidelobe structures. For phase-coded waveforms, this implies correlating the code sequence with another discrete-time sequence, not necessarily restricted in the amplitudes or phases of its coefficients, such that some metric of the sidelobe structure is optimized. Mismatched filters can be designed to minimize the output PSL, minimize the output ISL, or to shape the output sidelobe response, for instance to enforce particularly low near-in

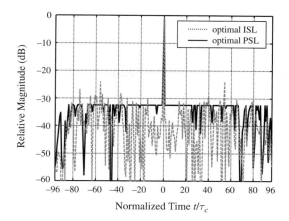

FIGURE 4.60 Autocorrelation of two $L = 130$ mismatched filters for the same $N = 64$ MPS code. Solid line: Optimal PSR filter response. Dotted line: Optimal ISR filter response. (After Keel and Baden, 2012.)

sidelobes at the expense of higher distant sidelobes. The filter order L is usually larger than the code length. In many cases the design of the mismatched filter coefficients can be formulated as the solution of a weighted least squares problem, for which many numerical algorithms are available. Other optimization techniques, such as L^1 minimization using convex optimization algorithms, can also be employed. It is also possible to use optimization techniques to seek code families that meet multiple simultaneous objectives on the autocorrelation, cross-correlation, and spectral properties (O'Donnell and Baden, 2018).

Figure 4.60 illustrates two examples of mismatched filter design. In both cases, the waveform phase code was the same $N = 64$ MPS biphase code. The peak sidelobe of the matched filter output for this length is 4, giving a PSL in dB of $20 \log_{10}(4/64) = -24.1$ dB. The ISL for the matched filter is -6.7 dB. The solid black line in the figure is the result of a mismatched filter of length $L = 130$ designed to minimize the PSL. The filter impulse response is normalized to have the same energy as the code and its matched filter impulse response, namely 64. This filter achieves a PSL of -31.4 dB, an 8.3-dB improvement compared to the matched filter. The ISL is -9.8 dB, an improvement of about 3.1 dB. However, there is now a loss in processing gain of 1.11 dB relative to the matched filter.

The dashed gray line is the result of an $L = 130$ filter designed to minimize the ISL and also normalized to an energy of 64. The PSL is now about -23.1 dB, 1 dB worse than the matched filter and 8.3 dB worse than the minimum-PSL filter. However, the ISL is now -12 dB, 5.3 dB better than the matched filter and 2.2 dB better than the minimum-PSL filter. The LPG is 0.85 dB relative to the matched filter, a 0.26-dB improvement compared to the minimum-PSL filter.

Research suggests that, at least for biphase codes, the SNR loss increases with the mismatched filter length but eventually levels off when the filter length is on the order of three times the code length. At the same time, for ISL-optimized filters the ISL continues to decline as the filter length increases (Levanon, 2005).

4.10 Continuous Wave Radar

All of the preceding discussion in this chapter is centered on pulsed radar. Pulsed radars are capable of very long-range application, can easily measure range and velocity, and can achieve fine-resolution imaging. However, pulsed radars require high peak powers in order to achieve good average power and suffer eclipsing and blind zones.

Continuous wave radars form another class of radar systems that transmit and receive continuously. Because transmission is continuous the average power equals the peak power, a situation more amenable to the use of solid-state or other peak-power-limited transmit sources. The capability for good average power without high peak powers is also helpful when a low probability of intercept is desired. Solid-state sources in turn enable the development of very low-cost radar systems, making the technology practical for cost-sensitive mass production applications. Another advantage of CW radar is that the transmitter is never off, so there is no inherent minimum range and blind ranges may not occur (depending on system design), making CW systems superior for short-range measurements. Some of the simpler CW radars do not require as complex a transceiver as do pulsed systems, but only measure velocity, not range. More complex CW systems measure both at the expense of transceiver complexities similar to pulsed radars.

Given these characteristics, CW radar is popular for a variety of low-power short-range applications, especially those involving velocity measurements. Common examples include police and sports "speed guns," radar altimeters and fuzes, missile seekers, and patient monitoring in health care. A particularly important and rapidly growing application area is automotive radar. The automotive industry is placing millions of radars into use each year, many or perhaps most of them CW. These radars implement such functions as lane keeping, cruise control, parking assist, collision avoidance, and autonomous operation. There are also many more complex or unusual applications such as short-range synthetic aperture imaging, radar cross section measurements, and even storage tank level measurements.

Like pulsed radar, CW radar can be operated with different waveforms, many of them analogous to the pulsed waveforms studied so far. These include constant frequency, linear and nonlinear FM, biphase and polyphase coding, and frequency coding, as well as techniques less common in pulsed radar such as frequency shift keying (FSK), sinusoidal modulation, and noise modulation. The most common CW waveform is linear FM. Of special interest in this text is the "fast chirp" variant of linear FMCW common in automotive radars, which is conducive to datacube-based CPI processing similar to many pulsed systems.

The ambiguity function can be used to analyze CW waveform resolution, sidelobes, and ambiguities, just as with pulsed waveforms. Because of the infinite duration of a periodically modulated CW waveform, the standard AF definition of Eq. (4.32) cannot be applied directly. Instead, the periodic ambiguity function of Eq. (4.81), or a variant, is used. However, unlike many pulsed radars, most FMCW radars do not use the explicit matched filter receiver structure assumed by the AF or PAF. Instead, the transmitted and received signal are mixed and the beat frequency signal is analyzed to obtain target range and velocity information in a manner similar to the deramp receiver for sawtooth linear FMCW described in Chap. 3 and the stretch processing of LFM pulses in Sec. 4.6.5. For this reason, ambiguity functions for CW waveforms are not discussed here. The PAF for a number of periodic waveforms is discussed in Levanon and Mozeson (2004).

A thorough overview of CW radar configurations, design, waveforms, and applications is available in Baker and Piper (2014). A comparative discussion of CW, FMCW, and FSK waveforms in the context of automotive radar is given in Rohling and Kronauge (2012).

4.10.1 Single-Frequency CW

The most basic continuous wave radar transmits a continuous signal $\bar{x}(t)$ at a constant RF frequency $\Omega_t = 2\pi F_t$. The baseband waveform $x(t)$ is simply the amplitude A:

$$\bar{x}(t) = \underbrace{A}_{x(t)} \exp(j\Omega_t t) \quad \forall t \tag{4.142}$$

Target and clutter reflections are generally much weaker than the transmitted signal and must somehow be separated from it if they are to be detected. Pulsed radars accomplish this via time multiplexing: echoes are detectable so long as they don't arrive while the radar is transmitting another pulse so that they are eclipsed. Relatively low duty cycles mean that targets are detectable at most ranges.

Because CW radars have a 100% duty cycle, there is no transmitter quiet time during which echoes can be detected.[14] CW radars instead separate echoes from transmissions using one or both of two techniques. The first is physical separation of the transmit and receive antennas. This generally takes the form of using a separate though co-located antenna for each function. A physical divider is sometime installed between the two antennas to improve isolation. The second method is to rely on frequency shifts of the received echo relative to the transmitted signal to create separation. This will occur through the Doppler shift if there is sufficient radar-target relative motion. It can also be created by periodically frequency modulating the CW waveform, as will be seen shortly. In this case, two antennas are not necessary.

The Doppler shift due to a target approaching the radar at a relative velocity of v m/s is given by the usual formula

$$F_D = +\frac{2v}{\lambda_t} = +\frac{2vF_t}{c} \text{ Hz} \quad (4.143)$$

A variety of receiver configurations are used to detect this shift of the CW echo frequency. Traffic control and sports radars often use the *autodyne* configuration, which is simple and inexpensive but has poor sensitivity. More elaborate *homodyne* and *heterodyne* receivers improve sensitivity at the cost of increased complexity (Baker and Piper, 2014).

4.10.2 Periodically Modulated CW

Simple CW systems are unable to measure range because the waveform provides no identifiable timing marks. This problem is solved by imposing a periodic modulation on the carrier. A periodically modulated CW signal at RF can be written as

$$\begin{aligned}
\bar{x}(t) &= \underbrace{\left\{\sum_{n=-\infty}^{\infty} x_p(t-nT)\right\}}_{x(t)} \exp(j\Omega_t t) \\
&= \underbrace{\left\{\sum_{n=-\infty}^{\infty} a(t-nT)\exp[j\varphi(t-nT)]\right\}}_{x(t)} \exp(j\Omega_t t)
\end{aligned} \quad (4.144)$$

where $x_p(t) = a(t)\exp[j\varphi(t)]$ is the finite length waveform prototype. The envelope $a(t)$ is of finite duration T; thus so is $x_p(t)$. As with pulsed waveforms, amplitude modulation is rare in CW radar so that $a(t)$ is usually constant amplitude as well. Equation (4.144) states that the prototype is simply repeated continuously with no gaps for all time to form the CW baseband signal, which is them modulated onto the RF carrier Ω_t. The baseband portion of Eq. (4.144) is the same as Eq. (4.54), describing the finite pulse train but with $M \to \infty$ and

---

[14]A variation called *interrupted FMCW* (IFMCW) turns off the transmitter to permit a quiet reception window, improving sensitivity and supporting longer-range performance. Strictly speaking, IFMCW is a pulsed radar. It is not further discussed here.

the pulse length $\tau \to T$. Setting $x_p(t) \equiv 1$ gives the basic unmodulated CW signal again. In this text, the focus is on phase- and especially frequency-modulated CW.

4.10.3 Linear Frequency-Modulated CW

In FMCW the amplitude of the prototype waveform is identically one, and all modulation is applied through the phase function $\varphi(t)$. For "sawtooth" linear FMCW the baseband prototype waveform is the linear FM sweep of Eq. (4.83) or (4.84) with the pulse length τ extended to the full period T. The complex form is also the same as Eq. (3.4) except that the latter includes the RF term as well. Section 3.2.2 showed how this waveform and a "deramp" receiver structure could be used to measure a range profile of stationary scatterers. Specifically, recall that for a swept bandwidth of β Hz over a repetition interval T seconds and a stationary target at range R_0 corresponding to a time delay t_0, the primary beat frequency F_{b1} was given by Eq. (3.6), repeated here for convenience:

$$F_{b1} = -\left(\frac{\beta}{T}\right)t_0 = -\left(\frac{t_0}{T}\right)\beta = -\left(\frac{2R_0}{cT}\right)\beta \text{ Hz}, \quad t_0 \le t \le T \tag{4.145}$$

The second version of F_{b1} is directly analogous to the stretch processor beat frequency of $-\beta(\delta t_b/\tau)$ discussed in Sec. 4.6.5. Figure 3.12, repeated here as Fig. 4.61 for convenience, illustrated the relationships between the transmitted and received instantaneous frequencies and the beat frequencies for a sawtooth linear FMCW waveform and a stationary target at delay t_0.

As was discussed earlier in stretch processing of pulsed LFM, the bandwidth of the beat frequency signal at the FMCW mixer output is usually much less than the swept bandwidth β, enabling a significant reduction in the required data sampling rate and storage capacity in the subsequent signal processing. Consider an automotive radar example with $F_t = 77$ GHz, $\beta = 300$ MHz (giving about 1-m range resolution), a range window extending from $R_{\min} = 10$ m to $R_{\max} = 300$ m, and a maximum relative velocity range of ± 60 m/s (± 134

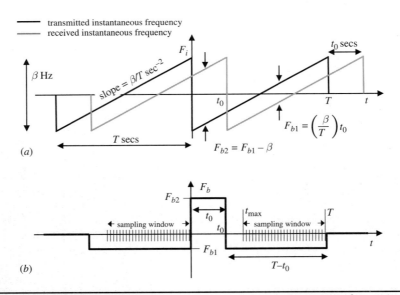

FIGURE 4.61 Repeat of Fig. 3.12. (a) Transmitted and received instantaneous frequency patterns for a sawtooth linear FMCW radar and a scatterer at delay t_0. (b) Beat frequency pattern and sampling window resulting from the signals in (a). The beat frequency is the received frequency minus the transmitted frequency.

mph). The maximum beat frequency magnitude at a sweep repetition rate of $RR = 10$ kHz ($T = 100$ μs) will be only 6 MHz, 50 times lower than the waveform bandwidth. This bandwidth reduction enabled by the deramp receiver structure is a major technology advantage of linear FMCW waveforms over many other wideband CW waveforms.

Figure 4.61 shows that the maximum time delay t_{max} corresponding to the far edge of the range window R_{max} limits the maximum observation time over which the beat frequency can be sampled to $T - t_{max}$ seconds. It was shown in Eq. (3.12), repeated here, that this effect limits the range resolution for sawtooth linear FMCW to

$$\Delta R = \left(\frac{T}{T - t_{max}}\right)\left(\frac{c}{2\beta}\right) \text{ m} \qquad (4.146)$$

However, t_{max} is normally a small fraction of the sweep time T so the range resolution is close to the usual $c/2\beta$.

Linear FMCW does not have an inherent minimum range, although in practice a minimum range may exist due to a limited ability to measure beat frequency magnitudes very near zero hertz due to internal phase noise or external clutter interference. However, there is a limit on the longest measurable target range determined by the fast-time sampling rate F_s. Beat frequency is measured by computing the IDFT of the sampled beat frequency signal. Range is non-negative so this spectrum can be interpreted as covering the interval $[-F_s, 0]$ Hz instead of $[-F_s/2, +F_s/2]$ Hz. The range corresponding to the maximum magnitude beat frequency F_s is then the unambiguous range

$$R_{ua} = \left(\frac{cT}{2\beta}\right) F_s \text{ m} \quad \text{(sawtooth linear FMCW)} \qquad (4.147)$$

Targets at longer ranges will have beat frequency magnitudes outside the $[0, F_s]$ interval. Their beat frequencies will therefore alias to a lower magnitude beat frequency, appearing as an aliased range measurement.

If the target, radar, or both are moving such that the range between them is decreasing at a rate of v m/s, there will be a Doppler shift of the echo waveform of $F_D = +2v/\lambda_t$ Hz. The primary beat frequency for a sawtooth linear FMCW system will become, in several equivalent versions,

$$\begin{aligned} F_{b1} &= -\left(\frac{2\beta}{cT}\right) R_0 + F_D \text{ Hz}, \quad t_0 \leq t \leq T \\ &= -\left(\frac{2\beta}{cT}\right)\left(R_0 - \frac{cT}{2\beta} F_D\right) = -\left(\frac{2\beta}{cT}\right)\left(R_0 - \frac{cT}{\lambda_t \beta} v\right) = -\left(\frac{2\beta}{cT}\right)\left(R_0 - \frac{T \cdot F_t}{\beta} v\right) \text{ Hz} \end{aligned} \qquad (4.148)$$

Equation (4.148) shows that sawtooth linear FMCW exhibits the same range-Doppler coupling as pulsed linear FM [Eqs. (4.106) and (4.116)], with a Doppler shift of F_D Hz resulting in an apparent range shift of $-(cT/2\beta)F_D$ meters.

Notice that F_{b1} depends on both the range and velocity of the target, so that a single beat frequency measurement cannot determine both; a second linearly independent measurement is needed. Measurements of both range and Doppler for a single target can be obtained by using a second beat frequency measurement having a different sweep rate. One common way to implement this is to alternate up and down chirp sawtooth sweeps, effectively creating the triangular FMCW waveform shown in Fig. 4.62. This version completes both the upchirp and downchirp in T seconds, making the effective duration of each chirp $T/2$ seconds. The received signal shows the effect of a positive Doppler shift of F_D Hz. This figure also illustrates the beat frequency patterns, sampling intervals, and other details. Measuring

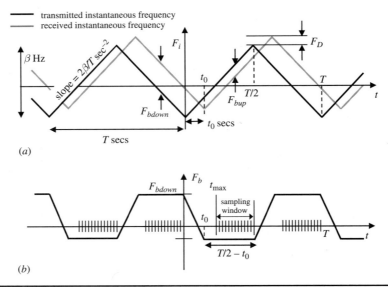

FIGURE 4.62 (a) Transmitted and received instantaneous frequency patterns for a triangular linear FMCW radar and a moving scatterer at delay t_0 with a radial velocity of v m/s causing a Doppler shift of $F_D = 2v/\lambda_t$ Hz. (b) Beat frequency pattern and sampling windows resulting from the signals in (a). The beat frequency is the received frequency minus the transmitted frequency.

both the upchirp and downchirp beat frequencies using this waveform gives two equations in the two unknowns R and v,

$$F_{bup} = \left(\frac{4\beta}{cT}\right)\left(-R + \frac{T \cdot F_t}{2\beta}v\right) \text{ Hz}, \quad F_{bdown} = \left(\frac{4\beta}{cT}\right)\left(+R + \frac{T \cdot F_t}{2\beta}v\right) \text{ Hz} \quad (4.149)$$

These equations can be solved for R and v on a sweep-by-sweep basis (see Prob. 32).

Several similar waveform variations can be used to measure range and velocity simultaneously. For example, they can be measured using two chirps having different sweep rates to provide independent equations. Another approach uses a linear FM segment to measure range, followed by an unmodulated CW segment to allow direct measurement of velocity (and correction of the range-Doppler coupling in the range measurement). This can also be viewed as a special case of the different-slopes approach with one of the slopes being set to zero. These approaches do not appear to be common.

Resolving the range and velocity of multiple targets with these methods requires waveforms with increasing numbers of different slopes, an approach that rapidly becomes unwieldy (Baker and Piper, 2014). The fast-chirp method of the next subsection provides a more effective way to deal with multiple targets.

4.10.4 "Fast Chirp" Linear Frequency-Modulated CW

A large number of modern linear FMCW radars in high-volume applications such as automotive radar and medical patient monitoring solve the problem of measuring both range and velocity for multiple targets by using the CPI-based approach described in Chap. 3. In this method, a sawtooth waveform is employed. Velocity is measured from the spatial Doppler phase history by collecting multiple sweeps of echo data and computing the slow-time spatial Doppler spectrum, while range is measured by forming the spectrum of the beat frequency signal at the mixer output. Consequently, the range-Doppler matrix can be obtained by a computationally efficient two-dimensional Fourier transform of the raw fast-time/slow-time CPI of data. For good range measurements it is desirable to design the waveform so that range-Doppler coupling is insignificant, meaning that the maximum Doppler shift is much less than the smallest beat frequency of interest:

$$\frac{2F_t|v|_{max}}{c} \ll \frac{2\beta R_{min}}{cT} \quad \Rightarrow \quad T \ll \left(\frac{\beta}{F_t}\right)\left(\frac{R_{min}}{|v|_{max}}\right) \text{ seconds} \quad (4.150)$$

Notice that smaller fractional bandwidths, closer range windows, and higher velocities all force shorter repetition intervals. Continuing the previous automotive radar example, the bound will be $T \ll 649$ μs, requiring the repetition rate $RR \gg 1540$ Hz and so suggesting a moderately fast RR of perhaps 15 kHz or more.

An alternative constraint on the range shift due to range-Doppler coupling is that it be less than the range resolution so that targets are shifted by less than a resolution cell. Using the maximum relative velocity and, for simplicity, the nominal (and finest) range resolution $c/2\beta$, this constraint becomes

$$\frac{cT}{2\beta}|F_D|_{max} < \frac{c}{2\beta} \quad \Rightarrow \quad T < \frac{1}{|F_D|_{max}} = \frac{\lambda_t}{2|v|_{max}} = \frac{c}{2|v|_{max}F_t} \text{ seconds} \quad (4.151)$$

Applying this constraint to the automotive example gives the bound $T < 32.5$ μs. The corresponding repetition rate constraint is $RR > 30.8$ kHz. At least in this example, both constraints lead to repetition rates in the low tens of kilohertz.

Minimization of range-Doppler coupling is not necessarily the only or even the tightest constraint on the repetition rate. Because the RR is the slow-time sampling rate, it must also satisfy the Nyquist criterion for the required unambiguous velocity interval. Suppose it is desired to have no velocity ambiguities over the entire ±60 m/s velocity interval in the automotive example. The Doppler bandwidth and therefore the required RR is 61.6 kHz, thus becoming the driving requirement on RR in this example.

Linear FMCW waveforms designed to have insignificant range-Doppler coupling and no velocity ambiguities in the scenarios of interest often have repetition rates in the tens of kilohertz, like the example above. Such waveforms are called *fast chirp* linear FMCW and are quite common in automotive and health applications.[15] The higher the RR, the less the range error due to range-Doppler coupling and the wider the unambiguous velocity interval. The primary limitation to increasing the RR is the degradation of the range resolution as T is decreased for a given range window extent t_{max}.

The range resolution and unambiguous range of the linear FMCW waveform were given in Eqs. (4.146) and (4.147). The Doppler or velocity resolution, like the pulse burst waveform of Sec. 4.5.4, are determined by the CPI duration MT and are

$$\begin{aligned}\Delta F_D &= \frac{1}{MT} \text{ Hz} \\ \Delta v &= \frac{\lambda_t}{2}\Delta F_D = \frac{\lambda_t}{2MT} \text{ m/s}\end{aligned} \quad (4.152)$$

Finally, the unambiguous Doppler interval, again like the pulse burst waveform, is determined by the slow-time sampling rate (the RR) as

$$\begin{aligned}F_{Dua} &= RR \quad \Rightarrow \quad F_D \in [-RR/2, +RR/2] \text{ Hz} \\ v_{ua} &= \frac{\lambda_t}{2}RR \quad \Rightarrow \quad v \in [-\lambda_t \cdot RR/4, +\lambda_t \cdot RR/4] \text{ m/s}\end{aligned} \quad (4.153)$$

---

[15]Another term sometimes seen is *chirp sequence* FMCW.

4.10.5 Sidelobe Control in Linear FMCW

Control of range and Doppler sidelobes in linear FMCW using a deramp receiver structure is straightforward. The range profile is computed as an IDFT of the beat frequency signal. Range sidelobes can be controlled by applying an appropriate window to the sampled fast-time data before that IDFT. The Doppler signature is computed as a DFT of the slow-time signal in a given range bin, and so can also be controlled independently by applying a window of choice in the slow time dimension prior to that DFT. Unlike the otherwise very similar stretch processing, most applications of linear FMCW do not exhibit range skew of targets at different ranges that must be tolerated or corrected. In both dimensions, applying a window will carry the usual costs (see App. B): degraded resolution, reduced peak gain, and reduced SNR.

4.10.6 Other CW Waveforms

There are many other CW waveforms of varying popularity. Examples include sinusoidal FM, phase-coded CW using various phase codes similar or even identical to those discussed here for pulsed radar, and frequency shift keying (FSK). Orthogonal frequency division multiplexing (OFDM), a technique borrowed from communications technology, is being actively explored as a possible waveform for advanced CW systems. However, few fielded systems are using OFDM at this writing (2021). More information on these waveforms is available in Levanon and Mozeson (2004), Baker and Piper (2014), and in the radar literature.

4.11 Frequency-Modulated versus Phase-Modulated Waveforms

As has been seen in this chapter, both frequency- and phase-modulated waveforms exist that are capable of large time-bandwidth products, fine resolution, and low sidelobes. How does one decide which is the better choice for a given system?

The answer is anything but obvious, involving a number of tradeoffs as well as a good bit of tradition. Metrics of interest include attainable time-bandwidth products; blind ranges; peak and integrated sidelobe levels; Doppler tolerance; spectral compactness; ease of generation and processing; duty cycle; electronic warfare (EW) susceptibility; and electromagnetic interference (EMI) generation and susceptibility.

In practice, the most widely used waveforms in existing fielded radars (in addition to the simple pulse, which still has its place) are LFM and biphase codes, especially Barker codes. Both are easy to generate and process with current technology. Recall that the BT product of a waveform determines the pulse compression gain. If only a small BT (perhaps less than 100) and modest sidelobe suppression are needed to meet performance specifications, then short biphase codes may be suitable. For example, Barker codes are fairly common in air-to-air applications, where there is often little sidelobe clutter. Higher BT products with low ISLs may be needed in air-to-ground applications such as ground moving target indication (GMTI) and imaging to support fine resolution and good distributed clutter rejection. Linear FM and perhaps nonlinear FM are often used in these cases, since they are easily designed and implemented for virtually any desired BT product and, through appropriate weighting, any desired PSL and ISL. However, greater bandwidth is needed to maintain resolution as sidelobes are pushed lower. For very fine resolution applications such as imaging or target classification, wideband LFM with stretch processing is common because of its ease of implementation and reduced processing loads.

A high degree of Doppler tolerance is often desirable in search and in GMTI. Doppler tolerance is very good with LFM provided that range-Doppler coupling is acceptable. Phase codes based on quadratic phase progressions may also provide fairly good Doppler tolerance, but many other phase codes do not. The same statements apply to spectral compactness, which is the degree to which the waveform spectrum is confined to the principal bandwidth. Waveforms with high spectral sidelobes, such as the 13-bit Barker spectrum of Fig. 4.46, have difficulty meeting spectrum allocation constraints and may require extra preprocessing such as precompression bandlimiting that degrades performance. Quadriphase code transformations provide an alternative means of controlling spectral sidelobes.

A drawback of high-BT pulsed systems is that the pulses tend to be long. Long pulses may introduce new problems. For instance, they imply large minimum ranges and blind zones, and so are unsuitable for short-range applications. They also imply larger duty cycles, which may exceed thermal limits of high-power pulsed transmitters. CW systems do not isolate the receiver for a portion of the repetition interval and so don't have a minimum range (at least for this reason). CW systems also need less peak power to achieve a given average power. Both these attributes make CW systems a better choice for short-range low-power applications. CW radars still offer waveform options similar to pulsed radars for fine resolution and low sidelobes in range and Doppler, Doppler tolerance, and so forth.

Electronic warfare (jamming) susceptibility is heavily influenced by the number of parameters it takes to specify a waveform. The fewer the number, the fewer parameters have to be estimated by a hostile emitter to jam or mimic the waveform. For instance, an LFM pulse train is fully determined by five parameters: nominal RF F_t, pulse length τ, swept bandwidth β, choice of upchirp or downchirp, and PRI T. Waveforms with so few degrees of freedom (DOF) also suffer more mutual EMI when multiple units are in close proximity, such as with automotive radar.

Longer phase codes have more DOF, which brings more capabilities. Long phase codes can have low sidelobes without sacrificing resolution, especially with the use of mismatched filters, and can be designed to have specialized sidelobe patterns such as a lower sidelobe level in certain range intervals to attenuate particular clutter scatterers or reduce nearby target masking. Long phase codes also have better spectral sidelobe control. The ability to have many different codes of a longer length with equivalent PSLs and ISLs allows better EW resistance through code diversity. Diversity also aids EMI reduction between nearby systems. For these reasons, phase codes may be gaining somewhat in popularity.

Finally, there are a number of hardware implementation issues that affect the performance of various waveforms, both pulsed and CW, that are mostly beyond the scope of this textbook. Examples include unintentional FM sweep nonlinearities, transmitter and receiver saturation and nonuniform frequency responses, phase noise, gain and phase errors in quadrature signal generation, and so forth. Information on some of these error sources is available in Scheer and Kurtz (1993).

4.12 Summary

The radar waveform determines or is a major contributor to many important characteristics of the system: resolution in range and Doppler, ambiguity intervals in range and Doppler, peak and integrated sidelobe levels, range-Doppler coupling and Doppler tolerance, and signal-to-noise ratio. The choice of waveform(s) is one of the most important decisions in designing a radar system.

Consider pulsed waveforms first. Simple pulses, while easy to generate and process, suffer a fundamental conflict between goals of high SNR (requiring long pulses) and fine range resolution (requiring short pulses). The solution to this conflict is the use of modulated waveforms. A key point is that modulated waveforms decouple resolution and SNR only when used in conjunction with their matched filter, a critical tool in radar signal processing. When processing a waveform through its matched filter, the SNR depends on the waveform length while the resolution depends not on the waveform shape, but on that of its autocorrelation. Good SNR and range resolution are obtained simultaneously using "pulse compression" waveforms that have narrowly concentrated autocorrelations, even though the waveform itself is long. The previous section compared common frequency and phase modulated pulse compression waveforms.

CW radars have much in common with pulsed waveforms but some essential differences as well. Linear FMCW achieves essentially the same resolution in range and Doppler as do pulsed waveforms, exhibits the same range-Doppler coupling, and uses similar techniques for sidelobe control. On the other hand, the implementation of range measurement with FMCW is very different from that for pulsed waveforms, and minimum range and blind range behavior are determined by different factors.

"Fast chirp" linear FMCW is an especially common CW waveform and is the primary one considered here. Like pulsed systems, it supports acquisition of data in the format of a radar datacube, as described in Chap. 3. This in turn provides a common basis for subsequent processing such as clutter rejection, imaging, and adaptive processing. These algorithms are the subject of the ensuing chapters.

References

Baker, C. J., and S. O. Piper, "Continuous Wave Radar," Chap. 2 in J. A. Scheer and W. L. Melvin (eds.), *Principles of Modern Radar: Radar Applications*. SciTech Publishing, Raleigh, NC, 2014.

Born, M., and E. Wolf, *Principles of Optics*. Pergamon Press, London, 1959.

Carrara, W. G., R. S. Goodman, and R. M. Majewski, *Spotlight Synthetic Aperture Radar*. Artech House, Boston, 1995.

Cook, C. E., and M. Bernfeld, *Radar Signals: An Introduction to Theory and Application*. Artech House, Boston, 1993.

Costas, J. P., "A Study of a Class of Detection Waveforms Having Nearly Ideal Range-Doppler Ambiguity Properties," *Proceedings of the IEEE*, vol. 72, no. 8, pp. 996–1009, Aug. 1984.

Coxson, G. E., J. C. Russo, and A. Luther, "Long Low-PSL Binary Codes by Multi-Thread Evolutionary Search," *Proceedings of 2020 IEEE International Radar Conference (RADAR)*, Washington, DC, USA, 2020, pp. 256–261.

De Witte, E., and H. D. Griffiths, "Improved Ultra-Low Range Sidelobe Pulse Compression Waveform Design," *Electronics Letters*, vol. 40, no. 22, pp. 1448–1450, 2004.

Felhauer, T., "Design and Analysis of New $P(n, k)$ Polyphase Pulse Compression Codes," *IEEE Transactions on Aerospace and Electronic Systems*, vol. 30, no. 3, pp. 865–874, Jul. 1994.

Gini, F., A. De Maio, and L. Patton (eds.), *Waveform Design and Diversity for Advanced Radar Systems*. Institution of Engineering and Technology (IET), London, 2012.

Jonsson, B. E., "A Survey of A/D Converter Performance Evolution," *Proceedings of 17th IEEE International Conference on Electronics, Circuits, and Systems (ICECS)*, pp. 766–769, 2010.

Jonsson, B. E., "A/D-Converter Performance Evolution," *Converter Passion*, Aug. 2012. Available at https://converterpassion.wordpress.com/articles/a-d-converter-performance-evolution.

Kay, S. M., *Fundamentals of Statistical Signal Processing, Vol. II: Detection Theory*. Prentice Hall, Upper Saddle River, NJ, 1998.

Keel, B. M., "Fundamentals of Pulse Compression Waveforms," Chap. 20 in M. A. Richards, J. A. Scheer, and W. A. Holm (eds.), *Principles of Modern Radar: Basic Principles*. SciTech Publishing, Raleigh, NC, 2010.

Keel, B. M., and J. M. Baden, "Advanced Pulse Compression Waveform Modulations and Techniques," Chap. 2 in W. L. Melvin and J. A. Scheer (eds.), *Principles of Modern Radar: Advanced Techniques*. SciTech Publishing, Edison, NJ, 2012.

Levanon, N., "Cross-Correlation of Long Binary Signals with Longer Mismatched Filters," *IEE Proceedings—Radar, Sonar and Navigation*, vol. 152, no. 6, pp. 377–382, Canton, MA, Dec. 2005.

Levanon, N., "The Periodic Ambiguity Function—Its Validity and Value," *Proceedings 2010 IEEE International Radar Conference*, pp. 204–208, Arlington, VA, 2010.

Levanon, N., and E. Mozeson, *Radar Signals*. Wiley, New York, 2004.

Lewis, B. L., and F. K. Kretschmer, Jr., "Linear Frequency Modulation Derived Polyphase Pulse Compression Codes," *IEEE Transactions on Aerospace and Electronic Systems*, vol. AES-18, no. 5, pp. 637–641, Sep. 1982.

Nathanson, F. E. (with J. P. Reilly and M. N. Cohen), *Radar Design Principles*, 2d ed. McGraw-Hill, New York, 1991.

O'Donnell, B., and J. M. Baden, "Fast Gradient Descent for Multi-Objective Waveform Design," *Proceedings 2016 IEEE Radar Conference* (RadarConf), Philadelphia, PA, pp. 1–5, 2016.

Oppenheim, A. V., and R. W. Schafer, *Discrete-Time Signal Processing*, 3rd ed. Pearson, Englewood Cliffs, NJ, 2010.

Peebles, P. Z., Jr., *Radar Principles*. Wiley, New York, 1998.

Price, R., "Chebyshev Low Pulse Compression Sidelobes via a Nonlinear FM," URSI National Radio Science meeting, Seattle, WA, June 18, 1979.

Raney R. K., "A New and Fundamental Fourier Transform Pair," *Proceedings of the IEEE 12th International Geoscience & Remote Sensing Symposium (IGARSS' 92)*, pp. 26–29, 106–107 May, 1992.

Richards, M. A., "Time and Frequency Domain Weighting of LFM Pulses," technical memorandum, Sep. 29, 2006. Available at http://radarsp.com.

Rihaczek, A. W., *Principles of High-Resolution Radar*. Artech House, Boston, MA, 1996.

Rohling, H., and M. Kronauge, "Continuous Waveforms for Automotive Radar Systems," Chap. 7 in F. Gini et al. (eds.), *Waveform Design and Diversity for Advanced Radar Systems*. Institution of Engineering and Technology (IET), London, 2012.

Scheer, J. A., and J. L. Kurtz, eds., *Coherent Radar Performance Estimation*. Artech House, Canton, MA, 1993.

Problems

1. Consider a stationary radar transmitting a simple square pulse (modulation only, not including the carrier term) of duration τ:

$$x(t) = \begin{cases} 1, & 0 \leq t \leq \tau \\ 0, & 0, \text{otherwise} \end{cases}$$

The receiver uses a causal matched filter with $T_M = \tau$, so $h(t) = x^*(\tau - t)$. The pulse is transmitted with the leading edge being emitted at time $t = 0$. An echo is received from a stationary target at a range of R meters. At what time t_{peak} will the peak output of the matched filter be observed? Show all work.

2. Consider the same pulse and matched filter used in the previous problem. Assume that now the target is at range R meters when the pulse hits it, but is moving with a radial velocity toward the radar of $k\lambda_t/2\tau$ m/s, where k is any integer (except $k \neq 0$). The received signal (again, after the carrier is removed) can be modeled as

$$r(t) = x\left(t - \frac{2R}{c}\right) \exp\left[j2\pi F_D\left(t - \frac{2R}{c}\right)\right]$$

where F_D is the Doppler shift in hertz. Find the output waveform of the causal matched filter. What is its value at $t = 2R/c + \tau$ seconds?

3. Suppose the ambiguity function of some waveform $x(t)$ of duration $\tau = 1$ ms is given by

$$A(t, F_D) = \exp\left\{-\left[\left(\frac{t}{3\tau}\right)^2 + (2F_D\tau)^2\right]\right\}$$

(Note: This is not a possible AF because it is not time-limited to $\pm \tau$ seconds duration in the delay coordinate, but it will do for this problem.) Suppose there are two targets in the radar's line of sight, one at $R = 10$ km and one at $R = 10.1$ km. Also assume that both have the same RCS and ignore the effect of the small range difference on the received echo power. The radar and the first target are stationary. The second target is traveling toward the radar at 100 m/s. The radar is operating at 1 GHz. What is the Doppler shift of the echo from the second target, in hertz? If the matched filter output is sampled at a time delay corresponding to the range to the first target ($= 2 \times (10 \text{ km})/c = 66.67$ μs), the sample will contain contributions from both the first and second targets. Use $A(t, F_D)$ to determine the relative amplitude of the contribution from the second target compared to that of the first target. Express the answer in dB.

4. Consider a simple pulse burst waveform with $M = 30$ pulses, each of 10 μs duration, and a PRI of $T = 100$ μs. Assuming no weighting functions are used, what are the range resolution, Doppler resolution, unambiguous range, and unambiguous Doppler shift of this waveform?

5. Consider a linear FM waveform that sweeps from 9.5 to 10.5 GHz over a pulse length of 20 μs. What is the bandwidth β? What is the time-bandwidth product? What will be the Rayleigh resolution (peak to first null) of the matched filter output in meters? What would be the Rayleigh resolution in meters of a square pulse of the same energy (assuming both have the same amplitude)?

6. Continuing with the same LFM waveform as in the previous problem, what will be the frequency in hertz of the first zero of the zero delay cut of the ambiguity function? (This will be the Doppler resolution, or Doppler sensitivity, of the pulse.)

7. Consider an LFM waveform of bandwidth $\beta = 1$ MHz and pulse length $\tau = 1$ ms. Suppose an echo is received from a target at a true range of 10 km that is Doppler shifted by 1 kHz. What will be the apparent range of the target, that is, what will be the range corresponding to the time at which the matched filter output peaks?

8. Consider an LFM pulse with $\beta = 50$ MHz and $\tau = 1$ ms. Compute the Doppler shift required to displace the matched filter output by three Rayleigh range resolution cells. No windowing for sidelobe control is used. At 10 GHz, compute the radial velocity associated with that value of Doppler shift. Compute the loss in peak amplitude due to the Doppler shift in dB.

9. Suppose a radar uses a simple rectangular pulse of duration τ seconds and processes it through the corresponding matched filter. Assume the matched filter output is sampled at a rate equal to its Rayleigh bandwidth. What is the worst-case straddle loss in dB? Repeat for an LFM waveform with a sufficiently large BT product so that its spectrum is well-approximated by a rectangle of width β Hz. Assume no weighting for sidelobe control is used with either waveform.

10. Consider an LFM pulse of duration $\tau = 1$ ms. Suppose that a range window of only 1.5 km extent is of interest, so it is decided to use stretch processing. The range window is centered on a nominal range of 100 km (think of this as "zooming in" on targets in the vicinity of 100 km). A range resolution of 1.5 m is required. What is the required bandwidth β? What will be the $\beta\tau$ product of the LFM pulse? What will be the bandwidth of the stretch mixer output?

11. Continue with the same scenario and LFM waveform as in the previous problem. Suppose that a beat frequency of 100 kHz is observed at the mixer output. What is the range of the target, relative to the 100 km center of the range window? Ignore any delay in the matched filter.

12. Consider a stationary X-band (10 GHz) radar transmitting a $\beta = 500$ MHz LFM waveform and using stretch processing in the receiver. The pulse length is $\tau = 10$ μs. A radar is often considered "narrowband" if the percentage bandwidth, defined as β divided by the RF frequency, is less than 10 percent; otherwise it is "wideband." Is this radar narrowband or wideband? What is the expected range resolution in meters?

13. Continuing with the same LFM waveform, suppose a Hamming window is applied to the signal at the output of the stretch mixer, before the FFT is performed. What will be the new value for the expected range resolution, based on the Rayleigh definition of resolution? (*Hint:* The peak-to-null width of the DTFT of a Hamming window of length τ seconds is $2/\tau$ Hz; for a rectangular window it is $1/\tau$ Hz.) What bandwidth β would be required to achieve 0.3 m resolution if the Hamming window is used to keep the range sidelobes low?

14. Continuing with the same radar and 500 MHz LFM pulse as in the previous two problems, suppose the stretch processor is set up for a nominal range (center of the range window) of $R_0 = 200$ km and a range window of 300 m (200 km $\pm$ 150 m). However, suppose the reference LFM signal is only $\tau = 10$ μs seconds long, that is, it is not lengthened to allow for signals arriving from the leading or trailing edges of the range window. The reference signal is timed to overlap exactly with the echo from a target at range R_0. What will be the duration at the mixer output of the beat frequency tone between the echo from a scatterer at the leading edge of the range window (200 km $-$ 150 m) and the LFM reference? Assuming a rectangular window (i.e., no Hamming window), what will be the range resolution at the leading edge of the window?

15. It was stated that range skew at the output of a stretch processor could be corrected with a filter having the frequency response $H(\Omega) = \exp(-j\Omega^2\tau/4\pi\beta)$, where Ω is in radian frequency units and β is in hertz. Show that the group delay function $d_g(\Omega)$ of this filter meets the stated requirement, namely $d_g(-2\pi\beta \cdot \delta t_b/\tau) = -\delta t_b$ seconds. Group delay in seconds is defined as $d_g(\Omega) \equiv -d\Phi(\Omega)/d\Omega$, where $\Phi(\Omega) = \arg[H(\Omega)]$.

16. Assuming a sampling rate of F_s samples per second at the stretch mixer output, convert the analog frequency response $H(\Omega)$ of the previous problem to an equivalent discrete-time frequency response $H(\omega)$. Also give the expression for $H(\omega)$ in the particular case when F_s is chosen to match the stretch mixer output bandwidth of Eq. (4.111).

17. Explicitly compute the loss in processing gain LPG and the processing loss PL as a function of K for a triangular window of odd length $K + 1$ (so K is even) defined according to

$$w[k] = \begin{cases} 2k/K, & 0 \leq k \leq K/2 \\ 2 - 2k/K, & K/2 \leq k \leq K \\ 0, & \text{otherwise} \end{cases}$$

Numerically evaluate the result for $K = 4$ and $K = 20$ and give the answers in dB. What are the asymptotic values in dB for LPG and PL as $K \to \infty$? The following facts may be useful (be careful about the limits):

$$\sum_{k=1}^{n} k = n(n+1)/2, \quad \sum_{k=1}^{n} k^2 = n(n+1)(2n+1)/6$$

(*Hint:* Sum just the first half of the triangle, then use symmetry to get the sum of the whole function. Be careful not to double-count any samples.)

18. Consider the array steering factor $E(\theta_0)$ of Eq. (4.123) and use the weights given in Eq. (4.124) with $|a_n| = 1$ for all n. Assume the phases of the weights are computed for a wavelength λ_t and steering angle θ_0, but the waveform bandwidth is approximately 10 percent of the nominal frequency so that the effective wavelength varies over the range of $(1 \pm 0.05)\lambda_t$. Derive an equation that gives the new angle θ at which $E(\theta)$ will be maximum in terms of λ_t, θ_0, and the actual wavelength λ. When the actual wavelength is 5 percent larger than λ_t and the design steering angle is $\theta_0 = 10°$, what will be the actual steering angle [angle of the maximum of $E(\theta)$]? Repeat for $\theta_0 = 30°$ and $70°$.

19. There are $2^4 = 16$ length-4 biphase codes. Select one that is *not* equivalent to one of the two length-4 Barker codes in Table 4.1. Compute its autocorrelation explicitly and show that it is not a Barker code.

20. Compute explicitly the autocorrelation of the length-5 Barker code in Table 4.1 and verify that it meets the definition of a Barker code.

21. Compute the integrated sidelobe ratios in linear units and in decibels for the Barker codes in Table 4.1. A computer may be used to do the calculations if desired.

22. Determine the chip length and pulse length of a biphase-coded waveform for a pulsed radar to meet the following requirements:

    a. Rayleigh range resolution $= 0.3$ m.
    b. Pulse compression gain >15 dB.
    c. Maximum allowable minimum range within first range ambiguity $= 50$ m.
    d. Less than one-quarter cycle of Doppler phase rotation across the pulse for Doppler shifts up to 2000 Hz.

    One or both parameters may have a range of allowable values. Give the full range if this is the case.

23. Consider designing an M-pulse biphase-coded burst waveform for detecting an aircraft using a ground-based radar. Assume the aircraft velocity relative to the radar is 300 m/s, and that the system can coherently integrate the returns from the M pulses. Assume a maximum target length of 20 m and a search mode of operation. Specify the waveform parameters of pulse length τ, chip length τ_c, code length N, PRF, and total signal processing gain (combined pulse compression and coherent integration) to meet all of the following requirements:

    a. To avoid sub-resolving the target into multiple range bins (bad for detection), the range resolution is required to be $\Delta R > 20$ m.
    b. To prevent masking of weaker targets by stronger targets, the waveform PSL relative to its peak is required to be less than -15 dB.
    c. The minimum range is required to be less than 360 m.
    d. Range migration over the CPI duration must be less than one resolution cell.

e. The unambiguous range must satisfy 50 km $\leq R_{ua} \leq$ 60 km.

f. The combined SNR processing gain (pulse compression gain and coherent pulse integration) must be at least 28 dB.

24. Suppose the radar in the preceding problem has an RF of 10 GHz. How many cycles of Doppler shift would there be across a pulse of the waveform that was designed if the radar did not compensate for the Doppler? Based on the "quarter cycle of Doppler across the pulse" criterion for Barker phase codes, would a significant reduction in pulse compression gain be expected?

25. Consider designing a pulse burst waveform for an air-to-air engagement between "blue" and "red" aircraft. The blue force pilot wants to employ a high PRF waveform to ensure that an approaching red force aircraft is not ambiguous in Doppler. The maximum speed of the blue aircraft is 300 m/s, and the maximum speed of the red force aircraft is 350 m/s. The RF is 10 GHz. Specify the number of pulses M, PRF, and CPI duration for a pulse burst waveform that meets all of the following requirements:

  a. Unambiguous velocity $\geq$ 1300 m/s to allow for both opening and closing targets at maximum speeds.

  b. Velocity resolution $\leq$ 20 m/s (measured by Rayleigh width with no windowing)

  c. Unambiguous range >1.5 km. This requirement is driven by other system constraints including the radar system's duty cycle.

26. Consider Barker, MPS, and pseudorandom biphase codes. State whether each code type can meet the requirements of the previous problem. If not, state the reason; if so, state at least one specific length that will work.

27. Compute explicitly the $N = 4 = 2^2$ Frank code. What is the sequence of phases ϕ_n in the code (expressed as an angle in radians, e.g., 0, $\pi/3$, etc.)? Compute the autocorrelation function of the detected code sequence $\exp(j\varphi_n)$ explicitly by hand. Sketch the magnitude of the autocorrelation function. What is the peak sidelobe level relative to the peak of the autocorrelation function in dB?

28. Repeat the previous problem for an $M = 4$ P4 code. (Be sure to use the complex autocorrelation function.)

29. Equation (4.135) expressed the continuous autocorrelation function $s_x(t)$ of a phase-coded waveform for $t = k\tau_c + \eta$ as a linear interpolation between the possibly complex discrete autocorrelation values $s_A[k]$ and $s_A[k+1]$ of the code sequence $s_A[k]$. Show that this linear interpolation of the complex values also linearly interpolates the real and imaginary parts of $s_A[k]$ and $s_A[k+1]$, but that the magnitude of the interpolated value is not the linear interpolation of the magnitudes of $s_A[k]$ and $s_A[k+1]$.

30. Some waveform/matched filter pairs are more sensitive to Doppler mismatch ("less Doppler tolerant") than others. Consider three different waveforms, all using pulses of length τ seconds: a single simple pulse, a single LFM pulse with $\beta\tau = 1000$, and a pulse burst composed of 30 simple pulses of length τ with a PRI of 10τ seconds. Denote the time of the peak matched filter output when there is no Doppler shift as t_{max}. Suppose a target with a Doppler shift of $F_D = 1/\tau$ Hz is present. The matched filter does not compensate for this Doppler shift. For each waveform, what will be the magnitude of the matched filter output waveform at $t = t_{max}$ compared to the value at t_{max} when there is no Doppler shift? Which of these waveforms is most Doppler tolerant in this case? Which is least Doppler tolerant?

31. It was stated that an unmodulated CW waveform can measure velocity, but not range, because there is no "timing mark" in the waveform. Consider an unmodulated CW radar and receiver viewing a stationary target at range R_0 m. The transmitted waveform is simply $A\exp(j\Omega_t t)$. What will be the received waveform? (Ignore amplitude factors.) The answer will depend on R_0. Why can't the difference in the two waveforms be used in practice to measure range, at least for stationary targets?

32. Consider an automotive radar using a triangular continuous wave waveform like that of Fig. 4.62. The transmit frequency is $F_t = 77$ GHz, the upchirp + downchirp combined sweep duration is $T = 50$ μs, and the swept bandwidth is $\beta = 300$ MHz. Suppose a target is detected exhibiting upchirp and downchirp beat frequencies of $F_{bup} = -7.9846$ MHz and $F_{down} = +8.0154$ MHz. What are the range and radial velocity of the target?

33. Suppose the waveform in Prob. 32 is modified to retain the same upchirp in the first 25 μs of the sweep interval, but to use a constant frequency (no chirp) CW waveform instead of a downchirp in the second 25 μs. Write the equations relating the primary beat frequencies observed to the range and velocity. What would be the measured values of the beat frequencies in the scenario of Prob. 32? Given the two beat frequencies that would be measured, is it still possible to compute both the target range and velocity?

34. Continuing Probs. 32 and 33, suppose the waveform is modified again to use a slightly different constant frequency in the first 25 μs of the sweep period, say 77 GHz + δF. Thus the modulation now consists of alternating every 25 μs between the frequencies 77 GHz and 77 GHz + δF. The receiver continues to demodulate the echo signal by mixing it with the outgoing transmitted signal. Can the range and radial velocity of the target still be determined from the two primary beat frequencies? If so, write the appropriate equations for range and velocity in terms of the two frequencies.

35. A fast-chirp linear FMCW 16-GHz radar with a sweep is to be designed to have 3-dB velocity resolution of 2 m/s and 3-dB range resolution of 1 m. A Hamming window will be used in the processing for sidelobe reduction in both range and velocity. (Table B.2 may be helpful.) The unaliased velocity range of interest is ±50 m/s and the maximum range of interest is 3 km.

   a. What is the maximum allowable repetition interval T? Assume this value is used going forward.
   b. What is the maximum time delay of interest, t_{max}?
   c. What swept bandwidth β is required to achieve a 3-dB range resolution of $\Delta R = 1$ m? Remember to allow for the effect of Hamming weighting.
   d. How many sweeps must be processed to achieve a 3-dB velocity resolution of $\Delta v = 2$ m/s, again allowing for Hamming weighting in the processing? What is the CPI duration?

36. Two different conditions were given for considering a linear FMCW radar to be a "fast chirp" system. Determine if the system in the previous problem is a "fast chirp" system according to either or both of these conditions. Assume $R_{min} = 100$ m. If the system is *not* fast chirp under either of the conditions, what changes are required to make it fast chirp? Describe how those changes would affect other aspects of the design, if any.

CHAPTER 5
Doppler Processing

5.1 Introduction

Doppler processing is the term applied to filtering or spectral analysis of the signal received from a fixed range or range interval over a period of time corresponding to several pulses or sweeps. The purpose is generally to suppress clutter returns and to enable the detection of targets in the presence of significant clutter.

Figure 5.1*a* shows a notional scenario where a down-looking stationary radar observes four moving targets in a ground clutter background. The gray dashed lines represent range bins. Receding targets are in bins 4, 11, and 18, while an approaching target shares bin 11. Figure 5.1*b* is a stylized representation of the range-Doppler power spectrum that might result from this scenario. The light gray background represents the receiver noise floor, which is spread uniformly throughout the range-Doppler map. The band of energy extending through all of the range bins represents the ground clutter echo. Because the radar is stationary, the clutter is centered at zero Doppler shift, and its power fades with range in accordance with the range equation. The four small ovals represent the returns from the moving targets. Their echo energy is located in the appropriate range bins. Their Doppler coordinates depend on the direction and speed of each target with respect to the radar.

Figure 5.1*c* highlights the notional Doppler spectrum of the slow-time signal from range bin 11, containing the two middle moving targets. Since the slow-time data is sampled at a rate equal to the repetition rate of the radar, this spectrum is periodic with a period equal to the repetition rate (RR), so only the principal period from $-RR/2$ to $+RR/2$ is shown. The portion of the spectrum where clutter is the dominant interference is often termed the *clutter region*. The width β_C in hertz of the clutter region is determined by the actual clutter motion, the radar frequency, and the pulse repetition frequency. The portion where noise is the dominant interference is called the *clear region*; note that the clear region is clear of clutter but not of all interference. Sometimes a *skirt region* is defined at the transition between the clutter and clear regions; in the skirt region, both noise and clutter are significant interference sources. The moving targets appear in the spectrum at Doppler shifts consistent with their radial velocities relative to the radar.

Doppler processing is most often of interest when the relative amplitudes of the clutter, target, and noise signals are as shown: the target returns are above the noise floor (signal-to-noise ratio $SNR \gg 1$), but weaker than the clutter (signal-to-clutter ratio $SCR \ll 1$). In this case, targets cannot be detected reliably based on amplitude in the slow-time domain alone because the presence or absence of the target makes little difference to the total signal power, which is dominated by the clutter. Doppler processing is used to separate the target and clutter signals in the frequency domain. The clutter can be filtered out, leaving the target return as the strongest signal present, or the spectrum can be computed explicitly so that targets in

252 Chapter Five

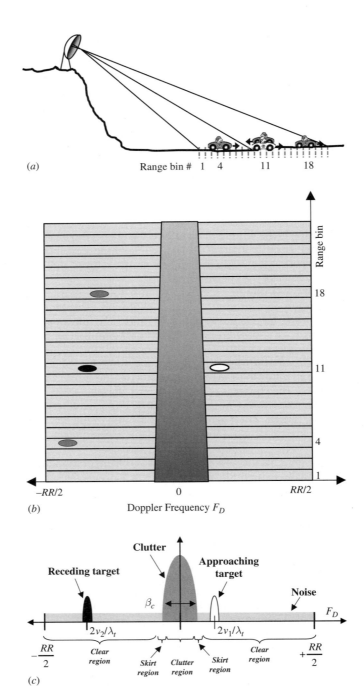

FIGURE 5.1 Notional range-Doppler spectrum for a down-looking stationary radar. Noise, clutter, and target components are shown. (*a*) Scenario. (*b*) Range-Doppler echo power distribution. (*c*) Doppler spectrum of range bin 11.

the clear region can be located by finding frequency components that significantly exceed the noise floor.

In this chapter, the two major classes of Doppler processing, *moving target indication* (MTI) and *pulse Doppler* (PD) processing, are described. In the terminology used here MTI refers to explicitly suppressing the clutter component of the slow-time signal using time domain filtering. Pulse Doppler processing refers to separating the target and clutter components of the signal using frequency domain processing.[1] As will be seen, MTI processing produces limited information at very low computational cost; pulse Doppler processing requires more computation but produces more information and greater SCR and signal-to-interference ratio (SIR) improvement. Only coherent Doppler processing using digital implementations is considered since this is the approach taken in essentially all modern radars. Good general references include Richards (2010) and Schleher (2010). Alternative systems using noncoherent Doppler processing and implementations based on analog technologies are described in Eaves and Reedy (1987), Nathanson (1991), and Schleher (2010).

5.2 Moving Platform Effects on the Doppler Spectrum

The notional Doppler spectrum of Fig. 5.1 represents a very simple case. While it is realistic for some scenarios, the Doppler spectrum for a given range bin can be greatly complicated by factors such as a moving radar platform or range and Doppler ambiguities caused by aliasing of the target signatures.

The effect of the RR in the Doppler dimension is straightforward. It establishes the width of one period of the Doppler spectrum as discussed above. Clutter or target signals having Doppler shifts outside of the range $\pm RR/2$ will alias into that interval. Figure 5.2 illustrates how the spectrum of Fig. 5.1c might look if the RR were reduced by 40 percent. The clutter spectrum is unchanged but now represents a larger percentage of the total spectrum width, while the clear region is now a smaller percentage. The target originally at velocity v_1 is still unaliased and so still appears at Doppler shift $2v_1/\lambda_t$. The Doppler shift of the target originally at the (negative) velocity v_2 was outside the new spectrum range $\pm RR'/2$ and so aliases to a new apparent Doppler shift of $2v_2/\lambda_t + RR'$.

Now consider the effect of platform motion on the range-Doppler map observed from an airborne or spaceborne platform. As described in Chap. 3, the Doppler shift observed for any target or clutter scatterer is increased by the platform-induced Doppler shift component of $F_D = 2v \cos\psi/\lambda_t$ Hz, where ψ is the *cone angle* between the platform velocity vector and the line of sight (LOS) to the scatterer. Figure 5.3a shows an aircraft in level flight with a forward but down-looking radar illuminating ground clutter. In this case, the cone angle ψ is the depression angle ψ_{MLC} shown. While most of the echo energy will be in the radar mainbeam, it is common to have measurable (above the noise floor) clutter echo through both forward- and rear-looking sidelobes as well. Ground clutter echoes received from different directions will appear at both different Doppler shifts due to different cone angles and at different ranges due to different distances to the ground. Target echoes will again appear at the appropriate Doppler shift and range bin, where the Doppler shift now is proportional to the total radial velocity due to both the platform motion and the target motion.

As shown in Fig. 5.3a, the mainlobe clutter (MLC) will occur at some positive Doppler shift and the appropriate range bins. Clutter echo received from below the aircraft will occur at a near range bin corresponding to the aircraft altitude and at zero Doppler shift. This

---

[1]Skolnik (2001) distinguishes MTI and pulse Doppler by defining pulse Doppler as a system that uses a PRF high enough to avoid blind speeds (see Sec. 5.3.4). In this text, the two terms are instead differentiated based on the processing approach used and the information obtained.

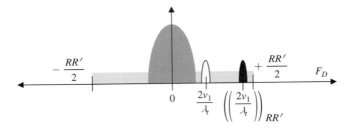

FIGURE 5.2 Effect of a 40 percent RR reduction on the Doppler spectrum of Fig. 5.1c.

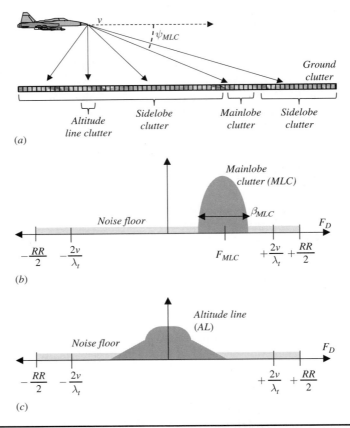

FIGURE 5.3 Notional Doppler spectrum of ground clutter observed from a moving radar platform. (a) Sources of clutter return. (b) Notional clutter spectrum in the boresight range bin. (c) Notional spectrum in the altitude line range bin.

clutter component is termed the "altitude line" (AL). Although it is transmitted and received through the radar sidelobes, the altitude line nonetheless tends to be relatively strong due to the relatively short vertical range and the high reflectivity of most clutter at normal incidence (recall Fig. 2.22). Other ranges may exhibit *sidelobe clutter* (SLC) seen through the antenna sidelobes. SLC is weaker than MLC because it is observed through the lower-gain

Doppler Processing   **255**

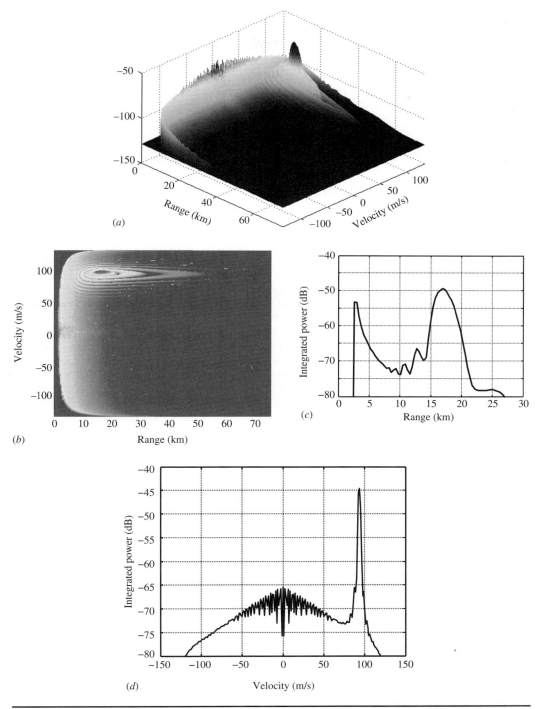

FIGURE 5.4 Simulated clutter range-velocity spectrum for an airborne radar viewing ground clutter. (a) Perspective view. (b) Overhead view. (c) Clutter power versus range, integrated over velocity. (d) Clutter power versus velocity, integrated over range. See text for details.

sidelobes. It is also weaker than the AL due to longer ranges and non-vertical incidence. SLC can potentially extend from Doppler shifts approaching $+2v/\lambda_t$ (directly ahead of the radar) to $-2v/\lambda_t$ (directly behind the radar), though the amplitude will drop rapidly due to increasing range at those maximum Doppler shifts.

Figure 5.3b is a sketch of the notional MLC spectrum at the range bin corresponding to the radar boresight, assuming the RR is high enough to avoid any aliasing. The MLC will be centered at $F_{MLC} = 2v\cos(\psi_{MLC})/\lambda_t$, where ψ_{MLC} is the depression angle from the horizontal velocity vector to the radar boresight vector. The width β_{MLC} of the MLC will be approximately the sum of the intrinsic clutter spectrum width β_C and the mainlobe widening due to platform motion of $\beta_D = 2v\theta\sin(\psi_{MLC})/\lambda_t$, where θ is the antenna azimuth beamwidth in radians. Figure 5.3c is a similar sketch for the altitude line, which would occur at a closer range bin. Range bins nearer than that of the AL can contain only noise.

Figure 5.4 is a more detailed but still idealized simulated range-Doppler clutter spectrum for a down-looking airborne radar viewing flat homogeneous ground clutter, not yet including any aliasing in range or Doppler due to the RR. The aircraft is traveling level at an altitude of 10,000 ft (3048 m) and a velocity of 300 mph (134.1 m/s). The RF is 10 GHz. The radar antenna has a 2.5° beamwidth, and its boresight is scanned 45° clockwise (right) of the velocity vector in azimuth and 10° down in elevation from level. The boresight will then intersect the ground at a range of 17.3 km. The component of velocity along the boresight (i.e., $v\cdot\cos\psi_{MLC}$) is 93.4 m/s, giving a Doppler shift at the center of the MLC of 6.23 kHz. The maximum clutter Doppler shift that could be observed is ±8.94 kHz, corresponding to relative velocities of ±134.1 m/s, the aircraft velocity. The simulation used a constant-γ model for clutter reflectivity; a sinc^2 two-way antenna power gain pattern; and accounted for receiver noise, clutter cell size, and power variation with range and other factors.

Figure 5.4a gives a three-dimensional view, while Fig. 5.4b shows a perhaps more useful two-dimensional view. It is readily seen that the maximum clutter power occurs at the mainlobe peak, centered at 93.4 m/s and 17.3 km. The high antenna sidelobes (no weighting was used) create visible rings of clutter power around the mainlobe. The shortest-range clutter occurs at zero velocity and just over 3 km; this is the altitude line and it is also relatively strong. The clutter energy spreads to the maximum relative velocities of ±134.1 m/s (Doppler shifts of ±89.4 kHz) as the range increases. In these figures, positive velocities (Doppler shifts) must come from clutter in front of the aircraft, while negative velocities come from clutter behind the aircraft. Figure 5.4c shows the total clutter power versus range, integrated over velocity. This illustrates the high AL return, the fall-off of the clutter power with range, and the large MLC return before the clutter falls off again. Figure 5.4d shows the power versus velocity, integrated over range. It illustrates the MLC, the relatively strong AL and the fall-off of clutter power at higher velocities, which implies smaller cone angles, longer ranges, and shallower grazing angles with attendant lower clutter reflectivity.

This example only hints at the complexity of the range-Doppler spectrum. The subject will be revisited in Sec. 5.6.4 in the discussion of PRF regimes. Despite the potential complexity, the simple spectrum of Fig. 5.1c has all the features needed to introduce MTI.

5.3 Moving Target Indication

Figure 5.5 illustrates a two-dimensional data matrix formed from the coherently demodulated baseband returns from a series of M range profiles (RPs) comprising one coherent processing interval (CPI). This matrix corresponds to one two-dimensional horizontal plane from the radar datacube of Fig. 3.19. A similar matrix exists for each phase center in the antenna system. In a single-aperture system, or at the point in an array where the data from

Doppler Processing

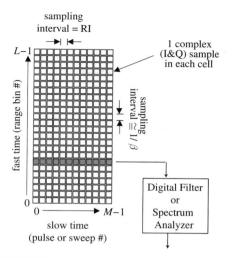

FIGURE 5.5 Notional two-dimensional data matrix. Each cell is one complex-valued echo voltage sample.

multiple phase centers have been combined, there is only a single two-dimensional data matrix as shown.[2]

The samples in each column are successive samples of the complex amplitude of the returns from a single pulse or sweep, that is, successive range bins. Collectively, they form an RP for one pulse or sweep. Each element of a column is one complex number representing the real and imaginary (I and Q) components of one range sample. Each row then represents the series of measurements from the same range bin over successive RPs. In a pulsed radar using standard matched filtering, the sampling rate in the fast time or range dimension (vertical column in Fig. 5.5) is at least equal to the transmitted pulse bandwidth and therefore is on the order of hundreds of kilohertz to tens or even hundreds of megahertz. For a pulsed radar using stretch processing or a fast-chirp frequency-modulated continuous wave (FMCW) system, the rate is usually much lower. The slow-time or pulse number dimension (horizontal row in Fig. 5.5) is sampled at the repetition interval (RI) of the radar. Thus the sampling rate in this dimension is the RR and is on the order of ones to tens, and sometimes hundreds of kilohertz. As indicated by the shading, Doppler filtering operates on rows of this matrix.

MTI processing applies a linear filter to the slow-time data sequence in order to suppress the clutter component. Figure 5.6 illustrates the process. The type of filtering needed can be understood by considering Fig. 5.7. In this figure, it is again assumed that knowledge of the platform motion and scenario geometry has been used to center the clutter spectrum at zero Doppler frequency. Clearly, some form of highpass filter, as suggested by the notional frequency response $|H(F_D)|$ shown, is needed to attenuate the clutter without filtering out moving targets in the clear region of the Doppler spectrum.

The output of the highpass MTI filter will be a new slow-time signal containing components due to noise and, possibly, one or more targets. This signal is passed to a detector

---

[2]Not all digital processors necessarily form a data matrix similar to Fig. 5.5 explicitly. MTI processors in particular can be implemented more simply. However, the data matrix is used explicitly in many other processors and is useful for illustration of Doppler filtering concepts.

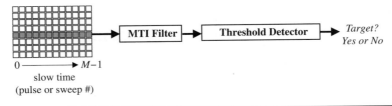

FIGURE 5.6 MTI filtering and detection process.

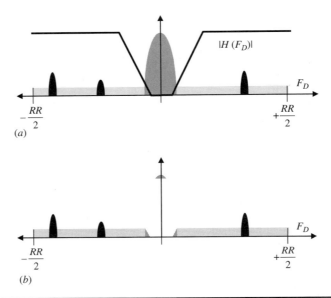

FIGURE 5.7 The concept of moving target indication filtering. (*a*) Doppler spectrum and MTI filter frequency response. (*b*) Doppler spectrum after filtering.

typically consisting of a matched filter followed by a threshold test. If the peak matched filter output exceeds the threshold (i.e., its energy is too great to likely be the result of noise alone), a target will be declared. Note that in MTI processing, the presence or absence of a moving target in the range bin of interest is the only information obtained. The filtering process of Fig. 5.6 does not provide any estimate of the Doppler frequency at which the target energy causing the detection occurred, or even of its sign; thus, it "indicates" the presence of a moving target but does not determine whether the target is approaching or receding, or at what radial velocity. Furthermore, it provides no indication of the number of moving targets present. If multiple moving targets are present in the slow-time signal from a particular range, the result will still be only a "target present" decision from the detector. On the other hand, MTI processing is very simple and computationally undemanding. Despite its simplicity, a well-designed MTI can improve the SCR by several decibels, and 20 or more decibels in some clutter conditions.

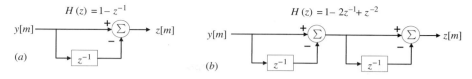

FIGURE 5.8 Flowgraphs and system functions of basic MTI cancellers. (a) Two-pulse canceller. (b) Three-pulse canceller.

5.3.1 Pulse Cancellers

The major MTI design decision is the choice of the particular MTI filter to be used. MTI filters are typically simple low-order designs. Some of the most common and traditional MTI filters are based on very simple heuristic design approaches. Suppose a fixed radar illuminates a moving target surrounded by perfectly stationary clutter. The clutter component of the echo signal from each pulse or sweep would be identical, while the phase of the moving target component would vary due to the changing range. Subtracting the echoes from the same range bin and successive pairs of pulses would cancel the clutter components completely. The target signal would not cancel in general due to the phase changes.

This observation motivates the two-pulse MTI canceller,[3] also referred to as the single or first-order canceller. Figure 5.8a illustrates the flowgraph of a two-pulse canceller. The input data are a sequence of baseband complex (I and Q) data samples from the same range bin over successive range profiles, forming a discrete-time sequence $y[m]$ with an effective sampling interval T equal to the RI. The discrete-time transfer function (also called the *system function*) of this linear finite impulse response (FIR, also called *tapped delay line* or *nonrecursive*) filter is $H(z) = 1 - z^{-1}$. The coefficients of this polynomial in z^{-1} are $[1, -1]$; they imply that the filter can be implemented with the difference equation

$$z[m] = y[m] - y[m-1] \tag{5.1}$$

where $y[m]$ is the slow-time input data and $z[m]$ is the filtered result.

The frequency response as a function of analog frequency F in hertz is obtained by setting $z = \exp(2\pi FT)$:

$$\begin{aligned} H(F) &= (1-z)^{-1}\big|_{z=\exp(j2\pi FT)} \\ &= 1 - \exp(-j2\pi FT) = \exp(-j\pi FT)[\exp(+j\pi FT) - \exp(-j\pi FT)] \\ &= 2j\exp(-j\pi FT)\sin(\pi FT) \end{aligned} \tag{5.2}$$

It is common to work with normalized frequency $f = FT$ cycles per sample or with the radian equivalent, $\omega = \Omega T = 2\pi FT$ radians per sample. In terms of normalized radian frequency, the frequency response of the two-pulse canceller is

$$H(\omega) = 2j\exp(-j\omega/2)\sin(\omega/2) \quad \text{radians per sample} \tag{5.3}$$

Recall that F ranges from $-1/2T$ to $+1/2T$, f ranges from -0.5 to $+0.5$, and ω from $-\pi$ to $+\pi$.

---

[3]The two-pulse canceller was developed for, and is most commonly used in pulsed radars, but can also be applied to fast-chirp linear FMCW. While it might be more generally called a two-RP canceller, the traditional "two-pulse" name is used here.

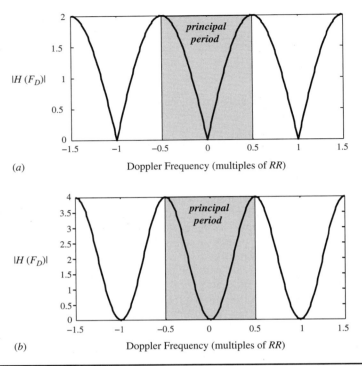

FIGURE 5.9 Frequency response of basic MTI cancellers. (a) Two-pulse canceller. (b) Three-pulse canceller.

Figure 5.9a plots the magnitude of the frequency response of the two-pulse canceller. The filter does indeed have a null at zero frequency to suppress the clutter energy. Spectral components representing moving targets may either be partially attenuated or amplified, depending on their location on the Doppler frequency axis. As with all discrete-time filters, the frequency response is periodic with a period of 1 in the normalized cyclical frequency variable f, corresponding to a period of 2π in the normalized frequency variable ω or a period of $1/T = RR$ in actual frequency in hertz. The shaded area highlights the principal period from $-RR/2$ to $+RR/2$; this is all that is normally plotted. Considering only this frequency range, it is clear that the frequency response is highpass in nature. The implications of the periodicity will be considered in Sec. 5.3.4.

The two-pulse canceller is a very simple filter. It is computationally very efficient: its implementation requires no multiplications and only one subtraction per output sample, as seen in Eq. (5.1). As Fig. 5.9a shows, however, it is a poor approximation to an ideal highpass filter for clutter suppression. The next traditional step up in MTI filtering is the three-pulse (second-order or double) canceller obtained by cascading two two-pulse cancellers. The flowgraph and frequency response are shown in Figs. 5.8b and 5.9b. The three-pulse canceller clearly improves the null breadth in the vicinity of zero Doppler, but it does not improve the consistency of response to moving targets at various Doppler shifts away from zero Doppler. It requires only two subtractions per output sample.

Despite their simplicity, the two- and three-pulse cancellers can be very effective against clutter with moderate-to-high sample-to-sample correlation. Figure 5.10 shows a simulated clutter sequence (gray line) formed by passing a white noise sequence through a filter with

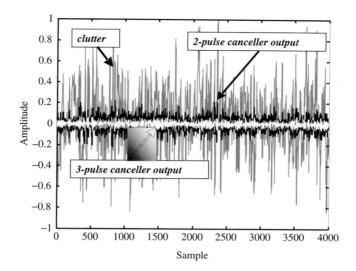

Figure 5.10 Example of clutter cancellation using two-and three-pulse cancellers. See text for details.

a Gaussian power spectrum having a standard deviation of $\sigma_f = 0.05$ on a normalized frequency scale; that is, the standard deviation is 5 percent of the full width of the principal period of the spectrum. Also shown are the results of passing this signal through the two-pulse canceller (black line) and three-pulse canceller (white line). The power in the two-pulse canceller output has been reduced by 13.4 dB relative to the unfiltered clutter sequence. For the three-pulse canceller, the reduction is 21.9 dB.

The idea of cascading two-pulse canceller sections to obtain higher-order filters can be extended to the N-pulse canceller obtained by cascading $N-1$ two-pulse canceller sections. The transfer function of the N-pulse canceller is therefore

$$H_N(z) = (1 - z^{-1})^{N-1} \tag{5.4}$$

The corresponding impulse response coefficients of the filter are given by the binomial series

$$h_N[m] = (-1)^m \binom{N-1}{m} = (-1)^m \frac{(-1)!}{m!(N-1-m)!}, \quad m = 0, \ldots, N-1 \tag{5.5}$$

Other types of digital highpass filters could also be designed for MTI filtering. For example, an FIR highpass filter could be designed using standard digital filter design techniques such as the window method or the Parks-McClellan algorithm (Oppenheim and Schafer, 2010). To be suitable as an MTI filter, the FIR filter frequency response should have a zero at $F = 0$. In terms of the four recognized classes of FIR digital filters (see Oppenheim and Schafer, 2010, Sec. 5.7), the MTI filter can be either type I (even order with symmetric impulse response) or type IV (odd order with antisymmetric impulse response). The transfer functions of type IV filters always have a zero at $z = 1$ so that the frequency response is zero at $f = 0$, ideal for an MTI filter. The two-pulse canceller (which has an *order* of 1) is an example of a type IV filter. Type I filters do not necessarily have a zero at $f = 0$, but can be made to have one by requiring that the sum of the impulse response coefficients $h[m]$ equal zero. The three-pulse canceller is an example of a type I filter that has been designed to be suitable as an MTI filter.

Type II and III filters are unsuitable because they always have a zero at $z = -1$, corresponding to a frequency response null at a normalized frequency of $f = 0.5$; this creates extra undesirable blind speeds (see Sec. 5.3.4). Alternatively, *infinite impulse response* (IIR) highpass filters could be designed. Many operational radar systems, however, use two- or three-pulse cancellers for the primary MTI filtering due to their computational simplicity.

5.3.2 Vector Formulation of the Matched Filter

The *N*-pulse cancellers described previously can be remarkably effective and have been widely used. Nonetheless, they are motivated by heuristic ideas. Can a more effective pulse canceller be designed? Since the goal of MTI filtering is to improve the signal-to-clutter ratio, it should be possible to apply the matched filter concept of Chap. 4 to this problem. To do so for discrete-time signals, it is convenient to first restate the matched filter using vector notation. This will also aid in generalizing the matched filter somewhat.

Consider a complex signal vector $\mathbf{y} = [y[m]\ y[m-1]\ \cdots\ y[m-N+1]]^T$ and a filter weight vector $\mathbf{h} = [h[0]\ h[1]\ \cdots\ h[N-1]]^T$. The superscript T represents matrix transpose so that $\mathbf{y}$ and $\mathbf{h}$ are N-element column vectors. A single output sample z of the filter is given by $z = \mathbf{h}^T\mathbf{y}$. The power in the output sample is given by

$$\text{Power in } z = |z|^2 = z^* z^T = \mathbf{h}^H \mathbf{y}^* \mathbf{y}^T \mathbf{h} \tag{5.6}$$

where a superscript H represents the Hermitian (complex) transpose.

The matched filter is obtained by finding the filter coefficient vector $\mathbf{h}$ that maximizes the SIR of the filtered data. Denote the desired target signal vector by $\mathbf{t}$ and the interference vector by $\mathbf{w}$, so that $\mathbf{y} = \mathbf{t} + \mathbf{w}$. The interference is a random process, but it is *not* assumed to be white or Gaussian, thus allowing modeling of both noise and clutter. The filtered signal and interference are, respectively, $\mathbf{h}^T\mathbf{t}$ and $\mathbf{h}^T\mathbf{w}$. The power in the signal component is therefore $\mathbf{h}^H\mathbf{t}^*\mathbf{t}^T\mathbf{h}$, and in the interference component it is $\mathbf{h}^H\mathbf{w}^*\mathbf{w}^T\mathbf{h}$. Because the interference power is a random variable (RV), its expected value is used to get meaningful results. The expected value of the matrix $\mathbf{w}^*\mathbf{w}^T$ is the interference covariance matrix $\mathbf{S}_I$:

$$\mathbf{S}_I \equiv E\{\mathbf{w}^*\mathbf{w}^T\} \tag{5.7}$$

It follows also that $\mathbf{S}_I^T = \mathbf{S}_I^*$, $\mathbf{S}_I = \mathbf{S}_I^H$ and $(\mathbf{S}_I^{-1})^T = (\mathbf{S}_I^{-1})^*$. With this definition the SIR becomes

$$\text{SIR} = \frac{\mathbf{h}^H \mathbf{t}^* \mathbf{t}^T \mathbf{h}}{\mathbf{h}^H \mathbf{S}_I \mathbf{h}} \tag{5.8}$$

As in Chap. 4, the filter $\mathbf{h}$ that maximizes Eq. (5.8) is found using the Schwarz inequality, which in a form suitable for vector-matrix manipulations is

$$|\mathbf{p}^H \mathbf{q}|^2 \leq \|\mathbf{p}\|^2 \|\mathbf{q}\|^2 \tag{5.9}$$

where $\|\mathbf{p}\| \equiv \sqrt{\mathbf{p}^H \mathbf{p}}$ and equality occurs if and only if $\mathbf{p} = k\mathbf{q}$ for some scalar constant k. To apply Eq. (5.9), first note that the matrix $\mathbf{S}_I$ will be positive definite so that it can be factored into the form $\mathbf{S}_I = \mathbf{A}^H\mathbf{A}$ for some matrix $\mathbf{A}$; that is, $\mathbf{A}$ is the "square root" of $\mathbf{S}_I$ in a sense.

Define $\mathbf{p} = \mathbf{Ah}$ and $\mathbf{q} = (\mathbf{A}^H)^{-1}\mathbf{t}^*$. This choice is contrived so that $\mathbf{p}^H\mathbf{q} = \mathbf{h}^H\mathbf{t}^*$ and therefore $|\mathbf{p}^H\mathbf{q}|^2 = \mathbf{h}^H\mathbf{t}^*\mathbf{t}^T\mathbf{h}$, which is the numerator of Eq. (5.8). The Schwarz inequality then gives

$$\mathbf{h}^H\mathbf{t}^*\mathbf{t}^T\mathbf{h} \leq \|\mathbf{Ah}\|^2 \|(\mathbf{A}^H)^{-1}\mathbf{t}^*\|^2 \\ = (\mathbf{h}^H\mathbf{S}_I\mathbf{h})(\mathbf{t}^T\mathbf{S}_I^{-1}\mathbf{t}^*) \tag{5.10}$$

Rearranging Eq. (5.10) to isolate the SIR of Eq. (5.8) shows that

$$\mathrm{SIR} \leq \mathbf{t}^T\mathbf{S}_I^{-1}\mathbf{t}^* \tag{5.11}$$

with equality only when $\mathbf{p} = k\mathbf{q}$. The optimal weight vector therefore satisfies $\mathbf{Ah}_{\mathrm{opt}} = k(\mathbf{A}^H)^{-1}\mathbf{t}^*$ or, with $k = 1$,

$$\mathbf{h}_{\mathrm{opt}} = \mathbf{S}_I^{-1}\mathbf{t}^* \tag{5.12}$$

Finally, the filtered data become, using $(\mathbf{S}_I^{-1})^T = (\mathbf{S}_I^{-1})^*$,

$$z = \mathbf{h}_{\mathrm{opt}}^T\mathbf{y} = \mathbf{t}^H(\mathbf{S}_I^{-1})^*\mathbf{y} \tag{5.13}$$

Equation (5.12) is of fundamental importance and great versatility in signal processing generally and radar signal processing in particular. It will be used in this text not only for clutter filtering and pulse Doppler processing, but also for space-time adaptive processing (STAP), detection, and ground moving target indication (GMTI).

5.3.3 Matched Filters for Clutter Suppression

The results of the previous section can now be applied to N-point (order $N-1$) MTI filters that are more optimal than the N-pulse canceller. Equation (5.12) shows that the linear filter that optimizes detection performance in the presence of additive interference is the FIR matched filter and that the coefficients of the filter are given by the matrix equation

$$\mathbf{h} \equiv \begin{bmatrix} h[0] \\ \vdots \\ h[N-1] \end{bmatrix} = \mathbf{S}_I^{-1}\mathbf{t}^* \tag{5.14}$$

where $\mathbf{h} = N \times 1$ column vector of filter coefficients
  $\mathbf{S}_I = N \times N$ covariance matrix of the interference
  $\mathbf{t} = N \times 1$ column vector representing the desired target signal to which the filter is matched

To determine the optimal filter coefficients $\mathbf{h}$, models are needed to describe the interference and target characteristics $\mathbf{S}_I$ and $\mathbf{t}$. For a simple example, consider the first-order (length $N = 2$) matched filter. Assume the interference $w[m]$ consists of the sum of zero mean stationary white noise $n[m]$ of power (variance) σ_n^2 and zero mean stationary colored clutter $c[m]$ of power σ_c^2,

$$w[m] = n[m] + c[m], \\ \mathbf{w} = [w[m] \ w[m-1]]^T \tag{5.15}$$

The clutter exhibits a correlation from one pulse to the next given by $E\{c[m]c^*[m+1]\} = E\{c^*[m]c[m-1]\} = s_c[1] = \sigma_c^2 \rho_c[1]$, where $s_c[k]$ is the clutter autocorrelation at lag k. The noise and clutter are uncorrelated with one another.

For this problem, $\mathbf{S_I}$ will be a 2×2 matrix, defined as

$$\mathbf{S_I} \equiv E\{\mathbf{w}^*\mathbf{w}^T\} = \begin{bmatrix} s_{11} & s_{12} \\ s_{21} & s_{22} \end{bmatrix} \tag{5.16}$$

Consider the s_{11} element. Using the fact that the noise and clutter are uncorrelated and that each is zero mean, it can be quickly concluded that

$$s_{11} = E\{(n[m] + c[m])^*(n[m] + c[m])\} = \sigma_n^2 + \sigma_c^2 \tag{5.17}$$

Next, consider the s_{12} element

$$\begin{aligned} s_{12} &= E\{(n[m] + c[m])^*(n[m-1] + c[m-1])\} \\ &= E\{n^*[m]n[m-1]\} + E\{n^*[m]c[m-1]\} + \\ &\quad E\{c^*[m]n[m-1]\} + E\{c^*[m]n[m-1]\} \end{aligned} \tag{5.18}$$

Again, the noise and clutter are zero mean and uncorrelated and the noise is white, so the expected values of the first three terms are zero. The last term becomes the pulse-to-pulse or sweep-to-sweep correlation of the clutter so that

$$s_{12} = 0 + 0 + 0 + E\{c^*[m]n[m-1]\} = \rho\sigma_c^2 \tag{5.19}$$

where $\rho_c[1]$ is denoted as ρ for simplicity. It is easy to see that $s_{22} = s_{11}$ and that $s_{21} = s_{12}^*$. Therefore,

$$\mathbf{S_I} = \begin{bmatrix} \sigma_c^2 + \sigma_n^2 & \rho\sigma_c^2 \\ \rho^*\sigma_c^2 & \sigma_c^2 + \sigma_n^2 \end{bmatrix} \tag{5.20}$$

so that

$$\mathbf{S_I}^{-1} = \frac{1}{(\sigma_c^2 + \sigma_n^2)^2 - |\rho|^2 \sigma_c^4} \begin{bmatrix} \sigma_c^2 + \sigma_n^2 & -\rho\sigma_c^2 \\ \rho^*\sigma_c^2 & \sigma_c^2 + \sigma_n^2 \end{bmatrix} = k \begin{bmatrix} \sigma_c^2 + \sigma_n^2 & -\rho\sigma_c^2 \\ -\rho^*\sigma_c^2 & \sigma_c^2 + \sigma_n^2 \end{bmatrix} \tag{5.21}$$

where k absorbs the constants resulting from the matrix inversion.

$\mathbf{S_I}$ represents the available information on the interference. The elements on the main diagonal will always be identical and equal to the total interference power, which is the sum of the independent interference source powers. The off-diagonal elements represent the correlation properties of the interference over one RI. Because the noise is white, it does not contribute to the off-diagonal elements, whereas the clutter does contribute provided $\rho = \rho_c[1] \neq 0$. More generally, an Nth order filter will require the $N \times N$ covariance matrix and will involve correlation coefficients up to $\rho_c[N-1]$.

To finish computing $\mathbf{h}$, a model is needed for the assumed target signal phase history $\mathbf{t}$. For a target moving at a constant radial velocity, the expected target signal is just a discrete complex sinusoid at the appropriate Doppler frequency F_D. Following the discussion in Sec. 2.7.3, assume the waveform is a train of M simple pulses with RI T and RF transmit frequency F_t. If the target is at a nominal range R_0 and is moving toward the radar at a radial velocity of v meters per second, the slow-time phase history will be of the form

$$y[m] = A \exp[+j2\pi F_D mT] \tag{5.22}$$

where $F_D = 2v/\lambda_t$ is the usual Doppler shift and all constants are absorbed into A at each step.

Only N samples at a time of $y[m]$ are of interest in analyzing an N-sample canceller. Assuming $N \leq M$ and referencing the results of App. B on vector representation of linear filtering, the series of N samples ending at $m = m_0$, $\{y[m_0], y[m_0-1], \ldots, y[m_0-N+1]\}$ can be represented in vector form as

$$\begin{aligned}\mathbf{t} &= A[\exp[j2\pi F_D m_0 T] \quad \exp[j2\pi F_D(m_0-1)T] \quad \cdots \quad \exp[j2\pi F_D(m_0-N+1)T]]^T \\ &= A[1 \quad \exp[-j2\pi F_D T] \quad \cdots \quad \exp[-j2\pi F_D(N-1)T]]^T\end{aligned} \tag{5.23}$$

where the phase terms due to the delay to the first sample of interest, m_0, have been absorbed into A and the signal has been renamed $\mathbf{t}$ to emphasize that it is only the target component. For the specific case $M = 2$ this becomes

$$\mathbf{t} = A[1 \quad \exp[-j2\pi F_D T]]^T \tag{5.24}$$

In practice, the target velocity and therefore Doppler shift are unknown; a target might be anywhere in the Doppler spectrum. The Doppler shift F_D is therefore modeled as a random variable with a uniform probability density function over $[-RR/2, +RR/2)$ and the expected value of $\mathbf{t}$ is computed. The expected value of the constant 1 is, of course, 1. The expected value of the second component of $\mathbf{t}$ is

$$\begin{aligned}E\{\exp(-j2\pi F_D T)\} &= \frac{1}{RR}\int_{-RR/2}^{+RR/2}\exp\{-j2\pi F_D T\}\,dF_D \\ &= \frac{1}{RR}\int_{-RR/2}^{+RR/2}\exp(-j2\pi F_D/RR)\,dF_D = 0\end{aligned} \tag{5.25}$$

The signal model then becomes simply

$$\mathbf{t} = [1 \quad 0]^T \tag{5.26}$$

Finally, combining Eqs. (5.21) and (5.26) in Eq. (5.14) gives the coefficients of the optimum two-pulse filter

$$\mathbf{h} = \left[\sigma_c^2 + \sigma_n^2 \quad -\rho^*\sigma_c^2\right]^T \tag{5.27}$$

In this equation the constant A has been dropped because it affects target, clutter, and noise equally and is therefore of no consequence.

To interpret this result, consider the case where the clutter is the dominant interference. Then σ_n^2 is negligible compared to σ_c^2 and $\mathbf{h} \approx [1 \quad -\rho^*]^T$. Now suppose the clutter is highly correlated over one RI so that ρ is close to one. Then $\mathbf{h} \approx [1 \quad -1]^T$, nearly the same as the two-pulse canceller. Despite its simplicity, the two-pulse canceller is therefore nearly a first-order matched filter for MTI processing when the clutter-to-noise ratio (CNR) is high and the clutter is highly correlated over one RI. In the limit of very high CNR and perfectly correlated clutter, the two-pulse canceller is exactly the first-order matched MTI filter.

The vector matched filter derivation of the optimum two-pulse MTI filter is easily extended to higher-order MTI filters. As the order increases, the corresponding N-pulse canceller becomes a poorer approximation of the matched filter (Schleher, 2010).

It is interesting to consider the form of the optimum filter when the dominant interference is noise rather than clutter, that is, $\sigma_c^2 \ll \sigma_n^2$. In this case, the optimum first-order MTI filter of Eq. (5.27) reduces to (ignoring overall scale factors again)

$$\mathbf{h} \approx [1 \ 0]^T \tag{5.28}$$

Equation (5.28) states that in the presence of completely *un*correlated interference and with no knowledge of the target velocity, the filter impulse response reduces to a single impulse, $h[m] = \delta[m]$. Since convolving any signal with $\delta[m]$ just returns the same signal, the filter does nothing. In the clutter-dominated case, the filter combined the two slow-time samples because, even though constructive interference of the target could not be guaranteed, the high correlation of the clutter did guarantee that the clutter signal would be suppressed. On average the overall effect was beneficial. In the noise-dominated case, there is still no guarantee that the target signal will be reinforced, and in addition there is now no guarantee that the noise will be suppressed. The filter therefore does not combine the two data samples at all.

The previous analysis assumes that the target Doppler shift is unknown and therefore considers all target Doppler frequencies equally likely. It is easy to modify the analysis to match the MTI filter to a specific Doppler shift or to the case where the target Doppler extends only over a portion of the Doppler spectrum. These alternative assumptions manifest themselves as alternate models for the desired signal vector $\mathbf{t}$. The second case is treated in Schleher (2010) and in Prob. 8; here the case of a known Doppler shift for the target and a two-pulse canceller is considered. The interference and signal models are exactly the same as given earlier except that now the target Doppler shift in $\mathbf{t}$ is not a random variable but a specific, fixed value. Therefore, it is not necessary to take an expected value of $\mathbf{t}$. The filter coefficient vector is

$$\mathbf{h} = \mathbf{S}_I^{-1}\mathbf{t}^* = \begin{bmatrix} \sigma_c^2 + \sigma_n^2 & -\rho\sigma_c^2 \\ -\rho^*\sigma_c^2 & \sigma_c^2 + \sigma_n^2 \end{bmatrix} \begin{bmatrix} 1 \\ \exp(+j2\pi F_D T) \end{bmatrix}$$

$$= \begin{bmatrix} (\sigma_c^2 + \sigma_n^2) - \rho\sigma_c^2 \exp(+j2\pi F_D T) \\ (\sigma_c^2 + \sigma_n^2)\exp(+j2\pi F_D T) - \rho^*\sigma_c^2 \end{bmatrix} \tag{5.29}$$

where again all overall constants have been dropped. While this result is easy to implement if the interference statistics are known, it is difficult to interpret. However, in the noise-limited case ($\sigma_n^2 \gg \sigma_c^2$), it reduces to

$$\mathbf{h} = [1 \quad \exp(+j2\pi F_D T)]^T \tag{5.30}$$

Equation (5.30) shows that in this case the optimum filter adds the two target samples together with a phase correction to the second so that they add in phase. In other words, the filter compensates for the sample-to-sample phase change and then performs a coherent integration of the two target samples.

5.3.4 Blind Speeds and Staggered PRFs

The frequency response of all discrete-time filters is periodic, repeating with a period of one in normalized cyclical frequency, corresponding to a period of $RR = 1/T$ Hz of Doppler shift.

Figure 5.9 illustrated this for the two- and three-pulse cancellers. Since MTI filters are designed to have a null at zero frequency, they will also have nulls at Doppler frequencies that are multiples of the repetition rate. Consequently, a target moving with a radial velocity that results in a Doppler shift equal to a multiple of the RR will be suppressed by the MTI filter. Velocities that result in these unfortunate Doppler shifts are called *blind speeds* because the target return will be suppressed; the system is "blind" to such targets. From a digital signal processing point of view, blind speeds represent target velocities that will be aliased to zero velocity. Equivalently, they correspond to Doppler shifts that will be aliased to zero frequency.

Blind speeds can be mitigated by making measurements at more than one RR. Different RRs correspond to different blind speeds, so any particular target velocity cannot be blind at all the RRs. This thought underlies the use of *staggered RRs*. Before continuing, note that the use of staggered RRs is common in pulsed radars but not in fast-chirp linear FMCW. The reason is that typical fast-chirp FMCW applications are usually designed for relatively short range and therefore short RIs. The corresponding RRs are high enough that blind speeds are often not a problem. Another reason is that using multiple RRs in FMCW significantly complicates the system. For example, changing the repetition interval T results in a change in range bin spacing (see Sec. 3.2.2) unless the fast-time sampling rate F_s is also changed. On the other hand, if the sweep length is to be kept constant to avoid this issue, a variable pause where no transmission occurs must be inserted at the end of each sweep to achieve different RIs. These problems do not arise in pulsed radar because the sweep duration is not tied directly to the RI. Given these considerations, the discussion of staggered RRs will be limited to pulsed systems, where it is referred to as staggered PRFs (or PRIs).

For a given PRF, the unambiguous range is

$$R_{ua} = \frac{c}{2PRF} \tag{5.31}$$

The first blind speed is

$$v_b = \frac{\lambda_t PRF}{2} = \frac{cPRF}{2F_t} \tag{5.32}$$

and is also called the unambiguous velocity v_{ua}. The corresponding blind Doppler shift F_{bs} simple equals the PRF. As the PRF is increased for a given RF, the unambiguous range decreases and the first blind speed increases. Figure 5.11 shows the unambiguous range-Doppler coverage regions that are possible. For example, each point on the line marked "1 GHz" represents a combination of R_{ua} and v_b corresponding to some PRF. The solid gray lines mark one example, corresponding to 400 m/s for the first blind speed and a 56.25-km unambiguous range. Equation (5.32) can be used to see that these values correspond to PRF = 2667 pulses per second.

As discussed in Chap. 3, there is some confusion in the terminology regarding the meanings of "unambiguous velocity" or "unambiguous Doppler shift" and "blind speed." Here, the blind speed as defined in Eq. (5.32) is the velocity interval between zeros of the MTI filter response. Generally, the range of velocities that are considered to be measurable without ambiguity is $[-v_b/2, +v_b/2]$. In some literature, the unambiguous velocity is then said to be $v_b/2$ because targets having a velocity magnitude greater than this will be aliased. However, in this text the unambiguous velocity is considered to be the full value v_b. Similarly, the blind Doppler shift is PRF Hz, allowing unambiguous measurements over the interval $[-PRF/2, +PRF/2]$, and the unambiguous Doppler shift interval is also considered to be PRF Hz. Because sources may differ on whether the "blind speed" is v_b or $v_b/2$, caution is needed in interpreting the literature. Again, unambiguous velocity and blind speed are synonymous in this text.

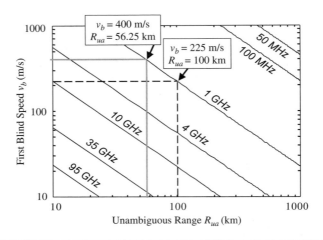

Figure 5.11 Ambiguity-free range-Doppler coverage regions.

The unambiguous range could be thought of as the range coverage interval of the radar for a given PRF, while the corresponding unambiguous Doppler interval could be considered the velocity or Doppler coverage interval. Equations (5.31) and (5.32) can be combined to show that for a given RF and single PRF the combined range-velocity coverage region is a constant, independent of the PRF. The combined range-Doppler coverage is also independent of the RF:

$$R_{ua}v_b = \frac{c}{2PRF} \cdot \frac{\lambda_t PRF}{2} = \frac{\lambda_t c}{4}$$
$$R_{ua}F_b = \frac{c}{2PRF} \cdot PRF = \frac{c}{2} \qquad (5.33)$$

Blind speeds could be avoided by choosing the PRF high enough so that any actual velocity likely to be observed for targets of interest falls within the first unambiguous velocity interval of $[-v_{ua}/2, +v_{ua}/2]$ m/s. Unfortunately, higher PRFs also correspond to shorter unambiguous ranges. It is frequently not feasible to operate at a PRF that allows unambiguous coverage of both the range and velocity intervals of interest. For example, suppose a designer requires an unambiguous range of at least 100 km and at least ±112.5 m/s of unambiguous velocity coverage, corresponding to a 225-m/s blind speed. The black dashed lines in Fig. 5.11 shows that the maximum RF at which this combination is possible is 1 GHz. If the radar is required to be at X band (10 GHz), the combination of 100 km unambiguous range coverage and a $[-112.5, +112.5]$ m/s unambiguous velocity interval is not obtainable at any PRF and some ambiguity must be accepted in range, Doppler, or both.

The use of *staggered PRFs* or *staggered PRIs* is an alternative approach that raises the blind speed significantly with only a modest reduction in the unambiguous range (Levanon, 1988; Schleher, 2010). PRF staggering can be performed on either a pulse-to-pulse or CPI-to-CPI basis. The latter case is common in airborne pulse Doppler radars and is deferred to Sec. 5.4.8. Pulse-to-pulse stagger varies the PRI from one pulse to the next within a single CPI or dwell. One common approach is to cycle through a set of P preselected PRIs from one pulse to the next, repeating when all of the PRIs have been used. Figure 5.12 illustrates the pulse timing sequence for a pulsed radar with $P = 2$. The resulting slow-time data for a given range bin is

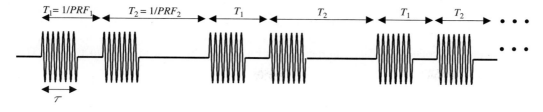

FIGURE 5.12 Pulse sequence timing for two staggered PRIs.

then passed through an MTI filter. As will be seen, this process has the advantage of achieving increased Doppler coverage with a single dwell.

One disadvantage of staggering is that the slow-time data are now a nonuniformly sampled sequence, making coherent Doppler filtering impractical and complicating analysis. Another is that ambiguous MLC can cause large pulse-to-pulse amplitude changes as the PRI varies since the range of the second-time-around clutter that folds into each range cell will change as the PRI changes. Consequently, pulse-to-pulse PRI stagger is generally used only in low PRF modes where there are no significant echoes from ranges beyond the unambiguous range.

Staggered PRI operation can be analyzed in terms of either the PRIs or the corresponding PRFs. The former is more direct and is used here. Consider a system using P staggered PRIs $\{T_0, T_1, \ldots, T_{P-1}\}$. The corresponding set of pulse repetition intervals is $\{PRF_p\} = \{1/T_p\}$. Assume that each of the PRIs is selected as an integer multiple of a base interval T_g,

$$T_p = k_p T_g, \quad p = 0, \ldots, P-1 \tag{5.34}$$

with corresponding PRFs, $PRF_p = 1/T_p$ and $F_g = 1/T_g$. This is reasonable for many radar systems, where T_g may correspond to the fast-time sampling interval and the set of integers $\{k_p\}$ to the number of range bins in each PRI. The $\{k_p\}$ are called the *staggers*[4] and the ratio $k_m:k_p$ of any two of them is called a *stagger ratio*. In many practical cases the staggers are chosen to be relatively prime integers.

For a given PRI, any MTI filter will exhibit blind Doppler frequencies at all integer multiples of the corresponding PRF. Consequently, the first true blind Doppler frequency F_{bs} of a system using staggered PRIs will be the lowest frequency that is blind at all of the corresponding PRFs, that is, the least common multiple (LCM) of the PRF set:

$$\begin{aligned} F_{bs} &= \text{lcm}(PRF_0, \ldots, PRF_{P-1}) = \text{lcm}(1/T_0, \ldots, T_{P-1}) \\ &= F_g \text{lcm}(1/k_0, \ldots, 1/k_{P-1}) \end{aligned} \tag{5.35}$$

A complete cycle through the set of PRIs takes a total period T_{tot} equal to the sum of each of the staggered PRIs

$$T_{\text{tot}} = \sum_{p=0}^{P-1} T_p = T_g \sum_{p=1}^{P-1} k_p \tag{5.36}$$

---

[4] Some authors work in terms of the PRFs instead of the PRIs and use the term "staggers" to refer to ratios of the PRF_n values.

How much is the blind Doppler of the staggered system increased relative to a comparable unstaggered system? A reasonable choice for an unstaggered system to serve as a baseline reference is one whose PRI equals the average PRI of the staggered system. The time required to collect N pulses will then be approximately the same for the two systems. The average PRI is

$$T_{\text{avg}} = \frac{T_{\text{tot}}}{P} = \frac{T_g}{P} \sum_{p=0}^{P-1} k_p \tag{5.37}$$

The blind Doppler frequency that would be observed in an unstaggered waveform with this average PRI is

$$F_b = \frac{1}{T_{\text{avg}}} \tag{5.38}$$

Using Eqs. (5.35), (5.37), and (5.38) and noting that $F_g T_g = 1$ gives an expression for the first blind Doppler frequency of the staggered PRI system in terms of the staggers $\{k_p\}$ and the blind Doppler of the reference unstaggered system

$$\frac{F_{bs}}{F_b} = \frac{1}{P} \text{lcm}(1/k_0, \ldots, 1/k_{P-1}) \sum_{p=0}^{P-1} k_p \tag{5.39}$$

For example, a two-PRI system with a stagger ratio of 3:4 would have a first blind Doppler that is 3.5 times that of a system using a fixed PRI equal to the average of the two individual PRIs. If a third PRF is added to give the set of staggers {3, 4, 5}, the first blind Doppler will be four times that of the comparable unstaggered system.

These equations simplify if all of the staggers are indeed mutually prime. In that event, the LCM of the set of inverse staggers $\{1/k_p\}$ equals 1 (see Prob. 14). The blind Doppler shift [Eq. (5.35)] of the staggered system then equals F_g and the factor by which the blind Doppler frequency is expanded relative to the unstaggered case [Eq. (5.39)] is just the average of the staggers.

If a pure sinusoid $A\exp(j\Omega t)$ is input to a *linear time-invariant* (LTI) system, the output will necessarily be another pure sinusoid at the same frequency but with possibly different amplitude and phase, $A'\exp(j\Omega t + \phi)$. However, if a pure sinusoid is sampled at nonuniform time intervals the resulting series of samples, if interpreted as a conventional discrete-time sequence, will *not* be equivalent to a uniformly sampled pure sinusoid at the appropriate frequency. Instead, the sampled signal will contain multiple frequency components. Any subsequent processing, even though itself LTI, will still result in an output spectrum containing multiple frequency components. Thus, a system utilizing nonuniform time sampling is not LTI and the frequency response of a pulse-to-pulse staggered system does not exist in a conventional sense. Instead, an approach based on first principles can be used to explicitly compute the effect of the MTI filter structure of interest on a complex sinusoid of arbitrary frequency and initial phase. Repeating for each possible sinusoid frequency, the effect of the combination of staggered sampling and MTI filtering can be determined for targets of different Doppler shifts (Roy and Lowenschuss, 1970; Levanon, 1988; Schleher, 2010).

A constant-PRI system with PRI T transmits pulse #m in an N-pulse sequence at time $t_m = mT$, $m = 0,\ldots, N-1$. In a P-stagger system, the sequence of sampling times $\{t_m\}$ is

$$t_m = \begin{cases} 0, & m = 0 \\ \sum_{p=0}^{m-1} T_{((p))p}, & m = 1,\ldots, N-1 \end{cases} \tag{5.40}$$

The notation $((\cdot))_P$ denotes evaluation of the argument modulo P. Note that the $\{T_p\}$ are the sampling time *increments*, not the absolute sampling times, and that $t_m - t_{m-1} = T_{((m-1))_P}$. The slow-time phase history of Eq. (5.22) for a constant PRI and constant-velocity target is easily generalized to the following form for a staggered PRI system:

$$y[m] = k\exp(+j2\pi F_D t_m) \tag{5.41}$$

Note that the initial phase of $y[m]$ is absorbed into the factor k. Now consider the two-pulse canceller network of Fig. 5.8a.[5] Using Eqs. (5.40) and (5.41), the output $z[m] = y[m] - y[m-1]$ can be written explicitly as

$$\begin{aligned} z[m] &= k\exp[j2\pi F_0 t_m] - k\exp[j2\pi F_0 t_{m-1}] \\ &= k\exp[j\pi F_0(t_m + t_{m-1})]\{\exp[+j\pi F_0(t_m - t_{m-1})] - \exp[-j\pi F_0(t_m - t_{m-1})]\} \\ &= k\exp[j\pi F_0(t_m + t_{m-1})]\{\exp[+j\pi F_0 T_{((m-1))_P}] - \exp[-j\pi F_0 T_{((m-1))_P}]\} \\ &= 2jk\exp[j\pi F_0(t_m + t_{m-1})]\sin(\pi F_0 T_{((m-1))_P}) \end{aligned} \tag{5.42}$$

The magnitude of the frequency response for the combined sampling and filter system can be defined as the square root of the ratio of the power of the filter output sequence to that of the input sequence. The power of each input sample is $|y[m]|^2 = |k|^2$. The power of the output samples $|z[m]|^2$ depends on the index m due to the varying PRIs. The average output power is therefore computed over one cycle of the staggered PRIs. The sum from $m=0$ to $P-1$ of $T_{((m-1))_P}$ is the same as the sum of T_m so the expression for the squared magnitude of the average two-pulse canceller filter frequency response becomes

$$|H_{2,P}(F_0)|^2 = \frac{\frac{1}{P}\sum_{m=0}^{P-1}|z[m]|^2}{|x[m]|^2} = \frac{4k^2\sum_{m=0}^{P-1}\sin^2(\pi F_0 T_m)}{Pk^2} \tag{5.43}$$

where the notation $|H_{N,P}(F_0)|^2$ indicates the power gain or attenuation of an N-pulse canceller using P staggers when the input is a sinusoid of frequency F_0. Generalizing the specific F_0 to an arbitrary frequency F and renaming the summation index gives the squared magnitude of the frequency response of the two-pulse canceller with staggered PRIs

$$|H_{2,P}(F)|^2 = \frac{4}{P}\sum_{p=0}^{P-1}\sin^2(\pi F T_p) = \frac{4}{P}\sum_{p=0}^{P-1}\sin^2(\pi F/PRF_p) \tag{5.44}$$

The actual frequency in hertz rather than normalized frequency is used in Eq. (5.44) because the nonuniform sampling rate invalidates the usual definition of normalized frequency. The response of more general MTI filters can be obtained using a similar approach.

Figure 5.13a compares the frequency response of a two-pulse canceller using two ($P=2$) PRIs versus the reference baseline single-PRF system. The staggered case uses PRIs of $4/3$ μs and 1 μs (thus PRFs of 750 and 1000 pulses per second). $T_g = 1/3$ μs, giving $F_g = 3$ kHz. The set of staggers k_p is $\{3, 4\}$. The first blind Doppler shift F_{bs} occurs at the least common multiple of 750 and 1000 Hz, which is also 3000 Hz. The reference unstaggered PRF and blind Doppler F_{bs}, which is the reciprocal of the average PRI $T_{avg} = 7/6$ μs, is 857.14 Hz. The

---

[5]The number of staggers used is independent of the canceller order selected. A two-pulse canceller can be used with two staggers or 10 staggers.

baseline unstaggered response collected using $PRI = T_{\text{avg}}$ shows blind Doppler frequencies at integer multiples of 857.14 Hz. Staggering the PRF has increased the blind Doppler frequency and thus the velocity and Doppler coverage by a factor of 3.5 ($= 3000/857.14$) consistent with Eq. (5.39).

It remains to determine the effect of the pulse stagger on the unambiguous range. The unambiguous range R_{uas} of the staggered-PRI system is simply the shortest of the unambiguous ranges for each individual PRI

$$R_{uas} = \frac{c}{2} \min[\{T_p\}] = \frac{cT_g}{2} \min[\{k_p\}] \tag{5.45}$$

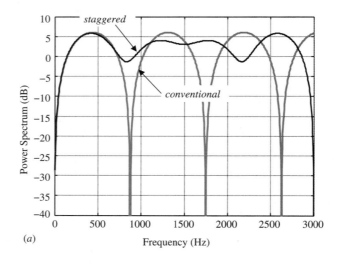

(a)

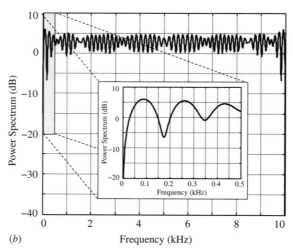

(b)

Figure 5.13 Comparison of two-pulse canceller frequency response with unstaggered waveform and 3:4 staggered waveform. (a) Two PRIs in a 3:4 stagger ratio. (b) Five PRIs in a 51:62:53:61:58 stagger ratio. The inset expands the gray-shaded portion of the graph near zero frequency.

The unambiguous range for the reference unstaggered system is $R_{ua} = cT_{avg}/2$. The ratio is

$$\frac{R_{uas}}{R_{ua}} = \frac{(cT_g/2)\min[\{k_p\}]}{(cT_{avg}/2)} = \frac{P\min[\{k_p\}]}{\sum_{p=0}^{P-1} k_p} \qquad (5.46)$$

which is 6/7 for this example, about a 14 percent reduction. The increase in the total range-Doppler coverage is the product of Eqs. (5.39) and (5.46),

$$\frac{R_{uas}F_{bs}}{R_{ua}F_b} = \min(\{k_p\})\,\mathrm{lcm}(\{1/k_p\}) \qquad (5.47)$$

For mutually prime staggers, $\mathrm{lcm}(\{1/k_p\}) = 1$ and this reduces to simply $\min[\{k_p\}]$, equal to a factor of three in the example above. Thus, the use of two PRIs with a 3:4 stagger ratio has reduced the unambiguous range by 14 percent but has expanded the Doppler coverage by 250 percent and the combined range-Doppler coverage by 200 percent.

Figure 5.13b illustrates the frequency response obtained using $T_g = 100$ μs and five mutually prime staggers in the ratio 51:62:53:61:58 with a two-pulse canceller.[6] Only the staggered response is shown; the equivalent unstaggered response would have nulls every 175.4 Hz. The insert expands the power spectrum around zero Doppler shift to make the clutter null more visible. The first blind Doppler shift is at 10 kHz, a 57 times increase as expected from Eq. (5.39). The unambiguous range reduction is 10.5 percent. The combined range-Doppler coverage is increased by 51 times.

Additional examples, special cases suited to weather radar, and the use of both infinite impulse response (IIR) filters and time-varying filters for MTI are discussed in Shrader and Gregers-Hansen (2008). An alternative design approach based on randomized PRIs to extend the blind speed is discussed in Vergara-Dominguez (1993).

5.3.5 MTI Figures of Merit

The goal of MTI filtering is to suppress clutter. In doing so, it also attenuates or amplifies the target return, depending on the particular target Doppler shift. The change in signal and clutter power then affects the system probabilities of detection and false alarm.

There are three principal MTI figures of merit in use. *Clutter attenuation* measures only the reduction in clutter power at the output of the MTI filter as compared to the input but is simplest to compute. *Improvement factor* quantifies the increase in signal-to-clutter ratio due to MTI filtering; as such, it accounts for the effect of the filter on the target as well as on the clutter. *Subclutter visibility* is a more complex measure that also takes into account the detection and false alarm probabilities and the detector characteristic. Because of its complexity, it is less often used. In this chapter, attention is concentrated on clutter attenuation CA and improvement factor I.

There are several ways to approach the calculation of the improvement factor. These include a frequency domain approach using clutter power spectra and MTI filter transfer functions, a time domain approach based on autocorrelation functions of the input and output of the MTI filter, and a vector method. Each will be illustrated in turn starting with the frequency domain approach, which is perhaps the most intuitive.

---

[6]The order of the PRIs makes no difference to the power spectrum. Alternating shorter and longer PRIs keeps the transmitter duty cycle more nearly constant.

Clutter attenuation is simply the ratio of the clutter power at the input of the MTI filter to the clutter power at the output,

$$CA = \frac{\sigma_{ci}^2}{\sigma_{co}^2} = \frac{\int_{-RR/2}^{RR/2} S_c(F)\,dF}{\int_{-RR/2}^{RR/2} S_c(F)|H(F)|^2\,dF} \qquad (5.48)$$

where
σ_{ci}^2 and σ_{co}^2 = clutter power at the filter input and output, respectively
$S_c(F)$ = sampled clutter power spectrum
$H(F)$ = discrete-time MTI filter frequency response

Since the MTI filter presumably reduces the clutter power, the clutter attenuation will be greater than one. In fact, it can be 13 dB or more in favorable conditions. However, CA also depends on the clutter itself through $S_c(F)$. The shape of the clutter power spectrum and its spread in hertz are determined by the RF and the physical phenomenology, such as terrain or weather conditions. The percentage of the Doppler spectrum width to which a given clutter spectrum is mapped depends on the PRF. Consequently, a change in RF, PRF, or clutter power spectrum will alter the achieved clutter cancellation.

Improvement factor I is defined formally as the signal-to-clutter ratio at the filter output divided by the signal-to-clutter ratio at the filter input, averaged over all target radial velocities of interest (IEEE, 2017). Considering for the moment only a specific target Doppler shift, the improvement factor can be written in the form

$$I = \frac{(S/C)_{out}}{(S/C)_{in}} = \left(\frac{S_{out}}{S_{in}}\right)\left(\frac{C_{in}}{C_{out}}\right) = G \cdot CA \qquad (5.49)$$

where G is the filter *gain*. Figure 5.9 makes clear that the effect of the MTI filter on the target signal is a strong function of the target Doppler shift. Thus, G is a function of target velocity while CA is not. The improvement factor is the signal processing gain G_{sp} in the radar range equation due to MTI filtering.

To reduce I to a single number instead of a function of target Doppler, the definition calls for averaging uniformly over all target Doppler shifts "of interest." If a target is known to be at a specific velocity the improvement factor can be obtained by simply evaluating Eq. (5.49) at the known target Doppler. It is more common to assume the target velocity is unknown a priori and use the average target gain over all possible Doppler shifts, which is just

$$G = \frac{1}{PRF} \int_{-RR/2}^{RR/2} |H(F)|^2\,dF \qquad (5.50)$$

An alternative expression for the gain that is often easier to compute for simple MTI filters follows from converting Eq. (5.50) back to normalized frequency units and applying Parseval's theorem:

$$G = \frac{1}{RR} \int_{-RR/2}^{RR/2} |H(F)|^2\,dF = \frac{1}{2\pi} \int_{-\pi}^{\pi} |H(\omega)|^2\,d\omega = \sum_{m=-\infty}^{\infty} |h[m]|^2 \qquad (5.51)$$

For example, a two-pulse canceller has only two nonzero coefficients, $+1$ and -1, giving immediately $G = 2$. Combining Eqs. (5.48) and (5.50) in (5.49) gives the expression for the improvement factor

$$I = G \cdot CA = \frac{\int_{-RR/2}^{RR/2} S_c(F) \, dF \int_{-RR/2}^{RR/2} |H(F)|^2 \, dF}{RR \int_{-RR/2}^{RR/2} S_c(F)|H(F)|^2 \, dF} \quad (5.52)$$

Equivalent expressions for improvement factor can be developed in terms of the autocorrelation function of the clutter and the MTI filter impulse response (Levanon, 1988; Nathanson, 1991). For low-order filters such as two- or three-pulse cancellers and clutter power spectra with either measured or analytically derivable autocorrelation functions, the resulting equations can be easier to evaluate than the frequency domain versions.

As an example of the autocorrelation approach, consider the output of the two-pulse canceller when the input is just clutter; this is $c'[m] = c[m] - c[m-1]$. The expected value of the filter output power is

$$\mathbf{E}\{|c'[m]|^2\} = \mathbf{E}\{|c[m]|^2 - 2\,\mathrm{Re}\{c[m]c^*[m-1]\} + |c[m-1]|^2\} \quad (5.53)$$

where $\mathrm{Re}\{\cdot\}$ denotes the real part of the argument.[7] Assuming that $c[m]$ is stationary,

$$\mathbf{E}\{|c'\{m\}|^2\} = 2s_c[0] - 2\,\mathrm{Re}\{s_c[1]\} \quad (5.54)$$

where

$$s_c[k] \equiv \mathbf{E}\{c[m]c^*[m+k]\} \quad (5.55)$$

is the autocorrelation function of $c[m]$. Note that $s_c[k] = s_c^*[-k]$. Also define the normalized autocorrelation function

$$\rho_c[k] \equiv \frac{1}{s_c[0]} s_c[k] \quad (5.56)$$

The clutter attenuation component of the improvement factor in Eq. (5.49) can now be written as

$$CA = \frac{\mathbf{E}\{|c[m]|^2\}}{\mathbf{E}\{|c'[m]|^2\}} = \frac{s_c[0]}{2(s_c[0] - \mathrm{Re}\{s_c[1]\})}$$
$$= \frac{1}{2(1 - \mathrm{Re}\{\rho_c[1]\})} \quad (5.57)$$

---

[7]Since the power spectrum $S_c(F)$ is real valued, the autocorrelation function $s_c[k]$ must be Hermitian symmetric. It can therefore be complex valued. However, clutter is usually modeled by a power spectrum that is also an even function of frequency (i.e., symmetric about $F = 0$), for example a Gaussian clutter spectrum with a zero mean. With this extra constraint, the autocorrelation function must be real-valued also, so the Re{} operator can be dropped in Eq. (5.53) and other equations in this section.

As noted above, the gain for a two-pulse canceller is $G = 2$; thus the improvement factor for the two-pulse canceller is

$$I = CA \cdot G = \frac{1}{1 - \text{Re}\{\rho_c[1]\}} \qquad (5.58)$$

A similar analysis can be used to derive the improvement factor for a three-pulse canceller; it is

$$I = CA \cdot G \frac{1}{1 - \frac{4}{3}\text{Re}\{\rho_c[1]\} + \frac{1}{3}\text{Re}\{\rho_c[2]\}} \qquad (5.59)$$

To see how these formulas are used, consider the case where the clutter spectrum is Gaussian with variance σ_ω^2 (in normalized radian frequency units), $S_c(\omega) = A \exp(-\omega^2/\sigma_\omega^2)$. Assuming $\sigma_w \ll \pi$ so that the continuous-time Fourier transform pair for Gaussian functions can be used to a good approximation, the normalized autocorrelation function for $c[m]$ at lag k is (Richards, 2006)

$$\rho_c[k] \approx \exp[-(\sigma_\omega k)^2/2] \qquad (5.60)$$

Using Eq. (5.60) in Eqs. (5.58) and (5.59) gives the improvement factor for a Gaussian clutter spectrum with a two- or three-pulse canceller:

$$I = \begin{cases} \dfrac{1}{1 - \exp\left[-\sigma_\omega^2/2\right]} & \text{(two pulse canceller)} \\[2mm] \dfrac{1}{1 - \frac{4}{3}\exp\left[-\sigma_\omega^2/2\right] + \frac{1}{3}\exp\left[-2\sigma_\omega^2\right]} & \text{(three pulse canceller)} \end{cases} \qquad (5.61)$$

Table 5.1 shows the improvement factor predicted for two- and three-pulse cancellers for the case of a Gaussian clutter power spectrum of various spectral widths using Eq. (5.61). If the clutter spectrum is narrow compared to the PRF, the improvement factor can be 13 dB or more even for the simple two-pulse canceller. If the clutter spectrum is wide, much of the clutter power will be in the passband of the MTI highpass filter and the improvement factor will be slight.

The third approach for computing the improvement factor uses the vector analysis techniques employed in determining the matched filter for MTI. For comparison with

Standard Deviation of Clutter Power Spectrum, Hz	Improvement Factor, dB Two-Pulse Canceller	Improvement Factor, dB Three-Pulse Canceller
RR/3	0.5	0.7
RR/10	7.5	12.5
RR/20	13.2	21.7
RR/100	24	51

TABLE 5.1 Improvement Factor for Gaussian Clutter Power Spectrum

the autocorrelation analysis given previously, consider the case where $\sigma_n^2 = 0$ (clutter only) and $\mathbf{h} = [1 \ -1]^T$ (two-pulse canceller). Improvement factor is the ratio of the signal-to-interference ratio at the filter output to the SIR at the filter input. While the optimum MTI filter was derived by averaging over possible target Doppler frequencies, in evaluating the improvement factor it is assumed that any specific target has a specific Doppler frequency. The improvement factor is calculated for that specific target Doppler frequency and then averaged over allowable Doppler frequencies. Since SIR at the input does not depend on Doppler frequency, it is sufficient to do the averaging on the output SIR.

Consider therefore the signal vector given by Eq. (5.24) and the clutter covariance matrix given by Eq. (5.20) (with $\sigma_n^2 = 0$). The input SIR is just $|A|^2/\sigma_c^2$. Equation (5.8) gave an explicit expression for the output SIR. The numerator of this expression is

$$\mathbf{h}^H \mathbf{t}^* \mathbf{t}^T \mathbf{h} = |A|^2 [1 \ -1] \begin{bmatrix} 1 & \exp(-j2\pi F_D T) \\ \exp(+j2\pi F_D T) & 1 \end{bmatrix} \begin{bmatrix} 1 \\ -1 \end{bmatrix}$$

$$= 2|A|^2 (1 - \text{Re}\{\exp(j2\pi F_D T)\}) \qquad (5.62)$$

This is the two-pulse canceller MTI filter output signal power for a target at Doppler shift F_D Hz. Averaging over all target Doppler shifts gives for the numerator

$$\mathbf{E}_{F_D} \{\mathbf{h}^H \mathbf{t}^* \mathbf{t}^T \mathbf{h}\} = 2|A|^2 \qquad (5.63)$$

The denominator of Eq. (5.8) is

$$\mathbf{h}^H \mathbf{S}_I \mathbf{h} = \sigma_c^2 [1 \ -1] \begin{bmatrix} 1 & \rho_c \\ \rho_c^* & 1 \end{bmatrix} \begin{bmatrix} 1 \\ -1 \end{bmatrix}$$

$$= 2\sigma_c^2 (1 - \text{Re}\{\rho_c\}). \qquad (5.64)$$

Dividing Eq. (5.63) by Eq. (5.64) gives the output SIR; further dividing that ratio by the input SIR gives the improvement factor for a two-pulse canceller operating against clutter only (no noise)

$$I = \frac{1}{1 - \text{Re}\{\rho_c\}} \qquad (5.65)$$

Since ρ_c in the matrix formulation is the same as $\rho_c[1]$, this is the same expression obtained using autocorrelation methods in Eq. (5.58). The same result could also have been obtained in the vector analysis by simply using the target model vector averaged over Doppler, $\mathbf{t} = [1 \ 0]^T$.

Additional MTI metrics can be defined. Improvement factor I is an average of the improvement in signal-to-clutter ratio over one Doppler period. At some Doppler shifts, the target is above the clutter energy, while at others it is below the clutter and therefore not detectable. I does not indicate over what percentage of the Doppler spectrum a target can be detected. The concept of *MTI visibility factor* or *target visibility* V has been proposed to quantify this effect (Kretschmer, 1986). V is the percentage of the Doppler spectrum over which the improvement factor for a target at a specific frequency is greater than or equal to the average improvement factor I. A related metric is the *usable Doppler space fraction* (UDSF), which in turn is determined by the *minimum detectable velocity* (MDV) or *minimum detectable Doppler* (MDD). These metrics are common in space-time adaptive processing, so their discussion is deferred to Chap. 9.

5.3.6 Limitations to MTI Performance

The basic idea of MTI processing is that repeated measurements of stationary clutter yield the same echo amplitude and phase; thus successive measurements, when subtracted from one another, should cancel. Any effect internal or external to the radar that causes the received echo from a stationary target to vary will cause imperfect cancellation, limiting the improvement factor.

Perhaps the simplest example is transmitter amplitude instability. Consider a two-pulse canceller and suppose that the amplitude of each pulse may differ in amplitude from the nominal amplitude by up to ± 5 percent (equivalent to $20\log_{10}(1.05/1) = 0.42$ dB). The signal resulting from subtracting two echoes from a perfectly stationary target can have an amplitude that is as large as 10 percent that of the nominal echo amplitude. Consequently, clutter attenuation may be as poor as $20\log_{10}(1/0.1) = 20$ dB even though the clutter is perfectly stationary. For a two-pulse canceller with an average signal gain G of 2 (6 dB), the maximum achievable improvement factor is 26 dB.

A more realistic analysis of the limitations due to amplitude jitter can be obtained by modeling the amplitude of the mth transmitted pulse as $A[m] = k(1 + a[m])$, where $a[m]$ is a zero mean, white random process with variance σ_a^2 that represents the percentage variation in transmitted amplitude, and k is a constant. The received signal will have a complex amplitude of the form $k'(1 + a[m]\exp(j\varphi))$, where φ is the phase of the received slow-time sample and the constant k' absorbs all the radar range equation factors. The average power of this signal, which is the input to the pulse canceller, is

$$E\{|y[m]|^2\} = k'^2 E\{1 + 2a[m] + a^2[m]\} = k'^2\left(1 + \sigma_a^2\right) \qquad (5.66)$$

The expected value of the two-pulse canceller output power will be

$$\begin{aligned}E\{|z[m]|^2\} &= E\{|y[m] - y[m-1]|^2\} = E\{|k'\exp(j\varphi)(a[m] - a[m-1])|^2\}\\ &= k'^2\left(E\{a^2[m]\} - 2E\{a[m]a[m-1]\} + E\{a^2[m-1]\}\right)\\ &= 2k'^2\sigma_a^2.\end{aligned} \qquad (5.67)$$

The achievable clutter cancellation is thus

$$\frac{\text{input power}}{\text{output power}} = \frac{k'^2\left(1 + \sigma_a^2\right)}{2k'^2\sigma_a^2} = \frac{1 + \sigma_a^2}{2\sigma_a^2} \qquad (5.68)$$

For example, an amplitude variance of 1 percent $(\sigma_a^2 = 0.01)$ limits two-pulse clutter cancellation to a factor of 50.5, or 17 dB. Because the average target gain of the two-pulse canceller is $G = 2$ (3 dB), the limit to the improvement factor I is $50.5 \times 2 = 101$, or 20 dB.

Another example is phase drift in either the transmitter or receiver. This can occur due to instability in coherent local oscillators used either as part of the waveform generator on the transmit side or in the demodulation chains on the receiver side. Consider the weighted coherent integration of M data samples $y[m]$ with a zero-mean stationary white phase error $\varphi[m]$,

$$Z = \sum_{m=0}^{M-1} a_m y[m]\exp(j\varphi[m]) \qquad (5.69)$$

where the $\{a_m\}$ are the integration weights. Assume that each data sample $y[m]$ is a (possibly complex) constant A. Then the power of the weighted coherent sum in the absence of phase error $\varphi[m]$ is

$$|Z|^2 = |A|^2 \left| \sum_{m=0}^{M-1} a_m \right|^2 \quad (5.70)$$

It can be shown that the integrated power when a white Gaussian phase error is present is (Richards, 2003)

$$|Z|^2 = |A|^2 \left\{ \sum_{m=0}^{M-1} |a_m|^2 + \exp(-\sigma_\varphi^2) \sum_{\substack{m=0 \\ m \neq k}}^{M-1} \sum_{k=0}^{M-1} a_m a_k^* \right\} \quad (5.71)$$

where σ_φ^2 is the variance of the phase noise in radians. This can be applied to the two-pulse canceller by letting $M = 2$ and $a_0 = 1$, $a_1 = -1$. Equation (5.71) then gives the power at the canceller output as $2|A|^2[1 - \exp(-\sigma_\varphi^2)]$. Since the power of a clutter sample before the canceller is $|A|^2$, the limitation on two-pulse cancellation due to the phase noise becomes $1/(2[1 - \exp(-\sigma_\varphi^2)])$.

A Gaussian probability density function (PDF) for phase error is reasonable when the error is small, but not when it becomes large because the PDF is not confined to the interval $[-\pi, \pi]$. An extension to this analysis uses instead the Tikhonov distribution for phase:

$$p_\varphi(\varphi) = \frac{1}{2\pi I_0(\alpha)} \exp(\alpha \cos \varphi) \quad (5.72)$$

This PDF is illustrated in Fig. 5.14. When the parameter α equals zero, the PDF is uniform over the interval $(-\pi, \pi]$. As $\alpha \to \infty$ the PDF approaches the Dirac impulse function $\delta_D(\varphi)$ at the origin, representing a fixed phase value of zero radians. Using this PDF produces very similar results to the Gaussian analysis, differing only in replacing the quantity $\exp(-\sigma_\varphi^2)$ in Eq. (5.71) and the two-pulse canceller improvement factor limit with the quantity $[I_1(\alpha)/I_0(\alpha)]^2$, where $I_1(\alpha)$ and $I_0(\alpha)$ are the modified Bessel functions of the first kind and orders one and zero, respectively (Richards, 2011).

Other sources of limitation due to radar system instabilities include instability in transmitter or oscillator frequencies, transmitter phase drift, coherent oscillator locking errors, PRI jitter, pulse width jitter, and quantization noise. Formulas can be developed to bound the achievable clutter attenuation due to each of these error sources. Still another source is PRI stagger. When used in a range-ambiguous scenario, PRI stagger results in clutter from different distant ranges aliasing to the same near ranges on different PRIs so that the clutter becomes less stationary and the clutter attenuation is degraded. Additional information and analysis on these and other issues is available in Shrader and Gregers-Hansen (2008), Nathanson (1991), and Schleher (2010).

External to the radar, the chief factor limiting MTI improvement factor is simply the width of the clutter spectrum itself. Wider spectra put more clutter energy outside of the MTI filter null so that less of the clutter energy is filtered out. This effect is evident in Eq. (5.48) and was illustrated numerically in Table 5.1. The effective clutter spectrum width can be increased by radar system instabilities or by measurement geometry and dynamics. For instance, a scanning antenna adds some amplitude modulation due to antenna pattern weighting to the clutter return. The power spectrum of the measured clutter is then the convolution in the frequency domain of the actual clutter power spectrum and the squared magnitude of the Fourier transform of the amplitude modulation caused by the antenna scanning. This convolution increases the observed spectral width somewhat. In some cases, the clutter power spectrum may not be centered on zero Doppler shift. A good example is

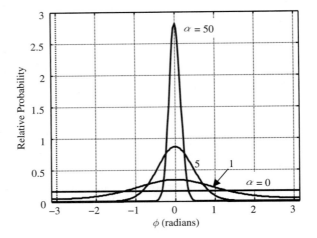

FIGURE 5.14 The Tikhonov PDF for phase.

rain clutter. Moving weather systems will have a nonzero average Doppler representing the rate at which the rain cell is approaching or receding from the radar system. Unless this average motion is estimated and compensated the MTI filter null will not be centered on the clutter spectrum and cancellation will be poor.

The largest source of clutter offset and spreading is radar platform motion. Recall from Eq. (3.15) that the motion-induced clutter bandwidth is

$$\beta_D \approx \frac{2v\theta_3}{\lambda_t}\sin\psi \tag{5.73}$$

where θ_3 is the 3-dB azimuth beamwidth and ψ is the cone angle. The offset in center frequency of the clutter spectrum can be as much as a few kilohertz for fast aircraft when forward looking, while the motion-induced spectral spread can be tens to a few hundreds of hertz for a sidelooking configuration. This clutter spreading adds to the intrinsic spread of the clutter spectrum due to internal motion and can often be the dominant effect determining the observed clutter spectral width and therefore determining the MTI performance limits.

5.4 Pulse Doppler Processing

Pulse Doppler processing is the second major class of Doppler processing. Despite the name, it is common in both pulsed and fast-chirp FMCW radars, and so RR and RI will now be used in preference to PRF and PRI. Recall that in MTI processing the fast-time/slow-time data matrix is highpass filtered in the slow-time dimension, yielding a new fast-time/slow-time data sequence in which the clutter components have been attenuated. Pulse Doppler processing differs in that filtering in the slow-time domain is replaced by explicit spectral analysis of the slow-time data for each range bin. Target detection is then performed directly on the resulting range-Doppler matrix of data. Because the range-Doppler matrix is the fundamental data quantity in pulse Doppler processing, the first step is to form it from the fast-time/slow-time CPI matrix by computing the one-dimensional spectrum of the slow-time signal in each range bin. The spectral analysis is most commonly by far performed using the *fast Fourier transform* (FFT) algorithm to compute the discrete Fourier transform, as shown in Fig. 5.15, but other techniques can also be used. The computed DFT is a

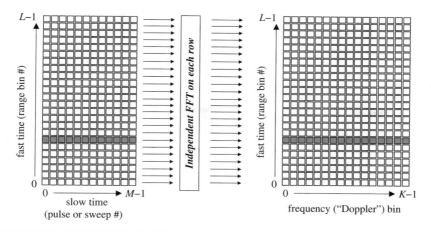

Figure 5.15 Conversion of the fast-time/slow-time data matrix to a range-Doppler matrix.

frequency-sampled version of the discrete-time Fourier transform (DTFT) of the slow-time signal. The figure emphasizes the fact that the number of frequency samples K does not have to equal the number of slow-time samples M; often $K > M$ as shown.

Consider the notional pulse Doppler spectrum for one range bin, shown in Fig. 5.16. The DTFT of the data is shown in a fashion similar to earlier figures. The white dots represent the samples of the DTFT computed by the DFT; these are the only actually available data. Assuming that the clutter has been centered at zero Doppler, spectral samples at or near zero frequency will be dominated by the strong clutter signal even though noise is also present. Spectral samples in the clear region have only thermal noise to interfere with signal detection. Each clear-region spectrum sample can be individually compared to a noise-based threshold to determine whether the signal at that range bin and Doppler frequency appears to be noise only or noise plus a target. If the sample crosses the threshold, it not only indicates the presence of a target in that range bin but also its approximate radial velocity, since the Doppler frequency bin is known. The samples that are clutter-dominated are often simply discarded on the grounds that the SIR will be too low for successful detection. However, other systems use a technique called clutter mapping, discussed in Sec. 5.7, to attempt detection of strong targets in the clutter region using a clutter-based threshold.

The advantages of pulse Doppler processing are that it provides at least a coarse estimate of the radial velocity component of a moving target and also a way to detect multiple targets in the same range bin, provided they are separated enough in Doppler to be resolved. The chief disadvantages are greater computational complexity of pulse Doppler processing as compared to MTI filtering and longer required dwell times due to the use of more pulses for the Doppler measurements. Thorough discussions of pulse Doppler processing are contained in Morris and Harkness (1996), Stimson (2014), and Alabaster (2012).

5.4.1 The Discrete-Time Fourier Transform of a Moving Target

To understand the issues in pulse Doppler processing, it is useful to understand the appearance of noise, clutter, and target signals in the range-Doppler map. Begin by again considering the Fourier spectrum of an ideal constant radial velocity moving point target and the effects of a sampled Doppler spectrum. The issues are the same as those discussed regarding the sampling of the Doppler spectrum in Chap. 3 and App. B. Consider a radar illuminating

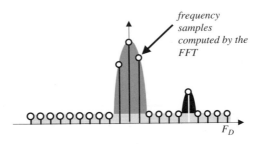

FIGURE 5.16 The computed pulse Doppler spectrum is a frequency-sampled version of an underlying discrete-time Fourier transform.

a moving target over a CPI of M pulses or sweeps and suppose a moving target is present in a particular range bin. If the target's velocity is such that the Doppler shift is F_D Hz, the slow-time received signal after quadrature demodulation is

$$y[m] = A\exp(j2\pi F_D mT), \quad m = 0, \ldots, M-1 \tag{5.74}$$

where T is the radar's repetition interval and is the effective sampling interval in slow time. The signal of Eq. (5.74) is the same signal considered in Chap. 3 [Eq. (3.31)], except for the change from normalized frequency ω_D in radians to analog frequency F_D in hertz; they are related according to $\omega_D = 2\pi F_D T$. Equation (3.32) gave the DTFT of this signal; converting to analog frequency gives

$$\begin{aligned} Y(F) &= \sum_{m=-\infty}^{\infty} y[m]\exp(-j2\pi F T m) \\ &= A\frac{\sin[\pi(F-F_D)MT]}{\sin[\pi(F-F_D)T]}\exp[-j\pi(M-1)(F-F_D)T] \end{aligned} \tag{5.75}$$

$Y(F)$ is an asinc function with its peak at $F = F_D$ as expected and a peak magnitude of MA. Its magnitude is illustrated in Fig. 5.17a for the case where $F_D = RR/4$, $M = 20$ samples, and $A = 1$. So long as $M \geq 4$, the Rayleigh (peak-to-null) mainlobe bandwidth is $1/MT$ Hz; this is also the 4-dB bandwidth. The width of the mainlobe at the -3 dB points is $0.89/MT$ Hz. The first sidelobe is 13.2 dB below the response peak. These mainlobe width measures determine the Doppler resolution of the radar system. They are all inversely proportional to MT, which is the total elapsed time of the set of pulses used to make the spectral measurement. Thus, Doppler resolution is determined by the observation time of the measurement as first discussed in Chap. 1. Longer observation times allow finer Doppler resolution.

Because of the high sidelobes, it is common to use a data window to weight the slow-time data samples $y[m]$ prior to computing the DTFT or DFT. The major effects of windowing are analyzed in App. B and are exactly the same as discussed in Chap. 4 for range sidelobe control. In brief, applying nonrectangular windows achieves major reductions in sidelobe levels and modest reductions in straddle losses at the cost of a loss of peak integration gain ("loss in processing gain," LPG) loss of SNR ("processing loss," PL), and degraded resolution (widened mainlobe). Figure 5.17b illustrates the effect of the window on the DTFT for the same data used in Fig. 5.17a. In fact, the asinc function of Fig. 5.17a is also just the Fourier transform of the rectangular window function (equivalent to no window).

As an example, for the Hamming window the loss in SNR is 1.75 dB for a very short ($M = 8$) window, decreasing asymptotically to about 1.35 dB for long windows. Table B.2

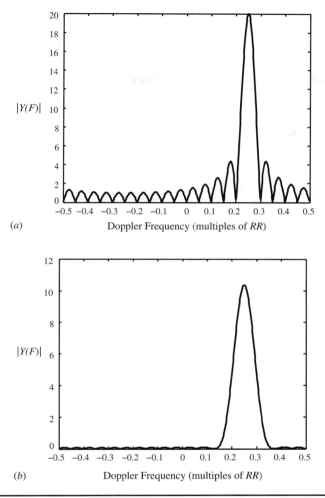

FIGURE 5.17 Magnitude of the discrete-time Fourier transform of an ideal moving target slow-time data sequence with $F_D = RR/4$ and $M = 20$ samples. (a) No window. (b) Hamming window.

summarizes the four key properties of 3-dB resolution, peak sidelobe level, loss in processing gain, and processing loss for several common windows at two moderate lengths. Also shown is worst-case straddle loss, to be discussed next. In general, as sidelobes are lowered the PL and LPG get worse and the mainlobe gets broader, degrading resolution. Straddle loss is actually reduced by windowing. A much more extensive table, including both more metrics and many more types of windows, is given in Harris (1978).

5.4.2 Sampling the DTFT: The Discrete Fourier Transform

In practice, the DTFT is not computed because its frequency variable is continuous. Instead, the discrete Fourier transform is computed:

$$Y[k] = \sum_{m=0}^{M-1} y[m]\exp(-j2\pi mk/K), \quad k = 0,\ldots, K-1 \qquad (5.76)$$

For a finite length data sequence $Y[k]$ is just $Y(F)$ sampled at frequencies $F = k/KT = k(RR/K)$ Hz. These samples are called *Doppler bins*. Thus, the DFT computes K samples of the DTFT evenly spaced across one period of the DTFT. The DFT is almost invariably computed using an FFT algorithm. The appearance of a plot of the Doppler spectrum computed using a DFT can be a strong function of the relation between the actual DTFT shape and the number and location of the DFT frequency samples.

In some situations, the number of data samples available can be greater than the desired DFT size, that is, $M > K$. This occurs when there is a need to reduce the DFT size for computational reasons, or when the radar timeline permits the collection of more pulses than the number of Doppler bins required and it is desirable to use the extra pulses to improve the SNR of the Doppler spectrum measurement. The data turning procedure described in Chap. 3 and Fig. 3.21 allows the use of a K-point DFT while taking advantage of all of the data.

Some caution is needed in applying a data window when the data are modified by zero padding or turning. In either case, a length M window should be applied to the data *before* it is either zero padded or turned. Applying a K-point window to the full length of a zero-padded sequence has the effect of multiplying the data by a truncated, asymmetric window (the portion of the actual window that overlaps the M nonzero data points), resulting in greatly increased sidelobes. Applying a shortened K-point window to a turned data sequence results in DFT samples that do not equal samples of the DTFT of the windowed M-point original data sequence.

Because the DFT is a sampled version of the DTFT, the peak value of the DFT obtained for a pure sinusoidal signal is greatest when the Doppler frequency coincides exactly with one of the DFT sample frequencies, and decreases when the signal peak is between DFT frequencies. This reduction in amplitude is called a Doppler *straddle loss*. The amount of loss depends on the particular window used and the ratio K/M. For a given signal length M, the straddle loss is always greatest for signal frequencies exactly halfway between DFT sample frequencies.

To calculate the worst-case straddle loss, assume the signal frequency is $F_D = 0$ so that $y[m] = w[m]$ and then evaluate Eq. (5.76) with $k = 0$ (DFT sample on the spectrum peak) and $k = 1/2$ (1/2 bin away from the sinusoid peak). To be explicit, consider the rectangular window case; then

$$|Y[k]| = \left| \sum_{m=0}^{M-1} e^{-j2\pi mk/K} \right| = \left| \frac{\sin(\pi M k/K)}{\sin(\pi k/K)} \right| \tag{5.77}$$

Assuming $K \geq M$ and evaluating at $k = 1/2$ gives

$$\left| Y\left[\frac{1}{2}\right] \right| = \left| \frac{\sin(\pi M/2K)}{\sin(\pi/2K)} \right| \approx \frac{2K}{\pi} \sin(\pi M/2K) \tag{5.78}$$

The last step was obtained by assuming that K is large enough to allow a small angle approximation to the sine function in the denominator. $Y[0]$ is obtained either by applying L'Hôpital's rule to the second form of Eq. (5.77) or computing it explicitly from the first form; the result is $Y[0] = M$. The maximum straddle loss for the DFT filterbank with no windowing is

$$\text{Maximum straddle loss} = \frac{\pi M}{2K \sin(\pi M/2K)} = \frac{1}{\text{sinc}(M/2K)} \tag{5.79}$$

This equation verifies that the loss depends on the ratio K/M. The worst case occurs when the sinc term is minimized; this happens when $K = M$. In decibels, this worst-case loss for a rectangular window is $\text{sinc}(1/2) = \sin(\pi/2)/(\pi/2) = 2/\pi$, equivalent to -3.92 dB.

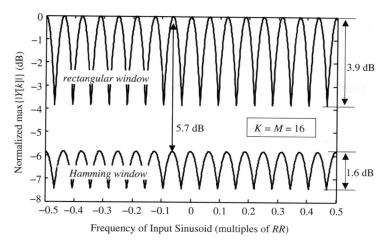

FIGURE 5.18 Variation of DFT output with complex sinusoid input frequency for two different data analysis windows. Window length = DFT length = 16.

The straddle loss for shaped windows varies somewhat with M as can be seen in Table B.2. A calculation for the Hamming window similar to that leading to Eq. (5.79) results in a smaller maximum straddle loss of 1.75 dB for very long windows ($M > 300$ or so), dropping to 1.5 dB for very short ($M = 8$) windows. Figure 5.18 shows the maximum DFT output amplitude as a function of the target Doppler shift for rectangular and Hamming windows for the case of $M = K = 16$. Observe the general reduction in peak amplitude for the Hamming-windowed data compared to the unwindowed data. On the other hand, the *variation* in amplitude is significantly less for the windowed data (1.6 dB versus 3.9 dB); that is, the amplitude response is more consistent, an underappreciated benefit of windowing.

5.4.3 The DFT of Noise

The target signals discussed in the previous sections compete with interference, primarily noise and clutter, so the characteristics of the DTFT and DFT of the interference are of interest. First consider noise. Assume the usual model of additive complex WGN with variance σ_w^2 in the M-sample slow-time signal $y[m]$. The value of the DTFT at any particular frequency or of any DFT sample is a weighted sum of the same M i.i.d. (independent identically distributed) complex Gaussian time samples and so is also a complex Gaussian RV. If no window is used, the weights are the DTFT or DFT kernel values $\exp(-j\omega m)$ or $\exp(-j2\pi km/K)$. Because these weights have unit magnitude, the resulting spectrum (DTFT or DFT) sample will have variance $M\sigma_w^2$. If the slow-time data is windowed, the variance becomes $E_w \sigma_w^2$, where $E_w = \sum_w |w[m]|^2$ is the energy of the window sequence. Because the DTFT and DFT samples are complex Gaussian, it also follows that the magnitude of a spectrum sample has a Rayleigh PDF, the magnitude-squared has an exponential PDF, and the phase has a uniform PDF over $[-\pi, \pi]$. The PDF and its parameters are important in setting detection thresholds as will be seen in Chap. 6. These and additional properties of the DTFT and DFT of noise are derived in Richards (2007).

DTFT and DFT values at different frequencies or indices for random inputs are, in general, correlated. In the unwindowed case the normalized autocorrelation function in frequency is an asinc function of peak amplitude 1 and a zero spacing of $2\pi/M$ radians per sample. For the windowed case, the normalized autocorrelation function follows the shape of the DTFT of the window function. In the case of an unwindowed DFT of size $K = M$, the DFT sample spacing matches the zero spacing of the asinc function so that the DFT samples will be uncorrelated. Because the complex samples are Gaussian, they are also independent in this case.

5.4.4 Pulse Doppler Processing Gain

The DTFT and DFT represent forms of coherent integration in the slow-time domain and result in a processing gain. Equation (5.75) shows that each sample of $y[m]$ is phase-adjusted by multiplication with the DTFT kernel and then summed ("integrated"). If no window is used, then at the value of F that matches the signal frequency, $F = F_D$, the kernel values exactly compensate the phase history of the data such that all of the signal samples add in phase. At this frequency $Y(F_D) = M \cdot A$. The variance of the noise component at the same frequency value will be $M\sigma_w^2$. The SNR is therefore $(M \cdot A)^2 / M\sigma_w^2 = M \cdot (A^2/\sigma_w^2)$ a processing gain factor of $G_{sp} = M$ compared to the SNR in the slow-time signal before the DTFT. The same result applies for the DFT. Note that in the DFT case, the processing gain is determined by M, the number of slow-time samples and not by K, the DFT size. Computing a DFT larger than the number of time domain samples, $K > M$, does not increase the processing gain. Rather, G_{sp} is limited to the number of samples of signal-plus-noise available to integrate.

Various issues can reduce G_{sp}. When a window is used with either a DTFT or DFT, G_{sp} is decreased by the processing loss PL described earlier and derived in App. B. In this case, the range equation signal processing gain due to Doppler processing becomes $G_{sp} = M/PL$. When a DFT is used, G_{sp} is also reduced by any straddle loss that may occur.

5.4.5 Matched Filter and Filterbank Interpretations of Pulse Doppler Processing with the DFT

Equation (5.14) defined the coefficients of the matched Doppler filter. In MTI filtering, it is assumed that the target Doppler shift is unknown. The resulting signal model of Eq. (5.26) leads to the pulse canceller as a near-optimum MTI filter for small order N. In contrast, DFT-based pulse Doppler processing attempts to separate target signals based on their particular Doppler shift. Assume that the signal is a pure complex sinusoid (constant radial velocity target) at a Doppler shift of F_D Hz. Based on Eq. (5.22), the model of the signal vector is then

$$\begin{aligned}\mathbf{t} &= [y[m] \quad y[m+1] \quad \cdots \quad y[m+M-1]]^T \\ &= A[1 \quad \exp[j2\pi F_D T] \quad \cdots \quad \exp[j2\pi F_D(M-1)T]]^T\end{aligned} \quad (5.80)$$

where the complex scalar A absorbs all constants. If the interference consists only of white noise (no correlated clutter) $\mathbf{S_I}$ reduces to $\sigma_n^2 \mathbf{I}_M$, where $\mathbf{I}_M$ is the Mth-order identity matrix. It follows that for an arbitrary data vector $\mathbf{y}$ the output $\mathbf{h}^T\mathbf{y}$ of the matched filter becomes

$$\mathbf{h}^T \mathbf{y} = A \sum_{m=0}^{M-1} y[m] \exp(-j2\pi F_D m T) \quad (5.81)$$

This is the DTFT of $y[m]$ to within the scale factor A. When F_D is a "DFT frequency," $F_D = k/KT = k \cdot RR/K$ for some integer k, Eq. (5.81) is the K-point DFT of the data sequence $y[m]$ to within the scale factor A. Consequently, the DFT is a matched filter to ideal constant radial velocity moving target signals, provided that the Doppler shift equals one of the DFT sample frequencies and the interference is white. This result is very closely related to the two-pulse canceller for a specific target Doppler and noise interference only considered in Sec. 5.3.3.

Since the K-point DFT computes K different outputs from each input vector, it effectively implements a bank of K matched filters at once, each tuned to a different Doppler frequency. The frequency response shape of each matched filter is the asinc function. To see this, denote the impulse response vector in Eq. (5.81) when $F_D = k/MT$ as $\mathbf{h}_k$. To within a scale factor,

$$\mathbf{h}_k = [1 \ \exp(-j2\pi k/K) \ \exp(-j4\pi k/K) \ \cdots \ \exp(-j2\pi(K-1)k/K)]^T \qquad (5.82)$$

The corresponding discrete-time frequency response $H_k(\omega)$ is

$$\begin{aligned} H_k(\omega) &= \sum_{m=0}^{K-1} h_k[m]\exp(-j\omega m) \\ &= \sum_{m=0}^{K-1} \exp[-j(\omega + 2\pi k/K)m] \end{aligned} \qquad (5.83)$$

This summation evaluates to the asinc function of Eq. (5.75) shifted to a center frequency of $\omega = -2\pi k/K$, which is equivalent to $\omega = 2\pi(K-k)/K$. The kth DFT sample therefore corresponds to filtering the data with a bandpass filter having a frequency response with an asinc function shape centered at the frequency of the $(K-k)$th DFT sample.

If the data are windowed before processing with a window function $w[m]$, Eq. (5.81) becomes (again for $F_D = k/KT$)

$$\mathbf{h}_k^T \mathbf{y} = A \sum_{m=0}^{M-1} w[m]y[m]\exp(-j2\pi mk/K) \qquad (5.84)$$

The impulse response vector and frequency response are then

$$\mathbf{h}_k = [w[0] \ w[1]\exp(-j2\pi k/K) \ \cdots \ w[M-1]\exp(-j2\pi(M-1)k/K)]^T \qquad (5.85)$$

$$H_k(\omega) = \sum_{m=0}^{M-1} w[m]\exp[-j(\omega + 2\pi k)/K] = W\left(\omega + \frac{2\pi k}{K}\right) \qquad (5.86)$$

The DFT still implements a bandpass filter centered at each DFT frequency, but the filter frequency response shape now becomes that of the window function.

The relation between the DFT and a bank of filters can be made more explicit. Consider a slow-time signal $y[m]$ obtained with a long series of pulses and an M-point window function $w[m]$. The window function can be slid along the data sequence to select a portion of the data for spectral analysis, as shown in Fig. 5.19. The DTFT of the resulting sequence $w[m-p] \cdot y[p]$ is, in terms of analog frequency F,

$$\begin{aligned} Y_m(F) &= \sum_{p=-\infty}^{\infty} w[m-p]y[p]\exp(-j2\pi FpT) \\ &= \exp(-j2\pi FmT) \sum_{p=-\infty}^{\infty} w[m-p]\exp[-j2\pi F(p-m)T]y[p] \\ &= \exp(-j2\pi FmT)\{[w[p]\exp(+j2\pi FpT)] * y[p]\}_{p=m} \end{aligned} \qquad (5.87)$$

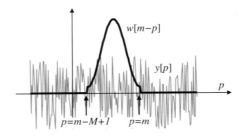

FIGURE 5.19 Relationship between data sequence x[m] and M-point sliding analysis window w[m].

Equation (5.87) shows that, aside from a phase factor, the DTFT at a particular frequency is equivalent to the convolution of the input sequence and a modulated window function, evaluated at time m. Furthermore, if $W(F)$ is the discrete-time Fourier transform of $w[m]$ (converted to an analog frequency scale), the DTFT of $w[p]\exp(+j2\pi F_D pT)$ is $W(F + F_D)$, which is simply the Fourier transform of the window shifted so that it is centered at Doppler frequency $-F_D$ Hz. This means that measuring the DTFT at a frequency F_D is equivalent to passing the signal through a bandpass filter centered at $-F_D$ and having a passband shape equal to the Fourier transform of the window function. Since the DFT evaluates the DTFT at K distinct frequencies at once, it follows that pulse Doppler spectral analysis using the DFT is equivalent to passing the data through a bank of bandpass filters.

Of course, it is possible to build a literal bank of bandpass filters, each one perhaps individually designed, and some systems are constructed in this way. For example, the zero-Doppler filter in the filterbank can be optimized to match the expected clutter spectrum or even made adaptive to account for changing clutter conditions. Most commonly, however, the DFT is used for Doppler spectrum analysis. This places several restrictions on the effective filterbank design. There will be K filters in the bank, where K is the DFT size; the filter center frequencies will be equally spaced, equal to the DFT sample frequencies; and all the passband filter frequency response shapes will be identical, determined by the window used and differing only in center frequency. The advantages to this approach are simplicity and speed with reasonable flexibility. Specifically, the FFT provides a computationally efficient implementation of the filterbank; the number of filters can be changed by changing the DFT size; the filter shape can be changed by choosing a different window; and the filter optimizes the output SNR for targets coinciding with a DFT filter center frequency in a noise-limited interference environment.

5.4.6 Fine Doppler Estimation

Peaks in the DFT output that are sufficiently above the noise level to cross an appropriate detection threshold are interpreted as responses to moving targets, that is, as samples of the peak of an asinc component of the form of Eq. (5.75). As has been emphasized, there is no guarantee that a DFT sample will fall exactly on the asinc function peak. Consequently, the amplitude and frequency of the DFT sample giving rise to a detection are only approximations to the actual amplitude and frequency of the asinc peak. In particular, the estimated Doppler frequency of the peak can be off by as much as one-half Doppler bin, equal to $RR/2K$ Hz.

If the DFT size K is significantly larger than the slow-time data sequence length M, several DFT samples will be taken on the asinc mainlobe and the largest may well be a good estimate of the amplitude and frequency of the asinc peak. However, frequently $K = M$ and sometimes, with the use of data turning, it is even true that $K < M$. In these cases, the

Doppler samples are far apart and a half-bin error may be intolerable. One way to improve the estimate of the true Doppler frequency F_D is to interpolate the DFT in the vicinity of the detected peak.

The most accurate way to interpolate the DFT is to zero pad the data and compute a larger DFT. In the absence of noise the interpolated values are exact. However, this approach is computationally expensive and interpolates all of the spectrum. If finer sampling is needed only over a small portion of the spectrum around a detected peak the zero padding approach is inefficient.

Computing a larger DFT is tantamount to interpolation using an asinc interpolation kernel. To see this, consider evaluating the DTFT at an arbitrary value of ω using only the available DFT samples. This can be done by computing the inverse DFT to recover the original time-domain data and then computing the DTFT from those samples:

$$\begin{aligned} Y(\omega) &= \sum_{m=0}^{M-1} y[m]\exp(-j\omega m) \\ &= \sum_{m=0}^{M-1} \left(\frac{1}{K} \sum_{m=0}^{K-1} Y[k]\exp(+j2\pi mk/K) \right) \exp(-j\omega m) \\ &= \frac{1}{K} \sum_{m=0}^{K-1} Y[k] \left\{ \sum_{m=0}^{M-1} \exp\left[-jm\left(\omega - \frac{2\pi k}{K}\right)\right] \right\} \end{aligned} \quad (5.88)$$

The term in braces is the interpolating kernel. It can be expressed in closed form as

$$\sum_{m=0}^{M-1} \exp\left[-jm\left(\omega - \frac{2\pi k}{K}\right)\right] = \exp\left[-j\left(\omega - \frac{2\pi k}{K}\right)(M-1)/2\right] \frac{\sin\left[\left(\omega - \frac{2\pi k}{K}\right)M/2\right]}{\sin\left[\left(\omega - \frac{2\pi k}{K}\right)/2\right]} \quad (5.89)$$

$$\equiv Q_{M,K}(\omega, k)$$

Combining these gives

$$Y(\omega) = \frac{1}{K} \sum_{m=0}^{K-1} Y[k] Q_{M,K}(\omega, k) \quad (5.90)$$

Equations (5.89) and (5.90) can be used to compute the DTFT at any single value of ω from the DFT samples. They can be applied to interpolate the values of the DFT over localized regions with any desired sample spacing and the result is exact in the absence of noise. However, it remains relatively computationally expensive.

A simpler but very serviceable technique for interpolating local peaks is illustrated in Fig. 5.20. For each detected peak in the magnitude of the DFT output, a second-order polynomial is fit through that peak and the two adjacent magnitude data samples. Once the polynomial coefficients are known, the amplitude and frequency of its peak are easily found by differentiating the formula for the polynomial and setting the result to zero. It can be shown that the second-order polynomial passing through the three samples can be expressed as

$$|Y[k_0 + \Delta k]| = \frac{1}{2}\{(\Delta k - 1)\Delta k|Y[k_0 - 1]| - 2(\Delta k - 1)(\Delta k + 1)|Y[k_0]| + \ldots \\ \ldots + (\Delta k + 1)\Delta k|Y[k_0 + 1]|\} \quad (5.91)$$

where Δk is the location of the interpolated peak relative to the index k_0 of the apparent peak so that the estimated peak location becomes $k' = k_0 + \Delta k$ (see Fig. 5.20); thus, $\Delta k \in [-0.5, +0.5]$.

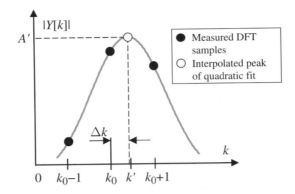

FIGURE 5.20 Refining the estimated target amplitude and Doppler shift by quadratic interpolation around the DFT peak.

Differentiating this equation with respect to Δk, setting the result to zero, and solving for Δk gives the estimated location of the polynomial peak relative to k_0 as

$$\Delta k = \frac{-(1/2)\{|Y[k_0+1]| - |Y[k_0-1]|\}}{|Y[k_0-1]| - 2|Y[k_0]| + |Y[k_0+1]|} \tag{5.92}$$

The amplitude of the estimated peak $A' = |Y[k_0 + \Delta k]|$ is found by computing Δk and then using that result in Eq. (5.91). Note that the formula for Δk behaves in intuitively satisfying ways. If the first and third DFT magnitude samples are equal, $\Delta k = 0$; the estimated peak is the middle sample. If the second and third samples are equal, $\Delta k = 1/2$, indicating the estimated peak is halfway between the two samples; a similar result applies if the first and second DFT magnitude samples are equal.

This interpolation technique is less effective when the width of the presumed mainlobe response that is to be interpolated is so narrow that the apparent peak and its two neighbors are not on the same lobe of the response. This occurs when the spectrum is sampled at the Nyquist rate in Doppler, $K = M$; no window is applied to the data; and the data frequency does not happen to fall on a DFT frequency sample (the very situation in which interpolation is most needed). Figure 5.21a illustrates this case. The data are $M = 20$ samples of a 1.35-kHz complex sinusoid sampled at 5 kHz. The spectrum was computed using a $K = 20$ point DFT. Without interpolation, the largest DFT magnitude sample has a value of 15.15. This apparent peak amplitude is in error by 24.3 percent from the true DTFT peak value of 20. The DFT samples occur every $5000/20 = 250$ Hz, so the apparent peak occurs at a frequency of 1500 Hz, an error of 150 Hz.

If the interpolation procedure is applied to these data, poor results will be achieved because the assumption that the three points are on an approximately quadratic curve segment is not valid. For these particular data, the interpolation technique will estimate the true frequency and amplitude of the spectral peak as 1295.4 Hz and 15.41, respectively. The amplitude estimate is improved only slightly, to a 23 percent error. The frequency error is reduced significantly to 54.6 Hz but is still large.

This problem can be avoided by ensuring that the sample set is dense enough to guarantee that the three samples are all on the mainlobe. One way to do this is to oversample in Doppler, that is, choose $K > M$. Another is to window the data. For most common windows, the expansion of the mainlobe that results is sufficient to guarantee that the apparent peak

Doppler Processing

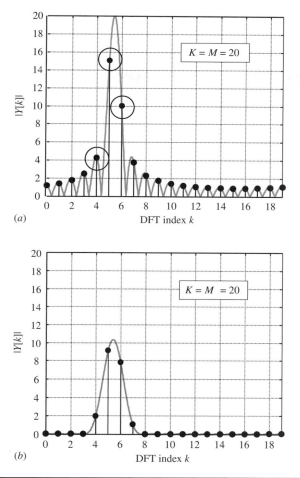

FIGURE 5.21 Ideal Doppler spectrum due to a moving target, sampled at the Nyquist rate. (*a*) No window. (*b*) Hamming window.

samples and its two neighbors fall on the same lobe, so that the basic assumption of an approximately quadratic segment is more valid. (It also helps that windowing reduces straddle loss by widening the mainlobe, as shown previously.) Figure 5.21*b* illustrates this effect by applying a Hamming window to the same data used for Fig. 5.21*a* and again applying a 20-point DFT. The peak DTFT amplitude is now 10.34 due to the effect of the Hamming window. Applying quadratic interpolation to this spectrum gives an estimated spectral peak frequency and amplitude of 1336.6 Hz and 9.676, respectively, errors of 6.4 percent in amplitude and 13.4 Hz in frequency, both about one-quarter of the errors in the unwindowed case.

A hybrid technique can be defined that combines attributes of the quadratic interpolation and the more exact asinc interpolation. The quadratic method is used to estimate the frequency of the peak, and then Eq. (5.90) is used to estimate the amplitude. This approach improves amplitude accuracy while avoiding the need to compute Eq. (5.90) more than once.

Figure 5.22 illustrates the noise-free frequency estimation performance of the quadratic interpolator both with and without Hamming windowing on a sinusoidal data sequence of length $M = 30$. Figure 5.22a illustrates the minimally sampled case $K = M$. The interpolated frequency estimates are best when the actual frequency is either very close to a sample frequency or exactly half way between two sample frequencies. If no window is used, the worst case error of 0.23 bin occurs when the actual frequency is 0.35 bins away from a sample frequency. A Hamming window reduces this maximum error to 0.067 bins at an offset of 0.31 bins. Figure 5.22b shows the performance when the spectrum sampling density is slightly more than doubled to $K = 64$. The maximum frequency estimation error is only 0.022 bins without the window and 0.014 bins with the Hamming window.

Figure 5.23 shows the amplitude estimation performance for the same cases, again with no noise. Figure 5.23a shows that the interpolator reduces the worst-case straddle loss from

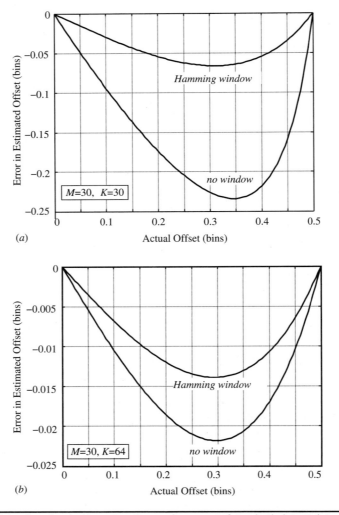

FIGURE 5.22 Noise-free frequency estimation performance of quadratic interpolator, $M = 30$. (a) Minimally sampled case ($K = 30$). (b) Oversampled case ($K = 64$).

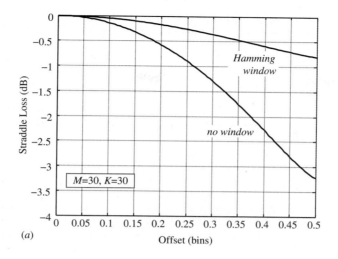

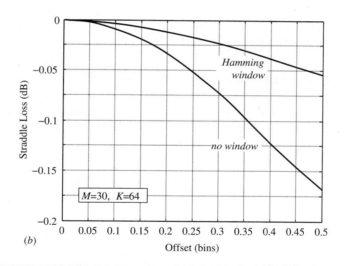

FIGURE 5.23 Noise-free amplitude estimation performance of quadratic interpolator, $M = 30$. (a) Minimally sampled case ($K = 30$). (b) Oversampled case ($K = 64$).

3.92 dB (recall Fig. 5.18) to 3.22 dB when no window is used, and from 1.61 to 0.82 dB when a Hamming window is used. The worst-case amplitude error with the oversampled spectrum using $K = 64$ is reduced to 0.17 dB without a window and only 0.05 dB with the Hamming window.

Many other interpolation-based estimators have been proposed for the DFT; several are described in Jacobsen and Kootsookos (2007) and MacLeod (1998). They include versions of Eqs. (5.91) and (5.92) that use the complex DFT data $Y[k]$ instead of its magnitude and achieve significantly better frequency estimation accuracy in noise-free data, as well as adjustments to the weighting coefficients in either the complex or magnitude version which improve accuracy when a window is used on the data. Another family of interpolators uses the apparent peak and only the larger of its two neighbors to compute Δk as

$$\Delta k = \frac{|Y[k_0]|}{|Y[k_0]| + |Y[k_\pm]|} \qquad (5.93)$$

where $|Y[k_\pm]|$ is the larger of $|Y[k_0 - 1]|$ and $|Y[k_0 + 1]|$. While this estimator seemingly uses only two DFT values explicitly, it implicitly uses three since two must be examined to determine which is $|Y[k_\pm]|$. Complex and magnitude-only versions of this estimator also exist. Its frequency estimation performance is surprisingly good in the absence of noise. Additional variations and performance analysis are given in Candan (2011) and Candan (2013).

These interpolation techniques are not limited to Doppler frequency estimation only. The same issues of sampling density and straddle loss arise at the output of the matched filter in fast time, for instance, and the same interpolation techniques can be applied to improve the range estimates. The application to time delay (range) estimation, along with simulation results for this and alternative techniques, is discussed in Chap. 7.

The above results are all for noise-free data, an unrealistic assumption. Estimation accuracy and the effect of interpolation algorithms will be revisited more in Chap. 7. It will be seen there the two-point interpolator is of little value in the presence of noise, but that the quadratic interpolator is still effective at high SNR. Other effects unrelated to interpolation dominate the estimation precision at mid to low SNRs.

5.4.7 Modern Spectral Estimation in Pulse Doppler Processing

So far, the DFT has been used exclusively to compute the spectral estimates needed for pulse Doppler processing. Other spectral estimators can be used. One that has been applied to radar is the *autoregressive* (AR) model, which models the actual spectrum $Y(\omega)$ of the slow-time signal with a spectrum of the form

$$\hat{Y}(\omega) \frac{\alpha}{1 + \sum_{p=1}^{P} a_p \exp(-j\omega p)} \qquad (5.94)$$

The algorithm finds the set of model coefficients $\{a_p\}$ that optimally fits $\hat{Y}(\omega)$ to $Y(\omega)$ for a given model order P. These coefficients are found by solving a set of *normal equations* (Hayes, 1996) derived from the autocorrelation of the slow-time data $y[m]$; the actual spectrum $Y(\omega)$ is not needed. The $\{a_p\}$ are then used to compute an estimated spectrum according to Eq. (5.94), which can be analyzed for target detection, pulse pair processing (Sec. 5.5), or other functions.

Modeling the spectrum, as shown in Eq. (5.94), is equivalent to modeling the slow-time signal $y[m]$ as the impulse response of an IIR filter with frequency response $\{1 + \sum_{p=1}^{P} a_p \exp(-j\omega p)\}^{-1}$. The inverse filter is an FIR filter with impulse response coefficients $h[m] = [1 \ a_1 \ a_2 \ a_P]$. If $y[m]$ is passed through this filter the output spectrum will be approximately constant provided that the actual signal spectrum is accurately modeled by Eq. (5.94). It follows that if the $\{a_p\}$ are chosen such that $|\hat{Y}(\omega)|^2$ is a good model of the power spectrum of random process data such as noise and clutter, then passing that data through the inverse filter will produce a new random process with an approximately flat power spectrum. Thus, the FIR filter designed from the model coefficients *whitens* the signal, removing any correlated signal components such as clutter.

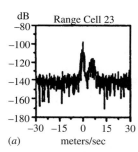 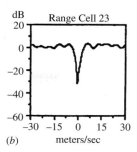 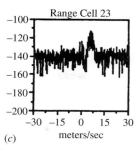

Figure 5.24 Clutter suppression and windshear detection using an autoregressive Doppler spectrum estimate. (*a*) Fourier spectrum of raw data. (*b*) Frequency response of clutter suppression filter derived from the AR coefficients. (*c*) Fourier spectrum of filtered data. (Figure courtesy of Dr. Byron M. Keel. Used with permission.)

Figure 5.24 illustrates the application of AR spectral estimation to design a clutter filter to enhance detection of windshear from an airborne radar (Keel, 1989). Figure 5.24*a* shows the Fourier spectrum of the slow-time data from one range bin. Two peaks are evident above the noise floor. The one at zero velocity is ground clutter. The smaller peak at approximately 8 m/s is due to windblown rain. The middle plot shows the frequency response of an optimal clutter filter implemented from the $\{a_p\}$. The third plot shows the Fourier spectrum of the slow-time data after processing with the clutter filter. The ground clutter has been significantly suppressed and the weather echo is now the dominant spectral feature.

Perhaps the most important advantage of modern spectral estimators is that, used in appropriate situations, they can achieve finer resolution than Fourier methods for the same amount of available data. This is possible because they assume a particular structural model of the data (e.g., that it is a sum of P sinusoids in white noise) and use that extra information to obtain enhanced resolution. Another good property is the ability to define optimal interference rejection filters as in the example above. They also suffer several disadvantages. Most importantly, if the model is not a good fit to the data a poor spectral estimate may result. In any event, the order of the model (P, for example) must be known or estimated. Also, the spectrum calculation is nonlinear, which tends to result in threshold effects where a small change in SNR may rapidly degrade the spectrum estimate. In contrast, Fourier methods are linear and tend to exhibit "graceful degradation" as the SNR decreases. Finally, while fast algorithms exist for many of these algorithms, none is as efficient as the FFT. In practice, the DFT is by far the most common means of estimating the Doppler spectrum.

5.4.8 CPI-to-CPI Stagger and Blind Zone Maps

In CPI-to-CPI PRF stagger, a coherent processing interval of M pulses is transmitted at a fixed PRF.[8] A second CPI is then transmitted at a different fixed PRF. Because the blind speeds are different for each PRF used, a target that falls in a blind speed of one PRF may be visible in the others. This concept is illustrated in Fig. 5.25, which shows a notional Doppler spectrum for two different PRFs. The plots are shown on the same frequency scale. First, consider the upper spectrum plot, which corresponds to data collected at PRF_1. A target whose Doppler equals PRF_1 will be aliased to zero Doppler shift, where it may be undetectable

---

[8]As with pulse-to-pulse staggered PRFs, it is assumed here that CPI-to-CPI staggered PRFs are used mostly with pulsed systems, so the pulsed radar terminology of PRFs and PRI will be used in this section instead of the more general RRs and RIs.

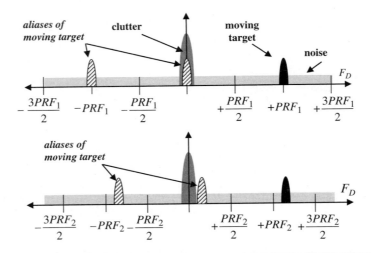

FIGURE 5.25 The use of two PRFs to avoid blind speeds in pulse Doppler radar. The target Doppler shift equals the PRF in the upper sketch, but not in the lower sketch.

if clutter is present. If the same target scenario is measured with a lower pulse repetition frequency PRF_2 as shown in the lower half of the figure, the Doppler shift of the target no longer matches the PRF. The target energy aliases to a nonzero Doppler, where it does not compete with the clutter and is still detectable.

In some systems, as many as eight PRFs may be used. The first velocity that is blind at all of the PRFs is the LCM of the individual blind speeds, which will usually be much higher than any one of them alone. Target detections are typically accepted and passed to subsequent processing only if they occur in some minimum fraction of the PRFs used, for example two out of five, or three out of eight PRFs. A major disadvantage of CPI-to-CPI stagger is that transmission of so many CPIs in each dwell consumes large amounts of the radar timeline and energy, impacting system-level search and track options.

Coherent pulse cancellers can be used within each CPI to suppress aliased or unaliased range clutter (Schleher, 2010). If a pulse canceller is not used, there is no highpass filter and therefore a target whose Doppler shift equals an integer multiple of the PRF will not be filtered out. Rather, the target energy will alias to the DC portion of the spectrum and combine with the clutter energy there. These targets will still go undetected, so the corresponding target velocities are still blind speeds.

For a given PRF there are periodic blind zones in Doppler spaced by the PRF. A target whose Doppler shift is such that it either is in the clutter region or aliases to the clutter region is unlikely to be detected. Equivalently, if the target falls into a clutter region caused by the periodically repeated clutter spectrum it will likely go undetected. The idea is illustrated in Fig. 5.26a, which indicates clutter at zero Doppler and its first two periodic repetitions at $\pm PRF$ Hz. The target shown just above PRF Hz but still within the clutter region is considered to be in the blind zone centered at $F_D = PRF$ Hz.

Recall that there are also blind zones in range or fast time in a monostatic radar as discussed in Sec. 3.3. The range blind zones due to pulse eclipsing are deliberately extended in some systems by keeping the receiver disconnected from the antenna for an additional period of time beyond the pulse length. The purpose is to avoid saturating the receiver with very strong near-in clutter returns, such as those from the clutter altitude line. Figure 5.26b illustrates range blind zones and eclipsing. Pulses of duration τ are transmitted every T seconds as shown in the upper line. The middle line shows three target echoes after each pulse and strong

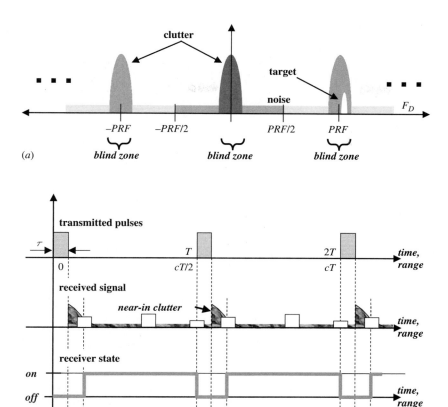

FIGURE 5.26 Blind zones in Doppler and range. (a) Blind zones in Doppler due to the periodic repetition of the mainlobe clutter spectrum. (b) Blind zones in range due to pulse transmission and near-in clutter in a monostatic radar. Partial eclipsing of two of the three target echoes is also shown.

near-in clutter returns. The bottom line shows that the receiver will be off during the time of transmission of the pulse and of reception of the near-in clutter echoes. These "off" periods represent the blind zones in range. In this example, the first target is close enough to the radar that its echo is 50 percent eclipsed because it arrives during the near-in clutter interval. The third echo is also 50 percent eclipsed because the target is far enough away that a portion of the echo is received during transmission of the next pulse. The middle echo is not eclipsed.

Blind zones in range and Doppler collectively combine to form a two-dimensional *blind zone map*, as shown in Fig. 5.27. Targets that fall within a range blind zone are undetectable no matter what their Doppler shift. Targets that fall within a Doppler blind zone are undetectable at any range. The spacing of the blind zones in both dimensions is determined by the PRF of the radar. If the PRF is increased, the Doppler blind zones spread further apart but the range blind zones contract to become closer together. These observations lead to the idea of using CPI-to-CPI stagger in combination with an "*M* out of *N*" detection logic to maximize the combinations of range and Doppler shift at which targets can be detected.

Figure 5.28 shows blind zone maps for a 10-GHz radar having a 10-μs pulse length and using two PRIs, 100 μs and 120 μs, on successive CPIs. The range blind zones are not lengthened

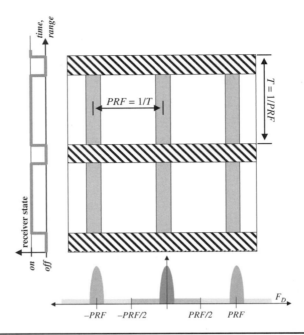

FIGURE 5.27 Blind zone map for one PRF formed by combined blind zones in Doppler and range.

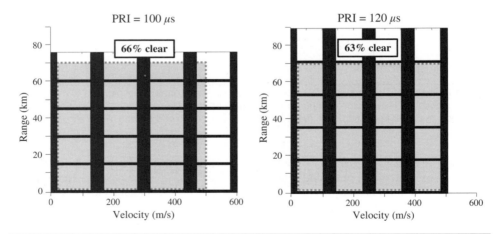

FIGURE 5.28 Blind zone maps for two different PRIs. The maps are drawn to the same velocity and range scales. Left: PRI = 100 μs. Right: PRI = 120 μs. The gray region represents the 70 km by 500 m/s range-Doppler detection region of primary interest. The clear percentage is of that region, excluding the first range and Doppler blind zones.

for near-in clutter. The horizontal axis is in velocity units rather than Doppler shift, and is shown only for positive velocity; the negative portion of the axis would be its mirror image. The clutter velocity spread is assumed to be ±20 m/s. Blind range-velocity cells due to sidelobe clutter are not considered; see Alabaster (2012) for more information. Both maps are shown to the same scale in range and velocity. Notice that the first-range blind zone and the zero-velocity blind zone are the same for any PRF and so always overlap in all PRFs used. This is not true for their repetitions in range and velocity due to the change in periodicity when the PRF is changed. For example, a target at a velocity of 300 m/s and range of 30 km is blind in both range and velocity at the 100 μs PRI but is not in a blind zone at the 120 μs PRI.

The gray shaded area represents the range-velocity search space of interest, in this case 70 km in range by 500 m/s in velocity. Ideally, the entire gray region, which does *not* include the first range and velocity blind zones, would be a clear region, suggesting that a target anywhere in that region is detectable so long as the SIR is adequate. The "percentage clear" shown on each blind zone map is the portion of the desired range-velocity region that is in the clear. A common goal is a 96 percent or higher clear percentage. As seen in this example, a single PRF often will not meet this goal, emphasizing the need for multiple PRF operation.

The left image of Fig. 5.29 results from overlaying the two blind zone maps in Fig. 5.28 on common intervals in range and velocity. Range-velocity coordinates that are in the blind zone at both PRFs are shown as black; those in only one of the two blind zones are shown in gray; and those in the clear on both PRFs are shown in white. For example, the target at 300 m/s and 30 km mentioned above falls in the gray area, suggesting that a radar collecting a CPI of data at each of these two PRIs would detect the target on only one of them (120 μs). The image on the right is the blind zone map that would result from the use of these two PRIs and a "1 of 2" detection logic. This means that a detection at particular range-velocity coordinates is accepted as being valid if it is detected on at least one of the CPIs; detection on both is not required. With this logic, a target at any range-velocity pair in the white or gray area of the left figure would be expected to be detectable (assuming adequate SIR). Only the coordinate pairs in the black area of the left image would be considered blind, leading to the blind zone map shown on the right. The clear percentage of the desired range-velocity region is greatly increased, from 63 or 66 percent to 92 percent.

A problem with a "1 of 2" rule is that a single false alarm would be accepted as a target. A more conservative "2 of 2" rule would mean a target was detectable only if it was in the clear on both PRIs. The effective blind zone map would be black in all of the black and gray regions of the left image in Fig. 5.29, with the result that very few combinations of range and velocity would correspond to detectable targets. Thus, there is a tradeoff between the range-velocity coverage area in which targets are detectable and the immunity to false alarms and other error sources. Using values of N larger than 2 offers more compromise options and better results. Figure 5.30 shows an exceptionally good blind zone map for a system using

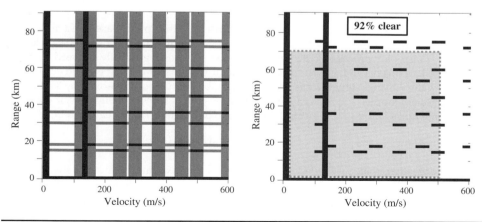

FIGURE 5.29 Blind zones using two PRIs and a "1 of 2" detection rule. Left: Overlay of the two blind zone maps of Fig. 5.28. Right: Resulting blind zone map using "1 of 2" detection logic.

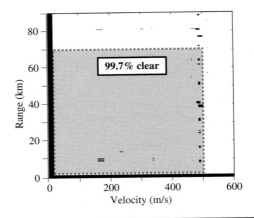

FIGURE 5.30 "3 of 8" blind zone map obtained using evolutionary algorithms to select the PRIs. See text for details.

eight PRFs with a "3 of 8" detection rule. The radar operates at 10 GHz with a 1-μs pulse length, blanking of 10 near-in clutter cells, a ± 17 m/s mainlobe clutter (MLC) spread, and a 10-μs CPI. An evolutionary optimization algorithm was used to select PRIs that maximized the detectable region within the range and velocity limits shown; the result was the set {51, 53, 60, 63, 67, 84, 89, 93} μs (Davis and Hughes, 2002). Nearly all (99.7%) of the desired range-velocity space, excluding the first range and velocity blind zones, is in the clear.

5.5 Pulse Pair Processing

Pulse pair processing (PPP) is a form of Doppler processing common in meteorological radar. Unlike the MTI and pulse Doppler techniques discussed so far in this chapter, the goal of pulse pair processing is weather detection and tracking, not clutter suppression to enable the detection of moving targets. In PPP it is assumed that the spectrum of the slow-time data consists of noise and a single Doppler peak, generally not located at zero Doppler (though it could be), that is due to echo from moving weather events such as wind-blown rain or other particulates. The goal of PPP is to estimate the parameters of that spectral peak. These are then used as the input to higher-level algorithms for estimating precipitation types and rates, forecasting severe weather, and so forth in both ground-based and airborne weather radars. In airborne radars, it is also one of the techniques used for windshear detection.

Pulse pair processing assumes the radar is looking generally upward if it is ground-based, or forward if it is airborne. Consequently, it is assumed that ground clutter competing with the weather signatures is small or negligible, or has been removed by MTI filtering. The notional Doppler spectrum $S_y(F)$ assumed by PPP is shown in Fig. 5.31. It consists only of white noise and a single spectral peak due to backscatter from weather-related phenomena,

$$S_y(F) = S_w(F) + S_n(F) \tag{5.95}$$

The weather peak $S_w(F)$ is often assumed to be approximately Gaussian shaped and is characterized by its amplitude, mean, and standard deviation. The total area of the $S_w(F)$ power spectrum component equals the power of the meteorological echo. PPP is used to estimate the power, mean Doppler shift F_0, and variance σ_F^2 (commonly called the *spectral width*) of the weather component. Each of these can be estimated using either time- or frequency-domain algorithms, all of which are included under the PPP rubric.

Doppler Processing

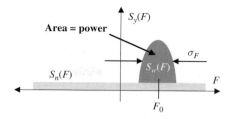

FIGURE 5.31 Notional slow-time power spectrum assumed in pulse pair processing.

Consider time-domain measurements first. The autocorrelation and power spectrum of the slow-time data sequence $y[m]$, $m = 0, \ldots, M-1$ obtained from M pulses sampled at a particular range bin are

$$s_y[k] = \sum_{m=0}^{M-k-1} y[m] y^*[m+k] \tag{5.96}$$

$$S_y(\omega) = \mathbf{F}\{s_y[k]\} = |Y(\omega)|^2 \tag{5.97}$$

The power in the slow-time signal can be estimated in the time domain from the peak of the autocorrelation function,

$$\hat{P}_y = \frac{1}{M} \sum_{m=0}^{M-1} |y[m]|^2 = \frac{1}{M} \sum_{m=0}^{M-1} y[m] y^*[m] = \frac{1}{M} s_y[0] \tag{5.98}$$

To see how to estimate the mean frequency, ignore noise for the moment and assume that the signal component is a pure sinusoid; the power spectrum of a finite segment would then be an asinc function squared. Now compute the first autocorrelation lag

$$y[m] = A \exp(j2\pi F_0 T m), \quad m = 0, \ldots, M-1$$

$$s_y[1] = \sum_{m=0}^{M-2} y[m] y^*[m+1]$$

$$= \sum_{m=0}^{M-2} A \exp(j2\pi F_0 T m) A^* \exp[-j2\pi F_0 T(m+1)] = |A|^2 \sum_{m=0}^{M-2} \exp(-j2\pi F_0 T) \tag{5.99}$$

$$= |A|^2 \exp(-j2\pi F_0 T) \sum_{m=0}^{M-2} (1) = |A|^2 (M-1) \exp(-j2\pi F_0 T)$$

The argument of the exponential, $-2\pi F_0 T$, is simply the negative of the amount of phase rotation in one sample period for a sampled sinusoid of original analog frequency F_0 Hz. The frequency can be estimated from Eq. (5.99) as

$$\hat{F}_0 = \frac{-1}{2\pi T} \arg\{s_y[1]\} \text{ Hz} \tag{5.100}$$

Multiplying $\hat{F}_0$ by $\lambda_t/2$ converts the result into units of velocity. Although derived for the pure sinusoid, this time-domain PPP frequency estimator works well for more general signals provided there is a single dominant frequency component with adequate SNR. The frequency estimate will be aliased if the Doppler frequency is outside the interval $\pm RR/2$.

Since the complex exponential inside the summation in Eq. (5.99) does not depend on m and so could be brought out of the sum, it would seem sufficient to simply compute $y[m]y^*[m+1]$ using only two slow-time samples instead of computing the full autocorrelation lag $s_y[1]$. In reality, noise is present in all of the samples and using all M available samples in the full summation averages the noise and improves the estimate quality.

To obtain a time-domain estimate of σ_F^2, assume that the Doppler power spectrum exhibits a Gaussian shape with standard deviation σ_F. The estimate of F_0 can be used to remove the mean Doppler component, giving a modified sequence $y'[m]$ with its Doppler spectrum centered at zero frequency. It is convenient to start with the continuous-time equivalent $y'(t)$. $S_{y'}(F)$ will be a zero-mean Gaussian function

$$S_{y'}(F) = \frac{|A|^2}{\sqrt{2\pi} \cdot \sigma_F} \exp\left(-F^2/2\sigma_F^2\right) \qquad (5.101)$$

It follows that the continuous-time autocorrelation function is also Gaussian

$$s_{y'}(z) = |A|^2 \exp(-2\pi^2 \sigma_F^2 z^2) \qquad (5.102)$$

where the variable z represents the autocorrelation lag. If the sampling interval T is chosen sufficiently small to guarantee that $S_{y'}(1/2T) \approx 0$, the discrete-time spectrum and autocorrelation will also form a Gaussian pair to a very good approximation (Richards, 2006). The sampled autocorrelation function will be

$$s_{y'}[k] = s_{y'}(z)|_{z=kT} = |A|^2 \exp(-2\pi^2 \sigma_F^2 k^2 T^2) \qquad (5.103)$$

and the corresponding DTFT, still on a hertz scale, is

$$S_{y'}(F) = \frac{|A|^2}{\sqrt{2\pi} \cdot \sigma_F T} \exp(-F^2/2\sigma_F^2) \qquad (5.104)$$

Because $s_{y'}[0] = |A|^2$, the first autocorrelation lag can be written

$$s_{y'}[1] = |A|^2 \exp\left(-2\pi^2 \sigma_F^2 T^2\right) = s_{y'}[0] \exp\left(-2\pi^2 \sigma_F^2 T^2\right) \qquad (5.105)$$

Equation (5.105) is easily solved to give an estimate for the spectrum standard deviation in terms of only $s_{y'}[0]$ and $s_{y'}[1]$,

$$\hat{\sigma}_F^2 = -\frac{1}{2\pi^2 T^2} \ln\left\{\frac{s_{y'}[1]}{s_{y'}[0]}\right\} \text{ Hz}^2 \qquad (5.106)$$

Equations (5.98), (5.100), and (5.106) are the time-domain pulse pair processing estimators. They can be computed from only two autocorrelation lags of the slow-time data in each range bin.

Equation (5.106) is sometimes simplified to avoid the natural logarithm calculation. Consider the following series expansion and approximation of $\ln(x)$,

$$\begin{aligned}\ln(x) &= \frac{x-1}{x} + \frac{1}{2}\left(\frac{x-1}{x}\right)^2 + \frac{1}{3}\left(\frac{x-1}{x}\right)^3 + \cdots \\ &\approx \frac{x-1}{x} = 1 - \frac{1}{x}\end{aligned} \qquad (5.107)$$

Applying Eq. (5.107) to Eq. (5.106) gives the simplified spectral width estimator

$$\hat{\sigma}_F^2 = -\frac{1}{2\pi^2 T^2}\left\{1 - \frac{s_{y'}[0]}{s_{y'}[1]}\right\} \text{ Hz}^2 \tag{5.108}$$

The basic PPP measurements of signal power, frequency, and spectral width can also be performed in the frequency domain. The power is obtained by applying Parseval's theorem to Eq. (5.98),

$$\hat{P}_y = \frac{1}{2\pi M}\int_{-\pi}^{+\pi} S_y(\omega)\,d\omega = \frac{1}{2\pi M}\int_{-\pi}^{+\pi}|Y(\omega)|^2\,d\omega \tag{5.109}$$

A practical calculation uses the DFT version of Parseval's theorem with the DFT $Y[k]$ of $y[m]$,

$$\hat{P}_y = \frac{1}{M^2}\sum_{k=0}^{M-1}|Y[k]|^2 \tag{5.110}$$

There are two frequency-based methods for estimating the mean frequency of the signal. The first is a direct analog to Eq. (5.100),

$$\hat{F}_0 = \frac{-1}{2\pi T}\arg\{s_y[1]\}$$
$$= \frac{-1}{2\pi T}\arg\left\{\frac{1}{2\pi}\int_{-\pi}^{+\pi}|Y(\omega)|^2\exp(j\omega)\,d\omega\right\} \text{ Hz} \tag{5.111}$$

The integrand in Eq. (5.111) is real except for the $\exp(j\omega)$ term. Noting that for a complex number z, $\arg\{z\} = \mathrm{atan}\{\mathrm{Im}(z)/\mathrm{Re}(z)\}$, Eq. (5.111) gives the estimator

$$\hat{F}_0 = \frac{-1}{2\pi T}\mathrm{atan}\left\{\frac{\int_{-\pi}^{+\pi}|Y(\omega)|^2\sin(\omega)\,d\omega}{\int_{-\pi}^{+\pi}|Y(\omega)|^2\cos(\omega)\,d\omega}\right\} \text{ Hz} \tag{5.112}$$

In practice, the DFT version is used:

$$\hat{F}_0 = \frac{-1}{2\pi T}\mathrm{atan}\left\{\frac{\sum_{k=0}^{K-1}|Y[k]|^2\sin(2\pi k/K)}{\sum_{k=0}^{K-1}|Y[k]|^2\cos(2\pi k/K)}\right\} \text{ Hz} \tag{5.113}$$

The other frequency-domain mean frequency and spectral width estimators result from viewing the signal spectrum of Fig. 5.31 as a probability density function. A valid PDF must be real and nonnegative, a condition met by the power spectrum. However, a PDF must also have unit area, so the power spectrum must be normalized to ensure this is the case. By Parseval's theorem, the integral of $|Y(\omega)|^2 = 2\pi E_y$, where E_y is the energy in

$y[m]$; this is the required normalization factor. For any arbitrary PDF $p_z(z)$, the mean and variance are given by

$$\bar{z} = \int_{-\infty}^{+\infty} z \cdot p_z(z)\, dz$$
$$\sigma_z^2 = \int_{-\infty}^{+\infty} (z - \bar{z})^2 p_z(z)\, dz \tag{5.114}$$

Applying the first of these definitions to the power spectrum gives an alternative mean frequency estimator

$$\hat{F}_0 = \frac{1}{2\pi T} \int_{-\pi}^{+\pi} \omega \left\{ |Y(\omega)|^2 / 2\pi E_y \right\} d\omega = \frac{1}{4\pi^2 TE_y} \int_{-\pi}^{+\pi} \omega\, |Y(\omega)|^2\, d\omega \quad \text{Hz} \tag{5.115}$$

Similarly, an estimator of the spectral width is

$$\hat{\sigma}_F^2 = \frac{1}{(2\pi T)^2} \int_{-\pi}^{+\pi} (\omega - \omega_0)^2 \left(|Y(\omega)|^2 / 2\pi E_y \right) d\omega$$
$$= \frac{1}{8\pi^3 T^2 E_y} \int_{-\pi}^{+\pi} (\omega - \omega_0)^2\, |Y(\omega)|^2\, d\omega \quad \text{Hz}^2 \tag{5.116}$$

Generally, the time-domain estimators are preferred if the SNR is low or the spectral width is very narrow (Doviak and Zrnic, 1993). In the latter case, the signal is closer to the pure sinusoid assumption that motivated the time-domain estimator. In addition, the time-domain methods are more computationally efficient because no Fourier transform calculations are required. Conversely, the frequency-domain estimators tend to provide better estimators at high SNR and large spectral widths. The frequency domain estimator also allows reduction of the noise before the estimates are calculated, reducing the estimate bias. This process, called *spectral subtraction*, is depicted in Fig. 5.32. The noise power spectrum $N(\omega)$ is estimated from a presumed clear region of the spectrum and then simply subtracted off to form a reduced-noise power spectrum,

$$S_y'(\omega) = S_y(\omega) - N(\omega) \tag{5.117}$$

Because of the statistical variations of any given data set, it is possible that $S_y'(\omega)$ may have some negative values; these are usually set to zero.

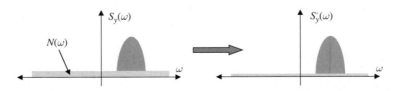

Figure 5.32 Spectral subtraction to reduce noise power and improve clutter parameter estimation.

Doppler Processing

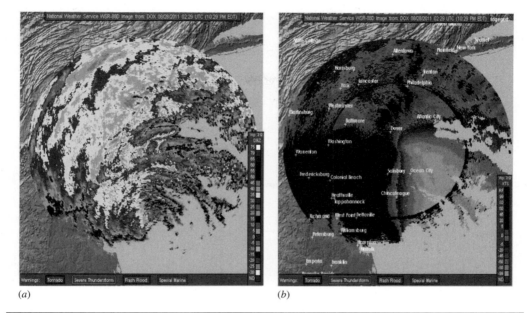

FIGURE 5.33 Sample weather radar images obtained with pulse pair processing. (a) Power image. (b) Velocity image. (Images courtesy of National Severe Storms Laboratory.)

Figure 5.33 shows two images of Hurricane Irene collected on August 28, 2011 by the DOX WSR-88D NEXRAD weather radar located at Dover Air Force Base, Delaware and operated by the U.S. Air Force and the U.S. National Weather Service. The radar images are overlaid on a topographical map of the area. While these images are much more easily interpreted in color than in grayscale, some features are visible. Figure 5.33a is a map of the power estimate in dBz, proportional to the volume reflectivity η as discussed in Chap. 2. Lighter grays represent areas of heavier rainfall. The circular shape of the rain bands is apparent. Figure 5.33b is a map of the radial velocity measured by the radar and thus of the radial wind speed component. The darker areas in the left part of the image are positive velocities relative to the radar; in other words, the areas where the wind is blowing away from the radar. The lighter grey regions on the right represent negative velocities (winds toward the radar).

5.6 Additional Doppler Processing Issues

5.6.1 Range Migration and the Keystone Transform

The basic moving target slow-time signal model of Eq. (5.74) presumes that the target motion is slow enough that all target samples remain in the same range bin over the CPI duration. This will be the case if

$$vTM \leq \delta R \text{ m} \quad \Rightarrow \quad v \leq \frac{\delta R}{MT} \text{ m/s} \qquad (5.118)$$

where δR is the range bin spacing. If this condition is met, the DFT of the slow-time signal will include the target signature in all M samples, yielding the full-resolution Doppler signature assumed in the foregoing discussion, and confining that Doppler signature to a single range bin.

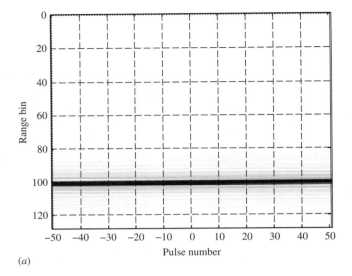

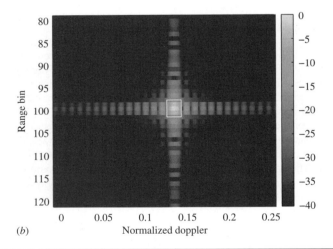

FIGURE 5.34 Effect of range migration on the range-Doppler signature of a moving target. (a) Magnitude of post-pulse compression fast-time/slow-time CPI of data for a target with no range migration. (b) Range-Doppler signature showing ideal resolution and sidelobes in both dimensions. (c) Faster target showing approximately 13 bins of range migration during the CPI. (d) Range-Doppler signature showing mainlobe smearing and resolution loss and sidelobe changes in both dimensions.

Some systems with relatively long CPIs and relatively fine resolution do not meet this constraint. In this case, the target signature is said to exhibit *range migration*: samples of the target echo slew across two or more range bins within a single CPI. Figure 5.34 illustrates the effect of range migration on the range-Doppler signature of a moving target. A constant radial velocity target observed by a 1-GHz radar was simulated. A linear frequency

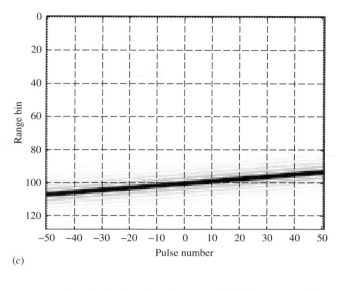

(c)

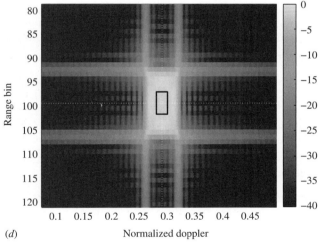

(d)

FIGURE 5.34 (Continued)

modulated (LFM) pulse with a bandwidth of 200 MHz was used without windowing, making the expected range resolution 0.75 m. However, the fast-time data was oversampled by 2.3 times, making the range bin spacing only 0.33 m. The CPI is 101 pulses long at a PRF of 10 kHz for a total duration of 10.1 ms, making the expected Doppler resolution 99 Hz or 0.0099 cycles per sample. Figure 5.34a shows the magnitude of the fast-time/slow-time data matrix after pulse compression when the target velocity is 20 m/s toward the radar. During the CPI the target moves 0.2 m, well under the 0.33 m range bin spacing. Therefore, the target stays in the same range bin (#100) for the entire CPI; there is no range migration. The light gray horizontal lines above and below the main response are the range sidelobes. Figure 5.34b shows the portion of the range-Doppler spectrum containing the signature of this target on a decibel scale, normalized to the peak amplitude. As expected, it is a sinc-like function in both dimensions. The white rectangle marks the theoretical null-to-null extent of the mainlobe, 0.75 m by 0.0099 cycles per sample in this case, showing that the actual signal mainlobe does exhibit the expected resolution.

Figure 5.34c shows the effect of significant range migration. In this example, the target velocity increased to 440 m/s (Doppler shift of 0.29 cycles per sample), making the range migration over the CPI 4.44 m or about 13.5 range bins. The slewing of the target response across the range bins is clearly evident. Figure 5.34d shows the badly degraded range-Doppler response that results. Because portions of the target data are spread over 13 to 14 range bins, the mainlobe of the response is smeared by the same amount in range; and because each of those range bins includes only a portion of the total target signature, the mainlobe of the DFT in each range bin is spread significantly.[9] Range migration has severely blurred the target response relative to the expected widths indicated by the dark rectangle. Although not evident from these normalized spectrum figures, the amplitude of the mainlobe is also reduced significantly.

The *keystone transform* (KT) is an algorithm for preprocessing the fast-time/slow-time data to correct for range migration. Its derivation and full explanation is beyond the scope of this textbook, but a detailed discussion including MATLAB® code is given in Richards (2020). Given a CPI of fast-time/slow-time data $y[l, m]$, the key steps are as follows:

1. Perform a DTFT in the *range* dimension for each pulse to get the range frequency/pulse number spectrum $Y_{Rd}(F, m)$. Note this is the opposite of the usual process of applying the DFT in the slow-time dimension. The range frequency variable F will be discrete in practice, but the analog notation is most convenient for this description.

2. In each range frequency bin, resample (interpolate) the slow-time data to form the new spectrum $Y_{Rd\_key}(F, m) = Y_{Rd}\left[F, \left(\frac{F_t}{F_t + F}\right)m\right]$, where F_t is the nominal RF. Note that the interpolation factor is different for each range frequency value F.

3. Compute the inverse DTFT in the range frequency dimension to obtain a modified fast-time/slow-time CPI of data $y_{key}[l, m]$.

4. Compute the usual slow-time DTFT in each range bin to obtain the final corrected range-Doppler spectrum $Y[l, F_D]$.

This flow is illustrated in Fig. 5.35, along with an example of the thumbnail images of the magnitude of the range frequency/slow-time spectrum just before and after the keystone resampling. These thumbnails illustrate why the algorithm is called the keystone transform. Prior to the resampling the spectrum occupies a rectangular region spanning the 101-sample slow-time duration of the CPI and the range frequency range of 900 MHz to 1.1 GHz. After the resampling, the spectrum takes on a trapezoidal "keystone" shape, so called because it resembles the shape of the keystone at the top of a masonry arch. Figure 5.36 shows the result of applying the KT to the example of Fig. 5.34c and d. The range migration has been corrected and the target is now stabilized in one range bin over the CPI, though there is some degradation of the pulse-compressed range profile in the first and last 5 percent of the CPI due to interpolation transients. The range-Doppler signature now shows a correctly located, full-resolution target response, though with some inconsequential sidelobe changes, especially in Doppler.

The KT has the excellent property that it can simultaneously correct the signatures of multiple targets at different velocities. Velocity-ambiguous targets require an additional

---

[9]The Doppler spread factor is less than 13 times because even when the mainlobe of the compressed pulse has migrated to another range bin, the range sidelobes still provide some attenuated target signature as they pass through a given range bin.

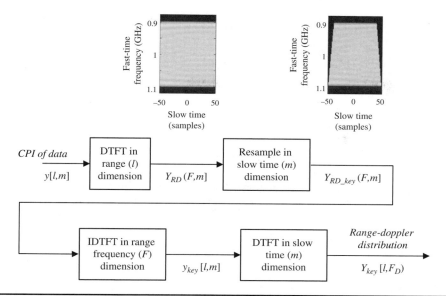

FIGURE 5.35 Flow diagram for keystone transform algorithm. The thumbnail images show the effect of the slow-time resampling of the range frequency/slow-time spectrum on the spectrum region of support.

simple phase compensation. Unfortunately, the compensation function requires that the number of Doppler foldovers be known in advance, or that a search process be carried out to estimate it. Furthermore, if there are multiple targets, all must have the same number of Doppler foldovers to be simultaneously corrected. Additional details of the keystone transform derivation, limitations, and usage are given in Richards (2020).

5.6.2 Combined MTI and Pulse Doppler Processing

It is not unusual to have both MTI filtering for gross clutter removal and pulse Doppler spectrum analysis for detailed examination of the Doppler spectrum. Since both operations are linear, the order in which they are applied would appear to make no difference to the final Doppler spectrum used for detection. However, differences in signal dynamic range can make their order significant in considering hardware effects when finite-wordlength hardware is used.

Clutter is often the strongest component of the signal; it can be several tens of decibels above the target signals of interest. If the DFT of the slow-time signal is computed prior to MTI filtering, the sidelobes of the response from the clutter around DC may swamp potential target responses at near-in velocities, masking these targets from possible detection. If the processor dynamic range is limited as well, the effect of the strong clutter signal on processor automatic gain control may drive the target signal amplitude below the minimum detectable signal of the processor, effectively filtering out the target.

For these reasons, the MTI filter is generally placed first if both processes are used. The MTI filter will attenuate the clutter component selectively so that the target signals become the dominant components. Subsequent finite wordlength processing will adapt the dynamic range to the targets rather than the now-suppressed clutter. In the now common floating point processors, dynamic range is less of an issue. However, strong clutter can still mask targets, so MTI filtering is still typically applied prior to Doppler spectrum formation.

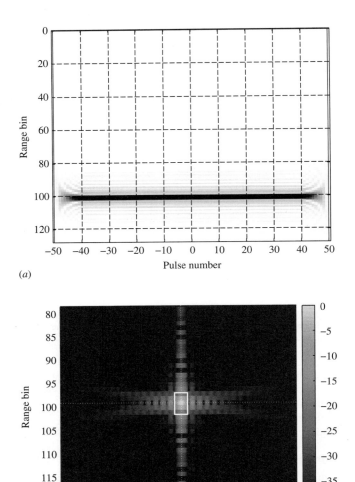

FIGURE 5.36 Effect of the keystone transform on the example of Fig. 5.34c and d. (a) Range migration is removed in the fast-time/slow-time data. (b) Full resolution in range and Doppler is restored, with some degradation of sidelobes.

5.6.3 Transient Effects

All of the discussion in this chapter has assumed a steady-state scenario in the sense that the clutter spectrum is stationary and that filter transient effects have been ignored. In range-ambiguous medium and high PRF modes the received signal sample in each range bin contains contributions from multiple ranges because of the multiple contributing pulses.[10]

---

[10]This is primarily a pulsed radar issue since range ambiguities are typically less of an issue with fast-chirp FMCW systems.

Whenever the radar PRF changes, several pulses, known as *clutter fill pulses*, must be transmitted before a steady-state situation is achieved. Here "steady state" means that the physical clutter intervals contributing to a given range bin are the same for each pulse so that the clutter statistics can be expected to be stationary from one pulse to the next. For example, suppose that in steady state each range gate contains significant contributions from $L = 4$ pulses (four range ambiguities). Then the clutter signature in each range bin will reach a steady-state condition only for the fourth and subsequent pulses in a CPI. This issue was first discussed in Chap. 4.

Steady-state operation of the digital filters used for MTI processing occurs when the output value depends only on actual data input values rather than any initial (typically zero-valued) samples used to initialize the processing. For FIR filters of length N, the first and last $N - 1$ samples of the complete convolution are transient in the sense that the filter impulse response does not fully overlap the finite data sequence. These transient output samples are often discarded.

These two effects are independent. To see how many pulses are needed in total to obtain an M-point non-transient steady-state sequence $y_{ss}[m]$, consider Fig. 5.37. Assume P total pulses are transmitted. This sketch assumes $L = 4$ range ambiguities and a three-pulse canceller ($N = 3$) MTI filter but is labeled for general L and N. The notional data sequence $y[m]$ is shown as ramping up in amplitude over the first L samples. While actual data would vary unpredictably depending on the clutter scenario, this represents the increasing number of range ambiguities present in each sample, stabilizing at four when $m = 3$ (the fourth sample). Recall that the convolution of h and y is given by $y_{ss}[n] = \sum y[m]h[n - m]$. The three-sample sequence in the figure represents the three-pulse canceller filter coefficients $h[n - m]$. (The actual coefficient values would be $\{+1, -2, +1\}$.) It can be seen from the figure that in general the first value of n for which the filter coefficients will overlap only with steady-state measured data occurs when $L - 1 = n - N + 1$, that is, $n = L + N - 2$ ($n = 5$ in this example). The last value for which this is true occurs when $n = P - 1$. The number of output samples in this interval is $M = (P - 1) - (L + N - 2) + 1$. Therefore, $P = M + L + N - 2$ pulses are needed to obtain M valid outputs for further processing.

For example, suppose 20 valid stationary pulses are needed for the pulse Doppler DFT. Also suppose a three-pulse canceller ($N = 3$) is used and the unambiguous range and radar sensitivity are such that $L = 4$ range ambiguities are present in the measured data. Then the CPI should collect 25 pulses of data, discard the first five outputs of the MTI filter (three for the range ambiguity buildup and two for the filter transient) and pass only the last 20 outputs to the pulse Doppler DFT or other processing. Additional pulses may be used to set the automatic gain control of the receiver and are also not used for Doppler processing.

5.6.4 PRF Regimes

As was seen in Chap. 4, measurements made with a pulse burst waveform can be ambiguous in range, Doppler, or both. Pulse Doppler radars in particular frequently operate in

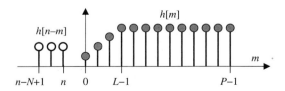

FIGURE 5.37 Determination of the relationship between number of range ambiguities L, MTI filter length N, and number of steady-state slow-time samples P needed to obtain M non-transient steady-state samples.

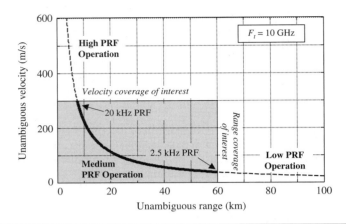

FIGURE 5.38 Low, medium, and high PRF regimes for a notional X-band radar.

scenarios that are ambiguous in one or both of the range and Doppler dimensions. Modern airborne pulse Doppler radars operate in a dizzying variety of modes having various range and Doppler span and resolution requirements. Pulse burst waveforms using a variety of constituent pulses, including simple pulses, LFM, and Barker phase codes at a minimum, are common. To meet the various mode requirements, PRFs ranging from several hundred hertz to 100 kHz or more are used.[11]

Pulse Doppler radar operation is commonly divided into three *PRF regimes* according to their ambiguity characteristics. Many airborne radars operate in all three regimes, depending on the requirements of the moment. The dividing lines are not standardized but depend on the mission requirements. Given an unambiguous range R_{ua} and unambiguous velocity interval $[-v_{ua}/2, +v_{ua}/2]$ of interest, the radar is considered to be in a *low PRF mode* if the PRF is sufficiently low to be unambiguous in range over the interval of interest, but is ambiguous in velocity, meaning targets of interest have velocities outside the range $\pm v_{ua}/2$. The *high PRF mode* is the opposite: the system is ambiguous in range but not in velocity. In a *medium PRF mode*, the radar is ambiguous in both. This tradeoff is summarized in Fig. 5.38. The line plots the achievable combinations of R_{ua} and v_{ua} at 10 GHz.

Suppose the desired range and velocity coverage are 60 km and 300 m/s (±150 m/s), respectively. The range limit might be set by the expected maximum detection range of the radar, while the velocity limit might be set by the maximum expected relative velocity of targets of interest. The shaded area indicates a range of PRFs (in this case, 2.5 to 20 kHz) that will result in ambiguities in both dimensions. PRFs below 2.5 kHz will be ambiguous in Doppler but not in range; those from 2.5 to 20 kHz will be ambiguous in both; and those above 20 kHz will be ambiguous in range but not in Doppler. Here, an "ambiguity" means a range ambiguity within the 0- to 60-km range interval of interest and/or a velocity ambiguity within the ±150 m/s velocity interval of interest.

The choice of PRF regime has major effects on the range-Doppler target and clutter spectrum. Consider again the radar and clutter scenario associated with Fig. 5.4 and suppose the maximum range and velocity intervals of interest are 75 km and ±150 m/s. Now consider viewing the same scenario with low, medium, and high PRFs of 2, 10, and 30 kHz, respectively.

---

[11] Again, pulsed waveform terminology is used in this section because these issues arise primarily in pulsed systems.

PRF, kHz	PRF Regime	Unambiguous, Range km	Unambiguous Doppler Shift Interval, kHz	Unambiguous Velocity Interval, m/s
2	Low	75	±1	±15
10	Medium	15	±5	±75
30	High	5	±15	±225

TABLE 5.2 Unambiguous Range, Doppler, and Velocity Intervals for Selected PRFs and a 10-GHz RF

Table 5.2 lists the unambiguous range, Doppler, and velocity intervals for each of these PRFs for reference.

Figure 5.39 compares the unaliased range-Doppler clutter power distribution of Fig. 5.4b to the aliased versions at each of these PRFs. In all cases the clutter echo out to a range of 75 km has been included in the computation. Figure 5.39a repeats the unaliased distribution for convenient comparison. The MLC is centered at 17.3 km and 93.4 m/s as noted earlier. The sidelobe clutter (SLC) spreads over ±134.1 m/s but is mostly confined to ranges of 15 km or less.

Figure 5.39b is the low PRF distribution. There is no clutter present in the first 3 km of range, corresponding to the platform altitude. The MLC is centered at its unambiguous range of 17.3 km, but the velocity has aliased from its actual value of 93.4 m/s to an ambiguous velocity of 3.4 m/s (93.4 − 3 × 30). The SLC is heavily aliased in velocity but fades with range so that most targets do not compete with it. If MTI filtering is applied to suppress the MLC there will be blind speeds every 30 m/s in the affected range bins.

Figure 5.39c is the medium PRF case. The MLC is ambiguous in both range and Doppler, having aliased to a range of 2.3 km (17.3 − 15) and a velocity of −56.6 m/s (93.4 − 150). Notice that the MLC and its sidelobes now wrap around in the range dimension and that the SLC wraps around in Doppler. SLC is now present at essentially all ranges and Dopplers, though in varying amounts and patterns in different range cells. The SLC also wraps in range, but this is less evident.

Figure 5.39d is the high PRF case. The MLC again wraps to the ambiguous range of 2.3 km (17.3 − 3 × 5) but is spread fairly uniformly throughout the short 5 km unambiguous range. It is located at its unambiguous velocity of 93.4 m/s. The narrow spread of the MLC in velocity allows it to be filtered out with a relatively narrow MTI or other notch filter with little risk of filtering out moving targets. The clutter out to 75 km has folded over 15 times to "fit" into the 5-km unambiguous range at this PRF. There is now significant and relatively constant SLC at all ranges, though the AL and other near-in clutter is still visible beginning at just over 2 km. On the other hand, the radar is now unambiguous in Doppler and the full SLC spread in velocity of ±134.1 m/s can be seen. In addition, there is now a clear region in the Doppler spectrum for positive and negative velocities having a magnitude between 134.1 and 225 m/s that was not present in the other figures, enabling noise-limited detection of targets at these high relative velocities.

Table 5.3 summarizes the major strengths and weaknesses of low, medium, and high PRF operation, especially from the viewpoint of an airborne radar. Broadly speaking, low PRF modes are very effective for ranging, mapping, and imaging modes but poor at detection of moving targets due to the lack of a sizable clear region. High PRF operation is complementary to low PRF operation in both its strengths and weaknesses. High PRF modes are good for detection of high-Doppler shift targets (e.g., rapidly closing aircraft or missiles) in high clutter due to the large clear region but poor at detection of low-Doppler targets

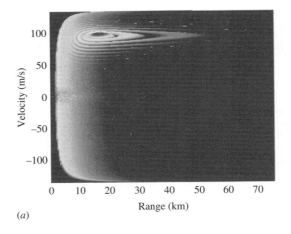

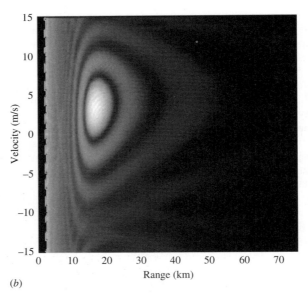

Figure 5.39 Effect of low, medium, and high PRFs of Table 5.2 on the ground clutter range-Doppler distribution. (*a*) Unaliased spectrum. (*b*) Viewed with a 2-kHz PRF. (*c*) Viewed with a 10-kHz PRF. (*d*) Viewed with a 30-kHz PRF. See text for details.

(slow-moving closing targets or opening targets) due to high sidelobe clutter and little or no range gating capability. Medium PRF operation is a compromise that retains most of the strengths of each without the inherited weaknesses becoming too severe. In-depth discussion of the properties and processing for all three regimes is given in Alabaster (2012), Morris and Harkness (1996), and Stimson (2014).

5.6.5 PRF Selection

The problem of picking the particular PRFs to be used in a given PRF regime and scenario remains. For a specified range-Doppler space, the general goals are to maximize the clear

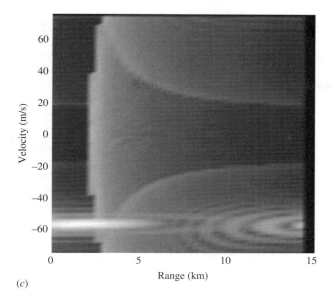

(c)

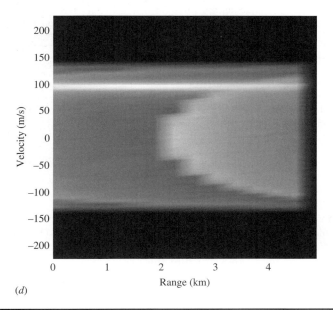

(d)

FIGURE 5.39 *(Continued)*

percentage of the space to provide high detectability, be capable of resolving range and/or Doppler ambiguities for a reasonable number of targets to support tracking, and minimize ghosting (to be discussed shortly) to avoid false detections and track, again for a reasonable number of targets. PRF selection can be particularly daunting for the medium PRF mode.

Assuming an *M*-of-*N* detection approach as described in Sec. 5.4.8, the first issue is choosing M and N. Suppose the maximum range and Doppler of interest are R_{max} and $F_{D\text{max}}$.

PRF Regime	Advantages	Disadvantages
Low	• No range ambiguities • Precise range measurement • Fine range resolution possible • Sidelobe clutter rejection via range gating • Simple processing	• Existence of blind speeds • Highly ambiguous Doppler, esp. at higher RFs • Poor detection performance in look-down modes • High peak power or pulse compression required
Medium	• Good detection over wide range of target Dopplers • Good rejection of both mainlobe and sidelobe clutter • Accurate ranging • Reduced eclipsing compared to high PRF operation	• Sidelobe clutter at all velocities • Large number of PRF and pulse width combinations • Complex range and Doppler ambiguity resolution processing • Poor performance for large targets in sidelobes
High	• High average power • Unambiguous Doppler • No blind speeds • Mainlobe clutter rejection without rejecting targets	• Highly ambiguous range • Increased eclipsing of targets • Complex, reduced-accuracy ranging • Reduced sensitivity to low-Doppler targets due to sidelobe clutter

TABLE 5.3 Partial List of Advantages and Disadvantages of PRF Regimes for Airborne Radar

It can be shown that the set of PRFs used for disambiguating range and velocity must satisfy the "decodability constraints" (Alabaster, 2012)

$$\operatorname{lcm}(PRI_1,\ldots,PRI_M) \geq \frac{2R_{\max}}{c} \quad \text{(range decodability constraint)},$$
$$\operatorname{lcm}(PRF_1,\ldots,PRF_M) \geq F_{D\max} \quad \text{(Doppler decodability constraint)} \tag{5.119}$$

These constraints ensure that the ambiguity resolution algorithms discussed next can provide a unique solution within that coverage area. Equation (5.119) must hold true for all $N!/[M!(N-M)!]$ selections of M of the N PRFs. For a 2-of-5 structure, there are 10 combinations to be tested; for a 3-of-8 structure, this rises to 56!

The fewer PRFs are used, the fewer decodability constraints must be met; this suggests using only two PRFs. However, as a practical matter, the range and Doppler coordinates can be disambiguated more reliably if three or more PRFs are used (Alabaster, 2012). M is therefore often chosen to be three. Turning to the choice of N, larger values increase the probability that at least M PRFs will be clear at any given range-Doppler coordinates, increasing the clear percentage. However, using more CPIs in a dwell reduces the time available per CPI, degrading Doppler resolution and reducing processing range. Alternatively, the dwell time can be lengthened at the cost of increasing search time for a given spatial region. A "3-of-8" rule is a common choice in airborne radar.

There are often additional constraints that must be satisfied by the PRFs. The minimum desired percentage of the Doppler spectrum that is outside the MLC region sets a minimum PRF. The maximum allowable duty cycle of the radar transmitter and the pulse length (derived from the required range resolution when a simple pulse is used) establish a maximum PRF. Candidate PRFs are often adjusted so that there are an integer number of range bins within the unambiguous range for each PRI; this eases ambiguity resolution but raises additional issues concerning duty cycle and range resolution variability.

A variety of techniques have been used to search for sets of particular PRF values that result in good blind zone maps within bounds on the maximum and minimum PRF. One traditional technique is the "major-minor method." It begins by picking at least two and more likely three "major" PRFs in the regime of interest. The PRFs are separated by at least the MLC Doppler width β_{MLC} to guarantee that the first repetitions of the MLC portion of the Doppler spectrum at each PRF do not overlap, thus making detection presumably possible on at least one of the PRFs for targets outside of the zero-Doppler clutter region. For each major PRF, two "minor" PRFs are then selected for use in range ambiguity resolution. This technique results in sets of six or nine PRFs in total. Another method considers the entire set of N PRFs, choosing values that are stepped by at least $\beta_{MLC}/(N - M)$ Hz. This choice ensures that the first Doppler blind zones for each PRF do not overlap for more than $N - M$ PRFs, meaning that any target Doppler shift in that vicinity will be in the clear on at least M PRFs.

Generally, the best PRF sets are obtained in modern systems using advanced constrained search algorithms rather than by ad hoc methods. Figure 5.30 was a particularly good example. Extensive discussion of PRF selection considerations and examples with an emphasis on airborne multimode radars is given in Alabaster (2012).

PRF selection may be less complex in radars with less varied missions and less complicated environments. For example, weather radars typically map weather conditions in their surrounding region using smaller and less complex PRF sets. In weather radar, volume clutter is the target of interest. Ground clutter is minimal because the system is not generally down-looking (although airborne weather radars may be). Weather radars typically require relatively long unambiguous ranges and therefore low PRFs to provide adequate area coverage. The WSR-88D radar used by the U.S. National Weather Service for long-range weather observation uses PRFs of 322 to 1282 Hz, giving unambiguous ranges of 466 to 117 km. The RF is approximately 3 GHz, so the corresponding unambiguous velocity intervals are ±8 m/s (about ±18 mph) and ±32 m/s (about ±72 mph). These are well short of the approximately ±100 mph velocity interval considered adequate by meteorologists. Consequently, ambiguous windspeed measurements are a basic problem in weather radar. Because weather radars are interested in continuous reflectivity fields (wind flows, storm cells, etc.) distributed in three dimensions instead of discrete targets, they can take advantage of special ambiguity resolution methods that rely on continuity of the measured velocities and other special features of the measured data; some of these techniques are described in Doviak and Zrnić (1993). Another approach combines pulse pair processing with two staggered PRFs to compute two autocorrelation values that can be combined to extend the unambiguous range. Another novel technique applies a pulse-to-pulse phase code similar to the LFM-like P3 and P4 codes discussed in Chap 4. Correlating with specific elements of the code on receive can emphasize a particular range foldover region at the expense of the others. These last two techniques are discussed in Zrnić (2008).

5.6.6 Ambiguity Resolution

Several techniques exist to resolve range and Doppler ambiguities when multiple-PRF data is available. Consider range ambiguity resolution first. Once a PRF is selected, it establishes an unambiguous range $R_{ua} = c/2PRF$. A target at an actual range $R_t > R_{ua}$ will be detected at an apparent range R_a that satisfies

$$R_t = R_a + kR_{ua} \text{ m} \qquad (5.120)$$

for some integer k. Equivalently

$$R_a = ((R_t))_{R_{ua}} \text{ m} \tag{5.121}$$

where the notation $((\cdot))_x$ denotes modulo x. Normalize the range measurements to the range bin spacing δR; for example, $n_a = R_a/\delta R$. Also suppose there are N range bins in the unambiguous range interval, $N = R_{ua}/\delta R$. Then Eqs. (5.120) and (5.121) become

$$n_t = n_a + kN \Rightarrow n_a = ((n_t))_N \tag{5.122}$$

The basic approach to resolving range ambiguities relies on multiple PRFs. Consider a set of r PRFs, $\{PRF_0, PRF_1, \ldots, PRF_{r-1}\}$. Assume the range bin spacing (range sampling interval) δR is kept the same for each PRF. The unambiguous range R_{ua_i} and number of range bins N_i must then satisfy $R_{ua_i} = N_i \cdot \delta R$, so that the number of range bins is different for each PRF. Then the actual target range bin and its apparent (aliased) range bin for the various PRFs must satisfy the equations

$$\begin{aligned} n_t &= n_{a_0} + k_0 N_0 = n_{a_1} + k_1 N_1 = \cdots = n_{a_{r-1}} + k_{r-1} N_{r-1} \Rightarrow \\ n_{a_i} &= ((n_t))_{N_i}, \quad 0 \leq i \leq r-1 \end{aligned} \tag{5.123}$$

The set of equations in Eq. (5.123) is called a set of *congruences*.

The set of congruences can be solved using the *Chinese remainder theorem* (CRT) (Trunk and Brockett, 1993). The CRT states that given a set of r relatively prime integers $N_0, N_1, \ldots, N_{r-1}$ and the set of congruences in Eq. (5.123), there exists a unique solution (modulo $N = N_0 N_1 \cdots N_{r-1}$) for n_t given by the equations

$$\begin{aligned} n_t &= k_0 \beta_0 n_{a_0} + k_1 \beta_1 n_{a_1} + \cdots + k_{r-1} \beta_{r-1} n_{a_{r-1}}, \\ k_i &= N/N_i = \prod_{j=0, j \neq i}^{r-1} N_j, \quad \beta_i = ((k_i^{-1}))_{N_i} \Rightarrow ((\beta_i k_i))_{N_i} = 1 \end{aligned} \tag{5.124}$$

To make the procedure clearer, consider the case of $r = 3$ PRFs. Then n_t satisfies

$$n_t = ((\alpha_0 n_{a_0} + \alpha_1 n_{a_1} + \alpha_2 n_{a_2}))_{N_0 N_1 N_2} \tag{5.125}$$

where

$$\alpha_i = \beta_i k_i = \beta_i \prod_{\substack{j=0 \\ j \neq i}}^{2} N_j \tag{5.126}$$

(for example, $\alpha_1 = \beta_1 N_0 N_2$) and the β_i are the smallest integers such that

$$((\beta_0 N_1 N_2))_{N_0} = 1, \quad ((\beta_1 N_0 N_2))_{N_1} = 1, \quad ((\beta_2 N_0 N_1))_{N_2} = 1 \tag{5.127}$$

Now suppose that the true range of a target, normalized to the range bin size, is $n_t = 19$. Further suppose that the three PRFs are chosen such that the number of range bins in the unambiguous range for each PRF are $N_0 = 11$, $N_1 = 12$, and $N_2 = 13$. On the first PRF, the target will be detected in the apparent range bin $n_{a_0} = ((19))_{11} = 8$. Similarly, $n_{a_1} = ((19))_{12} = 7$ and $n_{a_2} = ((19))_{13} = 6$. From Eq. (5.127), β_0 is the smallest integer that satisfies $((\beta_0 \times 12 \times 13))_{11} = 1$; that is, $156\beta_0 = 11k + 1$ for some integer k. The solution is $\beta_0 = 6$. In the same manner it is found that $\beta_1 = 11$ and $\beta_2 = 7$. Equation (5.126) then gives $\alpha_0 = 6 \times 12 \times 13 = 936$, $\alpha_1 = 1573$, and $\alpha_2 = 924$. Finally, Eq. (5.125) gives the estimate of the true range bin as

$$\hat{n}_t = ((\alpha_0 n_{a_0} + \alpha_1 n_{a_1} + \alpha_2 n_{a_2}))_{N_0 N_1 N_2} = 19 \qquad (5.128)$$

which is the correct result.

A serious problem with the CRT is its extreme sensitivity to errors induced by noise and range quantization. There is no guarantee that the actual range R_t will be an integer multiple of the range bin spacing δR as assumed previously; the target may in fact straddle range bins. In addition, noise in the measurements may cause the target to be located in an incorrect range bin. To illustrate the effect of such errors, repeat the previous example but suppose that n_{a_2} is for some reason measured to be 7 instead of the correct value of 6. Carrying out the previous calculations will give $\hat{n}_t = 943$ instead of 19, a huge error.

A "robust CRT" that controls the maximum error is given in Li et al. (2010). However, it is more common to introduce the *coincidence algorithm* for determining n_t. This technique is essentially a graphical implementation of the CRT (Hovanessian, 1976; Morris and Harkness, 1996; Alabaster, 2012). The method is best illustrated with an example. Again presume that $r = 3$ PRFs are used. Suppose that there are two targets "a" and "b" with true ranges corresponding to range bins $n_a = 6$ and $n_b = 11$. Further suppose that PRFs are such that the number of range bins in each unambiguous range interval are $N_0 = 7$, $N_1 = 8$, and $N_2 = 9$. This means that the first target is actually unambiguous at each PRF, while the second is ambiguous at each PRF. The measured data will be

$$\begin{aligned} n_{a_0} = n_{a_1} = n_{a_2} = 6 \\ n_{b_0} = 4, \quad n_{b_1} = 3, \quad n_{b_2} = 2 \end{aligned} \qquad (5.129)$$

This measurement scenario is illustrated in Fig. 5.40.

The graphical technique proceeds by taking the pattern of detections at each PRF and replicating it, as shown in Fig. 5.41. In essence, the replication implements the first version of Eq. (5.122), placing a detection at each value of $n_a + kN_0$ and $n_b + kN_0$ within the maximum detection range of the radar. These detections represent the plausible ranges for each target at each PRF. The algorithm then searches for a range bin that exhibits a detection at all three PRFs, indicating that that range bin is consistent with the measurements at all three PRFs. As shown in Fig. 5.41, this process correctly detects the true range bins $n_a = 6$ and $n_b = 11$ in this example.

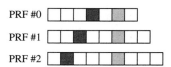

Figure 5.40 Notional measured data for illustrating coincidence algorithm for range ambiguity resolution.

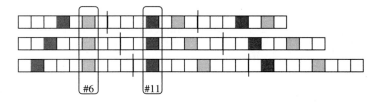

Figure 5.41 Coincidence detection of target ranges in replicated range data.

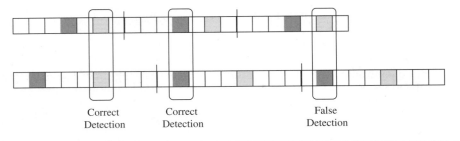

FIGURE 5.42 Formation of ghosts in range ambiguity resolution.

The graphical interpretation suggests various methods to reduce the sensitivity of the CRT to measurement errors. In one approach, exact coincidence is not required to declare a target. Instead, a tolerance N_T is established and a detection is declared if a detection occurs in all three PRFs at some range bin $n_t \pm N_T$. Depending on the range bin size and SNR, N_T will typically be only 1 or 2 range bins. A more sophisticated version of this basic idea is described in Trunk and Kim (1994). Their method combines a systematic approach to clustering plausible ranges from each PRF into candidate target ranges with a maximum likelihood calculation to recognize multiple target situations.

In the last example, three PRFs proved sufficient to resolve two different range-ambiguous targets. In general, N PRFs are required to successfully disambiguate $N-1$ targets. If the number of targets exceeds $N-1$, *ghosts* can appear (Morris and Harkness, 1996). Ghosts are false targets resulting from false coincidences of range-ambiguous data from different targets. The problem is illustrated in Fig. 5.42, which repeats the example of Fig. 5.41 using only two of the previous three PRFs. While targets will still be detected at the correct bins $n_a = 6$ and $n_b = 11$, a third coincidence occurs between detections from targets 1 and 2 at range bin $n_c = 20$, representing an apparent third target. Unless additional data such as tracking information is available, the signal processor has no way of recognizing that the last coincidence is among detections from different targets. Thus, the processor will declare the presence of three targets in this example, the two correct targets and one "ghost." Use of a third PRF, as in Fig. 5.41, eliminates this ghost.

Ghosting can also be caused by noise- or clutter-induced false alarms that happen to coincide with legitimate detections. More extensive discussion of ghosting in range and Doppler and of ghosting due to false alarms is given in Alabaster (2012), who also introduces several de-ghosting approaches.

In a medium or high PRF mode the radar may also suffer velocity ambiguities. This problem is identical to that of range ambiguities: given an apparent Doppler shift F_D, the actual Doppler shift must be of the form $F_D + k \cdot PRF$ for some integer k. Use of the DFT for spectral estimation results in quantization of the Doppler spectrum into Doppler bins (equivalently, velocity bins), analogous to range bins in the range dimension. The same techniques used for range disambiguation can therefore be used to resolve velocity ambiguities as well.

The coincidence algorithm can be readily extended to simultaneously disambiguate detections in both range and Doppler. Other extensions to deal with radar systems that do not use the same range resolution in each PRF (typically so as to maintain constant duty cycle) or the same velocity resolution, as well as references to other disambiguation algorithms are given in Alabaster (2012). Another approach using the emerging technique of sparse reconstruction has been proposed in Shaban and Richards (2013). It also works in range and Doppler simultaneously and appears tolerant of reasonable numbers of missed detections due to blind zones or other causes, false alarms, and other data inconsistencies.

5.7 Clutter Mapping

All of the MTI and pulse Doppler processing discussed so far has been focused on reducing the clutter power that interferes with the signature of a moving target so as to improve the SIR and ultimately the probability of detection. These techniques are not effective for targets with little or no Doppler shift, and that therefore are not separable from the clutter based on Doppler shift. *Clutter mapping* is a technique for detection of moving targets with zero or very low Doppler shift by stationary radars. It is intended for maintaining detection of targets on crossing paths, that is, passing orthogonal to the radar line of sight so that the radial velocity is zero; such targets are discarded by standard MTI and pulse Doppler processing. Clutter mapping can be effective if the target radar cross section (RCS) is relatively large and the competing clutter is relatively weak as depicted in Fig. 5.43. This situation can arise, for instance, in a ground-based air surveillance radar where the antenna is typically tilted upward. Then mainlobe ground clutter is not present to compete with the target echo (though weather clutter may be); the clutter is primarily from the sidelobes.

The concept of clutter mapping is shown in Fig. 5.44, which presumes that conventional pulse Doppler processing is applied to targets in the clear region of the Doppler spectrum. The output of the zero-Doppler bin and others in the clutter region is used to create a stored map of recent clutter echo power for each range-azimuth cell in the radar's search area. This map is updated continuously to adapt to clutter power variations due to weather and other environmental changes. On each scan, the received power in the clear region Doppler bins is tested against a conventional threshold based on the noise that dominates the interference in those bins. Instead of being discarded, the current received power in the clutter region Doppler bins for each range-azimuth cell is applied to a separate detector using a threshold based on the current stored clutter power estimate for that cell. The clutter map procedure is a form of *constant false alarm rate* (CFAR) detection but with the interference power estimated by averaging in time instead of in space. The details of threshold detection and CFAR are discussed in Chap. 6.

Instead of using the zero-Doppler output of a pulse Doppler processor (typically a DFT of the slow-time data), some clutter mapping systems pass the I/Q slow-time data through a separate "zero velocity" or "zero Doppler" filter, as shown in Fig. 5.45. The zero-velocity filter serves the opposite purpose of an MTI filter. It is a lowpass design, the output of which consists only of ground clutter and crossing target returns. The design of the zero-velocity filter can be optimized for the clutter environment at a specific radar site and can also be made adaptive to clutter changes, for instance due to changing weather in the area.

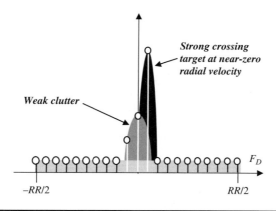

FIGURE 5.43 Pulse Doppler spectrum for a large RCS crossing target in weak clutter.

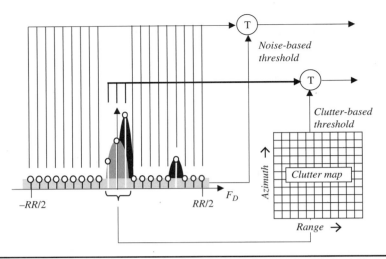

FIGURE 5.44 The concept of clutter mapping for detection of strong targets in clutter.

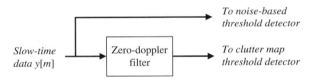

FIGURE 5.45 Zero-Doppler filter used to isolate low-Doppler targets and ground clutter.

5.8 The Moving Target Detector

The *moving target detector* (MTD) is a term applied to the Doppler processing system used in many airport surveillance radars. It is a relatively simple example of a system for stationary radars that combines most of the techniques discussed previously and others to achieve good overall moving target detection performance. Specifically, it incorporates MTI pulse cancellers, staggered PRFs, pulse Doppler spectrum formation and processing, Doppler domain windowing for sidelobe control, clutter mapping, and CFAR threshold detection.

A block diagram of the original MTD is shown in Fig. 5.46 (Nathanson, 1991). The upper channel begins with a standard MTI three-pulse canceller. The clutter-cancelled output is then applied to an eight-point FFT for pulse Doppler analysis. Two PRFs are used in a CPI-to-CPI stagger to extend the unambiguous velocity region. The "frequency domain weighting" is a computationally efficient implementation of time-domain windowing of the data for Doppler sidelobe control.[12] The individual FFT samples are then applied to a 16-range-bin cell-averaging CFAR threshold detector (to be discussed in Chap. 6) with thresholds selected separately for each frequency bin.

---

[12] Certain windows, the Hamming used in the MTD among them, can be most efficiently implemented as a convolution in the frequency domain with a three-point kernel instead of the usual time-domain data weighting.

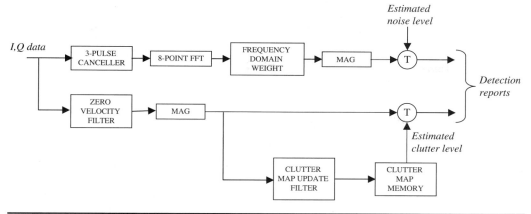

FIGURE 5.46 Block diagram of an early complete "moving target detector" system.

To provide some detection capability for crossing targets, the lower channel uses a site-specific zero-velocity filter to isolate the echo from clutter and low-Doppler targets. The output is applied to a clutter map threshold detector. The original MTD updated the clutter map using an eight-scan moving average, corresponding to 32 seconds of data history (Skolnik, 2001); however, a moving average or other mean estimator could also be used, as could other CFAR interference estimating methods.

The MTD design has progressed through several generations since the early implementation described here. Versions used in the ASR-9 and ASR-12 airport surveillance radars are described respectively in Taylor and Brunins (1985) and Cole et al. (1998). Additional discussion of the design and performance of the Doppler filterbank and zero-velocity filters is given in Shrader and Gregers-Hansen (2008).

5.9 MTI for Moving Platforms: Ground Moving Target Indication

MTI filtering and pulse Doppler processing provide an effective way to detect moving targets whose Doppler shift is in the clear region of the spectrum on at least one PRF. Airborne targets can generally be detected in this manner. However, slow-moving ground targets having actual Doppler shifts only slightly higher than the ground clutter or less will appear within the clutter spectrum or on its skirts at all PRFs and are therefore very difficult to detect. Recall that platform motion can substantially spread the ground clutter spectrum as described in Eq. (5.73). This spread exacerbates the problem, raising the minimum velocity at which slow-moving ground targets can be detected. The phenomenon is illustrated in Fig. 5.47, which shows the spreading of the MLC by the platform motion after the change in Doppler center frequency has been removed. Because of this spreading, clutter energy may compete directly with relatively slow-moving targets, typically surface targets such as vehicles on land and ships on the sea, making MTI processing less effective and detection difficult. Processing techniques intended to detect such "slow movers" from airborne or spaceborne platforms are referred to as *ground moving target indication* (GMTI) or *surface moving target indication* (SMTI).

There are a variety of GMTI techniques in use. Three introduced here are *displaced phase center antenna* (DPCA) processing, *along-track interferometry* (ATI), and *clutter suppression interferometry* (CSI). First consider DPCA and ATI. In their basic forms each requires an array antenna subdivided into two subapertures on receive. Each can operate in either the slow

FIGURE 5.47 Illustration of the effect of a moving radar platform on the Doppler spectrum and the detection of "slow movers." The change in Doppler center frequency has been removed.

time or Doppler domain, though normally the Doppler domain is preferred.[13] Each begins by aligning the data from the receive apertures in space and time, countering platform-induced clutter spectral spreading to improve the probability of detection for slow-moving targets. They differ in that DPCA applies an MTI-like pulse canceller to the time-aligned data, suppressing clutter and detecting moving targets based on amplitude, while ATI rejects clutter and detects targets based on interferometer-like phase difference calculations (Deming et al., 2014). CSI is a sequential combination of these two techniques using three antenna subapertures. DPCA is performed first on data from the first and second subapertures, and independently on data from the second and third apertures. ATI is then applied to the two resulting clutter-cancelled data streams to detect the ground movers.

5.9.1 Simplified GMTI Clutter and Target Models

GMTI processing attempts to compensate for platform motion by using multiple receive subapertures to create carefully controlled multiple virtual phase centers such that data received on one subaperture have the same virtual phase center as the data received on a *different* subaperture some time later. These two data streams can then be treated as if they were both collected from a stationary antenna at the *same* location and combined in a pulse canceller (DPCA) or subjected to a phase analysis (ATI) to detect moving targets.

Figure 5.48 illustrates the approach using an electronic antenna having two subapertures. A sidelooking configuration is assumed for simplicity, but GMTI is not restricted to this case. The entire antenna is used on transmission for maximum gain, so the phase center for transmission is the point **T** in the middle of the antenna. Each half of the antenna has its own independent receiver so there are two receive subapertures having respective phase centers **R1** and **R2**, each d_{pc} meters from the transmit phase center.

Suppose the antenna position is such that **T** is at x coordinate x_0, placing **R1** at $x_0 + d_{pc}$. Transmit a pulse using the full antenna and consider the data received on the fore subaperture. Recalling the concept of a virtual element from Sec. 1.3.4, the data received will be equivalent to the data that would be obtained by both transmitting and receiving at the **T-R1** path virtual element (VE) location $x_0 + d_{pc}/2$, halfway between the actual transmit and receive phase centers.

The idea of GMTI is to achieve effective clutter rejection by combining measurements made from the *same* phase center location in space, like a stationary MTI radar. Because the platform is moving forward, a second measurement from the **T-R1** VE location can be obtained when the **T-R2** VE passes through the same coordinate. If the pulse repetition

---

[13]Each can also be applied to synthetic aperture radar images; SAR GMTI is not further pursued in this text.

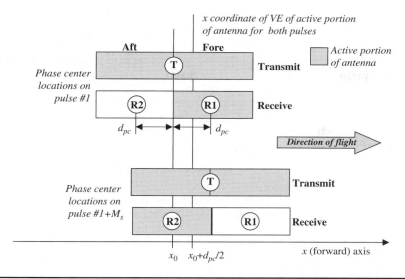

FIGURE 5.48 Relationship of transmit and receive aperture phase centers in DPCA processing.

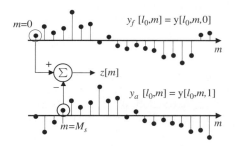

FIGURE 5.49 Illustration of two-pulse cancellation across two received data streams in DPCA for a time slip of approximately $M_s = 3$ PRIs.

interval is T and the platform velocity is v, this will occur d_{pc}/v seconds later. The required shift in pulses of the fore and aft data with respect to one another is then

$$M_s = \frac{d_{pc}}{vT} \text{ pulses} \qquad (5.130)$$

M_s is called the *time slip* or *time shift*. Equation (5.130) is known in the GMTI literature as the "DPCA condition." To the extent that the DPCA condition is met, the clutter component of the slow-time sequence in a given range bin observed at the output of the fore subaperture should be similar to that observed from the aft aperture M_s samples later, as suggested by Fig. 5.49 where $M_s = 3$. More precisely, the fore and aft subaperture clutter data sequences should be highly correlated except for a time shift of M_s samples.

In general, M_s will not be integer. For example, if $d_{pc} = 3$ m, $v = 200$ m/s and $T = 2$ μs, then $M_s = 3.75$ pulses. A typical GMTI implementation may continuously adjust platform velocity v and PRI T (d_{pc} is fixed once the antenna is built) to keep M_s as constant as possible and preferably close to an integer, but other constraints on v and T and a host of other issues make that difficult. If DPCA is being implemented in the Doppler domain, fractional-PRI time slips can be implemented by applying an appropriate linear phase slope to the Doppler spectrum. In practice additional adaptive processing is used to maximize correlation of the subaperture outputs.

To begin developing a simplified model for GMTI processing, transmit a CPI of $M + M_s$ pulses and consider a single range bin. Denote the clutter components of the slow-time data in the fore and aft receive apertures as $c_f[m]$ and $c_a[m]$, respectively. Assume the actual time slip is M_s pulses and that clutter fluctuations are fairly minimal over M_s PRIs. Then

$$c_a[m] \approx \alpha \cdot c_f[m - M_s] \tag{5.131}$$

The complex constant α allows for some degree of gain and phase mismatch between the receive subapertures. This equation applies to all of the range bins, but it is sufficient to model just one.

Processing is usually performed in the range-Doppler domain instead of fast time/slow time. Taking the DTFT of Eq. (5.131) gives a Doppler-domain model of the relationship between the fore and aft clutter signals:

$$C_a(\omega) \approx \alpha \cdot \exp(-j\omega M_s) C_f(\omega) \tag{5.132}$$

The signature of a moving target with complex amplitude A_t and normalized radian Doppler frequency ω_D in a particular range bin of the fore subaperture data will be

$$t_f[m] = A_t \exp(j\omega_D m) \tag{5.133}$$

$T_f(\omega)$ would take the form of an asinc function like that of Eq. (5.75) but converted to normalized frequency units. The signal observed from the same location in the aft subaperture data M_s PRIs later will be approximately the same as $t_f[m]$ but with the addition of a Doppler phase shift due to the target motion during the time slip:

$$\begin{aligned} t_a[m + M_s] &\approx \alpha \cdot \exp(j\omega_D M_s) t_f[m] \Rightarrow \\ t_a[m] &\approx \alpha \cdot \exp(j\omega_D M_s) t_f[m - M_s] \end{aligned} \tag{5.134}$$

The Doppler domain equivalent is

$$T_a(\omega) = \alpha \cdot \exp(-j\omega M_s) \exp(+j\omega_D M_s) T_f(\omega) \tag{5.135}$$

5.9.2 DPCA and ATI

Both DPCA and ATI start with Eqs. (5.132) and (5.135). Similar to MTI radars, DPCA seeks to suppress the clutter using the two subaperture data streams as inputs to a two-pulse canceller (Staudaher, 1990; Skolnik, 2001). It is a special case of the more general *space-time adaptive processing* (STAP). The connection between the two will be developed in Chap. 9.

Assume the Doppler spectrum in a particular range bin from the fore subaperture is composed of clutter, target, and noise components, $Y_f(\omega) = C_f(\omega) + T_f(\omega) + N_f(\omega)$, and similarly for the aft subaperture. Based on the clutter and target models above, the DPCA processor forms the new signal

$$Z(\omega) = Y_f(\omega) - \alpha^{-1} \cdot \exp(+j\omega M_s) Y_a(\omega) \tag{5.136}$$

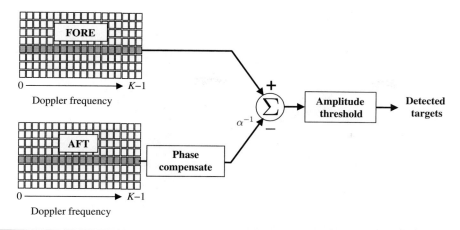

Figure 5.50 Processing flow for DPCA amplitude-based clutter rejection in the range-Doppler domain.

as shown in Fig. 5.50. The output clutter component will be

$$Z_c(\omega) = C_f(\omega) - \alpha^{-1} \cdot \exp(+j\omega M_s)C_a(\omega)$$
$$\approx C_f(\omega) - C_f(\omega) \approx 0 \tag{5.137}$$

To the extent that Eq. (5.132) holds and α and M_s can be estimated accurately, DPCA will cancel the clutter signal.

Applying the same operation in a range bin containing a target component gives the output target signal component

$$Z_t(\omega) = T_f(\omega) - \alpha^{-1} \cdot \exp(+j\omega M_s)T_a(\omega)$$
$$\approx [1 - \exp(+j\omega_D M_s)]T_f(\omega) \tag{5.138}$$

For a moving target, the magnitude of the target component will not be near zero in general but instead is proportional to $|1 - \exp(+j\omega_D M_s)|$ and therefore depends on the target Doppler frequency. For a stationary target ($\omega_D = 0$) the magnitude will be zero and the target will not be detected. When $\omega_D \neq 0$, the magnitude will be nonzero and, if sufficiently large, the target can be detected using standard threshold detection techniques to be discussed in Chap. 6.

Figure 5.51 shows an experimental airborne radar system used in some of the early development of the DPCA technique. The phased array antenna was mounted along the side of the fuselage to provide a sidelooking configuration. The antenna was divided into three subapertures on receive to support angle interferometry, to be discussed shortly. Figure 5.52 shows typical results. Figure 5.52a shows the range-Doppler spectrum obtained using a single subaperture, that is, no DPCA processing. Ground clutter forms a strong ridge extending through all range bins and about 50 percent of the Doppler spectrum. (The receiver noise expected in the clear region on the outskirts of the Doppler region is below the amplitude floor of the plot.) Figure 5.52b shows the same scene after DPCA processing. Nearly all of the clutter energy has been attenuated to a level below the plot floor, revealing a single slow-moving target having a radial velocity of about -6 knots (-3 m/s, -6.9 mph). Completely undetectable before the processing, this target should now be easily detected.

Figure 5.51 Experimental aircraft for early DPCA development. (*Source:* C. E. Muehe and M. Labitt, "Displaced Phase Center Antenna Technique," *Lincoln Laboratory Journal*, vol. 2, no. 2, pp. 281–296, 2000. Used with permission.)

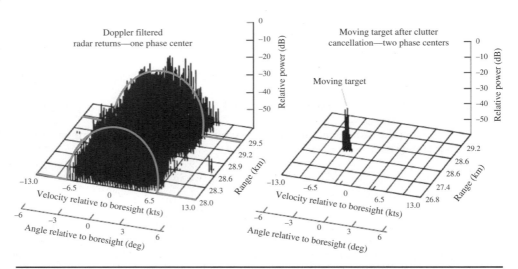

Figure 5.52 Example of DPCA processing effect on ground clutter and detection of a slow mover. See text for discussion. (*Source*: C. E. Muehe and M. Labitt, "Displaced Phase Center Antenna Technique," *Lincoln Laboratory Journal*, vol. 2, no. 2, pp. 281–296, 2000. Used with permission.)

Like conventional MTI, the DPCA process suffers blind speeds. The output magnitude will be zero whenever $\omega_D M_s$ in Eq. (5.138) is a multiple of 2π, which occurs when the Doppler shift in hertz is an integer multiple of v/d_{pc} (see Prob. 44).

DPCA is a clutter suppression technique, not intended for noise reduction. Applying Eq. (5.136) to the noise component does not produce a benefit. Assuming the usual white noise models, the noise in the two data sets will be uncorrelated i.i.d. processes. Weighting and subtracting them will increase the noise power rather than suppressing it. Consequently, DPCA should only be applied in the clutter portion of the Doppler spectrum.

DPCA detects moving targets based on an amplitude threshold. ATI, in contrast, applies a threshold to the phase differences between the fore and aft subaperture signals to distinguish targets from clutter. An ATI processor forms the output signal

$$Z(\omega) = \arg\{Y_f(\omega) \times [\alpha^{-1}\exp(+j\omega M_s)Y_a(\omega)]^*\} \tag{5.139}$$

as shown in Fig. 5.53. Applying Eq. (5.139) to the clutter component of the data gives the output

$$\begin{aligned}Z_c(\omega) &= \arg\{C_f(\omega) \times [\alpha^{-1}\exp(+j\omega M_s)C_a(\omega)]^*\} \\ &\approx \arg\{|C_f(\omega)|^2\} \approx 0\end{aligned} \tag{5.140}$$

This equation shows that the phase of the output signal should be approximately zero for clutter. For the target component, the result is

$$\begin{aligned}Z_t(\omega) &= \arg\{T_f(\omega) \times [\alpha^{-1}\exp(+j\omega M_s)T_a(\omega)]^*\} \\ &\approx \arg\{\exp(-j\omega_D M_s)\,|T_f(\omega)|^2\} \approx -\omega_D M_s\end{aligned} \tag{5.141}$$

Thus ATI can differentiate targets and clutter by applying a threshold to the output phase. Like DPCA, ATI is not effective in suppressing noise.

A complication in the analysis of ATI is that it is not a linear process. In particular, the product $Y_f(\omega) \times Y_a^*(\omega)$ has not only clutter, target, and noise components but cross-products among them all which complicate analysis. The cross-products are relatively negligible so long as one component (clutter, target, or noise) dominates the signal.

Both DPCA and ATI are vulnerable to false alarms in particular clutter scenarios. Mismatches in channel gain and phase responses in DPCA can cause echoes from particular strong clutter scatterers (often called *clutter discretes*) to cancel imperfectly and cause false alarms. ATI avoids this problem by rejecting clutter and detecting targets based on phase, not amplitude. However, ATI is more likely to exhibit false alarms in regions of low-reflectivity

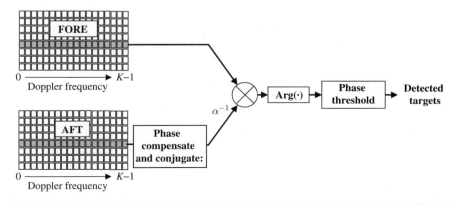

FIGURE 5.53 Processing flow for ATI phase-based clutter rejection in the range-Doppler domain.

clutter. In these regions the clutter-to-noise ratio will be relatively low, making the ATI phase noisy so that the phase measured for clutter cells is not necessarily near zero as expected. A detailed analysis of the strengths and weaknesses of each approach is given in Wang et al. (2020).

5.9.3 Clutter Suppression Interferometry

CSI combines Doppler-domain DPCA and ATI to provide better clutter suppression and detection performance than does either alone. The processing flow is shown in Fig. 5.54. A three-channel antenna is required. Denote the receiver outputs as Y_1, Y_2, and Y_3. DPCA processing is applied to the Y_1 and Y_2 data streams, producing a clutter cancelled output denoted Z_{12}. DPCA is applied independently to the Y_2 and Y_3 streams, producing a second clutter-cancelled output Z_{23}. However, threshold detection is *not* applied to the DPCA outputs. Instead, these two data streams are then combined in an ATI operation, $\arg\{Z_{12}Z_{23}^*\}$, and the result subjected to a phase threshold to form the final detection map (Deming et al., 2012; Greenspan, 2015). The DPCA step reduces the effect of clutter on the phase of any target signatures at its output, improving the performance of the ATI step.

CSI also usually includes estimation of the angle of arrival (AOA) of a signal using an angle interferometer. A radar interferometer consists of two displaced coherent receiver phase centers, as illustrated in Fig. 5.55, which shows a wavefront impinging on two receivers separated by a distance d_{pc}. The angle of arrival of the wavefront, measured from the normal to the line connecting the two receivers, is θ_a. As shown in the figure, the additional distance the wavefront must travel after arriving at the first receiver before it reaches the second receiver is $d_{pc}\sin\theta_a$. Consequently, at any given instant in time that the signal is present, the phase of the signal at the left phase center will be less than that of the signal at the right phase by

$$\Delta\phi = \frac{2\pi}{\lambda_t} d_{pc} \sin\theta_a \text{ rad} \tag{5.142}$$

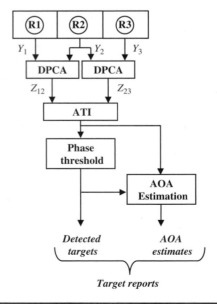

Figure 5.54 Processing flow for CSI for ground moving target indication, combining DPCA and ATI.

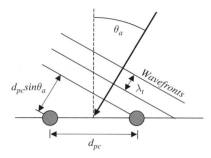

FIGURE 5.55 Geometry for relating angle of arrival to phase difference at two subaperture receive phase centers.

The AOA can be estimated by measuring the phase difference $\Delta\phi$ and inverting Eq. (5.142). For wideband subbands it may be necessary to adjust the wavelength for different subbands or Doppler bins in order to maintain good accuracy in estimating the AOA.

CSI provides a natural two-channel interferometer. The two DPCA outputs are the two channels. In a range-Doppler bin continuing a moving target, the phase of the signal in that bin is expected to differ by the sum of the Doppler-induced phase shift of $-\omega_D M_s$ from Eq. (5.141) and the AOA phase shift of Eq. (5.142). The ATI operation measures the phase shift for each range-Doppler bin. For those that pass the detection threshold the Doppler frequency ω_D is known, at least approximately, based on which Doppler bin contains the detected target. Consequently, the $-\omega_D M_s$ term can be subtracted from the total ATI phase, leaving an estimate of the AOA phase term $\Delta\phi$ that can in turn be used to estimate the AOA.

5.9.4 Analysis of Adaptive DPCA

While conventional bandlimited interpolation could be used to implement fractional-PRI timing adjustments, in practice there will also be mismatches between channels that will make it impossible to achieve high cancellation ratios even if the time alignment is perfect. Adaptive processing can be combined with the basic DPCA cancellation to minimize the clutter residue at the processor output and therefore maximize the improvement factor. This section provides a more detailed analysis of adaptive DPCA, modeled after the "suboptimum matched filter algorithm" in Shaw and McAulay (1983). It is presented here as an example of a more complex application of the vector matched filter discussed earlier for MTI, and of more detailed signal modeling than used above.

This algorithm assumes that an integer PRI delay of one channel with respect to the other is used to achieve coarse time alignment of the two channels to be combined. Each received signal channel is then divided into Doppler bins using a DFT. MTI cancellation is performed independently in each subband, allowing the adaptive cancellation weight to be optimized separately for each Doppler bin and improving overall performance.

The vector analysis approach will be used to model the signals and develop the adaptive filtering. Transmit a CPI of $M + M_s$ pulses and collect L range bins of data for each pulse and each of the $N = 2$ phase centers, resulting in an $L \times (M + M_s) \times 2$ datacube $\mathbf{y}[l, m, n]$. Advance the aft channel slow-time data ($n = 1$) to coarse-align it with the fore channel ($n = 0$) in each range bin and retain only the overlapped slow-time samples to obtain the $L \times M \times 2$ datacube

$$\tilde{\mathbf{y}}[l,m,n] = \begin{bmatrix} \mathbf{y}[l,m,0] \\ \mathbf{y}[l,m+M_s,1] \end{bmatrix} = \begin{bmatrix} y_f[l,m] \\ y_a[l,m+M_s] \end{bmatrix} \quad (5.143)$$

where $y_f[l,m]$ is the fore channel data plane, $y_a[l,m]$ is the aft channel data plane, and the dotted horizontal line represents vertical concatenation of datacube planes. The pulse number (slow time) index m is now in the range $0 \leq m \leq M-1$. The separation of the datacube into phase center planes is done for convenience because the DPCA filter will apply weighting in that dimension only. Now take the K-point slow time DFT of the data in each range bin to get the $L \times K \times 2$ range-Doppler datacube

$$\tilde{Y}[l,k,n] = \begin{bmatrix} Y_f[l,k] \\ \hdashline \exp(-j2\pi M_s k/K)Y_a[l,k] \end{bmatrix} \quad (5.144)$$

Equation (5.144) points out that the time slip adjustment adds a linear phase term to the aft channel Doppler data. This effect applies to both target and interference signal components. For a given range-Doppler bin $[l,k]$, $\tilde{Y}[l,k,n]$ is a 2×1 column vector.

The signals at each subaperture consist of clutter, noise, and (if present) target components. Because the weighting will be in the phase center dimension, a model of the covariance matrix $\mathbf{S_I} = E\{\tilde{Y}^*\tilde{Y}^T\}$ of the phase center interference data for each range-Doppler bin similar to that of Eq. (5.20) is needed. Begin with the clutter. It is not white in slow time, so its power spectrum is not flat and the clutter covariance $\sigma_c^2[l,k]$ is a function of the range index l and Doppler index k. The clutter at the same range-Doppler bin is correlated across the spatial phase center channels to some degree determined by the platform motion and time slip correction. That correlation is denoted by the normalized phase-center dimension correlation coefficient $\rho[k]$ that also varies with the Doppler bin. The thermal noise is assumed uncorrelated between channels and is white. Therefore, for a fixed range-Doppler bin $\mathbf{S_I}$ will take the 2×2 form

$$\mathbf{S_I}[l,k] = \mathbf{S_I}[k] = \begin{bmatrix} \sigma_c^2[k] + \sigma_n^2 & \rho[k]\sigma_c^2[k] \\ \rho^*[k]\sigma_c^2[k] & \beta[k]\left(\sigma_c^2[k] + \sigma_n^2\right) \end{bmatrix} \quad (5.145)$$

The coefficient $\beta[k]$ accounts for any Doppler-dependent mismatch in the gain and frequency response or subaperture antenna patterns of the two channels. This equation also assumes that the clutter power and correlation coefficient are constant over the range window of interest, removing the dependence on l.

Next, a model similar to Eq. (5.24) or (5.26) for the target data in the Doppler domain is needed. A CPI of fast-time/slow-time data for a moving point target in range bin l_t is modeled as

$$\mathbf{t}[l,m] = A_t \delta[l - l_t] \exp(j2\pi F_D mT) \begin{bmatrix} \gamma_f \\ \gamma_a \end{bmatrix} \quad (5.146)$$

where
γ_f and γ_a = complex scalar constants representing the unknown target phases and (possibly unequal) amplitudes in the fore and aft receive channels
A_t = target amplitude
$\delta[\cdot]$ = discrete impulse function

The target phases represented by γ_f and γ_a are determined by the absolute range and the angle of arrival as well as the electrical lengths of the receive paths.

Assuming $|\gamma_f| = |\gamma_a| = 1$ and letting R be the range from the scatterer to the fore aperture (rightmost in Fig. 5.48), γ_f and γ_a of Eq. (5.146) can be expanded as follows:

$$\begin{aligned} \gamma_f &= \exp\left[j\left(\frac{4\pi}{\lambda_k}R + \psi_f\right)\right] \\ \gamma_a &= \exp\left[j\left(\frac{4\pi}{\lambda_k}R + \psi_a - \Delta\phi\right)\right] \end{aligned} \quad (5.147)$$

where $\Delta\phi$ is the channel-to-channel phase difference due to the AOA of Eq. (5.142). The terms ψ_f and ψ_a represent the receiver phase shifts, which are different for each channel in general. The wavelength λ_k is that corresponding to Doppler bin k, that is, $\lambda_k = c/(F_t + F_{D_k})$, where F_{D_k} is the Doppler shift of the kth Doppler bin.

After the Doppler DFT, the target data model of Eqs. (5.146) and (5.147) results in the target range-Doppler domain signal model

$$\mathbf{t}[l,k] = MA_t \delta[l - l_t]\delta[k - k_t] \begin{bmatrix} \gamma_f \\ \exp(-j\pi M_s k / K) \cdot \gamma_a \end{bmatrix} \tag{5.148}$$

where $k_t = (F_D KT / 2\pi)$ is the target Doppler shift converted to an equivalent DFT bin number. The two impulse functions serve to confine the response to range bin l_t and Doppler bin k_t. If $K > M$, there will be multiple DFT samples on the target DTFT mainlobe and the assumption that the target response is essentially confined to one Doppler bin is less valid.

As before, the SIR can be maximized with a matched filter that computes the scalar quantity for each range-Doppler bin

$$\mathbf{z}[l,k] = \left(\mathbf{S}_\mathbf{I}^{-1}[k]\mathbf{t}^*\right)^T \mathbf{Y}[l,k] = \mathbf{t}^H[l,k]\left(\mathbf{S}_\mathbf{I}^{-1}[k]\right)^* \mathbf{Y}[l,k] \tag{5.149}$$

Since the target signal AOA is not known a priori the target signal vector is again averaged over all values of AOA in $[-\pi/2, \pi/2]$. Absorbing all common constants into the complex amplitude, the new target vector becomes simply

$$\mathbf{t}[l,k] = \mathbf{t} = M\hat{A}_t \begin{bmatrix} 1 \\ 0 \end{bmatrix} \tag{5.150}$$

The target location in range-Doppler space is not assumed known so the same target model is used in each range and Doppler bin.

An alternative to assuming an unknown AOA is to assume specific values for θ_a. For example, a series of values in an interval equal to the antenna mainlobe width and centered on the nominal steering angle used on transmit might be generated and used to search an angular region. However, the loss in SIR from using the much simpler Eq. (5.150) is very small (Shaw and McAulay, 1983).

The exact clutter and noise statistics are also not known a priori. Consequently, $\mathbf{S}_\mathbf{I}$ cannot be known exactly, but it can be estimated from the data. Since the clutter covariance is expected to have essentially the same form in every range bin, one way to estimate $\mathbf{S}_\mathbf{I}$ would be to compute a sample average over several range bins,

$$\hat{\mathbf{S}}_\mathbf{I}[k] = \frac{1}{L_2 - L_1 + 1} \sum_{l=L_1}^{L_2} \mathbf{Y}^*[l,k]\mathbf{Y}^T[l,k] \tag{5.151}$$

This estimate of $\hat{\mathbf{S}}_\mathbf{I}$ then replaces the actual $\mathbf{S}_\mathbf{I}$ in Eq. (5.149). Since the coefficients used to combine the fore and aft data streams are computed from the data itself, this is now an *adaptive* DPCA processor. This method for estimating $\mathbf{S}_\mathbf{I}$ is analogous to cell-averaging CFAR interference estimation to be discussed in Chap. 6 and revisited in Chap. 9.

Equation (5.151) implicitly assumes that the covariance matrices in the range bins adjoining bin k are all the same as the covariance matrix in bin k itself. Even if the physical clutter is the same over the averaging interval, this assumption also requires that a preprocessing gain control step compensate for the expected R^3 variation[14] in clutter power with range. Noise power does not vary with range or Doppler.

Combining Eqs. (5.144), (5.145), (5.146), and (5.150) gives the output of the DPCA system. Assuming that $\hat{\mathbf{S}}_I$ is a good approximation to $\mathbf{S}_I$ and absorbing all constants into a single constant α, it is

$$z[l,k] = \alpha\left\{\beta[k]\left(1+\frac{\sigma_n^2}{\sigma_c^2}\right)\mathbf{Y}_f[l,k] - \rho^*[k]\exp(+j2\pi M_s k/K)\mathbf{Y}_a[l,k]\right\} \quad (5.152)$$

While complicated, the structure of a two-pulse canceller is present in the subtraction of $\mathbf{Y}_a[l,k]$ from $\mathbf{Y}_f[l,k]$. If the interference is clutter-limited so that $\sigma_c^2 \gg \sigma_n^2$ and also highly correlated across phase centers so that $\rho[k] \to 1$ (implying that the coarse alignment was very successful), the output simplifies to

$$z[l,k] = \alpha\{\beta[k]\mathbf{Y}_f[l,k] - \exp(+j2\pi M_s k/K)\mathbf{Y}_a[l,k]\} \quad (5.153)$$

The two-pulse canceller structure is clearer here. The complex exponential in the Doppler index k is equivalent to a time-domain shift of M_s samples, in accordance with the DPCA condition discussed earlier. The factor $\beta[k]$ provides a Doppler-dependent weighting factor that can be optimized to maximize cancellation in each Doppler channel.

The matched filter design assumes that $\hat{\mathbf{S}}_I$ is an estimate of the covariance of the interference only, that is, it should not contain any target signal components. A practical system must take steps to ensure this is the case, perhaps by skipping range bins containing known targets such as those already in track, averaging over enough range bins to minimize any unknown target influences, prescreening the data for large amplitudes that might indicate a target, or other means. Many of the required techniques are similar to those used in constant false alarm rate detection, discussed in Chap. 6.

As mentioned earlier, DPCA is a clutter suppression technique, and as such it makes sense to apply it only on the Doppler bins in the clutter region. It is easy to see that if the interference is noise-limited the adaptive DPCA operation reduces to performing no operation (see Prob. 43). Consequently, DPCA should not be applied in the clear region of the Doppler spectrum.

5.10 Summary

The Doppler effect enables radars to localize targets not only in three spatial dimensions but in a fourth dimension of radial velocity as well. This capability improves the ability to separate targets from clutter and improves target tracking accuracy. This chapter focused on common strategies for processing Doppler information to achieve these ends, specifically concentrating on using differences in the Doppler signature of targets and clutter to aid target detection. All of the techniques discussed apply to pulsed radar; most also apply to linear FMCW.

Much of Doppler processing falls into one of two major categories: moving target indication (MTI) or pulse Doppler (PD) processing. The MTI strategy is to filter the slow-time data to remove clutter energy, so that the target energy competes only with noise. MTI filters are usually very low order and relatively simple. The traditional, heuristically motivated pulse cancellers were discussed in detail. The vector matched filter was introduced as a more

---

[14]This correction factor assumes resolution-limited ground clutter as described in Chap. 2. Other appropriate factors are used if this is not the case.

effective and flexible design (and to provide a gentle introduction to vector processing notation and optimum filtering). Pulse Doppler processing implements a very different strategy. Instead of filtering out the clutter, it forms and then acts on the Doppler spectrum in each range bin, relying on differences in target and clutter Doppler signatures to make targets detectable in the presence of stronger clutter interference.

Some common limitations and performance metrics for each technique were discussed. For MTI, improvement factor I was emphasized, and the effect on I of various system imperfections such as amplitude jitter or channel mismatches were described. The primary PD metrics are the processing gain G_{sp} along with various factors detracting from it such as straddle loss due to DFT sampling of the DTFT, and the processing loss PL and loss in processing gain LPG associated with the windowing usually needed to control sidelobes. A family of techniques for interpolating frequency estimates from DFT data were discussed to improve Doppler estimation accuracy.

A number of auxiliary issues were also identified. Pulse Doppler processing implies the use of multiple pulses or sweeps to create the slow-time signal needed, but such waveforms also create blind zones and ambiguities in both range and velocity. Basic techniques for resolving ambiguities and avoiding blind zones based on the use of multiple repetition rates were described, and methods for picking those rates discussed. Range migration, when present, can severely degrade the range-Doppler signature of a target. The relatively advanced technique of keystone transformation was briefly introduced to compensate.

Several moving target detection applications build on these basic techniques. Pulse pair processing, common in weather radars, takes advantage of application-specific signal and clutter models to efficiently produce large-area reflectivity and velocity maps. These in turn provide the basis for higher-level data products in common use such as "color Doppler radar" and severe storm warning and tracking algorithms. Clutter mapping, common in airport surveillance radars, combines MTI and PD processing with several other techniques to optimize detection of aircraft by a stationary radar, especially those with tangential crossing trajectories where targets and ground clutter do not separate well in Doppler. Finally, it was shown how multiple phase centers in an array antenna and the MTI vector matched filter can enable detection of slow moving ground targets from a fast moving aircraft or spacecraft.

References

Alabaster, C., *Pulse Doppler Radar: Principles, Technology, Applications.* SciTech Publishing, Edison, NJ, 2012.

Candan, C., "Analysis and Further Improvement of Fine Resolution Frequency Estimation Method From Three DFT Samples," *IEEE Signal Processing Letters*, vol. 18, no. 6, Jun. 2011.

Candan, C., "A Method for Fine Resolution Frequency Estimation From Three DFT Samples," *IEEE Signal Processing Letters*, vol. 20, no. 9, Sep. 2013.

Cole, E. L. et al., "ASR-12: A Next Generation Solid State Air Traffic Control Radar," *Proceedings of the 1988 IEEE Radar Conference*, Dallas, TX, pp. 9–13, May 1998.

Davis, P. G., and E. J. Hughes, "Medium PRF Set Selection Using Evolutionary Algorithms," *IEEE Transactions on Aerospace and Electronic Systems*, vol. 38, no. 3, pp. 933–939, Jul. 2002.

Deming, R., M. Best, and S. Farrell, "Simultaneous SAR and GMTI Using ATI/DPCA," *Proceedings SPIE 9093, Algorithms for Synthetic Aperture Radar Imagery XXI*, Jun. 13, 2014.

Deming, R., S. Macintosh, and M. Best, "Three-Channel Processing for Improved Geo-Location Performance in SAR-Based GMTI Interferometry," *Proceedings SPIE 8394—Algorithms for Synthetic Aperture Radar XIX*, May 7, 2012.

Doviak, D. S., and R. J. Zrnić, *Doppler Radar and Weather Observations*, 2d ed. Academic Press, New York, 1993.

Eaves, J. L., and E. K. Reedy (eds.), *Principles of Modern Radar*. Van Nostrand Reinhold, 1987.

Greenspan, M., "The Evolutionary Development of Airborne Surface Moving Target Detection," *Proceedings 2015 IEEE Radar Conference (RadarCon)*, Arlington, VA, 2015, pp. 1412–1416.

Harris, F. J., "On the Use of Windows for Harmonic Analysis with the Discrete Fourier Transform," *Proceedings of the IEEE*, vol. 68, no. 1, pp. 51–83, Jan. 1978.

Hayes, M. H., *Statistical Digital Signal Processing and Modeling*. Wiley, New York, 1996.

Hovanessian, S. A., "An Algorithm for Calculation of Range in a Multiple PRF Radar," *IEEE Transactions on Aerospace and Electronic Systems*, vol. 12, no. 2, pp. 287–290, Mar. 1976.

IEEE Standard Radar Definitions, IEEE Standard 686-2017. Institute of Electrical and Electronics Engineers, New York, Sep. 13, 2017.

Jacobsen, E., and P. Kootsookos, "Fast, Accurate Frequency Estimators," *IEEE Signal Processing Magazine*, pp. 123–125, May 2007.

Keel, B. M., "Adaptive Clutter Rejection Filters for Airborne Doppler Weather Radar," M.S. Thesis, Clemson University, Clemson, AL, 1989.

Kretschmer, F. F., Jr., "MTI Visibility Factor," *IEEE Transactions on Aerospace and Electronic Systems*, vol. AES-22, no. 2, pp. 216–218, Mar. 1986.

Levanon, N., *Radar Principles*. Wiley, New York, 1988.

Li, X., X. Xia and H. Liang, "A Robust Chinese Remainder Theorem with Its Applications in Moving Target Doppler Estimation," 2010 IEEE Radar Conference, 2010, pp. 1289–1294.

MacLeod, M. D., "Fast Nearly ML Estimation of the Parameters of Real of Complex Single Tones or Resolved Multiple Tones," *IEEE Transactions on Signal Processing*, vol. 46, no. 1, pp. 141–148, Jan. 1998.

Morris, G. V., and L. Harkness (eds.), *Airborne Pulse Doppler Radar*, 2d ed. Artech House, Boston, MA, 1996.

Muehe, C. E., and M. Labitt, "Displaced Phase Center Antenna Technique," *Lincoln Laboratory Journal*, vol. 2, no. 2, pp. 281–296, 2000.

Nathanson, F. E., *Radar Design Principles*, 2d ed. McGraw-Hill, New York, 1991.

Oppenheim, A. V., and R. W. Schafer, *Discrete-Time Signal Processing*, 3d ed. Prentice Hall, Englewood Cliffs, NJ, 2010.

Richards, M. A., "Coherent Integration Loss due to White Gaussian Phase Noise," *IEEE Signal Processing Letters*, vol. 10, no. 7, pp. 208–210, Jul. 2003.

Richards, M. A., "Discrete-Time Gaussian Fourier Transform Pair, and Generating a Random Process with Gaussian PDF and Power Spectrum," technical memorandum, Oct. 3, 2006. Available at http://radarsp.com.

Richards, M. A., "The Discrete-Time Fourier Transform and Discrete Fourier Transform of Windowed Stationary White Noise," technical memorandum, Oct. 27, 2007. Available at http://radarsp.com.

Richards, M. A., "Doppler Processing," Chap. 17 in M. A. Richards, J. A. Scheer, and W. A. Holm (eds.), *Principles of Modern Radar: Basic Principles*. SciTech Publishing, Raleigh, NC, 2010.

Richards, M. A., "A Slight Extension of 'Coherent Integration Loss due to White Gaussian Phase Noise," technical memorandum, Mar. 3, 2011. Available at http://radarsp.com.

Richards, M. A., "The Keystone Transformation for Correcting Range Migration in Range-Doppler Processing," technical memorandum, Feb. 2020. Available at http://radarsp.com.

Roy, R., and O. Lowenschuss, "Design of MTI Detection Filters with Nonuniform Interpulse Periods," *IEEE Transactions on Circuit Theory*, vol. CT-17, no. 4, pp. 604–612, Nov. 1970.

Schleher, D. C, *MTI and Pulse Doppler Radar*, 2d ed. Artech House, Boston, MA, 2010.

Shaban, F., and M. A. Richards, "Application of L¹ Reconstruction of Sparse Signals to Ambiguity Resolution in Radar," *Proceedings IEEE 2013 Radar Conference*, Ottawa, May 2013.

Shaw, G. A., and R. J. McAulay "The Application of Multichannel Signal Processing to Clutter Suppression for a Moving Platform Radar," *IEEE Acoustics, Speech, and Signal Processing (ASSP) Spectrum Estimation Workshop II*, Tampa, FL, Nov. 10–11, 1983.

Shrader, W. W., and V. Gregers-Hansen, "MTI Radar," Chap. 2 in M. I. Skolnik (ed.), *Radar Handbook*, 3d ed. McGraw-Hill, New York, 2008.

Skolnik, M. I., *Introduction to Radar Systems*, 3d ed. McGraw-Hill, New York, 2001.

Staudaher, F. M., "Airborne MTI," Chap. 16 in M. I. Skolnik (ed.), *Radar Handbook*, 2d ed. McGraw-Hill, New York, 1990.

Stimson, G. W. (author); H. D. Griffiths, C. J. Baker, and D. Adamy (eds.), *Introduction to Airborne Radar*, 3d ed. SciTech Publishing, Mendham, NJ, 2014.

Taylor, J. W., Jr., and G. Brunins, "Design of a New Airport Surveillance Radar (ASR-9)," *Proceedings of the IEEE*, vol. 73, no. 2, pp. 284–289, Feb. 1985.

Trunk, G., and S. Brockett, "Range and Velocity Ambiguity Resolution," *Record of the 1993 IEEE National Radar Conference*, pp. 146–149, Apr. 20–22, 1993.

Trunk, G., and M. W. Kim, "Ambiguity Resolution of Multiple Targets Using Pulse-Doppler Waveforms," *IEEE Transactions on Aerospace and Electronic Systems*, vol. 30, no. 4, pp. 1130–1137, Oct. 1994.

Vergara-Dominguez, L., "Analysis of the Digital MTI Filter with Random PRI," *IEEE Proceedings, Part F*, vol. 140, no. 2, pp. 129–137, Apr. 1993.

Wang, X. et al., "Performance Comparison and Assessment of Displaced Phase center Antenna and Along-Track Interferometry Techniques Used in Synthetic Aperture Radar-Ground Moving Target Indication," *Journal of Applied Remote Sensing*, vol. 8, 2014.

Zrnić, D., "Weather Radar—Recent Developments and Trends," *Proceedings Microwaves, Radar and Remote Sensing Symposium, 2008* (MRRS 2008), pp. 174–178, Kiev, Ukraine, Sep. 2008.

Problems

1. An aircraft has a 4° azimuth 3-dB beamwidth. The RF is 10 GHz and the antenna is steered to a squint angle ψ of 30°. If the aircraft flies at 100 m/s, what is the Doppler spread of the clutter echoes induced by the aircraft motion?

2. Suppose the aircraft in the previous problem has an RR of 10 kHz. Sketch a Doppler spectrum similar to that of Fig. 5.1c, but with noise and mainlobe clutter components only. What range of Doppler shifts lie in the mainlobe clutter region of the spectrum? What range of Doppler shifts lie in the clear region of the spectrum? What percentage of the total spectrum width from $-RR/2$ to $+RR/2$ is in the clear region?

3. If the RR in the previous problem is changed to 1 kHz, what percentage of the total spectrum width will lie in the clear region?

4. Consider a 10-GHz airborne radar traveling straight, level, and forward at 200 mph at an altitude of 30,000 ft. The antenna is pointed at an azimuth angle of 0° and an elevation angle of $-20°$, similar to Fig. 5.3a. What will be the radial velocity in meters per second of echoes from stationary scatterers in the following locations: (a) directly ahead of the aircraft, (b) at the point where the antenna boresight intercepts the ground, (c) directly below the aircraft, (d) on the ground directly to the left or right of the aircraft (azimuth angle of $\pm 90°$), and (e) directly behind the aircraft?

5. Verify all of the calculations regarding Fig. 5.4 given in the text. These include boresight range to the ground, component of velocity along the boresight, Doppler

shift at the middle of the mainlobe clutter, maximum Doppler shift of clutter, and maximum radial velocity of clutter.

6. For the aircraft in Prob. 4, sketch the approximate unaliased slant range-velocity distribution of the ground clutter (mainlobe + sidelobe) in a "bird's-eye" format similar to that of Fig. 5.39a. The slant range axis of the sketch should cover 0 to 100 km and the velocity axis should cover $\pm v_{max}$, where v_{max} is the maximum possible radial velocity in meter per second that could be observed from scatterers in front of the radar. It is not necessary to represent antenna gain effects; concentrate on indicating where the mainlobe clutter will be centered and the intervals in range and velocity where clutter energy will be seen.

7. Suppose the radar in the previous problem has an operating frequency of 10 GHz and a PRF of 3 kHz. What are the unambiguous slant range R_{ua} and unambiguous velocity interval v_{ua}? Sketch the approximate aliased range-velocity distribution of the ground clutter (mainlobe + sidelobe) in a "bird's-eye" format similar to that of Fig. 5.39c. The slant range axis of the sketch should cover 0 to R_{ua} and the velocity axis should cover $\pm v_{ua}/2$. Concentrate on indicating where the mainlobe clutter will be centered and the intervals in range and velocity where clutter energy will be seen.

8. Use the vector form of the matched filter to find the coefficients of an optimum two-pulse ($N = 2$) MTI filter under the assumptions that (a) the interference is white noise only, and (b) only approaching targets are of interest, that is, those with positive Doppler shifts; however, targets can be approaching with any positive velocity between 0 and $\lambda_t RR/4$ (corresponding to $F_D = RR/2$) being equally likely. Repeat for receding targets. Interpret the result: that is, explain how the particular form of the filter coefficients will maximize the SIR at the filter output for this target and interference model.

9. Suppose an MTI radar is placed on a moving platform such that the clutter spectrum, rather than being centered at normalized radian frequency $\omega = 0$, is instead centered on $\omega = \pi/2$. If the clutter observed by a stationary platform was $c[m]$, the new clutter process can be modeled crudely as $c'[m] = c[m]\exp(j\pi m/2)$. (This is an oversimplified model because it does not account for the broadening of the clutter spectrum caused by platform motion, but that effect is ignored in this problem for simplicity.) Assume the target can be at any velocity with equal likelihood. Ignore the noise, that is, assume $\sigma_n^2 = 0$. Assume the stationary clutter $c[m]$ has power σ_c^2 and first autocorrelation lag $\rho\sigma_c^2$. Use the vector approach to find the coefficients of the optimum two-pulse ($N = 2$) matched filter for this case. Compare to the $N = 2$ vector matched filter for the stationary radar. Explain how the shift in the clutter power spectrum center frequency changes the coefficients. Interpret the resulting coefficients as in the previous problem.

10. Use the vector form of the matched filter to find the coefficients of an optimum two-pulse MTI filter under the assumptions that (a) the interference is white noise plus a pure complex sinusoid at some specific normalized radian frequency ω_J, and (b) target velocity is uniform random over the entire spectrum. Specifically, the interference signal becomes

$$w[m] = n[m] + q[m]$$
$$q[m] = A_J \exp(j\omega_J m)$$

where $n[m]$ is a stationary white noise process with power σ_n^2. This could be a simple model for a jammer at frequency ω_J. What is the frequency response $H(\omega)$ of the filter having the resulting coefficients? Explain how this frequency response will affect the jammer and target signals.

11. Consider a pulse-to-pulse staggered PRF system using a series of $P = 3$ PRIs, namely $\{66.\overline{6}\ \mu s, 83.\overline{3}\ \mu s, 100\ \mu s\}$.

    a. What are the base interval T_g and the set of staggers $\{k_p\}$?
    b. What is the average PRI, T_{avg}?
    c. What is the first blind Doppler frequency, F_{bs}, of the staggered system?
    d. What is the ratio F_{bs}/F_b of the first blind Doppler of the staggered system to the first blind speed of a constant-PRI system having the average PRI T_{avg}?

12. Now consider the effect of the staggered PRIs of the previous problem on range coverage.

    a. What is the unambiguous range, R_{us}, corresponding to the average PRI from Prob. 11? (This would be the unambiguous range of a constant-PRI system that used the same amount of time to collect N pulses as the staggered-PRF system.)
    b. For the three PRFs in Prob. 11, what is the maximum unambiguous range R_{uas}?
    c. What is the factor by which the range coverage (unambiguous range) is reduced in the staggered PRF system of Prob. 11 relative to the unstaggered system?

13. Using the numerical results from the previous two problems, determine the percentage by which the total range-velocity coverage, defined as the product of the unambiguous range and first blind Doppler, increased in the staggered case as compared to the unstaggered case. In other words, compute $R_{uas}F_{bs}/R_{ua}F_b$. Show that the numerical results agree with the result predicted by Eq. (5.47).

14. Assume a set of integer staggers $\{k_p\}$ is mutually prime. Show that $\text{lcm}\{1/k_p\} = 1$. *Hint #1*: For any set of P numbers β, $\alpha \cdot \text{lcm}(\beta_1, \ldots, \beta p) = \text{lcm}(\alpha \beta_1, \ldots, \alpha \beta p)$. *Hint #2*: Use hint #1 with $\alpha = \beta_1 \cdot \beta_2 \cdot \ldots \cdot \beta p$ and the prime factorization method for finding the least common multiple.

15. Consider a three-pulse canceller; thus the filter coefficients are $h[m] = \{1, -2, 1\}$. Suppose the clutter $c[m]$ at the canceller input has an autocorrelation function given by

    $$s_c[k] = \sigma_c^2(\delta[k] + 0.5\delta[|k|-1] + 0.25\delta[|k|-2])$$

    a. Find the clutter power spectrum on a normalized radian frequency scale, $S_c(\omega)$.
    b. The improvement factor I was expressed as the product of "gain" and "clutter attenuation," $I = G \cdot CA$, in Eq. (5.49). Find the gain G for the three-pulse canceller using Eq. (5.50).
    c. Use the autocorrelation method to find the improvement factor I for the three-pulse canceller with this clutter process.

16. Repeat Prob. 15(c) using the frequency domain approach. The value of G from Prob. 15(b) can be reused.

17. Repeat Prob. 15(c) using the vector matched filter approach. Recompute both G and CA using the vector method.

18. Compute the clutter attenuation CA when using a two-pulse canceller, assuming the clutter Doppler power spectrum is of the form

$$S_c(F_D) = \begin{cases} \sigma_c^2, & |F_D| \leq F_\omega \\ 0, & F_\omega < |F_D| \leq RR/2 \end{cases}$$

Assume $F_{co} \leq RR/2$ and express the answer in terms of F_{co}.

19. Derive the three-pulse canceller version of Eq. (5.68) for the limit on clutter cancellation due to transmitted signal amplitude jitter.

20. Apply Eq. (5.71) to the three-pulse canceller case to obtain the limit on clutter cancellation due to transmitted signal phase jitter.

21. Consider an $M = 32$ pulse sequence of slow-time data collected with $RR = 10$ kHz. A radix 2 FFT algorithm is used to compute the Doppler spectrum of the data. If the Doppler frequency samples are to have a spacing of 100 Hz or less, what is the minimum FFT size K that should be used? What is the resulting spacing of the Doppler frequency samples in hertz?

22. Consider a C-band (5 GHz) radar using a pulse repetition frequency of PRF = 3500 pulses per second. The radar collects 30 pulses of data. For a given range, the slow-time data sequence is zero-padded and input to a 64-point DFT to compute the Doppler spectrum. What is the spacing of the DFT samples in normalized radian frequency (i.e., on the $-\pi$ to $+\pi$ scale)? What is the spacing in hertz? In meters per second? What is the Rayleigh resolution (peak-to-first null width) in Doppler, in hertz? In meters per second?

23. An X-band (10 GHz) pulse Doppler radar collects a fast-time/slow-time matrix of 30 pulses by 200 range bins per pulse. This is converted to a range-Doppler matrix by applying a Hamming window and then a 64-point fast Fourier transform to each slow-time row. Suppose that there is a target with a constant radial velocity of 30 m/s approaching the radar at a range corresponding to range bin #100. The PRF is 6000 samples per second. There is no ground clutter, and noise can be ignored as well. For which FFT sample index k_0 is $|Y[k_0]|$ the largest? (Remember that the DC sample is $k = 0$.) What velocity in meters per second does this sample correspond to? What is the error between the apparent velocity based on the largest FFT sample and the actual velocity?

24. Continuing Prob. 23: in terms of the window function $w[m]$, give an expression for the peak value of the DTFT (not DFT) of the windowed data in range bin #100, assuming that each slow-time sample has an amplitude of 1 before windowing. What is the numerical value of this peak? (MATLAB® or a similar computational tool can be used to compute this value.) Now suppose the peak value of the magnitude of the FFT of the data $|Y[k_0]| = 15.45$. What is the straddle loss in dB?

25. Continuing with Probs. 23 and 24, suppose also that $|Y[k_0 - 1]| = 11.61$ and $|Y[k_0 + 1]| = 14.61$. Use the amplitude-based quadratic interpolation technique of Eqs. (5.91) and (5.92) to estimate the velocity of the target and the peak amplitude of the DTFT. Compute the new values of velocity error and straddle loss and compare to those found in Probs. 23 and 24.

26. Consider a 10-GHz fast-chirp FMCW radar collecting a CPI of data at an RR of 2000 sweeps per second, with $M = 30$ sweeps per CPI. The fast-time/slow-time matrix is converted to a range-Doppler matrix by a $K = 30$ point slow-time DFT.

What is the pulsed Doppler processing gain G_{sp} due to the DFT, in linear and dB units? Repeat for $K = 64$.

27. How are the gains in the previous problem changed if a Hamming window is applied to the data prior to computing the Doppler DFT? Any window parameters needed can be estimated using the information in Table B.2.

28. Continuing the previous problem, now consider a sinusoidal signal with a frequency exactly halfway between two DFT samples (the worst case for straddle loss). What is G_{sp} for a signal at this frequency when no window is used and $K = 30$? Repeat for $K = 64$. Now compute the worst-case G_{sp} for the same worst-case straddle frequency when a Hamming window is used for $K = 30$ and 64. Does using a window incur less total worst-case loss (straddle loss plus processing loss) than does not using a window?

29. Consider two radars. The first is a 3-GHz weather radar having a desired unambiguous range of $R_{ua} = 300$ km and unambiguous velocity interval of ± 50 m/s (about ± 112 mph). The second is a 10-GHz airborne radar having a desired unambiguous range of $R_{ua} = 100$ km and unambiguous velocity interval of ± 250 m/s (about ± 560 mph). For each radar, is a 1-kHz PRF considered to be a low, medium, or high PRF?

30. What is the lowest PRF that would be considered "high" for each of the two radars in the previous problem?

31. A 1-GHz radar is designed to make an unambiguous velocity measurement over a radar-target radial velocity interval of width 300 m/s with a Doppler resolution of 10 Hz. The range bin spacing is 3 m. What are the minimum RR and the minimum number of pulses or sweeps M required in the CPI? Assuming a $+300$ m/s radar-target closing rate, how many range bins will the target signature migrate across during one CPI?

32. The key step in the keystone transform is the range frequency-dependent slow-time interpolation $Y_{Rd\_key}(F, m] = Y_{Rd}\left(F, \dfrac{F_t}{F_t + F}m\right]$. Suppose the KT is applied to the radar in the previous problem. What is the bandwidth of the waveform required to achieve 3-m range resolution (ignoring any windowing)? What are the minimum and maximum values of the interpolation scale factor $F_t/(F_t + F)$, and the total percentage change of that factor across the occupied range bins? When the slow-time DFT is applied to the range-transformed data $Y_{Rd}(F, m]$ to obtain $Y_{Rd\_key}(F, F_D)$ the result has a rectangular region of support in (F, F_D) space with F_D in hertz (F can be in either hertz or cycles/sample). Sketch the ROS of $Y_{Rd\_key}(F, m)$ that will result from applying the KT interpolation to $Y_{Rd}(F, m)$. Label significant values.

33. Consider range ambiguity resolution using three PRFs. Suppose the three PRFs correspond to $N_0 = 4$, $N_1 = 5$, and $N_2 = 7$ range cells. A single target is detected in the first range bin on the first PRF, the fourth range bin on PRF #2, and the second range bin on PRF #3; that is, $n_{a_0} = 1$, $n_{a_1} = 4$, $n_{a_2} = 2$. Assume the radar sensitivity is such that it could possibly detect targets out to the 15th range bin. Use the coincidence algorithm to determine the true range bin number for this target.

34. Repeat the previous problem using the Chinese remainder theorem approach.

35. Again consider range ambiguity resolution using three PRFs. Suppose the three PRFs correspond to $N_0 = 3$, $N_1 = 4$, and $N_2 = 5$ range cells. Two targets are detected in each PRF: in range bins 1 and 2 of the first PRF, bins 1 and 4 of the second PRF, and bins 2 and 4 of the third PRF. Assume the radar sensitivity is such that it could

possibly detect targets out to the 25th range bin. How many targets will be reported by the coincidence algorithm, and in which range bins?

36. Now assume that in the previous problem, one false alarm occurs on the third PRF in range bin #3. How many targets will be reported by the coincidence algorithm, and in which range bins?

37. Now suppose that in Prob. 35, the target in range bin #4 on the third PRF is not detected (a missed detection). How many targets will be reported by the coincidence algorithm, and in which range bins?

38. Finally, suppose only the second and third PRFs in Prob. 35 are used. How many targets will be reported by the coincidence algorithm, and in which range bins?

39. Suppose a radar has a pulse length of $\tau = 10$ μs and a PRF of 10 kHz. Assume that the clutter observed by the radar has a two-sided spectral width of 1 kHz (i.e., the clutter spectrum occupies the range from -500 to $+500$ Hz). Sketch the blind zone map for these operating conditions. For the vertical axis, use time in seconds from 0 to 400 μs; for the horizontal axis, use Doppler frequency in hertz from $-10,000$ to $+10,000$ Hz.

40. A weather radar has a PRF of 2 kHz. Using a series of 50 samples of data from a particular range bin and look direction, the following values of the autocorrelation function are computed: $s_y[0] = 50$, $s_y[1] = 30\exp(j\pi/3)$. Use the pulse-pair processing (PPP) time domain method to compute the estimated mean frequency of the echo in hertz.

41. Continuing with the same weather radar, the mean Doppler shift is now removed from the data to obtain a new sequence $y'[m]$ with autocorrelation lags $s_{y'}[0] = 50$, $s_{y'}[1] = 30$. Use the PPP time domain method to compute the estimated spectral width of the echo in hertz. Use the version of the spectral width estimator that contains the $\ln(\cdot)$ function. Remember that spectral width is the variance of the spectrum, not the standard deviation.

42. Repeat the spectral width calculation given in Eq. (5.108) using the version based on the series approximation to $\ln(x)$. What is the percentage error in this estimate compared to the estimate in the previous problem?

43. Assume the simplified target model of Eq. (5.150) and suppose the adaptive DPCA system is noise limited at a particular Doppler bin, $\sigma_n^2[l, k] \gg \sigma_c^2[l, k]$. Show that the output $z[l, k]$ of Eq. (5.152) reduces to simply $Y_f[l, k]$ to within a scale factor. That is, the processor does not combine the fore and aft channels in noise-limited Doppler bins.

44. Show that the target component of the output of the DPCA process given in Eq. (5.138) will be zero whenever the target Doppler shift in hertz is a multiple of v/d_{pc}.

CHAPTER 6
Detection Fundamentals

6.1 Introduction

The primary functions to be carried out by a radar signal processor are detection, tracking, and imaging. In this chapter, the concern is detection: deciding whether a given radar measurement is the result of an echo from a target or simply represents the effects of interference. If it is decided that the measurement indicates the presence of a target, further processing is usually undertaken. This additional processing might, for instance, take the form of tracking via precise range, angle, or Doppler measurements.

Detection decisions can be applied to signals present at various stages of the radar signal processing, from raw echoes to heavily preprocessed data such as Doppler spectra or even synthetic aperture radar images. In the simplest case, each range bin (fast-time sample) for each pulse or sweep can be individually tested to decide if a target is present at the range corresponding to the range bin, and the spatial angles corresponding to the antenna pointing direction for that measurement. Since the number of range bins can be in the hundreds or even thousands and pulse or sweep repetition rates can range from hundreds per second to tens of thousands per second or more, the radar can be making many thousands to millions of detection decisions each second.

It was seen in Chap. 2 that both the interference and the echoes from complex targets are best described by statistical signal models. Consequently, the process of deciding whether or not a measurement represents the influence of a target or only interference is a problem in statistical hypothesis testing. In this chapter, it will be shown how this basic decision strategy leads to the concept of threshold testing as the most common detection logic in radar. Performance curves will be derived for the most basic signal and interference models.

Clutter is sometimes interference and sometimes the target. If one is trying to detect a moving vehicle ground clutter is the interference, along with noise and possibly jamming; but if one is trying to image a region of the earth, this same terrain becomes the desired target and only noise and jamming are the interference.

An excellent concise reference for modern detection theory is given in Chap. 5 of Johnson and Dudgeon (1993). When greater depth is needed, another excellent modern reference with a digital signal processing point of view is Kay (1998). An important classical textbook in detection theory is Van Trees et al. (2013), while Meyer and Mayer (1973) provide a classical in-depth analysis and many detection curves specifically for radar applications. Recent developments in constant false alarm rate (CFAR) processing, especially for marine applications, are described in Weinberg (2017).

6.2 Radar Detection as Hypothesis Testing

For any radar measurement that is to be tested for the presence of a target, one of two hypotheses can be assumed to be true:

1. The measurement is the result of interference only.
2. The measurement is the combined result of interference and echoes from a target.

The first hypothesis is denoted as the *null hypothesis* H_0 and the second as the non-null hypothesis H_1. The detection logic therefore must examine each radar measurement to be tested and select one of the hypotheses as best accounting for that measurement. If H_0 best accounts for the data, the system declares that a target was not present at the range, angle, or Doppler coordinates of that measurement; if H_1 best accounts for the data, the system declares that a target was present.[1]

Because the signals are described statistically, the decision between the two hypotheses is an exercise in statistical decision theory. A general approach to this problem is described in many texts (e.g., Kay, 1998). The analysis starts with a probability density function (PDF) that describes the measurement to be tested under each of the two hypotheses. If the sample to be tested is denoted as y, the following two PDFs are required:

$p_y(y \mid H_0)$ = PDF of y given that a target was *not* present

$p_y(y \mid H_1)$ = PDF of y given that a target *was* present

Thus, part of the detection problem is to develop models for these two PDFs. In fact, analysis of radar performance is dependent on estimating these PDFs for the system and scenario at hand. Furthermore, a good deal of the radar system design problem is aimed at manipulating these two PDFs in order to obtain the most favorable detection performance. More generally, detection will be based on N samples of data y_n forming a column vector $\mathbf{y}$:

$$\mathbf{y} \equiv [y_0 \quad \cdots \quad y_{N-1}]^T \qquad (6.1)$$

The N-dimensional joint PDFs $p_\mathbf{y}(\mathbf{y} \mid H_0)$ and $p_\mathbf{y}(\mathbf{y} \mid H_1)$ are then used.

Assuming the two PDFs are successfully modeled, the following probabilities of interest can be defined:

Probability of detection, P_D: The probability that a target is declared (i.e., H_1 is chosen) when a target is in fact present.

Probability of false alarm, P_{FA}: The probability that a target is declared (i.e., H_1 is chosen) when a target is in fact *not* present.

Probability of miss, P_M: The probability that a target is *not* declared (i.e., H_0 is chosen) when a target is in fact present.

Because $P_M = 1 - P_D$, P_D and P_{FA} suffice to specify all of the probabilities of interest.[2] As P_{FA} and P_M imply, it is important to realize that because the problem is statistical, there will be a finite probability that the decisions will be wrong.

---

[1] In some detection problems, a third hypothesis is allowed: "don't know." Most radar systems, however, force a choice between "target present" and "target absent" on each detection test.

[2] A fourth probability can be defined, that of choosing H_0 and thus declaring a target not present when in fact the test sample is due to interference only. This probability, equal to $1 - P_{FA}$, is not normally of direct interest.

6.2.1 The Neyman-Pearson Detection Rule

The next step in making a detection decision is to define what constitutes an optimal choice between the two hypotheses. This is a rich field. The Bayes optimization criterion assigns a cost or risk to each of the four possible combinations of actual state (target present or not) and decision (select H_0 or H_1). In radar it is common to use a special case of the Bayes criterion called the *Neyman-Pearson criterion*. Under this criterion, the decision process is designed to maximize the probability of detection P_D under the constraint that the probability of false alarm P_{FA} does not exceed a set value. The achievable combinations of P_D and P_{FA} are affected by the quality of the radar system, the signal processor design, and the scenario. However, it will be seen that for a fixed system design, increasing P_D implies increasing P_{FA} as well. The radar system designer will generally decide what rate of false alarms can be tolerated based on the implications of acting on a false alarm, which may include using radar resources to start a track on a nonexistent target, or in extreme cases even firing a weapon! Recalling that the radar may make tens or hundreds of thousands, even millions of detection decisions per second, values of P_{FA} must generally be quite low. Values in the range of 10^{-4} to 10^{-8} are common, and yet may still lead to false alarms every few seconds. Higher-level logic implemented in downstream data processing, beyond the scope of this book, is often used to reduce the number or impact of false alarms.

Each vector of measured data values $\mathbf{y}$ can be considered to be a point in N-dimensional space. To have a complete decision rule, each point in that space (each possible combination of N measured data values) must be assigned to one of the two allowed decisions, H_0 or H_1. Then, when the radar measures a particular set of data values ("observation") $\mathbf{y}$, the system declares either "target absent" or "target present" based on the preexisting assignment of $\mathbf{y}$ to either H_0 or H_1. Denote the set of all observations $\mathbf{y}$ for which H_1 will be chosen as the region $\Re_1$. $\Re_1$ is not necessarily a connected region. General expressions can now be written for the probabilities of detection and false alarm as integrals of the joint PDFs over the region $\Re_1$ in N-dimensional space:

$$P_D = \int_{\Re_1} p_\mathbf{y}(\mathbf{y}|H_1) \, d\mathbf{y}$$
$$P_{FA} = \int_{\Re_1} p_\mathbf{y}(\mathbf{y}|H_0) \, d\mathbf{y} \qquad (6.2)$$

Because probability density functions are nonnegative, Eq. (6.2) proves a claim made earlier, namely that P_D and P_{FA} must rise or fall together. As the region $\Re_1$ grows to include more of the possible observations $\mathbf{y}$, either integral encompasses more of the N-dimensional space and therefore integrates more of the nonnegative PDF. The opposite is true if $\Re_1$ shrinks. That is, as $\Re_1$ grows or shrinks, *both* P_D and P_{FA} must decrease or increase.[3] In order to increase detection probability, the false alarm probability must be allowed to increase as well. Loosely speaking, to achieve a good balance of performance the points that contribute more probability mass to P_D than to P_{FA} are assigned to $\Re_1$. If the system can be designed so that $p_\mathbf{y}(\mathbf{y}|H_0)$ and $p_\mathbf{y}(\mathbf{y}|H_1)$ are as disjoint as possible, this task becomes easier and more effective. This point will be revisited later.

6.2.2 The Likelihood Ratio Test

The Neyman-Pearson criterion is motivated by the goal of obtaining the best possible detection performance while guaranteeing that the false alarm probability does not exceed some tolerable value. Thus, the Neyman-Pearson decision rule is to

$$\text{Choose } \Re 1 \text{ such that } P_D \text{ is maximized, subject to } P_{FA} \leq \alpha \qquad (6.3)$$

---

[3] The exception occurs if points are added to or subtracted from $\Re_1$ for which $p_\mathbf{y}(\mathbf{y}|H_0)$, $p_\mathbf{y}(\mathbf{y}|H_1)$, or both are zero. In that case, the corresponding probability is unchanged.

where α is the maximum allowable false alarm probability.[4] This optimization problem is solved by the method of Lagrange multipliers. Construct the Lagrangian

$$F \equiv P_D + \lambda(P_{FA} - \alpha) \tag{6.4}$$

The constant λ (not to be confused with the radar wavelength) is called the Lagrange multiplier. To find the optimum solution, maximize F and then choose α to satisfy the constraint criterion $P_{FA} = \alpha$. Substituting Eq. (6.2) into Eq. (6.4),

$$\begin{aligned} F &= \int_{\Re_1} p_\mathbf{y}(\mathbf{y}|H_1)\, d\mathbf{y} + \lambda\left[\int_{\Re_1} p_\mathbf{y}(\mathbf{y}|H_0)\, d\mathbf{y} - \alpha\right] \\ &= -\lambda\alpha + \int_{\Re_1} \{p_\mathbf{y}(\mathbf{y}|H_1) + \lambda p_\mathbf{y}(\mathbf{y}|H_0)\}\, d\mathbf{y} \end{aligned} \tag{6.5}$$

Remember that the design variable here is the choice of the region $\Re_1$. The first term in the second line of Eq. (6.5) does not depend on $\Re_1$, so F is maximized by maximizing the value of the integral over $\Re_1$. Since λ could be negative, the integrand can be either positive or negative, depending on the values of λ and the relative values of $p_\mathbf{y}(\mathbf{y}|H_0)$ and $p_\mathbf{y}(\mathbf{y}|H_1)$. The integral is therefore maximized by including in $\Re_1$ all the points, and only the points, in the N-dimensional space for which $p_\mathbf{y}(\mathbf{y}|H_1) + \lambda p_\mathbf{y}(\mathbf{y}|H_0) > 0$, that is, $\Re_1$ is all points $\mathbf{y}$ for which $p_\mathbf{y}(\mathbf{y}|H_1) > -\lambda \cdot p_\mathbf{y}(\mathbf{y}|H_0)$. This leads directly to the decision rule:

$$\frac{p_\mathbf{y}(\mathbf{y}|H_1)}{p_\mathbf{y}(\mathbf{y}|H_0)} \underset{H_0}{\overset{H_1}{\gtrless}} -\lambda \tag{6.6}$$

Equation (6.6) is known as the *likelihood ratio test* (LRT). Although derived from the point of view of determining what values of $\mathbf{y}$ should be assigned to decision region $\Re_1$, it in fact allows one to skip over explicit determination of $\Re_1$ and gives a rule for optimally guessing, under the Neyman-Pearson criterion, whether a target is present or not based directly on the observed data $\mathbf{y}$ and a threshold $-\lambda$ (which must still be computed). This equation states that the ratio of the two PDFs, each evaluated for the particular observed data $\mathbf{y}$, should be compared to a threshold. If that "likelihood ratio" exceeds the threshold, choose hypothesis H_1, that is, declare a target to be present. If it does not exceed the threshold, choose H_0 and declare that a target is not present. Under the Neyman-Pearson optimization criterion, the probability of a false alarm cannot exceed the original design value P_{FA}. Note again that models of $p_\mathbf{y}(\mathbf{y}|H_0)$ and $p_\mathbf{y}(\mathbf{y}|H_1)$ are required in order to carry out the LRT. Finally, realize that in computing the LRT, the data processing operations to be carried out on the observed data $\mathbf{y}$ are being specified. What exactly the required operations are depends on the particular PDFs. This point will be made clearer shortly.

The LRT is as ubiquitous in detection theory and statistical hypothesis testing as is the Fourier transform in signal filtering and analysis. It arises as the solution to the hypothesis testing problem under several different decision criteria, such as the Bayes minimum cost criterion, or maximization of the probability of a correct decision. Substantial additional detail is provided in Johnson and Dudgeon (1993) and Kay (1998). As a convenient shorthand, the LRT is often expressed in the following notation:

$$\Lambda(\mathbf{y}) \underset{H_0}{\overset{H_1}{\gtrless}} \eta \tag{6.7}$$

From Eq. (6.6), $\Lambda(\mathbf{y}) = p_\mathbf{y}(\mathbf{y}|H_1)/p_\mathbf{y}(\mathbf{y}|H_0)$ and $\eta = -\lambda$.

---

[4]Some subtleties that can arise if the PDFs are noncontinuous are being ignored. See Johnson and Dudgeon (1993) for additional detail.

Because the decision depends only on whether the LRT exceeds the threshold or not, any monotone increasing[5] operation can be performed on both sides of Eq. (6.7) without affecting the values of observed data **y** that cause the threshold to be exceeded, and therefore without affecting the performance (P_D and P_{FA}). A well-chosen transformation can sometimes greatly simplify the computations required to actually carry out the LRT. Most common is to take the natural logarithm of both sides of Eq. (6.7) to obtain the *log likelihood ratio test*:

$$\ln \Lambda(\mathbf{y}) \underset{H_0}{\overset{H_1}{\gtrless}} \ln(\eta) \qquad (6.8)$$

To make these procedures more concrete, consider what is perhaps the simplest example, detection of the presence or absence of a real-valued positive constant m in real-valued zero-mean white Gaussian noise (WGN). Let **w** be a vector of independent identically distributed (i.i.d.) zero mean Gaussian random variables (RVs) of variance σ_w^2. When the constant is absent (hypothesis H_0) the data vector $\mathbf{y} = \mathbf{w}$ follows an N-dimensional zero-mean normal distribution with a scaled identity covariance matrix. When the constant is present (hypothesis H_1), $\mathbf{y} = \mathbf{m} + \mathbf{w} = m\mathbf{1}_N + \mathbf{w}$ and the distribution is simply shifted to a nonzero positive mean[6]:

$$\begin{aligned} H_0 &: \mathbf{y} \sim N(\mathbf{0}_N, \sigma_w^2 \cdot \mathbf{I}_N) \\ H_1 &: \mathbf{y} \sim N(m \cdot \mathbf{1}_N, \sigma_w^2 \cdot \mathbf{I}_N) \end{aligned} \qquad (6.9)$$

where $m > 0$ and $\mathbf{0}_N$, $\mathbf{1}_N$, and $\mathbf{I}_N$ are, respectively, a vector of N zeros, a vector of N ones, and the identity matrix of order N. The model of the required PDFs is therefore

$$\begin{aligned} p(\mathbf{y}|H_0) &= \prod_{n=0}^{N-1} \frac{1}{\sqrt{2\pi\sigma_w^2}} \exp\left\{-\frac{1}{2}\left(\frac{y_n}{\sigma_w}\right)^2\right\} \\ p(\mathbf{y}|H_1) &= \prod_{n=0}^{N-1} \frac{1}{\sqrt{2\pi\sigma_w^2}} \exp\left\{-\frac{1}{2}\left(\frac{y_n - m}{\sigma_w}\right)^2\right\} \end{aligned} \qquad (6.10)$$

The likelihood ratio $\Lambda(\mathbf{y})$ and the log-likelihood ratio can be directly computed from Eq. (6.10), giving

$$\Lambda(\mathbf{y}) = \frac{\prod_{n=0}^{N-1} \exp\left\{-\frac{1}{2}\left(\frac{y_n - m}{\sigma_w}\right)^2\right\}}{\prod_{n=0}^{N-1} \exp\left\{-\frac{1}{2}\left(\frac{y_n}{\sigma_w}\right)^2\right\}} \qquad (6.11)$$

$$\begin{aligned} \ln[\Lambda(\mathbf{y})] &= \sum_{n=0}^{N-1}\left[-\frac{1}{2}\left(\frac{y_n - m}{\sigma_w}\right)^2 + \frac{1}{2}\left(\frac{y_n}{\sigma_w}\right)^2\right] \\ &= \frac{1}{\sigma_w^2}\sum_{n=0}^{N-1} m \cdot y_n - \frac{1}{2\sigma_w^2}\sum_{n=0}^{N-1} m^2 \end{aligned} \qquad (6.12)$$

---

[5] A monotone decreasing operation would simply invert the sense of the threshold test.

[6] All of the following development is fairly easily generalized for the case when m is negative or is of unknown sign. It will be seen later that radar detection generally involves working with the magnitude of the signal, thus it is sufficient to work with a positive value of m.

Because of its simpler form, the log-likelihood ratio will be used. Substituting Eq. (6.12) into Eq. (6.8) and rearranging gives the decision rule

$$\sum_{n=0}^{N-1} y_n \underset{H_0}{\overset{H_1}{\gtrless}} \frac{\sigma_w^2}{m} \ln(-\lambda) + \frac{Nm}{2} \equiv T \qquad (6.13)$$

Note that the right-hand side of the equation consists only of constants, though not all are yet known; these can be combined into a single constant T. Equation (6.13) thus specifies that the available data samples y_n be summed (*integrated*) and the integrated data compared to a threshold. This integration is an example of how the LRT specifies the data processing to be performed on the measurements. Note also that Eq. (6.13) does *not* require specifically evaluating the PDFs, let alone determining what exactly is the region in N-space comprising $\Re_1$ or whether the observation $\mathbf{y}$ is in it.

In many cases of interest, the specific form of the log-likelihood ratio can be further rearranged to isolate on the left-hand side of the equation only those terms explicitly including the data samples y_n, moving all other constants to the right-hand side. Equation (6.13) is such a rearrangement of Eq. (6.12). The quantity $\sum y_n$ is called a *sufficient statistic* for this problem, and is denoted by $\Upsilon(\mathbf{y})$. The sufficient statistic, if it exists, is a function of the data $\mathbf{y}$ that has the property that the likelihood ratio (or log-likelihood ratio) can be written as a function of $\Upsilon(\mathbf{y})$, that is, the data appear in the likelihood ratio *only* through $\Upsilon(\mathbf{y})$ (Van Trees et al., 2013). This means that in making a decision that is optimal under the Neyman-Pearson criterion, knowing the sufficient statistic $\Upsilon(\mathbf{y})$ is as good as knowing the actual data $\mathbf{y}$. In particular, the decision criterion in Eq. (6.8) can be expressed as

$$\Upsilon(\mathbf{y}) \underset{H_0}{\overset{H_1}{\gtrless}} T \qquad (6.14)$$

The idea of a sufficient statistic is quite rich. For example, it can be interpreted as a geometric coordinate transformation chosen to place all of the useful information in the first coordinate (Van Trees et al., 2013). Procedures for verifying that a statistic (a function of the data $\mathbf{y}$) is sufficient are given in Kay (1993), as is the Neyman-Fisher factorization theorem for identifying sufficient statistics. Detailed consideration of the properties of sufficient statistics is beyond the scope of this text; the reader is referred to the references for greater depth.

The specific value of the threshold $\eta = -\lambda$ that will ensure that $P_{FA} = \alpha$ as desired has not yet been found. The original expression for P_{FA} was given in Eq. (6.2), but this is not very useful since its evaluation requires the N-dimensional joint PDF of $\mathbf{y}$ and an explicit definition of the region $\Re_1$, which has been defined only implicitly as the points in N-space for which the LRT exceeds the still-unknown threshold. Since they are functions of the random data $\mathbf{y}$, Λ and Υ are also random variables and thus have their own probability density functions. An alternate approach to computing the LRT threshold is therefore to express P_{FA} in terms of Λ or Υ and then solve those expressions for η or equivalently for T. The required expressions are

$$P_{FA} = \int_{\eta = -\lambda}^{+\infty} p_\Lambda(\Lambda \mid H_0) \, d\Lambda = \alpha \qquad (6.15)$$

or

$$P_{FA} = \int_{T}^{+\infty} p_\Upsilon(\Upsilon \mid H_0) \, d\Upsilon = \alpha \qquad (6.16)$$

Because false alarms are the result of interference and do not involve targets, the result depends only on the PDF of the likelihood ratio [if using Eq. (6.15)] or the sufficient statistic [if using Eq. (6.16)] when a target is not present. Given a specific model of that PDF, a specific value can be computed for η (equivalently, λ) or T.

To illustrate, continue the real-valued "constant in WGN" example by finding the threshold and then evaluating the performance, working with the sufficient statistic $\Upsilon(\mathbf{y})$. In this case, $\Upsilon(\mathbf{y})$ is the sum of the individual data samples y_n. Under hypothesis H_0 (no target), the samples are i.i.d. It follows that $\Upsilon \sim N(0, N\sigma_w^2)$. A false alarm occurs anytime $\Upsilon > T$, so

$$\alpha = P_{FA} = \int_T^{+\infty} p_\Upsilon(\Upsilon|H_0)\, d\Upsilon$$

$$= \int_T^{+\infty} \frac{1}{\sqrt{2\pi N\sigma_w^2}} \exp\left[\frac{-\Upsilon^2}{2N\sigma_w^2}\right] d\Upsilon \tag{6.17}$$

Equation (6.17) is the integral of a Gaussian PDF, so the *error function* erf(x) will appear in the solution. The standard definition is (Olver et al., 2020)[7]

$$\mathrm{erf}(x) \equiv \frac{2}{\sqrt{\pi}} \int_0^x e^{-t^2}\, dt \tag{6.18}$$

Also define the *complementary error function* erfc(x) corresponding to erf(x),

$$\mathrm{erfc}(x) \equiv \frac{2}{\sqrt{\pi}} \int_x^{+\infty} e^{-t^2}\, dt = 1 - \mathrm{erf}(x) \tag{6.19}$$

One will generally be interested in finding the value of x that results in a certain value of erf(x) or erfc(x); thus the inverse error and complementary error functions, denoted by $\mathrm{erf}^{-1}(z)$ and $\mathrm{erfc}^{-1}(z)$, respectively, are of interest. It follows from Eq. (6.19) that the two are related by $\mathrm{erfc}^{-1}(z) = \mathrm{erf}^{-1}(1-z)$.[8]

With the change of variables $t = \Upsilon/\sqrt{2N\sigma_w^2}$, Eq. (6.17) can be written as

$$\alpha = P_{FA} = \frac{1}{\sqrt{\pi}} \int_{T/\sqrt{2N\sigma_w^2}}^{+\infty} \exp(-t^2)\, dt = \frac{1}{2}\left[1 - \mathrm{erf}\left(\frac{T}{\sqrt{2N\sigma_w^2}}\right)\right] \tag{6.20}$$

Finally, Eq. (6.20) can be solved to obtain the threshold T in terms of the inverse error function,

$$T = \sqrt{2N\sigma_w^2} \cdot \mathrm{erf}^{-1}(1 - 2P_{FA}) \tag{6.21}$$

Equations (6.20) and (6.21) show how to compute P_{FA} given T and vice versa.

---

[7] The definitions of Eqs. (6.18) and (6.19) are the same as those used in MATLAB®.
[8] The $\mathrm{erf}^{-1}(\cdot)$ function will often be used here, even when $\mathrm{erfc}^{-1}(\cdot)$ gives a slightly more compact expression, because of the wider availability of $\mathrm{erf}^{-1}(\cdot)$ functions than $\mathrm{erfc}^{-1}(\cdot)$ in MATLAB® and similar computational software packages.

All of the information needed to carry out the LRT in its sufficient statistic form of Eq. (6.14) is now available. $\Upsilon(\mathbf{y})$ is just the sum of the data samples, while the threshold T can be computed from the number N of samples, the variance σ_w^2 of the noise, which is assumed known, and the desired false alarm probability P_{FA}.

The performance of this detector is evaluated by constructing a *receiver operating characteristic* (ROC) curve. There are four interrelated variables of interest: P_D, P_{FA}, the noise power σ_w^2, and the constant m whose presence or absence is to be decided. The latter two are characteristics of the given signals, while P_{FA} is generally fixed as part of the system specifications at whatever level is deemed tolerable. Thus it is necessary only to determine P_D. The approach is identical to that used for determining P_{FA}: determine the probability density function of the sufficient statistic Υ under hypothesis H_1 and integrate the area under it from the threshold to $+\infty$.

Continuing the example, note that the only change under hypothesis H_1 is that the individual data samples y_n now each have mean m, so their sum Υ has mean Nm. Thus $\Upsilon(\mathbf{y}) \sim N(Nm, N\sigma_w^2)$ and

$$P_D = \int_T^{+\infty} p_\Upsilon(\Upsilon|H_1)\,d\Upsilon$$
$$= \int_T^{+\infty} \frac{1}{\sqrt{2\pi N\sigma_w^2}} \exp\left[\frac{-(\Upsilon-Nm)^2}{2N\sigma_w^2}\right] d\Upsilon \tag{6.22}$$

Again applying the definition of the error function in Eq. (6.18) leads to

$$P_D = \frac{1}{2}\left[1 - \operatorname{erf}\left(\frac{T-Nm}{\sqrt{2N\sigma_w^2}}\right)\right] \tag{6.23}$$

Since the primary concern is the relationship between the performance metrics P_D and P_{FA}, Eq. (6.21) can be used in Eq. (6.23) to eliminate the threshold T and arrive at

$$P_D = \frac{1}{2}\left[1 - \operatorname{erf}\left\{\operatorname{erf}^{-1}(1-2P_{FA}) - \frac{\sqrt{N}\cdot m}{\sqrt{2\sigma_w^2}}\right\}\right]$$
$$= \frac{1}{2}\operatorname{erfc}\left\{\operatorname{erfc}^{-1}(2P_{FA}) - \frac{\sqrt{N}\cdot m}{\sqrt{2\sigma_w^2}}\right\} \tag{6.24}$$

Nm is considered to be the signal voltage of interest in the sufficient statistic $\Upsilon(\mathbf{y})$. The corresponding signal power is $(Nm)^2$. The power of the noise component of $\Upsilon(\mathbf{y})$ is $N\sigma_w^2$. Thus, the term $\sqrt{N}\cdot m/\sigma_w$ is the square root of the signal-to-noise ratio (SNR) χ for this problem, and Eq. (6.24) can be rewritten as

$$P_D = \frac{1}{2}\left[1 - \operatorname{erf}\left\{\operatorname{erf}^{-1}(1-P_{FA}) - \sqrt{\chi/2}\right\}\right]$$
$$= \frac{1}{2}\operatorname{erfc}\left[\operatorname{erfc}^{-1}(2P_{FA}) - \sqrt{\chi/2}\right] \tag{6.25}$$

Figure 6.1 illustrates how the detection and false alarm probabilities follow from the PDFs under the two hypotheses and the threshold, and how their relative values depend on

Detection Fundamentals 351

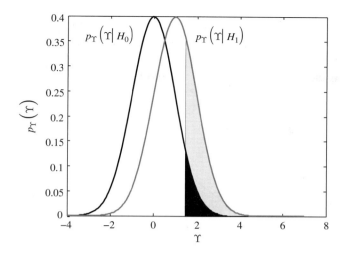

FIGURE 6.1 Gaussian probability density functions of the sufficient statistic when $\sigma_w^2 = 1/N$ and $m = 1/N$ under hypothesis H_0 (left) and H_1 (right).

the relation between the two PDFs. Two Gaussian PDFs with variance equal to one are shown. The leftmost has a zero mean, while the rightmost has a mean of 1. If $\sigma_w^2 = 1/N$ and $m = 1/N$, these fit the model of Eq. (6.9) and the subsequent analysis. P_D and P_{FA} are the areas under the right and left PDFs, respectively, from the threshold (shown as a vertical line at about $\Upsilon = 1.5$) to $+\infty$. The receiver design thus consists of adjusting the position of the threshold until the black area equals the acceptable false alarm probability. The detection probability is then the gray area (which includes the black area). This figure again makes it clear that P_D and P_{FA} must increase or decrease together as the threshold moves lower or higher. The achievable *combinations* of P_D and P_{FA} are determined by the degree to which the two distributions overlap.

Figure 6.2 illustrates the ROC for this problem with the SNR χ as a parameter. Figure 6.2a plots the ROC using linear scales for both P_{FA} and P_D. Several features are worth noting. First, $P_{FA} = P_D$ when $\chi = 0$ (implying $m = 0$). This is to be expected since in that case, the PDF of $\Upsilon(\mathbf{y})$ is the same under either hypothesis. For a given P_{FA} and $\chi > 0$, P_D increases as the SNR increases, a result which should be intuitively satisfying. Finally, note how abrupt the transition between near-zero and near-unity detection probabilities becomes as the SNR increases. This is a little misleading, since radars normally operate with very low values of P_{FA}; depending on the type of system, P_{FA} is typically no higher than 10^{-3} and very often is in the range of 10^{-6} to 10^{-8} or even lower. Figure 6.2b plots the same data on a logarithmic scale for P_{FA}, which better reveals the characteristics of the ROC for false alarm probabilities of interest in radar signal processing.

If the achievable combinations of P_D and P_{FA} do not meet the performance specifications, what can be done? Consideration of Fig. 6.1 suggests two answers. First, for a given P_{FA}, P_D can be increased by causing the two PDFs to move further apart when a target is present. That is, the presence of a target must cause a larger shift in the mean m of the distribution of the sufficient statistic. It was shown that the SNR equals m^2/σ_w^2, and that the way to improve the detection/false alarm tradeoff is to increase the SNR. This conclusion is borne out in Fig. 6.2. Figure 6.3a illustrates the effect of increased SNR on the PDFs and performance probabilities under the two hypotheses using the same threshold as Fig. 6.1.

Chapter Six

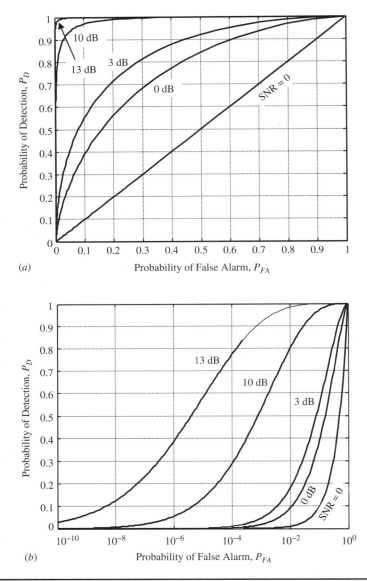

FIGURE 6.2 Receiver operating characteristic for the Gaussian example. (a) Displayed on a linear P_{FA} scale. (b) Displayed on a logarithmic P_{FA} scale.

The second way to improve the performance tradeoff is to reduce the overlap of the PDFs by reducing their variance. Reducing the noise power σ_w^2 will reduce the variance of both PDFs, leading to the situation shown in Fig. 6.3b (where the area corresponding to P_{FA} is too small to be seen) and again improving performance. As with the first technique of increasing m, reducing σ_w^2 again constitutes increasing the SNR. Thus, consistent with Eq. (6.25), improving the tradeoff between P_D and P_{FA} requires increasing the SNR χ. This is a fundamental result that will arise repeatedly.

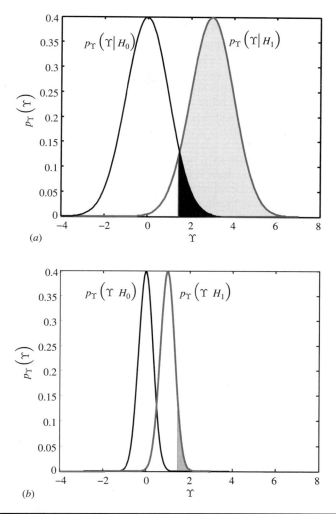

FIGURE 6.3 Two ways to modify the PDFs of Fig. 6.1 to improve the tradeoff between detection and false alarms. (a) Increasing the signal power. (b) Reducing the noise power.

Radar systems are designed to achieve specified values of P_D and P_{FA} subject to various conditions, such as specified ranges, target types, interference environments, and so forth. The designer can work with antenna design, transmitter power, waveform design, and signal processing techniques, all within cost and form factor constraints. The job of the designer is therefore to develop a radar system design which ultimately results in a pair of "target absent" and "target present" PDFs at the point of detection with a small enough overlap to allow the desired P_D and P_{FA} to be achieved. If the design does not do this, the designer must redesign one or more of these elements to reduce the variance of the PDFs, shift them further apart, or both until the desired performance is obtained. Thus, a significant goal of radar system design is controlling the overlap of the two PDFs analogous to those in Fig. 6.1, or equivalently, maximizing the SNR.

6.3 Threshold Detection in Coherent Systems

The real-valued Gaussian problem considered so far is useful to introduce and explain the major elements of Neyman-Pearson detection such as the likelihood ratio test, probabilities of detection and false alarm, receiver operating characteristics, and the major design tradeoffs that follow. The problem seems "radar-like": under one hypothesis, only Gaussian noise is observed; under the other, a constant was added to the noise, which could be interpreted as the echoes from a steady target. Figure 6.4 summarizes the design and analysis strategy that was used. Beginning with models of the PDF of the data under hypotheses H_0 and H_1, the LRT or log-LRT is written down and manipulated to isolate the terms involving the measured data. If necessary or useful, a simplified detector law is substituted (see Sec. 6.3.3). The sufficient statistic is then identified and its PDF under each hypothesis determined. The PDF under H_0 is integrated to get a relationship between the threshold T and the P_{FA} which is solved analytically or numerically to determine the threshold that gives the desired P_{FA}. Finally, that threshold value is used with the PDF of Υ under H_1 to get P_D. This same analysis strategy will be used repeatedly to develop the detector design and evaluate its performance for various data models. The only difference will be in the raw data PDFs $p_\mathbf{y}(\mathbf{y}|H_0)$ and $p_\mathbf{y}(\mathbf{y}|H_1)$ assumed in the first step. However, changes to these PDFs ripple

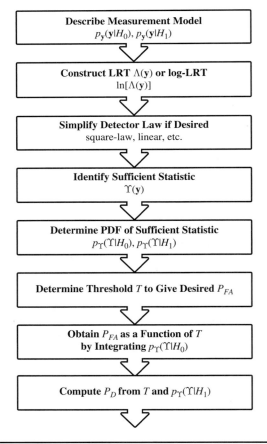

FIGURE 6.4 Strategy for radar detector design, analysis, and performance evaluation.

through the entire process, changing the likelihood ratio, then the sufficient statistic and thus the detector design, the threshold setting, and finally the detection performance.

The real constant-in-WGN example is not a good model for any actual radar detection problem due to at least three major limitations. First, coherent radar systems that produce complex-valued measurements are of most interest. The approach must therefore be extended to the complex case. Second, there are unknown parameters. The analysis so far has assumed that such signal parameters as the noise variance and target amplitude are known, when in fact these are not known a priori but must be estimated if needed. To complicate matters further, some parameters are linked. Specifically in radar, the (unknown) echo amplitude varies with the (unknown) echo arrival time according to the appropriate version of the radar range equation. Thus, the LRT must be generalized to develop a technique that can work when some signal parameters are unknown. Finally, as was seen in Chap. 2, there are a number of established models for radar signal phenomenology that must be incorporated. In particular, it is necessary to account for fluctuating targets, that is, statistical variations in the amplitude of the target components of the measured data when a target is present. Furthermore, while the Gaussian PDF remains a good model for noise, in many problems the dominant interference is clutter which may have one of the distinctly non-Gaussian PDFs discussed in Chap. 2. The next subsections begin addressing these shortcomings by extending the LRT to coherent systems.

6.3.1 The Gaussian Case for Coherent Receivers

An appropriate model for noise at the output of a coherent receiver was developed in Chap. 2. It was shown there that if the noise in the system prior to quadrature signal generation is a zero mean, white Gaussian process with power $\sigma_w^2 = kT$,[9] the I and Q channels will each contain independent, identically distributed zero-mean white Gaussian processes with power $kT/2 = \sigma_w^2/2$. That is, the noise power splits evenly but independently between the two channels. The expression for the joint PDF of N complex samples of the Gaussian random process is

$$p_{\mathbf{y}}(\mathbf{y}) = \frac{1}{\det\{\pi \mathbf{S}_\mathbf{y}\}} \exp\left\{-(\mathbf{y} - \mathbf{m})^H \mathbf{S}_\mathbf{y}^{-1}(\mathbf{y} - \mathbf{m})\right\} \qquad (6.26)$$

where $\mathbf{m}$ is the $N \times 1$ vector mean of the $N \times 1$ vector signal $\mathbf{y} = \mathbf{m} + \mathbf{w}$, $\mathbf{S}_\mathbf{y}$ is the $N \times N$ covariance matrix of $\mathbf{y}$,

$$\mathbf{S}_\mathbf{y} = E\{\mathbf{y} \cdot \mathbf{y}^H\} - \mathbf{m} \cdot \mathbf{m}^H \qquad (6.27)$$

and H is the Hermitian (conjugate transpose) operator. In most cases, the noise samples are i.i.d. so that $\mathbf{S}_\mathbf{y} = \sigma_w^2 \mathbf{I}_N$, which in turn means that $\det\{\pi \mathbf{S}_\mathbf{y}\} = \pi^N \sigma_w^{2N}$. Treatment of the case where the noise samples are not equal variance and the colored noise case, where $\mathbf{S}_\mathbf{y}$ is not diagonal is beyond the scope of this text. The reader is referred to Dudgeon and Johnson (1993) and Kay (1998) for these more complex situations.

Equation (6.26) now reduces to

$$p_{\mathbf{y}}(\mathbf{y}) = \frac{1}{\pi^N \sigma_w^{2N}} \exp\left\{-\frac{1}{\sigma_w^2}(\mathbf{y} - \mathbf{m})^H(\mathbf{y} - \mathbf{m})\right\} \qquad (6.28)$$

---

[9]Here T refers to receiver temperature, not the detection threshold. Which meaning of T is intended should be clear from context throughout this chapter.

The notation $\mathbf{y} \sim CN(\mathbf{m}, \sigma_w^2)$ will be used as shorthand for "$\mathbf{y}$ is distributed as an i.i.d. complex normal r.v. with mean $\mathbf{m}$ and variance σ_w^2," that is, its PDF is of the form of Eq. (6.28). Further simplifications occur when all of the means under H_1 are identical so that $\mathbf{m} = m\mathbf{1}_N$, where m can now be complex-valued. In this case, Eq. (6.28) reduces slightly further to

$$p_\mathbf{y}(\mathbf{y}) = \frac{1}{\pi^N \sigma_w^{2N}} \exp\left\{-\frac{1}{\sigma_w^2}(\mathbf{y} - m\mathbf{1}_N)^H (\mathbf{y} - m\mathbf{1}_N)\right\} \qquad (6.29)$$

The LRT for the coherent version of the previous Gaussian example can be obtained by repeating the steps in the example of Eqs. (6.10) through (6.25) using the PDF of Eq. (6.28), with $\mathbf{m} = \mathbf{0}_N$ under hypothesis H_0 and $\mathbf{m} \neq \mathbf{0}_N$ under H_1. The log-likelihood ratio is

$$\begin{aligned}\ln(\Lambda) &= \frac{1}{\sigma_w^2}[2\,\mathrm{Re}\{\mathbf{m}^H \mathbf{y}\} - \mathbf{m}^H \mathbf{m}] \\ &= \frac{2}{\sigma_w^2} \mathrm{Re}\left\{\sum_{n=0}^{N-1} m^* y_n\right\} - \frac{1}{\sigma_w^2} N|m|^2\end{aligned} \qquad (6.30)$$

where the second line of Eq. (6.30) applies only to the case where the means are identical ($\mathbf{m} = m\mathbf{1}_N$).

Some interpretation of Eq. (6.30) is in order. The term $\mathbf{m}^H \mathbf{y}$ is the dot product of the complex vectors $\mathbf{m}$ and $\mathbf{y}$. As seen in App. B, this dot product represents an FIR filtering operation evaluated at the particular instant when the equivalent impulse response $\mathbf{m}^H$ and the data vector $\mathbf{y}$ completely overlap. Furthermore, since the impulse response $\mathbf{m}^H$ of the filter is the conjugate transpose of the signal whose presence is to be detected under hypothesis H_1, namely $\mathbf{m}$, it is a matched filter. The same reasoning applies if the elements of $\mathbf{m}$ are the samples of a modulated waveform or any other function of interest.

The second term in the log LRT is the complex dot product of $\mathbf{m}$ with itself, which expands to $\mathbf{m}^H \mathbf{m} = \sum_{n=0}^{N-1} |m_n|^2$. This is just the energy E in $\mathbf{m}$. In the equal means case $E = N|m|^2$.

Finally, note the $\mathrm{Re}\{\cdot\}$ operator applied to the matched filter output $\mathbf{m}^H \mathbf{y}$. Because $\mathbf{m}$ and $\mathbf{y}$ are complex, one might be concerned that the dot product could be purely imaginary or nearly so, such that $\mathrm{Re}\{\mathbf{m}^H \mathbf{y}\} \approx 0$. The measured data $\mathbf{y}$ would then have little or no effect on the threshold test. For this example, $\mathbf{m} = \mathbf{0}_N$ under hypothesis H_0 and the $\mathrm{Re}\{\cdot\}$ operator is inconsequential. Under hypothesis H_1 each element of $\mathbf{m}$ is a complex number $m_n \exp(j\theta_n)$. If the target is actually present the elements of the measured data vector $\mathbf{y} = \mathbf{m} + \mathbf{w}$ will be of the form $m_n \exp(j\theta_n) + w_n$, where w_n is a zero mean complex Gaussian noise sample. It follows that

$$\begin{aligned}\mathbf{m}^H \mathbf{y} &= \mathbf{m}^H \mathbf{m} + \mathbf{m}^H \mathbf{w} \\ &= \sum_{n=0}^{N-1} |m_n|^2 + \sum_{n=0}^{N-1} w_n m_n \exp(-j\theta_n)\end{aligned} \qquad (6.31)$$

The first term is again the energy E in the signal $\mathbf{m}$; this is real-valued and therefore unaffected by the $\mathrm{Re}\{\cdot\}$ operator. The second term is weighted and integrated noise samples. The phase of this noise component and therefore the effect of the $\mathrm{Re}\{\cdot\}$ operator is random. Its effect on the phase of the sum will be large when the SNR is low, but minimal when the SNR is high.

It is evident by inspection of Eq. (6.30) that the sufficient statistic is now $\mathrm{Re}\{\mathbf{m}^H \mathbf{y}\}$. Expressing the LRT in its sufficient statistic form for the complex case gives

$$\Upsilon = \mathrm{Re}\{\mathbf{m}^H \mathbf{y}\} \underset{H_0}{\overset{H_1}{\gtrless}} \frac{\sigma_w^2}{2}\ln(-\lambda) + \frac{E}{2} = T \qquad (6.32)$$

Note that if $\mathbf{m} = m\mathbf{1}_N$, the term $\text{Re}\{\mathbf{m}^H\mathbf{y}\} = m\sum y_n$, and Eq. (6.32) is very similar to Eq. (6.13).

To complete consideration of the complex Gaussian case, its performance, that is, P_D and P_{FA}, must be determined. The sufficient statistic $\Upsilon = \text{Re}\{\mathbf{m}^H\mathbf{y}\} = \text{Re}\{\sum m_n^* y_n\}$ is a sum of Gaussian random variables and so will also be Gaussian. To determine the performance of the coherent detector, the PDF of Υ must be determined under each hypothesis. To do so, it is useful to first consider the quantity $\mathbf{z} = \mathbf{m}^H\mathbf{y}$, which will be a complex Gaussian r.v. First suppose hypothesis H_0 is true. In this case, the $\{y_n\}$ are zero mean and therefore so is z. Because the $\{y_n\}$ are independent, the variance of z is just the sum of the variances of the individual weighted samples:

$$\text{var}(z) = \sum_{n=0}^{N-1} \text{var}(m_n^* y_n) = \sum_{n=0}^{N-1} |m_n|^2 \sigma_w^2 = E\sigma_w^2 \tag{6.33}$$

Thus, under hypothesis H_0, $z \sim CN(0, E\sigma_w^2)$. Similarly, under hypothesis H_1, $\mathbf{y} = \mathbf{m} + \mathbf{w}$ and $z \sim CN(E, E\sigma_w^2)$. Note that the mean of z is real in both cases. The power of the complex Gaussian noise splits evenly between the real and imaginary parts of z. Since $\Upsilon = \text{Re}\{\mathbf{m}^H\mathbf{y}\}$, it follows that $\Upsilon \sim N(0, E\sigma_w^2/2)$ under H_0 and $\Upsilon \sim N(E, E\sigma_w^2/2)$ under H_1. Following the procedure used in Sec. 6.2.2, it can be seen that

$$P_{FA} = \frac{1}{2}\left[1 - \text{erf}\left(\frac{T}{\sqrt{E\sigma_w^2}}\right)\right] \tag{6.34}$$

Repeating the development of Eqs. (6.22) to (6.24) gives the probability of detection

$$\begin{aligned} P_D &= \frac{1}{2}\left[1 - \text{erf}\left\{\text{erf}^{-1}(1 - 2P_{FA}) - \sqrt{\frac{E}{\sigma_w^2}}\right\}\right] \\ &= \frac{1}{2}\text{erfc}\left\{\text{erfc}^{-1}(2P_{FA}) - \sqrt{\frac{E}{\sigma_w^2}}\right\} \end{aligned} \tag{6.35}$$

Note again that the last term in Eq. (6.35) is the square root of the energy in the signal $\mathbf{m}$, divided this time by the noise power σ_w^2; that is, the SNR. Thus Eq. (6.35) can be written as

$$\begin{aligned} P_D &= \frac{1}{2}\{1 - \text{erf}[\text{erf}^{-1}(1 - 2P_{FA}) - \sqrt{\chi}]\} \\ &= \frac{1}{2}\text{erfc}[\text{erfc}^{-1}(2P_{FA}) - \sqrt{\chi}] \end{aligned} \tag{6.36}$$

Finally, in the equal means case when $\mathbf{m} = m\mathbf{1}_N$, Eq. (6.35) is similar (but not identical) to Eq. (6.24). The coherent case includes the term $\sqrt{Nm^2/2\sigma_w^2}$ instead of $\sqrt{Nm^2/\sigma_w^2}$ because all of the signal energy competes with only half of the noise power.

Figure 6.5 shows the receiver operating characteristic for this example. It is identical in general form to the real-valued case of Fig. 6.2; however, for a given SNR the performance is better because in the coherent receiver the signal competes with only half the noise power. For example, in this coherent case an SNR of 13 dB produces $P_D = 0.94$ at a $P_{FA} = 10^{-6}$. Figure 6.2b shows that in the real case the same χ and P_{FA} produce a P_D of just under 0.39.

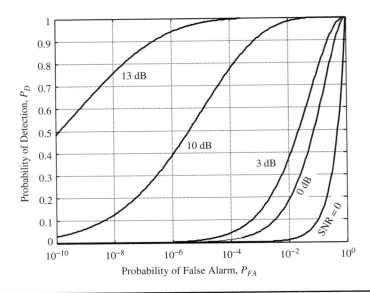

FIGURE 6.5 Performance of the coherent receiver on the complex Gaussian example.

6.3.2 Unknown Parameters and Threshold Detection

In general, perfect knowledge of each of the parameters of the PDFs $p_\Upsilon(\Upsilon|H_0)$ and $p_\Upsilon(\Upsilon|H_1)$ is required to carry out the LRT, which usually means having perfect knowledge of $p_\mathbf{y}(\mathbf{y}|H_0)$ and $p_\mathbf{y}(\mathbf{y}|H_1)$. For instance, in the Gaussian example it was assumed that the expected signal $\mathbf{y}$ is known under the various hypotheses, as well as the noise sample variance σ_w^2. This is not the case in the real world, where the PDFs that form the likelihood ratio may depend on one or more parameters ξ that are unknown. Depending on the available information three cases arise:

1. ξ is a random variable with a known probability density function.
2. ξ is a random variable with an unknown probability density function.
3. ξ is deterministic but unknown.

Different techniques are used to handle each of these cases. The first is the most important because it has the greatest effect on the structure of the optimal Neyman-Pearson detector.

To illustrate the approach for handling a random parameter with a known PDF, consider yet again the complex Gaussian case. The optimal detector implemented a matched filter operation $\mathbf{m}^H\mathbf{y}$ followed by the Re$\{\cdot\}$ operator. The success of the matched filter structure depended on knowing exactly the constant component $\mathbf{m}$ of $\mathbf{y} = \mathbf{m} + \mathbf{w}$ under hypothesis H_1, so that the filter coefficients could be set equal to $\mathbf{m}^H$ and the filter output would be real-valued. Recall that, when applied to radar, $\mathbf{y}$ under H_0 is considered to consist only of samples $\mathbf{w}$ of receiver noise, and under H_1 to consist of noisy samples $\mathbf{m} + \mathbf{w}$ of the echoes from a radar target over multiple pulses, or alternately successive fast-time samples of the waveform of one pulse echo from a target.

Claiming perfect knowledge of $\mathbf{m}$ implies knowing the range to the target very precisely, since a variation in one-way range of only $\lambda_t/4$ causes the received echo phase to change by 180°. A quarter-wavelength is only 30 cm at L band and 3.16 mm at 95 GHz. Because this precision is usually unrealistic, it is more reasonable to assume $\mathbf{m}$ is known only to within a phase factor $\exp(j\theta)$, where the phase angle θ is considered to be a random variable distributed uniformly over $[0, 2\pi)$ and independent of the random variables $\{m_n\}$. In other words, $\mathbf{m} = \tilde{\mathbf{m}} \exp(j\theta)$, where $\tilde{\mathbf{m}}$ is known exactly, but θ is a random phase. Note that the energy in $\tilde{\mathbf{m}}$ is the same as that in $\mathbf{m}$, that is, $\mathbf{m}^H \mathbf{m} = \tilde{\mathbf{m}}^H \tilde{\mathbf{m}}$. This "unknown phase" assumption cannot usually be avoided in radar. What is its effect on the optimal detector and its performance?

The goal remains to carry out the LRT, so it is necessary to return to its basic definition of Eq. (6.6) and determine $p_\mathbf{y}(\mathbf{y} \mid H_0)$ and $p_\mathbf{y}(\mathbf{y} \mid H_1)$, both of which now presumably depend on θ, and use the technique known as the *Bayesian approach* for random parameters with known PDFs (Kay, 1998).[10] Specifically, compute the PDF under H_i by averaging the conditional PDF $p_\mathbf{y}(\mathbf{y} \mid H_i, \theta)$ over θ:

$$p_\mathbf{y}(\mathbf{y} \mid H_i) = \int p_\mathbf{y}(\mathbf{y} \mid H_i, \theta) p_\theta(\theta) \, d\theta, \quad i = 0, 1 \qquad (6.37)$$

The unconditional PDFs $p_\mathbf{y}(\mathbf{y} \mid H_i)$ are then used to define the likelihood ratio in the usual way.

As an example of the Bayesian approach for random parameters, consider again the complex Gaussian case, but now with an unknown phase in the data, $\mathbf{m} = \tilde{\mathbf{m}} \exp(j\theta)$. The conditional PDF of the observations $\mathbf{y}$ becomes, under each of the two hypotheses,

$$\begin{aligned} p_\mathbf{y}(\mathbf{y} \mid H_0, \theta) &= \frac{1}{\pi^N \sigma_w^{2N}} \exp\left[-\frac{1}{\sigma_w^2} \mathbf{y}^H \mathbf{y}\right] \\ p_\mathbf{y}(\mathbf{y} \mid H_1, \theta) &= \frac{1}{\pi^N \sigma_w^{2N}} \exp\left\{-\frac{1}{\sigma_w^2} [\mathbf{y} - \tilde{\mathbf{m}} \exp(j\theta)]^H [\mathbf{y} - \tilde{\mathbf{m}} \exp(j\theta)]\right\} \end{aligned} \qquad (6.38)$$

Expanding the exponent in $p_\mathbf{y}(\mathbf{y} \mid H_1, \theta)$ gives

$$\begin{aligned} p_\mathbf{y}(\mathbf{y} \mid H_1, \theta) &= \frac{1}{\pi^N \sigma_w^{2N}} \exp\left[-\frac{1}{\sigma_w^2} (\mathbf{y}^H \mathbf{y} - 2\,\mathrm{Re}\{\tilde{\mathbf{m}}^H \mathbf{y} \exp(-j\theta)\} + E)\right] \\ &= \frac{1}{\pi^N \sigma_w^{2N}} \exp\left[-\frac{1}{\sigma_w^2} (\mathbf{y}^H \mathbf{y} - 2|\tilde{\mathbf{m}}^H \mathbf{y}|\cos(\phi - \theta) + E)\right] \end{aligned} \qquad (6.39)$$

where ϕ is the unknown but fixed phase of the inner product $\tilde{\mathbf{m}}^H \mathbf{y}$.

Notice that $p_\mathbf{y}(\mathbf{y} \mid H_0, \theta)$ does not depend on θ after all (not surprising since there is no target present in this case to present an unknown phase), so it is not necessary to apply Eq. (6.37). However, in $p_\mathbf{y}(\mathbf{y} \mid H_1)$ the dependence on θ is explicit. Assuming a uniform random phase, defining $\theta' = \phi - \theta$ and applying Eq. (6.37) under H_1 gives

$$p_\mathbf{y}(\mathbf{y} \mid H_1) = \frac{1}{\pi^N \sigma_w^{2N}} \exp[-(\mathbf{y}^H \mathbf{y} + E)/\sigma_w^2] \cdot \frac{1}{2\pi} \int_0^{2\pi} \exp\left[\frac{2}{\sigma_w^2} |\tilde{\mathbf{m}}^H \mathbf{y}| \cos \theta'\right] d\theta' \qquad (6.40)$$

Equation (6.40) is a standard integral. Specifically, integral 10.32.1 in Olver et al. (2020) is

$$\frac{1}{\pi} \int_0^\pi \exp(\pm z \cos \theta) \, d\theta = I_0(z) \qquad (6.41)$$

---

[10] An alternative approach called the *generalized likelihood ratio test* (GLRT), in which the unknown parameter(s) are replaced by their maximum likelihood estimates, is discussed in many detection theory texts (e.g., Kay, 1998).

where $I_0(z)$ is the modified Bessel function of the first kind. Using this result and properties of the cosine function, Eq. (6.40) becomes

$$p_\mathbf{y}(\mathbf{y}|H_1) = \frac{1}{\pi^N \sigma_w^{2N}} \exp[-(\mathbf{y}^H\mathbf{y} + E)/\sigma_w^2] I_0\left(\frac{2|\tilde{\mathbf{m}}^H\mathbf{y}|}{\sigma_w^2}\right) \quad (6.42)$$

The log-LRT now becomes

$$\ln(\Lambda) = \ln\left[I_0\left(\frac{2|\tilde{\mathbf{m}}^H\mathbf{y}|}{\sigma_w^2}\right)\right] - \frac{E}{\sigma_w^2} \underset{H_0}{\overset{H_1}{\gtrless}} \ln(-\lambda) \quad (6.43)$$

or, in sufficient statistic form,

$$\Upsilon = \ln\left[I_0\left(\frac{2|\tilde{\mathbf{m}}^H\mathbf{y}|}{\sigma_w^2}\right)\right] \underset{H_0}{\overset{H_1}{\gtrless}} \ln(-\lambda) + \frac{E}{\sigma_w^2} = T \quad (6.44)$$

Equation (6.44) defines the signal processing required for optimum detection in the presence of an unknown phase. It calls for taking the magnitude of the matched filter output $\tilde{\mathbf{m}}^H\mathbf{y}$, passing it through the memoryless nonlinearity $\ln[I_0(\cdot)]$, and comparing the result to a threshold. This result is appealing in that the matched filter is still applied to utilize the *internal* phase structure of the known signal and maximize the integration gain, but then a magnitude operation is applied because the absolute phase of the result cannot be known. Also, note that the argument of the Bessel function is the energy in the matched filter output divided by half the noise power, again an SNR. Only half of the noise power appears because the total noise power in the complex case is split between the real and imaginary channels.

As a practical matter, it is desirable to avoid having to compute the natural logarithm and Bessel function for every threshold test, since these might occur millions of times per second in some systems. Because the function $\ln[I_0(\cdot)]$ is monotonically increasing, the same detection results can be obtained by simply comparing its argument $2|\tilde{\mathbf{m}}^H\mathbf{y}|/\sigma_w^2$ to a modified threshold. Folding $2/\sigma_w^2$ into T to form a new threshold T', Eq. (6.44) then becomes simply

$$|\tilde{\mathbf{m}}^H\mathbf{y}| \underset{H_0}{\overset{H_1}{\gtrless}} T' \quad (6.45)$$

Figure 6.6 illustrates the optimal detector for the coherent detector with an unknown phase.

The performance of this detector will now be established. Let $z = |\tilde{\mathbf{m}}^H\mathbf{y}|$. The detection test becomes simply $z \gtrless T'$; thus the distribution of z under each of the two hypotheses is

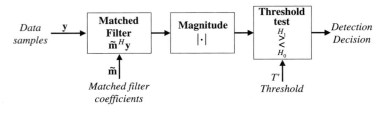

FIGURE 6.6 Structure of optimal detector when the signal phase is unknown.

needed. As in the known phase case, under hypothesis H_0 (target absent) the real and imaginary parts of $\tilde{\mathbf{m}}^H \mathbf{y}$ are independent of one another and each distributed as $N(0, E\sigma_w^2/2)$. It follows (see App. A) that z is Rayleigh distributed,

$$p_z(z|H_0) = \begin{cases} \dfrac{2z}{E\sigma_w^2} \exp\left(-\dfrac{z^2}{E\sigma_w^2}\right), & z \geq 0 \\ 0, & z < 0 \end{cases} \tag{6.46}$$

The probability of false alarm is

$$P_{FA} = \int_{T'}^{+\infty} p_z(z|H_0)\, dz = \exp\left(\dfrac{-T'^2}{E\sigma_w^2}\right) \tag{6.47}$$

This equation can be inverted to obtain the threshold setting in terms of P_{FA}:

$$T' = \sqrt{-E\sigma_w^2 \ln(P_{FA})} \tag{6.48}$$

Now consider hypothesis H_1, that is, target present. In this case $\tilde{\mathbf{m}}^H \mathbf{y} \sim CN(E, E\sigma_w^2)$. For a particular target and thus a particular value of θ, the real and imaginary parts of $\tilde{\mathbf{m}}^H \mathbf{y}$ are distributed, respectively, as $N(E, E\sigma_w^2/2)$ and $N(0, E\sigma_w^2/2)$. Regardless of the value of θ, the PDF of $z = |\tilde{\mathbf{m}}^H \mathbf{y}|$ is then

$$p_z(z|H_1) = \begin{cases} \dfrac{2z}{E\sigma_w^2} \exp\left[-\dfrac{1}{E\sigma_w^2}(z^2+E^2)\right] I_0\left(\dfrac{2z}{\sigma_w^2}\right), & z \geq 0 \\ 0, & z < 0 \end{cases} \tag{6.49}$$

where $I_0(z)$ is again the modified Bessel function of the first kind. Equation (6.49) is the Rician PDF. The probability of detection is obtained by integrating it from T' to $+\infty$.

In normalized form, the required integral is

$$Q_M(\alpha, \gamma) = \int_{\gamma}^{+\infty} t \cdot \exp\left[-\dfrac{1}{2}(t^2 + \alpha^2)\right] I_0(\alpha t)\, dt \tag{6.50}$$

The expression $Q_M(\alpha, \gamma)$ is known as *Marcum's Q function*. It arises frequently in radar detection calculations. A closed form for this integral is not known. Algorithms for evaluating $Q_M(\alpha, \gamma)$ are compared in Cantrell and Ojha (1987). The Communications Toolbox™ and Signal Processing Toolbox™ optional packages in MATLAB® include a marcumq function to evaluate $Q_M(\alpha, \gamma)$; another MATLAB® algorithm is given in Kay (1998).

By defining a change of variables the integral of Eq. (6.49) can be put into the form of Eq. (6.50). Specifically, choose $t = z/\sqrt{E\sigma_w^2/2}$ and $\alpha = \sqrt{2E}/\sigma_w$. Substituting into Eq. (6.49) and doing the integration gives

$$P_D = Q_M\left(\sqrt{\dfrac{2E}{\sigma_w^2}}, \sqrt{\dfrac{2T'^2}{E\sigma_w^2}}\right) \tag{6.51}$$

Finally, noting that E/σ_w^2 is the SNR χ and expressing the threshold in terms of the false alarm probability using Eq. (6.48) gives

$$P_D = Q_M\left(\sqrt{2\chi}, \sqrt{-2\ln(P_{FA})}\right) \quad (6.52)$$

It is usually the case that the energy E in $\mathbf{m}$ or $\tilde{\mathbf{m}}$ is not known. Fortunately, Eq. (6.52) does not depend on E or the noise power σ_w^2 explicitly, but only on their ratio χ, so that it is possible to generate the ROC without this information. However, actually implementing the detector requires a specific value of the threshold T' as given in Eq. (6.45), and this does require knowledge of both E and σ_w^2. One way to avoid this problem is to replace the matched filter coefficients $\tilde{\mathbf{m}}$ with a normalized coefficient vector $\hat{\mathbf{m}} = \tilde{\mathbf{m}}/E$. This choice simply normalizes the gain of the matched filter to 1. The energy in this modified sequence is $\hat{E} = 1$, leading to a modified threshold

$$\hat{T} = \sqrt{-\sigma_w^2 \ln(P_{FA})} \quad (6.53)$$

The modified matched filter gain and threshold result in no change to the ROC so that Eq. (6.52) remains valid. Setting of the threshold $\hat{T}$ still requires knowledge of the noise power σ_w^2; removal of this restriction is the subject of Sec. 6.5. The handling of unknown amplitude parameters is discussed in somewhat more detail in Sec. 6.3.4.

The performance of the envelope detector in this example is given in Fig. 6.7. The general behavior is very similar to the known phase coherent detector case of Fig. 6.5. Closer inspection, however, shows that for a given P_{FA}, the coherent detector obtains a higher P_D. To make this point clearer, Fig. 6.8 compares the detection curves for the coherent and envelope detectors (known and unknown phase, respectively), plotted two different ways. Figure 6.8a simply repeats the 10 dB curves from the two earlier figures. At $P_{FA} = 10^{-4}$ and $\chi = 10$ dB,

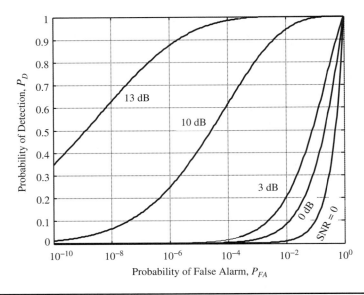

FIGURE 6.7 Performance of the linear envelope detector for the Gaussian example with unknown phase.

Detection Fundamentals 363

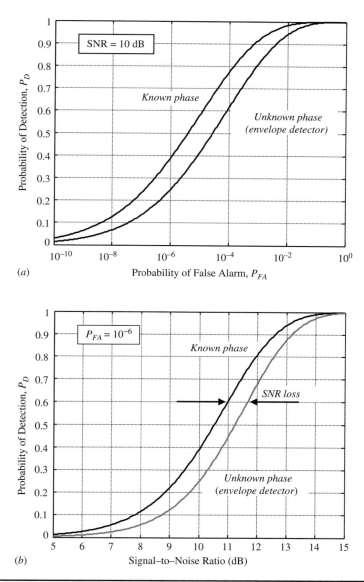

FIGURE 6.8 Performance difference between coherent and envelope detectors for the complex Gaussian example. (a) Difference in P_D for $\chi = 10$ dB. (b) Difference in P_D for $P_{FA} = 10^{-6}$.

for example, P_D is about 0.77 for the coherent detector. This figure drops to 0.62 when the envelope detector is used.

Figure 6.8b plots the detection performance as a function of χ with P_{FA} fixed (at 10^{-6} in this example). This figure shows that, to achieve the same probability of detection, the envelope detector requires about 0.6 dB higher SNR than the coherent detector at $P_D = 0.9$, and about 0.7 dB more at $P_D = 0.5$. The extra SNR required to maintain the detection performance of the envelope detector compared to the coherent case is called an *SNR loss*. SNR losses can result from many factors; this particular one is often called the *detector loss*. It represents extra SNR that must be obtained in some way if the performance of the envelope detector is to match that of the ideal coherent detector. Increasing the SNR in turn implies one or more of many radar system changes, such as greater transmitter power, a larger antenna gain, reduced range coverage, and so forth.

The phenomenon of detector loss illustrates a very important point in detection theory: the less that is known about the signal to be detected, the higher must be the SNR to achieve a given combination of P_D and P_{FA}. In this case, not knowing the absolute phase of the signal has cost about 0.6 dB. Inconvenient though it may be, this result is intuitively satisfying: the worse the knowledge of the signal details, the worse the performance of the detector will be.

6.3.3 Linear and Square Law Detectors

Equation (6.44) defines the optimal Neyman-Pearson detector for the Gaussian example with an unknown phase in the data. It was shown that the $\ln[I_0(x)]$ function could be replaced by its argument x without altering the performance. In Sec. 6.4.2 a simpler detector characteristic than $\ln[I_0(\cdot)]$ will again be desirable for noncoherent integration, but it will not be possible to simply substitute any monotonic increasing function. It is therefore useful to see what approximations can be made to the $\ln[I_0(\cdot)]$ function.

A standard series expansion for the Bessel function holds that

$$I_0(x) = 1 + \frac{x^2}{4} + \frac{x^4}{64} + \cdots \tag{6.54}$$

Thus for small x, $I_0(x) \approx 1 + x^2/4$. Furthermore, one series expansion of the natural logarithm has $\ln(1+z) = z - z^2/2 + z^3/3 + \cdots$. Combining these gives

$$\ln[I_0(x)] \approx \frac{x^2}{4} \qquad x \ll 1 \tag{6.55}$$

Equation (6.55) shows that if x is small, the optimal detector is well approximated by a matched filter followed by a so-called *square law detector*, that is, a magnitude squaring operation. The factor of 4 can be incorporated into the threshold in Eq. (6.44).

For large values of x, $I_0(x) \approx \exp(x)/\sqrt{2\pi x}$; then

$$\ln[I_0(x)] \approx x - \frac{1}{2}\ln(2\pi) - \frac{1}{2}\ln(x) \tag{6.56}$$

The constant term on the right of Eq. (6.56) can be incorporated into the threshold in Eq. (6.44), while the linear term in x quickly dominates the logarithmic term for $x \gg 1$. This leads to the *linear detector* approximation for large x

$$\ln[I_0(x)] \approx x, \qquad x \gg 1 \tag{6.57}$$

Figure 6.9 illustrates the fit between the square law and linear approximations and the exact $\ln[I_0(\cdot)]$ functions. The square law detector is an excellent fit for $x < 3$ dB, while the linear detector fits $\ln[I_0(x)]$ very well for $x > 10$ dB.

Finally, note that it is easy enough to compute the squared magnitude of a complex-valued test sample as simply the sum of the squares of the real and imaginary parts. The linear magnitude requires a square root and is less computationally convenient.

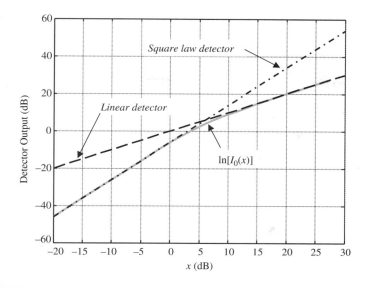

FIGURE 6.9 Approximation of the $\ln[I_0(\cdot)]$ detector characteristic by the square law detector when its argument is small, and the linear detector when its argument is large.

6.3.4 Other Unknown Parameters

The preceding sections have shown the effect of unknown phase of the received signal on the optimal detector. However, other parameters of the received signal are also unknown in practice. The amplitude of the echo depends on all of the factors in the radar range equation, including especially the unknown target radar cross section and, at least until it is successfully detected, its range. In addition, the target may be moving relative to the radar, so that the echo is modified by a Doppler shift.

The derivation of the magnitude-based detectors of Secs. 6.3.2 and 6.3.3 included an assumption that the received signal amplitude was known. Specifically, it was assumed that the received signal sample vector $\tilde{\mathbf{m}}$ was known except for its absolute phase. However, as noted the absolute amplitude is also unknown in general. To determine the effect of an unknown amplitude, assume that the received signal is $A \cdot \tilde{\mathbf{m}}$, where A is an unknown but deterministic scale factor.[11] The analysis of Sec. 6.3.2 can be repeated under this assumption. The detector output under hypothesis H_0 is unchanged as would be expected, since the target echo with its unknown amplitude is not present in this case. Under hypothesis H_1, the detector output is now $\text{Re}(\tilde{\mathbf{m}}^H \mathbf{y}) \sim N(A^2 E, E\sigma_w^2/2)$. Note that the detector still considers the quantity $\tilde{\mathbf{m}}^H \mathbf{y}$ rather than $A \cdot \tilde{\mathbf{m}}^H \mathbf{y}$ because the $\tilde{\mathbf{m}}$ arises from the matched filter applied to the data and thus does not include the unknown amplitude factor A of the signal echo. Also, the quantity $E = \tilde{\mathbf{m}}^H \mathbf{m}$ is now the energy of the matched filter reference signal, while the actual signal energy becomes $A^2 E$.

The equivalent of Eq. (6.51) is now

$$P_D = Q_M\left(\sqrt{\frac{2A^2 E}{\sigma_w^2}}, \sqrt{\frac{2T'^2}{E\sigma_w^2}}\right) \tag{6.58}$$

---

[11]For instance, the earlier discussion of unknown signal energy corresponds to having $A = 1/E$.

As before, the second argument of Eq. (6.58) can be written in terms of the probability of false alarm. Furthermore, because the actual signal energy is now A^2E, the first argument is still $\sqrt{2\chi}$. Thus, the detection performance is still given by Eq. (6.52). The unknown echo amplitude neither requires any change in the detector structure nor changes its performance.

Despite the unknown amplitude, the sufficient statistic was not changed. Furthermore, the probability of false alarm could be computed without knowledge of the amplitude. When both these conditions hold, the detection test is called a *uniformly most powerful* (UMP) test (Johnson and Dudgeon, 1993).

A UMP does not exist for the case where the not-yet-detected signal delay (range) is unknown, which again is the only realistic assumption that can be made in radar. It is therefore necessary to resort to a generalized likelihood ratio test (GLRT), in which the likelihood ratio is written as a function of the unknown signal delay Δ, and then the value of Δ that maximizes the likelihood ratio is found. Details are given in Johnson and Dudgeon (1993). The problem of estimating the time delay or range that maximizes the likelihood ratio is a major topic in Chap. 7. The result simply requires evaluation of the matched filter output to identify the range that produces the maximum output. In practice, each matched filter output sample is compared to a threshold. If the threshold is crossed, a detection is declared and the value of Δ at which the threshold crossing occurs is taken as an estimate of the target delay.

If the target is moving, an unknown Doppler shift will be imposed on the incident signal. The received echo will then be proportional not to $\tilde{m}$ but to a modified signal $\tilde{m}'$, where the samples of the reference signal $\tilde{m}$ have been multiplied by the complex exponential sequence $\exp(j\omega_D n)$, with ω_D the normalized Doppler shift. The required matched filter impulse response is now $\tilde{m}'$; if $\tilde{m}$ is replaced by $\tilde{m}'$ in the derivations of Sec. 6.2.2, the same performance results as before will be obtained. Because ω_D is unknown, however, it is necessary to test for different possible Doppler shifts by conducting the detection test for multiple possible values of ω_D, similar to the procedure used to test for unknown range. If a set of K potential Doppler frequencies uniformly spaced from $-RR/2$ Hz to $+RR/2$ Hz is to be tested, the matched filter can be implemented for all K frequencies at once using the pulse Doppler processing techniques described in Chap. 5.

6.4 Threshold Detection of Radar Signals

The results of the preceding sections can now be applied to some reasonably realistic scenarios for detecting radar targets in noise. These scenarios will almost always include unknown parameters of the signal to be detected (the target), specifically, its amplitude, absolute phase, time of arrival, and Doppler shift. Both detection using a single sample of the received signal and, when available, multiple samples are of interest. In the latter case, as discussed in Chap. 2, the target signal is often modeled as a random process, rather than a simple constant; the discussion in this chapter will be limited to the four Swerling models to illustrate both the approach and the classical, and still very useful, results obtained in these cases. Also, it will be seen that the idea of sample integration is needed in the case of multiple samples. Finally, a square law detector will be assumed, though one important approximation that applies to linear detectors will also be introduced. Figure 6.10 represents one possible taxonomy of the most common variations on the radar detection problem. Each of these will be discussed in turn in this section, with the exception of the Swerling 3 and 4 cases, for which the strategy is shown but the details are not carried out; the partially correlated case, which is not considered; and adaptive threshold-setting techniques (CFAR), which are the subject of Sec. 6.5.

6.4.1 Coherent, Noncoherent, and Binary Integration

The ability to detect targets is inhibited by the presence of noise, clutter, and jamming or other electromagnetic interference. All are modeled as random processes; the noise as uncorrelated

Detection Fundamentals

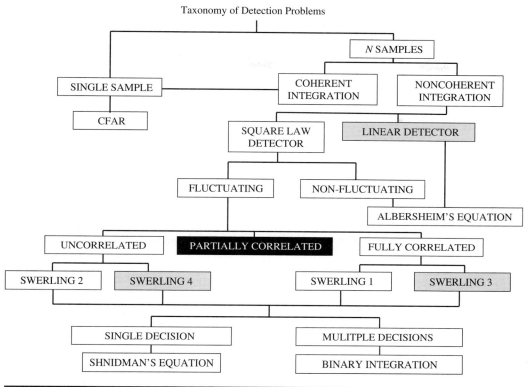

FIGURE 6.10 Taxonomy of detection problems considered. (Adapted from Levanon, 1988.)

from sample to sample, the clutter as partially correlated (including possibly uncorrelated) from sample to sample. Jamming can be either. The target is modeled as either nonfluctuating (i.e., a constant) or a random process that can be either completely correlated or completely uncorrelated from sample to sample (the Swerling models), or partially correlated from sample to sample. The signal-to-interference ratio (SIR) and thus the detection performance are often improved by *integrating* (adding) multiple samples of the target and interference, motivated by the idea that the interference signal can be "averaged down" and the target signal reinforced by adding multiple samples. This idea was first discussed in Chap. 1. Thus, in general, detection will be based on N samples of the target-plus-interference. Note that care must be taken to integrate only samples that represent the same range and Doppler resolution cells.

Integration is commonly applied to the data at three different stages in the processing chain:

1. After coherent demodulation, to the baseband complex-valued data (I and Q, or magnitude and phase). Combining complex data samples is referred to as *coherent integration*.

2. After envelope detection, to the magnitude (or squared magnitude or log magnitude) data. Combining magnitude samples after the phase information is discarded is referred to as *noncoherent integration*.

3. After threshold detection, to the target present/target absent decisions. This technique is called *binary integration* or *M-of-N processing*.

A system could elect to use none, one, or a combination of these techniques. The major cost of integration is the time and energy required to obtain multiple samples of the same range, Doppler, and/or angle cell (or multiple threshold detection decisions for that cell); this is time that cannot be spent searching for targets elsewhere, or tracking already-known targets, or imaging other regions of interest. Integration also increases the signal processing computational load. Modern systems vary as to whether this is a significant burden: the required operations are simple, but must be performed at a very high rate in many systems.

In coherent integration complex data samples y_n are combined to form a new complex variable z:

$$z = \sum_{n=0}^{N-1} y_n \tag{6.59}$$

As shown in Chap. 1, if the SNR of a single sample y_n is χ_1, the integrated data sample z has an SNR $\chi_N = N \cdot \chi_1$ provided that all of the samples add in phase. That is, coherent integration attains an *integration gain* of a factor of N. This is the signal processing gain G_{sp} in the radar range equation due to coherent integration. Detection results are then obtained by using single-sample ($N = 1$) results with χ_1 replaced by the integrated χ_N.

In noncoherent integration, phase information is discarded. Instead, the magnitudes or squared magnitudes of the data samples are integrated. (Sometimes another function of the magnitude, such as the log-magnitude, is used.) Most classical detection results have been developed for the square law detector, which bases detection on the quantity

$$z = \sum_{n=0}^{N-1} |y_n|^2 \tag{6.60}$$

Consideration will be largely restricted to the square law detector in this section.

The situation for noncoherent integration is more complicated. As shown in Chap. 1, the integrated signal z cannot be expressed as the sum of a target-only part and a noise-only part, so an integrated SNR cannot be defined directly. It will prove necessary to determine the actual probability density function of the integrated random variable z to compute detection results; this is done in the next subsection.

Binary integration takes place after an initial detection decision has taken place. That initial decision may be based on a single sample or on data that have already been coherently or noncoherently integrated. Whatever the processing before the threshold detection, after it the result is a choice between hypothesis H_0, "target absent," and H_1, "target present." Because there are only two possible outputs of the detector each time a threshold test is made, the output is said to be binary. Multiple binary decisions can be combined in an "M out of N" decision logic in an attempt to further improve the performance. This type of integration is analyzed in Sec. 6.4.8.

6.4.2 Nonfluctuating Targets

Consider detection based on noncoherent integration of N samples of a nonfluctuating target (sometimes called the "Swerling 0" or "Swerling 5" case) in white Gaussian noise. The amplitude and absolute phase of the target component are unknown. Thus, an individual data sample y_n is the sum of a complex constant $m = \tilde{m}\exp(j\theta)$ for some real amplitude $\tilde{m}$ and phase θ, and a complex white Gaussian noise sample w_n of power $\sigma_w^2/2$ in each of the I and Q channels (total noise power σ_w^2):

$$y_n = m + w_n \tag{6.61}$$

Under hypothesis H_0, the target is absent and $y_n = w_n$. The PDF of $z_n \equiv |y_n|$ is Rayleigh

$$p_{z_n}(z_n|H_0) = \begin{cases} \dfrac{2z_n}{\sigma_w^2} \exp(-z_n^2/\sigma_w^2), & z_n \geq 0 \\ 0, & z_n < 0 \end{cases} \qquad (6.62)$$

Under hypothesis H_1 z_n is a Rician voltage density

$$p_{z_n}(z_n|H_1) = \begin{cases} \dfrac{2z_n}{\sigma_w^2} \exp\left[-(z_n^2 + \tilde{m}^2)/\sigma_w^2\right] I_0\left(\dfrac{2\tilde{m}z_n}{\sigma_w^2}\right), & z_n \geq 0 \\ 0, & z_n < 0 \end{cases} \qquad (6.63)$$

For a vector $\mathbf{z}$ of N such i.i.d. samples the joint PDFs are, for each $z_n \geq 0$,

$$p_{\mathbf{z}}(\mathbf{z}|H_0) = \prod_{n=0}^{N-1} \frac{2z_n}{\sigma_w^2} \exp(-z_n^2/\sigma_w^2) \qquad (6.64)$$

$$p_{\mathbf{z}}(\mathbf{z}|H_1) = \prod_{n=0}^{N-1} \frac{2z_n}{\sigma_w^2} \exp\left[-(z_n^2 + \tilde{m}^2)/\sigma_w^2\right] I_0\left(\frac{2\tilde{m}z_n}{\sigma_w^2}\right) \qquad (6.65)$$

The LRT and log LRT become

$$\Lambda = \prod_{n=0}^{N-1} \exp(-\tilde{m}^2/\sigma_w^2) I_0\left(\frac{2\tilde{m}z_n}{\sigma_w^2}\right) = \exp(-N\tilde{m}^2/\sigma_w^2) \prod_{n=0}^{N-1} I_0\left(\frac{2\tilde{m}z_n}{\sigma_w^2}\right) \underset{H_0}{\overset{H_1}{\gtrless}} -\lambda \qquad (6.66)$$

$$\ln(\Lambda) = -\frac{N\tilde{m}^2}{\sigma_w^2} + \sum_{n=0}^{N-1} \ln\left[I_0\left(\frac{2\tilde{m}z_n}{\sigma_w^2}\right)\right] \underset{H_0}{\overset{H_1}{\gtrless}} \ln(-\lambda) \qquad (6.67)$$

Incorporating the term involving the ratio of signal power and noise power on the left-hand side into the threshold gives

$$\sum_{n=0}^{N-1} \ln\left[I_0\left(\frac{2\tilde{m}z_n}{\sigma_w^2}\right)\right] \underset{H_0}{\overset{H_1}{\gtrless}} \ln(-\lambda) + \frac{N\tilde{m}^2}{\sigma_w^2} \equiv \tilde{T} \qquad (6.68)$$

Equation (6.68) shows that, given N noncoherent samples of a nonfluctuating target in white noise, the optimal Neyman-Pearson detection test scales each sample by the quantity $2\tilde{m}/\sigma_w^2$, passes it through the monotonic nonlinearity $\ln[I_0(\cdot)]$, and then integrates the scaled samples and performs a threshold test. There are two practical problems with this equation. First, it is still desirable to avoid computing the function $\ln[I_0(\cdot)]$ possibly millions of times per second. Second, both the target amplitude $\tilde{m}$ and the noise power σ_w^2 must be known to perform the required scaling. The test can be simplified by using the results of Sec. 6.3.3. Applying the square law detector approximation of Eq. (6.55) to Eq. (6.68) gives the test

$$\sum_{n=0}^{N-1} \left(\frac{\tilde{m}^2 z_n^2}{\sigma_w^4}\right) \underset{H_0}{\overset{H_1}{\gtrless}} \tilde{T} \qquad (6.69)$$

Combining all constants into the threshold gives us the final detection rule:

$$z = \sum_{n=0}^{N-1} z_n^2 \underset{H_0}{\overset{H_1}{\gtrless}} \frac{\sigma_w^4 \tilde{T}}{\tilde{m}^2} \equiv T \qquad (6.70)$$

Equation (6.70) states that the squared magnitudes of the data samples are simply integrated and the integrated sum compared to a threshold to decide whether a target is present or not. The integrated variable z is the sufficient statistic Υ for this problem.

The performance of this detector must now be determined. Note that the performance of the actual detector of Eq. (6.70) is sought, not that of the ideal detector of Eq. (6.68). It is convenient to scale the z_n, replacing them with the new variables $z'_n = z_n/\sigma_w$ and thus replacing z with $z' = \sum(z'_n)^2 = z/\sigma_w^2$. Such a scaling does not change the performance, but merely alters the threshold value that corresponds to a particular P_D or P_{FA}. Specifically, a threshold value T applied to the unnormalized data z gives the same performance as a normalized threshold $T' = T/\sigma_w^2$ applied to the normalized data z'.

The PDF of z'_n is still either Rayleigh or Rician as in Eqs. (6.62) and (6.63), but now with unit noise variance:

$$p_{z'_n}(z'_n \mid H_0) = \begin{cases} 2z'_n \exp(-z'^2_n), & z'_n \geq 0 \\ 0, & z'_n < 0 \end{cases} \quad (6.71)$$

$$p_{z'_n}(z'_n \mid H_1) = \begin{cases} 2z'_n \exp\left[-(z'^2_n + \chi)\right] I_0(2z'_n \sqrt{\chi}), & z'_n \geq 0 \\ 0, & z'_n < 0 \end{cases} \quad (6.72)$$

where $\chi = \tilde{m}^2/\sigma_w^2$ is the single-sample SNR. Since a square law detector is being used, define $r_n = (z'_n)^2$; then $z' = \sum r_n$. The PDF of r_n is exponential under H_0 and a generalized noncentral chi-squared density under H_1:

$$p_{r_n}(r_n \mid H_0) = \begin{cases} \exp(-r_n), & r_n \geq 0 \\ 0, & r_n < 0 \end{cases} \quad (6.73)$$

$$p_{r_n}(r_n \mid H_1) = \begin{cases} \exp(-r_n + \chi) I_0(2\sqrt{\chi r_n}), & r_n \geq 0 \\ 0, & r_n < 0 \end{cases} \quad (6.74)$$

Since z' is the sum of N i.i.d. random variables r_n, the PDF of z' is the N-fold convolution of the PDF given in Eq. (6.73) or (6.74). This is most easily found using *characteristic functions* (CFs; see App. A). If $C_z(q)$ is the CF corresponding to a PDF $p_z(z)$, the CF of the N-fold convolution of the PDFs is the product $C_z^N(q)$ of their individual characteristic functions.

Under hypothesis H_0, the CF of r_n can be readily shown to be

$$C_{r_n}(q) = \frac{1}{1 - jq} \quad (6.75)$$

The characteristic function of z' is therefore

$$C_{z'}(q) = [C_{r_n}(q)]^N = \left(\frac{1}{1 - jq}\right)^N \quad (6.76)$$

The PDF of z' is obtained by inverting its characteristic function using the Fourier-like inverse CF transform:

$$p_{z'}(z' \mid H_0) = \frac{1}{2\pi} \int_{-\infty}^{\infty} C_{z'}(q) \exp(-jz'q)\, dq \quad (6.77)$$

Using Eq. (6.76) in Eq. (6.77) and referring to any good Fourier transform table (with allowance for the reversed sign of the Fourier kernel in the definition of the characteristic function), the Erlang density is obtained:

$$p_{z'}(z'|H_0) = \begin{cases} \dfrac{(z')^{N-1}}{(N-1)!}\exp(-z'), & z' \geq 0 \\ 0, & z' < 0 \end{cases} \quad (6.78)$$

This reduces to the exponential PDF when $N = 1$ as would be expected since in that case z' is the magnitude squared of a single sample of complex Gaussian noise.

The probability of false alarm is obtained by integrating Eq. (6.78) from the normalized threshold value T' to $+\infty$. It is convenient to manipulate that into the form

$$\begin{aligned} P_{FA} &= \int_{T'}^{\infty} p_{z'}(z'|H_0)\,dz = \int_{0}^{\infty} p_{z'}(z'|H_0)\,dz - \int_{0}^{T'} p_{z'}(z'|H_0)\,dz \\ &= 1 - \int_{0}^{T'} \frac{(z')^{N-1}}{(N-1)!} \exp(-z')\,dz' \end{aligned} \quad (6.79)$$

The normalized upper incomplete gamma function is defined as (Olver et al., 2020, §8.2)[12,13]

$$I(x, N) = \int_{0}^{x} \frac{\exp(-\tau)\tau^{N-1}}{(N-1)!}\,d\tau \quad (6.80)$$

Applying this definition to Eq. (6.79) gives, in both normalized and unnormalized threshold forms,

$$\begin{aligned} P_{FA} &= 1 - I(T', N) \\ &= 1 - I(T/\sigma_w^2, N) \end{aligned} \quad (6.81)$$

Using the incomplete gamma function property $I(x, 1) = 1 - \exp(-x)$, Eq. (6.81) reduces for a single sample ($N = 1$) to the especially simple results

$$\begin{aligned} P_{FA} &= \exp(-T') &\Rightarrow\quad T' &= -\ln(P_{FA}) \\ &= \exp(-T/\sigma_w^2) &\Rightarrow\quad T &= -\sigma_w^2 \cdot \ln(P_{FA}) \end{aligned} \quad (6.82)$$

Equation (6.81) (or (6.82) when $N = 1$) can be used to determine the probability of false alarm P_{FA} for a given threshold T or T' or, more likely, the required value of T or T' to achieve a desired P_{FA}.

Now the probability of detection P_D corresponding to the same threshold must be determined. Start by finding the PDF of the integrated and square-law-detected samples under

---

[12]"Normalized" so that $I(\infty, N) = 1$, like a probability density function.

[13]This definition is the same used in MATLAB®'s `gammainc(·,·)` function. Meyer and Mayer (1973) use "Pearson's form" of the incomplete gamma function, denoted here as $I_P(u, M)$. The two are related according to $I_P(u, M) = I(u\sqrt{M+1}, M+1)$.

hypothesis H_1. Each individual data sample r_n has a generalized noncentral chi-squared PDF [Eq. (6.74)]; the corresponding characteristic function is

$$C_{r_n}(q) = \frac{1}{1-jq}\exp\left(\frac{jx q}{1-jq}\right) \qquad (6.83)$$

The CF of the sum z' of N such samples is

$$C_{z'}(q) = \left(\frac{1}{1-jq}\right)^N \exp\left(\frac{jN\chi q}{1-jq}\right) \qquad (6.84)$$

and the PDF of z' is[14]

$$p_{z'}(z'|H_1) = \left(\frac{z'}{N\chi}\right)^{\frac{N-1}{2}} \exp(-z' - N\chi)\, I_{N-1}(2\sqrt{N\chi z'}) \qquad (6.85)$$

P_D is found by using Eq. (6.82) to obtain the required threshold T' for a specified PFA and then integrating Eq. (6.85) from T' to $+\infty$. One version of the result and the approach to deriving it is given in Meyer and Mayer (1973); it is[15,16]

$$\begin{aligned}P_D &= \int_{T'}^{\infty} p_{z'}(z'|H_1)\,dz' \\ &= Q_M(\sqrt{2N\chi},\sqrt{2T'}) + \exp(-T' - N\chi)\left\{\sum_{r=2}^{N}\left(\frac{T'}{N\chi}\right)^{\frac{r-1}{2}} I_{r-1}(2\sqrt{N\chi T'})\right\}\end{aligned} \qquad (6.86)$$

The summation term in the second line of Eq. (6.86) only contributes when $N \geq 2$. Equations (6.81) and (6.86) define the performance achieved for a nonfluctuating target with noncoherent integration using a square law detector.

Figure 6.11 shows the effect of the number of samples noncoherently integrated, N, on the receiver operating characteristic when $P_{FA} = 10^{-8}$. It shows that noncoherent integration reduces the single-sample SNR required to achieve a given P_D and P_{FA}, but by less than the factor N achieved with coherent integration. For example, consider the single-sample SNR required to achieve $P_D = 0.9$. For $N = 1$, this is 14.2 dB; for $N = 10$, it drops to 6.1 dB, a reduction of 8.1 dB, but less than the 10 dB that corresponds to the factor of 10 increase in the number of pulses integrated. This reduction in required single-sample SNR to achieve a specified P_D and P_{FA} is called the *noncoherent integration gain* G_{nc}. Unlike the coherent case, G_{nc} depends on the particular values of P_D and P_{FA} sought and the particular PDFs under hypotheses H_0 and H_1.

---

[14]$I_{N-1}(x)$ is the modified Bessel function of the first kind and order $N - 1$, not to be confused with the incomplete gamma function $I(x, N)$.
[15]Equations (6.84) through (6.86) are the same as Meyer and Mayer's (3-30), (3-33), and (3-37) with the following changes of variables: $Y \to z'$, $x \to \chi$, and $Y_b \to T'$.
[16]DiFranco and Rubin (1980) give an alternate expression in terms of the Toronto function; see https://en.wikipedia.org/wiki/Toronto_function.

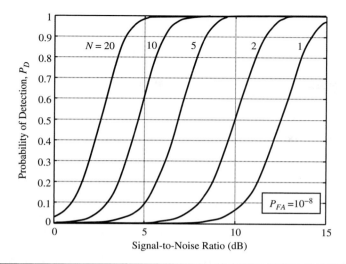

Figure 6.11 Effect of noncoherent integration on detection performance for a nonfluctuating target in complex Gaussian noise as a function of single-sample SNR.

6.4.3 Albersheim's Equation

The performance results for the case of a nonfluctuating target in complex Gaussian noise are given by Eqs. (6.81) and (6.86). While relatively easy to implement in a modern software analysis system such as MATLAB®, these equations do not lend themselves to manual calculation. Fortunately, there does exist a simple closed-form expression relating P_D, P_{FA}, and χ that can be computed with simple scientific calculators. This expression is known as *Albersheim's equation* (Albersheim, 1981; Tufts and Cann, 1983).

Albersheim's equation is an empirical approximation to the results in Robertson (1967) for computing the single-sample SNR χ_1 required to achieve a given P_D and P_{FA}. It applies under the following conditions:

- Nonfluctuating target in Gaussian (i.i.d. in I and Q) noise
- Linear (not square law) detector
- Noncoherent integration of N samples

The estimate is given by the series of calculations

$$A = \ln\left(\frac{0.62}{P_{FA}}\right), \quad B = \ln\left(\frac{P_D}{1 - P_D}\right),$$
$$\chi_1 = -5\log_{10}(N) + \left[6.2 + \left(\frac{4.54}{\sqrt{N + 0.44}}\right)\right] \cdot \log_{10}(A + 0.12AB + 1.7B) \text{ dB} \quad (6.87)$$

Note that χ_1 is in decibels. The error in the estimate of χ_1 is less than 0.2 dB for $10^{-7} \leq P_{FA} \leq 10^{-3}$, $0.1 \leq P_D \leq 0.9$, and $1 \leq N \leq 8096$, a useful range of parameters. For the special case of $N = 1$, Eq. (6.87) reduces to

$$A = \ln\left(\frac{0.62}{P_{FA}}\right), \quad B = \ln\left(\frac{P_D}{1 - P_D}\right),$$
$$\chi_1 = 10\log_{10}(A + 0.12AB + 1.7B) \text{ dB} \quad (6.88)$$

To illustrate, suppose $P_D = 0.9$ and $P_{FA} = 10^{-6}$ are required for a nonfluctuating target in a system using a linear detector. If detection is to be based on a single sample, what is the required SNR of that sample? This is a direct application of Albersheim's equation. Compute $A = \ln(0.62 \times 10^6) = 13.34$ and $B = \ln(9) = 2.197$. Equation (6.88) then gives $\chi_1 = 13.14$ dB; on a linear scale, this is 20.59.

If $N = 100$ samples are noncoherently integrated, it should be possible to obtain the same P_D and P_{FA} with a lower single-sample SNR. To confirm this, use Eq. (6.87). The intermediate parameters A and B are unchanged. χ_1 is now reduced to -1.26 dB, a reduction of 14.4 dB corresponding to a factor of 27.54 on a linear scale. This value closely matches that obtained using the exact expressions. It is much better than the $\sqrt{N}$ rule of thumb sometimes given for noncoherent integration, which would give a gain factor of only 10 for $N = 100$ samples integrated. Rather, the gain is approximately $N^{0.7}$ in this example. Albersheim's equation will be used shortly to develop an expression for estimating noncoherent integration gain.

Albersheim's equation is useful because it requires no function more exotic than the natural logarithm and square root for its evaluation so it can be used with any scientific calculator. If a somewhat larger error can be tolerated, it can also be used for square law detector results for the nonfluctuating target, Gaussian noise case. Specifically, square law detector results are within 0.2 dB of linear detector results (Robertson, 1967; Tufts and Cann, 1983). Thus, the same equation can be used for rough calculations over the range of parameters given previously with errors not exceeding 0.4 dB.

Equations (6.87) and (6.88) provide for calculation of χ_1 given P_D, P_{FA}, and N. It is possible to solve Eq. (6.87) for either P_D or P_{FA} in terms of the other and χ_1 and N, further extending its usefulness. For instance, the following calculations show how to estimate P_D given the other factors (χ_1 is in dB):

$$A = \ln\left(\frac{0.62}{P_{FA}}\right), \quad Z = \frac{\chi_1 + 5\log_{10}(N)}{6.2 + (4.54/\sqrt{N + 0.44})}, \quad B = \frac{10^Z - A}{1.7 + 0.12A} \quad (6.89)$$

$$\Rightarrow P_D = \frac{1}{1 + \exp(-B)}$$

In Eq. (6.89) A and B are the same values as in Eq. (6.87), though B cannot be computed in terms of P_D since P_D is now the unknown. A result similar to Eq. (6.89) can be derived for computing P_{FA} (see Prob. 11).

Albersheim's equation can also be used to write a relatively compact formula for estimating the noncoherent integration gain for nonfluctuating targets. G_{nc} is the reduction in single-sample SNR required to achieve a specified P_D and P_{FA} when N samples are combined and is the range equation signal processing gain G_{sp} for noncoherent integration. In dB, it is given by

$$\begin{aligned} G_{nc}(N)\,(\text{dB}) &= \chi_{1|1\text{ sample}} - \chi_{1|N\text{ samples}} \\ &= 5\log_{10}(N) - \left[6.2 + \left(\frac{4.54}{\sqrt{N + 0.44}}\right)\right] \cdot \log_{10}(A + 0.12AB + 1.7B) \\ &\quad + 10\log_{10}(A + 0.12AB + 1.7B) \text{ dB} \\ &= 5\log_{10}(N) - \left[\left(\frac{4.54}{\sqrt{N + 0.44}}\right) - 3.8\right] \cdot \log_{10}(A + 0.12AB + 1.7B) \text{ dB} \end{aligned} \quad (6.90)$$

On a linear scale this becomes

$$G_{nc}(N) = \frac{\sqrt{N}}{k^{f(N)}} \quad (6.91)$$

Detection Fundamentals

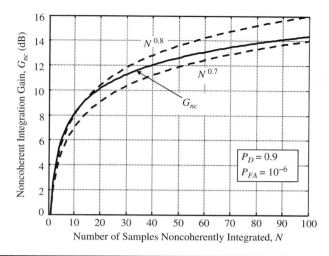

FIGURE 6.12 Noncoherent integrations gain G_{nc} for a nonfluctuating target, estimated using Albersheim's equation.

where

$$k = A + 0.12AB + 1.7B$$

$$f(N) = \left(\frac{0.454}{\sqrt{N+0.44}}\right) - 0.38 \qquad (6.92)$$

The constant k depends only on P_D and P_{FA}, while the term $f(N)$ is a slowly declining function only of N.

Figure 6.12 plots this estimate of G_{nc} in decibels for Albersheim's nonfluctuating, linear detector case as a function of N for $P_D = 0.9$ and $P_{FA} = 10^{-6}$. Also shown are curves corresponding to $N^{0.7}$ and $N^{0.8}$. The noncoherent gain is slightly better than $N^{0.8}$ for very few samples integrated ($N = 2$ or 3), with the effective exponent on N declining slowly as N increases. G_{nc} is bracketed by $N^{0.7}$ and $N^{0.8}$ to in excess of $N = 100$ samples integrated; the gain eventually slows asymptotically to become proportional to $\sqrt{N}$ for very large N as the $5\log_{10}(N)$ term dominates Eq. (6.90). This can be seen from Eqs. (6.91) and (6.92), which show that as $N \to \infty$, $f(N) \to -0.38$ and G_{nc} becomes proportional to $\sqrt{N}$.[17] Requiring a large value of N to achieve a given P_D and P_{FA} implies a very poor single-sample SNR so that a large amount of integration is needed, while small N implies a relatively large single-sample SNR. Noncoherent integration is therefore more efficient when the single-sample SNR is higher to begin with. In any event, the simplicity and robustness of noncoherent integration, requiring no knowledge of the phase, means it is widely used to improve the SNR before the threshold detector.

6.4.4 Fluctuating Targets

The analysis in the preceding section considered only nonfluctuating targets, sometimes called the "Swerling 0" or "Swerling 5" case. A more realistic model allows for target fluctuations. If one of the Swerling fluctuation models is chosen, the target RCS is drawn

---

[17] The general trend of the proportionality of G_{nc} to $\sqrt{N}$ holds also for fluctuating target models.

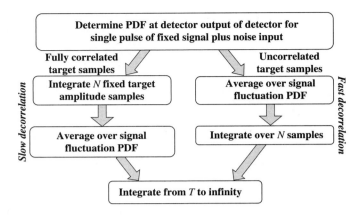

FIGURE 6.13 Strategy for computing P_D for fluctuating target models.

from either the exponential or chi-squared PDF and the RCS of a group of N samples follows either the uncorrelated or fully correlated model, as described in Chap. 2. Note that using a fluctuating target model has no effect on the probability of false alarm. P_{FA} is determined only by the PDF when no target is present; thus Eq. (6.81) still applies.

The strategy for determining the probability of detection depends on the target fluctuation model used. Figure 6.13 illustrates the approach. In all cases, the PDF of the magnitude-squared of a single sample of signal-plus-noise, normalized to the noise power σ_w^2, is a generalized noncentral chi-squared distribution so its CF is still given by Eq. (6.83). However, the SNR χ in that expression is now a random variable because the target RCS is a random variable.

In a slow decorrelation case such as Swerling 1 or 3 the target RCS is a fixed value for all N pulses integrated to form z', like the nonfluctuating case. The CF of z' is again the N-fold product of Eq. (6.83) with itself

$$C_{z'}(q; \chi, N) = \left(\frac{1}{1-jq}\right)^N \exp\left(\frac{jN\chi q}{1-jq}\right) \qquad (6.93)$$

This is the same expression as Eq. (6.84) except that now $C_{z'}$ is written explicitly as a function of all of q, χ, and N. The SNR χ is fixed for one detection test but varies according to the RCS PDF from one test to the next. Next take the expected value of the CF over the target SNR distribution:

$$\bar{C}_{z'}(q; \chi, N) = \int_0^\infty p_\chi(\chi) C_{z'}(q; \chi, N)\, d\chi \qquad (6.94)$$

For the Swerling 1 model the PDF of the target power, and therefore of the SNR, is exponential:

$$p_\chi(\chi) = \left(\frac{1}{\bar{\chi}}\right) \exp(-\chi/\bar{\chi}) \qquad (6.95)$$

where $\bar{\chi}$ is the mean value of the SNR. Using Eqs. (6.93) and (6.95) in Eq. (6.94) gives the characteristic function averaged over the target fluctuations:

$$\bar{C}_{z'}(q; \bar{\chi}, N) = \frac{1}{(1-jq)^{N-1}[1-j(1+N\bar{\chi})q]} \qquad (6.96)$$

The $N = 1$ and $N > 1$ cases are best handled separately. For $N = 1$, the CF simplifies to just $\bar{C}_{z'}(q; \bar{\chi}, N) = [1 - j(1 + N\bar{\chi})q]^{-1}$. Its inverse is the PDF of z' under hypothesis H_1 for Swerling case 1 and $N = 1$,

$$p_{z'}(z' | H_1) = \frac{1}{1 + \bar{\chi}} \exp\left(\frac{-z'}{1 + \bar{\chi}}\right) \quad (N = 1) \tag{6.97}$$

The inverse of the more general CF of Eq. (6.96) for any N is the PDF:

$$p_{z'}(z' | H_1) = \frac{1}{N\bar{\chi}} \left(1 + \frac{1}{N\bar{\chi}}\right)^{N-2} I\left[\frac{z'}{(1 + 1/N\bar{\chi})}, N - 1\right] \exp\left(\frac{-z'}{1 + N\bar{\chi}}\right) \quad (N \geq 1) \tag{6.98}$$

Integrating the PDF of Eqs. (6.97) and (6.98) from the normalized threshold T' to $+\infty$ leads to the following expression for the probability of detection in the Swerling 1 case (Meyer and Mayer, 1973):

$$\begin{aligned} N = 1: &\quad P_D = \exp[-T'/(1 + \bar{\chi})], \\ N \geq 1: &\quad P_D = 1 - I[T', N - 1] + \ldots \\ &\quad \ldots + \left(1 + \frac{1}{N\bar{\chi}}\right)^{N-1} I\left[\frac{T'}{(1 + (1/N\bar{\chi}))}, N - 1\right] \exp[-T'/(1 + N\bar{\chi})] \end{aligned} \tag{6.99}$$

A simplified expression for P_D that can be used in many cases is given in the next subsection.

In an uncorrelated fluctuation model such as the Swerling 2 or 4 cases, each of the N samples noncoherently integrated has a different value of SNR. Consequently, it is appropriate to average over the SNR in the single-sample CF first to get the "average single-sample CF"

$$\bar{C}_{r_n}(q; \bar{\chi}) = \left(\frac{1}{1 - jq}\right) \int_0^\infty p_\chi(\chi) \exp\left(j \frac{q\chi}{1 - jq}\right) d\chi \tag{6.100}$$

and then perform the N-fold multiplication to get the CF of the integrated data

$$\bar{C}_{z'}(q; \bar{\chi}, N) = [\bar{C}_r(q; \bar{\chi})]^N \tag{6.101}$$

For the Swerling 2 model specifically the exponential PDF can again be used for the SNR [Eq. (6.95)], applying it this time in Eq. (6.100) to arrive at

$$\bar{C}_{r_n}(q; \bar{\chi}) = \frac{1}{[1 - j(1 + \bar{\chi})q]} \tag{6.102}$$

so that

$$\bar{C}_{z'}(q; \bar{\chi}, N) = \frac{1}{[1 - j(1 + \bar{\chi})q]^N} \tag{6.103}$$

Inverse transforming Eq. (6.103) gives the PDF of z' under hypothesis H_1 for Swerling case 2:

$$p_{z'}(z' | H_1) = \frac{(z')^{N-1} \exp[-z'/(1 + \bar{\chi})]}{(1 + \bar{\chi})^N (N - 1)!} \tag{6.104}$$

Integrating Eq. (6.104) gives the probability of detection, which can be shown to be (Meyer and Mayer, 1973)

$$P_D = 1 - I\left[\frac{T'}{(1+\bar{\chi})}, N\right] \qquad (6.105)$$

When $N = 1$, correlation models are irrelevant. Because they are based on the same PDF for RCS, the Swerling 1 and 2 cases therefore produce the same outcome for $N = 1$.

Results for Swerling 3 and 4 targets can be obtained by repeating the previous analyses for the Swerling 1 and 2 cases, but with a chi-squared instead of exponential density function for the SNR:

$$p_\chi(\chi) = \frac{4\chi}{\bar{\chi}^2} \exp(-2\chi/\bar{\chi}) \qquad (6.106)$$

Derivations of the resulting expressions for P_D can be found in Meyer and Mayer (1973) and DiFranco and Rubin (1980) and many other radar detection texts. Table 6.1 summarizes one form of the resulting expressions. For $N = 1$, the Swerling 3 and 4 results can be shown to be identical.

Figure 6.14 compares the detection performance of the four Swerling model fluctuating targets and the nonfluctuating target for $N = 10$ samples as a function of the mean single-sample SNR for a fixed $P_{FA} = 10^{-8}$. Assuming that the primary interest is in relatively high (> 0.5) values of P_D, the upper half of the figure is of greatest interest. In this case, the nonfluctuating target is the most favorable in the sense that it achieves a given probability of detection at the lowest SNR. The worst case (highest required SNR for a given P_D) is the Swerling case 1, which corresponds to fully correlated samples and an exponential PDF of the target RCS. For instance, $P_D = 0.9$ requires $\chi \approx 6.1$ dB for the nonfluctuating case, but $\chi \approx 14.5$ dB for the Swerling 1 case, a difference of 8.4 dB.

At least two general conclusions can be drawn from Fig. 6.14. First, for values of $N\bar{\chi}$ in the mid-teens of decibels (certainly the case if detection is to be very likely), P_D is greatest for nonfluctuating targets. Evidently target fluctuations make detection more difficult, that is, require a higher SNR to achieve a given P_D and P_{FA}. Second, given that a target exhibits a fluctuating RCS, uncorrelated "fast" fluctuations of the target RCS (Swerling 2 and 4) aid target detectability compared to correlated "slow" fluctuations (Swerling 1 and 3). The last observation suggests that it is desirable to be able to force the data collected from a complex target and subsequently combined noncoherently to be fully decorrelated. Many radars use *frequency agility* to accomplish this. As discussed in Chap. 2, stepping the radar RF from one CPI or pulse to the next, as appropriate, will decorrelate successive target RCS measurements provided the frequency step size $\Delta F \geq c/2L_d$, where L_d is the target depth as viewed from the radar.

6.4.5 Simplified Equations for P_D for Some Swerling Cases

Properties of the normalized incomplete gamma function can be used to simplify the expressions for P_D in a few cases, producing better-known and more calculator-friendly expressions. These expressions are exact for some Swerling cases and values of N and approximate for others, as noted in the following discussion. The relevant properties of $I(\cdot,\cdot)$ are (Olver et al., 2020):

- For any c, $I(c,0) = 1$
- For any c, $I(c,1) = 1 - \exp(-c)$
- For any c, $I(c,2) = 1 - (1+c)\exp(-c)$
- For $d > 0$, $\lim_{c \to \infty} I(c,d) = 1$

Detection Fundamentals

Case	P_D	Comments
0 or 5	$Q_M(\sqrt{2N\bar{\chi}},\sqrt{2T'}) + e^{-(T'+N\bar{\chi})}\sum_{r=2}^{N}\left(\dfrac{T'}{N\bar{\chi}}\right)^{\frac{r-1}{2}} I_{r-1}(2\sqrt{N\bar{\chi}T'})$	Nonfluctuating case. Sum term applies only for $N \geq 2$.
1	$1 - I(T', N-1) + \ldots$ $+ \left(1+\dfrac{1}{N\bar{\chi}}\right)^{N-1} \exp\left(\dfrac{-T'}{1+N\bar{\chi}}\right) I\left(\dfrac{T'}{[1+(1/N\bar{\chi})]}, N-1\right)$	
2	$1 - I\left(\dfrac{T'}{1+\bar{\chi}}, N\right)$	
3	$N = 1 \text{ or } 2: \left(1+\dfrac{1}{(N\bar{\chi}/2)}\right)^{N-2} \times \ldots$ $\times \left[1 + \dfrac{T'}{1+(N\bar{\chi}/2)} - \dfrac{N-2}{(N\bar{\chi}/2)}\right]\exp\left[\dfrac{-T'}{1+(N\bar{\chi}/2)}\right]$ $N > 2: \dfrac{(T')^{N-1}e^{-T'}c}{(N-2)!} + \sum_{l=0}^{N-2}\dfrac{e^{-T'}(T')^l}{l!} + \dfrac{e^{-cT'}}{(1-c)^{N-2}} \times \ldots$ $\times \left[1 - \dfrac{(N-2)c}{(1-c)} + cT'\right]\left[1 - \sum_{l=0}^{N-2}\dfrac{e^{-(1-c)T'}(T')^l(1-c)^l}{l!}\right]$	$c \equiv \dfrac{1}{1+(N\bar{\chi}/2)}$
4	$c^N \sum_{k=0}^{N}\left\{\dfrac{N!}{k!(N-k)!}\left(\dfrac{1-c}{c}\right)^{N-k}\left[\sum_{l=0}^{2N-k-1}\dfrac{e^{-cT'}(cT')^l}{l!}\right]\right\} \quad T' > N(2-c)$ $1 - c^N \sum_{k=0}^{N}\left\{\dfrac{N!}{k!(N-k)!}\left(\dfrac{1-c}{c}\right)^{N-k}\left[\sum_{l=2N-k}^{\infty}\dfrac{e^{-cT'}(cT')^l}{l!}\right]\right\} \quad T' < N(2-c)$	$c \equiv \dfrac{1}{1+(\bar{\chi}/2)}$

$P_{FA} = 1 - I(T', N)$ in all cases.
T' is the threshold T normalized to the noise power σ_w^2, $T' = T/\sigma_w^2$.

$I(\cdot, \cdot)$ is the normalized incomplete gamma function.
$I_k(\cdot)$ is the modified Bessel function of the first kind and order k.
$Q_M(\cdot, \cdot)$ is Marcum's Q function.

TABLE 6.1 Probability of Detection for Swerling Model Fluctuating Targets with a Square-Law Detector

Swerling 1 Case: When $N = 1$, applying the property $I(c, 0) = 1$ immediately reduces the Swerling 1 expression in Table 6.1 to the much simpler but still exact expression

$$P_D = \exp\left(\dfrac{-T'}{1+\bar{\chi}}\right) \quad \text{(Swerling 1, } N = 1\text{)} \tag{6.107}$$

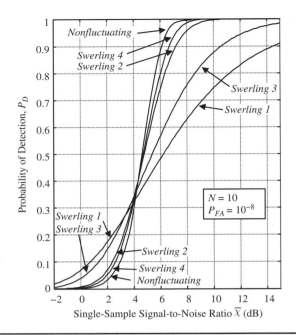

FIGURE 6.14 Comparison of detection performance for nonfluctuating and Swerling fluctuating target models using noncoherent integration of 10 pulses ($N = 10$) and a fixed probability of false alarm $P_{FA} = 10^{-8}$ with a square law detector.

Also in the $N = 1$ case, the property $I(c, 1) = 1 - \exp(-c)$ can be applied to the expression for P_{FA} to get

$$P_{FA} = \exp(-T') \quad (N = 1) \tag{6.108}$$

Eliminating T' from Eqs. (6.107) and (6.108) gives, for a single sample of a Swerling 1 target in noise,

$$P_D = P_{FA}^{1/(1+\bar{x})} \quad \text{(Swerling 1 or 2, } N = 1\text{)} \tag{6.109}$$

(The applicability to the Swerling 2 case will be explained shortly.)

Now assume that $P_{FA} \ll 1$ and $N\bar{x} > 2$. The quantity $N\bar{x}$ can be thought of crudely as the integrated SNR at the input to the square law detector. Both conditions must hold true if there is to be any chance of detecting a target while maintaining a small P_{FA} (DiFranco and Rubin, 1980, p. 390). Further, because $\lim_{c \to \infty} I(c, d) = 1$, the two incomplete gamma functions in the Swerling 1 exact result are both approximately equal to 1. Under these conditions, the exact expression in the table reduces to

$$P_D \approx \left(1 + \frac{1}{N\bar{x}}\right)^{N-1} \exp\left(\frac{-T'}{1 + N\bar{x}}\right) \quad \text{(Swerling 1; } P_{FA} \ll 1 \text{ and } N\bar{x} > 2\text{)} \tag{6.110}$$

This result is exact [and matches Eq. (6.107)] for $N = 1$.

Swerling 2 Case: Like the Swerling 1 case, applying the property $I(c,0) = 1$ when $N = 1$ reduces the Swerling 2 expression in Table 6.1 to

$$P_D = \exp\left(\frac{-T'}{1+\bar{\chi}}\right) \quad \text{(Swerling 2, } N = 1\text{)} \tag{6.111}$$

This result is identical to the Swerling 1, $N = 1$ result of Eq. (6.107). This means that Eq. (6.109) also applies in the Swerling 2, $N = 1$ case. For $N = 2$, the property $I(c,2) = 1 - (1+c)\exp(-c)$ gives

$$P_D = \left(1 + \frac{T'}{1+\bar{\chi}}\right)\exp\left(\frac{-T'}{1+\bar{\chi}}\right) \quad \text{(Swerling 2, } N = 2\text{)} \tag{6.112}$$

Swerling 3 Case: Using the same arguments as in the Swerling 1 case and the property $\lim_{c\to\infty} I(c,d) = 1$, it can be shown that the exact expression given in Table 6.1 for the Swerling 3 case with $N = 1$ or 2 is a reasonable approximation for larger N as well, provided again that $P_{FA} \ll 1$ and $N\bar{\chi} > 2$ (DiFranco and Rubin, 1980, p. 421). Therefore

$$P_D \approx \left(1 + \frac{1}{(N\bar{\chi}/2)}\right)^{N-2}\left[1 + \frac{T'}{1+(N\bar{\chi}/2)} - \frac{N-2}{N\bar{\chi}/2}\right]\exp\left[\frac{-T'}{1+(N\bar{\chi}/2)}\right] \tag{6.113}$$

(Swerling 3; $P_{FA} \ll 1$ and $N\bar{\chi} > 2$)

Again, this expression is exact for $N = 1$ and $N = 2$. Table 6.2 summarizes these simplified formulas.

6.4.6 Shnidman's Equation

Most of the analytic results in Table 6.1 are too complex for "back-of-the-envelope" calculations or even for calculation on programmable calculators. The simplified results of Table 6.2 help for some Swerling cases and values of N, but not all. Albersheim's equation provided a simple approximation for the nonfluctuating target case, but it is not applicable to fluctuating targets in general or the Swerling models in particular.

Fortunately, empirical approximations have also been developed for the Swerling cases. One example is *Shnidman's equation* (Shnidman, 2002). Similar to Albersheim's equation, this series of equations estimates the single-sample SNR χ_1 required to achieve a specified P_D and P_{FA} with noncoherent integration of N samples. Unlike Albersheim's equation, the results are for a square law detector. However, as noted previously the differences in the required SNR for linear and square law detectors are typically no more than 0.2 dB.

Shnidman's equation is given by the following series of calculations:

$$K = \begin{cases} \infty, & \text{nonfluctuating target ("Swerling 0 or 5")} \\ 1, & \text{Swerling 1} \\ N, & \text{Swerling 2} \\ 2, & \text{Swerling 3} \\ 2N, & \text{Swerling 4,} \end{cases} \tag{6.114}$$

$$\alpha = \begin{cases} 0, & N < 40 \\ 1/4, & N \geq 40 \end{cases}$$

Case	N	Formula	Other Conditions	Exact or Approximate
1	1	$P_D = \exp\left(\dfrac{-T'}{1+\bar{\chi}}\right)$, $P_D = P_{FA}^{1/(1+\bar{\chi})}$ $P_{FA} = \exp(-T')$		Exact
1	All N	$P_D = \left(1 + \dfrac{1}{N\bar{\chi}}\right)^{N-1} \exp\left(\dfrac{-T'}{1+N\bar{\chi}}\right)$	$P_{FA} \ll 1$, $N\bar{\chi} > 2$	Exact for $N=1$ Approximate for $N \geq 2$
2	1	$P_D = \exp\left(\dfrac{-T'}{1+\bar{\chi}}\right)$, $P_D = P_{FA}^{1/(1+\bar{\chi})}$ $P_{FA} = \exp(-T')$		Exact
2	2	$P_D = \left(1 + \dfrac{T'}{1+\bar{\chi}}\right)\exp\left(\dfrac{-T'}{1+\bar{\chi}}\right)$		Exact
3	1, 2	$\left(1 + \dfrac{1}{(N\bar{\chi}/2)}\right)^{N-2} \times \ldots$ $\times \left[1 + \dfrac{T'}{1+(N\bar{\chi}/2)} - \dfrac{N-2}{N\bar{\chi}/2}\right]\exp\left[\dfrac{-T'}{1+(N\bar{\chi}/2)}\right]$	$P_{FA} \ll 1$, $N\bar{\chi} > 2$	Exact for $N=1$ or 2 Approximate for $N \geq 3$

TABLE 6.2 Simplified Formulas for P_D and P_{FA} for Some Swerling Cases

$$\eta = \sqrt{-0.8\ln[4P_{FA}(1-P_{FA})]} + \text{sgn}(P_D - 0.5)\sqrt{-0.8\ln[4P_D(1-P_D)]}$$
$$X_\infty = \eta\left(\eta + 2\sqrt{\dfrac{N}{2} + \left(\alpha - \dfrac{1}{4}\right)}\right) \tag{6.115}$$

$$C_1 = \{[(17.7006P_D - 18.4496)P_D + 14.5339]P_D - 3.525\}/K$$
$$C_2 = \dfrac{1}{K}\left\{\exp(27.31P_D - 25.14) + (P_D - 0.08)\left[0.7\ln\left(\dfrac{10^{-5}}{P_{FA}}\right) + \dfrac{(2N-20)}{80}\right]\right\}$$
$$C_{dB} = \begin{cases} C_1, & 0.1 \leq P_D \leq 0.872 \\ C_1 + C_2, & 0.872 < P_D \leq 0.99 \end{cases}$$
$$C = 10^{C_{dB}/10} \tag{6.116}$$

$$\chi_1 = \frac{C \cdot X_\infty}{N}$$
$$\chi_1 \text{ (dB)} = 10 \log_{10}(\chi_1) \tag{6.117}$$

The function $\text{sgn}(x)$ is $+1$ if $x > 0$ and -1 if $x < 0$. Note that several of the equations simplify in the nonfluctuating case ($K = \infty$). Specifically, in that case $C_1 = C_2 = 0$ so that in turn $C_{\text{dB}} = 0$ and $C = 1$.

The accuracy bounds on Shnidman's equation are somewhat looser than those specified for Albersheim's equation. Except at the extreme values of P_D for the Swerling 1 case, the error in the estimate of χ_1 is less than 0.5 dB for $0.1 \leq P_D \leq 0.99$, $10^{-9} \leq P_{FA} \leq 10^{-3}$, and $1 \leq N \leq 100$. This is a much wider range for P_D than used in Albersheim's equation. The range of N is much smaller but still large enough for almost all problems of interest. Figure 6.15 illustrates the error in the estimate of the single-pulse SNR χ_1 using Shnidman's equation for the case $P_{FA} = 10^{-6}$, $N = 5$, and P_D over the specified range of 0.1 to 0.99.

A still more accurate approximation for the nonfluctuating and Swerling 1 cases is described in Hmam (2003). However, it is not applicable to all of the Swerling cases and the computations, while easy, are more extensive.

6.4.7 Detection in Clutter

All of the results so far in this chapter have been for white Gaussian noise interference. In this case, the density of the interference power is an exponential PDF. However, as discussed in Chap. 2, the statistics of clutter interference are often better modeled with longer-tailed PDFs such as the log-normal, Weibull, or K power densities. If clutter is the dominant interference, the analysis strategy used in the preceding sections can still be applied, provided that appropriate PDFs to describe the data under hypotheses H_0 and H_1 are used. For

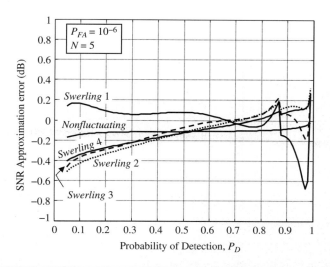

Figure 6.15 Example of error in estimating χ_1 via Shnidman's equation for $P_{FA} = 10^{-6}$ and $N = 5$.

instance, for $N = 1$ and a nonfluctuating target the H_0 (clutter-only) PDF might be a K power density. The PDF under H_1 will then be that of a constant plus the K-distributed clutter. If the target is fluctuating, then the PDF of clutter plus the now-random target component will be needed. For $N > 1$, the PDFs of the sum of a single clutter-only or clutter-plus-target sample must be obtained.

The effect of a long-tailed interference PDF on detection performance is easy to understand. A higher threshold will be required to achieve a given P_{FA}. The higher threshold will then require a stronger target echo, that is, a higher SIR, to obtain a given P_D. One example of this behavior is shown in Fig. 6.16, which compares the ROC curves for a single sample of a nonfluctuating target when the square-law-detected interference is exponential versus when it is distributed according to the K power PDF of Table 2.3. P_{FA} was set to 10^{-6}. The curve marked "exponential" is the same as the "unknown phase" case in Fig. 6.8b. Three different values of the K distribution shape parameter a are used; smaller values of a represent longer PDF tails and therefore higher thresholds to meet the P_{FA} specification. The scale parameter c was set equal to $2\sqrt{a}$, a choice which keeps the mean $\bar{\sigma} = 1$ as a changes. Clearly, the SIR must be several decibels higher to achieve a given P_D when the interference is clutter instead of noise.

The square law detector used in this example was developed based on Gaussian noise interference and is not necessarily the optimal LRT detector when the interference is clutter. A thorough discussion of optimal detection for coherent radar systems when the PDF of the dominant interference is a compound form such as Weibull or K is given in Sangston and Farina (2016). There they discuss several detector structures which approximate the LRT. In general, they tend to set a threshold that varies based on the instantaneous clutter power level, rather than using the fixed threshold of the Gaussian case. The reader is referred to Sangston and Farina (2016) for details.

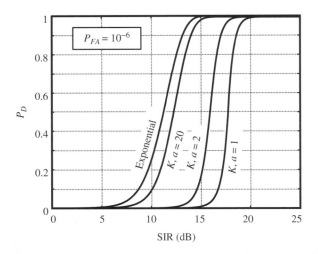

FIGURE 6.16 Effect of K power-distributed clutter interference with varying shape parameters on target detection performance. See text for details.

6.4.8 Binary Integration

As discussed in Sec. 6.4.1, any coherent or noncoherent integration is followed finally by comparing the integrated data to a threshold. The result is a choice between two hypotheses, "target present" or "target absent," so the output is binary in the sense that it takes one of only two possible outcomes. If the entire detection process is repeated N times for a given range, Doppler, or angle cell, N binary decisions will be available. Each decision of "target present" will have some probability P_D of being correct and a probability P_{FA} of being incorrect. To improve the reliability of the detection decision, the decision rule can require that a target be detected on some number M of the N decisions before it is finally accepted as a valid target detection. This process is called *binary integration*, "*M of N*" *detection*, or *coincidence detection* (Levanon, 1988; Skolnik, 2001).

To analyze binary integration, begin by assuming a nonfluctuating target so that the single-trial probability of detection P_D is the same for each of N threshold tests. Then the probability of *not* detecting an actual target (i.e., the probability of a miss) on one trial is $1 - P_D$. If they are independent, the probability of missing the target on all N trials is $(1 - P_D)^N$. Thus, the probability of detecting the target on at least one of N trials, denoted the *binary integrated probability* P_{BD}, is

$$P_{BD} = 1 - (1 - P_D)^N \qquad (6.118)$$

Table 6.3 shows the single-trial probability of detection required to achieve $P_{BD} = 0.99$ as a function of N. Clearly, a "1 of N" decision rule achieves a high binary integrated probability of detection with relatively low single-trial probabilities of detection. In other words, the "1 of N" rule increases the effective probability of detection. This has the effect of reducing the single-trial SNR required to achieve the final target value of P_D.

The trouble with the "1 of N" rule is that it "works" for the probability of false alarm also. The probability of at least one false alarm in N trials is the binary integrated probability of false alarm, P_{BFA},

$$P_{BFA} = 1 - (1 - P_{FA})^N \qquad (6.119)$$

Assuming that $P_{FA} \ll 1$, Eq. (6.119) can be approximated as

$$\begin{aligned} P_{BFA} &= 1 - (1 - P_{FA})^N \\ &= 1 - \left[1 - N \cdot P_{FA} + \frac{N(N-1)}{2} P_{FA}^2 - \ldots \right] \approx 1 - (1 - N \cdot P_{FA}) \\ &= N \cdot P_{FA} \end{aligned} \qquad (6.120)$$

N	1	2	4	10	20	100
P_D	0.99	0.90	0.68	0.37	0.2	0.045

TABLE 6.3 Single-Trial P_D Needed to Achieve $P_{BD} = 0.99$ Using Binary Integration

where the binomial series expansion was used to obtain the second line. Equation (6.120) shows that the "1 of N" rule increases P_{FA} by a factor of approximately N, an undesirable result. What is needed is a binary integration rule that increases P_{BD} compared to P_D while leaving P_{BFA} equal to or less than P_{FA}. An "M of N" strategy with $M > 1$ provides better results.

The integrated probability P_B of M threshold crossings in N trials when the probability of a crossing on a single trial is p is

$$P_B = \sum_{r=M}^{N} \binom{N}{r} p^r (1-p)^{N-r} \tag{6.121}$$

where

$$\binom{N}{r} \equiv \frac{N!}{(N-r)!r!} \tag{6.122}$$

Equation (6.121) can be applied to the probability of false alarm by letting $p = P_{FA}$ and to the probability of detection by letting $p = P_D$. Consider the specific example of a "2 of 4" rule, that is, $N = 4$ and $M = 2$. Using these parameters in Eq. (6.121) gives

$$\begin{aligned} P_B &= \sum_{r=2}^{4} \frac{24}{(4-r)!r!} p^r (1-p)^{4-r} \\ &= 6p^2(1-p^2) + 4p^3(1-p) + p^4 \end{aligned} \tag{6.123}$$

To determine the effect of this rule on the probability of false alarm, let $p = P_{FA}$. Assuming that $P_{FA} \ll 1$, Eq. (6.123) can be approximated by its first term to obtain

$$P_{BFA} \approx 6P_{FA}^2 \tag{6.124}$$

Thus, the "2 of 4" rule will result in a binary integrated false alarm probability that is less than the single-trial P_{FA} as desired.

Equation (6.123) cannot easily be approximated in a simple form similar to Eq. (6.124) for typical single-trial values of P_D. Table 6.4 shows the binary integrated probability obtained using a "2 of 4" rule for various values of the single-trial probability p. The three cases above the dotted line are appropriate for considering the effect on example single-trial probabilities of detection, while the two cases below the line are examples of the effect on example single-trial probabilities of false alarm. This table shows that the "2 of 4" rule not only reduces the probability of false alarms, but also increases the probability of detection so long as the single-trial P_D is reasonably high.

p	P_B
0.5	0.688
0.8	0.973
0.9	0.996
10^{-3}	5.992×10^{-6}
10^{-6}	6.0×10^{-12}

TABLE 6.4 Binary Integrated Probability Using a "2 of 4" Rule

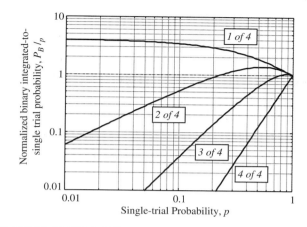

Figure 6.17 Ratio of the binary integrated probability to the single-trial probability for an "M of 4" binary integration rule.

Given a particular choice of M and N, Eq. (6.121) can be solved numerically to find the pre-binary integration P_D and P_{FA} required to achieve a specified P_{BD} and P_{BFA}. Continuing the "2 of 4" rule example so that Eq. (6.123) applies, suppose $P_{BD} = 0.99$ and $P_{BFA} = 10^{-8}$ is required. Then Eq. (6.123) can be solved for the required single-trial probabilities of $P_D = 0.86$ and $P_{FA} = 4.1 \times 10^{-5}$. [$P_{FA}$ can also be found more quickly from the approximation of Eq. (6.124).]

This example illustrates the characteristics required of an "M of N" rule. For small values of p, P_B should be less than or equal to p so that the rule reduces false alarm probabilities. For larger values of p, P_B should be greater than or equal to p so that detection probabilities are increased by binary integration. Figure 6.17 plots the ratio of P_B to p for $N = 4$ and all four possible choices of M. A ratio greater than 1 means P_B is greater than p; this should be the case for values of p appropriate to single-trial detection probabilities. Conversely, for small values of p appropriate to false alarm probabilities, the ratio should be less than 1. The figure shows that the ratio is greater than 1 for all values of p for the "1 of 4" rule, consistent with the earlier discussion. Similarly, the "4 of 4" rule results in a ratio that is always less than 1, good for false alarm reduction but bad for improving detection. The "2 of 4" and "3 of 4" rules both provide good false alarm reduction for small values of p and some detection improvement for large values of p. However, the "3 of 4" rule increases probabilities only for p equal to approximately 0.75 or higher, a relatively narrow range, and the increase is very slight. The "2 of 4" rule improves detection for values of p down to approximately 0.23, still well above any likely single-trial P_{FA}.

Another term sometimes used for the binary integrated probability is *cumulative probability*. However, that term is more commonly restricted to describe the probability of detecting a target at least once in N tries, for example, on N successive scans of a surveillance radar (IEEE, 2017). If the individual scan P_D is the same on each scan, the cumulative probability P_{CD} is the binary integrated probability for the "1 of N" case given in Eq. (6.118). If the target range changes significantly during the N scans, the individual scan SNR and thus P_D would be expected to change so that a more general formula would be required.

The discussion above suggests that the best choice of M for a given N is the one that maximizes the range of pre-threshold detection probabilities for which the post-threshold detection probability is increased. (Any reasonable pre-threshold false alarm probability will be decreased for any choice of M except $M = 1$). A more useful basis for evaluation would be to compute a binary integration gain G_{BI}, that is, the factor by which the required single-trial

388 Chapter Six

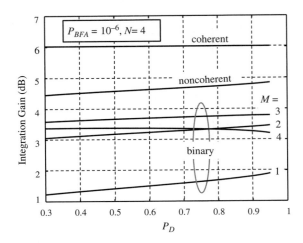

Figure 6.18 Binary integration gain G_{BI} for $N = 4$ and various values of M.

SNR needed to achieve a given P_D and P_{FA} is decreased when using binary integration, versus that required when detection is based on only a single trial. This is in exact analogy to the idea of both coherent and noncoherent integration. A simple numerical procedure for estimating G_{BI} for any M and N in the nonfluctuating target case is developed in Richards (2016). The result is illustrated in Fig. 6.18 for the "2 of 4" example with a binary integrated P_{BFA} of 10^{-6}, values of P_{BD} from 0.3 to 0.95, and all four possible values of M. Choosing $M = 3$ produces the highest integration gain of about 3.6 to 3.8 dB. However, except for $M = 1$, the variation with M is slight, about 0.6 dB or less between the $M = 2, 3,$ and 4 cases.

Figure 6.18 also includes the coherent and noncoherent integration gains that would be possible for integration of four predetection measurements. These curves should be taken with caution, since they represent integrating independent measurements before threshold detection while the binary integration curves represent the effect of combining four post-threshold decisions. Nonetheless, they emphasize that, in the sense of minimizing SNR required to achieve given detection and false alarm probabilities, coherent integration makes the most efficient use of data, followed by noncoherent integration, and finally binary integration. This observation is discussed further in the next subsection.

Recall that a nonfluctuating target has been implicitly assumed since the single-trial P_D was taken to be the same on each trial. The analysis can be extended to fluctuating targets (Weiner, 1991; Shnidman, 1998). One result of these analyses is that for a given Swerling model, P_{FA} specification, SNR, and number of trials N, the value M_{opt} of M that maximizes P_{BD} for a given P_{BFA}, N, and SNR can be estimated as

$$M_{opt} = 10^b N^a \tag{6.125}$$

where the parameters a and b are given in Table 6.5 for the various Swerling models with $P_D = 0.9$ and $10^{-8} \leq P_{FA} \leq 10^{-4}$ ("Swerling 0" is the nonfluctuating case) (Shnidman, 1998). M_{opt} must be rounded to the nearest integer. Applying this formula to the nonfluctuating $N = 4$ example above (even though $N = 4$ is slightly out of the parameter range in the table for the Swerling 0 case) gives $M = 2.9$, which rounds to the value of 3 in agreement with Fig. 6.18. The estimated value of M_{opt} for other target fluctuation models follows the same functional form as Eq. (6.125) but with different exponent values; see Shnidman (1998) for details.

Swerling Model	a	b	Range of N
0	0.8	−0.02	5–700
1	0.8	−0.02	6–500
2	0.91	−0.38	9–700
3	0.8	−0.02	6–700
4	0.873	−0.27	10–700

TABLE 6.5 Parameters for Estimating M_{opt}

Binary integration is often combined with multiple PRF data acquisition to achieve more than just detection. In particular, collecting N coherent processing intervals (CPIs) of data in a pulse Doppler radar, each at a different PRF; applying appropriate threshold detection processing to each CPI; and then applying binary integration across the CPIs in each range-Doppler cell not only achieves an integration gain for detection but can also be used to reduce or avoid range-Doppler blind zones and resolve range and Doppler ambiguities as discussed in Chap. 5.

6.4.9 Integration Summary

As has been seen, N independent measurements of the echo from a particular spatial or spatial and Doppler location can be combined in various ways to attempt to minimize the single-measurement SNR needed to achieve a given P_D and P_{FA}. Coherent integration offers the greatest integration gain G_c of a factor of N, but requires that the target components of the measurements be in phase, or be compensated to be in phase. Failure to maintain phase coherence can result in destructive combination of the target samples, degrading instead of enhancing the SNR. Noncoherent integration offers a lesser integration gain factor, $\sqrt{N} < G_{nc} < N$, but can be applied in more situations because it does not depend on maintaining coherence of the target echo components, which is sometimes difficult. Unlike coherent and noncoherent integration, binary integration takes place postdetection and does not use data values directly, instead combining N binary threshold crossing test results in a combinatorial "M of N" rule. It achieves an integration gain G_{BI} about 1 to 1.5 dB lower than noncoherent integration over a wide range of parameters and target fluctuation models.

Given multiple measurements of a nonfluctuating target in noise with the goal of minimizing the SNR required to achieve a specified P_D and P_{FA}, coherent integration should be performed first to the maximum extent possible. If coherent integration is not feasible or is limited to fewer than N samples at a time (perhaps due to radar-target motion-induced phase errors, transceiver phase instabilities, etc.), then there will still be samples available to be combined. These samples should be noncoherently integrated if possible. Finally, if noncoherent integration of the remaining samples is not feasible, binary integration can be applied with only a modest additional loss.

For example, suppose 32 measurements are collected, but the radar-target motion is such that the Doppler shift is reasonably constant only over about eight measurements. Coherently integrating more than eight measurements could result in destructive combining of the target components, degrading detection performance. A more effective approach might be to coherently combine groups of eight successive measurements and then noncoherently combine the four resulting integrated measurements, followed by a single threshold test. Alternatively, the four coherently integrated measurements could each be threshold tested and the results combined in a binary integration step using a "3 of 4" rule. These issues are discussed in more detail and a more detailed example is given in Richards (2014).

6.5 Constant False Alarm Rate Detection

Standard radar threshold detection as discussed in the preceding sections assumes that the interference level is known and constant. This in turn allows accurate setting of a threshold that guarantees a specified P_{FA}. In practice, interference levels are often variable. *Constant false alarm rate* (CFAR) detection, also frequently referred to as "adaptive threshold detection" or "automatic detection," is a set of techniques designed to provide predictable detection and false alarm behavior in more realistic interference scenarios.

6.5.1 The Effect of Unknown Interference Power on False Alarm Probability

In the preceding sections the detection and false alarm performance of a square law detector were considered for a target in complex white Gaussian interference as a function of the target fluctuation model and number of measurements noncoherently integrated. It was shown in Eq. (6.82) that for a single unnormalized data sample ($N = 1$) of a nonfluctuating target the false alarm probability and threshold were related according to

$$P_{FA} = \exp\left(-T/\sigma_w^2\right) \Rightarrow T = -\sigma_w^2 \ln(P_{FA}) \tag{6.126}$$

The threshold T is proportional to the interference power, $T = \alpha \sigma_w^2$, with the multiplier α a function of the desired false alarm probability.

To tune the square law detector for a particular radar system, an acceptable value of P_{FA} must be chosen. The threshold is then computed according to Eq. (6.126). The probability of detection that will be achieved is determined by the target SNR.

Accurate setting of the threshold requires accurate knowledge of the interference power σ_w^2. In some systems this is known, but in many it is not. When the interference is principally receiver noise, it is possible to measure σ_w^2 and calibrate the detector. In day-to-day operation, however, the receiver noise will vary over time due to factors such as temperature changes and component aging. Temperature compensation and periodic recalibration, if possible, can combat this problem. If the total interference power is significantly affected by external sources, the variability can be much more severe. In very low noise radar systems, a significant part of the noise power is cosmic noise. The total receiver interference then varies with the look direction and the time of day. In conventional radars, the total interference power can be affected by in-band *electromagnetic interference* (EMI). For example, UHF radars can be affected by television transmissions, while certain wireless communication services can compete with higher frequency radars, especially in urban areas. If the dominant interference is ground clutter, its power will vary radically with the type of terrain being illuminated and even the weather and seasons. For instance, open desert has a relatively low reflectivity, while refrozen snow can have a very high reflectivity. Finally, the dominant interference can be hostile electromagnetic emissions deliberately directed at the radar system (jamming). In this case, the interference power can be extremely high.

In any of these cases, the observed P_{FA} will vary from the intended value. To see how significant this variation might be, let P_{FA0} be the intended probability of false alarm when the actual interference power is the expected value of σ_{w0}^2; the threshold will then be set at $T = -\sigma_{w0}^2 \ln P_{FA0}$. Now suppose the actual interference power is σ_w^2. The actual P_{FA}, using Eq. (6.126) with the threshold designed assuming an interference power of σ_{w0}^2, will be

$$\begin{aligned} P_{FA} &= \exp\left[\frac{\sigma_{w0}^2 \ln(P_{FA0})}{\sigma_w^2}\right] = \exp\left\{\ln\left[P_{FA0}^{(\sigma_{w0}^2/\sigma_w^2)}\right]\right\} \\ &= P_{FA0}^{(\sigma_{w0}^2/\sigma_w^2)} \end{aligned} \tag{6.127}$$

Detection Fundamentals

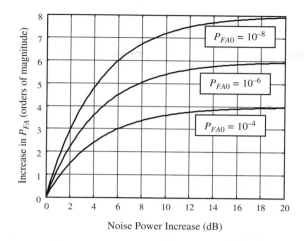

FIGURE 6.19 Increase in probability of false alarm for fixed threshold due to increase in noise power.

and the increase in false alarm probability will be a factor of

$$\frac{P_{FA}}{P_{FA0}} = (P_{FA0})^{\left[(\sigma_{w0}^2/\sigma_w^2)-1\right]} \tag{6.128}$$

Figure 6.19 plots Eq. (6.128) for three different values of the design false alarm probability. This figure shows that even modest increases of 2 dB can cause an unintended increase in P_{FA} of 1.5 to 3 orders of magnitude, with the largest changes occurring when the desired P_{FA} is lowest. When the increase is 3 dB (a factor of 2×), $P_{FA} = \sqrt{P_{FA0}}$. Clearly, such sensitivity to small changes in interference power or, equivalently, small errors in setting the threshold will have major impacts on radar performance.

The reason for the increase in P_{FA} observed in Fig. 6.19 is that the threshold T was based on an incorrect value for the interference power σ_w^2. More generally, as the interference power at the output of the radar receiver varies, the actual P_{FA} will vary widely. From a system point of view, this is highly undesirable. When the interference power rises, the number of false alarms will also rise, possibly by orders of magnitude. It might seem that the difference between a probability of false alarm of 10^{-8} and a rate of 10^{-6} may be insignificant, but consider the example of a simple radar system with a repetition rate of 10 kHz and 200 range bins. If each range bin is tested, this system makes $(10,000)(200) =$ two million detection decisions per second. With $P_{FA} = 10^{-8}$, false alarms occur on average only once every 50 seconds. If P_{FA} rises to 10^{-6}, the system is confronted with an average of two false alarms every second. How much of a concern this increase is depends on the impact of a false alarm in the overall radar system. This could include increased demand on radar or signal processor resources to confirm or reject the false alarm or to start unneeded tracks, increased cluttering of an operator display, or reduced time for search and tracking of other targets.

If the interference power drops below that assumed when calculating the threshold, the false alarm probability will drop. This may seem inconsequential or even desirable, but a reduced P_{FA} represents a threshold that is higher than necessary to achieve the system design goals. Since P_{FA} and P_D always increase or decrease together as discussed earlier, this means that the probability of detection is less than could be achieved with a correctly set threshold.

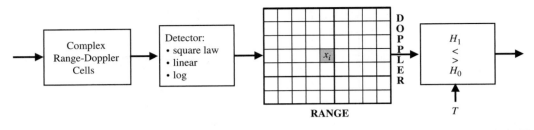

FIGURE 6.20   Generic detection processor.

6.5.2 Cell-Averaging CFAR

In order to obtain predictable and consistent performance the radar system designer would usually prefer to maintain a constant false alarm rate. To achieve this, the actual interference power must be estimated from the data in real time so that the detector threshold can be adjusted to maintain the desired P_{FA}. A detection processor that can maintain a constant P_{FA} is called a CFAR processor.

Figure 6.20 shows a generic radar CFAR processor. The detector shown is for a system using range-Doppler processing, but other systems might consider only a one-dimensional vector of range cells in making a decision. Still other systems might perform the detection process on a radar image, so that the individual cells are pixels in a two-dimensional image. Whatever the form of the data, the detector will test each available data sample for the presence or absence of a target. The current *cell under test* (CUT), denoted by x_i in Fig. 6.20, is compared against a threshold determined by the interference power. If the value of the data in the test cell exceeds the threshold, the processor declares a target present at the range and velocity (or range, or image location, as appropriate) corresponding to the CUT. The next cell is then tested and so forth until a target present/target absent decision has been made for all cells of interest.

To set the threshold for testing cell x_i, the interference power *in the same cell* is required. Since it may be variable, it must be estimated from the data. The approach used in CFAR processing is based on two major assumptions:

- The neighboring cells contain interference with the same statistics as the CUT (called *homogeneous* interference), so that they are representative of the interference that is competing with the potential target.
- The neighboring cells do *not* contain any targets; they contain interference only.

Under these conditions, the interference statistics in the CUT can be estimated from the measured samples in the adjoining cells.

The statistics that must be estimated are determined by the statistics needed to implement the threshold test. For Gaussian interference and linear or square law detectors, the interference will be Rayleigh or exponential distributed, respectively. In either case, the interference PDF has only one free parameter, the mean interference power. Thus, the CFAR processor must estimate the mean interference power in the CUT by using the measured data in the adjoining cells.

For a more specific example, consider the square law case. Assuming the interference is i.i.d. WGN in the I and Q signals with power $\sigma_w^2/2$ in each (total power σ_w^2), the PDF of the power observed in a cell x_i is

$$p_{x_i}(x_i) = \frac{1}{\sigma_w^2} \exp\left(-x_i/\sigma_w^2\right) \tag{6.129}$$

As seen in Eq. (6.121), knowledge of σ_w^2 is needed to set the threshold. When exact knowledge of σ_w^2 is not available, it must be estimated.

Assume that N cells in the vicinity of the cell under test are used to estimate σ_w^2 and that the interference in each is i.i.d. WGN. The joint PDF of a vector $\mathbf{x}$ of N such samples is

$$p_\mathbf{x}(\mathbf{x}) = \frac{1}{\sigma_w^{2N}} \prod_{i=1}^{N} \exp\left(-x_i/\sigma_w^2\right)$$
$$= \frac{1}{\sigma_w^{2N}} \exp\left[-\left(\sum_{i=1}^{N} x_i\right)/\sigma_w^2\right] \equiv \ell(\mathbf{x}) \qquad (6.130)$$

Equation (6.130) is the *likelihood function* ℓ (see Chap. 7) for the observed data vector $\mathbf{x}$. The maximum likelihood estimate of σ_w^2 is obtained by maximizing $\ell(\mathbf{x})$ with respect to σ_w^2 while $\sum x_i$ is held constant (Kay, 1993). It is equivalent and more convenient to maximize the log-likelihood function

$$\ln(\ell) = -N \ln(\sigma_w^2) - \frac{1}{\sigma_w^2}\left(\sum_{i=1}^{N} x_i\right) \qquad (6.131)$$

Setting the derivative of Eq. (6.131) with respect to σ_w^2 equal to zero gives

$$\frac{d[\ln(\ell)]}{d(\sigma_w^2)} = 0 = -N\left(\frac{1}{\sigma_w^2}\right) - \left(-\frac{1}{(\sigma_w^2)^2}\right)\sum_{i=1}^{N} x_i \qquad (6.132)$$

Solving Eq. (6.132) for σ_w^2 gives the unsurprising result that the maximum likelihood estimate is just the average of the available data samples:

$$\widehat{\sigma_w^2} = \frac{1}{N}\sum_{i=1}^{N} x_i \qquad (6.133)$$

The required threshold is then estimated as a multiple of the estimated interference power:

$$\hat{T} = \alpha \cdot \widehat{\sigma_w^2} \qquad (6.134)$$

Because the interference power and thus the threshold are estimated from an average of the power in the cells adjoining the test cell, this CFAR approach is referred to as *cell-averaging CFAR* (CA CFAR). Because the interference power is estimated rather than known exactly, the scale factor α will not have the same value as in Eq. (6.126); it will be derived in the next subsection.

Equation (6.133) states that the parameter of the exponential PDF describing the square-law-detected data should be estimated from an average of N adjoining data samples. Figure 6.21 shows two examples of how the samples to be averaged are selected. Figure 6.21a shows a one-dimensional data vector of range cells with the CUT, x_i, in the middle. The data in the grey cells to either side, representing data from ranges nearer and farther from the radar than the CUT, are averaged to estimate the noise parameter. These cells are called the *reference cells*. The diagonally-striped cells immediately adjacent to the CUT, called *guard cells*, are excluded from the average. The reason is that a target, if present, might straddle range cells. In that case, the energy in the cell adjacent to x_i would contain both interference and target energy and would therefore not be representative of the interference alone. The extra energy from the target would tend to raise the estimate

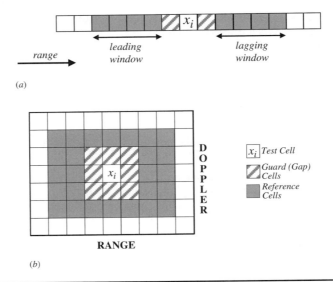

FIGURE 6.21 CFAR windows. (a) One-dimensional window for range-only processor. (b) Two-dimensional window for range-Doppler processor.

of the interference parameter. For instance, in the square-law-detector case the estimate of σ_w^2 would be too high, resulting in a threshold that was too high and a lower P_{FA} and P_D than intended. If the system range resolution is such that anticipated targets could extend over multiple range cells, more than one guard cell would be skipped on each side of the CUT. The combined reference cells, guard cells, and cell under test are referred to as the *CFAR window*.

Figure 6.21b shows a typical two-dimensional equivalent to the one-dimensional case, in this case applied to a range-Doppler matrix after detection. Both the guard region and the reference window are now two dimensional. A two-dimensional CFAR window could also be applied to synthetic aperture radar imagery, in which case the two dimensions would be simply the range and cross-range spatial dimensions. In the range-Doppler case, the cell-averaging CFAR might be applied only over certain range and Doppler cells because ground clutter renders the interference nonhomogeneous in Doppler and possibly in range.

6.5.3 Analysis of Cell-Averaging CFAR

The intent of the adaptive calculation of the threshold is to provide a constant P_{FA} despite varying interference power levels. In this subsection, the detector performance is analyzed for the case of a square law detector to see if this goal has been achieved. The threshold computed according to Eq. (6.134) will be a random variable, and therefore so will be the probability of false alarm. The detector will be considered to be CFAR if the expected value of P_{FA} does not depend on the actual value of σ_w^2.

Combining Eqs. (6.133) and (6.134) gives an expression for the estimated threshold

$$\hat{T} = \frac{\alpha}{N} \sum_{i=1}^{N} x_i \qquad (6.135)$$

Define $z_i = (\alpha/N)x_i$; thus $\hat{T} = \sum_{i=1}^{N} z_i$. Standard results from probability theory and Eq. (6.129) give the PDF of z_i as

$$p_{z_i}(z_i) = \frac{N}{\alpha \sigma_w^2} \exp(-Nz_i/\alpha\sigma_w^2) \tag{6.136}$$

The PDF of $\hat{T}$ is the Erlang density

$$p_{\hat{T}}(\hat{T}) = \begin{cases} \left(\dfrac{N}{\alpha\sigma_w^2}\right)^N \dfrac{\hat{T}^{N-1}}{(N-1)!} \exp(-N\hat{T}/\alpha\sigma_w^2), & \hat{T} > 0 \\ 0, & \hat{T} < 0 \end{cases} \tag{6.137}$$

The P_{FA} observed with the estimated threshold will be $\exp(-\hat{T}/\sigma_w^2)$. This is now also a random variable; its expected value is

$$\begin{aligned}\bar{P}_{FA} &= \int_0^\infty \exp(-\hat{T}/\sigma_w^2) p_{\hat{T}}(\hat{T}) d\hat{T} \\ &= \left(\frac{N}{\alpha\sigma_w^2}\right)^N \frac{1}{(N-1)!} \int_0^\infty \hat{T}^{N-1} \exp\{-[(N/\alpha)+1]\hat{T}/\sigma_w^2\} d\hat{T}\end{aligned} \tag{6.138}$$

Completing this standard integral and performing some algebraic manipulations gives the final result:

$$\bar{P}_{FA} = \left(1 + \frac{\alpha}{N}\right)^{-N} \tag{6.139}$$

For a given desired $\bar{P}_{FA}$, the required threshold multiplier is obtained by solving Eq. (6.139) to obtain

$$\alpha = N\left[(\bar{P}_{FA})^{-1/N} - 1\right] \tag{6.140}$$

Note that while $\bar{P}_{FA}$ depends on the number N of neighboring cells averaged and the multiplier, it does not depend on the actual interference power σ_w^2. Thus, the cell-averaging technique exhibits CFAR behavior.

Now that a rule for selecting the CA CFAR threshold has been determined, the detection performance can be determined. Equation (6.99) shows that for a single sample of a Swerling 1 or 2 target with the threshold $\hat{T}$, $P_D = \exp[-\hat{T}/(1+\bar{\chi})]$, where $\bar{\chi}$ is the mean SNR. The expected value of P_D is obtained by averaging over the threshold:

$$\bar{P}_D = \int_0^\infty \exp[-\hat{T}/(1+\bar{\chi})] p_{\hat{T}}(\hat{T}) d\hat{T} \tag{6.141}$$

This integral is the same general form as Eq. (6.138); the result is

$$\bar{P}_D = \left[1 + \frac{\alpha}{N(1+\bar{\chi})}\right]^{-N} \tag{6.142}$$

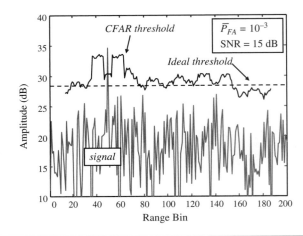

FIGURE 6.22 Example of cell-averaging CFAR threshold behavior.

Note that this also does not depend on the interference power. However, this result is specific to the assumptions of complex WGN, a square law detector, a Swerling 1 or 2 target, and a single test sample.

Figure 6.22 illustrates the operation of cell-averaging CFAR. The simulated data correspond to additive complex WGN with power $10\log_{10}(\sigma_w^2) = 20$ dB. A single nonfluctuating target with a power of 35 dB is present in range bin 50; the SNR is thus 15 dB. If the desired $P_{FA} = 10^{-3}$, Eq. (6.126) gives the ideal $T = 691$, equal to 28.4 dB. This threshold level is indicated on the plot. Note that the ideal threshold is a multiple of $-\ln(P_{FA}) = 6.91$ times the true interference power; equivalently, the threshold is 8.4 dB above the interference power level.

Now consider a CA CFAR with leading and lagging windows of 10 cells each after skipping a three-cell guard region to each side of the CUT. Thus $N = 20$ cells are averaged to estimate the interference power.[18] From Eq. (6.140), the multiplier α will be 8.25, placing the threshold about 9.2 dB above the *estimated* mean power. The line labeled "CFAR threshold" shows the computed threshold as the reference window slides across the data. Except in the vicinity of the target, the estimated threshold tracks the ideal threshold well, staying within 2 dB across most of the data. Note that the data exceed the CFAR threshold only at range bin 50. In this example the CFAR detector works very well: a detection would correctly be declared when the CFAR test cell is located at range bin 50, but there are no false alarms (threshold crossings) at any other range bins.

The elevated threshold to either side of the target location is characteristic of cell-averaging CFAR. For the particular CFAR window configuration used here the cell containing the target will be in the leading reference window and will be included in the estimate of the interference power when the test cell is between range bins 37 and 46. The estimated power $\widehat{\sigma_w^2}$ and, in turn, the computed threshold $\hat{T}$ will be significantly raised. This phenomenon repeats when the test cell is between bins 54 and 63 so that the target is in the lagging reference window. When a target is in the reference window, the assumption that all of the reference cells share the same interference statistics as the test cell is violated and the estimate of the interference

---

[18]Unless otherwise stated, this same arrangement of lead, lag, and guard cells is used in all examples in this chapter.

power is unreliable. However, when the test cell is located at the cell containing the target, the reference windows contain only noise samples and the threshold falls to an appropriate level, allowing target detection. The extent of the elevated threshold regions to either side of a target equals the extent of the leading and lagging windows. The extent of the region of normal threshold level between the two elevated regions equals the total number of guard cells plus one (for the test cell).

As the number of reference cells N becomes large, the estimate σ_w^2 should converge to the true value σ_w^2 and the average probabilities of detection and false alarm should also converge to the values obtained in Sec. 6.4. To see that this occurs, it is easier to work with $\ln(\bar{P}_{FA})$ than with $\bar{P}_{FA}$ itself:

$$\ln(\bar{P}_{FA}) = \ln\left\{\left(1 + \frac{\alpha}{N}\right)^{-N}\right\}$$
$$= -N \cdot \ln\left(1 + \frac{\alpha}{N}\right) = -N\left\{\frac{\alpha}{N} - \frac{1}{2}\left(\frac{\alpha}{N}\right)^2 + \ldots\right\} \quad (6.143)$$

Taking the limit as $N \to \infty$,

$$\lim_{N \to \infty} \{\ln(\bar{P}_{FA})\} = -N\left(\frac{\alpha}{N}\right) \Rightarrow \lim_{N \to \infty} (\bar{P}_{FA}) = \exp(-\alpha) \quad (6.144)$$

Similarly,

$$\lim_{N \to \infty} \{\bar{P}_D\} = \exp[-\alpha/(1+\bar{\chi})] \quad (6.145)$$

Combining Eqs. (6.144) and (6.145) gives the relation

$$\bar{P}_D = (\bar{P}_{FA})^{1/(1+\bar{\chi})} \quad (6.146)$$

which, except for the use of the expected values of the now-random probabilities of detection and false alarm, is identical to Eq. (6.109) for a Swerling 1 target, no noncoherent integration, and known interference power.

All of the previous discussion has been for a square law detector. Similar analyses can be carried out for a linear detector, but the results are more difficult to obtain in closed form. Suppose that the measurements $\{w_i\}$ are the output of a linear detector. The threshold will be set according to

$$\hat{T} = \kappa\left(\frac{1}{N}\sum_{i=1}^{N} w_i\right) \quad (6.147)$$

A formula can be found relating $\bar{P}_{FA}$ and $\hat{T}$ that must be solved iteratively and is numerically difficult (Raghavan, 1992). The exact results show excellent agreement with the approximation (Di Vito and Moretti, 1989):

$$\kappa = \sqrt{N[(\bar{P}_{FA})^{-1/N} - 1][c - (c-1)\exp(1-N)]} \quad (6.148)$$

where $c = 4/\pi$. Notice the similarity to Eq. (6.140). The square root is a consequence of using a linear rather than square law detector. For $N > 4$, the $(c-1)\exp(1-N)$ term is negligible and since $c \approx 1.27$, $\kappa \approx 1.13\sqrt{\alpha}$. Furthermore, although the square law detector performs marginally better than the linear detector for some parameter choices, its performance is virtually identical for parameter values of practical interest.

In the example of Fig. 6.22 it was noted that for the parameters given the ideal threshold would be 8.4 dB above the mean power if the interference power were known exactly, but if the power had to be estimated from the data with $N = 20$, the threshold would be 9.2 dB above the estimated power level. This higher threshold compared to the interference power level is necessary to compensate for the imperfectly known interference power and guarantee the desired $\bar{P}_{FA}$. Because the threshold multiplier is increased in CFAR, the average probability of detection for a target of a given SNR will be decreased relative to the known interference case. Alternately, to achieve a specified $\bar{P}_D$ for a given $\bar{P}_{FA}$, a higher SNR will be required than would be were the interference power known exactly. This increase in SNR required to achieve specified detection probabilities when using CFAR techniques is called the *CFAR loss*.

To quantify the CFAR loss in the case of a CA CFAR, combine Eqs. (6.139) and (6.142) to eliminate the multiplier α and solve for the value of SNR required to achieve a specified combination of $\bar{P}_{FA}$ and $\bar{P}_D$. The result is a function of the number of samples averaged and is denoted by $\bar{\chi}_N$:

$$\bar{\chi}_N = \frac{(\bar{P}_D/\bar{P}_{FA})^{1/N} - 1}{1 - (\bar{P}_D)^{1/N}} \qquad (6.149)$$

As $N \to \infty$, the estimate of interference power converges to the true value and so $\bar{P}_{FA}$ and $\bar{P}_D$ will converge to the values given by Eqs. (6.144) and (6.145). Similarly combining these two equations gives the value of SNR, denoted by $\bar{\chi}_\infty$, required to achieve the specified probabilities when the interference estimate is perfect:

$$\bar{\chi}_\infty = \frac{\ln(\bar{P}_{FA}/\bar{P}_D)}{\ln(\bar{P}_D)} \qquad (6.150)$$

The CFAR loss is then simply the ratio (Hansen and Sawyers, 1980; Levanon, 1988)

$$\text{CFAR loss} = \frac{\bar{\chi}_N}{\bar{\chi}_\infty} \qquad (6.151)$$

Figure 6.23 plots Eq. (6.151) for a $\bar{P}_D$ of 0.9 and three values of $\bar{P}_{FA}$. The loss is greatest for lower values of $\bar{P}_{FA}$ and decreases as expected when the number of reference cells increases. For small ($N < 20$) reference windows, the CFAR loss can be several dB. High losses make values of N less than 10 unacceptable in most cases. Although not shown here, the CFAR loss also increases with increasing $\bar{P}_D$ for a given $\bar{P}_{FA}$ and N. Also, although these results were derived for a Swerling 1 or 2 target, the literature shows that the CFAR loss is roughly the same for all of the Swerling target fluctuation models and the nonfluctuating case (Nathanson, 1991).

6.5.4 CA CFAR Limitations

The cell-averaging CFAR concept relies on two major assumptions:

1. Targets are isolated; specifically, targets are separated by enough cells to guarantee that whenever there is a target in the CUT there will not simultaneously be another in the reference window.
2. All of the reference window interference samples are independent and identically distributed, and that distribution is the same as that of the interference component in the CUT; in other words, the interference is homogeneous.

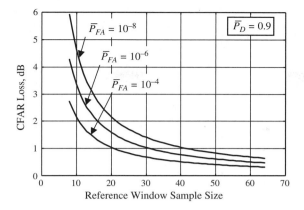

FIGURE 6.23 Cell-averaging CFAR loss for Swerling 1/2 target in complex WGN with $\bar{P}_D = 0.9$.

While useful in many situations, either or both of these conditions are frequently violated in real-world scenarios. The second assumption is particularly likely to be untrue when the dominant interference is clutter, that is, echo from terrain, rather than thermal noise. In this section, the effect on cell-averaging CFAR of violating these assumptions is discussed, and then some modifications that combat these effects are described.

Target *masking* occurs when two or more targets are present such that, when one target is in the test cell, one or more targets are located among the reference cells. Assuming that the power of the target in the reference cell exceeds that of the surrounding interference, its presence will raise the estimate of the interference power and thus of the CFAR threshold. The target(s) in the reference window can "mask" the target in the test cell because the increased threshold will cause a reduction in the probability of detection, that is, the detection is more likely to be missed. Equivalently, a higher SNR will be required to achieve a specified $\bar{P}_D$.

Figure 6.24 is an example of target masking. As before, the interference level is 20 dB, the target in range bin 50 has an SNR of 15 dB, and the threshold is computed using 20 reference cells and a desired $\bar{P}_{FA}$ of 10^{-3}. However, a second target with an SNR of 20 dB in range bin

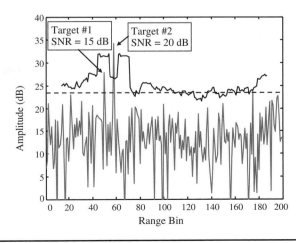

FIGURE 6.24 Illustration of target masking. See text for details.

58 elevates the estimated interference power when the first target is in the test cell. This increase in threshold is sufficient to prevent detection of the first target in this case. On the other hand, the 15-dB target does not affect the threshold enough to prevent detection of the second, stronger target.

Precise analysis of the effect of a target in the reference cells is conceptually simple but somewhat complicated in practice. However, a relatively simple approximation that illustrates the effect of an interfering target can be derived. Consider a single interfering target, possibly fluctuating, with power γ_i that contaminates only one of the N CFAR reference cells. The SNR of this interferer is $\chi_i = \bar{\gamma}_i / \sigma_w^2$. The expected value of the new threshold will be

$$E\{\hat{T}'\} = E\left\{\frac{\alpha}{N}\left(\gamma_i + \sum_{i=0}^{N-1} x_i\right)\right\}$$
$$= \frac{\alpha \bar{\gamma}_i}{N} + \alpha \sigma_w^2 = \alpha\left(1 + \frac{\bar{\chi}_i}{N}\right)\sigma_w^2 \qquad (6.152)$$

Thus, $E\{\hat{T}'\}$ is again a multiple of the interference power σ_w^2 as in Eq. (6.134) but with a multiplier α' given by

$$\alpha' \equiv \alpha\left(1 + \frac{\bar{\chi}_i}{N}\right) \qquad (6.153)$$

The elevated threshold will decrease both the P_D and P_{FA}. Using Eqs. (6.153) and (6.140) in Eq. (6.142) gives an expression for the new value of $\bar{P}_D$:

$$\bar{P}_D' = \left\{1 + \left[(\bar{P}_{FA})^{-1/N} - 1\right]\left[\frac{1 + (\bar{\chi}_i/N)}{1 + \bar{\chi}}\right]\right\}^{-N} \qquad (6.154)$$

Figure 6.25a plots this expression as a function of interferer SNR for a target SNR of 15 dB, $\bar{P}_{FA}$ of 10^{-3}, and N equal to 20 or 50. In this example, the approximate detection probability is significantly degraded even when the interferer SNR is still several decibels below the target SNR. Note also that if $\bar{\chi}_i \to 0$ (no interfering target) or $N \to \infty$ (interfering target influence becomes negligible), then $\bar{P}_D' \to \bar{P}_D$. The probabilities of detection without the interferer for these two cases are about 0.78 and 0.8, respectively.

Another way to characterize the effect of an interfering $\alpha' \to \alpha$ and target is by the increase in SNR required to maintain the original value of $\bar{P}_D$. Let $\bar{\chi}'$ be the value of SNR required to attain the original $\bar{P}_D$ using the elevated threshold $\hat{T}'$. Equation (6.142) expressed $\bar{P}_D$ in terms of the original value of $\bar{\chi}$ and threshold multiplier α. Approximately the same relationship will determine the detection probability $\bar{P}_D'$ attained with the new threshold multiplier α' and SNR $\bar{\chi}'$. $\bar{P}_D'$ will equal $\bar{P}_D$ if

$$\frac{\alpha}{N(1 + \bar{\chi})} = \frac{\alpha'}{N(1 + \bar{\chi}')} \qquad (6.155)$$

Using Eq. (6.153) in Eq. (6.155) leads to

$$\bar{\chi}' = \left(1 + \frac{\bar{\chi}_i}{N}\right)(1 + \bar{\chi}) - 1 \qquad (6.156)$$

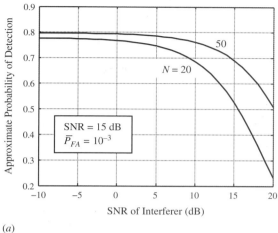

(a)

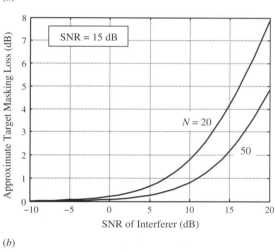

(b)

FIGURE 6.25 Approximate effect of interfering target on cell-averaging CFAR. Threshold set for $P_{FA} = 10^{-3}$. (a) Reduction in P_D. (b) Equivalent masking loss.

Figure 6.25b plots the approximate "target masking loss" $\bar{x}'/\bar{x}$ in decibels for the same conditions as in Fig. 6.25a. In this example, the masking loss is on the order of 3 dB when the interferer SNR equals the target SNR of 15 dB.

The results given in Eqs. (6.154) and (6.156) are only approximations. A more careful analysis would mimic the basic CA CFAR analysis of Sec. 6.5.3 by finding the PDF of the threshold in the presence of an interfering target, then using that PDF to find the expected values of P_D and P_{FA}. This approach is complicated by the fact that the interfering target changes the PDF of the cell containing it. For instance, if the interferer is nonfluctuating, the PDF of the power in its cell will be a generalized noncentral chi-squared, while all of the remaining cells will still be exponentially distributed. The PDF of the threshold will be a mixture of the noncentral chi-squared and exponential PDFs. To avoid calculating this PDF, the expected value of the threshold was used in the expressions for the case of no interfering target. This gives a simple approximation that behaves correctly in the limits of large and small mean interferer SNR $\bar{x}_i$.

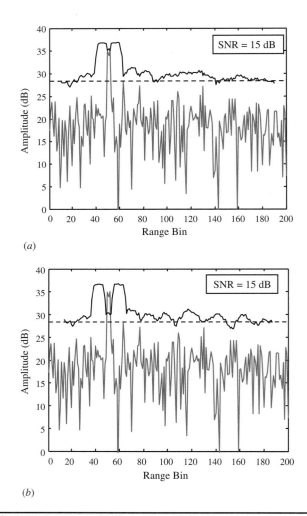

FIGURE 6.26 Self-masking and guard cells in CA CFAR. (a) Threshold using no guard cells. (b) Threshold using three guard cells to each side of the test cell. See text for details.

Figure 6.26 illustrates a related phenomenon, self-masking by a distributed target. The interference, detector, and target characteristics are the same as in Fig. 6.22, with the exception that the physical extent of the target is now greater than a range bin so that the target signature is spread over three consecutive cells. Figure 6.26a shows the effect when no guard cells are used. When one of the three target cells is the test cell, the other two contaminate the interference estimate, raising the threshold just enough to prevent detection. This effect is the reason for using guard cells. Their impact is illustrated in Fig. 6.26b, where three guard cells are used to each side of the test cell. This lengthens the total CFAR window slightly but ensures that when the test cell is centered on the target, the adjoining target cells do not contaminate the interference estimate and the target is now detected.

If the dominant interference is surface clutter rather than thermal noise or jamming, the interference will often be highly heterogeneous. The radar may collect data over a terrain region that is, for instance, part open field and part forested, or part land and part water. When the test cell is at or near the boundary between two clutter regions having different

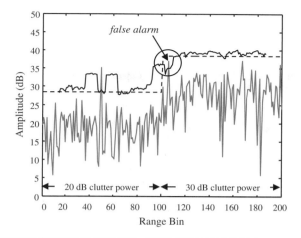

FIGURE 6.27 False alarms at a clutter edge.

reflectivities, the statistics in the leading and lagging window will not be the same. Such *clutter edges* can cause both false alarms at the edge and masking of targets near the edge and in the lower-reflectivity region.

Figure 6.27 shows the false alarm effect of a clutter edge. The first 100 bins have a mean interference power of 20 dB. The clutter power rises suddenly by 10 dB to a mean of 30 dB in the last 100 bins, simulating a change in terrain type, perhaps from open field to a wooded area. One target is present at bin 50. Two ideal thresholds corresponding to $P_{FA} = 10^{-3}$ are shown, one for each of the two clutter regions. The threshold estimated by the cell-averaging CFAR tracks each region, but with a transition region in the range of bins 87 through 113. In this example, the clutter happens to include a high-amplitude fluctuation near the clutter edge. Because the CFAR threshold does not rise to the correct level for the new clutter level until several cells after the transition, the clutter spike crosses the CFAR threshold and a false alarm occurs. However, the target at bin 50 is detected normally.

Figure 6.28 illustrates the second effect of clutter edges. The clutter regions are the same as in Fig. 6.27, but the particular clutter sequence is different. A target with an SNR of 15 dB

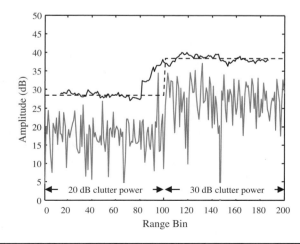

FIGURE 6.28 Target masking at a clutter edge. Clutter parameters are the same as in Fig. 6.27.

is located at bin 95, five bins away from the clutter edge. The CFAR window uses 10 cells in both the leading and lagging windows, as well as three guard cells on either side of the test cell, for a total CFAR window length of 27 cells. Thus, when the test cell is centered over the target, the leading reference window is mostly filled by clutter from the high-reflectivity region, elevating the threshold above the target and causing a missed detection. (Note that this example does not also suffer a clutter-edge false alarm.)

6.5.5 Extensions to Cell-Averaging CFAR

The performance limitations caused by nonhomogeneous clutter and interfering targets have led to the development of numerous extensions to the cell-averaging CFAR concept, each designed to combat one or more of the deleterious effects. These techniques are often heuristically motivated and can be difficult to analyze exhaustively due to the many variations in clutter non-homogeneity, target and interfering target SNR, CFAR window size, and CFAR detection logic. Additional information on many of the techniques described here is available in Keel (2010).

One common CFAR extension is the *smallest-of cell-averaging CFAR* (SOCA CFAR); the method is also known as the *least-of cell-averaging CFAR*. This technique is intended to combat the masking effect caused by an interfering target among the CFAR reference cells seen in Fig. 6.24. In an N-cell SOCA approach, the lead and lag windows are averaged separately to create two independent estimates σ_{w1}^2 and σ_{w2}^2 of the interference mean, each based on $N/2$ reference cells. The threshold is then computed from the smaller of the two estimates in a manner similar to Eq. (6.134):

$$\hat{T} = \alpha_{SO} \cdot \min\left(\sigma_{w1}^2, \sigma_{w2}^2\right) \tag{6.157}$$

If an interfering target is present in one of the two windows, it will raise the interference power estimate in that window. Thus, the lesser of the two estimates is more likely to be representative of the true interference level and should be used to set the threshold.

Because the interference power is estimated from $N/2$ cells instead of N cells, the threshold multiplier α required for a given design value of $\bar{P}_{FA}$ will be increased. It is tempting to suppose that the threshold multiplier α_{SO} for SOCA CFAR could be calculated using Eq. (6.140) with N replaced by $N/2$. A more careful analysis shows that the required multiplier is the solution of the equation (Weiss, 1982)

$$\bar{P}_{FA} = 2\left(2 + \frac{\alpha_{SO}}{(N/2)}\right)^{-\frac{N}{2}} \left\{ \sum_{k=0}^{(N/2)-1} \binom{N/2 - 1 + k}{k} \left[2 + \frac{\alpha_{SO}}{(N/2)}\right]^{-k} \right\} \tag{6.158}$$

This equation must be solved iteratively. As an example, for $\bar{P}_{FA} = 10^{-3}$ and $N = 20$, $\alpha_{SO} = 11.276$. In contrast, the CA CFAR multiplier is $\alpha = 8.25$ for the same conditions.

Figure 6.29a compares the behavior of conventional CA CFAR and SOCA CFAR on simulated data containing two closely spaced targets of mean SNRs of 15 and 20 dB, a 10-dB clutter edge, and a third 15-dB target near the clutter edge. As before, the lead and lag windows are both 10 samples (thus $N = 20$) and there are three guard cells to each side of the test cell. The ideal threshold shown is based on $\bar{P}_{FA} = 10^{-3}$. The threshold multipliers are $\alpha = 8.25$ for CA CFAR and $\alpha_{SO} = 11.276$ for SOCA CFAR as discussed previously.

The CA CFAR masks the weaker of the two closely spaced targets and also fails to detect the target near the clutter edge, but does not exhibit a false alarm at the clutter edge in this instance. The SOCA threshold, in contrast, easily allows detection of both targets; the half of

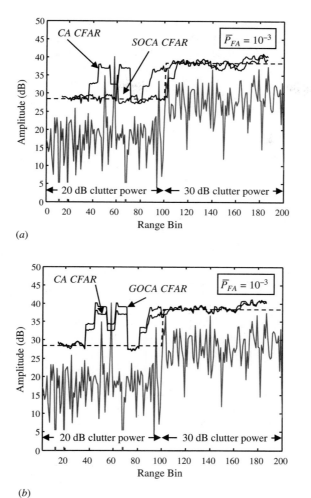

FIGURE 6.29 Comparison of conventional, "smallest-of," and "greatest of" cell-averaging CFAR with multiple targets and a clutter edge. (a) CA and SOCA CFAR. (b) CA and GOCA CFAR.

the CFAR window contaminated by the other target is simply ignored by the SOCA logic. Similarly, the SOCA CFAR detects the target near the clutter edge, again because the half of the window containing the higher-power clutter is ignored.

Figure 6.29a also shows the principal failing of the SOCA method. Although the CA CFAR did not exhibit a false alarm at the clutter edge, the SOCA CFAR does. This is a natural consequence of the SOCA logic. As the CFAR window crosses a clutter edge, there will be a region in which the test cell is in the higher interference power region, while one of the lead or lag windows is filled mostly or entirely with samples of the lower power interference. The SOCA logic ensures that the threshold is then based on the lower interference power, significantly raising the probability that the clutter in the test cell will cross the threshold.

For systems and environments in which closely spaced targets are unlikely but the clutter is highly nonhomogeneous, clutter-edge false alarms may be of much more concern than target masking. In this case, the observations above suggest that a *greater-of cell-averaging CFAR* (GOCA CFAR) logic be used. As with the SOCA technique, the lead and lag windows are averaged separately, but now the threshold is based on the larger of the two averages:

$$\hat{T} = \alpha_{GO} \cdot \max\left(\sigma_{w1}^2, \sigma_{w2}^2\right) \qquad (6.159)$$

Similar to the SOCA case, the GOCA threshold multiplier is the solution of the equation (Weiss, 1982)

$$\bar{P}_{FA} = 2\left(1 + \frac{\alpha_{GO}}{(N/2)}\right)^{-\frac{N}{2}} - \left(2 + \frac{\alpha_{GO}}{(N/2)}\right)^{-\frac{N}{2}} \left\{ \sum_{k=0}^{(N/2)-1} \binom{N/2 - 1 + k}{k} \left[2 + \frac{\alpha_{GO}}{(N/2)}\right]^{-k} \right\} \qquad (6.160)$$

For $N = 20$ and $\bar{P}_{FA} = 10^{-3}$, $\alpha_{GO} = 7.24$.

Figure 6.29*b* illustrates the performance of the GOCA CFAR logic on the same example used in Fig. 6.29*a*. The GOCA threshold is now equal to or higher than the CA CFAR threshold. Not surprisingly, the GOCA logic successfully avoids the false alarm at the clutter edge. However, the strong target masks the weaker target. Furthermore, the weaker target even makes detection of the stronger target more marginal, although in this case the detection is successful. The GOCA CFAR also misses the target near the clutter edge due to the masking effect of the elevated clutter. Additional analysis of the GOCA CFAR for the case of a linear detector is provided in Pace and Taylor (1994).

The SOCA and GOCA CFARs estimate the threshold using only half of the reference cells. Consequently, they exhibit a higher CFAR loss above than the conventional CA CFAR. This additional loss is less than 0.3 dB for the GOCA CFAR over a wide range of parameters (Hansen and Sawyers, 1980). The additional loss for the SOCA CFAR is greater, especially for small values of N. It is necessary to use $N > 32$, approximately, to ensure that the additional loss is less than 1 dB over a wide range of $\bar{P}_{FA}$ values (Weiss, 1982). A mitigating influence for either approach is that the use of a split window may allow a larger value of N than would normally be used in a conventional cell-averaging CFAR. The reason is that the window size in CA CFAR is often limited by concern over nonhomogeneous clutter. Since only half the window will actually be used, a larger value of N can be tolerated.

Still another way to combat the target masking problem is *censored* or *trimmed mean* CFAR (Ritcey, 1986). In these techniques, the M reference cells ($M < N$) having the highest power are discarded and the interference power is estimated from the remaining $N - M$ cells. In some versions of trimmed mean CFAR both the highest and lowest power reference cells are discarded. Consider an example where $M = 2$. If an interfering target is present, but is confined to only one or two cells (or if two interferers are present, each confined to one cell), the censoring process will completely eliminate their elevating effect on the estimate of interference power. There will, however, be a small additional CFAR loss due to the use of only $N - M$ cells instead of N cells (Ritcey and Hines, 1989). Proper selection of M requires some knowledge of the maximum number of interferers to be expected, as well as whether they will be confined to one cell or will be distributed over multiple cells. Typically, one-quarter to one-half of the reference window cells are discarded (Nathanson, 1991). In addition, implementation of the technique requires logic to rank order the reference cell data, sometimes a significant implementation consideration at the speeds at which real-time CFAR calculations often must be done.

Many additional variations on the approaches described previously can be used. For example, censoring can be combined with any of the CA, SOCA, or GOCA techniques. A more elaborate approach attempts to examine the behavior of the interference in the lead and lag windows and then choose an appropriate CFAR algorithm. One version of these ideas computes the mean and the variance in each of the lead and lag windows. If the variance in a window exceeds a certain threshold, it is assumed that the data in that window are not homogeneous Rayleigh interference, most likely due to target contamination. A series of logical decisions then determines whether to combine the windows for a CA CFAR using the data from both windows, use CA CFAR using only one window of data, or use GOCA or SOCA CFAR (Smith and Varshney, 2000). For example:

- If the means differ by less than a specified threshold and the variances in each window are less than the variance threshold, homogeneous interference is assumed and conventional CA CFAR is selected to set the detection threshold.
- If instead the means do differ by more than that threshold, a clutter edge is assumed and GOCA CFAR is applied.
- If the variance in one window exceeds the variance threshold, but in the other window does not, CA CFAR based only on the low-variance window is applied.
- If both windows have variances exceeding the variance threshold, SOCA CFAR is applied.

Another recent attempt to develop a CFAR algorithm that provides good performance in the presence of clutter edges and target masking while maintaining performance near that of CA CFAR in homogeneous clutter is the *switching CFAR* (S-CFAR) (Van Cao, 2004). In this approach, the CFAR reference window is divided into two groups, not necessarily contiguous: those cells above a threshold set as a fraction of the test cell value, and those below. If the number of cells in the low amplitude group exceeds some threshold N_t, typically set to about one half of the total number N of reference cells, all N cells are used in a cell-averaging calculation. If the number of low amplitude cells is less than N_t, the threshold is set based only on the low amplitude cells. The principal advantage appears to be reduced losses compared to order statistic CFAR (OS CFAR, described in the next section) due to masking targets and somewhat improved clutter-edge performance, while avoiding the need for sorting required by OS CFAR.

Yet another approach that has been proposed to combat masking is the use of alternate detector laws (i.e., not linear or square law). By far the most common is log CFAR, which applies conventional cell-averaging CFAR logic to the logarithm of the received power samples.[19] There appears to be no simple closed form analysis equivalent to Eqs. (6.129) through (6.134) for determining the relationship between an average of the log-detected data and the interference power σ_w^2. However, motivated by considering the logarithm of the threshold computation for a square law detector seen in Eq. (6.135), the log CFAR threshold is computed by adding an offset to the averaged logarithmic data:

$$\hat{T}_{\log} = \frac{1}{N} \sum_{i=1}^{N} 10 \log_{10}(x_i) + \alpha_{\log} \qquad (6.161)$$

In general, applying a logarithmic transformation to the data compresses its numerical dynamic range. This was an important implementation advantage in older systems built using analog or fixed-point digital hardware, but is less of a consideration with more modern processors. However, averaging the logarithmic data has the additional advantage that isolated interferers in the reference window do not have as great an influence on the numerical value of the estimated interference mean, thus reducing target masking effects. This effect is clearly seen in Fig. 6.30, which shows the same data set used previously in Fig. 6.29 containing closely spaced targets with 15 and 20 dB signal-to-clutter ratios and a 10-dB clutter edge. The

---

[19]The base of the logarithm affects the specific offset needed to set the threshold but is otherwise unimportant. In Eq. (6.161) it is assumed that the log data are on a decibel scale.

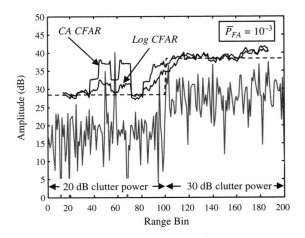

FIGURE 6.30 Comparison of CA CFAR and log CFAR on the same data as Fig. 6.28.

two targets are easily detected. Unfortunately, log CFAR exhibits poor performance at clutter edges, in particular an increased vulnerability to false alarms at clutter edges. In this example, not only does a false alarm occur at the clutter edge, but the target near the edge is not detected.

No explicit expression is known for finding the required threshold offset $\alpha_{\log}$ as a function of $\bar{P}_{FA}$. Results have been obtained for $\bar{P}_D$ and $\bar{P}_{FA}$ that can be solved numerically to find suitable values of the threshold (Novak, 1980). To create Fig. 6.30, a Monte Carlo simulation was instead used to determine the required value by trial and error. The result for $N = 20$ reference cells and $\bar{P}_{FA} = 10^{-3}$ is $\alpha_{\log} = 11.85$ dB.

The CFAR loss of a log CFAR detector can also be estimated using Monte Carlo simulation techniques (Hansen and Ward, 1972). The log CFAR increases the CFAR loss relative to the linear detector. In homogenous clutter the number of log CFAR reference window cells $N_{\log}$ required to achieve the same CFAR loss as an N-cell conventional CA CFAR using a linear detector is approximately

$$N_{\log} \approx 1.65N - 0.65 \tag{6.162}$$

Thus, the use of the log detector increases the required CFAR window size by about 65 percent to avoid increasing the CFAR loss. Furthermore, for $N > 8$ the log CFAR loss in decibels is about 65 percent more than the loss for a CA CFAR with the same value of N.

6.5.6 Order Statistic CFAR

An alternative to cell-averaging CFAR is the class of *rank-based* or *order statistic* CFARs (OS CFAR). Proposed primarily for combating masking degradations, OS CFAR retains the one-dimensional or two-dimensional sliding window structure of CA CFAR, including guard cells if desired, but does away entirely with averaging of the reference window contents to explicitly estimate the interference level. Instead, OS CFAR rank orders the reference window data samples $\{x_1, x_2,\ldots, x_n\}$ to form a new sequence in ascending numerical order, denoted by $\{x_{(1)}, x_{(2)}, \ldots, x_{(N)}\}$. The kth element of the ordered list is called the kth *order statistic*. For example, the first-order statistic is the minimum, the Nth order statistic is the maximum, and the $(N/2)$th order statistic is the median of the data $\{x_1, x_2,\ldots, x_N\}$. In OS CFAR, a value of k is pre-selected and the kth order statistic is selected as representative of the interference level and a threshold is set as a multiple of this value:

$$\hat{T} = \alpha_{OS} x_{(k)} \tag{6.163}$$

The interference is thus estimated from only one actual data sample, instead of an average of all of the data samples. Nonetheless, the threshold in fact depends on all of the data since all of the samples are required to determine which will be the kth largest.

It will be shown that this algorithm is in fact CFAR (i.e., does not depend on the interference power σ_w^2), and the threshold multiplier required to achieve a specified $\overline{P}_{FA}$ will be determined. The analysis follows Levanon (1988). To simplify the notation, consider the square-law-detected output x_i normalized to its mean, $y_i = x_i/\sigma_w^2$; this will have an exponential PDF with unit mean. The rank-ordered set of reference samples $\{y_i\}$ are denoted by $\{y_{(i)}\}$. For a given threshold T, the probability of false alarm will be

$$P_{FA}(T) = \int_T^{+\infty} \exp(-y)\, dy = \exp(-T) \tag{6.164}$$

The average P_{FA} will be computed as

$$\overline{P}_{FA} = \int_0^{+\infty} P_{FA}(T) p_T(T)\, dT \tag{6.165}$$

where $p_T(T)$ is the PDF of the threshold. Because T is proportional to the kth-ranked reference sample $y_{(k)}$, it is necessary to find the PDF of $y_{(k)}$, the kth largest of N i.i.d. random variables. The result can be found in many textbooks and is

$$p_{y_{(k)}}(y) = k \binom{N}{k} P_{y_i}^{k-1}(y) [1 - P_{y_i}(y)]^{N-k} p_{y_i}(y) \tag{6.166}$$

where $P_{y_i}(y)$ is the *cumulative distribution function* (CDF) of y_i. For complex WGN, the PDF and CDF of a single square-law-detected and normalized reference sample y_i are

$$\begin{aligned} p_{y_i}(y) &= \exp(-y), \\ P_{y_i}(y) &= \int_0^y p_{y_i}(y')\, dy' = 1 - \exp(-y) \end{aligned} \tag{6.167}$$

Using Eq. (6.167) in Eq. (6.166) gives the PDF of the kth ranked sample:

$$p_{y_{(k)}}(y) = k \binom{N}{k} [\exp(-y)]^{N-k+1} [1 - \exp(-y)]^{k-1} \tag{6.168}$$

The threshold is $\hat{T} = \alpha_{OS} y_{(k)}$, so the PDF of $\hat{T}$ is $p_{\hat{T}}(\hat{T}) = (1/\alpha_{OS}) p_{y_{(k)}}(\hat{T}/\alpha_{OS})$, giving

$$p_{\hat{T}}(\hat{T}) = \frac{k}{\alpha_{OS}} \binom{N}{k} [\exp(-\hat{T}/\alpha_{OS})]^{N-k+1} [1 - \exp(-\hat{T}/\alpha_{OS})]^{k-1} \tag{6.169}$$

Inserting this result into Eq. (6.165) gives

$$\overline{P}_{FA} = \int_0^\infty \exp(-\widehat{T})\frac{k}{\alpha_{OS}}\binom{N}{k}[\exp(-\widehat{T}/\alpha_{OS})]^{N-k+1}[1-\exp(-\widehat{T}/\alpha_{OS})]^{k-1}d\widehat{T} \quad (6.170)$$

$$= \frac{k}{\alpha_{OS}}\binom{N}{k}\int_0^\infty \exp[-(\alpha_{OS}+N-k+1)\widehat{T}/\alpha_{OS}][1-\exp(-\widehat{T}/\alpha_{OS})]^{k-1}d\widehat{T}$$

With the change of variable $T' = T/\alpha_{OS}$, this becomes the slightly more convenient form:

$$\overline{P}_{FA} = k\binom{N}{k}\int_0^\infty \exp[-(\alpha_{OS}+N-k+1)T'][1-\exp(-T')]^{k-1}dT' \quad (6.171)$$

Utilizing integral 3.312(1) in Gradshteyn and Ryzhik (1980) gives

$$\overline{P}_{FA} = k\binom{N}{k}B(\alpha_{OS}+N-k+1,k)$$
$$= k\binom{N}{k}\frac{\Gamma(\alpha_{OS}+N-k+1)\Gamma(k)}{\Gamma(\alpha_{OS}+N+1)} \quad (6.172)$$

where $B(\cdot,\cdot)$ is the beta function which in turn can be expressed in terms of the gamma function $\Gamma(\cdot)$ as shown. For integer arguments, $\Gamma(n) = (n-1)!$ and Eq. (6.172) reduces for integer α_{os} to

$$\overline{P}_{FA} = k\frac{N!}{k!(N-k)!}\frac{(k-1)!(\alpha_{OS}+N-k)!}{(\alpha_{OS}+N)!}$$
$$= \frac{N!(\alpha_{OS}+N-k)!}{(N-k)!(\alpha_{OS}+N)!} \quad (\alpha_{OS} \text{ integer}) \quad (6.173)$$

Figure 6.31 plots $\overline{P}_{FA}$ as a function of α_{OS} for two choices of OS windows, one with $N = 20$ and one with $N = 50$. In the first case, the $k = 15$th order statistic is chosen to set the

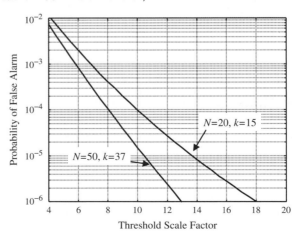

FIGURE 6.31 $\overline{P}_{FA}$ versus threshold scale factor α_{OS} in order statistics CFAR. The selected order statistic is chosen as approximately $0.75N$.

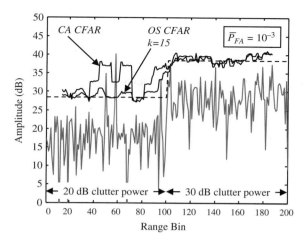

FIGURE 6.32 Comparison of CA CFAR and OS CFAR on the same data as Fig. 6.29.

threshold, while in the second the $k = 37$th order statistic is selected. Plots such as these can be used to determine the threshold multiplier needed to achieve a specified $\bar{P}_{FA}$ for a given OS CFAR window configuration. For example, with $N = 20$ and $k = 15$, a multiplier of $\alpha_{OS} = 6.857$ gives $\bar{P}_{FA} = 10^{-3}$.

Figure 6.32 shows the performance of the OS CFAR on the data set of Fig. 6.29 again using $N = 20$ again and choosing the $k = 15$th order statistic to set the threshold. The use of the ordered statistic instead of a mean estimate makes the detector almost completely insensitive to masking by closely spaced targets so long as the number of cells contaminated by interfering targets does not exceed $N - k$. In this example, both closely spaced targets are readily detected. Although they were used in Fig. 6.32 for consistency, guard cells are less important in an OS CFAR since the rank ordering process will not be affected by targets spreading outside of the test cell. The effect of the window length and choice of order statistic k on the behavior at clutter edges are discussed in Rohling (1983). If $k \leq N/2$, there will be extensive false alarms at clutter edges. Thus, k is usually chosen to satisfy $N/2 < k < N$. Typically, k is on the order of $0.75N$ (Nathanson, 1991).

To determine the OS CFAR loss, it is necessary to determine the SNR required to obtain a specified $\bar{P}_D$ for a given $\bar{P}_{FA}$ and compare that result to the ideal threshold or to CA CFAR in homogeneous interference. By repeating the analysis above but using the Swerling 1 or Swerling 2 model (Rayleigh amplitude PDF) for the test cell, it can be shown that the average probability of detection is also given by Eq. (6.172), but with α_{OS} replaced by (Levanon, 1988)

$$\alpha_{OS}^D = \frac{\alpha_{OS}}{1 + \bar{\chi}} \qquad (6.174)$$

where $\bar{\chi}$ is the average SNR of the target. By varying α_{OS} or α_{OS}^D, curves of $\bar{P}_D$ versus $\bar{P}_{FA}$ can be developed and the CFAR loss calculated. An alternative analysis approach is given in Blake (1988). In the absence of interfering targets, the OS-CFAR suffers a small additional loss over CA CFAR. The value depends on both k and N but is typically on the order of 0.3 to 0.5 dB. If interfering targets are present, the OS CFAR loss increases only very slowly until the number of interferers exceeds $N - k$, the number of "ignored" high-rank cells. In contrast, the loss in CA CFAR increases very rapidly in this case due to the elevated estimate of the interference power (Blake, 1988). Thus, OS CFAR losses are lower than CA CFAR losses in the presence of interferers. Additional results on the performance of OS CFAR, including the effect of noncoherent integration and its behavior in Weibull clutter, are available in Shor and Levanon (1991). A recent example of a fast algorithm for implementing OS CFAR is given in Colena et al. (2020).

6.5.7 Adaptive CFAR

"Adaptive CFAR" is the name given to a growing class of CFAR algorithms designed to improve performance in nonhomogeneous clutter.[20] Generally, they dispense with the fixed CFAR window structure (usually with half of the cells in each of the lead and lag windows). Instead they typically construct a statistical test to determine if the reference cells span one clutter field or two, that is, whether or not the reference cell data are homogeneous. If not homogeneous, the algorithms estimate not only the clutter statistics in each field but also at which cell the transition from one to the other occurs and therefore which type of clutter is competing with the target in the CUT.

The basic approach to adaptive CFAR for nonhomogeneous clutter was described in Finn (1986). The algorithm assumes a CFAR window of total length N cells, including the CUT, that spans two clutter fields, that is, two regions with different statistical parameters. The clutter edge is presumed to occur between samples M and $M + 1$; however, M is not known. Initially, it is also assumed that the clutter follows the usual square-law-detected exponential distribution. Let $\sigma_{w1}^2(M)$ denote the estimate of σ_w^2 obtained by averaging the M samples 1 through M forming the first clutter region, and $\sigma_{w2}^2(M)$ be the average obtained from the $N - M$ samples $M + 1$ through N forming the second region. The algorithm starts by setting $M = 1$ and computing $\sigma_{w1}^2(1)$, which is the "average" of only cell #1, and $\sigma_{w2}^2(1)$, the average of the remaining $N - 1$ cells. This process is repeated for $M = 2, ..., N = 1$. Thus, a pair of sample means $\sigma_{w1}^2(M)$ and $\sigma_{w2}^2(M)$ are computed for each of the $N - 1$ possible transition points between the two-clutter regions.

The next step is to choose the most likely transition point M_t. The maximum likelihood estimate of this transition point is the value M_t of M that maximizes the log-likelihood function (Finn, 1986):

$$\ln[\ell(M)] = -\left\{ M \cdot \ln\left[\sigma_{w1}^2(M)\right] - (N - M) \cdot \ln\sigma_{w2}^2(M) \right\} \quad (6.175)$$

Once M_t is identified, it is also known if the CUT is in the first or second clutter region. Standard CA CFAR can then be applied using the appropriate mean estimate and the number of cells with which it is estimated. For example, if the CUT is in the first region the threshold would be set according to

$$\alpha = M_t\left[(\overline{P}_{FA})^{-1/M_t} - 1\right]$$
$$\widehat{T} = \alpha \cdot \sigma_{w1}^2(M_t) \quad (6.176)$$

Note that this procedure is effectively SOCA CFAR when the CUT is in the low clutter region and GOCA CFAR when it is in the high clutter region, assuming the transition point is correctly located.

---

[20]This terminology is unfortunate because all CFARs are adaptive in that they estimate the detection threshold from the measured data. "Adaptive CFARs" are those for which one or more of the parameters of the estimation algorithm itself are varied depending on data characteristics.

The discussion above does not allow for the possibility that the clutter is uniform. To address this an additional likelihood test is conducted to compare $\ln[\ell(M)]$ as computed using Eq. (6.175) with $M = M_t$ to the corresponding metric under the assumption of uniform clutter, $\ln[\ell(0)] = -N\ln[\sigma_{w2}^2(0)] = -N\ln(\sigma_{w2}^2)$. If $\ln[\ell(0)] > \ln[\ell(M_t)]$, the clutter is assumed homogeneous and conventional CA CFAR is applied.

When the two-region clutter hypothesis is accepted, the estimate of the transition point M_t can, of course, be incorrect. If the target is, in fact, in the low clutter region but is incorrectly determined to be in the high clutter region, the high clutter statistics will be used to set the threshold. There will then be an increased probability that the high clutter will mask the target. If the target is in the high clutter region but is incorrectly determined to be in the low clutter region, the low clutter statistics will be used to set the threshold, which will then be too low. While the target will have an enhanced probability of detection, the false alarm probability will rise, possibly dramatically. For this reason, the adaptive CFAR algorithm is modified to bias the decision in favor of the hypothesis that the target is in the high clutter region. While this will increase masking effects somewhat, it avoids the generally more damaging problem of large increases in the false alarm rate. Details are given in Finn (1986).

Many of the extensions to CA CFAR can also be applied to the adaptive CFAR. It can be applied to log normal or Weibull clutter by computing both sample means and variances in each region. The data in each region can be censored prior to estimating the statistics. Order statistic rather than cell-averaging rules can be used to set the threshold. Many such variations are available in the literature, as are algorithms building on the basic adaptive concept but applying different statistical estimators and decision logics.

6.5.8 CFAR for Two-Parameter PDFs

All of the results discussed in this chapter so far have assumed exponential power (equivalently, Rayleigh voltage) interference, which is the appropriate model when the primary interference is WGN, whether the source is low-power receiver noise or high-power noise jamming. Only one parameter, the mean power σ_w^2, is required to completely specify the PDF. However, as discussed in Chap. 2, many types of clutter are best modeled by more complicated PDFs such as the log-normal, Weibull, or K PDF. Unlike the exponential PDF, these are two-parameter distributions and estimates of both the mean and variance (or a related parameter such as skewness) must be estimated in order to characterize them. Any threshold control mechanism must be based on estimates of both parameters if it is to exhibit CFAR behavior.

An example of a CFAR algorithm for log-normal clutter is given in Schleher (1977). The receiver uses a log detector so the detected samples $\{x_i\}$ are normally distributed. The CFAR structure is a conventional cell-averaging approach on the log data. The threshold is computed as follows:

$$\hat{\mu} = \frac{1}{N}\sum_{i=1}^{N} x_i, \quad \hat{s} = \sqrt{\frac{1}{N}\sum_{i=1}^{N}(x_i - \hat{\mu})^2},$$
$$\hat{T} = \hat{\mu} + \alpha\hat{s} \tag{6.177}$$

This CFAR threshold calculation could clearly be combined with many of the embellishments discussed earlier for CA CFAR, such as SOCA or GOCA rules or censoring.

Because of the need to estimate two parameters, the CFAR loss is greater with two parameter distributions than with the exponential distribution and in fact can be very large, especially for small numbers of reference cells. For example, with $\bar{P}_D = 0.9$, $\bar{P}_{FA} = 10^{-6}$, and $N = 32$ reference cells, the CFAR loss using Eq. (6.177) in log-normal interference is approximately 13 dB (Schleher, 1977). From Fig. 6.23, a conventional CA CFAR in exponential interference with the same detection statistics and window size has a CFAR loss of just under 1 dB.

The same calculations are used to set the CFAR threshold in Weibull clutter, though the specific values of α needed differ from the log-normal case. Two proposed Weibull detectors, the so-called log-t detector and another based on maximum likelihood estimates of the Weibull PDF parameters, have been shown to be equivalent to Eq. (6.177) (Gandhi et al., 1995).

Order statistic CFARs have also been proposed for two-parameter clutter. One example combines OS CFAR in each of the lead and lag windows with a greatest-of logic to estimate the interference mean, and then uses the single parameter Eq. (6.159) to set the threshold. Since the second (skewness) parameter of the PDF is not estimated implicitly or explicitly, the multiplier α must be made a function of the skewness, implying in turn that the skewness parameter must be known to correctly set the threshold. Performance results again suggest that choosing the order statistic k to be about $0.75N$ provides the best performance against interferers and uncertainty in the skewness parameter (Rifkin, 1994).

6.5.9 Temporal CFAR

In Chap. 5 the technique of clutter mapping was discussed for detection of stationary or slowly moving targets by ground-based fixed-site radars when the competing zero-Doppler clutter was not too strong. Airport surveillance radars are a common application. As with other CFAR algorithms, the threshold for each range-angle cell was computed as a multiple of the estimated clutter power in the same cell. However, in practice it is often the case that the typical spatial or spatial-Doppler averaging to estimate the CUT clutter power is not effective for these systems because the clutter is not sufficiently homogeneous in the spatial dimensions (typically range and angle). For example, the radar coverage area may include heavily developed urban areas in some directions, open terrain in others, busy roads, and so forth. Consider a radar sited at the Los Angeles International Airport. When looking to the west, the dominant clutter is the ocean, while to the east it is urban business and residential neighborhoods.

Figure 6.33 shows a notional range-angle radar grid having different types of clutter (highways, urban areas, forested areas) in different look directions. Figure 6.33a shows a two-dimensional spatial CFAR reference region around the cell under test. The clutter statistics needed for the particular CFAR algorithm are estimated from the data in the reference window cells. For some CUT locations the surrounding terrain might be relatively homogenous so this spatial averaging within a single scan works well. For others, that will not be the case because different clutter types will be present within the reference window.

An alternative is to estimate the CUT interference power using temporal averaging rather than spatial averaging. Figure 6.33b illustrates this idea. The clutter power in a given spatial resolution cell is estimated from the measured values in the *same* cell from current and previous scans. A common technique is to estimate the power using a simple first-order recursive filter of the form:

$$\hat{x}[n] = (1-\gamma)\hat{x}[n-1] + \gamma \cdot x[n] \qquad (6.178)$$

where $\hat{x}[n]$ is the estimate of the clutter power in the CUT at time n (n usually indexes complete radar scans of an area) and $x[n]$ is the currently measured clutter echo sample at time n. The factor γ varies between zero and one and controls the relative weight of the current measurement versus the preceding measurements. This equation is applied separately to each spatial cell of interest.

The detection threshold is set for each spatial cell as

$$\hat{T}[n] = \alpha \hat{x}[n-1] \qquad (6.179)$$

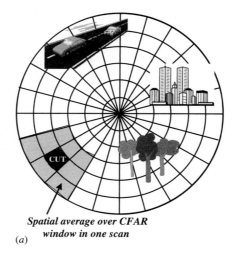

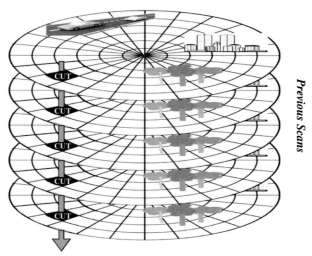

FIGURE 6.33 Temporal versus spatial averaging for estimation of CFAR parameters.
(a) Conventional spatial averaging around the CUT in spatially inhomogeneous clutter.
(b) Temporal averaging over multiple scans avoids spatial variations but requires temporal stationarity.

where α is the CFAR multiplier. Note that the threshold for testing for a target on the current scan is based on the clutter estimate from the previous scan. The current data are *not* included because, if the CUT does contain a target, it would distort the clutter measurement and raise the threshold, creating a self-masking effect. Even with this precaution, self-masking can cause CFAR losses of several dB if slow-moving targets are present, such that they persist in a map cell for more than one scan (Lops and Orsini, 1989). If the target persists in the map cell for a number of scans approaching the number of scans averaged to form the clutter statistics estimate, detectability is essentially lost entirely.

The first-order difference equation [Eq. (6.178)] corresponds to an infinite impulse response (IIR) filter with the impulse response:

$$h[n] = \gamma(1-\gamma)^n u[n] \quad (6.180)$$

where $u[n]$ is the unit step function. Consequently, the output of the filter can also be expressed as a convolution with $h[n]$:

$$\hat{x}[n] = x[n] * h[n]$$
$$= \gamma \sum_{m=0}^{\infty} (1-\gamma)^m x[n-m] \quad (6.181)$$

This equation shows that the threshold value will be based on an infinite, weighted average of previous clutter power samples rather than the finite sums of spatial-Doppler CFAR. However, the relative weight of previous samples declines as $(1-\gamma)^m$, so the choice of γ controls the number of scans that effectively contribute to the average.

The average probabilities of false alarm and detection when the threshold is computed with the recursive scheme are (Nitzberg, 1986)

$$\overline{P}_{FA} = \prod_{m=0}^{\infty} [1 + \alpha\gamma(1-\gamma)^m]^{-1} \quad (6.182)$$

$$\overline{P}_D = \prod_{m=0}^{\infty} \left[1 + \left(\frac{\alpha}{1+\bar{\chi}}\right)\gamma(1-\gamma)^m\right]^{-1} \quad (6.183)$$

These formulas are slow to converge in practice; a more rapidly converging variation is given in Levanon (1988). Again, α can be varied to generate curves of $\overline{P}_D$ versus $\overline{P}_{FA}$ and CFAR loss can then be determined by comparing these curves to the case where the interference is known exactly. (The case $\gamma = 0$ in fact corresponds to that ideal case.) This analysis can be used to show that for every increase of 0.2 in γ, the CFAR loss increases approximately 3 dB (Nitzberg, 1986).

Spatial CFAR assumes that clutter samples taken at the same time be homogeneous in space and/or Doppler, depending on the type of data. The various CA CFAR extensions discussed previously were all motivated by real-world violations of this assumption. In temporal CFAR spatial homogeneity is not required but homogeneity in time is. As a result, short-term variations in clutter sources such as weather or automobile traffic patterns in the radar coverage area limit the allowable time duration of the averaging to a few minutes in many cases. Both types of CFAR require that the clutter samples used be uncorrelated for the analyses given here to be valid.

It is also possible to estimate the clutter mapping threshold using a simple average of a finite number of previous measurements. One version of the *moving target detector* (MTD) discussed in Chap. 5 is said to have used an eight scan average covering about 32 seconds of data (Skolnik, 2001). Another implementation used a recursive estimator with a feedback coefficient of $\gamma = 7/8$ (Nathanson, 1991). A rational value of γ with a denominator that is a power of two was particularly amenable to the fixed point implementations used in early versions of the MTD, since the division by the denominator value could be implemented by simply right-shifting the binary data.

Some clutter map systems combine multiple range cells at a given azimuth direction to form a single, larger map cell. This allows the introduction of any of several standard CFAR techniques to combine the range cells and improve the map cell clutter estimate. Either cell-averaging or order statistics CFAR can be applied to the range cells to form the basic clutter power estimate, and the basic CA or OS approaches can be extended as appropriate for the environment with any of the techniques discussed previously for conventional CFAR: censoring, guard cells, smallest-of or greatest-of (SOCA and GOCA) detectors, log detectors, and two-parameter estimation algorithms (Lops, 1996; Conte and Lops, 1997).

6.5.10 Distribution-Free CFAR

The CFAR processors discussed so far assume a specific form of the PDF of the interference in order to determine the value of the threshold (equivalently, the value of the threshold multiplier α). For instance, the particular form of Eq. (6.139) and the formula for α in Eq. (6.140) is a result of having assumed an exponential distribution for the square-law detected interference-only samples. If noise is the dominant interference source, this is not a significant constraint. However, for a system operating in a clutter limited environment or in an environment where the dominant interference varies between clutter of various distributions, noise, and jamming, a threshold setting algorithm based on a particular interference PDF may produce large errors in the threshold setting when another interference PDF dominates. For this reason, threshold-setting algorithms that do not depend on the particular PDF of the interference are of interest. Such techniques are called *distribution free* or *nonparametric* CFAR algorithms (DF CFAR).

Historically, DF CFAR has been based most often on a two stage "double-threshold" approach, in which the first-stage threshold converts the raw data into binary detection/no detection decisions. This decision is repeated over multiple pulses, sweeps, or scans, and the individual detection decisions for a given resolution cell are combined using an "M out of N" rule as described in Sec. 6.4.8. The PDF of the output of the first stage is binomial since the data are binary at that point, independent of the input PDF. Four variations on this idea are discussed in Barrett (1987). Two of them, the "double-threshold detector" and the "rank order detector," use conventional cell-averaging or OS CFAR in the first stage to set a specific $\bar{P}_{FA}$ and therefore require knowledge of the interference PDF to set the first stage threshold. These are therefore not truly distribution free.

The "modified double-threshold detector" replaces the deterministically computed first-stage threshold with a feedback circuit that monitors $\bar{P}_{FA}$ at the first-stage output and adjusts the threshold to approximate the desired value. This technique requires large amounts of data to estimate the observed $\bar{P}_{FA}$ but will work for any input distribution. In the *rank sum double quantizer*, the first stage does not actually threshold the data but instead computes the rank of the test cell compared to the reference cells and passes the rank number, instead of a binary detection/no detection decision, to the second stage. The second stage integrates this rank number over multiple pulses, sweeps, or scans, producing a random variable with a distribution that depends only on the correlation between the data samples from successive pulses or scans (Barrett, 1987). A two-parameter cell-averaging CFAR computation is applied to this variable to obtain the final detection decision.

A more modern approach based on rank order ideas is described in Sarma and Tufts (2001). Consider a set of N reference samples $\{y_i\}$ and the kth order statistic for this set, $y_{(k)}$. The PDF for $y_{(k)}$ was given in Eq. (6.168). The *coverage* C is defined as the probability that a reference sample y_i is greater than $y_{(k)}$. Note that $0 \leq C \leq 1$. Since the cumulative distribution function $\bar{P}_{y_i}(y)$ is the probability that y_i is *less* than some value y, it follows that

$$C = 1 - P_{y_i}(y_{(k)}) \tag{6.184}$$

and the PDF of C becomes

$$p_C(C) = k \binom{N}{k}(1-C)^{k-1}C^{N-k} \tag{6.185}$$

Furthermore, the expected value of C is

$$\mathrm{E}\{C\} = \int_0^1 C \cdot p_C(C)\, dC = \frac{N+1-k}{N+1} \qquad (6.186)$$

Now consider a system that uses the kth rank order statistic of the reference cell data as the threshold value for the CUT. This differs from a conventional OS CFAR that computes a threshold as a multiple of the kth rank order statistic, where the multiplier α_{OS} is a function of the PDF of the interference data. If instead $y_{(k)}$ is itself the threshold, no multiplier α is required. Under these conditions, C will be the probability of false alarm. Equations (6.185) and (6.186) show that the PDF of C, and thus also its expected value $\bar{C} = \bar{P}_{FA}$, do not depend on the PDF of the raw data $\{y_i\}$. Thus, the detector is a distribution-free CFAR.

A limitation of this DF CFAR is that only certain values of $\bar{P}_{FA}$ are achievable, and they depend on the number N of reference cells. $\bar{P}_{FA}$ takes on the discrete set of values given by Eq. (6.186) as k varies from 1 to N. The minimum possible value is obtained with $k = N$ and is

$$\bar{P}_{FA\,\min} = \frac{1}{N+1} \qquad (6.187)$$

Equation (6.187) illustrates another limitation of this technique: small values of $\bar{P}_{FA\,\min}$ require large values of N. More realistically, practical limitations on the reference window size N limit this method to relatively high values of $\bar{P}_{FA}$. For a given value of N, the design value of the probability of false alarm, say $\bar{P}_{FAd}$, must be chosen to be greater than or equal to $\bar{P}_{\min}$. Assuming this condition is satisfied, the rank order to be used as a threshold is the one that produces a value of $\bar{P}_{FA}$ as close to $\bar{P}_{FA\,\min}$ as possible without exceeding it. That rank is given by

$$k = (N+1)(1 - \bar{P}_{FAd}) \qquad (6.188)$$

While the false alarm probability does not depend on the PDF of the interference, the detection probability does. For exponentially distributed interference and targets (Swerling 1 target in complex WGN), the average probability of detection is (Sarma and Tufts, 2001)

$$\bar{P}_D = \prod_{i=0}^{k-1} \frac{N-i}{N-i+(1+\bar{\chi})^{-1}} \qquad (6.189)$$

As with the other CFAR detectors, Eqs. (6.188) and (6.189) can be used to determine the CFAR loss of the DF CFAR. The additional loss over a CA CFAR for the Swerling 1 case is typically less than about 0.4 dB.

6.6 System-Level Control of False Alarms

It has been seen in this chapter that achieving good detection performance (high $\bar{P}_D$ and low $\bar{P}_{FA}$) requires a signal-to-interference ratio on the order of 15 dB or better at the point of detection. For a given target RCS, the SIR is determined in part by basic radar system design choices reflected in the radar range equation: transmitter power, antenna gain, operating frequency, and noise figure. Furthermore, the fundamental goal of many of the techniques of radar signal processing discussed in other chapters of this text is to improve the SIR before the point of detection. Examples include matched filtering, pulse compression, MTI, pulse Doppler processing, and space-time adaptive processing. Once the SIR has been maximized, the detector, whether fixed or adaptive threshold, sets the actual threshold value and thus the false alarm probability. The SIR then determines the detection probability.

In some cases the detection probability may still be lower than required. In this event, the threshold may be lowered, increasing $\hat{P}_D$ but also increasing $\bar{P}_{FA}$. Additional techniques can then be applied at other stages of the overall system processing in order to reduce $\bar{P}_{FA}$ back to an acceptable level. Several options are discussed in Nathanson (1991); their applicability depends on the particular system involved. If jamming is present, a sidelobe blanker or sidelobe canceller can be used to further improve the SIR before the detector (assuming more advanced techniques such as STAP have not already been applied). After the detector, a clutter map may be used in some systems to reject false alarms due to fixed clutter discretes or known *radio frequency interference* (RFI) sources. If valid targets can be expected to extend over more than one range, azimuth, or Doppler cell, apparent detections that occupy only a single cell can be rejected as false alarms after the detector, lowering the system $\bar{P}_{FA}$. Finally, an apparent target can be tracked to make sure it recurs over multiple scans; if it does not, it is rejected as a false alarm. If it does, its kinematics can be tracked over multiple scans. If the target track violates reasonable bounds on velocity and acceleration, it can be assumed to be a false alarm, quite possibly due to the presence of electronic countermeasures, and the track can be rejected. Thus, control of the overall system false alarm rate can be spread over virtually all stages of the system.

6.7 Summary

Detection of a target echo signal in the present of interference is the most fundamental task of radar signal processing. The most common detection strategy is to fix the probability of false alarm at a tolerable level based on system concerns and then maximize the probability of detection. This strategy leads to threshold detection as the key operation.

Details of the specific optimal threshold detection algorithm are determined by the particular target and interference signal statistical models of Chap. 2. Common nonfluctuating or Swerling target models in WGN interference lead to optimal algorithms that may be computationally burdensome and so are usually approximated in practice with simple linear or square law detectors. Detailed examples for a square law detector and nonfluctuating, Swerling 1 and Swerling 2 targets illustrated the process for determining the receiver operating characteristic.

Integration, the process of combining multiple measurements in the hope of reinforcing the target signal (if present) and "averaging down" the noise, is the most basic signal processing tool for improving detection performance for a given single-sample SNR. Three types were discussed. Coherent integration of N complex samples, when feasible, increases the SNR by the factor N and provides the greatest improvement in performance. Noncoherent integration avoids the risk of destructive interference that can occur with unsuccessful coherent integration, but this greater robustness comes at the cost of a lesser detection improvement factor. Binary integration combines multiple threshold crossing decisions to improve the detection statistics, at a still lesser effectiveness.

The formulas for detection performance can be difficult to compute, often involving special functions such as the Bessel, incomplete gamma, and Marcum's Q functions. Simplified expressions were developed for some cases. More comprehensively, Albersheim's equation and Shnidman's equation were introduced as simple calculator-friendly tools for estimating detection performance for nonfluctuating targets and all four Swerling target models with noncoherent integration.

Choosing the threshold T to achieve a given P_{FA} requires knowledge of the interference power. This information is often not available with adequate accuracy, especially when clutter, which may be highly variable in time and space, is the primary interference. Unfortunately, small errors in the interference power result in very large errors in P_{FA}. To avoid this problem, CFAR processing is used to estimate the interference power from the measured data in most radars. The most common variant is cell-averaging CFAR. While quite effective and widely used, it is prone to a variety of errors due to clutter edges and closely spaced targets that are combatted with various heuristic extensions such as greater-of, smaller-of, and censoring CA CFAR. Another class of CFAR algorithms that substitutes sorting operations for averaging are the order statistic CFARs, which have also come into common use and have various extensions to cope with real-world scenarios.

References

Albersheim, W. J., "Closed-Form Approximation to Robertson's Detection Characteristics," *Proceedings of IEEE*, vol. 69, no. 7, p. 839, Jul. 1981.

Barrett, C. R., Jr., "Adaptive Thresholding and Automatic Detection," Chap. 12 in J. L. Eaves and E. K. Reedy (eds.), *Principles of Modern Radar*. Van Nostrand Reinhold, New York, 1987.

Blake, S., "OS-CFAR Theory for Multiple Targets and Nonuniform Clutter," *IEEE Transactions on Aerospace and Electronic Systems*, vol. AES-24, no. 6, pp. 785–790, Nov. 1988.

Cantrell, P. E., and A. K. Ojha, "Comparison of Generalized Q-function Algorithms," *IEEE Transactions on Information Theory*, vol. IT-33, no. 4, pp. 591–596, Jul. 1987.

Colena, C. I., M. J. Russell and S. A. Braun, "Minesweeper: A Novel and Fast Ordered-Statistic CFAR Algorithm," 2020 IEEE High Performance Extreme Computing Conference (HPEC), Waltham, MA, USA, pp. 1–6, 2020.

Conte, E., and M. Lops, "Clutter-Map CFAR Detection for Range-Spread Targets in Non-Gaussian Clutter, Part I: System Design," *IEEE Transactions on Aerospace and Electronic Systems*, vol. AES-33, no. 2, pp. 432–442, Apr. 1997.

Di Vito, A., and G. Moretti, "Probability of False Alarm in CA-CFAR Device Downstream From Linear-law Detector," *Electronics Letters*, vol. 25, no. 5, pp. 1692–1693, Dec. 1989.

DiFranco, J. V., and W. L. Rubin, *Radar Detection*. Artech House, Dedham, MA, 1980.

Finn, H. M., "A CFAR Design for a Window Spanning Two Clutter Fields," *IEEE Transactions on Aerospace and Electronic Systems*, vol. AES-22, no. 2, pp. 155–169, Mar. 1986.

Gandhi, P. P., E. Cardona, and L. Baker, "CFAR Signal Detection in Nonhomogeneous Weibull Clutter and Interference," *Record of the IEEE International Radar Conference*, pp. 583–588, 1995.

Gradshteyn, I. S., and I. M. Ryzhik, *Tables of Integrals, Series, and Products*, A. Jeffrey (ed.). Academic Press, New York, 1980.

Hansen, V. G., and J. H. Sawyers, "Detectability Loss due to 'Greatest of Selection in a Cell-Averaging CFAR,'" *IEEE Transactions on Aerospace and Electronic Systems*, vol. AES-16, no. 1, pp. 115–118, Jan. 1980.

Hansen, V. G., and H. R. Ward, "Detection Performance of the Cell-Averaging LOG/CFAR Receiver," *IEEE Transactions on Aerospace and Electronic Systems*, vol. AES-8, no. 5, pp. 648–652, Sep. 1972.

Hmam, H., "Approximating the SNR Value in Detection Problems," *IEEE Transactions on Aerospace and Electronic Systems*, vol. AES-39, no. 4, pp. 1446–1452, Oct. 2003.

IEEE Standard Radar Definitions. IEEE Std. 686-2017, Institute of Electrical and Electronics Engineers, Sep. 13, 2017.

Johnson, D. H., and D. E. Dudgeon, *Array Signal Processing*. Prentice Hall, Englewood Cliffs, NJ, 1993.

Kay, S. M., *Fundamentals of Statistical Signal Processing, Vol. I: Estimation Theory*. Prentice Hall, Upper Saddle River, NJ, 1993.

Kay, S. M., *Fundamentals of Statistical Signal Processing, Vol. II: Detection Theory*. Prentice Hall, Upper Saddle River, NJ, 1998.

Keel, B. M., "Constant False Alarm Rate Detectors," Chap. 16 in M. A. Richards, J. A. Scheer, and W. A. Holm (eds.), *Principles of Modern Radar: Basic Principles*. SciTech Publishing, Raleigh, NC, 2010.

Levanon, N., *Radar Principles*. Wiley, New York, 1988.

Lops, M., "Hybrid Clutter-Map/L-CFAR Procedure for Clutter Rejection in Nonhomogeneous Environment," *IEE Proceedings of Radar, Sonar, and Navigation*, vol. 143, no. 4, pp. 239–245, Aug. 1996.

Lops, M., and M. Orsini, "Scan-by-Scan Averaging CFAR," *IEE Proceedings*, Part F, vol. 136, no. 6, pp. 249–253, Dec. 1989.

Meyer, D. P., and H. A. Mayer, *Radar Target Detection*. Academic Press, New York, 1973.

Nathanson, F. E. (with J. P. Reilly and M. N. Cohen), *Radar Design Principles*, 2d ed. McGraw-Hill, New York, 1991.

Nitzberg, R., "Clutter Map CFAR Analysis," *IEEE Transactions on Aerospace and Electronic Systems*, vol. AES-22, no. 4, pp. 419–421, Jul. 1986.

Novak, L. M., "Radar Target Detection and Map-Matching Algorithm Studies," *IEEE Transactions on Aerospace and Electronic Systems*, vol. AES-16, no. 5, pp. 620–625, Sep. 1980.

Olver, F. W. J. et al., editors, *NIST Digital Library of Mathematical Functions*, version 1.0.28, Sept 15, 2020. National Institute of Standards and Technology, U.S. Dept. of Commerce, and Cambridge University Press. https://dlmf.nist.gov/.

Pace, P. E., and I. L. Taylor, "False Alarm Analysis of the Envelope Detection GO-CFAR Processor," *IEEE Transactions on Aerospace and Electronic Systems*, vol. AES-30, no. 3, pp. 848–864, Jul. 1994.

Raghavan, R. S., "Analysis of CA-CFAR Processors for Linear-Law Detection," *IEEE Transactions on Aerospace and Electronic Systems*, vol. AES-28, no. 3, pp. 661–665, Jul. 1992.

Richards, M. A., "Notes on Noncoherent Integration Gain," technical memorandum, Jul 17, 2014. Available at http://radarsp.com.

Richards, M. A., "Binary Integration Gain," technical memorandum, Sep. 2016. Available at http://radarsp.com.

Rifkin, R., "Analysis of CFAR Performance in Weibull Clutter," *IEEE Transactions on Aerospace and Electronic Systems*, vol. AES-30, no. 2, pp. 315–329, Apr. 1994.

Ritcey J. A., "Performance Analysis of the Censored Mean-Level Detector," *IEEE Transactions on Aerospace and Electronic Systems*, vol. AES-22, no. 4, pp. 443–454, Jul. 1986.

Ritcey, J. A., and J. L. Hines, "Performance of Max-Mean-Level Detector with and without Censoring," *IEEE Transactions on Aerospace and Electronic Systems*, vol. AES-25, no. 2, pp. 213–223, Mar. 1989.

Robertson, G. H., "Operating Characteristic for a Linear Detector of CW Signals in Narrow Band Gaussian Noise," *Bell System Technical Journal*, vol. 46, no. 4, pp. 755–774, Apr. 1967.

Rohling, H., "Radar CFAR Thresholding in Clutter and Multiple Target Situations," *IEEE Transactions on Aerospace and Electronic Systems*, vol. AES-19, no. 4, pp. 608–620, Jul. 1983.

Sangston, K. J., and A. Farina, "Coherent Radar Detection in Compound-Gaussian Clutter: Clairvoyant Detectors," IEEE *Aerospace and Electronic systems Magazine*, pp. 42–63, Nov. 2016.

Sarma, A., and D. W. Tufts, "Robust Adaptive Threshold for Control of False Alarms," *IEEE Signal Processing Letters*, vol. 8, no. 9, pp. 261–263, Sep. 2001.

Schleher, D. C., "Harbor Surveillance Radar Detection Performance," *IEEE Journal of Oceanic Engineering*, vol. OE-2, no. 4, pp. 318–325, Oct. 1977.

Shnidman, D. A., "Binary Integration for Swerling Target Fluctuations," *IEEE Transactions on Aerospace and Electronic Systems*, vol. AES-34, no. 3, pp. 1043–1053, Jul. 1998.

Shnidman, D. A., "Determination of Required SNR Values," *IEEE Transactions on Aerospace and Electronic Systems*, vol. AES-38, no. 3, pp. 1059–1064, Jul. 2002.

Shor, M., and N. Levanon, "Performance of Order Statistics CFAR," *IEEE Transactions on Aerospace and Electronic Systems*, vol. AES-27, no. 2, pp. 214–224, Mar. 1991.

Skolnik, M. I., *Introduction to Radar Systems*, 3d ed. McGraw-Hill, New York, 2001.

Smith, M. E., and P. K. Varshney, "Intelligent CFAR Processor Based on Data Variability," *IEEE Transactions on Aerospace and Electronic Systems*, vol. AES-36, no. 3, pp. 837–847, Jul. 2000.

Tufts, D. W., and A. J. Cann, "On Albersheim's Detection Equation," *IEEE Transactions on Aerospace and Electronic Systems*, vol. AES-19, no. 4, pp. 643–646, Jul. 1983.

Van Cao, T.-T., "A CFAR Thresholding Approach Based on Test Cell Statistics," *Proceedings of the 2004 IEEE Radar Conference*, pp. 349–354, Apr. 26–29, 2004.

Van Trees, H. L., K. L. Bell, and Z. Tian, *Detection, Estimation, and Modulation Theory, Part I: Detection, Estimation, and Linear Modulation Theory*, 2d ed. Wiley, New York, 2013.

Weinberg, G., *Radar Detection Theory of Sliding Window Processes*. CRC Press, New York, 2017.

Weiner, M. A., "Binary Integration of Fluctuating Targets," *IEEE Transactions on Aerospace and Electronic Systems*, vol. AES-27, no. 1, pp. 11–17, Jan. 1991.

Weiss, M., "Analysis of Some Modified Cell-Averaging CFAR Processors in Multiple-Target Situations," *IEEE Transactions on Aerospace and Electronic Systems*, vol. AES-18, no. 1, pp. 102–114, Jan. 1982.

Problems

1. Consider a detection problem where under hypothesis H_0 the PDF of the signal x is $p_x(x|H_0) = \alpha \cdot \exp(-x/\alpha)$, $0 \le x \le \infty$, while under hypothesis H_1 the PDF of the signal x is $p_x(x|H_1) = \beta \cdot \exp(-x/\beta)$, $0 \le x \le \infty$, with $\beta > \alpha$. What are the likelihood ratio and log likelihood ratio for this problem?

2. Consider Neyman-Pearson detection of a constant $A = 0.5$ in uniform (not Gaussian) white noise. Specifically, the PDF of the noise $w[n]$ is

$$p_w(w) = \begin{cases} 1, & -0.5 < w < +0.5 \\ 0, & \text{otherwise} \end{cases}$$

The measured signal is $x[n]$. Under hypothesis H_0 (noise only), $x[n] = w[n]$. Under hypothesis H_1, $x[n] = A + w[n]$.

  a. On one graph, sketch carefully the PDF of $x[n]$ under each hypothesis, $p_x(x|H_0)$ and $p_x(x|H_1)$. Label all significant values.

  b. What is the likelihood ratio $\Lambda(x)$ (not the log-likelihood ratio or the likelihood ratio test) for this problem? It may be necessary to express this in more than one region, that is, "$\Lambda(x) = $ expression 1 for $a < x < b$," etc.

  c. Regardless of the LR found in part (b), suppose the rule $x \underset{H_0}{\overset{H_1}{\gtrless}} T$ is chosen as the detection test. Sketch the resulting curves for P_{FA} versus T and P_D versus T as T varies from -1.5 to $+1.5$. Label all important values.

Problems 3 through 6 are a related group exploring how changes in the values of m, σ_w^2, and T affect detection and false alarm performance in the real constant-in-WGN example.

3. Consider detection of a real-valued constant in zero-mean real-valued Gaussian noise. Let the noise variance $\sigma_w^2 = 2$, the number of samples $N = 1$, and the constant $m = 4$. What is the SNR χ for this case? Sketch the distributions $p(y|H_0)$ and $p(y|H_1)$; label appropriate numerical values on the axes. Write the likelihood ratio and log-likelihood ratio for this problem. Simplify the expressions.

4. Continuing with the same parameters given in the previous problem, suppose $P_{FA} = 0.01$ (1 percent) is required. What is the required value of the threshold T? What is the resulting value of P_D? Lookup tables or MATLAB® can be used to calculate the values of functions such as $\text{erf}(\cdot)$, $\text{erfc}(\cdot)$, $\text{erf}^{-1}(\cdot)$, or $\text{erfc}^{-1}(\cdot)$ that may be needed.

5. Suppose m in Prob. 3 is increased so as to double the SNR; $\sigma_w^2 = 2$ and $N = 1$ still. What is the new value of m? Sketch and label the distributions $p(y|H_0)$ and $p(y|H_1)$ with this new value of m. If the same threshold T found in Prob. 4 is retained, does P_{FA} change and, if so, what is the new value? If that same threshold T is retained, does P_D change and, if so, what is the new value?

6. Go back to the case of $m = 4$, but now reduce the noise variance to $\sigma_w^2 = 1$. $N = 1$ still. What is the SNR χ now? Sketch and label the distributions $p(y|H_0)$ and $p(y|H_1)$ with this value of m and σ_w^2. If the threshold value used in Probs. 4 and 5 is still retained, what is the value of P_{FA}? What is the value of P_D?

7. Consider detection of a real constant in *complex* Gaussian noise. Let the total noise variance $\sigma_w^2 = 2$, the number of samples $N = 1$, and the constant $m = 4$. What is the SNR χ for this case? Suppose $P_{FA} = 0.01$ (1 percent) is required. What is the required value of the threshold T? What is the resulting value of P_D? Lookup tables or MATLAB® can be used to calculate the values of functions such as erf(·), erfc(·), erf^{-1}(·), or erfc^{-1}(·) that may be needed.

8. Compute the threshold T and probability of detection P_D for the case of a constant with unknown phase (and so complex-valued) in zero-mean complex Gaussian noise. Use $m = 4$, $\sigma_w^2 = 2$, $N = 1$, and $P_{FA} = 0.01$ again. It will be necessary to numerically evaluate the Marcum Q function Q_M. MATLAB® can be used with either the marcumq.m function available in the Signal Processing Toolbox™ or Communications Systems Toolbox™ or the marcum.m function available in the "MATLAB Supplements" area of the website http://www.radarsp.com.

9. Use Albersheim's equation to estimate the single-sample SNR χ_1 required to achieve $P_{FA} = 0.01$ and P_D equal to the same value obtained in the previous problem. Compare to the actual SNR used in Prob. 8.

10. Repeat Prob. 9 using Shnidman's equation in place of Albersheim's equation.

11. Rearrange Albersheim's equation to derive a set of equations for P_{FA} in terms of P_D, N, and single-pulse SNR in dB, χ_1.

12. Use Shnidman's equation to estimate the single-pulse SNR χ_1 in dB required to achieve $P_{FA} = 10^{-8}$ and $P_D = 0.9$ when noncoherently integrating $N = 10$ samples. Do this for all four Swerling cases and for the nonfluctuating case. Figure 6.14 can be used to check the results.

Problems 13 through 16 are a related group comparing the effectiveness of coherent and noncoherent integration of measurements and showing how to compute noncoherent integration gain.

13. Coherently integrating N samples of signal-plus-noise produces an integration gain of N on a linear scale; that is, if the SNR of a single sample y_i is χ, the SNR of $z = \sum_{i=1}^{N} y_i$ is $N\chi$. In Probs. 13 through 16, Albersheim's equation will be used to investigate the relative efficiency of noncoherent integration for one example case. Throughout these problems, assume $P_D = 0.9$ and $P_{FA} = 10^{-6}$ are required, and that a linear detector is used. Start by assuming detection is to be based on a single sample, $N = 1$. Use Albersheim's equation to determine the SNR χ_1 in dB needed for this single sample to meet the specifications above.

14. Now consider noncoherent integration of 100 samples to achieve the same P_D and P_{FA} as in the previous problem. Each individual sample can then have a lower SNR. Use Albersheim's equation again to determine the SNR χ_{nc} of each sample in dB needed to achieve the required detection performance.

15. Now consider coherent integration of 100 pulses. What is the SNR χ_c in dB required for each sample such that the coherently integrated SNR will be equal to the value χ_1 found in Prob. 13?

16. Finally, the noncoherent integration gain is the ratio χ_1/χ_{nc}. Find α such that $\chi_1/\chi_{nc} = N^\alpha$. What would be the value of α for coherent integration? Which is more efficient (obtains more gain for the same number of samples integrated), coherent or noncoherent integration?

17. Use Fig. 6.16 to estimate the SIR required to achieve $P_D = 0.5$ and $P_{FA} = 10^{-6}$ when the interference is white Gaussian noise, and again when it is K power-distributed clutter with $a = 1$. (The increase in required SIR in clutter is a loss due to non-Gaussian interference.)

18. Consider 3-out-of-5 ($M = 3$, $N = 5$) binary integration. Determine the required values of the single-trial probabilities P_D and P_{FA} such that the cumulative probabilities are $P_{CFA} = 10^{-8}$ and $P_{CD} = 0.99$. A small-probability approximation can be used to solve for P_{FA}, but finding P_D will require some numerical trial-and-error; the estimate of P_D should be accurate to two decimal places. (*Hint*: The correct answer lies in the range $0.87 \le P_D \le 0.92$.)

19. A single noncoherently detected sample of a nonfluctuating target in complex Gaussian noise with power $\sigma_w^2 = 1$ is to be tested for the presence of a target. A square law detector is used. Assuming the interference power level is known exactly, what ideal value of threshold T is required to obtain $P_{FA} = 10^{-4}$? If the SNR is $\chi = 10$ dB, what is P_D? MATLAB® and one of the computer functions mentioned in Prob. 8 or their equivalent will be needed to evaluate the Marcum Q function Q_M.

20. Now assume that the interference level is not known a priori, so a cell-averaging CFAR is used to perform the detection test. Choose $N = 30$ reference cells. What will be the threshold multiplier α such that the average false alarm probability $\bar{P}_{FA}$ remains at 10^{-4}? It turns out that if the SNR is $\chi = 10$ dB, the value of P_D using the ideal threshold in the previous problem is 0.616. Assuming the SNR is still $\chi = 10$ dB, what will be the average detection probability $\bar{P}_D$ using the CA CFAR?

21. Suppose the threshold in a standard (non-CFAR) threshold detector designed using the Neyman-Pearson approach is chosen to give $P_{FA} = 10^{-6}$. If the interference power level increases by 6 dB, what will be the new value of P_{FA}?

22. Consider a detector designed to give an average false alarm probability of $\bar{P}_{FA} = 10^{-8}$. What is the SNR χ_∞ in dB required to achieve $\bar{P}_D = 0.9$ when using the ideal Neyman-Pearson threshold? What SNR χ_N in dB is required when using a cell-averaging CFAR with $N = 16$ reference cells? What is the CFAR loss for this case, in dB?

23. Consider a CA CFAR with a single interfering target (the "target masking" problem). Assume the SNR χ of the target of interest in the cell under test is 15 dB. What is the approximate "target masking loss" in dB if the SNR χ_i of the interferer is 10 dB? Repeat for $\chi_i = 15$ dB. Assume the number of averaging cells is $N = 20$. Compute the results numerically and show all work. Figure 6.25 can be used as an approximate check of the results.

24. Consider an order statistic CFAR in exponentially distributed interference with $N = 20$ and a threshold scale factor of $\alpha_{OS} = 10$. Compute the average false alarm probability $\bar{P}_{FA}$ for the order statistic $k = 15, 16, \ldots, 20$.

25. For the OS-CFAR in the previous problem with $k = 15$, how many outliers (e.g., other targets) could be tolerated in the lead and lag windows without serious degradation of the detection and false alarm performance?

CHAPTER 7

Measurements and Introduction to Tracking

It has been seen in earlier chapters that resolution in range, angle, and Doppler shift is determined by the temporal and spatial bandwidth and the coherent processing interval (CPI) duration of the radar; but what about the *precision* of location in these dimensions? To clarify the question, consider the notional output from the radar receiver as the system scans in angle past a single, isolated point target. Assume a high repetition rate relative to the antenna scan rate so that the angle samples are closely spaced. In the absence of noise, one would expect to measure an output voltage proportional to the two-way antenna voltage pattern as illustrated in Fig. 7.1a for a sinc-squared two-way voltage pattern and a linear detector. The angular position of the target can be determined exactly simply by finding the angle at which the peak output voltage occurs. Thus, the target is located in angle to a precision much finer than the resolution, which is proportional to the mainlobe width.

Now add receiver noise to the problem. The receiver output will consist of the sum of the target echo weighted by the antenna pattern and the noise. The noise may cause the observed peak to occur at an angle other than the true target location, as seen in Fig. 7.1b; the actual peak in this realization (expanded in the inset) is at −0.033 Rayleigh widths. The larger the noise variance, the greater the likely deviation of the measurement from the noise-free case.

Because of the noise, the measured peak location is now a random variable. If the peak measurement is repeated on noisy data many times, the probability density function (PDF) of that random variable (RV) can be estimated. Figure 7.2 shows two histograms for the observed peak location when complex Gaussian noise at two values of peak signal-to-noise ratio (SNR) is added to the received signal prior to the detector. The black curves are zero-mean Gaussian PDFs with the same variance as the simulated error data. In Fig. 7.2b, the SNR is 20 dB lower (a factor of 100× in power) than in Fig. 7.2a, resulting in a wider angle error distribution than in the higher SNR case. In this example, the variance of the distribution in Fig. 7.2b is 9.54 times that of the distribution in Fig. 7.2a. This factor is approximately the square root of the 100× change in SNR.

Measurement precision limitations due to additive noise are the primary focus in this chapter, but noise is not the only limitation on measurement accuracy. Others include sampling density, target reflectivity fluctuations, other interference sources such as clutter and jamming, and hardware limitations. For instance, the scanning antenna's received power would be measured only at discrete angles determined by the radar's repetition rate (RR) and the antenna scan rate rather than on the dense sampling grid used in the preceding example. In a search mode, there might be only one to two samples per beamwidth and the peak estimator used here would not be practical. The peak estimator is also a poor choice because it takes no advantage of multiple samples to average out the effects of the noise. More practical estimators will be considered shortly.

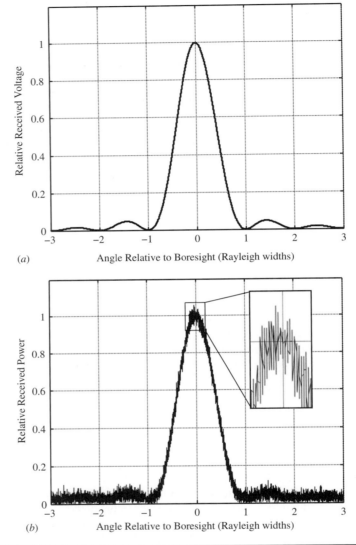

FIGURE 7.1 Received voltage from an angle scan of a single point target: (a) no noise, (b) 30 dB signal-to-noise ratio. The inset expands the area around the boresight angle.

7.1 Estimators

7.1.1 Estimator Properties

Suppose that one has a vector of N measured data samples $\mathbf{x} = \{x_i, i = 0,\ldots, N-1\}$ that depends on a deterministic but unknown parameter Θ. Θ might be the actual target angle, time delay (equivalent to range), or Doppler shift. An *estimator* $f(\mathbf{x})$ is an algorithm for computing an estimate $\hat{\Theta}$ of the actual value of Θ from the data $\mathbf{x}$:

$$\hat{\Theta} = f(\mathbf{x}) \tag{7.1}$$

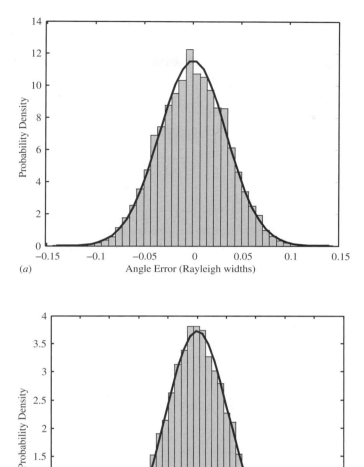

FIGURE 7.2 Figure histogram of angle error using peak power method. (*a*) 30 dB SNR at boresight. (*b*) 10 dB SNR at boresight. Note the wider spread of errors at lower SNR.

If the data is noisy, $\hat{\Theta}$ will be an RV with its own PDF and moments. Because the quality of different estimators will vary, it is reasonable to ask "What properties characterize a good estimator?"

Before answering that question, it is useful to define two important metrics of estimator quality, *accuracy* and *precision*. Figure 7.3 illustrates these two ideas heuristically with the example of target shooting. The hits on the left target are centered roughly on the bullseye (center of the target) but are widely dispersed. Because the average location is near the center

428 Chapter Seven

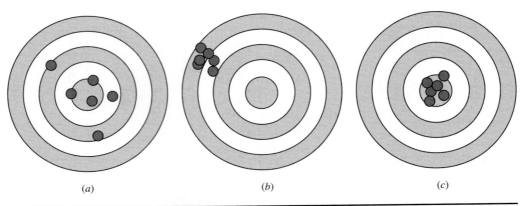

FIGURE 7.3 Distinction between precision and accuracy: (a) hits on the target are accurate but not precise, (b) precise but not accurate, (c) both accurate and precise.

of the target, this set of hits is said to exhibit good accuracy. However, because they are widely dispersed, they are said to exhibit poor precision. In the middle target, the hits are tightly clustered but the center of the cluster is far from the bullseye. These hits are precise but inaccurate. In the target on the right, the hits are both tightly clustered and centered on the bull's-eye: both precise and accurate.

To make these ideas more formal, define the accuracy of an estimate $\hat{\Theta}$ of a parameter Θ as the *bias* of the estimator, that is, the expected value of the difference between the mean of the estimate and the actual parameter value:

$$\text{Accuracy} = B_{\hat{\Theta}}(\hat{\Theta}) \equiv \mathbf{E}(\hat{\Theta}) - \Theta = \mathbf{E}(\hat{\Theta} - \Theta) \tag{7.2}$$

The precision is defined as the standard deviation of the estimate:

$$\text{Precision} = \sigma_{\hat{\Theta}} = \mathbf{E}\left\{\sqrt{\left(\hat{\Theta} - \bar{\hat{\Theta}}\right)^2}\right\} \tag{7.3}$$

(The overbar also indicates expected value.) In most cases, it will be convenient to work with the variance of $\hat{\Theta}$ (square of precision), $\sigma_{\hat{\Theta}}^2$.

Three desirable properties of an estimator are that it be *unbiased*, *consistent*, and *minimum variance*. The first two of these are defined by the behaviors

$$\begin{aligned}\mathbf{E}\{\hat{\Theta}\} &= \Theta \quad \text{(unbiased)} \\ \lim_{N \to \infty}\{\sigma_{\hat{\Theta}}^2\} &\to 0 \quad \text{(consistent)}\end{aligned} \tag{7.4}$$

The unbiased property states that the estimate matches the true parameter value "on average." The consistency property states that as more data is made available for the estimate, the precision improves and asymptotically approaches zero (perfect precision). The third property, minimum variance, expresses the goal that of all possible unbiased estimators, the one having the minimum variance (finest precision) should be selected. An estimator having both the first and third properties is called a *minimum variance unbiased* (MVU) estimator.

As an example of these properties, consider a simple estimation problem. The data **x** are N independent observations of a real constant A in additive real stationary white Gaussian noise (WGN):

$$x[n] = A + w[n], \quad w[n] \sim N(0, \sigma_w^2) \Rightarrow x[n] \sim N(A, \sigma_w^2) \tag{7.5}$$

The goal is to estimate the constant A from the noisy data. In this example, A is the parameter Θ on which the measured data $\mathbf{x} = [x[0] \ x[1] \ \cdots \ x[N-1]]^T$ depend. An obvious estimator is the sample mean:

$$\begin{aligned}\hat{A} = f(\mathbf{x}) &= \frac{1}{N}\sum_{n=0}^{N-1} x[n] \\ &= \frac{1}{N}\sum_{n=0}^{N-1}(A + w[n]) = A + \frac{1}{N}\sum_{n=0}^{N-1} w[n]\end{aligned} \tag{7.6}$$

Clearly the estimate is unbiased, since

$$E\{\hat{A}\} = E\left\{A + \frac{1}{N}\sum_{i=0}^{N-1} w[n]\right\} = A + \frac{1}{N}\sum_{i=0}^{N-1} E\{w[n]\} = A \tag{7.7}$$

Consistency is tested by considering the variance of $\hat{A}$:

$$\begin{aligned}\sigma_{\hat{A}}^2 &= E\left\{\left(\hat{A} - \bar{\hat{A}}\right)^2\right\} = E\{(\hat{A} - A)^2\} = E\left\{\left(\frac{1}{N}\sum_{m=0}^{N-1} w[m]\right)\left(\frac{1}{N}\sum_{n=0}^{N-1} w[n]\right)\right\} \\ &= \frac{1}{N^2} E\left[\sum_{m=0}^{N-1}\sum_{n=0}^{N-1} w[m]w[n]\right] = \frac{1}{N^2}\sum_{m=0}^{N-1}\sum_{n=0}^{N-1} E\{w[m]w[n]\}\end{aligned} \tag{7.8}$$

Because the noise samples are independent, $E\{w[m]w[n]\} = E\{w[m]\} \cdot E\{w[n]\}$. Because the noise is white, the mean of $w[n]$ is necessarily zero. Therefore, $E\{w[m]w[n]\}$ equals zero whenever $m \neq n$ but equals σ_w^2 when $m = n$. The latter case occurs N times during the double summation. Equation (7.8) reduces to

$$\sigma_{\hat{A}}^2 = \frac{1}{N^2}(N\sigma_w^2) = \frac{\sigma_w^2}{N} \Rightarrow \lim_{N \to \infty} \sigma_{\hat{A}}^2 = 0 \tag{7.9}$$

so the sample mean estimator is also consistent.

7.1.2 The Cramèr-Rao Lower Bound

Equation (7.9) established the variance of the sample mean estimator, but is it the estimator with the minimum possible variance? This question is answered by the famous *Cramèr-Rao Lower Bound* (CRLB). Denote the joint PDF of **x** given Θ as $p_\mathbf{x}(\mathbf{x}|\Theta)$. The CRLB states that the variance of any unbiased estimate $\hat{\Theta}$ of Θ is lower bounded to

$$\sigma_{\hat{\Theta}}^2 \geq \frac{1}{E[\{\partial \ln[p_\mathbf{x}(\mathbf{x}|\Theta)]/\partial \Theta\}^2]} \tag{7.10}$$

An alternate form of the CRLB is also common. If $p_x(x|\Theta)$ is twice differentiable and obeys some other mild regularity conditions, then it can be shown that

$$\sigma_{\hat{\Theta}}^2 \geq \frac{-1}{E[\partial^2 \ln\{p_x(x|\Theta)\}/\partial\Theta^2]} \tag{7.11}$$

The choice between Eq. (7.10) or (7.11) is a matter of convenience. The CRLB is derived in App. A, which also details its extensions to data dependent on multiple parameters and to complex data and parameters. The CRLB can be generalized to biased estimators, but the unbiased case is sufficient here.

Because the sample mean estimator proved to be unbiased, the CRLB can be applied to evaluate its variance. Identifying $\Theta = A$, the required PDF $p_x(x|\Theta)$ is

$$p_x(x|A) = \prod_{n=0}^{N-1} \frac{1}{\sqrt{2\pi\sigma_w^2}} \exp\left[-(x[n] - A)^2/2\sigma_w^2\right] \tag{7.12}$$

Considering the CRLB in the form of Eq. (7.10), the partial derivative with respect to A of $\ln\{p_x(x|A)\}$ is

$$\begin{aligned}\frac{\partial}{\partial A}\ln\{p_x(x|A)\} &= \frac{\partial}{\partial A}\left\{-\frac{N}{2}\ln(2\pi\sigma_w^2) - \sum_{n=0}^{N-1}(x[n] - A)^2/2\sigma_w^2\right\} \\ &= 2\sum_{n=0}^{N-1}(x[n] - A)/2\sigma_w^2 = \frac{1}{\sigma_w^2}\sum_{n=0}^{N-1}(x[n] - A) \\ &= \frac{1}{\sigma_w^2}\sum_{n=0}^{N-1} w[n] \end{aligned} \tag{7.13}$$

The expected value of the square of this quantity is

$$\begin{aligned}E\left[\left(\frac{\partial}{\partial A}\ln\{p_x(x|A)\}\right)^2\right] &= E\left[\left(\frac{1}{\sigma_w^2}\sum_{n=0}^{N-1} w[n]\right)^2\right] = \frac{1}{\sigma_w^4} E\left[\left(\sum_{m=0}^{N-1} w[m]\right)\left(\sum_{n=0}^{N-1} w[n]\right)\right] \\ &= \frac{1}{\sigma_w^4} E\left[\sum_{n=0}^{N-1} w^2[n]\right] = \frac{1}{\sigma_w^4}(N\sigma_w^2) = \frac{N}{\sigma_w^2}\end{aligned} \tag{7.14}$$

The second line follows because the noise is white so that the expected value of the cross-products is zero. Finally, using Eq. (7.14) in Eq. (7.10) gives the CRLB for this problem:

$$\sigma_{\hat{A}}^2 \geq \frac{\sigma_w^2}{N} \tag{7.15}$$

which is exactly the variance of the sample mean estimator obtained in Eq. (7.9). The CRLB guarantees that the sample mean estimator can be used with the assurance that no other unbiased estimator will have a lower variance. An unbiased estimator that achieves the CRLB is called an *efficient* estimator.

The case of a signal in additive white Gaussian noise (AWGN) is important enough to warrant further attention. Assume the measurement vector x is composed of N real-valued signal + noise samples:

$$x[n] = s[n;\Theta] + w[n], \quad n = 0,\ldots, N-1 \tag{7.16}$$

where Θ is the parameter to be estimated and the variance of $w[n]$ is σ_w^2. The following result is derived in App. A by applying Eq. (7.11) to the PDF for **x**:

$$\sigma_{\hat\Theta}^2 \geq \frac{\sigma_w^2}{\sum_{n=0}^{N-1}\left(\frac{\partial s[n;\Theta]}{\partial \Theta}\right)^2} \quad \text{(real-valued signal in AWGN)} \qquad (7.17)$$

Applying this to the constant-in-AWGN example, the parameter $\Theta = A$ and the signal $s[n; \Theta] = s[n; A] = A$ also, so $\partial s[n;\Theta]/\partial\Theta = \partial A/\partial A = 1$, again giving the result of Eq. (7.15).

If the signal and noise are complex-valued and the parameter to be estimated is real, the CRLB becomes (see App. A)

$$\sigma_{\hat\Theta}^2 \geq \frac{\sigma_w^2}{2\sum_{n=0}^{N-1}\left|\frac{\partial s[n;\Theta]}{\partial \Theta}\right|^2} \quad \text{(complex signals in AWGN), real parameter} \qquad (7.18)$$

The factor of two in the denominator of Eq. (7.18) is significant. It means that the real-valued CRLB is not a special case of the complex CRLB. Rather, the complex CRLB is smaller by a factor of 2.

The case where the parameter to be estimated is itself complex can be handled by treating it as two real parameters, $\Theta = \Theta_R + j\Theta_I$. Equation (7.18) establishes the CRLB for each of the real and imaginary parts. It is easy to show that the variance of the complex parameter Θ is the sum of the variances of Θ_R and Θ_I (see Prob. 9). The CRLB for complex Θ is then the sum of the two component CRLBs and therefore is also given by Eq. (7.17) or (7.18).

7.1.3 The CRLB and Signal-to-Noise Ratio

A relationship between the CRLB and SNR can be found for the case of a signal in AWGN. Consider a complex sampled signal $s[n]$ that is dependent on some real parameter Θ (not necessarily its amplitude) and write it as the product of its real-valued peak amplitude A and a normalized function $\tilde{s}[n]$ having a peak magnitude of one,

$$s[n] = A \cdot \tilde{s}[n] \qquad (7.19)$$

The peak SNR for the noisy signal $x[n] = s[n] + w[n]$ is $\chi = A^2/\sigma_w^2$. Suppose the parameter of interest, Θ, is *not* A. Then

$$\left|\frac{\partial s[n]}{\partial \Theta}\right|^2 = A^2\left|\frac{\partial \tilde{s}[n]}{\partial \Theta}\right|^2 \qquad (7.20)$$

and Eq. (7.18) becomes

$$\sigma_{\hat\Theta}^2 \geq \frac{\sigma_w^2}{2A^2\sum_{n=0}^{N-1}\left|\frac{\partial \tilde{s}[n]}{\partial \Theta}\right|^2} \qquad (7.21)$$

The quantity $\sum_{n=0}^{N-1} |\partial \tilde{s}[n]/\partial \Theta|^2$ depends on the signal shape $\tilde{s}$, but for a given signal it is some scalar k so that Eq. (7.21) is of the form:

$$\sigma_{\hat{\Theta}}^2 \geq \begin{cases} \dfrac{1}{2k\chi} & \text{(complex-valued signals)} \\ \dfrac{1}{k\chi} & \text{(real-valued signals)} \end{cases} \qquad (7.22)$$

The second line of Eq. (7.22) would result for the real-valued case beginning with Eq. (7.17). Equation (7.22) states that the CRLB is inversely proportional to SNR and that the constant of proportionality depends on the rate of change of $\tilde{s}$ with respect to Θ. A waveform that changes more rapidly with Θ produces a smaller CRLB.

The case where the parameter of interest is the peak amplitude A must be handled separately. In this event $\Theta = A$, giving

$$\frac{\partial s[n]}{\partial A} = \frac{\partial (A \cdot \tilde{s}[n])}{\partial A} = \tilde{s}[n] \qquad (7.23)$$

The CRLB is then

$$\sigma_{\hat{A}}^2 \geq \frac{\sigma_w^2}{2\sum_{n=0}^{N-1} |\tilde{s}[n]|^2} = \frac{\sigma_w^2}{2E_{\tilde{s}}} = \frac{1}{2(E_{\tilde{s}}/\sigma_w^2)} \qquad (7.24)$$

where the sum term in the denominator of Eq. (7.24) has been recognized as the energy in $\tilde{s}$, resulting in a CRLB that is one-half the inverse of the energy SNR based on the normalized signal.[1] However, when estimating the amplitude, the variance relative to the actual amplitude is more likely to be of interest than the absolute variance. It is easy to see that the CRLB for the normalized amplitude estimation variance is the same as Eq. (7.24) but with the energy of the unnormalized signal s. The result is (including the equivalent real-valued result as well)

$$\sigma_{\hat{A}/A}^2 \geq \begin{cases} \dfrac{1}{2(E_s/\sigma_w^2)} & \text{(complex-valued signals)} \\ \dfrac{1}{(E_s/\sigma_w^2)} & \text{(real-valued signals)} \end{cases} \qquad (7.25)$$

7.1.4 Maximum Likelihood Estimators

The CRLB establishes the minimum variance of an unbiased estimator. As discussed in App. A, it is sometimes but not always possible to find the minimum variance unbiased estimator. If the data can be modeled as a linear function of the parameters Θ with additive WGN, the MVU estimator can be found in a straightforward fashion and is further guaranteed to be efficient. In other cases, one may have to settle for a suboptimal estimator. Kay (1993) gives a thorough discussion of the hierarchy of techniques and achievable results in classical estimation.

---

[1] Energy SNR was defined in Chap. 4.

Maximum likelihood estimation (MLE, also maximum likelihood estimator or estimate) is by far the most common approach to finding practical estimators, and for several good reasons:

- The MLE can be found for most problems, including many complicated ones, by a straightforward procedure.
- While it is not the MVU estimator in general, as the number of data samples $N \to \infty$ it becomes asymptotically unbiased and efficient. The PDF of $\hat{\Theta}$ also becomes Gaussian.
- If an efficient estimator in fact exists, the maximum likelihood procedure will produce it.
- The MLE of a function $\Phi = g(\Theta)$ of Θ can be found by applying that same function to the MLE $\hat{\Theta}$ of Θ, $\hat{\Phi} = g(\hat{\Theta})$.
- Finally, various numerical methods exist for finding the MLE when a closed form expression cannot be obtained.

The MLE of Θ is that estimate $\hat{\Theta}$ that maximizes the likelihood function for the problem. The likelihood function is just the PDF of the data **x** given Θ, but viewed as a function of Θ with **x** fixed (see App. A):

$$\hat{\Theta} = \arg\max_{\Theta}\{\ell(\Theta|\mathbf{x})\} = \arg\max_{\Theta}\{p_\mathbf{x}(\mathbf{x}|\Theta)\} \qquad (7.26)$$

The maximization is performed only over allowed values of Θ, for example, $\Theta \geq 0$ if the parameter is the power of some signal. Note that maximizing a monotonically increasing function of the likelihood is the same as maximizing the likelihood itself. When the data are independent identically distributed (i.i.d.) and the noise is Gaussian, it is often the case that the likelihood function is the product of N scalar PDFs, each containing exponential terms. Consequently, it is common to maximize the log-likelihood function because the algebra is greatly simplified in these cases:

$$\hat{\Theta} = \arg\max_{\Theta}\{\ln[\ell(\Theta|\mathbf{x})]\} = \arg\max_{\Theta}\{\ln[p_\mathbf{x}(\mathbf{x}|\Theta)]\} \qquad (7.27)$$

To illustrate the procedure, consider yet again the real constant in real additive WGN example with N samples of data. The PDF is an N-dimensional joint Gaussian, so the likelihood function is

$$\ell(A|\mathbf{x}) = \frac{1}{(2\pi\sigma_w^2)^{N/2}} \exp\left[-\frac{1}{2\sigma_w^2}\sum_{n=0}^{N-1}(x[n]-A)^2\right] \qquad (7.28)$$

The log-likelihood function is

$$\ln[\ell(A|\mathbf{x})] = -\frac{1}{2\sigma_w^2}\sum_{n=0}^{N-1}(x[n]-A)^2 - \frac{N}{2}\ln(2\pi\sigma_w^2) \qquad (7.29)$$

Taking its derivative with respect to A and setting the result equal to zero to find the maximum gives

$$\frac{\partial}{\partial A}\{\ln[\ell(A|\mathbf{x})]\} = \frac{1}{\sigma_w^2}\sum_{n=0}^{N-1}(x[n]-A) = \frac{1}{\sigma_w^2}\left[-NA + \sum_{n=0}^{N-1}x[n]\right] = 0$$

$$\Rightarrow \hat{A} = \frac{1}{N}\sum_{n=0}^{N-1}x[n] \qquad (7.30)$$

The MLE for this problem is the sample mean estimator. It was seen earlier that this estimator is unbiased and achieves the CRLB. Thus, the MLE did produce the efficient estimator for this problem.

7.2 Range, Doppler, and Angle Estimators

7.2.1 Range Estimators

CRLB for Time Delay and Pulsed Radar Range Estimation

In this section, the CRLB for time delay is developed for a pulsed radar; this is trivially scaled to provide the CRLB for range for a pulsed radar. Because linear frequency-modulated continuous wave (FMCW) radar measures range by measuring the frequency of the beat signal, discussion of the range CRLB for a linear FMCW radar is deferred to Sec. 7.2.2 on frequency estimators.

Consider a complex-valued continuous-time signal $x_i(t)$ at the input to a radar receiver that is the sum of an echo of a complex transmitted signal $s_t(t)$ and complex additive WGN $w_r(t)$ with a power spectral density (PSD) of $\sigma_{w_r}^2$ W/Hz. The echoed signal component is delayed by an unknown amount t_0 to form the received signal component $s_r(t - t_0) = \alpha s_t(t - t_0)$, where α is a complex scalar whose magnitude represents attenuation due to the radar range equation and whose phase represents the unknown $-4\pi R/\lambda_t$ radian phase shift. It is assumed that demodulation has already removed the carrier so that $x_r(t)$ is at baseband. The receiver input signal is therefore $x_r(t) = s_r(t - t_0)$, as shown in Fig. 7.4. The goal is to estimate the unknown real parameter $\Theta = t_0$. The result can be scaled by $c/2$ to estimate the corresponding target range R_0.

The pulse $s_t(t)$, and therefore its scaled echo $s_r(t - t_0)$, is assumed to be τ seconds long and effectively bandlimited to β Hz. The radar receiver's frequency response is assumed to be a unit-gain bandpass characteristic over $\pm \beta/2$ Hz. Consequently, the pulse echo component of the output is essentially identical to the input pulse, while the total receiver filter output noise power is $\sigma_{w_r}^2 \beta$ W. This will be the variance of the output noise signal $w_o(t)$. The receiver output signal is $x_o(t) = s_r(t - t_0) + w_o(t)$.

The receiver output is sampled at the Nyquist rate of β samples per second to produce the observed data $x[n] = s[n - n_0] + w[n]$, where $n_0 \approx t_0/T_s$ and $T_s = 1/\beta$ is the sampling interval. The maximum time delay of interest is T; this might be the radar's pulse repetition interval, for example. It is assumed that $T > 2\tau$, the duration of a matched filter output for a single pulse at its input. $N = \lfloor T/T_s \rfloor$ is the total number of samples required to cover the maximum time interval of interest and $M = \lfloor \tau/T_s \rfloor$ is the number required to cover the signal duration τ. Let n_0 be the first sample that occurs within the received pulse so that $n_0 T_s \approx t_0$. The discrete-time data is

$$x[n] = \begin{cases} w[n], & 0 \le n \le n_0 - 1 \\ s_i(nT_s - t_0) + w[n] \equiv s[n - n_0] + w[n], & n_0 \le n \le n_0 + M - 1 \\ w[n], & n_0 + M \le n \le N - 1 \end{cases} \quad (7.31)$$

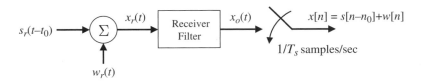

FIGURE 7.4 Receiver model for time delay and range estimation. See text for details.

Measurements and Introduction to Tracking

The sampled noise will be white (see App. A). The PSD of the sampled noise and the variance of $w[n]$ will both be $\sigma_w^2 = \sigma_{w_o}^2 = \sigma_{w_r}^2 \beta$.

Before discussing a specific estimator, consider the CRLB for this problem. Applying the form for a complex signal in additive complex Gaussian noise in Eq. (7.18) gives

$$\sigma_{\hat{t}_0}^2 \geq \frac{\sigma_w^2}{2\sum_{n=0}^{N-1}\left|\frac{\partial s[n;t_0]}{\partial t_0}\right|^2} = \frac{\sigma_w^2}{2\sum_{n=0}^{N-1}\left|\frac{\partial s(nT_s - t_0)}{\partial t_0}\right|^2} = \frac{\sigma_w^2}{2\sum_{n=n_0}^{n_0+M-1}\left|\frac{\partial s(nT_s - t_0)}{\partial t_0}\right|^2}$$

$$= \frac{\sigma_w^2}{2\sum_{n=n_0}^{n_0+M-1}\left|\frac{ds(t)}{dt}\right|_{t=nT_s-t_0}^2} = \frac{\sigma_w^2}{2\sum_{n=0}^{M-1}\left|\frac{ds(t)}{dt}\right|_{t=nT_s+n_0T_s-t_0}^2} \approx \frac{\sigma_w^2}{2\sum_{n=0}^{M-1}\left|\frac{ds(t)}{dt}\right|_{t=nT_s}^2} \tag{7.32}$$

The last step used $n_0 T_s \approx t_0$. Recall that the integral of a function $f(t)$ can be approximated by a sum using the rectangle rule, $\int_0^\tau f(t)dt \approx \sum_{n=0}^{\lfloor\tau/T_s\rfloor} f(nT_s)T_s$. Applying this to Eq. (7.32) and using $T_s = 1/\beta$ gives the CRLB for time delay estimation in complex signals as

$$\sigma_{\hat{t}_0}^2 \geq \frac{\sigma_w^2}{\frac{2}{T_s}\int_0^\tau \left|\frac{ds(t)}{dt}\right|^2 dt} = \frac{\sigma_w^2/\beta}{2\int_0^\tau \left|\frac{ds(t)}{dt}\right|^2 dt} \approx \frac{\sigma_{w_r}^2}{2\int_0^\tau \left|\frac{ds(t)}{dt}\right|^2 dt} \quad \text{(complex signals)} \tag{7.33}$$

Equation (7.33) can be put in a more useful form that reveals the role of the waveform bandwidth. The root-mean-square (RMS) bandwidth of $s(t)$ is defined as

$$\beta_{rms} \equiv \sqrt{\frac{\int_{-\infty}^{\infty} F^2 |S(F)|^2 \, dF}{\int_{-\infty}^{\infty} |S(F)|^2 \, dF}} \quad \text{Hz} \tag{7.34}$$

The derivative property of Fourier transforms states that the transform of $ds(t)/dt$ is $j2\pi F \cdot S(F)$. Combining this with Parseval's theorem, the numerator can be replaced by $(1/4\pi^2)\int |ds(t)/dt|^2 \, dt$. Therefore

$$\beta_{rms}^2 = \frac{\frac{1}{4\pi^2}\int_0^\tau \left|\frac{ds(t)}{dt}\right|^2 dt}{\int_0^\tau |s(t)|^2 dt} = \frac{1}{4\pi^2 E_s}\int_0^\tau \left|\frac{ds(t)}{dt}\right|^2 dt \tag{7.35}$$

where the integral in the denominator is the energy E_s of $s(t)$. Equation (7.33) can now be rewritten as

$$\sigma_{\hat{t}_0}^2 \geq \frac{\sigma_{w_r}^2}{8\pi^2 E_s \beta_{rms}^2} = \frac{1}{8\pi^2 (E_s/\sigma_{w_r}^2)\beta_{rms}^2} \quad \text{sec}^2$$
$$\sigma_{\hat{R}_0}^2 = \left(\frac{c}{2}\right)^2 \sigma_{\hat{t}_0}^2 \geq \frac{(\Delta R_{rms})^2}{8\pi^2 (E_s/\sigma_{w_r}^2)} \quad \text{m}^2 \quad \text{(complex signals)} \tag{7.36}$$

The first line of Eq. (7.36) shows that the lower bound on time delay estimation precision (square root of variance) becomes smaller when the mean-square signal bandwidth or the energy SNR $E_s/\sigma_{w_r}^2$ is increased. The second line scales the time delay variance by $(c/2)^2$ and defines $\Delta R_{\text{rms}} = c/2\beta_{\text{rms}}$, showing that the precision is proportional to RMS range resolution divided by the square root of the energy SNR.

The RMS bandwidth is not commonly used in radar signal processing. If it exists for a given waveform, it is proportional to more common metrics such as the Rayleigh or 3-dB bandwidth. For example, both the Rayleigh and 3-dB bandwidths of a rectangular spectrum are simply the spectrum width β. It is easy to compute from Eq. (7.34) that β_{rms} for this spectrum is $\beta/\sqrt{12} \approx 0.29\beta$. The RMS bandwidth does not exist for a rectangular pulse because the defining integral Eq. (7.34) does not converge; see Prob. 23.

Practical Range Estimators

The above results establish the minimum variance of a time delay or range estimator, but they do not show how to actually obtain an estimate, let alone one that meets the bound. Two approaches that might come to mind based on earlier chapters are to estimate the time delay by detecting when the leading edge of the received pulse echo crosses some threshold, or finding the time at which the peak of the matched filter output occurs. The results of this chapter, however, suggest starting by seeking the MLE.

The signal model of Eq. (7.31) shows that the appropriate likelihood function is a zero-mean complex Gaussian for the noise-only samples, and a nonzero mean complex Gaussian for the signal-plus-noise samples. Because the noise is i.i.d., the resulting joint likelihood function is

$$\ell(n_0|\mathbf{x}) = p_\mathbf{x}(\mathbf{x}|n_0) = \left\{\prod_{n=0}^{n_0-1} \frac{1}{\sqrt{2\pi\sigma_w^2}} \exp\left(-\frac{1}{2\sigma_w^2}|x[n]|^2\right)\right\} \times \cdots$$
$$\times \left\{\prod_{n=n_0}^{n_0+M-1} \frac{1}{\sqrt{2\pi\sigma_w^2}} \exp\left(-\frac{1}{2\sigma_w^2}|x[n]-s[n-n_0]|^2\right)\right\} \times \cdots$$
$$\times \left\{\prod_{n=n_0+M}^{N-1} \frac{1}{\sqrt{2\pi\sigma_w^2}} \exp\left(-\frac{1}{2\sigma_w^2}|x[n]|^2\right)\right\} \quad (7.37)$$
$$= \frac{1}{(2\pi\sigma_w^2)^{N/2}} \exp\left(-\frac{1}{2\sigma_w^2}\sum_{n=0}^{N-1}|x[n]|^2\right) \times \cdots$$
$$\left\{\prod_{n=n_0}^{n_0+M-1} \exp\left[-\frac{1}{2\sigma_w^2}(-2\operatorname{Re}\{x[n]s^*[n-n_0]\}+|s[n-n_0]|^2)\right]\right\}$$

Because n_0 appears only in the product of exponentials in the last line of Eq. (7.37), maximizing $\ell(n_0|\mathbf{x})$ with respect to n_0 is equivalent to minimizing the exponent of that term, which can be rewritten as

$$\exp\left[-\frac{1}{2\sigma_w^2}\sum_{n=n_0}^{n_0+M-1}(-2\operatorname{Re}\{x[n]s^*[n-n_0]\}+|s[n-n_0]|^2)\right] \quad (7.38)$$

With the change of variables $n' = n - n_0$ the sum involving $|s|^2$ is recognized as $\sum_{n'=0}^{M-1}|s[n']|^2 = E_s$, which does not actually depend on n_0. Therefore, minimizing the exponent of Eq. (7.38) with respect to n_0 reduces to maximizing the quantity $\operatorname{Re}\left\{\sum_{n=n_0}^{n_0+M-1} x[n]s^*[n-n_0]\right\}$. Because $s[n-n_0]$ is zero outside of the summation interval, this summation is the same as the correlation $z[n_0] = \sum_{n=-\infty}^{\infty} x[n]s^*[n-n_0]$, which is simply the output of the (noncausal) matched filter with impulse response $h[n] = s^*[-n]$ when the input is $x[n]$. The MLE estimate of n_0 becomes

$$\widehat{n_0} = \arg\max_{n_0}\left\{\text{Re}\left[\sum_{n=-\infty}^{\infty} x[n]s^*[n-n_0]\right]\right\} = \arg\max_{n_0}\{z[n_0]\} \qquad (7.39)$$

Equation (7.39) states that the maximum likelihood estimate of time delay is obtained by passing the sampled data at the receiver bandpass output through a filter matched to the received waveform and then finding the sample at which the real part of the output is maximum. To interpret this result further, recall that $x[n] = s[n-n_0] + w[n]$ and evaluate Eq. (7.39) at an estimated delay of $\widehat{n_0}$:

$$\text{Re}\left\{\sum_{n=-\infty}^{\infty} x[n]s^*\left[n-\widehat{n_0}\right]\right\} = \text{Re}\left\{\sum_{n=-\infty}^{\infty} s[n-n_0]s^*\left[n-\widehat{n_0}\right] + w[n]s^*\left[n-\widehat{n_0}\right]\right\} \qquad (7.40)$$

When $\widehat{n_0}$ equals the true delay n_0 this becomes

$$\text{Re}\left\{\sum_{n=-\infty}^{\infty} x[n]s^*\left[n-\widehat{n_0}\right]\right\}\bigg|_{\widehat{n_0}=n_0} = \text{Re}\left\{\sum_{n=-\infty}^{\infty} |s[n-n_0]|^2 + w[n]s^*[n-n_0]\right\}$$

$$= E_s + \text{Re}\left\{\sum_{n=-\infty}^{\infty} w[n]s^*[n-n_0]\right\} \qquad (7.41)$$

Equation (7.41) shows that at the true time delay, the output will consist of a real-valued component from the echoed signal, plus the real part of the filtered noise. This is consistent with the discussion of matched filters for waveforms in Chap. 4, where it was seen that the peak of the matched filter output occurred at the time delay of the target echo (increased by the delay T_M if a causal version of the impulse response is used) and that the SNR is maximized at that time. Thus, in the absence of noise the estimation rule of Eq. (7.39) will correctly identify the time delay n_0. The presence of the noise can perturb the location of the peak real output, giving rise to the uncertainty in estimating delay quantified by the CRLB. Nonetheless, when the SNR is reasonably high, it is likely that the matched filter output will be at its maximum or near-maximum at $n = n_0$ and will be primarily real-valued so that the time delay is estimated with good precision. When the SNR is not large, other effects dominate as will be seen shortly.

The analysis to this point assumes that the *received* pulse $s_r(t)$ is known and is used to define the matched filter. In practice, only the *transmitted* pulse $s_t(t)$ will be known, so the matched filter impulse response must be selected as $h[n] = s_t^*[-nT_s] = s[n]/\alpha$. The difference is the complex scale factor α. The magnitude of α is of no concern because it scales the signal and noise components equally and will not affect where the maximum occurs. However, the phase of α is the two-way propagation phase shift $-4\pi R/\lambda_t$ and must be considered uniform random over $(-\pi, \pi]$. It affects only the signal $x[n]$ in Eq. (7.41). Consequently, it will rotate the signal component E_s of Eq. (7.41) to the unknown non-real value $E_{\tilde{s}} = E_s/\alpha$ so that the Re$\{\cdot\}$ operator is no longer appropriate. In practice, Eq. (7.39) is modified to produce the final time delay estimation rule:

$$\widehat{n_0} = \arg\max_{n_0}\left\{\left|\sum_{n=-\infty}^{\infty} x[n]s^*[n-n_0]\right|\right\} = \arg\max_{n_0}\{|(x*h)[n]|_{n=n_0}\} \qquad (7.42)$$

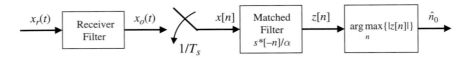

FIGURE 7.5 Maximum likelihood estimator time delay and range estimation.

Figure 7.5 modifies Fig. 7.4 to illustrate this structure by adding the matched filter and maximization of the magnitude.

It was shown in Chap. 4 that the peak SNR at the output of a matched filter is the input signal energy divided by the input noise PSD, $\chi_{out} = E_s/\sigma_{w_i}^2$. Equation (7.36) can therefore be rewritten as

$$\sigma_{\hat{t}_0}^2 \geq \frac{1}{8\pi^2 \chi_{out} \beta_{rms}^2} \quad \sec^2 \quad \text{(complex-valued signals)} \tag{7.43}$$

Equation (7.43) was obtained assuming that the signals involved were complex-valued, a case that includes real-valued signals in complex-valued noise. If both signal and noise are real-valued, then as seen in App. A the CRLB will be twice the value begun with in Eq. (7.32). In this case Eq. (7.43) is replaced by

$$\sigma_{\hat{t}_0}^2 \geq \frac{1}{4\pi^2 \chi_{out} \beta_{rms}^2} \quad \sec^2 \quad \text{(real-valued signals)} \tag{7.44}$$

The complex case is normally of interest in coherent radars.

An equivalent equation applies in discrete-time units. Converted to units of samples, Eq. (7.43) becomes

$$\sigma_{\hat{n}_0}^2 \geq \frac{1}{8\pi^2 \chi_{out} \beta_{rms}^2 T_s^2} \quad \text{samples}^2 \quad \text{(complex-valued signals)} \tag{7.45}$$

Using a tilde to distinguish quantities in discrete-time units from those in continuous-time units, the sampled signal energy, noise PSD, matched filter output SNR, and RMS bandwidth are related to the corresponding analog quantities according to

$$E_s = \int_{-\infty}^{\infty} |s(t)|^2 \, dt \approx T_s \sum_{n=-\infty}^{\infty} |s(nT_s)|^2 = T_s \sum_{n=-\infty}^{\infty} |s[n]|^2 = T_s \tilde{E}_s \Rightarrow \tilde{E}_s = \frac{1}{T_s} E_s$$

$$\sigma_{\tilde{w}}^2 = \frac{1}{T_s} \sigma_{w_r}^2 \quad \text{(effect of sampling on spectrum)} \tag{7.46}$$

$$\Rightarrow \tilde{\chi}_{out} = \tilde{E}_s/\sigma_{\tilde{w}}^2 = E_s/\sigma_{w_r}^2 = \chi_{out}$$

$$\tilde{\beta}_{rms} = T_s \beta_{rms} \quad \text{(conversion to normalized frequency in cycles per sample)}$$

Applying these conversions to Eq. (7.45) and considering the equivalent real-valued case gives

$$\sigma_{\hat{n}_0}^2 \geq \frac{1}{4\pi^2 \tilde{\chi}_{out} \tilde{\beta}_{rms}^2} \quad \text{samples}^2 \quad \text{(real-valued signals)}$$

$$\sigma_{\hat{n}_0}^2 \geq \frac{1}{8\pi^2 \tilde{\chi}_{out} \tilde{\beta}_{rms}^2} \quad \text{samples}^2 \quad \text{(complex-valued signals)} \tag{7.47}$$

which is exactly the same form as Eqs. (7.43) and (7.44).

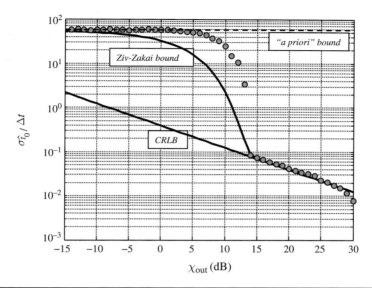

FIGURE 7.6 Behavior of simulated time delay maximum likelihood estimation precision in comparison to various bounds for a complex LFM pulse in complex noise. See text for details.

As an example, consider a complex-valued linear FM pulse with a time-bandwidth product of $\beta\tau = 100$. Because the spectrum will be approximately rectangular, $\beta_{\text{rms}} \approx \beta/\sqrt{12}$ as noted earlier. The signal is sampled at a very high rate of 20 times the Nyquist bandwidth β, which is necessary if it is to be possible to measure peak location errors that are a small fraction of the pulse width. The data record has a total length of $T = 2\tau$ seconds (the duration of the matched filter output). The gray circles in Fig. 7.6 plot the precision (standard deviation) of the error in the observed peak location,[2] normalized to the Rayleigh time resolution of $\Delta t = 1/\beta$ seconds, versus the matched filter output SNR. The precision at each value of SNR is estimated from 1000 random trials. At an SNR of 20 dB the precision in this example is about 4 percent of the time resolution $\Delta t = 1/\beta$ seconds.

Also shown in the figure is the CRLB computed from Eq. (7.43). For SNRs of 14 dB or larger, the observed precision follows the CRLB very closely,[3] confirming that the matched filter output peak is in fact a minimum variance estimator for large SNR. However, below 14 dB the estimate is much worse than suggested by the CRLB. Heuristically, reasonable performance of the MLE for time delay requires that the matched filter output peak be well above the filtered noise level. The value of the peak SNR at the output peak location is $E_s/\sigma_{w_i}^2$. High values of $E_s/\sigma_{w_i}^2$, say 20 dB or more, ensure that the peak of the signal component of the matched filter output is well above the filtered noise peaks so that the peak output, which is the MLE of delay, is in fact at the location of the signal peak or very close to it with high probability. For low values of $E_s/\sigma_{w_i}^2$, say 0 dB or less, the signal+noise peak is no larger than the noise fluctuations, so the peak output is likely to be anywhere in the complete output signal interval of $[0, N-1]$ samples or $[0, T]$ seconds with approximately equal

---

[2]In this and subsequent plots of estimate variance or precision in this chapter, the quantity plotted is the variance or standard deviation of the estimate error, rather than of the estimate itself. The result is the same, but computing and plotting the error variance has some simulation advantages.

[3]The measured precision falls below the CRLB for SNR = 29 and 30 dB because the time delay sampling grid in the simulation is not fine enough to reliably measure such small variations.

probability. In this event, the matched filter output peak conveys no information. The upper bound on delay estimation variance becomes simply the a priori (low SNR) bound of $N^2/12$ samples squared, equivalent to $N^2 T_s^2/12 = T^2/12$ seconds^2. This result applies to both real and complex signals. For the example of Fig. 7.6 where the record length T is only $2\tau = 200\Delta t$, the precision $\sigma_{\hat{t}_0}$ is about 58 times the linear frequency modulated (LFM) time resolution or 58 percent of the pulse length since $\beta\tau = 100$. For a record length of $100\tau = 10{,}000\Delta t$ the "precision" would be bounded at 2890 times the resolution! The low-SNR error can be more tightly constrained if a tracking gate or other a priori knowledge is available to limit the range of plausible time delays.

An early attempt to combine the CRLB performance bound for high SNR with the a priori upper bound for low SNR to obtain a more complete description of estimator performance was the "correlator performance estimate" described in Scarbrough et al. (1983) and attributed to Ianniello (1982). A number of other estimator bounds have been derived in the literature which also capture this behavior, extending the CRLB. The Ziv-Zakai bound (ZZB), applied to time delay estimation in the particular form given in Bell et al. (1997) and with the definitions used here for SNR and the erfc($\cdot$) function, is given by

$$\sigma_{\hat{t}_0}^2 \geq \frac{T^2}{12}\,\mathrm{erfc}\left(\sqrt{\frac{\chi_{\mathrm{out}}}{2}}\right) + \frac{1}{8\pi^2 \chi_{\mathrm{out}}\beta_{\mathrm{rms}}^2}\, I\!\left(\frac{\chi_{\mathrm{out}}}{2\sqrt{2}}, \frac{3}{2}\right) \quad \text{(complex signals)}$$

$$\sigma_{\hat{t}_0}^2 \geq \frac{T^2}{12}\,\mathrm{erfc}\left(\sqrt{\frac{\chi_{\mathrm{out}}}{2}}\right) + \frac{1}{4\pi^2 \chi_{\mathrm{out}}\beta_{\mathrm{rms}}^2}\, I\!\left(\frac{\chi_{\mathrm{out}}}{4\sqrt{2}}, \frac{3}{2}\right) \quad \text{(real signals)}$$

(7.48)

where erfc($\cdot$) is the complementary error function in the form defined in Chap. 6 and $I(\cdot,\cdot)$ is the incomplete gamma function.[4] The ZZB is also shown in Fig. 7.4. It matches the CRLB at high SNR and the a priori upper bound at low SNR and thus provides a better representation of achievable precision than the CRLB alone. The Weiss-Weinstein analysis provides another set of bounds that incorporate the a priori and Cramèr-Rao bounds and also provide tighter bounds in the transition regions between them (Weiss, 1986).

Comparison of Eq. (7.48) to Eqs. (7.43) and (7.44) and to the discussion above of the a priori bound shows that Eq. (7.48) can be rewritten as

$$\sigma_{\hat{t}_0}^2 \geq APB \cdot \mathrm{erfc}\left(\sqrt{\frac{\chi_{\mathrm{out}}}{2}}\right) + CRLB \cdot I\!\left(\frac{3}{2}, \frac{\chi_{\mathrm{out}}}{2\sqrt{2}}\right) \quad \text{(complex signals)}$$

$$\sigma_{\hat{t}_0}^2 \geq APB \cdot \mathrm{erfc}\left(\sqrt{\frac{\chi_{\mathrm{out}}}{2}}\right) + CRLB \cdot I\!\left(\frac{3}{2}, \frac{\chi_{\mathrm{out}}}{4\sqrt{2}}\right) \quad \text{(real signals)}$$

(7.49)

where APB and $CRLB$ are the applicable a priori and Cramèr-Rao bounds, respectively. The erfc($\cdot$) and $I(\cdot,\cdot)$ functions provide the weightings that transition between these two bounds as the output SNR changes. This observation will be used to quickly write the ZZB for estimating sinusoid parameters in the next section.

Figure 7.7 compares the CRLB and simulated delay estimation precision for two different waveforms. The gray curve and bound are for a complex-valued LFM pulse with $\beta\tau = 100$ as considered above. The black curve and bound are for a real-valued trapezoidal pulse with an arbitrary but fixed complex phase shift and 20 percent rise and fall times (see Fig. 7.9a) of the same length. The RMS bandwidth of the LFM waveform is then 49.1 times

---

[4]The definitions of both functions in this equation are consistent with those used in Chap. 6 and in MATLAB®.

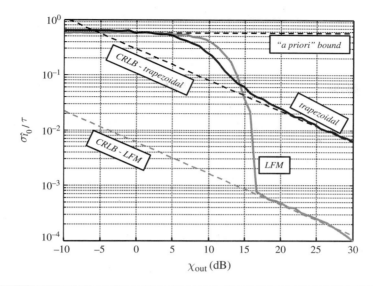

FIGURE 7.7 Comparison of maximum likelihood delay estimation error for a real trapezoidal pulse and a complex LFM pulse of the same length in complex noise. The time-bandwidth product of the LFM pulse is 100.

that of the trapezoidal pulse. The noise is complex in both cases. In this case, both are normalized to the pulse length, making the normalized variance 100 times smaller for the LFM pulse than in Fig. 7.6, but allowing direct comparison of the two waveforms. The ZZBs are not shown to avoid cluttering the graph. Consistent with the CRLB, the precision achieved with the LFM for a given SNR in the high-SNR region exceeds that of the trapezoidal pulse by about 50 times due to the greater RMS bandwidth.

A limiting factor at high SNR that has not yet been considered is the quantization of the time delay axis. In the SNR regime where it is effective, the MLE locates the target time delay to sample n_0, equivalent to time $n_0 T_s$. Even if n_0 is the correct location of the highest amplitude sample of the signal component at the filter output, the implied time estimate $n_0 T_s$ can differ from the actual peak location t_0 by up to $\pm T_s/2$. Modeling this difference as an independent uniformly distributed error source establishes a lower *sampling bound* (SB) of $1/12$ samples^2 or $T_s^2/12$ seconds^2 on the error variance. This will generally be higher than the CRLB unless the data is highly oversampled or the estimator is an analog (continuous time) implementation. Suppose the matched filter output is oversampled by a factor k_{os} compared to the Nyquist rate so that $T_s = 1/k_{os}\beta$. The SB on time delay estimation variance for real or complex signals is then

$$\sigma_{\hat{t}_0}^2 \geq \frac{1}{12 k_{os}^2 \beta^2} \quad \text{seconds}^2 \qquad (7.50)$$

Equating (7.50) with Eq. (7.43) or (7.44) shows the degree of oversampling needed to achieve the CRLB is $k_{os} = \pi(\beta_{\text{rms}}/\beta)\sqrt{\chi_{\text{out}}/3}$ in the real case and $\pi(\beta_{\text{rms}}/\beta)\sqrt{2\chi_{\text{out}}/3}$ in the complex case. For a complex-valued LFM waveform the required oversampling factor is about 7.4 at an SNR of 20 dB and 23.4 at 30 dB.

Figure 7.8 illustrates this effect for the same complex LFM pulse, MLE, and normalization to the time resolution used for Fig. 7.6. The matched filter output is oversampled by a

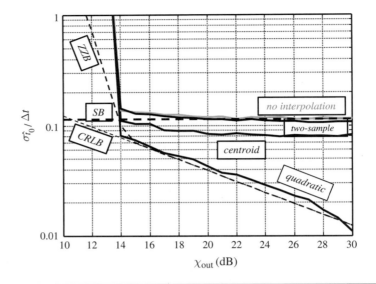

FIGURE 7.8 Effect of sampling rate limitations and interpolation techniques on high-SNR time delay estimation error. See text for details.

factor $k_{os} = 2.5$ so that the sampling bound on precision is $0.116/\beta = 0.116 T_s$. The analysis above shows (and the figure confirms) that the SB dominates the CRLB for $\chi_{out} > 10.1$ dB, but as can also be seen from the figure, in this example the SB does not dominate the ZZB until $\chi_{out} > 14$ dB. For values of SNR in this regime, the gray MLE curve labeled "no interpolation" hews closely to the SB.

Time delay measurement accuracy could be improved by oversampling to a greater degree, but the rate required to achieve the CRLB at SNRs greater than about 20 dB can be quite large, as seen above. A common alternative approach is to interpolate between the samples in the vicinity of a measured peak to refine the estimated peak location, as discussed previously in Sec. 5.4.6 for reducing discrete Fourier transform (DFT) straddle errors. For the LFM waveform, the peak of the magnitude of the signal component at the matched filter output closely approximates a sinc function, which in turn can be approximated near its peak at $t = t_0$ by a parabola. Denote the magnitude of the matched filter output as $y(t)$ with a peak occurring at time t_0. The largest sample actually measured is at sample n_0 corresponding to time $n_0 T_s$. Denote the value y at that time as $y[n_0] = y_0$, and the value of the two adjacent samples as $y[n_0 - 1] = y_{-1}$ and $y[n_0 + 1] = y_{+1}$. The actual peak is at location $t_0 = (n_0 + \Delta n)T_s$, where Δn is a fractional sample value between -1 and $+1$. Applying the interpolator of Eq. (5.92) gives

$$\Delta n = \frac{1}{2} \cdot \frac{y_{+1} - y_{-1}}{2y_0 - y_{-1} - y_{+1}}$$
$$\hat{t}_0 = (n_0 + \Delta n)T_s \quad (7.51)$$

The curve marked "quadratic in Fig. 7.8 shows the result of applying this technique to the coarsely sampled data that produced the "SB" (sampling bound) curve. The interpolation provides precision close to the CRLB up to at least 30 dB with only 2.5 times oversampling.

Other interpolation techniques can be used. One example is the centroid of the apparent peak and its two nearest neighbors, which is given by

$$n_0 + \Delta n = \frac{\sum_{n=n_0-1}^{n_0+1} n \cdot y[n]}{\sum_{n=n_0-1}^{n_0+1} y[n]} = \frac{(n_0-1)y_{-1} + n_0 y_0 + (n_0+1)y_{+1}}{y_{-1} + y_0 + y_{+1}} \quad (7.52)$$

The result of centroid interpolation is also shown in Fig. 7.7. While it improves the precision below the sampling bound, it does not approach the CRLB. Still another is the two-sample estimator, which uses only the peak sample and the larger of its two nearest neighbors (Macleod, 1998):

$$n_0 + \Delta n = n_0 + \text{sgn}(y_{+1} - y_{-1}) \left[\frac{\max(y_{+1}, y_{-1})}{y_0 + \max(y_{+1}, y_{-1})} \right] \quad (7.53)$$

where sgn(·) is the sign or signum function. The two-sample estimator is surprisingly accurate on noise-free data but is ineffective on noisy data as seen in Fig. 7.8.

There are many additional issues to be considered in using peak interpolators. For example, parabolic interpolator coefficients should be modified to obtain the best performance if the data is windowed for range sidelobe control (which alters the peak shape) or if interpolation is applied to complex data instead of real data (Agrež, 2002; Jacobsen and Kootsookos, 2007; Lyons, 2011). In addition, both the quality of interpolation generally and the relative effectiveness of various interpolators varies widely with oversampling rate. Another factor is that these techniques are designed for isolated peaks not contaminated by other nearby peaks. Also, most interpolators are nonlinear and introduce bias (degraded accuracy) in the estimate, although the bias is generally small in the high-SNR region where interpolation is useful. The performance of several of these estimators is compared in Candan (2011) and an extension to remove bias in one particular estimator is described in Candan (2013). Taken as a whole, interpolation procedures can be effective in improving precision but must be carefully selected and their details optimized for each particular data acquisition protocol.

Another common time delay estimator frequently used with approximately rectangular pulses is a leading edge threshold detector. This is a noncoherent technique that operates on the magnitude of the receiver output and is suitable only for high-SNR conditions. Consider a rectangular pulse filtered through a receiver filter of nominal bandwidth β Hz. The filter output can be approximated by a trapezoidal pulse with a rise time of $t_r \approx 1/\beta$ seconds, as shown in Fig. 7.9a. In a high-SNR environment, time delay could be estimated by measuring the time at which the received noisy pulse first crosses an amplitude threshold such as 50 percent of the peak amplitude. The addition of the filtered noise to the filtered pulse will perturb the time at which the threshold crossing occurs.

The behavior of the noise-free and noisy output pulses of Fig. 7.9a is an example. The energy SNR is 20 dB. The correlation interval of the noise will approximately equal the pulse rise time t_r, which is expressed as a fraction α of the total pulse length, $t_r = \alpha \tau$; in this example, $\alpha = 0.1$. This shape is the result of convolving a square pulse of original duration $(1 - 2\alpha)\tau$ with a filter having a rectangular impulse response of width $\alpha \tau$ seconds, and therefore a bandwidth of approximately $1/\alpha \tau$ Hz.

Suppose the threshold crossing time in the absence of noise is t_0. As suggested in the expanded view of the region around the amplitude threshold (here set at 50 percent of peak) in Fig. 7.9b, the slope of the noisy pulse is similar to that of the noise-free pulse so that the

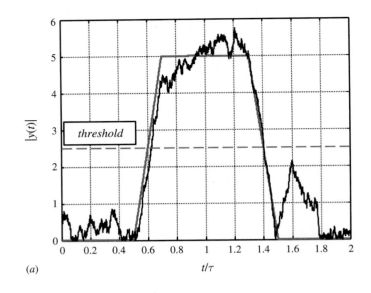

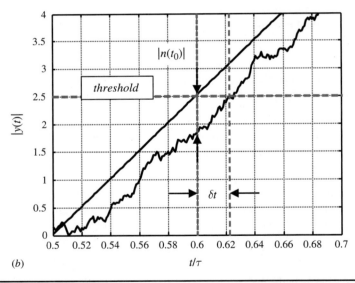

FIGURE 7.9 Leading edge detection for a trapezoidal pulse. (a) Trapezoidal pulse with rise time $t_r = 20$ percent of pulse length τ; threshold is set at 50 percent of peak noise-free pulse amplitude. (b) Close-up of threshold crossing region for noise-free and noisy pulse.

change in threshold crossing time, Δt, is related to the noise amplitude at t_0 approximately according to

$$\frac{|n(t_0)|}{\delta t} = \frac{A}{t_r} \cdot \frac{A}{\alpha \tau} \quad \Rightarrow \quad \delta t = \frac{\alpha \tau}{A} |n(t_0)| \tag{7.54}$$

If the bias in the estimate can be assumed small, the mean-square error in the time delay estimate, $E\{(\delta t)^2\}$, can be taken as an estimate of the variance of the delay error:

$$\sigma_{\delta t}^2 \approx \frac{\alpha^2 \tau^2}{A^2} \sigma_w^2 = \frac{\alpha^2 \tau^2}{\chi_{\text{out}}} \qquad (7.55)$$

Equation (7.55) is referred to as the *mean-square error* (MSE) *bound*. The SNR χ_{out} is the peak output voltage SNR A^2/σ_w^2, more appropriate than the waveform energy SNR since a matched filter is not used here.

It is interesting to compare the MSE bound to the general complex signal CRLB for time delay estimation of Eq. (7.43). It can be shown (see Prob. 22) that the RMS bandwidth of the trapezoidal pulse is $\left(\pi\tau\sqrt{2\alpha(1-4\alpha/3)}\right)^{-1}$. It follows that the ratio of the complex CRLB precision to the MSE bound is a factor of $\sqrt{(1-4\alpha/3)/4\alpha}$, which varies from 1.47 to 0.41 as α varies from 0.1 (nearly square pulse) to its maximum of 0.5 (triangular pulse). Thus in some cases, the CRLB is *larger* than the MSE bound! (For the value $\alpha = 0.2$ used in these examples, the two are nearly equal, with the MSE bound slightly smaller at 96 percent of the CRLB.) Since the CRLB is the minimum variance of any *un*biased estimator, the leading edge estimator must be biased. This can be confirmed by considering the exact PDF of the threshold crossing time. In the notation used here, the MSE and bias in estimating t_0 can be shown to be (Bar-David and Anaton, 1981)

$$E\{(\delta t)^2\} = \frac{\alpha^2 \tau^2}{\chi_{\text{out}}}, \qquad E\{\delta t\} = -\frac{\alpha \tau}{\chi_{\text{out}}} \qquad (7.56)$$

The MSE matches the earlier heuristic MSE bound of Eq. (7.55). Equation (7.56) confirms that the estimate is biased and shows that the bias asymptotically approaches zero at high SNR.

Figure 7.10 shows the precision, normalized to pulse length τ, for a simulated leading edge estimator with $\alpha = 0.1$ and the threshold at 50 percent of the noise-free peak amplitude. The total data record length was 3τ seconds. Also shown are the a priori bound, the MSE bound of Eq. (7.55), and the CRLB. The leading edge estimate approaches the MSE bound for SNRs of 20 dB or more. The MSE bound is below the CRLB for small α as discussed above. Also shown again as the solid gray line is the MLE for this problem, which approaches the CRLB but does not achieve the MSE bound.

One might expect that at low SNR neither the CRLB or MSE bound would apply because the a priori bound would become dominant. Figure 7.10 shows that the a priori bound does not control the low-SNR behavior of the leading-edge estimator. As the SNR decreases below 20 dB, the variance of the leading edge estimate starts to rise as in earlier examples, but then decreases rapidly. This behavior occurs because at low SNR the noise alone is likely to cross the amplitude threshold in the first few moments of the data record, regardless of the actual pulse position. The "arrival time" is then consistently reported as very near to the beginning of the data. While this estimate is badly biased, it varies little so the variance becomes small. This behavior can be avoided if the approximate position of the echo is known from a tracking loop so that such anomalous results can be ignored. Changing the level of the threshold alters somewhat the SNR at which the transition to CRLB-limited performance occurs. A higher threshold moves the "hump" in the data to the left by a few decibels, for instance. It is clear, however, that this estimator is only suitable for high-SNR signals.

Another common class of time delay or range estimators are various *split-gate* or *early-late gate* techniques. These attempt to find an estimated delay such that the energy in a finite window to either side of the estimated delay is approximately equal. As an example, for a rectangular pulse the matched filter output is a triangle. If the estimated delay coincides

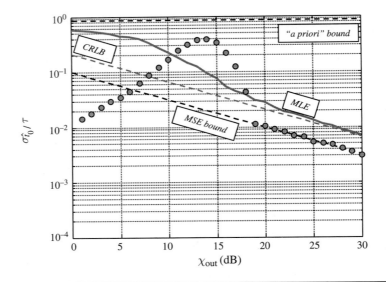

FIGURE 7.10 Time delay estimation error for leading-edge detection of a trapezoidal pulse. Circles indicate leading edge estimate. The solid gray line is the maximum likelihood estimate. See text for details.

with the center of the triangle, then the sum of the noisy triangle voltages (or their squares) over one pulse length to either side is likely to be nearly equal. An analysis of various implementations of this idea is given in Peebles (1998).

7.2.2 Doppler Signal Estimators

CRLB for Sinusoid Parameters

A general uniformly sampled complex sinusoid $s[n]$ in additive complex WGN $w[n]$ of power σ_w^2, using normalized cyclical frequency units is of the form

$$\begin{aligned} x[n] &= s[n] + w[n] \\ &= A\exp[j(2\pi f_0 n + \varphi)] + w[n] \\ &= \tilde{A}\exp(j2\pi f_0 n) + w[n], \quad 0 \le n \le N-1 \end{aligned} \quad (7.57)$$

where $\tilde{A} = A\exp(j\varphi)$ is the complex amplitude of the sinusoid. This signal is a good model for the slow-time signal produced by a target with a constant radial velocity and thus constant Doppler shift. Note that the SNR for this signal is $\chi = A^2/\sigma_w^2$.

There are three simultaneous parameters to be estimated: A, f_0, and φ. Finding the CRLB for each requires computing the Fisher information matrix in the form of Eq. (A.85). The (i,j)th element of the matrix is

$$[\mathbf{I}(\Theta)]_{ij} = \frac{2}{\sigma_w^2}\text{Re}\left\{\sum_{n=0}^{N-1}\left[\frac{\partial s[n;\Theta]}{\partial \Theta_i}\right]^*\left[\frac{\partial s[n;\Theta]}{\partial \Theta_j}\right]\right\} \quad (7.58)$$

where Θ is the three-element parameter vector $[A, f_0, \varphi]^T$. The derivatives needed are easily obtained:

$$\frac{\partial s[n;\Theta]}{\partial A} = \frac{\partial}{\partial A}\{A\exp[j(2\pi f_0 n + \varphi)]\} = \exp[j(2\pi f_0 n + \varphi)]$$

$$\frac{\partial s[n;\Theta]}{\partial f_0} = j2\pi n A \exp[j(2\pi f_0 n + \varphi)] \qquad (7.59)$$

$$\frac{\partial s[n;\Theta]}{\partial \varphi} = jA \exp[j(2\pi f_0 n + \varphi)]$$

The (1, 1) element of $\mathbf{I}(\Theta)$ is, from Eq. (7.58),

$$[\mathbf{I}(\Theta)]_{11} = \frac{2}{\sigma_w^2} \text{Re}\left\{\sum_{n=0}^{N-1}\left[\frac{\partial s[n;\Theta]}{\partial A}\right]^*\left[\frac{\partial s[n;\Theta]}{\partial A}\right]\right\}$$

$$= \frac{2}{\sigma_w^2}\sum_{n=0}^{N-1}\left|\frac{\partial s[n;\Theta]}{\partial A}\right|^2 = \frac{2}{\sigma_w^2}\sum_{n=0}^{N-1}|\exp[j(2\pi f_0 n + \varphi)]|^2 = \frac{2N}{\sigma_w^2} \qquad (7.60)$$

It is easy to see that the (1, 2), (1, 3), (2, 1), and (3, 1) elements are all zero because of the Re{·} operator in Eq. (7.58) and the fact that the derivatives with respect to f_0 and φ are purely imaginary. The remaining elements are easily found from the derivatives above, giving the complete matrix as

$$[\mathbf{I}(\Theta)]_{i,j} = \frac{2}{\sigma_w^2}\begin{bmatrix} N & 0 & 0 \\ 0 & A^2\sum_n(2\pi n)^2 & A^2\sum_n 2\pi n \\ 0 & A^2\sum_n 2\pi n & NA^2 \end{bmatrix}$$

$$= \frac{2}{\sigma_w^2}\begin{bmatrix} N & 0 & 0 \\ 0 & \left(\frac{2\pi^2 A^2}{3}\right)N(N-1)(2N-1) & \pi A^2 N(N-1) \\ 0 & \pi A^2 N(N-1) & NA^2 \end{bmatrix} \qquad (7.61)$$

The last step uses the identities

$$\sum_{n=0}^{N-1} n = \frac{N(N-1)}{2}, \quad \sum_{n=0}^{N-1} n^2 = \frac{N(N-1)(2N-1)}{6} \qquad (7.62)$$

The CRLB for each parameter is the corresponding diagonal element of $\mathbf{I}^{-1}(\Theta)$. The calculation is straightforward, if tedious, and gives the results

$$\sigma_{\hat{A}}^2 \geq \frac{\sigma_w^2}{2N} \longrightarrow \sigma_{\hat{A}/A}^2 \geq \frac{\sigma_w^2}{2NA^2} = \frac{1}{2N\chi} \quad \text{(dimensionless)}$$

$$\sigma_{\hat{f}_0}^2 \geq \frac{6}{(2\pi)^2 \chi N(N^2-1)} \xrightarrow{N\text{ large}} \frac{6}{(2\pi)^2 N^3 \chi} \quad \text{(cycles/sample)}^2 \qquad (7.63)$$

$$\sigma_{\hat{\varphi}}^2 \geq \frac{2N-1}{\chi N(N+1)} \xrightarrow{N\text{ large}} \frac{2}{N\chi} \quad \text{rad}^2$$

As usual, the CRLBs for frequency, phase, and relative amplitude are inversely proportional to SNR as shown above. The "large N" approximations shown are accurate to within 10 percent for $N \geq 10$ (frequency) and $N \geq 13$ (phase). It is interesting to note that the CRLBs for amplitude and phase decrease asymptotically as $1/N$ while that for frequency improves at the much faster rate of $1/N^3$ as the amount of data increases.

Two alternative forms for the frequency CRLB merit mention. The frequency in hertz is $F_0 = f_0/T_s$, where T_s is the sampling interval of interest. The data duration in seconds is NT_s, so the expected frequency resolution of the discrete-time Fourier transform (DTFT) in hertz is $\Delta F = 1/NT_s$. Applying these in the asymptotic form of Eq. (7.63) gives

$$\sigma_{\hat{F}_0}^2 \geq \frac{6}{(2\pi)^2 N^3 T_s^2 \chi} = \frac{6(\Delta F)^2}{(2\pi)^2 N \chi} \text{ Hz}^2 \quad (7.64)$$

The last form makes it clear that the N^3 improvement in the frequency estimation CRLB is due largely to the implied improvement in frequency resolution, which accounts for an N^2 term. The remaining factor of N represents the usual SNR improvement due to coherent integration that is also seen in the phase and amplitude CRLBs.

Having multiple simultaneously unknown parameters can result in CRLBs that are larger than they would be if all but one of the parameters is known. The interaction between parameters is manifested in nonzero off-diagonal elements of the Fisher information matrix. For the sinusoidal parameter estimation problem, it can be seen (see Prob. 30) that if the other two parameters are known, the CRLB of the remaining unknown parameter is equal to or tighter than the results of Eq. (7.63). Specifically, the CRLB for amplitude is the same, while in the limits of large N the CRLBs for frequency and phase are each four times smaller.

MLE for Sinusoid Parameters

The next step is to develop the MLE for the sinusoid parameters. This requires maximizing the likelihood function

$$\ell(A, \varphi, f_0 | \mathbf{x}) = \ell(\tilde{A}, f_0 | \mathbf{x}) = \frac{1}{(\pi \sigma_w^2)^N} \exp\left[-\frac{1}{\sigma_w^2} \sum_{n=0}^{N-1} |x[n] - \tilde{A} \exp(j2\pi f_0 n)|^2\right] \quad (7.65)$$

with respect to all three parameters. (In Eq. (7.65) the complex amplitude $\tilde{A}$ has been reintroduced for convenience.) Equivalently, the exponent term

$$J(\tilde{A}, f_0) \equiv \sum_{n=0}^{N-1} |x[n] - \tilde{A} \exp(j2\pi f_0 n)|^2 \quad (7.66)$$

must be minimized.

Begin by minimizing J with respect to $\tilde{A} = \tilde{A}_R + j\tilde{A}_I$ with f_0 temporarily assumed known. The partial derivative with respect to $\tilde{A}_R$ is (leaving $x[n]$ and $\exp(j2\pi f_0 n)$ in complex form)

$$\begin{aligned}\frac{\partial J}{\partial \tilde{A}_R} &= \sum_{n=0}^{N-1} \frac{\partial}{\partial \tilde{A}_R}\{[x[n] - (\tilde{A}_R + j\tilde{A}_I)\exp(j2\pi f_0 n)][x^*[n] - (\tilde{A}_R - j\tilde{A}_I)\exp(-j2\pi f_0 n)]\} \\ &= \sum_{n=0}^{N-1}\{[x[n] - (\tilde{A}_R + j\tilde{A}_I)\exp(j2\pi f_0 n)][-\exp(-j2\pi f_0 n)] + \cdots \\ &\quad \cdots + [x^*[n] - (\tilde{A}_R - j\tilde{A}_I)\exp(-j2\pi f_0 n)][-\exp(j2\pi f_0 n)]\}\end{aligned} \quad (7.67)$$

Setting Eq. (7.67) equal to zero leads to

$$0 = \frac{\partial J}{\partial \tilde{A}_R} = \sum_{n=0}^{N-1} \{2\tilde{A}_R - x[n]\exp(-j2\pi f_0 n) - x^*[n]\exp(j2\pi f_0 n)\} \quad (7.68)$$
$$= 2N\tilde{A}_R - \sum_{n=0}^{N-1} 2\operatorname{Re}\{x[n]\exp(-j2\pi f_0 n)\}$$

and finally, along with a similar development for minimization with respect to $\hat{A}_I$ (see Prob. 32),

$$\left.\begin{array}{l}\widehat{\tilde{A}_R} = \operatorname{Re}\left\{\dfrac{1}{N}\sum_{n=0}^{N-1} x[n]\exp(-j2\pi f_0 n)\right\} \\ \widehat{\tilde{A}_I} = \operatorname{Im}\left\{\dfrac{1}{N}\sum_{n=0}^{N-1} x[n]\exp(-j2\pi f_0 n)\right\}\end{array}\right\} \Rightarrow \hat{\tilde{A}} = \dfrac{1}{N}\sum_{n=0}^{N-1} x[n]\exp(-j2\pi f_0 n) \quad (7.69)$$

where $\hat{\tilde{A}}$ denotes the estimated value of $\tilde{A}$. Notice that if f_0 is indeed known, then the MLE for $\tilde{A}$ is simply the DTFT of the data evaluated at f_0, a very satisfying result.

However, with f_0 not known, J must be minimized with respect to f_0. Expanding the definition of J in Eq. (7.66) gives

$$J(\tilde{A}, f_0) = \sum_{n=0}^{N-1} [x[n] - \hat{\tilde{A}}\exp(j2\pi f_0 n)][x^*[n] - \hat{\tilde{A}}\exp(-j2\pi f_0 n)]$$
$$= \sum_{n=0}^{N-1} |x[n]|^2 - (\hat{\tilde{A}})^* \sum_{n=0}^{N-1} x[n]\exp(-j2\pi f_0 n) \cdots \quad (7.70)$$
$$\cdots - \hat{\tilde{A}} \sum_{n=0}^{N-1} x^*[n]\exp(+j2\pi f_0 n) + \sum_{n=0}^{N-1} |\hat{\tilde{A}}|^2$$

Recognizing the sums in the two middle terms as $N\hat{\tilde{A}}$ and $N\hat{\tilde{A}}^*$ [see Eq. (7.69)] reduces this to

$$J(\tilde{A}, f_0) = \sum_{n=0}^{N-1} |x[n]|^2 - N|\hat{\tilde{A}}|^2$$
$$= \sum_{n=0}^{N-1} |x[n]|^2 - \frac{1}{N}\left|\sum_{n=0}^{N-1} x[n]\exp(-j2\pi f_0 n)\right|^2 \quad (7.71)$$

The first term of Eq. (7.71) does not depend on f_0. Therefore, J is minimized by maximizing the second term, which is just the magnitude-squared of the DTFT of the data.[5] Again very sensibly, $\widehat{f_0}$ is found by computing the DTFT of the data and finding the frequency at which its squared magnitude is largest.

To summarize, the MLEs of the amplitude, frequency, and phase of a complex sinusoid in complex WGN are found by the following sequence of operations:

1. Compute the DTFT $X(f)$ of the data $x[n]$.

2. $\widehat{f_0}$ is the frequency at which the peak magnitude of the DTFT occurs, $\widehat{f_0} = \arg\max_f \{|X(f)|^2\}$.

---

[5]This function is often called the *periodogram* of the data $x[n]$.

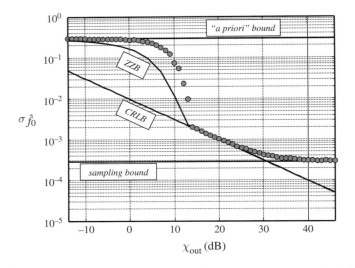

FIGURE 7.11 Behavior of simulated maximum likelihood normalized frequency estimation error. A $K = 1000$ point DFT was applied to sinusoids with $M = 40$ time samples.

3. The MLE of the complex amplitude is $\hat{A} = X(\hat{f_0})/N$. The MLEs $\hat{A}$ and $\hat{\varphi}$ of the magnitude and phase are the magnitude and angle of $\hat{\tilde{A}}$.

Because the MLE for frequency estimation using N samples is a DFT of size $K = N$, it is now known that there will be a corresponding integration gain of a factor of N.[6] The SNR at the DFT output will be $\chi_{\text{out}} = N\chi$.[7] The CRLBs of Eq. (7.63) can be rewritten [also using Eq. (7.64)] in the possibly more useful forms

$$\sigma^2_{\hat{A}/A} \geq \frac{\sigma^2_w}{2NA^2} = \frac{1}{2\chi_{\text{out}}} \quad \text{(dimensionless)}$$

$$\sigma^2_{\hat{f_0}} \geq \frac{6}{(2\pi)^2 \chi_{\text{out}}(N^2-1)} \xrightarrow{N \text{ large}} \frac{6}{(2\pi)^2 N^2 \chi_{\text{out}}} \quad \text{(cycles/sample)}^2 \quad (7.72)$$

$$\sigma^2_{\hat{F_0}} \geq \frac{6}{(2\pi)^2 N^2 T_s^2 \chi_{\text{out}}} = \frac{6(\Delta F)^2}{(2\pi)^2 \chi_{\text{out}}} \quad \text{Hz}^2$$

$$\sigma^2_{\hat{\varphi}} \geq \frac{2N-1}{\chi_{\text{out}}(N+1)} \xrightarrow{N \text{ large}} \frac{2}{\chi_{\text{out}}} \quad \text{rad}^2$$

Figures 7.11 to 7.13 illustrate the behavior of the error in the MLEs as a function of the DFT output SNR for frequency, phase, and relative amplitude, respectively. All three figures are based on simulations of the DFT-based MLE applied to complex sinusoids of length $N = 40$ samples with random frequencies and initial phases. A large DFT of size $K = 1000$

---

[6]Straddle loss will reduce this gain factor.
[7]As discussed in Chap. 5, the output SNR will still be $N\chi$ if the DFT size K is chosen to be greater than N ("zero padded").

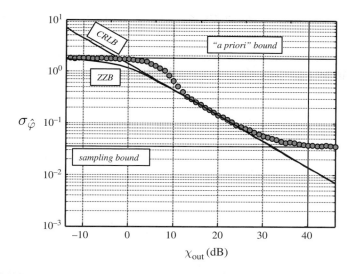

FIGURE 7.12 Behavior of simulated maximum likelihood phase estimation error in radians. A $K = 1000$ point DFT was applied to sinusoids with $N = 40$ time samples.

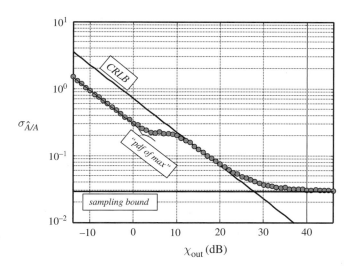

FIGURE 7.13 Behavior of simulated maximum likelihood relative amplitude estimation error. A $K = 80$ point DFT was applied to sinusoids with $N = 40$ time samples.

was used with the frequency and phase estimates, while $K = 80$ for the amplitude estimation in order to have a sampling bound high enough to have an impact within the SNR range shown. As with range estimation, in all three cases the measurement precision follows the CRLB only at relatively high SNRs. At low SNR an a priori bound dominates the estimates as the signal becomes essentially all noise, while at high SNR the precision is limited by a sampling bound based on sampling density.

A Priori and Sampling Bounds for Sinusoid Parameters Using the MLE

In the low-SNR case the signal peak will be swamped by the noise, so the MLE for frequency will return simply the location of the peak of the DFT of the noise. The peak is equally likely to occur anywhere in the normalized frequency interval $[-0.5, +0.5]$. Modeling the estimated normalized frequency $\hat{f}_0$ as a uniform random variable over that interval, the a priori bound on its variance is simply $1/12$ and the precision is the square root of this, $1/\sqrt{12}$. Similarly, the phase at the DFT peak is uniform over the full range $[-\pi, \pi]$ so that the a priori bound on phase precision is $2\pi/\sqrt{12}$.

Unlike frequency and phase, there is no hard limit to the values that the estimated amplitude can take on as the noise becomes large relative to the signal. In this low-SNR region the input signal is essentially all noise, so the MLE becomes the peak of the DFT of white noise. This peak is not guaranteed to occur at the frequency bin corresponding to the weak sinusoidal signal; rather, it is equally likely to occur at any frequency bin.

One estimate of relative amplitude precision in the low-SNR regions begins by noting that each sample of the K-point DFT of $x[n]$ is the sum of N samples of complex WGN with variance σ_w^2 weighted by the DFT kernel samples $\exp(-j2\pi k/K)$. Because the magnitudes of those weights are all unity, the DFT samples are also complex Gaussian with variance $N\sigma_w^2/2$. The magnitude of the DFT samples is therefore a Rayleigh-distributed RV with variance $(1-\pi/4)N\sigma_w^2$. The MLE scales the DFT by $1/N$ so the variance of the relative amplitude of a DFT sample becomes $(1-\pi/4)\sigma_w^2/N$. Even though it is not known which DFT sample will be the peak, for a white noise input the random peak location does not affect the amplitude variance.

The MLE variance value above could be used as a crude estimate of the low-SNR relative amplitude variance, but in practice it predicts too high a variance. Because the MLE is the maximum of the K DFT samples, the estimated amplitude variance will be the variance of that quantity. Denote the PDF of the magnitude of a scaled DFT sample as $p_{|X|}(|X|)$ and the corresponding cumulative distribution function as $P_{|X|}(|X|)$. For the scaled DFT of complex WGN these are (Richards, 2013)

$$p_{|X|}(|X|) = \frac{2N|X|}{\sigma_w^2} \exp\left[-\frac{N|X|^2}{\sigma_w^2}\right], \quad |X| > 0$$
$$P_{|X|}(|X|) = 1 - \exp\left[-\frac{N|X|^2}{\sigma_w^2}\right], \quad |X| > 0 \tag{7.73}$$

Now let z be the maximum of the K samples of the DFT of the N-point input. The PDF of the maximum of K independent samples of $|X|$ is (Levanon, 1988, pp. 258–260)

$$p_z(z) = K \cdot p_{|X|}(|X|)\left[P_{|X|}(|X|)\right]^{K-1}, \quad |X| > 0 \tag{7.74}$$

For $K = N$, the DFT samples are uncorrelated. It is easy to show that DFT samples are correlated when $K > N$, even for a white noise input (see Prob. 33). In fact, for $K > N$, the effective number of statistically independent samples remains approximately equal to N. Consequently, a good estimate of the low-SNR PDF of the MLE is obtained by setting $K = N$ and using Eq. (7.73) in Eq. (7.74) to compute $p_z(z)$. Unfortunately, this does not result in a simple formula for the variance of z, but it can be computed numerically from $p_z(z)$ to establish a bound on amplitude precision. The solid line labeled "pdf of max" in Fig. 7.13 is a rough numerical estimate of the precision of the MLE estimate of relative amplitude, showing that it matches the actual low-SNR behavior well.[8]

---

[8]Numerical difficulties limit the pdf-of-max bound calculation to the low SNR region only.

Note that the "pdf of max" bound is tighter than the CRLB. This implies that the MLE is biased, significantly so in the low-SNR region. Also, while there is no upper limit to the bound as the SNR decreases because the noise simply gets larger and larger, in practice there will be a limit based on the maximum representable signal value. That value will be determined by the number of bits and their format (fixed or floating point) used in the signal processor.

When the SNR is sufficiently high and the DFT size is relatively small, the DFT frequency sampling interval limits the precision of all three parameter estimates. Most obvious is the sampling bound on frequency estimation precision. A K-point DFT samples the normalized frequency axis in intervals of $1/K$ cycles per sample. For large SNR, this frequency quantization becomes the dominant factor. The frequency error is then modeled as a uniform RV over an interval of width $1/K$, so the sampling bound on the precision of the frequency estimate is simply $1/\sqrt{12}K$. This bound is also shown in Fig. 7.11.

The errors in the amplitude and phase estimates at high SNR are induced by the frequency estimation error. Recall that the DTFT of an N-point sinusoid like that of Eq. (7.57) with frequency f_0 and initial phase φ_0 in the absence of noise is

$$X(f) = A \exp(j\varphi_0) \frac{\sin[2\pi(f - f_0)N/2]}{\sin[2\pi(f - f_0)/2]} \exp[-j\pi(f - f_0)(N - 1)] \qquad (7.75)$$

When evaluated at the actual frequency f_0, $X(f_0) = NA\exp(j\varphi_0)$. Dividing by N and taking the magnitude and phase of $X(f_0)$ returns the correct amplitude and phase of the original sinusoid. However, if the DTFT is evaluated at a frequency $f_0 + \delta f$, a different amplitude and phase will be measured. The range of δf is $\pm 1/2K$ (one DFT bin). The resulting range of phase variations is $\mp\pi(N-1)/2K$, and for amplitude is $-A[N - \sin(\pi N/2K)/\sin(\pi/2K)]/N$. The phase error is expected to be uniform over its interval because it arises from a linear mapping of frequency error to phase error. This is not the case for the amplitude error, but when K is reasonably large compared to N, the amplitude error can be approximately modeled as uniformly distributed over its interval. The sampling bounds on phase and amplitude estimation precision are then $\pi(N-1)/K\sqrt{12}$ and $A[N - \sin(\pi N/2K)/\sin(\pi/2K)]/N\sqrt{12}$. The sampling bound for relative amplitude precision is $[N - \sin(\pi N/2K)/\sin(\pi/2K)]/N\sqrt{12}$.

The problem of estimating sinusoid frequency and the MLE that resulted is very similar in nature to that of maximum likelihood estimation of time delay. Both involve finding the peak of a relatively narrow waveform in noise. Consequently, the ZZB in the generalized form of Eq. (7.49) can be applied to this problem as well. The equivalent of χ_{out} for the frequency estimation MLE is $N\chi$. The Ziv-Zakai bound computed using this change and the CRLBs and a priori bounds computed above for frequency estimation is plotted on Figs. 7.11 and 7.12. It is not shown in Fig. 7.13 because the particular arguments in the erfc$(\cdot)$ and $I(\cdot,\cdot)$ weighting functions of Eq. (7.49) do not match well to the transition between the CRLB and the a priori "pdf of max" bounds in the amplitude estimation problem.

As with time delay estimation, interpolation techniques can be used to try to improve precision beyond the sampling bound at high SNR. In fact, many of the interpolation techniques have been developed primarily for the purpose of frequency estimation. Once an improved frequency estimate is obtained, the same logic used above to establish the sampling bounds can be used to infer improved phase and amplitude estimates. Figure 7.14 shows the results of quadratic interpolation for the case $N = 40$ and $K = 120$. All three parameters show significant precision improvements. The two-sample and centroid interpolation results are not shown because, similar to the time delay case in Fig. 7.8, they improve precision only slightly. These two interpolators are more effective for more coarse sampling, for example, $K = N$, but even then the quadratic interpolator produces better results. All of the previously discussed caveats on interpolation concerning the effects of windowing, real

454 Chapter Seven

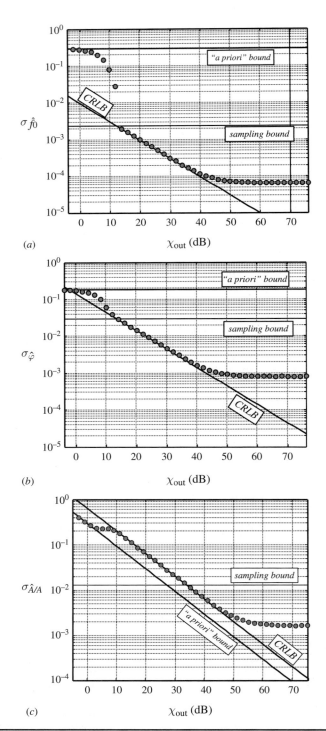

FIGURE 7.14 Improvement in precision using quadratic interpolation. $K = 120$ point DFT and $N = 40$. (a) Normalized frequency, (b) phase (radians), (c) relative amplitude.

versus complex data, oversampling rates, and bias apply in sinusoid parameter estimation as well.

Notice that both the time delay estimators of Sec. 7.2.1 and the sinusoid frequency, phase, and amplitude estimators of this section show the same basic CRLB behavior with respect to the SNR at the MLE output. Specifically, the precision of the estimator is limited by an a priori bound or other considerations at output SNRs of about zero decibels or less. As the output SNR reaches about 3 dB the precision starts to move toward the CRLB, following it closely when χ_{out} is around 13 dB or greater. At some relatively high SNR, a sampling bound prohibits further improvement in the precision. This suggests that the CRLB is a good guide only for MLE output SNRs between about 13 dB and the applicable sampling bound-limited region and that, perhaps surprisingly, this same rule applies to all the parameters considered so far.

CRLB for Range in Linear FMCW

In Chap. 3 it was shown that a sawtooth linear FMCW system measures range to a scatterer by computing the DFT of the primary beat frequency and finding its peak. Consequently, the preceding measurement precision analysis for the parameters of a sinusoid using the MLE are directly applicable, requiring only a rescaling from beat frequency to range. Recall that the relationship for linear FMCW between scatterer range R and the primary beat frequency F_b is $R = (cT/2\beta)F_b$. Equation (7.72) gave the CRLB for sinusoid frequency F_0 in hertz in terms of frequency resolution ΔF. The rescalings of both $\sigma_{F_0}^2$ and ΔF to meters cancel one another out, giving the CRLB on range as

$$\sigma_{\hat{R}}^2 \geq \frac{6(\Delta R)^2}{(2\pi)^2 \chi_{\text{out}}} \quad \text{m}^2 \tag{7.76}$$

This result is consistent with that in Scherr et al. (2012).

7.2.3 Angle Estimators

The remaining parameters to be estimated in order to locate a target in three-dimensional space and Doppler shift are the elevation and azimuth angles. It is sufficient to consider estimation of one angle; the results apply to both. There are two fundamental methods for estimating target angle relative to the antenna orientation in radar. The first uses phase measurements at multiple antenna phase centers in what is essentially phase interferometry, while the second uses multiple amplitude measurements in a process known as *lobing* or *lobe switching*. The choice between them depends on the type of antenna available and the data collection protocol. Some techniques are blends of the two basic approaches.

Phase-Based Angle Measurement

Consider the horizontally oriented uniform linear array shown in Fig. 7.15. Each of the gray triangles represents an antenna phase center having its own independent receiver. The phase centers could be either subarrays or individual elements of a phased array antenna. The number of phase centers N is assumed to be at least two. They are separated by a distance d. An electromagnetic plane wave of wavelength λ_t (frequency $\Omega_t = 2\pi/\lambda_t$ radians per second) arrives at the array from an angle of arrival (AOA) of θ radians relative to the array normal. Two successive planar phase fronts are shown, spaced by λ_t meters in the direction of propagation.

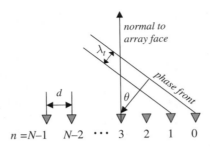

FIGURE 7.15 Plane wave impinging on a uniform linear array.

Consider a phase front arriving at phase center #0 at some time t. That phase front must travel $d\sin\theta$ meters in the direction of propagation in order to arrive at phase center #1, and that will take $d\sin\theta/c$ seconds. If the voltage signal observed at phase center #0 is of the form $\bar{y}_0(t) = A\exp[j(\Omega_t t + \varphi_0)]$ the signal at phase center #1 will be $\bar{y}_1(t) = A\exp(j[\Omega_t(t - d\sin\theta/c) + \varphi_0])$. More generally, the signal at the nth phase center will be

$$\bar{y}_n(t) = A\exp\{j[\Omega_t(t - nd\sin\theta/c) + \varphi_0]\}, \quad 0 \le n \le N-1 \tag{7.77}$$

Now consider the set of N voltage samples obtained by measuring the voltage at each phase center simultaneously at some time $t = t_0$ and assembling the results into a column vector **y**:

$$\begin{aligned} y[n] &\equiv \bar{y}_n(t_0) = A\exp\{j[\Omega_t(t_0 - nd\sin\theta/c) + \varphi_0]\} \\ &= \hat{A}\exp(-j\Omega_t nd\sin\theta/c) \\ &= \hat{A}\exp(-j2\pi nd\sin\theta/\lambda_t), \quad n = 0, \dots, N-1 \\ \mathbf{y} &= [y[0] \quad y[1] \quad \cdots \quad y[N-1]]^T \end{aligned} \tag{7.78}$$

The spatial phase history vector **y** is called a *spatial snapshot* of the signal arriving at the array. Define the normalized spatial frequency $k_\theta \equiv 2\pi d\sin\theta/\lambda_t$. $k\theta$ is the spatial frequency in radians per meter projected onto the face of the array, $(2\pi/\lambda_t)\sin\theta$, times the array sampling interval (the element spacing d), and has dimension of radians per sample. The spatial snapshot becomes

$$\mathbf{y} = \hat{A}[1 \quad \exp(-jk_\theta) \quad \exp(-j2k_\theta) \quad \cdots \quad \exp(-j(N-1)k_\theta)]^T \tag{7.79}$$

Equation (7.79) shows that the snapshot is a sampled complex sinusoid with normalized radian spatial frequency $-k_\theta$.

The radar's operating wavelength λ_t and the phase center spacing d are presumably known, so the AOA θ can be determined if k_θ can be measured; and since the snapshot is a sampled sinusoid, the results of the previous section on frequency estimation can be applied. Specifically, scaling Eq. (7.63) for a normalized radian frequency instead of the cyclical frequency f_0, the CRLB for estimating k_θ given a spatial snapshot with complex AWGN is

$$\sigma_{\hat{k}_\theta}^2 \ge \frac{6}{\chi N(N^2-1)} \xrightarrow{N \text{ large}} \frac{6}{N^3 \chi} \tag{7.80}$$

The CRLB for the AOA can be obtained using the transformation of parameters rule in Sec. A.3.2. Strictly speaking, a generalized version of this rule is needed for estimation of a transformed vector parameter; see Kay (1993) for details. However, in this sinusoid estimation problem the transformation from a CRLB on k_θ to one on θ simplifies to the scalar transformation case given in Eq. (A.72). Defining $\alpha = \lambda_t/2\pi d$ to temporarily simplify the notation,

$$\sigma_{\hat\theta}^2 \geq \left[\frac{\partial \theta}{\partial k_\theta}\right]^2 \sigma_{\hat{k_\theta}}^2 = \left[\frac{\partial}{\partial k_\theta}\sin^{-1}(\alpha k_\theta)\right]^2 \sigma_{\hat{k_\theta}}^2$$

$$= \frac{\alpha^2}{1-(\alpha k_\theta)^2}\sigma_{\hat{k_\theta}}^2 = \frac{\alpha^2}{1-\sin^2\theta}\sigma_{\hat{k_\theta}}^2 = \frac{\alpha^2}{\cos^2\theta}\sigma_{\hat{k_\theta}}^2 \qquad (7.81)$$

$$= \frac{\lambda_t^2}{(2\pi)^2 d^2 \cos^2\theta}\sigma_{\hat{k_\theta}}^2$$

Combining Eqs. (7.80) and (7.81) gives the CRLB on AOA,

$$\sigma_{\hat\theta}^2 \geq \frac{6\lambda_t^2}{(2\pi)^2 d^2 \cos^2(\theta)\chi N(N^2-1)} \xrightarrow{N\text{ large}} \frac{6\lambda_t^2}{(2\pi)^2 d^2 \cos^2(\theta)\chi N^3} \qquad (7.82)$$

Note that the accuracy depends on the AOA. In particular, signals arriving normal to the array ($\theta = 0$) are estimated with the finest precision, while the "precision" of the estimate of the AOA of those arriving from the end of the array ($\theta = \pm\pi/2$) becomes infinite. In this case, there is no difficulty estimating k_θ with the variance given in Eq. (7.80). However, the derivative of the mapping $\theta = \sin^{-1}(\lambda_t k_\theta/2\pi d) \to \infty$ as the argument tends to ± 1 so that a finite precision in k_θ becomes arbitrarily large in θ.

In practice, the AOA and therefore the precision cannot become unbounded. Ignoring signals arriving from behind the antenna, the AOA must be in the interval $[-\pi/2, +\pi/2]$. The worst-case variance is then $\pi^2/12$. This is then the a priori bound for low-SNR angle estimation.

Several substitutions put Eq. (7.82) in a more useful form. First, define the length of the array $D = (N-1)d$. Next, recall that the 3-dB beamwidth in the direction normal to the face of the array is $k\lambda_t/D$ for some k in the range of 0.89 to perhaps 2 or 3, depending on aperture weighting for sidelobe control. For an electronically scanned array antenna, the 3-dB beamwidth in a direction θ from the array normal then becomes $\theta_3 = k\lambda_t/D\cos\theta$ due to projection of the array length to the shorter effective length $D\cos\theta$ in the direction θ. Finally, note that χ is the SNR at an individual phase center; the coherently integrated SNR will be $\chi_{\text{out}} = N\chi$. With these substitutions, Eq. (7.80) becomes

$$\sigma_{\hat\theta}^2 \geq \frac{6\theta_3^2}{(2\pi)^2 \chi_{\text{out}}\left(\frac{N+1}{N-1}\right)k^2} \xrightarrow{N\text{ large}} \frac{6\theta_3^2}{(2\pi)^2 \chi_{\text{out}} k^2} \qquad (7.83)$$

This equation shows that the precision (square root of $\sigma_{\hat\theta}^2$) in estimating AOA is a fraction, determined by the integrated SNR, of its angular resolution. That fraction is independent of AOA. However, it must be remembered that θ_3 depends on the angle of arrival. The precision still becomes unbounded in theory for end-fire AOAs because the beamwidth $k\lambda_T/D\cos\theta$ becomes unbounded in theory, but in practice the beamwidth is limited to π radians (2π if allowing for signals from behind the array) and the estimate is still limited to $\pm\pi/2$.

Figure 7.16 plots the precision bound relative to the minimum 3-dB beamwidth (at AOA $\theta = 0°$) for a case where $k = 1.34$, $d = 5\lambda_t$ (implying a subarrayed antenna), $N = 10$, and the single-subarray (nonintegrated) SNR $= 0$ dB so that the integrated SNR is 10 dB. In this case, the minimum 3-dB beamwidth is $k\lambda_t/(N-1)d \approx 30$ mrad or 1.7°. For signals arriving within about $\pm 55°$ of the array normal, the precision is better than two-tenths of the minimum 3 dB beamwidth. The low-SNR precision bound is $\sqrt{\pi^2/12} \approx 0.91$ radian or 52°. This is 30.6 times the minimum beamwidth and is well above the range of the plot.

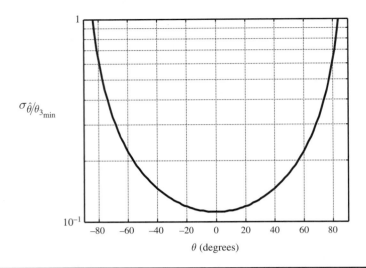

FIGURE 7.16 CRLB for relative precision of angle of arrival measured with a subarrayed antenna. See text for details.

If the phase center spacing $d > \lambda_t/2$, the Nyquist spatial sampling criterion of Chap. 3 will not be met. This will be the case if each phase center is a multielement subarray. In this event, $|k_\theta| > \pi$ for some range of AOAs, resulting in aliasing of the array factor portion of the antenna pattern (grating lobes) and making the estimate of θ ambiguous. For example, if $d = 2\lambda_t$ then AOAs of $0°$, $\pm 30°$, and $\pm 90°$ result in $k_\theta = 0$, $\pm 2\pi$, and $\pm 4\pi$, all of which alias to $k_\theta = 0$. If $\widehat{k_\theta} = 0$, the AOA could be any of those five angles. This concern is eased, if not eliminated, by the subarray antenna pattern, which attenuates some of the ambiguous lobes; see Bailey (2010) for more information.

Lobing-Based Angle Measurement

Consider the scenarios shown in Fig. 7.17. The radar antenna is pointed at an angle θ_0 relative to some reference axis. Now suppose the radar reports a target detection at some range. It is tempting to declare the angle to the target is θ_0 as shown in Fig. 7.17a, but in fact the target might actually be at some other angle $\theta_0 + \delta\theta$, as shown in Fig. 7.17b. It can reasonably be assumed that $\delta\theta$ is in the range of $\pm\theta_3/2$ so that the target is in the mainbeam if it is to have sufficient SNR to be detected. A method is sought to determine the target location within the mainbeam so that the angle can be measured to an accuracy significantly finer than the beamwidth.

This problem can be approached by noting that the target echo strength (power or voltage) measured in these two situations will not be the same because of the different gains of the two-way antenna pattern on and off the boresight. Suppose that once a target is detected at radar pointing angle θ_0, the procedure illustrated in Fig. 7.18 is followed. First, the antenna is steered to one side of the boresight by some amount $\Delta\theta$ radians and the target echo amplitude is measured again. The antenna is then steered $\Delta\theta$ radians to the opposite side of the original boresight and the echo amplitude measured once more. Either mechanical or electronic steering can be used, as appropriate for the particular antenna. The offset $\Delta\theta$ should be limited so that the target remains in the mainlobe in most cases. This issue is discussed in slightly more detail shortly.

By examining the pattern of target amplitude versus angle, the direction of the target relative to the original boresight can be determined. In the example of Fig. 7.18a, the target is to the

Measurements and Introduction to Tracking 459

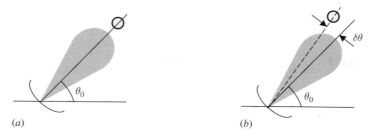

FIGURE 7.17 Effect of antenna gain pattern on received target echo amplitude. (a) Target on boresight of antenna. (b) Target off boresight.

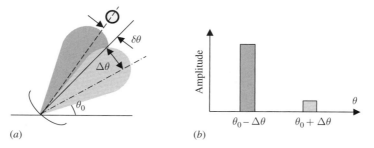

FIGURE 7.18 Sequential lobing concept. (a) New target measurements are taken with the antenna displaced from the nominal look direction. (b) Relative amplitude of resulting measurements.

left of the original boresight. When the antenna is shifted counterclockwise, the target is nearer the boresight and is therefore viewed with a higher antenna gain, resulting in a higher amplitude measurement. When the antenna is shifted clockwise, the antenna boresight moves away from the target, resulting in a lower amplitude measurement. This pattern of measurements, shown in Fig. 7.18b, indicates that the target must be counterclockwise from the original look direction. With knowledge of the antenna pattern, analysis of the relative amplitudes can be used to estimate by how much. It should be clear that if the target were clockwise from the original boresight, the pattern of the two measurements would be reversed, and if it was exactly on the boresight, the two measurements would produce equal amplitudes.

This process of making multiple measurements of target strength at offset angles relative to the nominal pointing angle and analyzing the results to indicate target angle is called *lobing*. When the measurements are made sequentially in time as described above it is called *sequential lobing*. Alternatively, antennas can be constructed that form multiple beams at once, providing multiple output signals corresponding to the original centered beams and to pairs of squinted beams in two orthogonal angle coordinates. These antennas are called *monopulse antennas* because they can form all of the required signals for target angle measurement on a single pulse or sweep. A brief introduction to monopulse antennas is given in Bailey (2010). A thorough discussion of monopulse radar, covering a variety of antenna structures and processing methods, is given in Sherman (1984).

Essentially the same approach can be used for analyzing the bias and precision of two-beam lobing for both the sequential and monopulse cases. An important modeling detail is whether the antenna pattern is the same for the squinted beams and the original beam, differing only in the boresight direction. This depends on antenna type and scanning mechanism. If squinted beams are formed by mechanically tilting a conventional antenna, all the beam shapes are the same. If a phased array antenna is electronically scanned to form squinted beams, they are not because the pattern broadens as the scan angle from normal increases. Monopulse antennas typically form identical squinted beams and then form the unsquinted beam from their sum, so that the unsquinted pattern is not the same as the squinted patterns. In this analysis it is assumed that the sum pattern is formed from two identically shaped squinted beams either in a monopulse or sequential lobing approach. The results obtained will still be generally applicable to other designs, though some details may change. The reader is referred to Sherman (1984) and Howard (2008) for much more information on antenna patterns for lobing systems.

Denote the complex amplitudes of the voltages after demodulation and matched filtering for a range bin containing a single point target of interest in each of the two beams of Fig. 7.18a as v_L and v_R. The symbols L and R refer to the antenna being shifted counterclockwise ("left" or L) and clockwise ("right" or R) by $\Delta\theta$ radians from the nominal pointing angle θ_0, respectively. The nominal pointing direction of the antenna can be set to $\theta_0 = 0$ for the time being without loss of generality.

Ignore for now any effects of noise or target fluctuations. Normalize the gain on the antenna boresight to $G(0) = 1$. Let $G(\theta - \Delta\theta)$ be the one-way antenna voltage pattern when the beam is squinted $\Delta\theta$ radians clockwise (right). The two-way "right" voltage pattern is then $G^2(\theta - \Delta\theta)$. Similarly, the two-way "left" voltage pattern is $G^2(\theta + \Delta\theta)$. The complex amplitude of the target L and R echo voltages will be $v_R = A\exp(j\varphi)G^2(\delta\theta - \Delta\theta)$ and $v_L = A\exp(j\varphi)G^2(\delta\theta + \Delta\theta)$ when the target is located at an angle $\delta\theta$ relative to the original pointing direction. Another important detail of the modeling is whether for a stationary nonfluctuating point target in the absence of noise the phases of v_L and v_R are identical. The answer depends on the design of the antenna and, in the non-monopulse case, the scanning mechanism used to form the squinted beams. In this analysis, it is assumed that they are identical for a single nonfluctuating point target with no noise. This is the intended result for many designs, although it is not always realized in practice.

The most common way by far to utilize v_L and v_R begins by computing the sum and difference of the received voltages. The left-minus-right *difference signal* is

$$v_\Delta(\Delta\theta, \delta\theta) = v_L - v_R = A\exp(j\varphi)[G^2(\delta\theta + \Delta\theta) - G^2(\delta\theta - \Delta\theta)]$$
$$\equiv A\exp(j\varphi)G_\Delta^2(\Delta\theta, \delta\theta) \quad (7.84)$$

Because this depends on the target amplitude and not just its angular location, a *sum signal* is also defined as

$$v_\Sigma(\Delta\theta, \delta\theta) = v_L + v_R = A\exp(j\varphi)[G^2(\delta\theta + \Delta\theta) + G^2(\delta\theta - \Delta\theta)]$$
$$\equiv A\exp(j\varphi)G_\Sigma^2(\Delta\theta, \delta\theta) \quad (7.85)$$

The Δ/Σ ratio is then independent of target amplitude:

$$v_{\Delta/\Sigma}(\Delta\theta, \delta\theta) = \frac{v_\Delta(\Delta\theta, \delta\theta)}{v_\Sigma(\Delta\theta, \delta\theta)} = \frac{G_\Delta^2(\Delta\theta, \delta\theta)}{G_\Sigma^2(\Delta\theta, \delta\theta)} \equiv G_{\Delta/\Sigma}^2(\Delta\theta, \delta\theta) \quad (7.86)$$

The arguments $(\Delta\theta, \delta\theta)$ in all three of these voltages emphasize that they depend on both the target location relative to the original boresight and the squint angle used in the lobing measurement system. For brevity, these arguments will usually be dropped in the remainder of this section.

Measurements and Introduction to Tracking 461

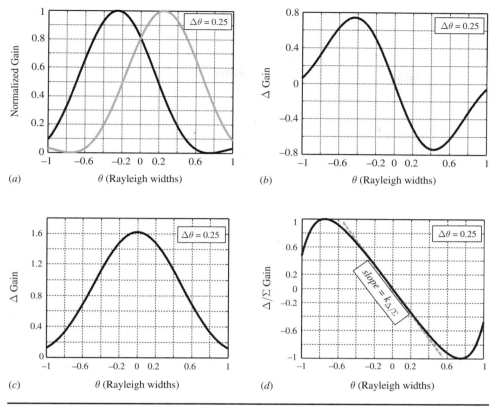

FIGURE 7.19 Two-way antenna voltage patterns for sequential lobing. (a) pattern $G^2(\theta)$ of Fig. 7.1a shifted left and right by $\Delta\theta = 0.25$ Rayleigh widths in each direction. (b) $G_\Delta^2 = L - R$ difference pattern. (c) $G_\Sigma^2 = L + R$ sum pattern. (d) Normalized difference pattern $G_{\Delta/\Sigma}^2$. The gray dashed line is a linear approximation to the central region with slope $k_{\Delta/\Sigma}$.

Figure 7.19 illustrates the antenna lobing patterns involved when the basic two-way antenna voltage pattern is the sinc² pattern of Fig. 7.1a. Figure 7.19a shows the central portion of the L and R patterns when each is shifted by 25 percent of the Rayleigh beamwidth away from the original boresight at $\theta = 0$. Figure 7.19b shows the difference pattern $G_\Delta^2 = L - R$, while Fig. 7.19c shows the sum pattern $G_\Sigma^2 = L + R$. Figure 7.19d is the pattern $G_{\Delta/\Sigma}^2$ of Eq. (7.86).

Notice that the normalized difference pattern is approximately linear in the angular region corresponding to the central 40 percent or so of the original antenna mainlobe. This suggests a simple algorithm for estimating the target angle $\delta\theta$ relative to the original boresight. For a given antenna, the Δ/Σ pattern $G_{\Delta/\Sigma}^2$ can be measured and its central portion approximated by a straight line with slope $k_{\Delta/\Sigma}$, $G_{\Delta/\Sigma}^2(\theta) \approx k_{\Delta/\Sigma} \cdot \theta$, as shown by the gray dashed line in Fig. 7.19d. When a target is detected, the lobing measurements v_L and v_R are collected and the quantity $v_{\Delta/\Sigma}$ computed. The estimated target angular position then becomes

$$\hat{\theta} = \theta_0 + \widehat{\delta\theta} = \theta_0 + \frac{1}{k_{\Delta/\Sigma}} \cdot v_{\Delta/\Sigma} \qquad (7.87)$$

where θ_0 is the boresight angle of the sum pattern and also the axis of odd symmetry of the difference pattern.

Now suppose that the noise-free values of v_L and v_R are each contaminated by i.i.d. complex WGN samples having variance σ_w^2. Because they are the sum and difference of v_L and v_R, v_Δ and v_Σ will each be contaminated with noise having a variance of $2\sigma_w^2$. The Δ/Σ voltage ratio will be

$$\frac{v_\Delta + n_\Delta}{v_\Sigma + n_\Sigma} \approx \frac{v_\Delta + n_\Delta}{v_\Sigma} = v_{\Delta/\Sigma} + \frac{n_\Delta}{v_\Sigma} \qquad (7.88)$$

The second step assumes high SNR so that $v_\Sigma \gg n_\Sigma$. The assumption that the phases of v_L and v_R are identical implies that the phases of v_Δ and v_Σ also have the same value, so that $v_{\Delta/\Sigma}$ is real. Consequently, the target angle information is contained entirely in the real part of the Δ/Σ ratio under ideal conditions; the imaginary part contains only noise. For this reason, the estimation algorithm is modified to its final form

$$\hat{\theta} = \theta_0 + \widehat{\delta\theta} = \theta_0 + \frac{1}{k_{\Delta/\Sigma}} \mathrm{Re}\{v_{\Delta/\Sigma}\} \qquad (7.89)$$

This is the form most commonly used in practice.

While this estimation approach seems reasonable, it is so far only heuristically based. A more systematic approach again derives the CRLB for angle estimation and the MLE given the measurements v_L and v_R. Begin with the CRLB. The signals v_L and v_R will be nonzero mean (due to the target component) independent complex Gaussians. They will not have the same mean due to the different L and R antenna weights, but will have the same variance σ_w^2. The sum and difference voltages v_Δ and v_Σ will also be nonzero mean complex Gaussians with variance $2\sigma_w^2$ but different means. The target angle $\delta\theta$ is real, so Eq. (7.18) applies. A slight extension is required since the antenna patterns for v_L and v_R differ due to the different squints. The result is

$$\sigma_{\widehat{\delta\theta}}^2 \geq \frac{\sigma_w^2}{2A^2 \left\{ \left|\frac{\partial G^2(\theta + \Delta\theta)}{\partial \theta}\right|^2 + \left|\frac{\partial G^2(\theta - \Delta\theta)}{\partial \theta}\right|^2 \right\}}$$
$$= \frac{1}{2\chi \left\{ \left|\frac{\partial G^2(\theta + \Delta\theta)}{\partial \theta}\right|^2 + \left|\frac{\partial G^2(\theta - \Delta\theta)}{\partial \theta}\right|^2 \right\}} \qquad (7.90)$$

Further progress requires assuming a specific antenna pattern.

Derivation of the MLE proves difficult. Several versions for various assumptions regarding the relative phase of the left and right signals, whether multiple samples are integrated, and so forth are given in Hofstetter and DeLong (1969). For the simple case assumed here, the MLE $\widehat{\delta\theta}$ of $\delta\theta$ is the value that satisfies

$$\frac{G^2(\widehat{\delta\theta} + \Delta\theta)}{G^2(\widehat{\delta\theta} - \Delta\theta)} = \frac{v_L}{v_R} \qquad (7.91)$$

that is, the value of $\delta\theta$ such that the ratio of the left and right squinted patterns at that angle matches the ratio of the measured left and right voltages. Simple algebraic manipulations

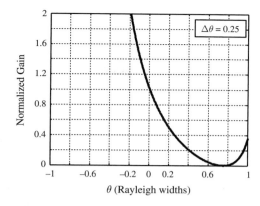

FIGURE 7.20 Left/right ratio pattern $G^2_{L/R}(\theta)$ for same antenna used in Fig. 7.19.

can restate $\widehat{\delta\theta}$ in terms of sum and difference voltages and patterns as the value that satisfies

$$\frac{G^2(\widehat{\delta\theta}+\Delta\theta)-G^2(\widehat{\delta\theta}-\Delta\theta)}{G^2(\widehat{\delta\theta}+\Delta\theta)+G^2(\widehat{\delta\theta}+\Delta\theta)}=\frac{v_L-v_R}{v_L+v_R} \Rightarrow G^2_{\Delta/\Sigma}(\Delta\theta,\widehat{\delta\theta})=\text{Re}\{v_{\Delta/\Sigma}\} \quad (7.92)$$

The Re{·} operator was again added in the last step because the Δ/Σ pattern is ideally real, so that in the presence of noise the imaginary part of the Δ/Σ ratio can be ignored.

If the linear approximation to the central portion of the difference pattern as illustrated in Fig. 7.19d holds, $G^2_{\Delta/\Sigma}(\theta)\approx k_{\Delta/\Sigma}\cdot\theta$, then $\widehat{\delta\theta}=v_{\Delta/\Sigma}/k_{\Delta/\Sigma}$ and Eq. (7.89) is in fact the maximum likelihood estimate of the target angle.

Equation (7.91) implies that the target angle could be estimated based on the ratio of v_L and v_R rather than the ratio of v_Δ and v_Σ. This is possible but is usually avoided because it would involve use of the pattern $G^2_{L/R}(\Delta\theta,\widehat{\delta\theta})\equiv G^2(\widehat{\delta\theta}+\Delta\theta)/G^2(\widehat{\delta\theta}-\Delta\theta)$ which is much less linear than $G^2_{\Delta/\Sigma}$, not symmetric around $\theta=0$, and not usable over as wide a range of θ. This pattern is shown in Fig. 7.20 for the same basic sinc² antenna pattern and ±0.25 beamwidth squint used earlier.

The PDF of the ratio voltage $v_{\Delta/\Sigma}$ is needed to determine the mean and bias of the estimator of Eq. (7.92). It has been derived in Kanter (1977), but the derivation is difficult and does not result in a tractable form. The result shows that the MLE exhibits a bias of the form $v_{\Delta/\Sigma}\exp(-\chi)$. This bias is very small for a target near θ_0 for any SNR and decreases rapidly with increasing SNR for any target angle. For SNR values adequate to provide quality tracking it is usually not significant.

The variance of the MLE is infinite in theory, but a simple approximation applies for high SNR, the main case of interest. Again suppose that noise-free values of v_Δ and v_Σ are each contaminated by i.i.d. complex WGN samples n_Δ and n_Σ having variances $2\sigma_w^2$. The error in the estimate of $v_{\Delta/\Sigma}$ will be

$$\begin{aligned}\varepsilon_{\Delta/\Sigma}&=\frac{v_\Delta+n_\Delta}{v_\Sigma+n_\Sigma}-\frac{v_\Delta}{v_\Sigma}=\frac{n_\Delta-v_{\Delta/\Sigma}n_\Sigma}{v_\Sigma+n_\Sigma}\\&\approx\frac{n_\Delta-v_{\Delta/\Sigma}\,n_\Sigma}{v_\Sigma}=\frac{n_\Delta}{v_\Sigma}-v_{\Delta/\Sigma}\frac{n_\Sigma}{v_\Sigma}\end{aligned} \quad (7.93)$$

In this equation, v_Σ represents the measured signal in the absence of noise so only the n_Δ and n_Σ terms are random.

Because the noise components are zero mean, the mean of $\varepsilon_{\Delta/\Sigma}$ in this analysis is zero. The variance is

$$\text{var}(\varepsilon_{\Delta/\Sigma}) = \overline{\varepsilon^2_{\Delta/\Sigma}} = \frac{\overline{n^2_\Delta}}{|v_\Sigma|^2} + v^2_{\Delta/\Sigma}\frac{\overline{n^2_\Sigma}}{v^2_\Sigma}$$
$$= \frac{2\sigma^2_w}{|v_\Sigma|^2}\left(1 + v^2_{\Delta/\Sigma}\right) = \frac{1}{\chi_\Sigma}\left(1 + v^2_{\Delta/\Sigma}\right) \quad (7.94)$$

where the sum-channel SNR has been defined as $\chi_\Sigma \equiv |v_\Sigma|^2/2\sigma^2_w$. The ideal lobing angle estimation processor selects the real part of $v_{\Delta/\Sigma}$. Ideally, this contains all of the target power but only half of the noise power. The expected variance of the real part of the error is then

$$\text{var}(\text{Re}\{\varepsilon_{\Delta/\Sigma}\}) = \frac{1}{2\chi_\Sigma}\left(1 + v^2_{\Delta/\Sigma}\right) = \frac{1}{2\chi_\Sigma}\left(1 + k^2_{\Delta/\Sigma} \cdot \theta^2\right) \quad (7.95)$$

This is the *lobing bound* on the variance of $v_{\Delta/\Sigma}$. The variance of the angle error is probably of more direct interest. Using $v_{\Delta/\Sigma} = G^2_{\Delta/\Sigma}(\theta) \approx k_{\Delta/\Sigma} \cdot \theta$ gives $\varepsilon_\theta = \varepsilon_{\Delta/\Sigma}/k_{\Delta/\Sigma}$. Equation (7.95) then becomes the lobing bound on angle precision in radians:

$$\text{var}(\text{Re}\{\varepsilon_\theta\}) = \frac{1}{2k^2_{\Delta/\Sigma} \cdot \chi_\Sigma}\left(1 + k^2_{\Delta/\Sigma} \cdot \theta^2\right)$$
$$= \frac{\theta^2_3}{2k^2_m \cdot \chi_\Sigma}\left[1 + k^2_m\left(\frac{\theta}{\theta_3}\right)^2\right] \quad (7.96)$$

Equation (7.96) is the desired result. In the second line, a normalized Δ/Σ slope $k_m \equiv k_{\Delta/\Sigma}\theta_3$ has been introduced. This or a closely similar definition is used in many common references, so it is included here for convenience of comparison. Several extensions to Eq. (7.96) to account for unequal L and R noise powers or correlated noise (both possibly due to jamming) and higher-order approximations are given in Sherman (1984).

Equation (7.96) is a reasonably good model of the error statistics for high SNR. However, because of the ratio calculation the error can occasionally become very large. In the absence of noise, v_L and v_R are in phase, but when significant noise is present, their relative phase can take on any value. When it happens that v_L and v_R are nearly equal in magnitude but close to 180° out of phase with one another, v_Σ can become very small while v_Δ does not. The ratio voltage $v_{\Delta/\Sigma}$ can then become very large, resulting in large outliers well outside the range $\pm\pi/2$ for the estimated angle and in turn resulting in large error variances. This behavior is consistent with the infinite variance of the exact theoretical PDF mentioned previously.

Unrealistically large angle errors can be essentially eliminated by one of several techniques, such as clipping the allowed magnitude of $v_{\Delta/\Sigma}$ to some limit or simply discarding those measurements that exceed a limit. The limit could be based, for instance, on the sum pattern beamwidth or the width of the linear region of the Δ/Σ pattern.

Figure 7.21 shows the precision observed in a simulation using Eq. (7.89) to estimate the target angle with the sinc² antenna patterns. Figure 7.21a shows the variation in precision versus SNR when the beams are squinted by ±0.2 Rayleigh beamwidths and the actual target angle is +0.1 Rayleigh beamwidth from the nominal boresight. Also shown are the lobing bound and CRLB of Eqs. (7.96) and (7.90), respectively, as well as the a priori bound for

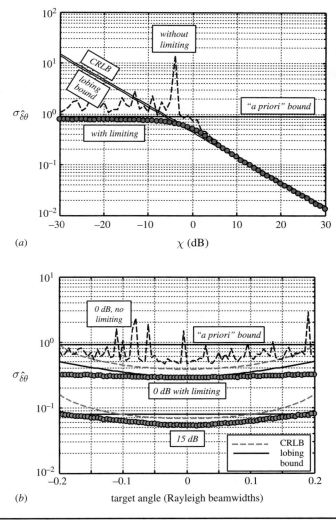

FIGURE 7.21 Angle estimation precision using lobing. (a) Precision versus SNR. (b) Precision versus target angle. See text for details.

a uniform random phase over the interval $\pm\pi/2$. Note that the lobing bound is very close to, but slightly less than, the CRLB. This is possible only because the estimator is biased.

Two curves are shown for the simulated precision. The dashed line shows the results obtained without any clipping of the magnitude of $v_{\Delta/\Sigma}$, while the gray dots show the result when it is (very loosely) clipped to a value 20 times that corresponding to the beam squint angle. This is sufficient to result in estimated angles that approach the a priori bound at low SNR while giving much more consistent behavior. Both curves follow the lobing bound and CRLB closely at high SNR.

In Fig. 7.21b the variation of precision versus target angle and the various bounds are shown for SNR = 0 dB and 15 dB. For the 0 dB case, the result is shown both with and without clipping of the $v_{\Delta/\Sigma}$ measurement. The lobing bound in the 15 dB case is obscured by the data. Again, the lobing bounds are slightly less than the CRLB, but this figure shows that the difference is greater at larger target angles and that neither bound is particularly tight at low SNR and larger angles.

There is a tradeoff between precision and width of the linear region around the boresight when choosing the squint angle. Equation (7.96) makes clear the effect of the Δ/Σ slope on estimate precision. Since $k_{\Delta/\Sigma} \cdot \theta \approx v_{\Delta/\Sigma}$ is less than 1, often significantly so, the dominant term in the angle measurement precision is $\left(k_{\Delta/\Sigma} \cdot \sqrt{2\chi_\Sigma}\right)^{-1}$, which is inversely proportional to the Δ/Σ slope. Consequently, a larger slope results in a finer precision for a given SNR. For a given basic squint antenna pattern $G^2(\theta \pm \Delta\theta)$, the slope increases if the squint angle $\Delta\theta$ is increased. On the other hand, larger slopes maintain the approximate linearity of $G^2_{\Delta/\Sigma}$ over a smaller angular region about the boresight. Furthermore, if the squint angle is too large the sum pattern loses both gain and resolution near the boresight. The gain reduction decreases χ_Σ, degrading the precision. All of these effects are evident in Fig. 7.22, which shows the sum pattern and normalized difference pattern for various squint angles.

The optimum squint depends on the antenna patterns and what is considered "optimum." The angle precision near boresight will be finest when the quantity $k_{\Delta/\Sigma} \cdot \sqrt{2\chi_\Sigma}$ is largest. For the sinc^2 pattern used here this occurs for $\Delta\theta = 0.42$ Rayleigh beamwidths or about $0.65\theta_3$. The cost of this finest precision is a relatively narrow linear region of about ± 0.15 Rayleigh beamwidths in the difference pattern.

The error analysis so far has considered only nonfluctuating targets and a single measurement. If N independent angle measurements are averaged, the error variance will decline by the factor $1/N$. Target fluctuations make the SNR a random variable. Their effect on the variance depends on the fluctuation model and whether or not a threshold is used to eliminate lower-SNR returns; the result can be either an increase or decrease in variance, depending on threshold setting, for correlated target echoes. For uncorrelated echoes, the variance approaches the nonfluctuating case with N measurements if N is sufficiently large. More information on both issues is available in Sherman (1984).

Angle measurement suffers an important additional source of error known as *glint*. Glint occurs when there are multiple point targets in the antenna beam or with multiple-scatterer targets. Consider two point targets in the tracking antenna beam at angles θ_1 and θ_2 relative to boresight. By superposition, the output of the L and R channels, and therefore of the sum and difference channels, is just the complex sum of the respective outputs. In the absence of noise, $v_{\Sigma 1}$ will be in phase with $v_{\Delta 1}$ and $v_{\Sigma 2}$ will be in phase with $v_{\Delta 2}$

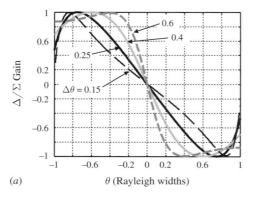
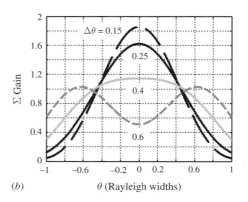

FIGURE 7.22 Effect of lobing squint angle $\Delta\theta$ on the sum and difference patterns. (*a*) Normalized difference pattern $G^2_{\Delta/\Sigma}(\theta)$. (*b*) Sum pattern $G^2_\Sigma(\theta)$.

as discussed previously. However, $v_{\Sigma 1}$ and $v_{\Sigma 2}$ will not be of equal amplitude nor in phase with one another in general. Write $v_{\Sigma 2} = \alpha \cdot v_{\Sigma 1}$ for some complex scalar $\alpha = \rho \exp(j\varphi)$. Assuming operation in the linear slope region, $v_{\Delta 1} = k_{\Delta/\Sigma} \cdot v_{\Sigma 1} \cdot \theta_1$ and $v_{\Delta 2} = k_{\Delta/\Sigma} \cdot v_{\Sigma 2} \cdot \theta_2 = \alpha k_{\Delta/\Sigma} \cdot v_{\Sigma 1} \cdot \theta_2$. The estimated angle becomes

$$\begin{aligned}\hat{\theta} &= \operatorname{Re}\left\{\frac{1}{k_{\Delta/\Sigma}} \frac{v_{\Delta 1} + v_{\Delta 2}}{v_{\Sigma 1} + v_{\Sigma 2}}\right\} \\ &= \operatorname{Re}\left\{\frac{\theta_1 + \alpha\theta_2}{1+\alpha}\right\} = \frac{\theta_1 + (\theta_1 + \theta_2)\rho\cos\varphi + \rho^2\theta_2}{1 + 2\rho\cos\varphi + \rho^2}\end{aligned} \qquad (7.97)$$

One reasonable definition of "ideal" behavior is that $\hat{\theta}$ always lies between θ_1 and θ_2, is halfway between them when $\rho = 1$ and is closer to the stronger of the two targets for other values of ρ.

It is easy to see from Eq. (7.97) that in fact $\hat{\theta} \to \theta_1$ or θ_2 when $\rho \to 0$ or ∞, corresponding to target 1 or 2 dominating the other. When $\rho = 1$, indicating equal-strength targets, $\hat{\theta}$ will be the average of the two angles, $(\theta_1 + \theta_2)/2$. However, for some values of α it is possible for $\hat{\theta}$ to fall outside of the angular interval defined by θ_1 and θ_2. The problem occurs when φ is close to 180° and ρ is close but not equal to 1 (see Prob. 37). Figure 7.23 shows the estimated angle $\hat{\theta}$ when $\theta_1 = -2°$ and $\theta_2 = -1°$. The two gray dashed lines mark these two angles, so it is desirable for $\hat{\theta}$ to fall between them. The figure confirms that the estimate equals the average for $\rho = 1$ and tends to θ_1 or θ_2 as appropriate when ρ differs from 1 by a large factor, that is, $\rho = 0.1$ or 10. However, when ρ is relatively close to 1 and φ is close to 180°, $\hat{\theta}$ can fall well outside of θ_1 and θ_2. If the two scatterers represent two scattering centers on a single target, this means the estimated AOA will be outside of the extent of the target! The closer ρ is to 1 without actually equaling it, the larger the error. If $\rho = 1.1$ in this example, $\hat{\theta}$ will be 9°, a full 10° beyond θ_2 and 10.5° from the average.

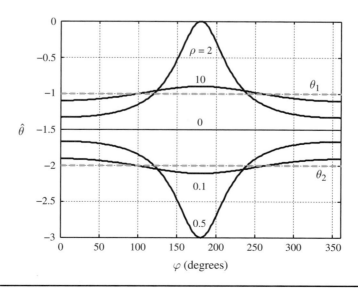

Figure 7.23 Estimated angle for two unresolved scatterers.

Real complex targets have many scattering centers, not just two. Statistical analyses suggest that a reasonable rule of thumb for the standard deviation of $\hat{\theta}$ due to glint is of the form

$$\sigma_{\hat{\theta}_{\text{glint}}} \approx k\theta_w = k\frac{L_w}{R} \tag{7.98}$$

where θ_w is the angular width in radians of the target as viewed from the radar and the scale factor k is variously reported to be between 0.15 and 0.35. If the width of the target as viewed from the radar is L_w, its angular extent will be $\theta_w = L_w/R$. Glint error is therefore a function of range. The combined effect of glint and noise on the angle precision is obtained by adding the variances or, equivalently, computing the RMS combination of the two precisions. Because glint error decreases with range while noise error increases with range due to declining SNR, their combination tends to produce an optimum range for a given system where the total error is minimum.

The discussion of angle measurement by lobing has been confined so far to one dimension, but angle tracking requires estimation of the target location in two angular dimensions: azimuth and elevation. Monopulse antennas provide difference outputs in both planes, while sequential lobing antennas can usually generate offset angles in both planes as well. The processing is the same in each plane.

Another means of 2D angle tracking is *conical scan* or *conscan*. In this technique, an antenna feed offset from a reflector antenna axis is mechanically rotated around that axis, producing a squinted beam that continually rotates around the antenna nominal boresight. This is suggested by Fig. 7.24a, which shows a series of beam positions at various points in the rotation. Two point targets are also shown, one on the centerline of the rotating squinted beam and displaced vertically (conscan rotation angle of 90°), the other much nearer to the nominal boresight (one-tenth of the centerline radius) and at a conscan rotation angle of 225°. Analysis of the variation in amplitude and the angular location of the peak response reveals the location of the target. For example, the pattern in the upper half of Fig. 7.24b suggests a target halfway between 80° and 100°, while the case in the lower half suggests a target close to 220°. Estimating the degree of squint from the amplitude variation requires use of the two-way voltage antenna pattern (see Prob. 38).

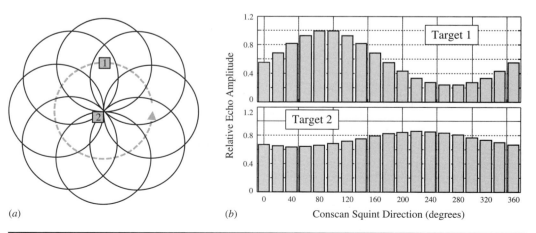

Figure 7.24 Conical scan. (a) Eight beam positions in a continuous conical scan. The gray dashed line shows the loci of the beam centers, rotating counterclockwise. Two target locations are shown. (b) Relative amplitude of signals received for each target.

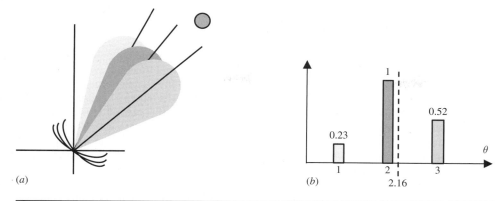

FIGURE 7.25 Centroiding of scanning radar angle measurements: (a) the target is detected on three successive beam positions, (b) relative power of the signals received at each position and their centroid.

More measurements would enable finer precision in both rotation angle and degree of squint. The number of pulses per rotation is determined by the rotation rate and the radar RR; 18 samples are shown in Fig. 7.24 before the cycle repeats on the 19th pulse or sweep, but real systems may have more or less than this.

The lobing techniques discussed so far are primarily for use in dedicated tracking radars or tracking modes of multipurpose radars. Scanning search radars may also utilize angle estimation algorithms as part of the search process. Consider a rotating search radar. As the beam passes over a target, that target may be detected in several sequential beam positions as suggested in Fig. 7.25a, and the resulting relative power measurements $P[n]$ may appear similar to those in Fig. 7.25b. The centroid of the measurements is

$$C_P = \frac{\sum_{n=n_1}^{n_2} n \cdot P[n]}{\sum_{n=n_1}^{n_2} P[n]} \quad (7.99)$$

For the measured powers and beam indices shown, $C_P = 2.16$. This value is then translated into an estimated AOA based on the angular orientation and spacing of the measurements. Other variations on this basic idea, such as a thresholding and binary-weighted centroiding techniques having some similarities to the split-gate range tracking method, are discussed in Blair et al. (2010).

7.3 Introduction to Tracking

The result of each measurement of the coordinates of a target in range, angle, or Doppler shift is an estimate of the true position of the target, corrupted due to the noise in the measurement process. The estimate therefore has an associated uncertainty. More information can be obtained by collecting a series of estimates over time and combining them with some (hopefully reasonable) assumptions about the target's kinematic motion to provide an improved estimate of the true coordinates of the target over time. The true coordinates of the target are elements of its *state*. The state is usually considered to also include the velocity and sometimes the acceleration in each spatial coordinate and may also include other aspects of the target, such as its pose relative to the radar. Usually the state components will be real-valued. *Tracking* is the process of developing the best estimate in some sense of the target state over time based on noisy measurements.

Several issues in addition to measurement noise make tracking a challenging problem. False alarms can cause tracks to deviate from the true target trajectory. Missed detections can allow the uncertainty in the track to grow excessively. If multiple targets are present in the same vicinity, unresolved measurements or crossing paths may make it difficult to associate new measurements with the correct track. In addition, radar naturally measures position in spherical coordinates relative to the antenna boresight, whereas it is usually desirable to track a target in Cartesian coordinates relative to the radar position or a fixed reference frame such as an earth-centered coordinate system. The nonlinear coupled transformation between the two coordinate systems complicates processing.

Tracking is a subject that supports many complete texts on its own. Only a brief treatment of the basic issues is sketched here. First, a simple problem of optimally combining two noisy measurements foreshadows a recursive estimator structure that will arise repeatedly. With that introduction, the fairly simple idea of sequential least-squares estimation (LSE) is then developed, borrowing heavily from Kay (1993). Next the common α-β filter for radar tracking is presented heuristically. Both the sequential least squares estimator and the α-β filter will be seen to share a similar prediction-correction structure. The Kalman filter (KF) is then presented as the optimal sequential least-mean-square estimator for a dynamically evolving system. While the extensive details required for its derivation are left to other texts, its structure is seen to be a generalization of the simpler sequential LSE problem, while the α-β filter is shown to be a special case.

Two excellent, concise introductory general references for tracking applied specifically to radar are Blair (2010) and Ehrman (2013). Another good tutorial text is Mahafza (2008). More in-depth radar-specific treatments are given in Bar-Shalom and Fortmann (1988) and Blackman (1986).

7.3.1 Optimal Combination of Two Noisy Measurements

Based on the discussion so far, it is expected that each radar measurement of position, velocity, or amplitude can be modeled as the true value corrupted by additive noise. The noise is typically modeled as zero mean. For example, consider a stationary target at a distance R from the radar. A first measurement will yield a range estimate

$$z[1] = R + w[1] \tag{7.100}$$

where $w[1]$ is a sample of stationary WGN with variance $\sigma_w^2[1]$. The index on σ_w^2 allows for the measurement noise power to vary from one measurement to the next. With only this one measurement available, the best estimate of R is $\hat{R}[1] = z[1]$. Now suppose a second measurement is made, $\hat{z}[2] = R + w[2]$. How should $\hat{R}[1]$ be updated with $\hat{z}[2]$ to provide the best estimate of R? For instance, should $z[1]$ and $z[2]$ simply be averaged?

The answer depends on the definition of "best." One reasonable definition is that the estimate should be unbiased and the error variance be minimized, subject to the unbiased constraint. Consider estimating the range as a linear combination of the two measurements,

$$\hat{R}[2] = k \cdot z[2] + p \cdot \hat{R}[1] \tag{7.101}$$

where k and p are constants. The mean of $\hat{R}[2]$ is

$$\overline{\hat{R}[2]} = k \cdot \overline{z[2]} + p \cdot \overline{\hat{R}[1]} = kR + pR \tag{7.102}$$

An unbiased estimate ($\overline{\hat{R}[2]} = R$) therefore requires that $p = 1 - k$. Equation (7.102) becomes

$$\begin{aligned} \hat{R}[2] &= k \cdot z[2] + (1-k)\hat{R}[1] \\ &= \hat{R}[1] + k(z[2] - \hat{R}[1]) \end{aligned} \qquad (7.103)$$

and the error in the estimate after the second measurement is

$$\begin{aligned} \varepsilon[2] &= \hat{R}[2] - R = \hat{R}[1] + k(z[2] - \hat{R}[1]) - R \\ &= (1-k)\hat{R}[1] - R + k(R + w[2]) \end{aligned} \qquad (7.104)$$

Notice that the mean of the error is zero. Its variance is

$$\text{var}\{\varepsilon[2]\} = (1-k)^2 \, \text{var}\{\varepsilon[1]\} + k^2 \sigma_w^2[2] \qquad (7.105)$$

To minimize the variance, set the derivative of Eq. (7.105) with respect to k to zero. After rearrangement, this gives

$$k = \frac{\text{var}\{\varepsilon[1]\}}{\text{var}\{\varepsilon[1]\} + \sigma_w^2[2]} \qquad (7.106)$$

Collecting results, the optimal update of the range estimate when the new measurement becomes available is

$$\begin{aligned} \hat{R}[2] &= \hat{R}[1] + k[2](z[2] - \hat{R}[1]) \\ k[2] &= \frac{\text{var}\{\varepsilon[1]\}}{\text{var}\{\varepsilon[1]\} + \sigma_w^2[2]} \end{aligned} \qquad (7.107)$$

k is called the *gain* of the estimator. In general, it will vary as new measurements arrive if their error variances are not constant, hence the introduction of the index $[n]$ on k. This structure will arise repeatedly in the more complex estimation problems discussed in subsequent sections.

The estimate update in the first line of Eq. (7.107) shows that the gain k determines the weight given to the difference between the new measurement and the previous estimate. If $k[2]$ is small, the estimate is not changed much and the updated value will be close to $\hat{R}[1]$. The second line of Eq. (7.107) shows that that will occur if $\sigma_w^2[2]$ is large compared to $\text{var}\{\varepsilon[1]\}$, that is, the second measurement is much noisier than the estimate based on the first. If the noise in the second measurement is much less than $\text{var}\{\varepsilon[1]\}$ then $k[2] \to 1$ and the updated value will be close to the new measurement $z[2]$, discounting the noisier previous estimate. If the measurement noise is the same on each measurement, then $k[2] = 1/2$ and the measurements are simply averaged.

7.3.2 Sequential Least Squares Estimation

Now consider a more general signal model and multiple available measurements in the form of a deterministic one-dimensional data sequence $x[n]$ dependent on some parameter Θ. Suppose a set of N possibly perturbed observations $z[n]$ of $x[n]$ is available. The perturbation could be additive noise in the data but could also be due to other sources such as inaccuracies in the assumed data model. The *least-squares estimate* (LSE) $\hat{\Theta}$ of Θ given $z[n]$, $n = 0, 1, \ldots, N-1$ is the value of Θ that minimizes

$$\varepsilon^2(\Theta) = \sum_{n=0}^{N-1} (z[n] - x[n])^2 \qquad (7.108)$$

ε^2 depends on Θ through $x[n]$. If $x[n] = \Theta$, straightforward minimization of ε^2 by differentiation with respect to Θ leads to the sample mean as the general LSE (see Prob. 39).

The linear LSE problem is of special interest. It assumes that the unperturbed data x are linearly related to Θ, $x[n] = f[n] \cdot \Theta$ for some known system function $f[n]$ (which could be just a constant). Using this model in Eq. (7.108) and minimizing ε^2 gives (see Prob. 40)

$$\hat{\Theta} = \frac{\sum_{n=0}^{N-1} z[n] f[n]}{\sum_{n=0}^{N-1} f^2[n]} \qquad (7.109)$$

with minimum error

$$\begin{aligned}\varepsilon^2(\hat{\Theta}) &= \sum_{n=0}^{N-1} (z[n] - \hat{\Theta} \cdot f[n])^2 \\ &= \sum_{n=0}^{N-1} z[n](z[n] - \hat{\Theta} \cdot f[n]) - \hat{\Theta} \sum_{n=0}^{N-1} f[n](z[n] - \hat{\Theta} \cdot f[n]) \\ &= \sum_{n=0}^{N-1} z^2[n] - \hat{\Theta} \sum_{n=0}^{N-1} f[n] z[n]\end{aligned} \qquad (7.110)$$

The last step relies on the fact that the last sum in the second step is identically zero, as can be verified by substituting $\hat{\Theta}$ from Eq. (7.109) into the sum.

As a trivial example, consider least squares estimation of a scalar constant A so that $\Theta = A$. Assume $f[n] = 1$ for all n so that the observations $z[n]$ are just perturbed measurements of A, most commonly due to additive noise. From Eq. (7.109), the LSE becomes the sample mean $\hat{A} = \sum_{n=0}^{N-1} z[n]/N$.

The procedure described so far is called *batch processing*. All of the measurements are collected and then the parameter estimate is generated in a single calculation. In tracking, new measurements are expected to occur on a regular basis and it is desirable to generate and update estimates of the target parameters as data arrives rather than wait until when (if ever) all of the measurements are collected. A procedure that allows this is *sequential least squares* estimation.

Continue the example above of estimating a constant. Let $\hat{A}[N-1]$ be the estimate based on all observations $z[n]$ up to and including time $N-1$, $\hat{A}[N-1] = \sum_{n=0}^{N-1} z[n]/N$. Now observe a new measurement $z[N]$. The new LSE of A is

$$\begin{aligned}\hat{A}[N] &= \frac{1}{N+1} \sum_{n=0}^{N} z[n] \\ &= \frac{N}{N+1} \hat{A}[N-1] + \frac{1}{N+1} z[N] \\ &= \hat{A}[N-1] + \frac{1}{N+1} (z[N] - \hat{A}[N-1])\end{aligned} \qquad (7.111)$$

Measurements and Introduction to Tracking

The estimate can be computed recursively using either the second or third form of Eq. (7.111). The third form is particularly interesting. It shows that the updated estimate is the previous estimate augmented with a correction term based on the difference between that previous estimate and the new observation. The weight accorded the correction term declines with successive measurements as the previous estimate incorporates more data.

The LSE for this problem can also be updated recursively. Using Eq. (7.111) in the definition of ε^2,

$$\begin{aligned}
\varepsilon^2(\hat{A}|N) &= \sum_{n=0}^{N}(z[n] - \hat{A}[N])^2 \\
&= \sum_{n=0}^{N-1}\left\{z[n] - \hat{A}[N-1] - \frac{1}{N+1}(z[N] - \hat{A}[N-1])\right\}^2 + (z[N] - \hat{A}[N])^2 \\
&= \varepsilon^2(\hat{A}|N-1) - \frac{2}{N+1}\sum_{n=0}^{N-1}(z[n] - \hat{A}[N-1])(z[N] - \hat{A}[N-1]) + \cdots \\
&\quad \cdots + \frac{N}{(N+1)^2}(z[N] - \hat{A}[N-1])^2 + (z[N] - \hat{A}[N])^2 \\
&= \varepsilon^2(\hat{A}|N-1) + \frac{N}{N+1}(z[N] - \hat{A}[N-1])^2
\end{aligned} \qquad (7.112)$$

The last line takes advantage of the fact that the term $\sum_{n=0}^{N-1}(z[n] - \hat{A}[N-1])$ is zero. The LSE grows with time because as more data is obtained the number of squared error terms increases.

No assumptions have yet been made about the nature of the perturbations in the observations $z[n]$, but going forward it will be assumed that the observations are linear functions of the parameter Θ corrupted by additive WGN, $z[n] = f[n] \cdot \Theta + w[n]$. In particular, $z[n] = A + w[n]$ for the problem of estimating a constant in noise.

A useful extension to the sequential LSE estimator is to allow weighting of the data in computing the MSE. Weighting would be appropriate for the constant-in-WGN model if the noise samples are uncorrelated but may have a time-varying variance $\sigma^2[n]$. Choose the weight on the nth term of the MSE defined in Eq. (7.108) as $w_n = 1/\sigma^2[n]$. This choice places greater weight on the least noisy measurements. Under these conditions it can be shown that the variance of the estimate $\hat{A}$ is

$$\text{var}(\hat{A}[N]) = \frac{1}{\sum_{n=0}^{N} w_n} = \frac{1}{\sum_{n=0}^{N}(1/\sigma^2[n])} \qquad (7.113)$$

The estimate update for this problem is

$$\begin{aligned}
\hat{A}[N] &= \hat{A}[N-1] + \frac{1/\sigma^2[N]}{\sum_{n=0}^{N}(1/\sigma^2[n])}(z[N] - \hat{A}[N-1]) \\
&= \hat{A}[N-1] + K[N](z[N] - \hat{A}[N-1])
\end{aligned} \qquad (7.114)$$

where the *gain* $K[N]$ is

$$K[N] = \frac{1/\sigma^2[N]}{\sum_{n=0}^{N}(1/\sigma^2[n])} = \frac{\text{var}(\hat{A}[N-1])}{\text{var}(\hat{A}[N-1]) + \sigma^2[N]} \qquad (7.115)$$

It follows that

$$\text{var}(\hat{A}[N]) = (1 - K[N])\,\text{var}(\hat{A}[N-1]) \tag{7.116}$$

Finally, the LSE update can be put in the form

$$\varepsilon^2(\hat{A}|N) = \varepsilon^2(\hat{A}|N-1) + \frac{(z[N] - \hat{A}[N-1])^2}{\text{var}(\hat{A}[N-1]) + \sigma^2[N]} \tag{7.117}$$

Equations (7.114) to (7.117) define the sequential LSE for a constant in additive WGN with nonstationary power. In particular, Eqs. (7.114) and (7.115) generalize Eq. (7.111), and Eq. (7.117) generalizes (7.112).

To execute this estimator, it is necessary to specify the initial values $\hat{A}[0]$ and $\text{var}(\hat{A}[0])$. The iteration can be initialized by choosing $\hat{A}[0] = z[0]$, which is the MLE of A based on the single measurement. The initial variance estimate is chosen as $\text{var}(\hat{A}[0]) = \sigma^2[0]$, implying that the initial noise power is known. Equations (7.115), (7.114), and (7.116) are then exercised in that order to update to time step $N = 1$, and the process repeated. If no estimate of $\sigma^2[0]$ is known, $\text{var}(\hat{A}[0])$ can be set to a "large" value. This in effect states that the confidence in the initial value $\hat{A}[0]$ is low so that the estimator places little weight on it. At $N = 1$, this will result in $K[1] \approx 1$, $\hat{A}[0] = z[1]$, and a reduced variance estimate going forward. Another initialization method is to use the first several observations in a batch process to develop an estimate of $\hat{A}$ and $\text{var}(\hat{A})$ and then switch to the sequential estimator for subsequent updates.

Figure 7.26 illustrates the behavior of this sequential LSE for $A = 10$, $\sigma^2[N] = 1$ for all N, $\hat{A}[0] = z[0]$, and $\text{var}(\hat{A}[0]) = 1$. It is easily seen in this case (see Prob. 41) that $K[N] = 1/(N+1)$ and $\text{var}(\hat{A}[N]) = \sigma^2/(N+1) = 1/(N+1)$ also. Figure 7.26a shows the gain and estimate variance versus the time step N. The curves overlay one another. Figure 7.26b shows the estimate $\hat{A}[N]$ of the actual value $A = 10$.

Figure 7.27 presents a similar example but with time-varying noise power. Specifically, the same specific sequence of random noise samples is used, except that the noise samples are multiplied by 10 (a 100 times variance increase) from time steps 80 to 160 and by $\sqrt{1/10}$ (a 10 times variance decrease) from time steps 240 through 320. Equation (7.115) shows that the sudden increase in noise variance will decrease the gain, while a decrease in noise variance will increase the gain. These effects are very evident in Fig. 7.27a. During the period of low estimator gain, the estimate variance changes very little; essentially, the new data is being discounted because it is so noisy. From steps 161 to 239, the noise in the data returns to its original value, the gain increases, and the estimate variance resumes its decrease. At step 240, the data noise is decreased and the gain increases to take advantage, allowing the estimate variance to decrease more rapidly. Figure 7.27b shows the actual estimate. It is clear that the estimate is relatively unchanging in periods of low gain and is more variable in periods of higher gain.

The generalization of the sequential LSE for a $P \times 1$ vector parameter is

$$\begin{aligned}
\hat{\Theta}[n] &= \hat{\Theta}[n-1] + \mathbf{k}[n](z[n] - \mathbf{f}^T[n]\hat{\Theta}[n-1]) \\
\mathbf{k}[n] &= \frac{\mathbf{M}[n-1]\mathbf{f}[n]}{\mathbf{f}^T[n]\mathbf{M}[n-1]\mathbf{f}[n] + \sigma^2[n]} \\
\mathbf{M}[n] &= (\mathbf{I} - \mathbf{k}[n]\mathbf{f}^T[n])\mathbf{M}[n-1]
\end{aligned} \tag{7.118}$$

Here $\mathbf{M}$ is the $P \times P$ covariance matrix of the estimate $\hat{\Theta}$. With the notation change $n \to N$ and the equivalences $\hat{\Theta} \to \hat{A}$, $\mathbf{f} \to 1$, $\mathbf{k}[n] \to K[n]$, and $\mathbf{M}[n] \to \text{var}(\hat{A}[N])$, the similarity

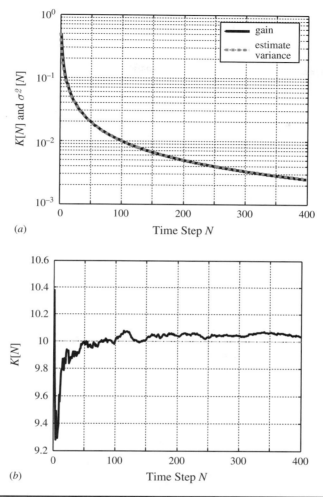

FIGURE 7.26 Sequential least squares estimation of a constant in additive WGN with constant variance. (a) Estimator gain and variance. (b) Estimated value.

between Eq. (7.118) and Eqs. (7.114) and (7.116) is obvious. Similar to the scalar example above, one simple initialization is to choose $\hat{\Theta}[0] = \mathbf{0}$ and $\mathbf{M}[0] = \kappa \mathbf{I}$ for some large value of κ, effectively discounting the initial covariance estimate.

7.3.3 The α-β Filter

Track filtering is readily viewed as an estimation problem. The radar makes noisy coordinate measurements at a series of discrete times t_n. The track filter has two goals: to *smooth* the measurements to provide an improved estimate of the coordinates at time n, and to use those smoothed estimates to *predict* the target position at the time t_{n+1} of the next planned measurement. A variety of filters are commonly used in the radar tracking community: the α-β filter, the α-β-γ filter, the Kalman filter (KF) and the extended Kalman filter (EKF). These are all closely related to the sequential LSE methods described above. New techniques such as "particle filters" are continuously emerging.

Chapter Seven

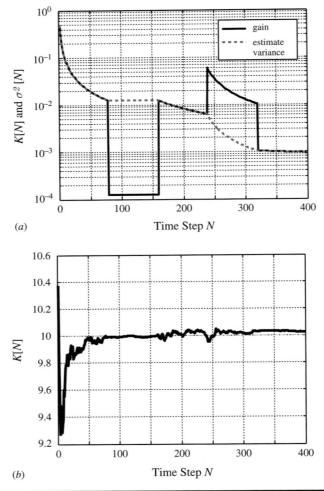

FIGURE 7.27 Sequential least squares estimation of a constant in additive WGN with time-varying variance. (a) Estimator gain and variance. (b) Estimated value. See text for details.

Two further generalizations of the sequential LSE results given so far are important in developing effective sequential estimation procedures for track filtering. The first is that the target model is dynamic, that is, the target parameters (position, velocity, etc.) change with time according to a defined model such as constant-velocity or constant-acceleration motion. A more fundamental generalization is to model the target parameters at a given time as random variables with a prior PDF instead of as deterministic quantities. The actual target parameters over time are then a single realization of a random process rather than a deterministic time series. This approach allows for uncertainty in the target model. The goal of the estimator is to minimize the expected value of squared error between the estimate and the realization, averaged over the joint PDF of the data and the target parameters. This modeling approach, called *Bayesian estimation*, is subtly different from the *classical estimation* considered so far. In either viewpoint the data **z** are random, typically due to additive noise in the measurements. In classical estimation the target parameters are not random, while in

Bayesian estimation they are. More discussion and examples of the difference in philosophies are given in Kay (1993).

Before stating the general solution to the sequential Bayesian LSE problem, consider the problem of tracking a target in a single dimension x with the simplest commonly used track filter, the α-β filter. The α-β filter is designed for a constant velocity target. Denoting the actual target position and velocity at time n as $x[n]$ and $\dot{x}[n]$, a true constant velocity model is

$$x[n] = x[n-1] + T \cdot \dot{x}[n-1] \tag{7.119}$$

where T is the interval between track update times.[9] In practice, it is useful to include a *process noise* $u[n]$ with variance σ_u^2 that allows for deviations from perfectly constant-velocity motion due primarily to target maneuvers, but also for such effects as turbulence or cross-winds affecting an aircraft. Process noise is not a measurement noise; rather, it is part of the model of the target state evolution. A typical model is to consider $u[n]$ to be a piecewise-constant random acceleration between each time step. The contribution of a constant acceleration to the position and velocity then gives the more complete "constant velocity" model:

$$\begin{aligned} x[n] &= x[n-1] + T \cdot \dot{x}[n-1] + \frac{T^2}{2} u[n] \\ \dot{x}[n] &= \dot{x}[n-1] + T \cdot u[n] \end{aligned} \tag{7.120}$$

The available measurements are the observations of only the target position,

$$z[n] = x[n] + w[n] \tag{7.121}$$

where $w[n]$ is the measurement noise with variance σ_w^2. The noises $u[n]$ and $w[n]$ are usually modeled as independent zero mean WGN processes.

Define the following estimated quantities:

$\hat{x}[n|n-1]$ Predicted position at time n given all measured data up through time $n-1$. Does not include knowledge of the new measurement $z[n]$.

$\hat{x}[n|n]$ Estimated (smoothed or corrected) position at time n given all measured data up through time n. This is an updated estimate incorporating the new information $z[n]$.

$\hat{\dot{x}}[n|n-1]$ Predicted velocity at time n given all measured data up through time $n-1$.

$\hat{\dot{x}}[n|n]$ Estimated (smoothed or corrected) velocity at time n given all measured data up through time n.

The α-β filter equations can be grouped in two stages: the *prediction* stage and the *innovation* stage. The prediction stage uses data through time $n-1$ and the assumed kinematic model to predict the target position at time n. The innovation stage then

---

[9]Constant update intervals are assumed here, but many of the results can be generalized to nonconstant update intervals.

applies the new measurement $z[n]$ at time n to correct or smooth the prediction. The equations are

Prediction stage:
$$\hat{x}[n|n-1] = \hat{x}[n-1|n-1] + T \cdot \hat{\dot{x}}[n-1|n-1]$$
$$\hat{\dot{x}}[n|n-1] = \hat{\dot{x}}[n-1|n-1]$$

Correction stage:
$$i[n] = z[n] - \hat{x}[n|n-1]$$
$$\hat{x}[n|n] = \hat{x}[n|n-1] + \alpha \cdot i[n]$$
$$= (1-\alpha)\hat{x}[n|n-1] + \alpha \cdot z[n]$$
$$\hat{\dot{x}}[n|n] = \hat{\dot{x}}[n|n-1] + \beta\frac{i[n]}{T}$$
$$= (1-\beta)\hat{\dot{x}}[n|n-1] + \beta\left(\frac{z[n] - \hat{x}[n-1|n-1]}{T}\right)$$
(7.122)

The constant velocity assumption is evident in the first equation of the prediction stage. Notice that the first form of the correction equation for $\hat{x}[n|n]$ has the same structure as the sequential LSE of Eq. (7.114) with a constant gain α taking the role of the more general time-varying gain $K[n]$. A typical practical initialization collects the first two measurements, $z[0]$ and $z[1]$, forms

$$\hat{x}[1|1] = z[1]$$
$$\hat{\dot{x}}[1|1] = (z[1] - z[0])/T$$
(7.123)

and begins iterating Eq. (7.122) at $n = 2$.

The quantity $i[n]$ is called the *residual* or the *innovation*. It represents new information available as a result of the new measurement $z[n]$. The last form of each of the last two equations makes the roles of α and β clear. α controls the relative weight assigned to new data versus prediction from prior data in updating the target position. Smaller values of α place more weight on the prediction and less on the new data and are appropriate when the new data has higher noise. β plays a similar role for updating velocity, though there is a subtle difference in the new data term. The parameters are usually confined to the ranges $0 \leq \alpha \leq 1$ and $0 \leq \beta \leq 2$. A complete analysis of the parameter choices on stability, convergence, and noise propagation is given in Tenne and Singh (2002).

Figure 7.28 illustrates the α-β filter in one dimension. The target is presumed to start at $x = 1000$ m at time $n = 0$ and move in the $+x$ direction with a constant velocity $\dot{x} = 50$ m/s for the time steps $n = 0, \ldots, 30$. For the next nine steps, the target decelerates at a constant rate such that at $n = 39$ the velocity reaches zero. The target then remains at $x = 2750$ m and $\dot{x} = 0$ m/s through $n = 69$. While the deceleration violates a strict constant-velocity model, it is permitted by the introduction of acceleration process noise in Eq. (7.120).

The noisy measurements $z[n]$ are shown in Fig. 7.28a by the diamond shapes and the actual position by the gray line in the background. The measurement noise is stationary additive WGN with $\sigma_w = 100$ m. The filter was initialized with $x[1|1] = z[0]$ and $\hat{\dot{x}}[1|1] = (z[1] - z[0])/T$. The black dashed line is the smoothed track $\hat{x}[n|n]$ obtained with $\alpha = 0.15$ and $\beta = 0.1$, while the solid black line is the track $\hat{x}[n|n]$ obtained by filtering the same data samples with the same $\beta = 0.1$ but $\alpha = 0.85$. As expected, small values

Measurements and Introduction to Tracking 479

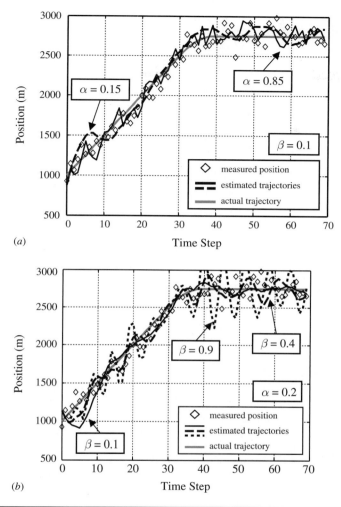

FIGURE 7.28 Target tracking in one dimension with an α-β filter. (a) Effect of varying α on position estimate. (b) Effect of varying β. See text for details.

of α de-emphasize the current measurement in favor of the prediction, producing smoother tracks but requiring longer to correct deviations. The larger value of α puts more emphasis on the current measurement, producing a noisier "smoothed" track but correcting more quickly.

Figure 7.28b shows the effect of β on the position estimate for a fixed $\alpha = 0.2$. Small values of β put more emphasis on the previous estimate of velocity, smoothing the velocity estimate more strongly and contributing to a smoother track. Larger values of β put more emphasis on the velocity estimate based on the innovation. Combined with a relatively small α, this tends to produce oscillatory behavior in the position estimate and is generally to be avoided. Although not shown here, the estimates of velocity become less smooth as β increases.

It is useful to express the α-β filter in a state space representation. Define the state vector for tracking in one dimension as $\mathbf{x}[n] = [x[n] \ \dot{x}[n]]^T$. The target kinematic and measurement models of Eqs. (7.120) and (7.121) are then

$$\begin{aligned}
\mathbf{x}[n] &= \mathbf{F} \cdot \mathbf{x}[n-1] + \mathbf{g} \cdot u[n] \\
&= \begin{bmatrix} 1 & T \\ 0 & 1 \end{bmatrix} \mathbf{x}[n-1] + \begin{bmatrix} T^2/2 \\ T \end{bmatrix} u[n] \\
z[n] &= \mathbf{h}^T \mathbf{x}[n] + w[n] \\
&= \begin{bmatrix} 1 \\ 0 \end{bmatrix}^T \mathbf{x}[n] + w[n]
\end{aligned} \quad (7.124)$$

The filter equations [Eq. (7.122)] become

Prediction stage:
$$\hat{\mathbf{x}}[n|n-1] = \mathbf{F} \cdot \hat{\mathbf{x}}[n-1|n-1]$$

Correction stage:
$$\begin{aligned}
\hat{\mathbf{x}}[n|n] &= \hat{\mathbf{x}}[n|n-1] + \mathbf{k}(z[n] - \mathbf{h} \cdot \hat{\mathbf{x}}[n|n-1]) \\
&= \hat{\mathbf{x}}[n|n-1] + \begin{bmatrix} \alpha \\ \beta/T \end{bmatrix} (z[n] - \mathbf{h} \cdot \hat{\mathbf{x}}[n|n-1])
\end{aligned} \quad (7.125)$$

In these equations, z, u, and w are scalars; $\mathbf{x}$, $\mathbf{h}$, $\mathbf{v}$, $\hat{\mathbf{x}}$, $\mathbf{g}$, and $\mathbf{k}$ are $P \times 1$ vectors; and $\mathbf{F}$ is a $P \times P$ matrix. Again, the similarity between the first version of the correction stage in Eq. (7.125) and the correction operation in Eq. (7.118) is clear.

7.3.4 The Kalman Filter

The new element in the α-β filter as compared to the sequential LSE is the prediction stage, which allows the state variables to evolve over time according to the kinematic model instead of being fixed as in the constant-in-WGN example. However, it is still a fixed-gain filter. The idea of sequential least squares estimation can be extended to apply to a dynamic model of the target motion by introducing a time-varying gain that minimizes the MSE at each step. The result is the *Kalman filter* (KF). The KF derivation details are beyond the scope of this text, but for a scalar observation the KF equations are

Kinematic and observation models:
$$\begin{aligned}
\mathbf{x}[n] &= \mathbf{F} \cdot \mathbf{x}[n-1] + \mathbf{g} \cdot u[n] \\
z[n] &= \mathbf{h}^T[n] \cdot \mathbf{x}[n] + w[n]
\end{aligned}$$

Prediction, and prediction MSE:
$$\begin{aligned}
\hat{\mathbf{x}}[n|n-1] &= \mathbf{F} \cdot \hat{\mathbf{x}}[n-1|n-1], \\
\mathbf{M}[n|n-1] &= \mathbf{F} \cdot \mathbf{M}[n-1|n-1] \cdot \mathbf{F}^T + \mathbf{g} \cdot \mathbf{g}^T \sigma_u^2
\end{aligned} \quad (7.126)$$

Kalman gain:
$$\mathbf{k}[n] = \frac{\mathbf{M}[n|n-1]\mathbf{h}[n]}{\sigma_w^2 + \mathbf{h}^T[n]\mathbf{M}[n|n-1]\mathbf{h}[n]}$$

Correction, and estimate MSE:
$$\begin{aligned}
\hat{\mathbf{x}}[n|n] &= \hat{\mathbf{x}}[n|n-1] + \mathbf{k}[n](z[n] - \mathbf{h}^T[n]\hat{\mathbf{x}}[n|n-1]) \\
\mathbf{M}[n|n] &= (\mathbf{I}_P - \mathbf{k}[n]\mathbf{h}^T[n])\mathbf{M}[n|n-1]
\end{aligned}$$

In addition to a time-varying gain, notice that the KF also allows for a time-varying observation noise by allowing $\mathbf{h}$ to vary. $\mathbf{M}$ is a $P \times P$ matrix that describes the MSE of the predicted and corrected signals, and $\mathbf{I}_P$ is the $P \times P$ identify matrix. The filter can be initialized by choosing $\hat{\mathbf{x}}[0|0] = 0$ and $\mathbf{M}[0|0] = \kappa \mathbf{I}$ for some large κ, or by using the same procedure described above for the α-β filter with the additional step of setting an initial value for $\mathbf{M}$,

$$\hat{\mathbf{x}}[0|0] = \begin{bmatrix} z[0] \\ 0 \end{bmatrix}, \qquad \hat{\mathbf{x}}[1|1] = \begin{bmatrix} z[1] \\ (z[1] - z[0])/T \end{bmatrix}$$
$$\mathbf{M}[1|1] = \sigma_w^2 \begin{bmatrix} 1 & 1/T \\ 1/T & 2/T^2 \end{bmatrix} \qquad (7.127)$$

This choice for $\mathbf{M}$ results from noting that the variance of $z[n]$ is σ_w^2 for all n, but that values of $z[n]$ at different values of n are uncorrelated. Computing the covariance matrix of $\hat{\mathbf{x}}[1|1]$ then gives $\mathbf{M}[1|1]$ as shown.

The scalar observation KF has a number of interesting properties. These include

- No matrix inversions are required for its execution. This would not be the case for batch estimation.
- The KF is a time-varying linear filter.
- The KF provides its own performance metric in the form of the MSE matrix $\mathbf{M}[n|n]$. Furthermore, computation of $\mathbf{M}$ does not depend on the actual observations so it can be precomputed, as can the Kalman gain $\mathbf{k}[n]$. The availability of $\mathbf{M}$ is helpful in gating and association, to be discussed in the next section.
- The KF exhibits the same prediction-correction structure seen in both the nondynamic sequential estimator and the α-β filter.
- If the driving noises u and w are stationary, the KF will asymptotically approach a steady state in which it becomes a time-invariant linear filter and can be viewed as a whitening filter.
- As will be seen, the α-β filter is an example of a steady-state KF.
- The same KF equations apply if the system model is extended to allow time-varying versions of $\mathbf{F}$, $\mathbf{g}$, and σ_w^2. This makes it possible to adapt to variable update times, missed updates, and other practical complications.

The equations above apply to a scalar observation, that is, a one-dimensional measurement. Normally a target is tracked in two or three dimensions using measurements of position in each. The KF can be further generalized to allow for an $M \times 1$ vector observation $\mathbf{z}[n]$ and an $R \times 1$ vector process noise $\mathbf{u}[n]$. The equations are

Kinematic and observation models:
$$\mathbf{x}[n] = \mathbf{F} \cdot \mathbf{x}[n-1] + \mathbf{G} \cdot \mathbf{u}[n]$$
$$\mathbf{z}[n] = \mathbf{H}[n] \cdot \mathbf{x}[n] + \mathbf{w}[n]$$

Prediction, and prediction MSE:
$$\hat{\mathbf{x}}[n|n-1] = \mathbf{F} \cdot \hat{\mathbf{x}}[n-1|n-1]$$
$$\mathbf{M}[n|n-1] = \mathbf{F} \cdot \mathbf{M}[n-1|n-1] \cdot \mathbf{F}^T + \mathbf{G}\mathbf{S}_\mathbf{u}[n]\mathbf{G}^T \qquad (7.128)$$

Kalman gain:
$$\mathbf{K}[n] = \mathbf{M}[n\|n-1]\mathbf{H}^T[n](\mathbf{S}_\mathbf{w}[n] + \mathbf{H}[n]\mathbf{M}[n|n-1]\mathbf{H}^T[n])^{-1}$$

Correction, and estimate MSE:
$$\hat{\mathbf{x}}[n|n] = \hat{\mathbf{x}}[n|n-1] + \mathbf{K}[n](\mathbf{z}[n] - \mathbf{H}[n]\hat{\mathbf{x}}[n|n-1])$$
$$\mathbf{M}[n|n] = (\mathbf{I} - \mathbf{K}[n]\mathbf{H}[n])\mathbf{M}[n|n-1]$$

F and **M** are $P \times P$; **G** is $P \times R$; **H** is $M \times P$; **K** is $P \times M$; **x** and $\hat{\mathbf{x}}$ are $P \times 1$; and **z** and **w** are $M \times 1$. $\mathbf{S_u}$ and $\mathbf{S_w}$ are the $R \times R$ and $M \times M$ covariance matrices of the noise processes **u** and **w**, which are assumed to each be zero mean uncorrelated Gaussian random processes, independent of one another. They have also been generalized to allow for time-varying noise. All of the previously listed properties still apply, except that a matrix inversion is now required to compute **K**.

Return now to the scalar observation case of Eq. (7.126). Over time, the KF will asymptotically approach a steady-state condition in which the gain **k** is constant. The prediction MSE $\mathbf{M}[n|n-1]$ and smoothed estimate MSE $\mathbf{M}[n|n]$ will also approach constants, with the former larger than the latter. The values of the steady-state gain can be determined in terms of the *tracking index* Γ (Kalata, 1984), also called the *maneuverability index*:

$$\Gamma \equiv \frac{\sigma_u T^2}{\sigma_w} \tag{7.129}$$

Γ is a measure of the target position uncertainty due to the acceleration noise relative to that due to the measurement noise. It can be shown that for the single-coordinate case considered here, the steady-state KF gains $\mathbf{k} = \begin{bmatrix} k_x & k_{\dot{x}} \end{bmatrix}$ obey the following relationships:

$$\begin{aligned} \Gamma^2 &= \frac{k_{\dot{x}}^2}{1 - k_x} \\ k_{\dot{x}} &= 2(2 - k_x) - 4\sqrt{1 - k_x} \end{aligned} \tag{7.130}$$

These equations can be solved to find the values of k_x and $k_{\dot{x}}$ for a given value of Γ (Blair, 2010):

$$\begin{aligned} k_x &= -\frac{\Gamma^2}{8} - \Gamma + \frac{1}{8}(\Gamma + 4)\sqrt{\Gamma^2 + 8\Gamma} \\ k_{\dot{x}} &= +\frac{\Gamma^2}{4} + \Gamma - \frac{1}{4}\Gamma\sqrt{\Gamma^2 + 8\Gamma} \end{aligned} \tag{7.131}$$

Not surprisingly, the α-β filter is a special case of the KF, obtained by choosing the gain vector to be a fixed value $\mathbf{k} = [\alpha \ \beta/T]^T$. A common way to choose α and β is to use the steady-state Kalman gains, that is, $\alpha = k_x$ and $\beta = k_{\dot{x}} T$.

Figures 7.29 and 7.30 show an example similar to the one of Fig. 7.28. Both a KF initialized according to Eq. (7.127) and an α-β filter were run on the same data samples. The process noise standard deviation was set at $\sigma_u^2 = a^2/2$, where a is the maximum acceleration (-5.5566 m/s^2 in this example). The resulting value is $\sigma_u = 3.93$ m/s^2. The values of $\alpha = 0.2443$ and $\beta = 0.0342$ were chosen based on Eq. (7.131).

Figure 7.29a compares the smoothed trajectory estimates from the two filters, while Fig. 7.29b shows the velocity estimates. Both filters track the first two measurements exactly due to the initialization process. This particular example was chosen to emphasize a disadvantage of the fixed-gain α-β filter. The first two samples happen to produce an initial velocity estimate that differs greatly from the actual target velocity. Consequently, the α-β estimate exhibits very large position errors over the first 20 time steps. The KF's variable gain recovers much more quickly and exhibits much lower errors. This is also seen in the velocity estimates, which track the actual values of 50 or zero in the appropriate regions. However, the KF converges toward the initial velocity of 50 m/s much sooner than does the α-β filter. If another random realization was used where the first two observations produced an initial

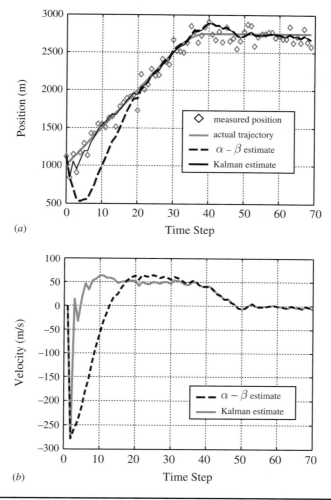

FIGURE 7.29 Comparison of α-β and Kalman filters on same data. (a) Smoothed position estimate. (b) Smoothed velocity estimate. See text for details.

velocity estimate closer to the true value, the results of the two filters would be much more comparable.

As seen in Fig. 7.30a, by time step 20 the KF gains have reached steady state. The fixed gains α and β were chosen to match these steady-state values, so both filters behave virtually identically from this point on, including the overshoot and recovery during the target deceleration from about time steps 35 to 45. Finally, Fig. 7.30b shows the variation of $\mathbf{M}[1, 1]$, which is the KF computation of the variance of the estimated position, for the first 30 time steps needed to attain steady-state behavior. Half-integer indices show the value of the prediction MSE $\mathbf{M}[n \mid n-1]$, while integer indices are the values of the smoothed MSE $\mathbf{M}[n \mid n]$. As expected, the MSE is always increased by the prediction step and then decreased by the correction step when new data is incorporated.

The failure of the KF to respond to the deceleration in steps 31 to 39 more quickly than the α-β filter can be remedied by a number of techniques that combine multiple filters

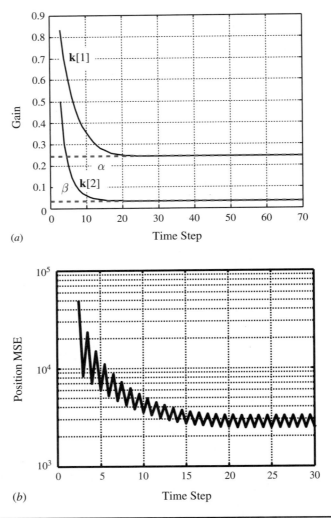

FIGURE 7.30 (a) α-β and Kalman filters gains for example of Fig. 7.29. (b) Kalman filter mean-square error estimate for position, $M[1, 1]$. See text for details.

designed for different target dynamics. One popular approach is *interacting multiple models* (IMM). Multiple model techniques run multiple track filters in parallel, each with different target dynamics represented by different process noise variances. At each step, the model yielding the smallest innovation is used to update the track. IMM is a particular technique in this class that attempts to assign relative probabilities to each model and blend their track updates. An introduction to the IMM approach and others is given in Ehrman (2013).

Just as the choice of α and β plays an important role in the tradeoff of smoothing and convergence time in the α-β filter, the choice of the process noise variance σ_u^2 plays a similar role in the KF. Figure 7.31 illustrates this effect on the position gain $\mathbf{k}[1] = k_x$ for an example similar to those preceding. As σ_u^2 increases, the steady-state gain rises and the filter converges more quickly, resulting in less smoothing and more rapid adaptation, the same effect observed as α is increased in the α-β filter. Selection of the process noise variance is discussed in Blair (2012).

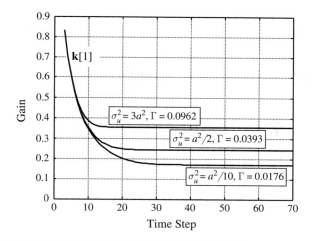

FIGURE 7.31 Effect of process noise variance on position gain. See text for details.

A technique called *gain scheduling* can be applied to improve the performance of the α-β filter during the startup transient without resorting to the more complex KF. Let α_{ss} and β_{ss} be the steady-state values of α and β obtained from the steady-state KF gains of Eq. (7.131). Similar to the $1/(N+1)$ gain of the least squares estimator of a constant in stationary WGN (Fig. 7.26 and Prob. 41), the gains for least squares estimation of position and velocity for the constant-velocity target model as a function of the number of data samples n can be computed and used until the steady-state value is reached (Blair, 2010):

$$\alpha[n] = \max\left\{\frac{2(2n+1)}{(n+1)(n+2)}, \alpha_{ss}\right\}$$
$$\beta[n] = \max\left\{\frac{6}{(n+1)(n+2)}, \beta_{ss}\right\} \quad (7.132)$$

Notice that $\alpha[n]$ is approximately proportional to $1/n$, similar to the earlier sequential LSE analysis, while $\beta[n]$ is approximately proportional to $1/n^2$.

The KF is based on a linear model for both the target dynamics and the sensor observation. As noted in Chap. 1, many radars naturally measure three-dimensional target position in the modified spherical coordinate system relative to the radar antenna of range, azimuth angle θ, and elevation angle ϕ. Some radars, particularly planar phased array systems, measure range and two positions along the *sine space* coordinates u and v, where $u = \cos\phi \cdot \cos\theta$ and $v = \cos\phi \cdot \sin\theta$ are the direction cosines from the antenna center to the target. On the other hand, targets are usually tracked in a stabilized Cartesian x-y-z system referenced to the radar system or possibly to a fixed coordinate system such as earth-centered. Because the transformation from spherical to Cartesian coordinates is nonlinear, the observation equation, state equation, or both can be nonlinear.

Consider a two-dimensional example where the radar measures the polar coordinates of range R and azimuth angle θ. It is desired to track the target in the Cartesian coordinates x and y, so the state vector is $\mathbf{x} = [x\ y]^T$. At the nth time step the radar makes noisy measurements:

$$\hat{R}[n] = R[n] + w_R[n] = \sqrt{x^2[n] + y^2[n]} + w_R[n]$$
$$\hat{\theta}[n] = \theta[n] + w_\theta[n] = \arctan(y[n]/x[n]) + w_\theta[n] \quad (7.133)$$

Chapter Seven

Clearly the measurement vector $\mathbf{z} = [\hat{R}\ \hat{\theta}]^T$ cannot be written as a linear function $\mathbf{z} = \mathbf{F} \cdot \mathbf{x} + \mathbf{w}$ of the state vector, so the KF cannot be applied as described so far.

What if the target were tracked in the more natural polar coordinates? In this case, the measurement equation will be linear. Assume the target travels at constant velocity with components v_x and v_y and ignore any process noise. The target motion in x-y coordinates satisfies

$$x[n] = x[n-1] + v_x T$$
$$y[n] = y[n-1] + v_y T \qquad (7.134)$$

The range becomes

$$R[n] = \sqrt{x^2[n-1] + y^2[n-1] + 2v_x T \cdot x[n-1] + 2v_y T \cdot y[n-1] + (v_x^2 + v_y^2)T^2}$$
$$= \sqrt{R^2[n-1] + 2R[n-1]T\{v_x \cos(\theta[n-1]) + v_y \sin(\theta[n-1])\} + (v_x^2 + v_y^2)T^2} \qquad (7.135)$$

The state update equation is now a nonlinear function of R and θ, so again the KF cannot be applied.

The *extended Kalman filter* (EKF) is a common technique for dealing with this problem. The basic idea of the EKF is to linearize a nonlinear state update equation around the current state estimate, and linearize a nonlinear observation equation around the current state prediction. The linear KF is then applied to the resulting approximate system equations.

As an example of linearization, consider the two-dimensional Cartesian state vector $\mathbf{x} = [x\ y]^T$. The position measurement model of Eq. (7.133), without the measurement noise and with the time index n suppressed to simplify the notation, is

$$\mathbf{z} = \begin{bmatrix} \hat{R} \\ \hat{\theta} \end{bmatrix} = \begin{bmatrix} \sqrt{x^2 + y^2} \\ \tan^{-1}\left(\dfrac{y}{x}\right) \end{bmatrix} \neq \mathbf{Fx} \qquad (7.136)$$

The inequality states that $\mathbf{z}$ is not a linear function of the Cartesian track state variable $\mathbf{x}$. Instead, $\mathbf{z}$ is a more general nonlinear function of $\mathbf{x}$,[10]

$$\mathbf{z} = F(\mathbf{x}) = \begin{bmatrix} \sqrt{x^2 + y^2} \\ \tan^{-1}\left(\dfrac{y}{x}\right) \end{bmatrix} \qquad (7.137)$$

The EKF linearizes the measurement model by computing a Taylor series approximation of $F(\cdot)$ in each state variable, evaluated at the current state. For this example, the linearized model becomes

$$\mathbf{z} = \begin{bmatrix} \dfrac{\partial R}{\partial x} & \dfrac{\partial R}{\partial y} \\ \dfrac{\partial \theta}{\partial x} & \dfrac{\partial \theta}{\partial y} \end{bmatrix}_{x,y} \cdot \mathbf{x} = \begin{bmatrix} \dfrac{\partial}{\partial x}\left(\sqrt{x^2+y^2}\right) & \dfrac{\partial}{\partial y}\left(\sqrt{x^2+y^2}\right) \\ \dfrac{\partial}{\partial x}\left(\tan^{-1}\left(\dfrac{y}{x}\right)\right) & \dfrac{\partial}{\partial y}\left(\tan^{-1}\left(\dfrac{y}{x}\right)\right) \end{bmatrix}_{x,y} \cdot \mathbf{x}$$

$$= \begin{bmatrix} \dfrac{x}{\sqrt{x^2+y^2}} & \dfrac{y}{\sqrt{x^2+y^2}} \\ \dfrac{-y}{x^2+y^2} & \dfrac{x}{x^2+y^2} \end{bmatrix}_{x,y} \cdot \mathbf{x} \qquad (7.138)$$

---

[10] In this and succeeding equations, the four-quadrant arctangent must be used.

Equation (7.138) is a linear relationship between z and x which can be utilized in the KF equations. This is a dynamic linearization that must be repeated at every time step as the state estimates evolve.

The EKF has no optimality properties, but is nonetheless a common technique as well as the basis for numerous other KF extensions. More information on the EKF is available in the references cited at the beginning of this section.

Many other approaches have been proposed to address nonlinear Bayesian estimation problems, including the "unscented" KF, "particle filters," nonlinear recursive filters, batch filtering methods, and numerical solution of the Fokker-Planck equation. A good overview of their philosophies, advantages, and disadvantages and a good initial bibliography is given in Daum (2005).

7.3.5 The Tracking Cycle

Track filtering is just one element of the overall target tracking cycle depicted in Fig. 7.32. The radar produces a new set of detections at every CPI or dwell. A "detection" generally consists of the estimated position in range, angle, and possibly velocity; an estimate of the measurement error in each dimension; and a time stamp. Other metadata may also be included, such as the estimated SIR of the detection. Since more than one target may be detected and tracked at a time, the first step is to *associate* detections from the current CPI or dwell with existing tracks, that is, determine which detections are believed to be due to which targets under track, so that the correct detection is used to update each track. In dense target scenarios this can be a daunting task. It will be discussed shortly, after consideration of the other elements of the tracking cycle.

Any detections that cannot be associated with existing tracks are candidates for new targets and are therefore sent to a *track initiation* process. An electronically scanned radar will typically transmit a confirmation CPI or dwell to verify the detection and ensure it is not a false alarm. This is generally not practical with rotating radars. In either case, an *M*-of-*N* logic is applied over the next *N* dwells to decide whether to promote a tentative track to a confirmed track. More sophisticated sequential probability analysis or detection pattern

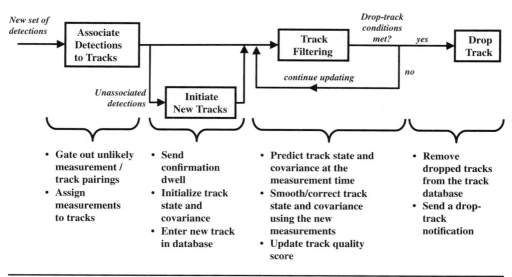

FIGURE 7.32 The tracking cycle.

recognition can also be used for this function (Blackman, 1986). Once a new track is established, it is entered into the radar database and its initial state and covariance established using an initialization procedure appropriate for the track filter that will be used.

The measurement data associated with each track is next passed into the track filter for that target. The filter first predicts the state at the current measurement time and then smoothes it using the innovation based on the new measurement. A track quality score is often maintained for each track based on the measurement-to-prediction distance for each association, normalized by the covariance. If the score exceeds a preestablished upper threshold it may indicate incorrect association or a poor target model. If it falls below a lower threshold, it may indicate that the target model process noise is too large.

Tracks must eventually be dropped, for instance when a target moves outside the detection range of the radar. For a single missed detection the track filter can be used to *coast* the track by simply performing another update without the correction step, or recomputing a single update from the last detection using a longer update time T. In practice, the track filter may be allowed to coast for a small number of misses. However, failure to associate a new detection with a track for some minimum number of updates will trigger deletion of the track. A more sophisticated approach than a hard threshold on the number of misses would again apply sequential probability techniques or threshold the track quality metric. When a track is deleted the track database must be updated. A notification may be sent to the end user of the track data as well.

Before discussing the data association stage, it is useful to introduce the idea of statistical distance. Figure 7.33 illustrates the statistics of two measurements that might be associated with a predicted value of $x = 30$. One has a mean $\mu_{x1} = 20$ and a standard deviation $\sigma_{x1} = 5$, while the other has $\mu_{x2} = 50$ and $\sigma_{x2} = 20$. Which is closer to the predicted value and should be used to smooth the prediction? μ_{x1} is only 10 units away from the prediction; that distance would be considered the innovation i_1 if the first measurement was used to update the predicted location. The second measurement μ_{x2} is 20 units away. However, the first measurement is $d_1 = \mu_{x1}/\sigma_{x1} = 2$ standard deviations away from the prediction while the second is only $d_2 = \mu_{x2}/\sigma_{x2} = 1$ standard deviation away. Consequently, the likelihood that the predicted value of 30 could have arisen from the estimated distribution of the first

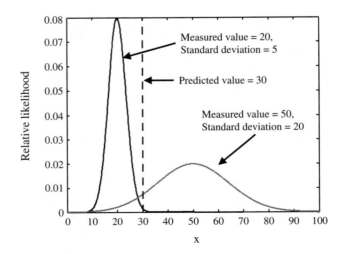

Figure 7.33 Statistical distance.

measurement is much less than the likelihood that it could have arisen from the second. The second measurement is said to be *statistically* closer to the track and should probably be associated with the track.

A generalization of this simple concept applicable to tracking in multiple dimensions is the *Mahalanobis distance* $d_M[n]$ applied to the innovation $\mathbf{i}[n]$ in the vector observation KF, given by the calculations:

$$\begin{aligned} \mathbf{i}[n] &= \mathbf{z}[n] - \mathbf{H}[n]\hat{\mathbf{x}}[n|n-1] \\ \mathbf{S_i}[n] &= \mathbf{S_w}[n] + \mathbf{H}[n]\mathbf{M}[n|n-1]\mathbf{H}^T[n] \\ d_M^2[n] &= \mathbf{i}^T[n](\mathbf{S_i}[n])^{-1}\mathbf{i}[n] \end{aligned} \quad (7.139)$$

In the scalar observation case this reduces to the heuristic measure used above, $d_M[n] = i[n]/\sigma_i[n]$. Notice that the expression for $\mathbf{S_i}[n]$ is the "denominator" of the Kalman gain calculation in Eq. (7.128). It combines a term $\mathbf{S_w}[n]$ for the covariance of the measurements with the covariance of the state prediction $\mathbf{M}[n|n-1]$ as observed through $\mathbf{H}[n]$.

Another distance measure is the log-likelihood function for the hypothesis that the prediction and measurement stem from the same object. Assuming Gaussian statistics, this is given by

$$\ln \Lambda = \frac{1}{2}\ln[2\pi \det(\mathbf{S_i})] + \frac{1}{2}\mathbf{i}^T\mathbf{S_i}^{-1}\mathbf{i} = \frac{1}{2}\ln[2\pi \det(\mathbf{S_i})] + \frac{1}{2}d_M \quad (7.140)$$

where the time index n has been temporarily dropped from the notation for clarity. This measure is said to be more robust against "track stealing" by young tracks, which tend to exhibit pessimistic (overly large) covariances when only a few updates have been made (Ehrman, 2013). The determinant term increases Λ for large covariances, reducing this effect.

Consider an arbitrary covariance matrix $\mathbf{S}$ in a multidimensional space. Suppose the eigenvectors of the $P \times P$ covariance matrix $\mathbf{S}$ are $\{\mathbf{e}_p\}$ with corresponding eigenvalues $\{\lambda_p\}$. Assuming Gaussian statistics, these describe a P-dimensional joint Gaussian PDF. Because $\mathbf{S}$ will be symmetric, the eigenvectors will be mutually orthogonal. The surface describing the standard deviation of the PDF is an ellipsoid having its axes in the directions $\{\mathbf{e}_p\}$ with one-sided lengths $\{\sqrt{\lambda_p}\}$. Figure 7.34a illustrates this for $P = 2$ dimensions. Figure 7.34b is a notional illustration of these *covariance ellipses* of a predicted state and the measurements corresponding to two different detections. The innovation is the vector connecting the prediction to whichever measurement is associated with it. The statistical distance between the measurement and a track prediction depends on the orientation and size of the covariance ellipsoids of both the measurements and the track prediction. In this notional example, both measurements are within the 1-σ ellipse of the prediction. The prediction is within the 1-σ ellipse of $\mathbf{z}_1[n]$ but not of $\mathbf{z}_2[n]$. Consequently, $\mathbf{z}_1[n]$ is statistically closer to $\hat{\mathbf{x}}[n|n-1]$ than is $\mathbf{z}_2[n]$. The Mahalanobis distance formalizes this calculation.

Statistical distance ideas are important in developing the algorithms for *gating* and *assignment* in the "associate detections" stage of the track cycle. Different classes of radars operating in different scenarios may have a number of detections at a given time step that is less than (sometimes much less than), equal to, or greater than the number of active tracks. The first step in determining which detections will be used to update which tracks (assignment) is usually gating, which seeks to reject highly implausible pairings so as to reduce the number of combinations that must be considered in the assignment process. Gating rejects any pairing for which the measurement exceeds some threshold distance from the track update, but accepts any measurements within that distance as potentially valid. These measurements

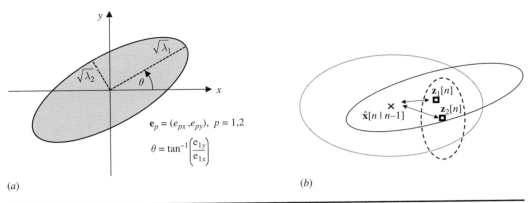

FIGURE 7.34 (a) Two-dimensional ellipsoid defined by the eigenstructure of a 2D covariance matrix, (b) covariance ellipsoids for measurements and prediction determine the statistical distance.

are said to be *validated* for that track. Common approaches are spherical, rectangular, and ellipsoidal gates:

Spherical gate:
$$\sqrt{(x_p - x_m)^2 + (y_p - y_m)^2 + (z_p - z_m)^2} < d_{spherical}$$

Rectangular gate:
$$|R_p - R_m| < d_R, \text{ or } \begin{array}{l} |x_p - x_m| < d_x, \text{ and} \\ |y_p - y_m| < d_y, \text{ and} \\ |z_p - z_m| < d_z \end{array} \quad (7.141)$$

Ellipsoidal gate:
$$\ln \Lambda < d_{statistical}$$

In this equation, x, y, and z are the three spatial components of the state vector $\mathbf{x}$; R is range. The subscripts m and p refer to measured and predicted values. Gating is often performed hierarchically. The surveillance space can be partitioned into regions so that a measurement falling in one region need not be tested against tracks in other regions. Similarly, it may be possible to group tracks within a region into mutually exclusive partitions, again reducing the number of comparisons to be made. The next step is often a relatively coarse but computationally efficient rectangular gating. The radar may gate only in range, or in all three spatial coordinates. Finally, a tighter ellipsoidal gating may be applied using either the Mahalanobis distance or the log-likelihood function.

The last major step in data association is assignment of the validated detections to the tracks. This process begins by constructing a measurement-to-track cost matrix. The elements of the matrix are some measure of the statistical cost of assigning a particular measurement to a particular track, often the negative of the log-likelihood of Eq. (7.140). One of a variety of algorithms is then used to decide the final assignments based on the cost matrix. Common algorithms include

- *Nearest neighbor (NN)*: Assigns the lowest cost validated measurement to each track, independent of the other tracks. As a result, one measurement could be used to update more than one track in some situations. A common modification called *sequential NN* removes each measurement from consideration when it is assigned to one track so that it cannot also be assigned to another.

- *Strongest neighbor* (SN): Assigns to each track the validated measurement with the highest SIR. Like NN, SN can assign the same measurement to more than one track unless it is modified to disallow this.
- *Global nearest neighbor* (GNN): Assigns validated measurements to tracks such that the total cost is minimized, subject to the constraint that each measurement can be assigned to only one track (or to no track).
- *Auction methods:* These globally optimum methods typically compare a prediction to the two or more closest validated measurements at each step. A bidding process assigns the closest measurement to each track. Costs are adjusted based on assignments made so far, allowing for reassignment of some measurements. A popular, computationally efficient version is the JVC algorithm (Malkoff, 1997).
- *Probabilistic data association* (PDA): Updates tracks with a weighted average of all validated measurements. The weights are based on an estimated probability that a given measurement is the correct one and are related to the likelihood function.

Figure 7.35a is a notional example of a cost matrix for an environment in which two tracks are being maintained, but three measurements have passed the gating stage and been validated as candidates to update the existing tracks. The numbers in each cell represent the log-likelihood cost of the corresponding assignment. The cost matrix includes a column for the possibility that a track receives no update, and a row for the possibility that a measurement is assigned to no track.

Figure 7.35b shows the results of a simple NN assignment. The algorithm considers each track in sequence. In this example, measurement 1 is the lowest cost candidate for both track 1 (cost = 5) and track 2 (cost = 10). There being no other tracks to assign measurements 2 and 3 to, both are assigned to "none," that is, not used, at a cost of 20 each. These measurements

(a)

Track	Measurement 1	2	3	None
1	5	10	15	20
2	10	30	25	20
None	20	20	20	∞

(b)

Track	Measurement 1	2	3	None
1	**5**	10	15	20
2	**10**	30	25	20
None	20	20	20	∞

(c)

Track	Measurement 1	2	3	None
1	**5**	10	15	20
2	10	30	25	**20**
None	20	**20**	20	∞

(d)

Track	Measurement 1	2	3	None
1	5	**10**	15	20
2	**10**	30	25	20
None	20	20	**20**	∞

FIGURE 7.35 (a) Track-measurement assignment cost matrix. (b) Result of simple NN assignment, cost = 55. (c) Sequential NN assignment starting with track 1, cost = 65. (d) Sequential NN assignment starting with track 2, and GNN assignment, cost = 40.

may be used to initiate new tracks. The total cost of the set of assignments is $5 + 10 + 20 + 20 = 55$.

One flaw highlighted by this example is that two tracks have been updated with the same measurement, so that one of the updates is probably wrong. Applying the *sequential NN* approach and starting with track 1 assigns measurement 1 to track 1; measurement 1 is then no longer eligible for assignment to any other track. No measurement is assigned to track 2, meaning that it will be coasted to the next measurement update. Measurements 2 and 3 are again not used. This result, shown in Fig. 7.35c, incurs a higher total cost of 65 but avoids dual use of measurement 1.

A flaw in this approach is that the results depend on the order in which the tracks are considered. Repeating the modified NN approach but starting with track 2 will produce the result shown in Fig. 7.35d, with measurement 1 being assigned to track 2, measurement 2 to track 1, and measurement 3 yet again not being used. The total cost is now only 40, lower than sequential NN starting with track 1 while still avoiding dual use of measurement 1.

Finally, in this particular example the solution shown in Fig. 7.35d happens to also be the global optimum, that is, the lowest possible total cost solution. The GNN algorithm would find this solution. The JVC algorithm would also be expected to find this solution.

Tracking, especially in multi-target scenarios, is a rich area of ongoing research. When there is little difference in assignment costs, *feature-aided tracking* may be used to augment the decision variables with estimates of target amplitude, pose, or image information such as high range resolution profiles. In especially dense scenarios, *multiple hypothesis tracking* methods may be used to improve track integrity. Introductory comments and references for these techniques are given in Ehrman (2013).

7.4 Summary

Once a target is detected, the radar system will usually be interested in estimating, as accurately as practicable, its position in three dimensions and possible its Doppler shift (radial velocity) and amplitude. The ability to do so is limited by interference signal such as noise, jamming, and clutter, and by system limitations such as sampling density, channel matching, and quantization errors. A good estimator is considered to be one that provides an unbiased (mean error equal to zero), accurate (low error variance), and consistent (error tends to zero if enough data is available) estimate of a parameter. The radar will also often be required to track the evolution of these parameters over time to understand the target's motion relative to the radar and predict its future position.

The Cramèr-Rao lower bound was introduced as a standard for the minimum error variance achievable with an unbiased estimator when the interference is additive white Gaussian noise. Effectively, the CRLB predicts the measurement error due to noise. It was seen that the CRLB is useful primarily in relatively high SNR situations, on the order of 13 dB or higher SNR at the point of estimation of the parameter. (This is the same rule of thumb introduced in Chap. 6 for achieving "reasonable" detection performance.)

The CRLB was seen to predict estimator performance closely only over a limited SNR range. At low SNR, the target fails to rise above the noise and the measurement variances become large, limited only by the allowable measurement window. In digital processors, sample spacing and amplitude quantization eventually limits estimate accuracy at very high SNRs. Computationally simple three- and two-point interpolators are often used to improve estimation accuracy beyond the limits imposed by sampling density.

Maximum likelihood estimators were introduced as practical algorithms for achieving the CRLB, at least asymptotically, in many problems. The MLEs for a problem are often quite intuitive. For instance, the MLE for estimating time delay or range consisted of matched filtering followed by peak detection, while the MLE for frequency or phase of a sinusoid in noise is a Fourier transform followed by peak detection in the resulting spectrum.

Two different approaches to angle estimation were introduced: phase-based and lobing-based. The phase-based technique is essentially angle interferometry and is well-suited to phased array antennas. A variety of lobing techniques tailored to various antenna configurations exist, but many use the same basic idea of refining the nominal angle to a target using a Δ/Σ correction term.

Once the parameter estimates are made, the tracking process attempts to smooth the sequence of measurements to reduce the errors and model the target's movement over time. The techniques presented all seek to minimize the squared error between the track and the actual position. It was seen that the optimal combination of new data with target state estimates smoothed from previous data is a predictor-corrector structure that appears in a variety of common track filters.

Those track filters include the α-β filter and the Kalman filter, both widely used in radar and the main focus of the tracking discussion here. The α-β filter is a simple and intuitive prediction-correction filter used in many older or simpler systems. The KF extends it to provide an optimal, multi-parameter, sequential least-squares estimation algorithm. In steady state, the KF reduces to an α-β filter with fixed gains. An important complication is that radars measure 3D position in spherical coordinates but usually track in Cartesian coordinates. The nonlinear transformation between these coordinates leads to the EKF, which linearizes the coordinate transformation at each step.

Track filtering is just one part of the complete tracking process, which also includes logics for track initiation, track deletion, and track updating. The latter comprises not only the track filtering but also gating and association of detections to tracks. Examples were given of a few simple association methods. Tracking remains an area of very active research due to its many complications such as false alarms, dense target environments, and close and unresolved tracks.

References

Agrež, D., "Weighted Multipoint Interpolated DFT to Improve Amplitude Estimation of Multifrequency Signal," *IEEE Transactions Instrumentation and Measurement*, vol. 51, no. 2, pp. 287–292, Apr. 2002.

Bailey, C. D., "Radar Antennas," Chap. 9 in M. A. Richards, J. A. Scheer, and W. A. Holm (eds.), *Principles of Modern Radar: Basic Principles*. SciTech Publishing, Raleigh, NC, 2010.

Bar-David, I., and D. Anaton, "Leading Edge Estimation Errors," *IEEE Transactions Aerospace and Electronic Systems*, vol. AES-17, no. 4, pp. 579–584, Jul. 1981.

Bar-Shalom, Y., and T. E. Fortmann, *Tracking and Data Association*. Academic Press, Orlando, Florida, 1988.

Bell, K. L. et al., "Extended Ziv-Zakai Lower Bound for Vector Parameter Estimation," *IEEE Transactions Information Theory*, vol. 43, no. 2, pp. 624–637, Mar. 1997.

Blackman, S. S., *Multiple-Target Tracking with Radar Applications*. Artech House, Dedham, MA, 1986.

Blair, W. D., "Radar Tracking Algorithms," Chap. 19 in M. A. Richards, J. A. Scheer, and W. A. Holm (eds.), *Principles of Modern Radar: Basic Principles*. SciTech Publishing, Raleigh, NC, 2010.

Blair, W. D., "Design of Nearly Constant Velocity Filters for Radar Tracking of Maneuvering Targets," *Proceedings IEEE 2012 Radar Conference*, pp. 1008–1013, Atlanta, GA, 2012.

Blair, W. D., M. A. Richards, and D. G. Long, "Radar Measurements," Chap. 18 in M. A. Richards, J. A. Scheer, and W. A. Holm (eds.), *Principles of Modern Radar: Basic Principles*. SciTech Publishing, Raleigh, NC, 2010.

Candan, C., "A Method for Fine Resolution Frequency Estimation from Three DFT Samples," *IEEE Signal Processing Letters*, vol. 20, no. 9, Sep. 2013.

Candan, C., "Analysis and Further Improvement of Fine Resolution Frequency Estimation Method from Three DFT Samples," *IEEE Signal Processing Letters*, vol. 18, no. 6, Jun. 2011.

Daum, F., "Nonlinear Filters: Beyond the Kalman Filter," *IEEE Aerospace and Electronic Systems Magazine*, vol. 20, no. 8, Part 2: Tutorials II, pp. 57–69, 2005.

Ehrman, L. M., "Multitarget, Multisensor Tracking," Chap. 15 in W. L. Melvin and J. A. Scheer (eds.), *Principles of Modern Radar: Advanced Topics*. SciTech Publishing, Raleigh, NC, 2013.

Hofstetter, E. M., and D. F. DeLong, Jr., "Detection and Estimation in an Amplitude-Comparison Monopulse Radar," *IEEE Transactions Information Theory*, vol. IT-15, no. 1, pp. 22–30, Jan. 1969.

Howard, D. D., "Tracking Radar," Chap. 9 in M. I. Skolnik (ed.), *Radar Handbook*, 3d ed. McGraw-Hill, New York, 2008.

Ianniello, J. P., "Time Delay Estimation via Cross-Correlation in the Presence of Large Estimation Errors," *IEEE Transactions Acoustics, Speech, and Signal Processing*, vol. ASSP-30, no. 6, pp. 998–1003, Dec. 1982.

Jacobsen, E., and P. Kootsookos, "Fast, Accurate Frequency Estimators," *IEEE Signal Processing Magazine*, pp. 123–125, May 2007.

Kalata, P. R., "The Tracking Index: A Generalized Parameter for α-β and α-β-γ Target Trackers," *IEEE Transactions Aerospace and Electronic Systems*, vol. AES-20, no. 2, pp. 174–182, Nov. 1984.

Kanter, I., "The Probability Density Function of the Monopulse Ratio for N Looks at a Combination of Constant and Rayleigh Targets," *IEEE Transactions Information Theory*, vol. IT-23, no. 5, pp. 643–648, Sep. 1977.

Kay, S. M., *Fundamentals of Statistical Signal Processing, Vol. 1: Estimation Theory*. Prentice-Hall, Upper Saddle River, New Jersey, 1993.

Levanon, N., *Radar Principles*. Wiley, New York, 1988.

Lyons, R. G., *Understanding Digital Signal Processing*, 3d ed., Sec. 13.15. Prentice-Hall, New York, 2011.

Macleod, M. D., "Fast Nearly ML Estimation of the Parameters of Real or Complex Single Tones or Resolved Multiple Tones," *IEEE Transactions Signal Processing*, vol. 46, no. 1, pp. 141–148, Jan. 1998.

Mahafza, B. R., *Radar Systems Analysis and Design Using MATLAB*, 3d ed. Chapman and Hall/CRC Press, Boca Raton, FL, 2008.

Malkoff, D. B., "Evaluation of the Jonker-Volgenant-Castanon (JVC) Assignment Algorithm for Track Association," *Proceedings SPIE 3068, Signal Processing, Sensor Fusion, and Target Recognition VI*, pp. 228–239, Jul. 28, 1997.

Peebles, P. Z., Jr., *Radar Principles*. Wiley, New York, 1998.

Richards, M. A., "The DTFT and DFT of Noise," technical memorandum, Nov. 24, 2013. Available at http://radarsp.com.

Scarbrough, K., R. J. Tremblay, and G. C. Carter, "Performance Predictions for Coherent and Incoherent Processing Techniques of Time Delay Estimation," *IEEE Transactions Acoustics, Speech, and Signal Processing*, vol. ASSP-31, no. 5, pp. 1191–1196, Oct. 1983.

Scherr, S., et al., "Accuracy Limits of a K-Band FMCW Radar with Phase Evaluation," *Proceedings 9th European Radar Conference*, pp. 246–249, Amsterdam, Oct. 2012.

Sherman, S. M., *Monopulse Principles and Techniques*. Artech House, Norwood, MA, 1984.

Tenne, D., and T. Singh, "Characterizing Performance of α-β-γ Filters," *IEEE Transactions Aerospace and Electronic Systems*, vol. 38, no. 3, pp. 1072–1087, Jul. 2002.

Weiss, A. J., "Composite Bound on Arrival Time Estimation Errors," *IEEE Transactions Aerospace and Electronic Systems*, vol. AES-22, no. 6, pp. 751–756, Nov. 1986.

Problems

1. Re-derive the CRLB of Eq. (7.15) for the real-valued constant-in-AWGN problem starting from the special form of the CRLB for AWGN given in Eq. (7.17).

2. Use the CRLB for transformed parameters (see App. A) and Eq. (7.15) to determine the CRLB for the signal power A^2 for the real-valued constant-in-AWGN problem.

3. Since the sample mean was an efficient estimator for a constant A in additive WGN, it seems plausible that an efficient estimator for A^2 might be the square of the sample mean, $\widehat{A^2} = (\hat{A})^2 = \left(\sum_{n=0}^{N-1} x[n]\right)^2$. Show that this estimator is not efficient even though the estimator for A was efficient. *Hint:* Consider the estimator bias.

4. Show that as $N \to \infty$ the estimator of Prob. 3 becomes efficient. *Hint:* Note that the estimate $\hat{A}$ is Gaussian with mean A and variance σ_w^2/N and use known results for the moments of a Gaussian RV.

5. Determine whether an alternative power estimator for the real-valued constant-in-AWGN problem that forms the sample mean of $x^2[n]$, $\widehat{A^2} = (1/N)\sum_{n=0}^{N-1} x^2[n]$, is efficient. If not, is it asymptotically efficient?

6. Suppose N samples are available of the real-valued signal $x[n] = A\cos(2\pi f_0 n + \varphi) + w[n]$, where $w[n]$ is i.i.d. white Gaussian noise with variance σ_w^2. Starting with Eq. (7.17), find the CRLB for the variance $\sigma_{\hat{A}}^2$ of an estimate $\hat{A}$ of A. *Hint #1:* $\cos^2(2\pi f_0 n + \varphi) = \frac{1}{2} + \frac{1}{2}\cos(4\pi f_0 n + 2\varphi)$. *Hint #2:* Assume $f_0 \ne 0$ or $\pi/2$ and that N is "reasonably large." In that event it can be shown that $\sum_{n=0}^{N-1} \cos(4\pi f_0 n + 2\varphi) \approx 0$.

7. Consider a complex signal $x[n] = s[n] + w[n]$. $w[n]$ is i.i.d. complex WGN with variance σ_w^2. $s[n]$ is the complex echo series from a Swerling 2 target, which means that it is also i.i.d. complex Gaussian but with variance σ_s^2. Write the joint PDF of N samples of $x[n]$. Use this and Eq. (7.11) to find the CRLB for estimating the signal power σ_s^2.

8. Verify that the same result is obtained for the CRLB as in the previous problem if the CRLB form of Eq. (7.10) is used instead. *Hint:* Use standard results for the moments of $N(0, \sigma^2)$ RVs.

9. Show that the variance of a complex parameter $\Theta = \Theta_R + j\Theta_I$ is the sum of the variances of the real and imaginary parts, $\sigma_\Theta^2 = \sigma_{\Theta_R}^2 + \sigma_{\Theta_I}^2$.

10. Starting with Eq. (7.17), verify the real-valued signal form (the second line) of Eq. (7.22).

11. Consider a signal $x[n] = s[n; \theta] + w[n] = Ar^n + w[n]$, $n = 0, \ldots, N-1$. $w[n]$ is white Gaussian noise with variance σ_w^2. The decay rate $r > 0$ is known. All signals and parameters are real-valued. Find the CRLB for estimating A. Use the geometric sum formula $\sum_{n=0}^{N-1} \alpha^n = (1 - \alpha^N)/(1 - \alpha)$ to put the result in a closed form.

12. If $r \to 1$, in the previous problem the data model becomes just a constant A in AWGN. Show that the CRLB found in that problem approaches the CRLB for a constant in AWGN in the limit as $r \to 1$.

13. Find the log-likelihood function for the data in Prob. 11. Use the fact that the noise is WGN and the signal portion Ar^n is a nonrandom constant for a given n to write the PDF of a single sample of $x[n]$. Then write the joint PDF of all N samples and take the natural logarithm to get the log-likelihood function.

14. Find the maximum likelihood estimator of A in the scenario of Prob. 11. To do this, find the value $\hat{A}$ for A that maximizes the log-likelihood function found in Prob. 13. Put any geometric sums in closed form.

15. Show that the MLE for A in Prob. 14 is unbiased.

16. For the same signal model used in Prob. 11, now assume that A is known but r is not. Find the CRLB for estimating r. *Hint:* It is not necessary (or advisable) to convert geometric sums or similar terms to a closed form in this problem; they can be left in summation form.

17. Consider the signal model $x[n] = A + Bn + w[n]$, $n = 0,\ldots, N - 1$. $w[n]$ is white Gaussian noise with variance σ_w^2. All signals and parameters are real-valued. Finding A and B amounts to fitting a straight line to the data. Find the Fisher information matrix for this problem [see Eq. (7.58) or App. A]. (The identities of Eq. (7.62) may be helpful) here as well.

18. Find the CRLBs for the slope B and intercept A in the previous problem. Which parameter estimate improves more rapidly as the amount of data N is increased? The following identities may be helpful:

$$\sum_{n=0}^{N-1} n = \frac{N(N-1)}{2}, \quad \sum_{n=0}^{N-1} n^2 = \frac{N(N-1)(2N-1)}{6}, \quad \begin{bmatrix} a & b \\ c & d \end{bmatrix}^{-1} = \frac{1}{ad-bc}\begin{bmatrix} d & -b \\ -c & a \end{bmatrix}$$

19. Suppose N i.i.d. samples are available from a random process described by the Gaussian PDF

$$p_x(x;\mu) = \frac{1}{\sqrt{2\pi}} \exp\left[-\frac{1}{2}(x-\mu)^2\right]$$

where the mean μ is an unknown parameter. Find the maximum likelihood estimator of μ.

20. Find the maximum likelihood estimator for the parameter λ in the exponential PDF

$$p_x(x;\lambda) = \begin{cases} \lambda \exp(-\lambda x) & x > 0 \\ 0, & x < 0 \end{cases}$$

Discuss how this estimator relates to the mean of the random variable x having this PDF.

21. Using the definition of Eq. (7.34), compute the RMS bandwidth of the following ideal lowpass spectrum $S(F)$:

$$S(F) = \begin{cases} 1, & -B/2 < F < B/2 \\ 0, & \text{otherwise} \end{cases}$$

22. Compute the RMS bandwidth of a trapezoidal pulse defined in the time domain as follows:

$$s(t) = \begin{cases} t/\alpha\tau, & 0 \leq t \leq \alpha\tau \\ 1 & \alpha\tau \leq t \leq (1-\alpha)\tau \\ (1-t/\tau)/\alpha, & (1-\alpha)\tau \leq t \leq \tau \\ 0, & \text{otherwise} \end{cases}$$

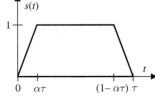

Hint: Use Eq. (7.35) instead of Eq. (7.34). How will time delay estimation precision vary with the steepness of the pulse edges?

23. Use the result from the previous problem to show that the RMS bandwidth of a rectangular pulse is unbounded.

24. Show that the time delay estimate of Eq. (7.42) is unchanged when an unknown gain and phase α are added to the data, that is, $x[n] \to \alpha x[n]$.

25. Consider a coherent (therefore complex-valued signals) radar using an LFM pulse with a "large" (greater than 100) BT product. What is the minimum SNR required to achieve a time delay estimation precision of 5 percent of the Rayleigh resolution in time? Figure 7.6 can be used as an approximate check on the result. If the bandwidth is $\beta = 60$ MHz, what is the range estimation precision in meters?

26. Suppose the output of the matched filter for a complex time delay MLE is oversampled by the factor $k_{os} = 10$. At what output SNR χ_{out} will the sampling bound become the limiting factor in the measurement precision? Assume the pulse is a large-BT product LFM so that $\beta_{rms} \approx \beta/\sqrt{12}$.

27. A time delay MLE produces a peak value of $y_0 = 22.47$ at sample $n_0 = 14$. The two neighboring output samples are $y_{-1} = 21.42$ at sample 13 and $y_{+1} = 16.47$ at sample 15. The SNR for this data was $\chi_{out} = 30$ dB. The true peak location in the absence of noise corresponds to sample 13.8. Estimate the peak location $n_0 + \Delta n$ using each of the techniques in Eqs. (7.51) to (7.53) and compare.

28. Verify that the (1, 2) and (1, 3) elements of the multi-parameter complex sinusoid estimation Fisher information matrix of Eq. (7.58) are zero.

29. Verify the CRLBs of Eq. (7.63) by explicitly computing the diagonal elements of the inverse of $\mathbf{I}(\Theta)$ of Eq. (7.61).

30. Compute the CRLB for estimation of the frequency of a complex sinusoid in AWGN when the amplitude and phase are presumed known. Verify that this CRLB is smaller by a factor of 4 in the limit of large N than the case when all three parameters were unknown given in Eq. (7.63).

31. What are the minimum *input* SNR values χ in dB required to estimate the parameters of a complex sinusoid to the following precision? In each case, give the SNR for $N = 40$ and $N = 400$ samples.

    a. Relative amplitude: 10 percent

    b. Normalized frequency: 0.001 cycles per sample

    c. Phase: 0.01 radians (0.573°)

    Some of the results may be approximately checked using Figs. 7.11 to 7.13.

32. Verify the result given for $\hat{A}_I$ in Eq. (7.69).

33. Suppose $w[n]$ is N samples of a complex white Gaussian noise process with variance σ_w^2 and $W[k]$ is its K-point DFT. Compute the correlation $E\{W[k_1]W^*[k_2]\}$ of two arbitrary DFT samples $W[k_1]$ and $W[k_2]$ as a function of N and K. Use the result to show that two different DFT samples ($k_1 \neq k_2$) are correlated if $K > N$, and also that they are uncorrelated if $K = N$.

34. Consider angle of arrival estimation with a phased array radar having $d = \lambda_t/2$ and $N = 10$ elements. What is the single-element SNR χ required to obtain a precision of 1° in estimating the AOA when the actual AOA $\theta = 0°$ (broadside) and $\theta = 45°$? Repeat for $N = 100$ elements.

35. What is the integrated SNR χ_{out} required to obtain a precision of 10 percent of the 3-dB beamwidth of a phased array radar? What is the equivalent input SNR χ if the array consists of $N = 10$ elements? Repeat for $N = 100$ elements?

36. Derive the MLE for lobing-based angle measurement in terms of the Σ and Δ signals [Eq. (7.92)] from the MLE in terms of the L and R signals [Eq. (7.91)].

37. Verify that in Eq. (7.97) the angle estimate $\hat{\theta} \to \theta_1$ as $\rho \to 0$ and $\hat{\theta} \to \theta_2$ as $\rho \to \infty$. Develop an expression for $\hat{\theta}$ when $\phi = 180°$ and $\rho = 1 + \varepsilon$ and show that the result can be well outside the range of $[\theta_1, \theta_2]$ when ε is small.

38. Suppose the two-way antenna pattern of a conical scan tracking system is modeled as a circularly symmetric Gaussian shape, $E^2(\theta) = G\exp(-\theta^2/\theta_0^2)$, where θ is the angular displacement from the boresight in any direction. Assume that the conical scan boresight locus is displaced from the center of the rotation pattern by the 3 dB point of the pattern, which is approximately $\theta_3 = 0.59\theta_0$ radians off the boresight. Derive an expression for estimating the angular displacement θ_t of a target in terms of the maximum and minimum echo amplitudes during the scan, G_{max} and G_{min}, and θ_3. Ignore noise and assume that the maximum and minimum response occur 180° apart in the rotation. Consider both $\theta_i < \theta_0$ and $\theta_i > \theta_0$.

39. When the signal $x[n] = \Theta$, the parameter to be measured, show that $\varepsilon^2(\Theta)$ of Eq. (7.108) is minimized by choosing the estimate $\hat{\Theta}$ as the sample mean of the data $z[n]$.

40. Show that $\varepsilon^2(\Theta)$ of Eq. (7.108) is minimized in the linear model by choosing the estimate $\hat{\Theta}$ as the weighted sample mean of the data $z[n]$ given in Eq. (7.109).

41. Using the initial conditions stated for Fig. 7.26, verify that the sequential LSE gain and variance obey $K[N] = 1/(N+1)$ and $\text{var}(\hat{A}[N]) = 1/(N+1)$.

42. Using the **F**, **g**, and **h** that define the α-β filter and assuming the process and observation noise variances are $\sigma_u^2 = 1$ and $\sigma_w^2 = 5$, compute the prediction MSE $\mathbf{M}[n\,|\,n]$ and Kalman gain $\mathbf{k}[n]$ for $n = 2, \ldots, 6$. Initialize according to Eq. (7.127) with $T = 1$. MATLAB® or other computational tools may be used to aid in the calculation.

43. With $T = 1$, what is the tracking index for the previous problem? What will be the steady-state Kalman gain for the previous problem? Do the computed gains of the previous problem appear to be converging to this value?

44. Verify the extended Kalman filter model of Eq. (7.138) by explicitly carrying out the computations to obtain the matrix **F**.

CHAPTER 8
Introduction to Synthetic Aperture Imaging

When it was first developed, radar had two primary functions: detection and tracking. To these have been added fine-resolution radar imaging in two and, more recently, three dimensions. The technique of fine-resolution two-dimensional radar imaging is called *synthetic aperture radar* (SAR). SAR is most often applied to imaging of static ground scenes; thus, the "target" in SAR operation is the ground clutter. Civilian applications of radar imagery include cartography, land use analysis, oceanography, forestry, agriculture, natural disaster assessment, and more (Henderson and Lewis, 1998). Equally numerous military applications include reconnaissance, surveillance, battle damage assessment, ground target classification and identification, navigation, and more. SAR maps are routinely generated from both airborne and spaceborne platforms with resolutions ranging from several tens of meters down to a few inches.

Figure 8.1a is an example of a SAR image produced in the mid-1990s. Collected by a Sandia National Laboratories K_u-band radar, this image obtains a resolution of 3 m at ranges of tens of kilometers. Figure 8.1b is an aerial photograph of the same scene; close examination reveals many similarities as well as many significant differences in the appearance of the scene at radar and visible wavelengths. Figure 8.2 is another example, a SAR image of the National Mall area of Washington, DC.

Despite the impressive quality of the SAR images, a human observer would likely prefer photographs for purposes of understanding and analyzing the scene. Though printed here in black and white, the original photograph in Fig. 8.1b is in color, whereas the SAR image is monochrome since SAR measures only the scalar quantity of reflectivity.[1] The photograph has finer resolution than the SAR image. The SAR image exhibits a granular speckle sometimes referred to as "salt-and-pepper" noise that is typical of coherent imaging systems (including holograms, for instance) but absent in the noncoherent optical image. Close inspection reveals differences in phenomenology that can confound image analysis. For example, in the bottom center of the photograph there is a large concrete pad area on which there are three rectangular buildings; the concrete appears light in color, the buildings dark. In the SAR image, the contrast is reversed and the building outlines are indistinct on the bottom side of the image. Another example is the painted stripes visible in the photograph at the ends of the runways (top center and right); these are entirely absent in the SAR image.

Why then is SAR of interest? The answer becomes apparent if the comparison of Fig. 8.1 is repeated on a cloudy night. The photograph would become a solid black, since the ground would not be visible through the clouds and the sun would not be present to provide

---

[1] "False color" or "pseudocolor" SAR imagery is often produced by combining multiple SAR images of the same scene collected at different polarizations and/or radar frequencies.

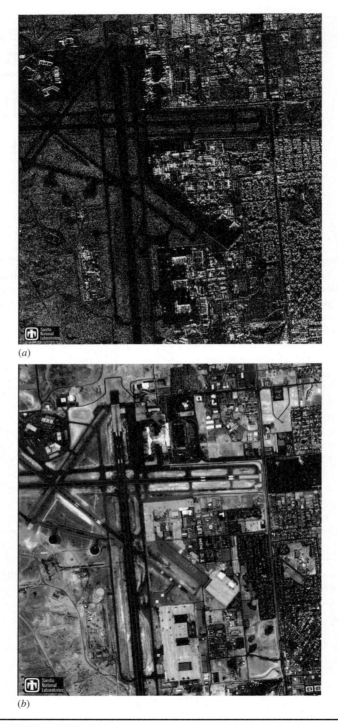

FIGURE 8.1 Comparison of optical and SAR images of the Albuquerque airport. (a) K_u-band (15 GHz) SAR image, 3-m resolution. (b) Aerial photograph. (Images courtesy of Sandia National Laboratories.)

FIGURE 8.2 Synthetic aperture radar image of the Capitol Mall area in Washington, DC. (Image courtesy of Sandia National Laboratories.)

illumination. The SAR image would be unchanged because the SAR is an active system that provides its own illumination, and because microwaves pass through clouds and other weather with little attenuation. Radar provides a means for surveillance at any time of day or night and in a much wider range of weather conditions. Figure 8.3 compares two images taken from the space shuttle of the Manhattan and Long Island areas of New York. Both were taken from an altitude of approximately 233 km in April 1994 at 3:00 A.M. New York time, though on different days. The lower image is a SAR map formed from three bands of radar data collected by the *shuttle imaging radar-C* (SIR-C) instrument.[2] The upper image is a photograph of the Manhattan and Long Island areas of New York city. The overlaid polygonal outline shows the portion of Long Island covered by the SAR image in the lower half of the figure. In the photograph, only those portions of the island that are well illuminated at 3:00 A.M. are visible. This does not include the northern coastal areas of the island. The SAR image, in contrast, successfully images all of the land mass. Though not illustrated, most SARs also easily penetrate most clouds and precipitation, producing imagery that looks exactly the same as if it were collected on a clear day. In fact, Fig. 8.2 is a portion of a larger image of the Washington, DC, area that is said to have been collected at night in a snowstorm. The lack of illumination and the weather have no noticeable effect on the resulting image.

The history of SAR is described briefly in Sherwin et al. (1962) and Wiley (1985). The SAR concept was first described and demonstrated by Carl Wiley of Goodyear Aircraft in 1951. The technique discussed in his patent (Wiley, 1965) would now be categorized as *Doppler beam sharpening* (DBS) (see Sec. 8.3.1). Since then, SAR has undergone several significant technology development phases. The late 1950s and early 1960s developed the

---

[2]This image is a grayscale representation of a false color image generated by assigning red to the L-band horizontal polarization transmit/horizontal polarization receive ("HH") data; green to the L-band horizontal transmit/vertical receive (HV) data; and blue to the C-band HV data.

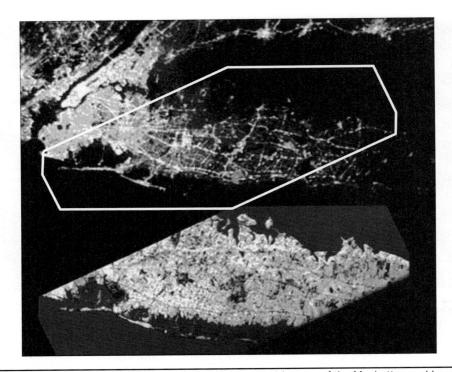

FIGURE 8.3 Comparison of optical (*top*) and radar (*bottom*) images of the Manhattan and Long Island, New York area as viewed from the space shuttle at 3:00 A.M., showing that no external illumination is needed by the radar to image the island land mass. (Image from NASA/JPL-CalTech.)

original concept and implementations of what is now known as *stripmap SAR*. In these pre-Moore's law days, SAR data were collected on photographic film and image formation was performed using remarkably elegant optical analog processing systems (Cutrona et al., 1966; Brown and Porcello, 1969; Harger, 1970; Ausherman, 1980; Elachi, 1988). The 1960s saw the development of *spotlight SAR*, generally credited to Jack Walker of the Environmental Research Institute of Michigan (ERIM) (Ausherman et al., 1984).[3] In the 1970s, digital processing for SAR image formation was developed (Kirk, 1975), while the range-Doppler algorithm (Wu et al., 1982) significantly improved the attainable resolution and image size. In the 1970s, David Munson of the University of Illinois published the connection between spotlight SAR imaging and certain forms of computerized tomography ("CAT scanning") (Munson et al., 1983). This observation was the first of several that significantly expanded the capabilities of SAR algorithms by moving them beyond the range-Doppler viewpoint adopted in their early development and adapting techniques from other fields, such as tomography and seismic prospecting (Munson and Visentin, 1989; Cafforio et al., 1991).

SAR is the first of two advanced radar signal processing techniques to which this text provides an introductory overview. (The other, adaptive array processing, is the subject of Chap. 9.) Beginning in the mid-1990s, a number of excellent textbooks on SAR became

---

[3]Subsequently part of Veridian Corporation, then General Dynamics Corporation, and now Lockheed Martin Corporation.

available (Curlander and McDonough, 1991; Carrara et al., 1995; Jakowatz et al., 1996; Franceschetti and Lanari, 1999; Soumekh, 1999; Cumming and Wong, 2005). The reader is referred to these for in-depth discussion of SAR systems and processing. Modern, concise introductory references are provided in the SAR chapters in the *Principles of Modern Radar* series (Showman, 2010; Cook, 2013; Richards, 2013; Showman, 2013). This chapter begins with a heuristic overview of the SAR concept from two points of view: that of the synthetic antenna aperture beamwidth, and of the Doppler resolution. These are sufficient to derive many of the fundamental equations describing SAR resolution, coverage, and sampling requirements and to describe the nature of the SAR data set. The signal processing required for SAR image formation is then addressed more directly, describing a basic SAR data model and introducing five of the most common algorithms for SAR image formation. A brief discussion of combined SAR and MTI extends the results in Chap. 5. Finally, the concept of the interferometric SAR approach to three-dimensional radar imaging is introduced.

SAR imaging developed in pulsed radars, and the discussion in this chapter, is limited to pulsed systems. However, today SAR imaging is also performed with many frequency-modulated continuous wave (FMCW) radars, again usually relying on the fast chirp variants and utilizing many of the same basic image formation algorithms. An example of short-range FMCW SAR imaging is found in Charvat (2014).

8.1 Fundamental SAR Concepts and Relations

8.1.1 Cross-Range Resolution in Radar

To be useful, a radar map must provide adequate resolution for its intended use. Resolution requirements may range from tens of meters to fractions of a meter. Furthermore, this resolution should be available in both ground dimensions, since there is no reason to prefer one over the other in most operational scenarios. Finally, the resolution should be maintained throughout the imaged scene. Sufficient range resolution for radar mapping is relatively easy to achieve using the high bandwidth waveforms discussed in Chap. 4. Comparable cross-range[4] resolution, however, is not possible in conventional operation, often termed "real-beam imaging."

Figure 8.4 illustrates cross-range resolution in a real-beam forward-looking radar. The two-dimensional scenario shown is equivalent to viewing the three-dimensional scenario in the *slant plane* formed by the boresight vector and the ground plane. The antenna scans in azimuth angle. It has an azimuth beamwidth of θ_{az} radians; thus, at a range R_0 the width of the beam is $R_0\theta_{az}$ meters to a good approximation. The cross-range dimension is the direction orthogonal to range. As discussed in Chap. 2, the receiver output for a fixed range as a function of azimuth scan angle is the range-averaged reflectivity convolved with the two-way antenna voltage pattern. In Fig. 8.4a, the two scatterers are separated in cross range by less than one beamwidth, so the receiver output will blur the response to the two scatterers together (see Fig. 2.27). In Fig. 8.4b, they are separated by more than the beamwidth so that the receiver output for the appropriate range bin as a function of scan angle will show two distinct peaks. By convention, the two scatterers are therefore considered just resolvable if

---

[4]"Cross-range" is the direction orthogonal to the radar range direction, and thus to the radar antenna boresight. It differs from azimuth, in that azimuth specifies an angular displacement relative to the boresight, while cross range specifies a displacement in an orthogonal Cartesian coordinate. If the radar is sidelooking and the platform is not crabbed, the cross-range direction is parallel to the platform velocity vector. Thus the cross-range dimension is sometimes referred to as the *along-track* dimension.

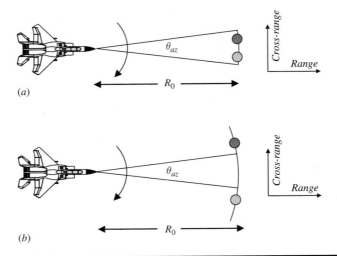

FIGURE 8.4 Resolution in cross range of two scatterers at the same range. (*a*) Unresolved in cross range. (*b*) Resolved in cross range.

they are separated by the width of the antenna beam. Assuming narrow azimuth beamwidths, the cross-range resolution ΔCR is well approximated by

$$\Delta CR = 2R_0 \sin\left(\frac{\theta_{az}}{2}\right) \approx R_0 \theta_{az} \quad \text{m} \tag{8.1}$$

The beamwidth θ_{az} in this equation is usually taken as the two-way 3-dB beamwidth, but it is sometimes taken as the Rayleigh (peak-to-null) or even the null-to-null beamwidth.

As shown in Chap. 1, the azimuth beamwidth of a conventional antenna is of the form

$$\theta_{az} = k\frac{\lambda_t}{D_{az}} \quad \text{rad} \tag{8.2}$$

where D_{az} is the width of the antenna in the azimuth dimension. The scale factor k depends on the antenna design. It is as little as 0.89 for an ideal aperture antenna with uniform illumination, but more often is on the order of 1.2 to 3.0 for practical antenna designs. In general, the lower the antenna sidelobes, the higher is k. Here it is sufficient to say that k is on the order of one and ignore it in subsequent calculations for compactness. Combining Eqs. (8.2) and (8.1) gives the cross-range resolution of a real-beam radar at range R as

$$\Delta CR_{\text{real beam}} = \frac{\lambda_t R}{D_{az}} \quad \text{m} \tag{8.3}$$

The resolution indicated by Eq. (8.3) is not acceptable for imaging purposes. Unlike range resolution, the cross-range resolution degrades in proportion to range instead of being constant throughout the image. Far more important, the cross-range resolution is too coarse for useful images. Consider some typical numbers. An airborne tactical X-band radar (10 GHz) with a 1-m antenna width would achieve a cross-range resolution of 300 m at 10 km range, or roughly three American football fields. A satellite in *low earth orbit* (LEO, for example around 770 km altitude) operating in C band (5 GHz) with a 10-m antenna

FIGURE 8.5 Relative size of a fighter aircraft and a phased array antenna large enough to achieve $\Delta CR = 3$ m at X band and 10 km range.

would exhibit a cross-range resolution of 4.6 km. These numbers are inadequate for useful imagery.

Equation (8.3) does suggest that cross-range resolution can be improved by restricting the operating range, using higher frequencies, or using larger antennas. Considering the airborne example, a change of two orders of magnitude is required to improve the resolution from 300-m to the 3-m resolution of Fig. 8.1a. This requires a change to either a 1-THz radar frequency, limitation to only 100 m operating range, an increase in the antenna size to 100 m, or some combination of less drastic but still very large changes in these parameters. Such large changes are impractical. For instance, Fig. 8.5 compares the approximate relative size of a 100-m phased array antenna and a typical fighter aircraft. It seems unlikely that such a large antenna could be constructed and flown on that aircraft.

8.1.2 The Synthetic Aperture Viewpoint

In fact, Fig. 8.5 does suggest a way in which fine cross-range resolution could be achieved. Rather than constructing a large physical phased array antenna to meet the requirements of Eq. (8.3), consider implementing only a single array element of the antenna, and then utilizing the platform motion to move that element through successive element positions to form the complete array. At each element position a pulse is transmitted and the fast-time data collected. When the element has traversed the length of the complete array, the data from each position is coherently combined in the signal processor to create the effect of a large phased array antenna with elements at each of the positions. The individual "element" can be the conventional antenna on the platform. In effect, the usual combining of phased array element signals in microwave hardware is performed instead in the signal processor. The system thus "synthesizes" a large phased array antenna aperture by operating a single element from multiple locations in space; hence the name, "synthetic aperture radar" (in some older literature, "synthetic array radar"). Put another way, the phased array antenna data is collected serially, one element at a time, rather than in parallel, all at once. This process is suggested by Fig. 8.6, which shows an aircraft that has collected data at four positions along the array; data are still to be collected at several more positions. While good range resolution is obtained via pulse compression, the cross-range resolution of the fast-time samples is wide and increases with range. After processing, the data have fine resolution in both range and cross range. Because data from multiple pulses are combined to form the effective fine-resolution beam, the scene being imaged should not change while the data are collected so that each pulse represents data from the same scenario. This again emphasizes that SAR is intended primarily for imaging static scenes.

In practice the radar does not usually collect data over just a single synthetic aperture length sufficient to obtain the desired narrow effective beamwidth. Instead, it operates continuously during flight, producing an ongoing sequence of fast-time data vectors from successive positions along the flight path. The effective synthetic aperture size D_{SAR} is determined by selecting the number of spatial positions from which data will be combined

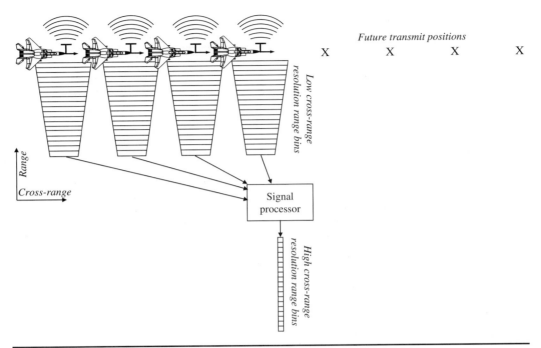

FIGURE 8.6 The concept of synthetic aperture operation.

to form a narrow effective beam. The amount of data combined is expressed in terms of D_{SAR} or equivalently the *aperture time* T_a, which is related to D_{SAR} by the platform velocity:

$$D_{SAR} = vT_a \text{ m} \tag{8.4}$$

By combining a sliding window of data collected over the most recent T_a seconds or D_{SAR} meters of the flight path, a series of narrow effective beams can be formed at successive cross-range positions. At each such position the signal processor forms a set of range bins with fine resolution in both range and cross-range centered at the cross-range position corresponding to the center of the synthetic aperture as illustrated in Fig. 8.7.[5] This series of range traces forms a two-dimensional radar image of the scene. So long as the radar remains in this mode and the signal processor can keep up with the data influx, the system can generate a continuous strip of imagery, rather like unrolling a long scroll.

The mode of operation implied by Fig. 8.7 is called *sidelooking stripmap SAR*. Because the synthetic aperture is formed by the forward motion of the radar platform, the array face is naturally oriented perpendicular to the flight path. The effective antenna pattern is then oriented orthogonal to the velocity vector, a configuration referred to as sidelooking radar. In stripmap SAR, the physical antenna that serves as the array element is *not* actively scanned. It is instead locked into the sidelooking position, and its antenna pattern therefore moves across the ground as the aircraft flies forward. Conventional phase steering combined

---

[5]Successive synthetic apertures are usually overlapped, with offsets ranging from as little as one pulse position to 50 percent or more of D_{SAR}. The two data subsets in Fig. 8.7 are shown nonoverlapping for clarity.

Introduction to Synthetic Aperture Imaging

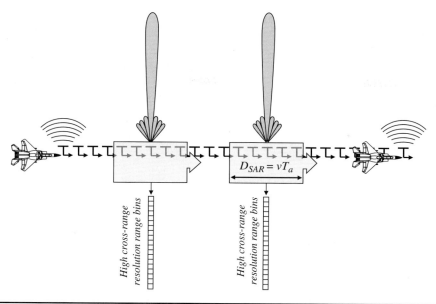

FIGURE 8.7 Forming multiple synthetic apertures by combining sliding subsets of continuously collected data.

with mechanical steering of the physical antenna can be used to orient the effective SAR beam at an angle to the velocity vector other than 90°; this is referred to as a *squinted SAR*. For simplicity, only the sidelooking case is considered in this chapter.

In a conventional phased array antenna, all elements are present and active at once on both transmit and receive for each pulse, so the phase center of the antenna is in the middle of the physical structure for both transmit and receive. For a synthetic phased array, this is not the case: only one element at a time is active, so the phase center for transmit and receive moves across the synthetic aperture as the data are collected. The resulting antenna pattern for a synthetic array differs somewhat from that of a physical array. To see this, consider the geometry of Fig. 8.8, which shows a scatterer at a range R and angle θ from the center of an array. Assuming that R is large compared to the array size, the range from the nth element is well approximated as

$$R_n = R - nd\sin\theta \text{ m} \tag{8.5}$$

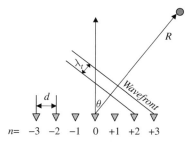

FIGURE 8.8 Geometry of synthetic array.

where d is the element spacing. Suppose the signal $\exp(j\Omega_t t)$ is transmitted from the nth element. The received radar frequency (RF) echo will be, ignoring scale factors,

$$\bar{y}_n(t) = \exp\left\{j\Omega_t\left[t - \frac{2}{c}(R - nd\sin\theta)\right]\right\} \tag{8.6}$$

In synthetic array operation, each element is operated separately and the outputs combined. The total RF output is

$$\bar{y}(t) = \sum_{n=-M}^{+M} \bar{y}_n(t) = \sum_{n=-M}^{+M} \exp\left\{j\Omega_t\left[t - \frac{2}{c}(R - nd\sin\theta)\right]\right\}$$

$$= \exp(j\Omega_t t)\exp(-j4\pi R/\lambda_t)\sum_{n=-M}^{+M}\exp(j2\Omega_t nd\sin\theta/c) \tag{8.7}$$

$$= \exp(j\Omega_t t)\exp(-j4\pi R/\lambda_t)\left\{\frac{\sin[(2M+1)\Omega_t d\sin\theta/c]}{\sin(\Omega_t d\sin\theta/c)}\right\}$$

The term in brackets is the array factor of the synthetic array. It determines the amplitude of the received signal and therefore specifies the two-way voltage pattern of the antenna. The Rayleigh beamwidth is the angle to the first null, which occurs when the argument of the sine function in the numerator equals π. Defining the total aperture size as $D_{SAR} = (2M+1)d$, this occurs when

$$\sin\theta_{SAR} \approx \theta_{SAR} = \frac{\lambda_t}{2D_{SAR}} \tag{8.8}$$

In comparison, the array factor of a conventional phased array, given in Eq. (1.14), gives a Rayleigh beamwidth of $\theta_R = \lambda_t/D$. Thus, the synthetic array has a beamwidth half that of a conventional array of the same aperture size.[6]

It is now possible to determine the cross-range resolution obtained by the synthetic array. Combining Eqs. (8.4) and (8.8) gives

$$\Delta CR = R\theta_{SAR} = \frac{\lambda_t R}{2D_{SAR}} = \frac{\lambda_t R}{2vT_a} \text{ m} \tag{8.9}$$

Equation (8.9) is a fundamental result that relates the amount of data combined, represented through the aperture time T_a or size D_{SAR}, and the SAR cross-range resolution. Solving for T_a or D_{SAR} gives the design equation

$$T_a = \frac{\lambda_t R}{2v \cdot \Delta CR} \text{ sec} \quad \text{or} \quad D_{SAR} = \frac{\lambda_t R}{2 \cdot \Delta CR} \text{ m} \tag{8.10}$$

Equation (8.10) also gives the first hint of one of the major complications in SAR signal processing: the aperture time required to obtain a constant cross-range resolution is proportional to range, implying that the required processing is different at different ranges. SAR is a linear process but is not shift-invariant in general.

---

[6] As another explanation of this effect, it will be seen in Chap. 9 that the virtual array corresponding to a synthetic aperture size D is twice that of the virtual array corresponding to a physical aperture of size D, giving a SAR beamwidth one-half that of the physical array beamwidth.

Beam Mode	Nominal Swath Length, km	Approximate Resolution, m Range	Approximate Resolution, m Cross Range
Ultra fine	20	3	3
Multi-look fine	50	8	8
Fine quad-pol	25	12	8
Standard quad-pol	25	25	8
Fine	50	8	8
Standard	100	25	26
Wide	150	30	26
ScanSAR narrow	300	50	50
ScanSAR wide	500	100	100
Extended high	75	18	26
Extended low	170	40	26

Data from http://www.asc-csa.gc.ca/eng/satellites/radarsat2/inf_data.asp.

TABLE 8.1 Resolution Modes in RADARSAT-2

One of the major advantages of SAR operation over a fixed physical antenna is that the effective aperture size is determined by the amount of data along the flight path that is combined to form any one pixel in the SAR image. The cross-range resolution can be changed by changing T_a (equivalently, D_{SAR}) in the signal processor according to Eq. (8.9). Consequently, a single radar can achieve several different resolution modes with the same antenna hardware. As an example, Table 8.1 lists the various resolution modes available in the Canadian RADARSAT-2 spaceborne SAR. Available resolutions vary by a factor of over 30×, while *swath lengths* (depth of the image in the range dimension) vary by 25×. Notice that the range resolution in most, though not all, modes is equal or nearly equal to the cross-range resolution, a condition referred to here as "square pixels." Table 8.1 also shows that finer resolution is generally associated with shorter swath lengths, implying lesser area coverage rates. The reason for this will be discussed in Sec. 8.1.4.

Equation (8.10) also suggests that arbitrarily fine resolution can be obtained, at least in principle, by letting T_a or D_{SAR} become large. However, the maximum practical value of aperture time is limited by the physical antenna on the platform. Consider Fig. 8.9. When the aircraft is at the position on the left, the target is just entering the mainbeam of the physical antenna; when the aircraft is at the position on the right, the target is just exiting the mainbeam. For aircraft positions before or after this interval, the target is not in the physical antenna mainbeam and the data collected outside of this interval will have no significant contribution from the target. Consequently, any one scatterer contributes significantly to the SAR data only over a maximum synthetic aperture size equal to the travel distance between the two points shown, which equals the width of the physical antenna beam at the range of interest, namely $R\theta_{az}$. The corresponding maximum effective aperture time is $R\theta_{az}/v$. Inserting this result into Eq. (8.9) and using Eq. (8.2) with $k = 1$ gives a lower bound on sidelooking stripmap SAR cross-range resolution of

$$\Delta CR_{min} = \frac{D_{az}}{2} \text{ m} \qquad (8.11)$$

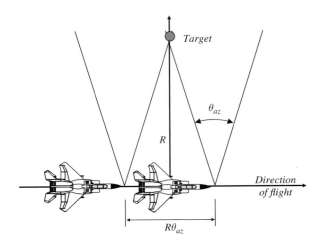

FIGURE 8.9 Limitation of the aperture time by the physical antenna beamwidth.

Equation (8.11) is a remarkable result. It states that the lower bound on stripmap SAR resolution is independent of range, a desirable result for imaging, and more importantly, is much smaller than the real-beam resolution of Eq. (8.3). Considering the same two examples given previously, the lower bound on cross-range resolution becomes 0.5 m instead of 300 m for the airborne case, and 5 m instead of 4.6 km for the spaceborne case.[7] Furthermore, the lower bound on cross-range resolution does not depend on the wavelength. Finally, note that Eq. (8.11) states that to improve the lower bound on cross-range resolution, the physical antenna size should be reduced! It is rare that improved performance calls for *reducing* the antenna size. In SAR, this occurs because a smaller physical antenna will have a broader beamwidth θ_{az}, allowing a larger maximum synthetic aperture size. Of course, shrinking the antenna size reduces the gain and SNR and, as will be seen, also reduces the maximum area imaging rate. The real implication of Eq. (8.11) is that fine cross-range resolution requires a large synthetic aperture, not a large physical antenna. Finally, note that in many cases the lower bound of Eq. (8.11) will be finer than required. In this case, the aperture time is simply limited to whatever value is required to obtain the desired resolution, as given by Eq. (8.10).

Resolutions of 1 to 10 m at ranges of 10 to 50 km, cruise velocities of 100 to 200 m/s, and frequencies from 10 to 35 GHz are typical of many operational airborne SARs, and thus might be considered "mainstream" parameters. Figure 8.10a is a representative plot of aperture time versus radar frequency with ΔCR as a parameter using Eq. (8.10). While low-frequency fine-resolution SARs can demand very long aperture times, for most systems in their mainstream modes T_a is in the range of a few tenths of a second to 1 or 2 seconds. Much finer resolutions are possible but require much longer aperture times, for example, about 10 seconds for 4-in resolution at 10 GHz. For spaceborne systems, frequencies are most often in the range of 1 to 5 GHz, with some newer systems operating at 10 GHz. In LEO a typical velocity and range are about 7500 m/s and 770 km. Figure 8.10b repeats the calculation of T_a for this scenario. Again, "mainstream" aperture times are on the order of a few tenths of a

---

[7]In practice, these resolution values will be degraded by 50 to 100 percent by the use of windows for sidelobe control in the processing.

Introduction to Synthetic Aperture Imaging 511

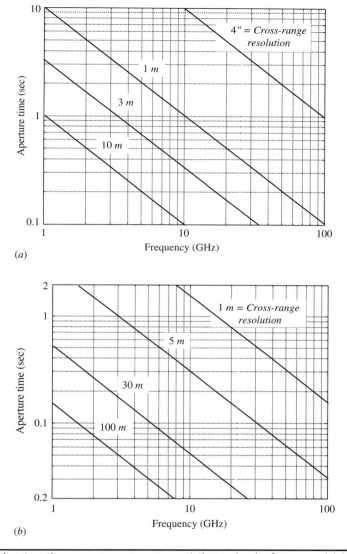

FIGURE 8.10 Aperture time versus cross-range resolution and radar frequency. (a) Airborne platform, $v = 150$ m/s and $R = 10$ km. (b) Spaceborne platform in low earth orbit, $v = 7500$ m/s and $R = 770$ km.

second to 1 or 2 seconds. Finer resolutions of 1 or 2 m are available with longer aperture times of a few seconds.

While the stripmap resolution lower bound of Eq. (8.11) is finer than needed in many cases, in some fine-resolution applications it may not be good enough. The solution is to construct a longer synthetic aperture. To keep a given scatterer within the physical antenna mainbeam, it then becomes necessary to abandon the restriction that the antenna is not scanning. As the radar flies the synthetic aperture, inertial navigation data are used to actively scan the antenna so as to keep its boresight pointed at the center of a *region of interest* (ROI)

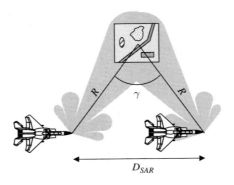

FIGURE 8.11 Spotlight mode SAR operation.

on the ground. In so doing, the ability to map a continuous strip is sacrificed for a finer-resolution map of the ROI. Once enough data have been collected to image the ROI, the antenna can be re-steered to image another discrete region. This mode of operation, termed *spotlight SAR*, is illustrated in Fig. 8.11. Because the radar essentially rotates around the region being imaged, spotlight SAR resolution is often expressed in terms of the rotation angle γ of the radar boresight vector as the platform traverses the synthetic aperture length. Clearly, $D_{SAR} = 2R\sin(\gamma/2) \approx R\gamma$ for small γ. Using this equivalence in Eq. (8.9) gives the alternate cross-range resolution expression

$$\Delta CR = \frac{\lambda_t}{2\gamma} \text{ m} \tag{8.12}$$

Figure 8.12 is a 1-m resolution spotlight SAR image of the Pentagon in Washington, DC, clearly showing remarkable details of the five rings of the building, individual trees, and the surrounding roads.

8.1.3 Doppler Viewpoint

The fundamental SAR resolution formula of Eq. (8.9) was derived above from a basic antenna properties point of view. It can also be derived from a Doppler processing point of view. In fact, this approach is more consistent with the original conception of SAR and serves as a better starting place for considering some SAR image formation algorithms. It also allows an easy generalization to squinted SAR.

Consider two scatterers at range R separated in cross range by ΔCR meters. The angular separation $\Delta\theta$ of the two scatterers satisfies

$$\Delta CR = 2R\sin\left(\frac{\Delta\theta}{2}\right) \approx R \cdot \Delta\theta \quad \Rightarrow \quad \Delta\theta \approx \frac{\Delta CR}{R} \text{ rad} \tag{8.13}$$

The small angle approximation is accurate within 1 percent so long as $\Delta\theta < 14°$. In Chap. 3 it was seen that two scatterers separated in angle by $\Delta\theta$ radians around a nominal squint angle ϕ from forward looking have a difference in Doppler shift of $(4v/\lambda_t)\sin(\Delta\theta/2)\sin\phi \approx (2v\Delta\theta/\lambda_t)\sin\phi$ Hz. Using this result in Eq. (8.13) gives the Doppler difference between two scatterers separated by ΔCR meters at a squint angle ϕ as

$$\Delta F_D = \frac{2v \cdot \Delta CR \sin\phi}{\lambda_t R} \text{ Hz} \tag{8.14}$$

Introduction to Synthetic Aperture Imaging   **513**

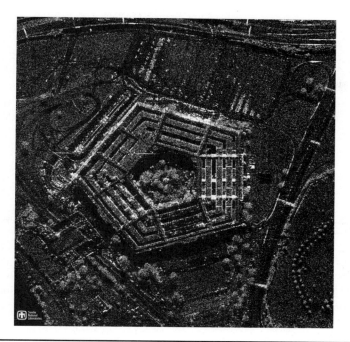

FIGURE 8.12 One-meter resolution spotlight SAR image of the Pentagon. (Courtesy of Sandia National Laboratories.)

These two scatterers could be resolved by Doppler processing of the slow-time data provided that the Doppler resolution is no larger than ΔF_D Hz. As discussed in Chap. 1 and App. B, frequency resolution is inversely proportional to signal duration in time, $\Delta F = 1/T$. In this scenario the signal duration is the aperture time T_a. Thus, two scatterers separated in cross range by ΔCR meters can just be resolved if $T_a = 1/\Delta F_D$. This gives

$$T_a = \frac{\lambda_t R}{2v \cdot \Delta CR \sin \phi} \text{ sec} \qquad (8.15)$$

which generalizes Eq. (8.10) to include the squinted case. In the sidelooking case $\phi = 90°$ and Eq. (8.15) is identical to Eq. (8.10).

The Doppler viewpoint suggests a starting point for SAR image formation algorithms. Figure 8.13 illustrates a two-dimensional view of a sidelooking stripmap SAR. A scatterer **P** is located at range and cross-range coordinates R_p and x_p relative to the antenna and is within the physical antenna mainbeam. Note that R_p is the distance to the range bin containing **P** or, alternatively, the range to **P** at the point of closest approach; the total range R to **P** is $\sqrt{x_p^2 + R_p^2}$. **P** is therefore at a squint angle $\psi_p = \tan^{-1}(x_p/R_p) \approx x_p/R_p$, where the approximation is accurate within 1 percent for $\theta \leq 9.9°$. This scatterer will produce a Doppler-shifted echo. Pulse Doppler processing of the slow-time data in the range bin corresponding to R_p will produce a peak in the Doppler spectrum at $F_{Dp} = (2v/\lambda_t)\cos(\psi_p) = (2v/\lambda_t)\sin(\theta_p)$ hertz. (The squint angle ψ_p of the scatterer from the forward-looking velocity vector has been used in previous chapters, but the angle θ_p measured from sidelooking is more

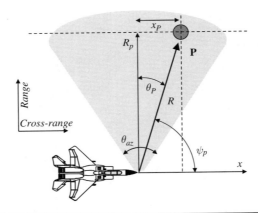

FIGURE 8.13 Geometry relating cross-range position to Doppler shift.

relevant here.) The corresponding cross-range position x_p is approximately R_p times the angular offset from sidelooking:

$$x_p = R_p \sin^{-1}\left(\frac{\lambda_t F_{Dp}}{2v}\right) \approx \frac{\lambda_t R_p F_{Dp}}{2v} \quad \text{m} \qquad (8.16)$$

Equation (8.16) provides a mapping of the Doppler axis to cross-range position. Note that the mapping is range-dependent and therefore different for each range bin. Some systems having a range swath that is a small fraction of the nominal range use only a single mapping based on the nominal range at the center of the imaged swath, since the percentage range change across the image is small.

An image is formed by collecting a fast-time/slow-time data set that covers the range window of interest and provides an aperture time adequate for the desired resolution. The Doppler spectrum is computed at each range of interest. The Doppler axis is then remapped to cross-range position using Eq. (8.16), resulting in a range/cross-range image. If the same aperture time (slow-time data set size) is used in each range bin, this particular SAR imaging algorithm is called *Doppler beam sharpening* (DBS). If the aperture time is increased proportionally to range so as to maintain constant cross-range resolution, it becomes a simple form of the *range-Doppler algorithm*. There are several effects that limit the scene size and resolution obtainable with this simple algorithm, primarily the problems of *range migration* and *quadratic phase errors*. Before discussing these, it is useful to consider some additional aspects of SAR operation.

8.1.4 SAR Coverage and Sampling

To determine the amount of terrain which a radar can image, the size of the image in two dimensions must be considered. In stripmap SAR, the along-track extent of the image is unlimited; so long as the radar platform continues collecting and processing data, the image is extended in cross range. What determines the swath length (range extent) of the image?

Consider Fig. 8.14, which illustrates a sidelooking SAR scenario viewed in the along-track direction. The nominal grazing angle is δ radians. Scatterers outside of the mainbeam elevation beamwidth θ_{el} will not produce significant echoes. Thus, the swath length L_s is upper bounded by the projection of the elevation beam onto the ground plane. This is the

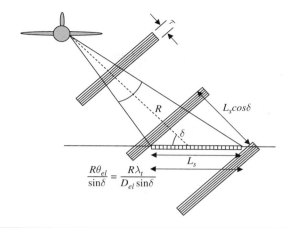

Figure 8.14 Illustration of swath length and unambiguous range interval.

same geometry considered in discussing the beam-limited resolution cell size in Chap. 2; the maximum swath length was seen to be

$$L_{s_{max}} = \frac{R\theta_{el}}{\sin\delta} = \frac{R\lambda_t}{D_{el}\sin\delta} \text{ m} \tag{8.17}$$

The swath length can be made less than this by simply collecting fast-time samples over a more limited extent. In the illustration, the swath length L_s is indicated by the row of range samples within the elevation beam and is shown at its maximum extent, $L_s = L_{max}$. Since the radar platform moves forward at v meters per second while imaging a swath L_s meters deep, the area coverage rate is simply

$$A = vL_s \text{ m}^2/\text{s} \tag{8.18}$$

A technique called *scanSAR* can be used to increase the area coverage rate by increasing the swath length beyond the elevation beamwidth limit, though at the cost of degraded cross-range resolution. The antenna is periodically switched between two elevation angles to create two contiguous swaths. Careful attention to timing is required to avoid leaving gaps in the imagery in either swath. The method, which can be extended to multiple swaths, is described in detail in Lacomme et al. (2001) and Cumming and Wong (2005).

To image the full swath L_s with the intended resolution, two conditions must be met by the received data. The first is that the full pulse echo must be received from all scatterers within the swath on each pulse. Any eclipsing will degrade the resolution of the eclipsed scatterers. The second is that clutter from other ranges must not be allowed to fold over onto the swath of interest. Clutter outside the mainbeam footprint is usually assumed insignificant, so this requirement applies only over the range footprint of the antenna. Also, it is not necessary that the radar be unambiguous out to the full range R. Rather, the requirement is that echoes from the far edge of the mainbeam footprint on one pulse be received before echoes of the next pulse from the near edge of the swath of interest.

The eclipsing constraint requires that the time interval from receipt of the leading edge of the echo from the near edge of the swath to the trailing edge of the echo from the far edge be less than the time $T - \tau$ between pulses when the receiver is on. At a nominal grazing angle of δ radians, the difference in delay between the swath edges is approximately $(2/c)L_s\cos\delta$, so the duration of the data from the swath including the tail of the pulse from the far edge is $(2/c)L_s\cos\delta + \tau$. The resulting requirement is the first relation in Eq. (8.19). The ambiguity constraint is tightest when the swath occupies the full range footprint of the antenna. In that case, it requires that the trailing edge of the clutter from the far edge of the swath not overlap with the leading edge echo from the near end and the next pulse. This results in the second relation in Eq. (8.19) and is seen to be slightly more lax than the eclipsing constraint. The shaded bands within the antenna beam in Fig. 8.14 indicate successive pulses in flight, with a spacing just large enough to meet this requirement. In many but not all SAR systems the swath length is much larger than the equivalent length of the pulse, so that either constraint results in the simplified approximate upper bound on pulse repetition frequency (PRF) shown in the third relationship in Eq. (8.19):

Eclipsing constraint:
$$\frac{2}{c}L_s\cos\delta + \tau < T - \tau \quad\Rightarrow\quad PRF < \frac{1}{\frac{2}{c}L_s\cos\delta + 2\tau}$$

Range ambiguity constraint:
$$\frac{2}{c}L_s\cos\delta + \tau < T \quad\Rightarrow\quad PRF < \frac{1}{\frac{2}{c}L_s\cos\delta + \tau} \qquad (8.19)$$

Approximation for $L_s \gg c\tau/2$:
$$PRF < \frac{c}{2L_s\cos\delta}$$

Because it is the slow-time sampling rate of the radar, a second constraint on the PRF is that it exceed the Doppler bandwidth β_D, as discussed in Chap. 3. For the moment, it is not assumed that β_D necessarily is the maximum sidelooking Doppler bandwidth of $2v\theta_{az}/\lambda_t$ [Eq. (3.15)]. β_D is related to the cross-range resolution of a sidelooking SAR as follows. A resolution of ΔCR meters is equivalent to a resolution in time of $\Delta_t = \Delta CR/v$ seconds. In the absence of weighting for sidelobe control, the time resolution is the inverse of the processed bandwidth, $\Delta t = 1/\beta_D$. The PRF lower bound in terms of cross-range resolution is therefore

$$PRF \geq \beta_D = \frac{v}{\Delta CR} \qquad (8.20)$$

Typical sidelooking SAR Doppler bandwidths vary from a few tens of hertz to a few kilohertz, while typical stripmap range swaths vary from a few ones of km to 100 km. The combination of these constraints results in SAR PRFs of several hundred to a few thousand pulses per second in most cases. SAR is therefore a low PRF radar mode.

Combining the last form of Eq. (8.19) with Eq. (8.20) establishes a range on the allowable PRF,

$$\frac{v}{\Delta CR} < PRF < \frac{c}{2L_s\cos\delta} \qquad (8.21)$$

Equation (8.21) is the basis for several interesting constraint equations in SAR processing. The first constraint follows directly:

$$\frac{L_s\cos\delta}{\Delta CR} < \frac{c}{2v} \qquad (8.22)$$

Notice that the left-hand side is proportional to the number of range resolution cells in the swath length, assuming square pixels ($\Delta CR = \Delta R$). The quantity $c/2v$ is on the order of

20,000 for LEO satellites and the space shuttle and can become a significant constraint. For airborne SAR, it is much larger at 300,000 to 750,000 and much less likely to constrain the system design. If a system is already operating near this limit, Eq. (8.22) shows that further improving resolution requires that the swath length and therefore the area coverage rate [Eq. (8.18)] must be reduced proportionally. Thus, there is a potential conflict between fine resolution and high area coverage rate.

Now suppose that the finest possible sidelooking stripmap resolution is desired so that the full Doppler bandwidth $\beta_D = 2v\theta_{az}/\lambda_t$ is processed. As seen earlier, with $\theta_{az} = \lambda_t/D_{az}$ this results in $\Delta CR = D_{az}/2$. Inserting this value into Eq. (8.22) and rearranging gives a bound on stripmap swath length,

$$L_s \leq \frac{cD_{az}}{4v\cos\delta} \text{ m} \tag{8.23}$$

Equation (8.23) shows that for a given platform velocity, the swath length is limited by the physical antenna width. It also shows that a wider antenna is needed to achieve a given swath length on a faster platform. This is one major reason that spaceborne SARs have much wider antennas than airborne SARs.

Using Eq. (8.23) in Eq. (8.18) gives a closely related area coverage rate constraint of

$$A \leq \frac{cD_{az}}{4\cos\delta} \text{ m}^2/\text{s} \tag{8.24}$$

While good SAR resolution encourages antennas that are narrow in the azimuth dimension [Eq. (8.11)], this equations states that high area coverage rates encourage wider azimuth antenna extents, again reflecting the conflict between fine resolution and high area coverage rates mentioned earlier.

Finally, if maximum area coverage is sought by setting the swath length to the full extent of the antenna range footprint while still maintaining minimum resolution, Eqs. (8.17) and (8.23) can be combined to derive the inequality

$$D_{az}D_{el} \geq \frac{4vR\lambda_t}{c\tan\delta} \text{ m}^2 \tag{8.25}$$

This result specifies a lower bound on the antenna area, not just its width. It shows that spaceborne SARs must have larger area antennas than airborne SARs due to their much larger velocities and ranges. It also states that lower RFs tend to require larger antennas.

Equations (8.23) and (8.24) are often referred to as the stripmap SAR *swath constraint* and *mapping rate constraint* equations, while Eq. (8.25) is called the *antenna area constraint*. A number of variations and extensions of these ideas for other cases are described in Freeman et al. (2000).

The previous equations apply to stripmap SAR. It is straightforward to estimate the mapping rate for spotlight SAR operation as well. From Eq. (8.12), an image with resolution ΔCR requires that the antenna LOS rotate through $\lambda_t/2\Delta CR$ radians. Assuming a straight flight path, the platform must travel $\lambda_t R/2\Delta CR$ meters; this will require $\lambda_t R/2v\Delta CR$ seconds. This is the time required for one spotlight SAR image. The number of images per unit time, N_{spot}, is therefore upper bounded by

$$\begin{aligned} N_{spot} &\leq \frac{2v \cdot \Delta CR}{\lambda_t R} \text{ images per second} \\ &= \frac{7200v \cdot \Delta CR}{\lambda_t R} \text{ images per hour} \end{aligned} \tag{8.26}$$

In SAR, the PRF sets the cross-range sampling interval $\delta x = vT = v/PRF \leq \Delta CR$, representing one pulse per cross-range pixel. Failure to use a high enough PRF results in cross-range ambiguities in SAR imaging. In order to minimize this problem, the maximum Doppler bandwidth $\beta_D = 2v\theta_{az}/\lambda_t$ is typically used as the lower bound on PRF even if the Doppler bandwidth needed for the desired cross-range resolution is less. Furthermore, θ_{az} is often interpreted conservatively as the null-to-null beamwidth, effectively increasing the PRF by about a factor of two compared to the result obtained if the 3-dB or Rayleigh beamwidth is used. Choosing a PRF greater than required by Eq. (8.20) tightens the swath length and area coverage constraints by the same factor.

8.2 Stripmap SAR Data Characteristics

Up to the point where the complex image is finally passed through a detector to convert it to pixel data suitable for display, the SAR signal processing system is nominally linear. It will be seen that stripmap SAR processing is shift-invariant in cross range but not in range. Consequently, the behavior of the SAR system, including such characteristics as resolution, sidelobes, and the response to complex multi-scatterer scenes, can be understood by considering the response to a single isolated point scatterer, parameterized as a function of range. In the nomenclature of linear systems analysis, the data set generated by a single point scatterer is called the *point scatterer response* (PSR) or the SAR system *impulse response*. The former term is preferred because the latter term is most often associated with shift-invariant systems.

8.2.1 Stripmap SAR Geometry

Figure 8.15 defines the geometry for sidelooking stripmap SAR data acquisition. The *ground plane*, which is the surface to be imaged, is defined by the x and y_g axes. The dimension y_s defines the *slant range* dimension, and the plane defined by the x and y_s axes is called the *slant plane*. Because the radar measures delay, this is the more natural plane in which to discuss imaging. Of course, real terrain varies in height. This aspect is discussed in Sec. 8.7. For the present, it is assumed that the ground plane scene is two dimensional.

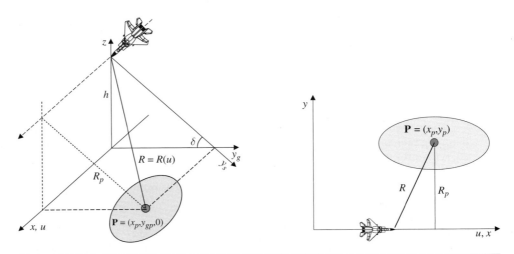

FIGURE 8.15 Geometry for stripmap SAR data acquisition. (*a*) Three dimensions. (*b*) Two-dimensional slant plane.

Consider a sidelooking radar platform flying at constant altitude h and velocity v along the $+x$ direction such that the antenna phase center is located at coordinates $(x = u = vt, 0, h)$ at time t. The antenna is therefore above the ground plane origin at $t = 0$. A scatterer **P** is located in the ground plane at coordinates $(x_p, y_{gp}, 0)$. Assume the antenna is steered such that **P** is in the center of the antenna beam azimuth extent at $t = 0$. If the aperture time required to achieve the desired resolution is T_a seconds corresponding to a synthetic aperture size of D_{SAR} meters, the time interval of interest in imaging **P** is $t \in [-T_a/2, T_a/2]$, corresponding to platform positions along the x axis (the synthetic aperture) of $u \in [-D_{SAR}/2, D_{SAR}/2]$. This geometry implies that if $x_p = 0$ the radar is sidelooking, while if $x_p \neq 0$ it is squinted. Figure 8.15a shows a case having the radar squinted forward.

The range from the radar to **P** is the Euclidean distance

$$R = R(u) = \sqrt{(u - x_p)^2 + y_{gp}^2 + h^2} = \sqrt{(u - x_p)^2 + R_p^2}$$
$$= R_p \sqrt{1 + \frac{(u - x_p)^2}{R_p^2}} \quad \text{m} \tag{8.27}$$

R_p is the range to the scatterer at the point of closest approach of the radar. Working in terms of R_p is also equivalent to letting $h = 0$ to reduce the problem to two dimensions, or to working in the slant plane, as shown in Fig. 8.15b.

Equation (8.27) shows that in a stripmap SAR the range between the radar and an arbitrary scatterer varies hyperbolically as the radar moves along the synthetic aperture. The form of the range variation is invariant to the scatterer's along-track position x_p, in the sense that the x_p dependence of $R(u)$ involves only the position of the aircraft relative to the scatterer ($u - x_p$). More explicitly, if the range variation for a scatterer at x_{p1} is denoted by $R_1(u)$, the range variation for a scatterer at position $xp_2 = x_{p1} + \Delta x$ is $R_2(u) = R_1(u - \Delta x)$. In contrast, the range variation does depend on the absolute slant or ground range of the target, R_p (or y_{gp}, which appears through R_p).

In the analysis of low- to medium-resolution SAR systems, it is common to further simplify Eq. (8.27) by using the series expansion of the square root $\sqrt{1 + x} = 1 + \frac{1}{2}x - \frac{1}{8}x^2 + \cdots$ and keeping only the first two terms, giving

$$R(u) \approx \left[1 + \frac{(u - x_p)^2}{2R_p^2}\right] R_p = R_p + \frac{(u - x_p)^2}{2R_p}$$
$$= R_p + \frac{u^2}{2R_p} - u\frac{x_p}{R_p} + \frac{x_p^2}{2R_p} \quad \text{m} \tag{8.28}$$

This simplification requires $|u - x_p|/R_p \ll 1$. Since the maximum value of $|u - x_p|$ of interest is $D_{SAR}/2$, this is equivalent to stating that the synthetic aperture size is short compared to the nominal range, implying a limit to the SAR cross-range resolution. Equation (8.28) shows that the range from the radar to the target varies approximately quadratically as the data set is collected. Since the received phase of the target echo is shifted from the transmitted phase by an amount proportional to range, namely $\phi = -(4\pi/\lambda_t)R$ radians, it follows that the phase of the received echoes will also vary approximately quadratically. In Chap. 4, it was seen that a quadratic phase function corresponds to linear frequency modulation (LFM). LFM was used to obtain fine resolution in fast time by spreading the waveform bandwidth. Equation (8.28) shows that the slow-time data from a given scatterer will also have an approximately quadratic phase modulation that will spread the slow-time bandwidth and enable fine cross-range resolution. This slow-time modulation is induced by the changing geometry due to platform motion relative to the imaged scene.

It is useful to determine the equivalent of the LFM time-bandwidth product for this cross-range chirp. For a given scatterer position x_p, the received phase will vary with aperture position u as $\phi(u) = -(4\pi/\lambda_t)R(u)$. The instantaneous frequency corresponding to this phase modulation is obtained as usual as

$$K_u = \frac{d\phi(u)}{du} = -\frac{4\pi}{\lambda_t R_p}(u - x_p) \text{ rads/m} \tag{8.29}$$

Since the independent variable u is a spatial coordinate in meters instead of time in seconds, K_u is a spatial frequency. As u varies over the synthetic aperture length of D_{SAR} meters, the spatial frequency bandwidth will sweep over $(4\pi/\lambda_t R_p)D_{SAR}$ radians per meter $= 2D_{SAR}/\lambda_t R_p$ cycles per meter. Defining the latter quantity as the spatial bandwidth β_u, the "space-bandwidth" product of the cross-range chirp is

$$\beta_u D_{SAR} = \frac{2D_{SAR}^2}{\lambda_t R_p} \tag{8.30}$$

Just as matched filtered LFM chirps have a time resolution equal to their duration divided by their BT product, it should be possible by proper processing to achieve a cross-range resolution in SAR that is the synthetic aperture size divided by this space-bandwidth product. In fact, dividing D_{SAR} by Eq. (8.30) does indeed agree with Eq. (8.9). Also, recall that the BT product $\beta\tau$ is the radar range equation signal processing gain for fast-time processing (pulse compression). Similarly, the signal processing gain for slow-time (SAR) processing will be the slow-time space-bandwidth product $\beta_u D_{SAR}$. The overall SAR signal processing gain G_{sp} in the SNR of a point scatterer will be the product

$$G_{sp} = (\beta\tau)(\beta_u D_{SAR}) \tag{8.31}$$

The variation in range to a point scatterer over the synthetic aperture is called *range migration*. In SAR processing it is often broken into two components, *range walk* ΔR_w and *range curvature* ΔR_c, as discussed in Chap. 5. Range walk is the difference between the range to **P** at the beginning and end of the synthetic aperture:

$$\begin{aligned}\Delta R_w &= R(-D_{SAR}/2) - R(+D_{SAR}/2) \\ &= -[(-D_{SAR}/2) - (+D_{SAR}/2)]\frac{x_p}{R_p} \\ &= \frac{D_{SAR}}{R_p}x_p = \frac{vT_a}{R_p}x_p \text{ m}\end{aligned} \tag{8.32}$$

Note that this is just the change over the aperture time in the linear term in u of Eq. (8.28). Range curvature is the variation in the quadratic term of $R(u)$, which has its maximum at either extreme of the aperture position, $u = \pm D_{SAR}/2$, and its minimum at $u = 0$:

$$\Delta R_c = R(\pm D_{SAR}/2) - R(0) = \frac{D_{SAR}^2}{8R_p} = \frac{v^2 T_a^2}{8R_p} \text{ m} \tag{8.33}$$

Both the range walk and range curvature depend on R_p and D_{SAR} or T_a. If a constant aperture time is used, range walk and curvature both decrease as $1/R_p$. However, if constant cross-range resolution is desired, T_a must increase proportionally to range. In this case, ΔR_w is constant over range for a given x_P, while ΔR_c increases in proportion to R_p.

Range migration is significant only if it exceeds the range bin size. If so, echo samples from a single scatterer will start in different range bins on different pulses, complicating the process of combining them to form a fine-resolution image. The degree of both range walk and range curvature can be predicted using only parameters known to the radar system: ownship velocity, aperture time, and range.

8.2.2 Stripmap SAR Data Set

The fast-time/slow-time (or fast-time/aperture position) complex baseband data set used to form the SAR image is the two-dimensional SAR phase history. Figures 8.16 through 8.18

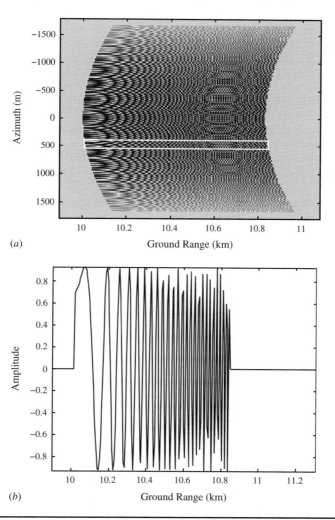

FIGURE 8.16 (a) Real part of the received data from a point scatterer located at $x_P = 0$, $y_{gP} = 10$ km. (b) Fast-time slice at azimuth position $u = 500$ m. See text for radar parameters.

show simulated received data from a single point scatterer located at $x_p = 0$, $y_{gp} = 10$ km. The data correspond to a 5-GHz radar at an altitude of 5 km. The pulse is an LFM upchirp of bandwidth $\beta = 10$ MHz and duration $\tau = 5$ μs, sampled at a fast-time rate of 30 megasamples per second. The PRF is 500 Hz and the platform velocity is 250 m/s. The aperture time is $T_a = 13.5$ seconds. With these parameters, the expected resolutions are 15 m in range and only 0.1 m in cross range. While not very realistic, this example results in a data set with exaggerated curvatures for convenience in illustrating some general characteristics of stripmap SAR data.

Figure 8.16a shows the real part of the complete two-dimensional PSR. The range curvature is clearly evident. The complicated pattern of the data within the PSR reflects the changing amplitude of the real part of the chirp pulse. Each fast-time slice through the data set is simply an echo of the transmitted chirp. For instance, the slice of data along the centerline of the thin rectangular box at azimuth +500 m is shown in Fig. 8.16b; the upchirp is clear. All other fast-time slices are identical except for their start time and initial phase offset, determined by the differing ranges to the target from each point along the radar platform's synthetic aperture.

Figure 8.17 shows a cross-range slice of the data in Fig. 8.16a, taken at ground range 10.4 km. Figure 8.17a shows the real part of the complete cross-range slice, emphasizing the attenuation caused by the physical antenna pattern at the edges of the data set when the point scatterer is near the edge of the antenna mainbeam. The segment inside the rectangular box is expanded in Fig. 8.17b, where it can be seen that the motion-induced phase modulation forms a chirp that passes through zero frequency at $u = x_p = 0$ (broadside from the scatterer). Thus, the PSR is approximately a two-dimensional chirp function on a curved *region of support* (ROS).

The size of the PSR data set which must be processed by a stripmap SAR can be quite large. For a swath length L_s, range resolution ΔR, and pulse time-bandwidth product $\beta\tau$, the number of fast-time samples required is

$$L = \frac{L_s}{\Delta R} + (\beta\tau - 1) \text{ samples} \tag{8.34}$$

This result assumes a fast-time sampling rate equal to the bandwidth β; if oversampling is used, L increases accordingly. The first term is simply the number of range bins in the swath; the second term allows for the samples of the tail of the pulse echo from the last range bin in the swath. The number of slow-time samples equals the number of pulses within the aperture time:

$$M = T_a \text{PRF}$$
$$= \left(\frac{\lambda_t R}{2v \cdot \Delta CR}\right)\left(\frac{2v\theta_{az}}{\lambda_t}\right) = \frac{R\theta_{az}}{\Delta CR} \text{ samples} \tag{8.35}$$

Notice that this is just the number of cross-range resolution cells in the physical beamwidth $R\theta_{az}$. Equation (8.35) assumes that the PRF equals the clutter bandwidth and that the synthetic aperture size is the full antenna azimuth footprint. This also means that the cross-range resolution is the minimum $D_{az}/2$. Again, M increases accordingly if oversampling is used.

As an example, consider the SIR-C, which operates at L band. A typical stripmap SAR mode may have a swath length of 15 km from an orbital range of 250 km, range and cross-range resolution of 15 m, and a modulated 17-μs pulse with a 10-MHz bandwidth. The antenna size is 10 m, suggesting a beamwidth θ_{az} of about 0.03 radians = 1.72° and therefore a cross-range antenna footprint of 7500 m and a minimum sidelooking cross-range resolution of 5 m. The actual cross-range resolution of 15 m, three times greater than the minimum, implies a processed synthetic aperture of one-third the maximum or $D_{SAR} = 2500$ m. The

Introduction to Synthetic Aperture Imaging   523

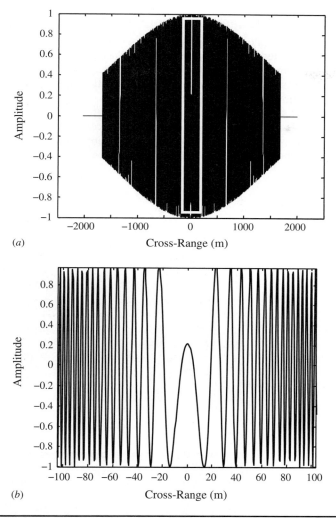

Figure 8.17 Cross-range slice of Fig. 8.16a at ground range $y_{gP} = 10.4$ km. (a) Complete slice. (b) Central portion of the slice.

number of pulses in the synthetic aperture will then be one-third of the value from the full-aperture value of Eq. (8.35), giving $M = 500/3 = 167$. The pulse parameters above give $L = 170$ fast-time samples per pulse from a single point scatterer and $L = 1169$ samples to collect the entire range swath. Putting these numbers together, the PSR representing the response from a single point scatterer will be $\beta\tau = 170$ fast-time samples by 166.7 slow-time samples. This large number of pulses contributing to the image of each point target differentiates SAR from the much simpler but related pulse Doppler processing. In pulse Doppler, a CPI is typically a few tens of pulses. Returning to the SIR-C example, the space-bandwidth product is found from Eq. (8.30) to be 166.7. The total available integration gain is then $(170)(166.7) = 28{,}339$, or about 44.5 dB. This is consistent with the gain expected from coherent integration of the 170×167 PSR samples.

8.3 Stripmap SAR Image Formation Algorithms

For a general scene, the stripmap SAR data set is the superposition of a large number of weighted and shifted PSRs. Recall that the PSR is, in general, a function of range but is independent of cross range. The goal of any SAR image formation algorithm is to compress the PSR of each scatterer into an impulse-like function with the appropriate location and amplitude, as shown notionally in Fig. 8.18.

A variety of stripmap SAR image formation algorithms exist. They vary in resolution capability, scene size capability, and computational complexity. In general, finer resolution, longer standoff range, lower radar frequency, squint geometries, and greater scene size require more elaborate and computationally expensive algorithms. In this section, four basic

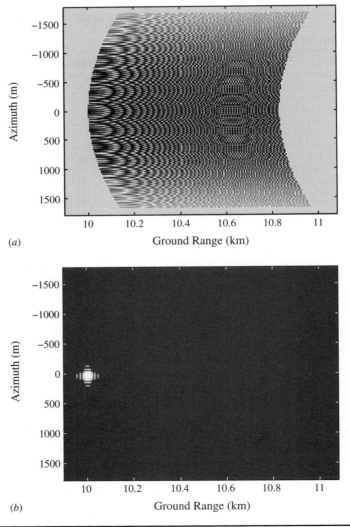

FIGURE 8.18 The goal of SAR image formation algorithms is to transform the two-dimensional PSR in (a) into the point target image in (b).

imaging algorithms are introduced: Doppler beam sharpening, *range-Doppler*, *range migration* (also known as the *ω–k algorithm*), and *backprojection*. These and other advanced techniques such as the *chirp scaling algorithm* are described in detail in Bamler (1992), Gough and Hawkins (1997), Franceschetti and Lanari (1999), Soumekh (1999), Cumming and Wong, (2005), and Showman (2013).

8.3.1 Doppler Beam Sharpening

Doppler beam sharpening is the original form of SAR envisioned by Carl Wiley (1965). It uses a constant aperture time for all ranges so that the cross-range resolution is proportional to range. It is the simplest SAR algorithm, and is suitable only for relatively coarse resolution imagery. Nonetheless, it provides a substantial improvement over real beam cross-range resolution and has relatively low computational requirements.

DBS is the algorithm implied by the Doppler viewpoint of Sec. 8.1.3. It presumes that the PSR is compressed in the fast-time dimension by conventional pulse compression, whether implemented by direct convolution, fast convolution, or stretch processing. Recall from Chap. 4 that the fast-time output of the matched filter is the autocorrelation of the transmitted baseband waveform $x(t)$ with a delay and phase shift proportional to the scatterer range (assuming the filter delay is removed for convenience, i.e., $T_M = 0$). For the wideband frequency- or phase-modulated waveforms used in radar imaging, the shape of the echo will be similar to that of Fig. 4.25 (LFM) or Fig. 4.49 (phase code), with a narrow peak and relatively low fast-time sidelobes. It is useful for the analyses in this and upcoming sections to ignore the sidelobes and model the matched filter response shape as a Dirac impulse function $\delta_D(\cdot)$ located at the appropriate range. Thus, the fast-time output of the matched filter for a fine-resolution pulse transmitted from aperture position u is modeled as

$$y(t) = A \exp[-j(4\pi/\lambda_t)R(u)]\delta_D(t - 2R(u)/c) \\ = A \exp[-j(2\Omega_t/c)R(u)]\delta_D(t - 2R(u)/c) \tag{8.36}$$

where Ω_t and λ_t are the RF frequency in rad/s and the wavelength, and the constant A absorbs all amplitude factors. The two equivalent expressions of the phase shift are interchangeable depending on which is more convenient. A subtlety of this model is that the Dirac impulse has infinite bandwidth, whereas the actual waveform has some finite, if large, bandwidth β that will be taken into account when necessary.

Using Eq. (8.28) to expand $R(u)$, the slow-time phase variation becomes

$$\phi(u) \approx -\left(\frac{4\pi}{\lambda_t}\right)\left(R_p + \frac{u^2}{2R_p} - u\frac{x_p}{R_p} + \frac{x_p^2}{2R_p}\right) \\ \approx -\left(\frac{4\pi}{\lambda_t}\right)\left(R_p - u\frac{x_p}{R_p} + \frac{x_p^2}{2R_p}\right) \text{ rads} \tag{8.37}$$

where the last step assumes that the quadratic phase term in u can be neglected; this assumption will be revisited in Sec. 8.3.2. The instantaneous cross-range spatial frequency K_u is

$$K_u = \frac{d\phi(u)}{du} \approx \frac{4\pi}{\lambda_t} \cdot \frac{x_p}{R_p} \text{ rads/m} \tag{8.38}$$

so that

$$x_p = \frac{\lambda_t R_p}{4\pi} K_u \text{ m} \tag{8.39}$$

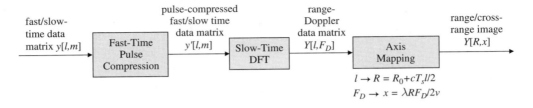

FIGURE 8.19 Block diagram of Doppler beam sharpening algorithm.

Equation (8.39) relates scatterer cross-range position to spatial frequency. Alternatively, x_p can be expressed in terms of Doppler (temporal) frequency. The Doppler frequency F_D in hertz due to a scatterer at (x_p, R_p) is $(1/2\pi)(d\phi/dt)$. Using $u = vt$ in Eq. (8.37) and repeating the calculation of Eqs. (8.38) and (8.39),

$$x_p = \frac{\lambda_t R_p}{2v} F_D \text{ m} \qquad (8.40)$$

a relation that is more natural to the DBS name. This is the same as Eq. (8.16) presented previously.

The algorithm implied by Eq. (8.40) is diagrammed in Fig. 8.19. It requires only one discrete Fourier transform (DFT) per range bin, followed by a rescaling of the axes to express the result in range/cross-range coordinates. Typically, windowing for sidelobe suppression would be included in the processing in both dimensions.

Figure 8.20 is the result of applying a simple DBS algorithm to simulated data from a test array. The magnitude of the DBS image in decibels is shown. Point scatterers were simulated at all combinations of x_p and R_p equal to -1000, 0, or $+1000$ m relative to the central reference point (CRP) at the center of the scene, which was located at a nominal range of 20 km from

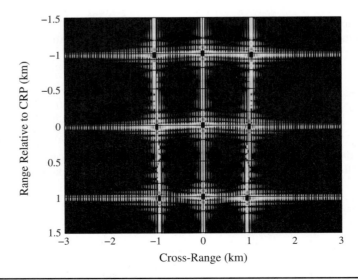

FIGURE 8.20 Magnitude in dB of the DBS image of a test array. See text for parameters.

the radar. The position of the radar is above the image, with range increasing from top to bottom. The aperture time was 40 ms and the velocity was 150 m/s. The radar frequency was 10 GHz, and an LFM pulse with $\beta = 3$ MHz and $\tau = 5\,\mu s$ ($\beta\tau = 15$) was used at a PRF of 7.5 kHz. The image resolution is $\Delta R = \Delta CR = 50$ m. The Doppler frequency axis was scaled to cross-range position using Eq. (8.38) with R_p set equal to the nominal range $R_{CRP} = 20$ km. Each of the scatterers is well-focused with clearly defined range and cross-range sidelobes. Scatterers off the center line of $x = 0$, however, are imaged as slightly further away from the radar than those on the center line because of the contribution of their nonzero cross-range coordinate to the total range. In addition, the columns of scatterers bow inward slightly as range increases. This is because the spectrum in each range bin was remapped from Doppler frequency to cross-range position using a scale factor based on the same central range of 20 km. That factor is too small at longer ranges, and too large at shorter ranges.

These geometric distortions are readily corrected after the basic image formation. The cross-range displacement is corrected by simply updating the cross-range scale for display of each row with the appropriate range R_p for each range bin, instead of using R_{CRP} for all of the range bins. The result of this operation is shown in Fig. 8.21a. The scatterers are now aligned vertically at the correct cross-range coordinates of $-1, 0$, or $+1$ km. Interpolation is required to provide samples on a common grid for display. However, the range shift for scatterers off the center line is still present. This can be corrected by computing for each cross-range bin the actual range $R = \sqrt{R_p^2 + x_p^2}$ to that bin, and then shifting that cross-range column by the difference of $R - R_p$ meters. Interpolation will again be required to accommodate partial range bin shifts. Figure 8.21b shows the result of this operation; the nine scatterers are now accurately centered on correct ground coordinates. The curvature of the cross-range sidelobes nicely reflects the range curvature that was present in the original DBS image and has now been corrected.

The DBS algorithm assumes that the echo samples from a given scatterer remain in the same range bin over the aperture time so that the DFT of the slow-time data will integrate all of the scatterer samples. This can be ensured by requiring that the range walk ΔR_w not exceed some fraction (typically one-half) of a range bin over the aperture. Using Eq. (8.32), this condition becomes

$$\Delta R_w = \frac{D_{SAR}}{R_p} x_p = \frac{vT_a}{R_p} x_p \leq \frac{\Delta R}{2} \quad \text{m} \tag{8.41}$$

Substituting from Eq. (8.15) for T_a gives the constraint

$$x_p \leq \frac{\Delta R \cdot \Delta CR}{\lambda_t} = \frac{(\Delta CR)^2}{\lambda_t} \quad \text{m} \tag{8.42}$$

where the last step assumes "square pixels," that is, $\Delta R = \Delta CR$. Equation (8.42) is a constraint on the maximum value of x and thus the maximum scene width in Doppler beam sharpening. The total allowable scene width is $2x_p$. A plot of sample values of this constraint as a function of resolution and RF is deferred to Fig. 8.24b.

The resolution performance of DBS is sometimes quantified in a Doppler *beam sharpening ratio* (BSR), defined as the ratio of the real-beam cross-range resolution to the DBS cross-range resolution. Using Eqs. (8.3) and (8.9), the BSR is

$$\begin{aligned}\text{BSR} &= \frac{\Delta CR_{\text{real beam}}}{\Delta CR_{\text{DBS}}} = \frac{(\lambda_t R_p / D_{az})}{(\lambda_t R_p / 2vT_a)} \\ &= \frac{2vT_a}{D_{az}} = 2\frac{D_{SAR}}{D_{az}}\end{aligned} \tag{8.43}$$

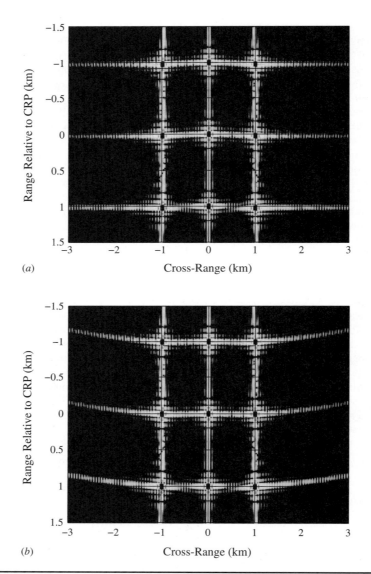

FIGURE 8.21 Correction of DBS geometric distortion. (*a*) Application of range-varying cross-range scale factor. (*b*) Application of cross-range-varying range shift.

where the last step used the relation $D_{SAR} = vT_a$. Thus, the BSR is just twice the ratio of the synthetic and real aperture sizes. The factor of two reflects the difference in synthetic and real array patterns discussed in Sec. 8.1.2.

8.3.2 Quadratic Phase Error Effects

The range curvature term $u^2/2R_p$ was neglected in simplifying Eq. (8.37). Inclusion of this term adds a quadratic phase component to the slow-time signal. Because the DBS algorithm associates a frequency component with a cross-range coordinate, the instantaneous frequency sweep implied by the quadratic phase will tend to smear the cross-range response of the processor. The ideal response of the DBS processor to a point scatterer is simply the discrete-time Fourier transform (DTFT) of a sinusoid. Figure 8.22a shows the DTFT of the signal $y[m] = \exp(j\phi_{max} m^2/M^2)$, $-M \leq m \leq +M$, where ϕ_{max} is the peak phase error. This sequence can be viewed as the product of a quadratic phase sequence and an ideal sinusoid of frequency zero, that is, a vector of ones. In the absence of quadratic phase error, the ideal spectrum would be an asinc function. Increasing the maximum value of the quadratic component attenuates and spreads the mainlobe. This causes a loss of brightness and resolution, respectively, in the DBS image of a scatterer (Richards, 1993). As the maximum of the quadratic component approaches π radians, the well-defined mainlobe/sidelobe structure

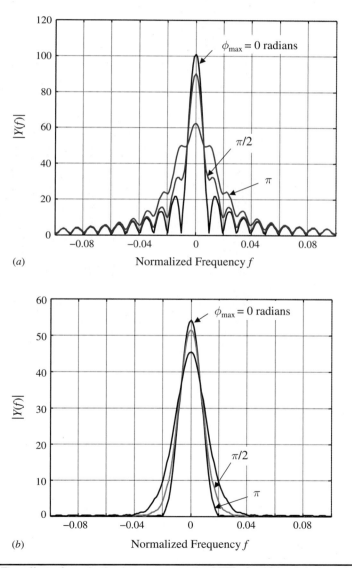

Figure 8.22 The effect of quadratic phase error on the DTFT of a 101-point sinusoid. (a) No window. (b) Hamming window.

of the response is breaking down. Figure 8.22b repeats the experiment with a Hamming window applied to the data before the DTFT. The Hamming window greatly moderates the effects of the phase error. The gain and resolution *variations* are reduced, and the general shape of the response is better maintained. This increased robustness of response is yet another benefit of windowing data.

Figure 8.23 plots the loss in peak amplitude and the increase in 3-dB mainlobe width of the DTFT as the peak quadratic phase increases. The moderating effect of the Hamming window is again evident. The degree of degradation is a mild function of sequence length. These plots are for a signal length of 101 samples. For shorter signals, the degradations are greater; for longer signals, somewhat less. For the example shown,

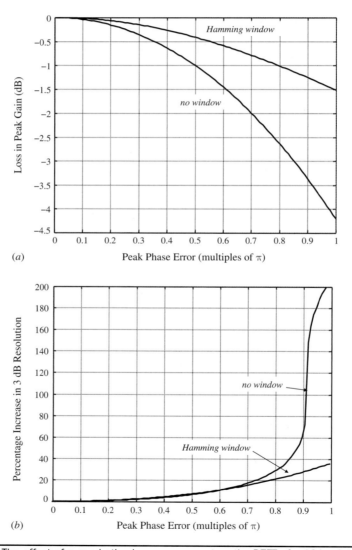

FIGURE 8.23 The effect of a quadratic phase component on the DTFT of a 101-point sinusoid. (a) Reduction in peak amplitude. (b) Percentage increase in 3-dB resolution.

limiting the maximum quadratic phase error to $\phi_{max} \leq \pi/2$ limits the loss in amplitude to 1 dB and the resolution increase to 7 percent, even without a window. This suggests that an uncompensated quadratic phase component can be ignored so long as it does not exceed $\pi/2$ radians over the aperture time. Tighter limits can be adopted if more stringent standards are desired.

The assumption that the $u^2/2R_p$ term in Eq. (8.37) can be neglected can now be revisited. This term contributes a quadratic phase component to $\phi(u)$. The maximum value of u of interest is $D_{SAR}/2 = vT_a/2$. Requiring that this phase term be limited to $\pi/2$ gives

$$\frac{4\pi}{\lambda_t} \cdot \frac{v^2 T_a^2}{8R_p} \leq \frac{\pi}{2} \tag{8.44}$$

Using $T_a = \lambda_t R_p/2v \cdot \Delta CR$ gives the constraint

$$\Delta CR \geq \frac{1}{2}\sqrt{\lambda_t R_p} \ \text{m} \tag{8.45}$$

The quadratic phase term can be ignored in DBS processing so long as this constraint is observed. Figure 8.24a plots the constraint due to range curvature of Eq. (8.45) for four values of range appropriate to airborne radars. Figure 8.24b plots the DBS scene size constraint due to range walk of Eq. (8.42). Taken together, these two constraints show that while DBS processing can achieve resolutions much better than real beam systems, it is most appropriate for imaging relatively small scenes at moderate-to-low resolution, and is best used at higher radar frequencies.

Figure 8.25a shows a portion of the DBS image for the same target array, but imaged with a resolution goal of 10 m. The standoff range is now 50 km, while the radar frequency remains 10 GHz. This scenario meets the range migration constraint of Eq. (8.42) (see Fig. 8.24b) but does not meet the quadratic phase error constraint of Eq. (8.45) (see Fig. 8.24a). The slow-time quadratic phase term represents a cross-range spatial frequency chirp. Because spatial frequency maps to cross-range position, the scatterer response is smeared in cross range. (This is the same effect illustrated in Fig. 8.22.) Range resolution is not affected.

This smearing can be corrected by compensating the data to remove the quadratic phase term. The required correction is implemented as a range-bin-dependent phase multiplication in each slow-time row of the raw data $y[l,m]$,

$$\begin{aligned} y'[l,m] &= y[l,m]\exp\left[j\left(\frac{4\pi}{\lambda_t}\right)\left(\frac{u^2}{2R_p}\right)\right] \\ &= y[l,m]\exp\left\{j\left(\frac{4\pi}{\lambda_t}\right)\frac{\left[v\left(m - \frac{M-1}{2}\right)T\right]^2}{2\left(R_0 + \frac{c(l-1)T_f}{2}\right)}\right\} \end{aligned} \tag{8.46}$$

where
T_f = fast-time sampling interval
T = slow-time sampling interval (pulse repetition interval)
R_0 = range corresponding to the first range bin
M = number of slow-time samples

Figure 8.25b shows the effect of this compensation, called *azimuth dechirp*, on the same data. The full cross-range resolution is restored, extending the usefulness of DBS.

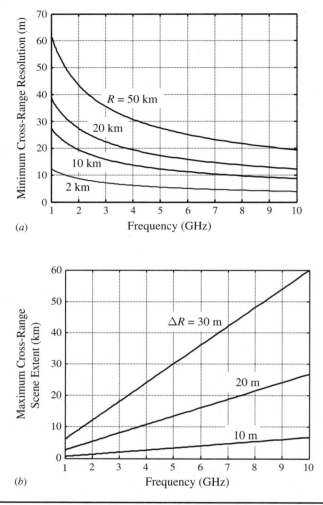

FIGURE 8.24 Limitations on the DBS algorithm to maintain focus quality. (*a*) Cross-range resolution constraint due to quadratic phase error. (*b*) Scene size constraint due to range walk.

Because of the short aperture times required for its modest resolution, DBS is often used in an actively scanning mode, unlike finer-resolution radar imaging modes (Stimson, 1998). As long as the region of interest stays within the area illuminated by the mainbeam over the aperture time the scanning is of little consequence. In keeping with the scanning operation, DBS is often used in squint mode rather than in a sidelooking configuration. In this event, range walk becomes large and must be compensated. Additional enhancements to DBS can also be implemented. For instance, *range migration correction* can be implemented to compensate for fractional-bin range curvature, and *secondary range compression* can improve focusing of targets at large cross-range displacements from the line of sight. These extensions are discussed in Schleher (1991) and Showman (2013).

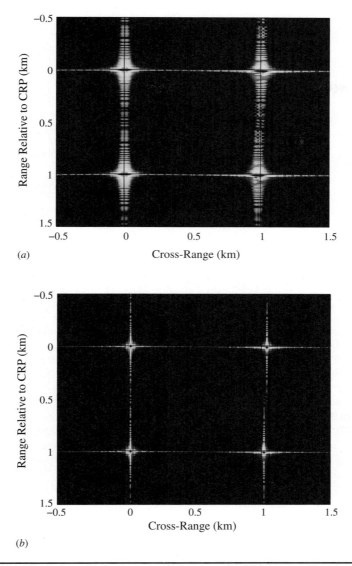

FIGURE 8.25 (a) Effect of quadratic phase on DBS image. (b) DBS image with azimuth dechirp.

8.3.3 Range-Doppler Algorithms

As resolution becomes finer, standoff range shorter, or radar frequency lower, both range curvature and azimuth quadratic phase become more pronounced. For example, Fig. 8.26 illustrates the amount of range curvature as a function of cross-range resolution for an L-band (1 GHz) radar at a standoff range $R_0 = 50$ km and velocity $v = 150$ m/s. If it is assumed that the range resolution is set approximately equal to the cross-range resolution (square pixels), the range curvature is completely insignificant when $\Delta CR = 50$ m and is still less than 1/5th of a range bin when $\Delta CR = 10$ m. Augmented DBS algorithms are effective for these cases. However, when ΔCR is reduced to 3 m the curvature rises to over five range

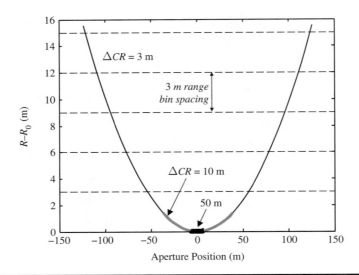

FIGURE 8.26 Increase in range curvature as cross-range resolution becomes finer.

bins, assuming $\Delta_R = 3$ m also. DBS algorithms are unsuitable in this situation. To achieve well-focused finer-resolution imagery, it is necessary to develop an algorithm that can address both significant range migration and quadratic phase. The *range-Doppler* (RD) algorithm is a family of algorithms in widespread use that offer this capability while maintaining separable range and cross-range processing and taking advantage of 2D fast Fourier transform (FFT) computational efficiency.

The baseband, pulse-compressed signal is again modeled by the phase-shifted impulse of Eq. (8.36). The responses from the series of pulses follow the range migration curve of Eq. (8.27). Because the shape of that curve is invariant to the scatterer cross-range location, that location can be set to zero without loss of generality. The response of the stripmap SAR system to a scatterer at some location $(0, R_p)$ after pulse compression is therefore (ignoring amplitude factors)

$$y(u,t) = \exp[-j(2\Omega_t/c)R(u)]\delta_D(t - 2R(u)/c)$$
$$= \exp\left[-j(2\Omega_t/c)\sqrt{u^2 + R_p^2}\right]\delta_D\left(t - \frac{2}{c}\sqrt{u^2 + R_p^2}\right) \tag{8.47}$$

In its most general form, the RD algorithm is implemented in the frequency domain, taking advantage of FFT efficiency. The two-dimensional Fourier transform of $y(u, t)$ is therefore of interest. It is

$$Y(K_u, \Omega) = \int_{-\infty}^{\infty}\int_{-\infty}^{+\infty} y(u,t)\exp(-j\Omega t)\exp(-jK_u u)\, dt\, du$$
$$= \int_{-\infty}^{\infty} \exp\left[-j\left(\frac{2\Omega_t}{c}\right)\sqrt{u^2 + R_p^2}\right] \times \cdots \tag{8.48}$$
$$\cdots \times \left(\int_{-\infty}^{+\infty} \delta_D\left[t - \frac{2}{c}\sqrt{u^2 + R_p^2}\right]\exp(-j\Omega t)\, dt\right)\exp(-jK_u u)\, du$$

Introduction to Synthetic Aperture Imaging

K_u is cross-range spatial frequency in radians per meter and Ω is temporal frequency in radians per second, not to be confused with the specific transmit frequency Ω_t.

The integral over t is easily completed, giving

$$Y(K_u, \Omega) = \int_{-\infty}^{+\infty} \exp\left\{-j\left[K_u u + \left[\frac{2(\Omega_t + \Omega)}{c}\right]\sqrt{u^2 + R_p^2}\right]\right\} du \qquad (8.49)$$

The remaining integral in Eq. (8.49) can be completed using the principle of stationary phase, discussed in Chap. 4 (Bamler, 1992; Raney, 1992); the resulting 2D spectrum for a scatterer in range bin R_p at cross-range $x_p = 0$ is

$$Y(K_u, \Omega) \approx \exp\left\{-jR_p\sqrt{\left[\frac{2(\Omega_t + \Omega)}{c}\right]^2 - K_u^2}\right\} \qquad (8.50)$$

This result ignores a frequency-dependent amplitude factor that is nearly constant because most SAR systems, despite their fine resolution, are still relatively narrowband, with a waveform bandwidth β that is usually 10 percent or less of the nominal RF Ω_t. A detailed derivation of Eq. (8.50) follows closely the similar derivation in Richards (2008).

Equation (8.50) is the frequency response for data collected from a scatterer in an arbitrary range bin R_p and cross-range coordinate $x_p = 0$. Because the PSR is shift-invariant in x, its spectrum can be generalized to an arbitrary x_p using the shift theorem of Fourier transforms, giving

$$Y(K_u, \Omega) \approx \exp(-jK_u x_p)\exp\left\{-jR_p\sqrt{\left[\frac{2(\Omega_t + \Omega)}{c}\right]^2 - K_u^2}\right\} \qquad (8.51)$$

The range-Doppler algorithm constructs a matched filter for a scatterer at $x_p = 0$ and a specific reference range R_0, usually the middle of the range swath. Setting $(x_p, R_p) = (0, R_0)$ in Eq. (8.51) gives the PSR $H(K_u, \Omega)$ for a scatterer at this CRP. Conjugating gives the range-Doppler algorithm matched filter frequency response

$$H^*(K_u, \Omega) \approx \exp\left\{+jR_0\sqrt{\left[\frac{2(\Omega_t + \Omega)}{c}\right]^2 - K_u^2}\right\} \qquad (8.52)$$

The image of a scatterer in an arbitrary range bin R_p can now be formed efficiently by computing the two-dimensional Fourier transform $Y(K_u, \Omega)$ of the data $y(u, t)$, applying the RD matched filter, and inverse transforming and rescaling:

$$\begin{aligned} d(u,t) &= \mathbf{F}^{-1}\{D(K_u, \Omega) = H^*(K_u, \Omega)Y(K_u, \Omega)\} \\ d(x,y) &= d(u,t)|(u,t) = (x, ct/2) \end{aligned} \qquad (8.53)$$

The image spectrum will be

$$\begin{aligned} D(K_u, \Omega) &= H^*(K_u, \Omega)Y(K_u, \Omega) \\ &= \exp(-jK_u x_p)\exp\left\{-j(R_p - R_0)\sqrt{\left[\frac{2(\Omega_t + \Omega)}{c}\right]^2 - K_u^2}\right\} \end{aligned} \qquad (8.54)$$

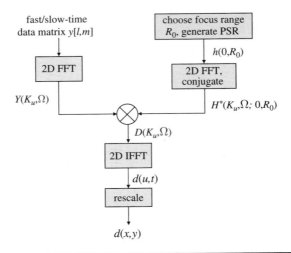

FIGURE 8.27 Block diagram of range-Doppler algorithm.

The spectrum of a scene containing multiple scatterers in varying range bins will be a superposition of terms like Eq. (8.54).

Note that if $R_p = R_0$, then $D(K_u, \Omega) = \exp(-jK_u x_p)$ over the region of support. The image of the scatterer would then be a product of sinc functions in the u and R_p directions centered at $(x_p, 0)$ with mainlobe widths determined by the bandwidths of the ROS.[8] Although not indicated explicitly, the support of D will be limited to $\Omega \in [\pm\pi\beta]$ rad/s by the waveform bandwidth β and $K_u \in [\pm\beta_u/2] = [\pm 2\pi D_{SAR}/\lambda_t R_p]$ rad/m by the synthetic aperture size D_{SAR}. Any other scatterer range will exhibit a spectrum with nonlinear phase terms, resulting in defocus.

The range-Doppler algorithm is diagrammed in Fig. 8.27. Its major advantages are its ability to process the entire range swath (provided both the swath and the synthetic aperture are small compared to the standoff range), compensate for both range migration and quadratic phase, and utilize the computational efficiency of the two-dimensional FFT. Since this version operates in the spatial frequency/temporal frequency domain, however, it is not clear why the technique is called the range-Doppler algorithm. In fact, the name range-Doppler derives from a version of the algorithm obtained by additional simplifications of Eq. (8.52).

The argument of the complex exponential in H can be approximated as follows:

$$\begin{aligned} R_0 \sqrt{\left[\frac{2(\Omega_t + \Omega)}{c}\right]^2 - K_u^2} &= R_0 \frac{2(\Omega_t + \Omega)}{c} \sqrt{1 - \frac{K_u^2}{[2(\Omega_t + \Omega)/c]^2}} \\ &\approx R_0 \frac{2(\Omega_t + \Omega)}{c}\left(1 - \frac{K_u^2}{2[2(\Omega_t + \Omega)/c]^2}\right) \\ &= \frac{2(\Omega_t + \Omega)}{c} R_0 - \frac{cK_u^2 R_0}{4(\Omega_t + \Omega)} \, m \end{aligned} \qquad (8.55)$$

---

[8]Because the PSR is referenced to the center of the swath, range is measured relative to R_0. A peak at $(x_p, 0)$ therefore corresponds to an actual position of (x_p, R_0).

The second line was obtained by expanding the square root in a binomial series and retaining only the first two terms. This requires $K_u \ll 2(\Omega_t + \Omega)/c$, normally the case in SAR systems. Now consider the $cK_u^2 R_0/[4(\Omega_t + \Omega)]$ term and recall that usually $\Omega \ll \Omega_t$. Again using the binomial series expansion, this time on the $(\Omega_t + \Omega)^{-1}$ term, gives

$$\frac{cK_u^2 R_0}{4(\Omega_t + \Omega)} = \frac{cK_u^2 R_0}{4\Omega_t(1 + \Omega/\Omega_t)} \approx \frac{cK_u^2}{4\Omega_t}\left(1 - \frac{\Omega}{\Omega_t}\right) = \frac{cK_u^2 R_0}{4\Omega_t} - \frac{cK_u^2 R_0}{4\Omega_t^2}\Omega \qquad (8.56)$$

Using Eqs. (8.55) and (8.56) in Eq. (8.52), the matched filter frequency response becomes

$$H^*(K_u, \Omega) \approx \exp\left[j\left(\frac{2\Omega_t}{c}\right)R_0\right]\exp\left[j\left(\frac{2R_0}{c}\right)\Omega\right]\exp\left(-j\frac{cR_0}{4\Omega_t}K_u^2\right)\exp\left(+j\frac{cK_u^2 R_0}{4\Omega_t^2}\Omega\right) \qquad (8.57)$$

The approximate matched filter frequency response of Eq. (8.57) implies two distinct operations for performing the RD cross-range compression. For a given cross-range spatial frequency K_u, the fourth term is a linear phase in the fast-time baseband frequency Ω, corresponding to a shift in the fast-time dimension of $cK_u^2 R_0/4\Omega_t^2$ seconds. This cross-range dependent shift in the data "straightens out" the range curvature in the PSR. Because it is a function of K_u rather than u itself, the shift must be applied after a slow-time DFT, that is, to the range-Doppler data. The third phase term compensates the slow-time quadratic phase modulation of the data, essentially performing an LFM pulse compression in slow time. The second term is a fast-time shift of $-2R_0/c$ seconds, effectively shifting the reference range to zero. Since the amount of shift is not dependent on K_u, the shift can be performed in either the fast-time domain or the frequency domain. Finally, the first phase term is the usual $(2\Omega_t/c)R_0 = (4\pi/\lambda_t)R_0$ phase shift for a scatterer at range R_0. Figure 8.28 is a purely notional illustration of this version of the range-Doppler algorithm, beginning with the pulse-compressed fast-time/slow-time data matrix in step 1 and continuing to the compressed point spread response in step 4.

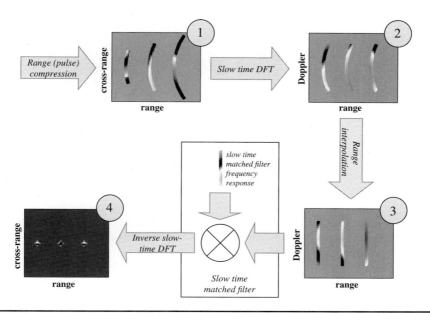

FIGURE 8.28 Notional illustration of sequence of operations in classic range-Doppler algorithm.

The fast-time shift can be implemented as a frequency domain multiplication as implied by Eq. (8.57) or as an explicit shift and interpolation in the fast-time domain. In the latter case, the data are operated on in the time domain in one dimension (fast time, or range) and the frequency domain in the other (spatial frequency, or Doppler), hence the name "range-Doppler" algorithm.

Figure 8.29 illustrates the performance of the range-Doppler algorithm on two simulated scenarios using the full RD point spread response of Eq. (8.52). In Fig. 8.29a, the RD

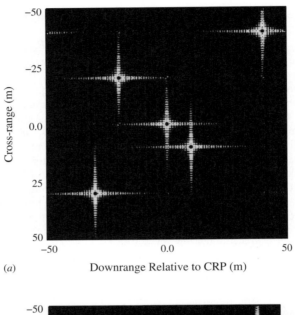

(a)

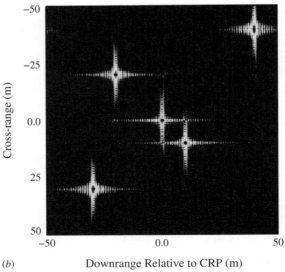

(b)

Figure 8.29 Magnitude in dB of the range-Doppler image of a test array. (a) X band, long standoff range, small scene size. (b) L band, short standoff range, larger scene size. See text for parameters. (Images courtesy of Dr. Gregory A. Showman.)

algorithm is applied to a collection of point scatterers in a 50 × 50-m area sensed with an X-band (9.5 GHz) radar from a standoff range of 7.5 km. While the range and cross-range resolutions are only 0.5 m, this example is relatively unchallenging due to the high RF frequency and long standoff range compared to the small scene size. All five scatterers are well-focused, with no visible degradations in the scene. Figure 8.29b is a more challenging case. Although the resolution is relaxed to 1.0 m in both dimensions, the image area is larger (100 × 100 m), the standoff range is shorter (4.1 km), and the RF is lowered to L band (1.5 GHz). The latter three factors increase the variability of the range curvature as a function of range. Although differences in sidelobe structure are evident, the scatterer at the center reference range R_0 remains well focused. However, the cross-range resolution of the other scatterers degrades in proportion to their distance from the scene center.

Many variations of the range-Doppler algorithm exist. One of the most common includes a third term in the series expansion of $(\Omega_t + \Omega)^{-1}$ leading to Eq. (8.56), resulting in another phase correction term in H^* that has the effect of compensating a fast-time quadratic phase that increases with K_u, causing defocusing in the range dimension at higher Doppler shifts (wider cross-range positions). This correction is called *secondary range compression* (Showman, 2013).

8.3.4 Depth of Focus

The exact stripmap PSR of Eq. (8.47) is range dependent. The range-Doppler algorithm uses the PSR for a scatterer at the particular range R_0. However, if applied over a deep enough swath, the variation in the PSR with R will become too significant to ignore. Failure to address this variation will result in increasingly poor focusing at increasing ranges from R_0 because of the increasing mismatch between a scatterer's actual PSR and the PSR at R_0. The swath length over which the RD algorithm can be applied without significant defocusing is called the *depth of focus* (DOF).

The DOF is determined by finding the change in range R_0 which will cause a specified change in range curvature over the aperture time. The tolerable variation is often taken as $\lambda_t/8$, corresponding to a two-way variation of $\lambda_t/4$ and thus a phase change of $\pi/2$ radians. It was seen in Sec. 8.3.2 that a quadratic phase error limited to this amount would cause minimal degradation in the DFT of a sinusoid, a model that is applicable to cross-range matched filtering once the primary quadratic phase is removed.

Equation (8.33) gave a formula for range curvature. Differentiating with respect to range gives the change in curvature as nominal range changes; setting that quantity equal to $\lambda_t/8$ gives the maximum change in range as (see Prob. 20)

$$\delta R_{max} = \frac{\lambda_t R_0^2}{D_{SAR}^2} = \frac{\lambda_t R_0^2}{(vT_a)^2} \text{ m} \tag{8.58}$$

Since this range change can occur in either direction from R_0, the DOF is twice this,

$$DOF = 2 \cdot \delta R_{max} = \frac{2\lambda_t R_0^2}{D_{SAR}^2} = \frac{2\lambda_t R_0^2}{(vT_a)^2} = \frac{8(\Delta CR)^2}{\lambda_t} \text{ m} \tag{8.59}$$

Consider SAR imaging of a scene with the swath center at $R_0 = 10$ km. If imaged using an X-band (10 GHz) radar with a cross-range resolution of 3 m, the depth of focus is 2.4 km, less than the swath length. One method to address this limited DOF is to break the desired swath into N subswaths, each less than DOF m deep and centered on ranges $R_{01}, \ldots, R_{0N}$. The PSR is updated and the RD algorithm applied independently in each subswath. The resulting images are then mosaicked together to form the complete image. In this example, the swath should be broken into at least four and possibly more subswaths.

If the scene is imaged with an L-band (1 GHz) radar at 1 m resolution, the depth of focus is only 240 m, requiring at least 42 subswaths to use the RD algorithm. The large number of subswaths needed indicates that the RD algorithm is poorly suited to this scenario and a more advanced algorithm such as the range migration algorithm, subject of the next section, should be used instead.

As another example, the scenario of Fig. 8.29a has a DOF of 63.33 m = ±31.67 m. This exceeds the maximum 20 m range offset from the scene center to the most distant scatterer, so all scatterers are focused well. Figure 8.29b has a DOF of 40 = ±20 m, only half the maximum range displacement of approximately 40 m in that scenario. Scatterers at ranges greater than 20 m from the scene center are significantly smeared in cross range.

8.3.5 Range Migration Algorithm

The major shortcoming of the range-Doppler algorithm is its limited depth of focus. As Eq. (8.59) makes clear, the DOF can become small at low RF, fine resolution, and short ranges, severely limiting focused scene size. The breakdown of the RD algorithm is manifested as increased cross-range smearing as distance from the scene center increases beyond the DOF.

The *range migration* (RM) *algorithm* (Bamler, 1992; Gough and Hawkins, 1997) directly accounts for the variation in the PSR as scatterer range changes, fully focusing scatterers anywhere in the scene. Equivalently, it achieves an infinite DOF. It is also known as the "omega-K" or "Ω-K" algorithm. The basic RM algorithm operates on the two-dimensional Fourier transform of the data, applying a 2D matched filter and a frequency domain coordinate remapping to form the image. An inverse 2D Fourier transform then forms the focused image. The efficiency of the frequency domain operations when using the FFT and the unrestricted scene size has made RM the dominant algorithm for many stripmap systems, especially under the challenging conditions mentioned above. It can also be adapted to spotlight SAR with either appropriate modifications to the data collection protocol or a preprocessing step; see Carrara et al. (1995).

The range migration algorithm itself migrated from the seismic community to SAR in an excellent example of cross-field fertilization. Cafforio, Prati, and Rocca published several papers developing its application to SAR; Cafforio et al. (1991) is a good starting point.

Figure 8.30 is a flowgraph of the RM algorithm. As in the range-Doppler algorithm analysis, set the center of the swath at range R_0. The frequency response $H^*(K_u, \Omega)$ of the matched filter for a target at that range and cross-range coordinate $x = 0$ is Eq. (8.52). This filter is applied to the 2D Fourier transform $Y(K_u, \Omega)$ of the data CPI to produce the spectrum $D(K_u, \Omega)$, a step often called *bulk compression*. To this point, this is simply the basic RD algorithm.

It is convenient now to convert fast-time frequency to *two*-way range spatial frequency in radians/meter (see App. B), $[2(\Omega_t + \Omega)/c] = K'_R$. Doing so puts the bulk compression output spectrum of Eq. (8.54) in the form

$$D(K_u, K'_R) \approx \exp(-jK_u x_p)\exp\left(-j(R_p - R_0)\sqrt{K'^2_R - K^2_u}\right) \qquad (8.60)$$

To progress further, consider the form that D would take for an ideal image of a scatterer at arbitrary coordinates (x_p, R_p). Expressing the final output in (x, y) coordinates and recalling that $R_p = y_p$, the ideal infinite-bandwidth spatial domain image and its Fourier transform would be

$$\begin{aligned}f(x,y) &= \delta_D(x - x_p, y - y_p) \Rightarrow \\ F(K_x, K_y) &= \exp(-jK_x x_p)\exp(-jK_y y_p)\end{aligned} \qquad (8.61)$$

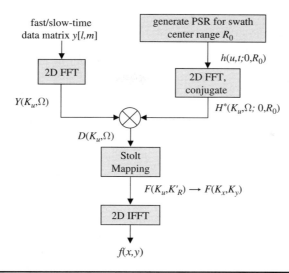

Figure 8.30 Flowchart of the full range-migration algorithm.

The goal of the RM algorithm will be to generate a practical finite-bandwidth approximation of this ideal as the product of sinc functions in the x and y dimensions having resolutions consistent with the waveform bandwidth and synthetic aperture size. It is also necessary that the algorithm work for scatterers at any coordinates and for all scatterers at once, so that full focus is obtained over the entire scene.

The key insight of the range migration algorithm is that the bulk compression output spectrum of Eq. (8.60) can be put in the desired form of Eq. (8.61) by remapping the spatial frequency variables according to the *Stolt mapping* or *Stolt interpolation* (Stolt, 1978),[9]

$$K_u \to K_x, \quad \sqrt{K_R'^2 - K_u^2} \to K_y \tag{8.62}$$

Applying this transformation converts Eq. (8.60) to

$$F(K_x, K_y) = D\left(K_x, \sqrt{K_x^2 + K_y^2}\right) = \exp(-jx_p K_x)\exp[-j(y_p - R_0)K_y] \tag{8.63}$$

By superposition, each scatterer in a complex scene will contribute an appropriately weighted term like that of Eq. (8.63) to the spectrum to form the complete image.

Equation (8.63) is the desired result. The Stolt mapping converts the RD algorithm output spectrum into a product of linear phase terms in K_x and K_y for each scatterer. An inverse 2D Fourier transform then produces the image of the scatterers, $f(x,y)$. As with the RD algorithm, the range (y) variable in Eq. (8.63) is obtained relative to the reference range R_0. Crucially, the details of the Stolt interpolation do not depend on the scatterer location.

---

[9]If the cross-range resolution is very fine so that K_u can become relatively large, and the RF is very low (perhaps UHF) so that K_R' can be relatively small, some spectral sample locations (K_u, K_R') may have $K_R'^2 - K_u^2 < 0$, making K_y imaginary. Any such spectral samples are excluded from the Stolt transformation to (K_x, K_y) space. In most real systems $K_R' \gg K_u$ over the ROS of the spatial spectrum so that this issue is not a concern.

Therefore, it converts the spectrum contributions of all scatterers to the form of Eq. (8.63) regardless of their location, focusing the entire scene at once.

The region of support of $F(K_x, K_y)$ remains to be specified. In the absence of weighting for sidelobe suppression, achieving design resolutions in x and y of ΔR and ΔCR requires spatial bandwidths of $\beta_x = 2\pi/\Delta CR$ and $\beta_y = 2\pi/\Delta R$ rad/m. An inverse Fourier transform of Eq. (8.63) over these bandwidths will produce, to within an amplitude scale factor,

$$f(x,y) = \text{sinc}[(x - x_p)/\Delta CR] \cdot \text{sinc}[(y - y_p + R_0)/\Delta R] \qquad (8.64)$$

which is the desired bandlimited image of the scatterer with the origin referred to ground coordinates $(0, R_0)$. Recalling that $\beta_u = 4\pi D_{SAR}/\lambda_t R_p$ rad/m, the required synthetic aperture size is $D_{SAR} = \lambda_t R_p/(2 \cdot \Delta CR)$ as usual. Similarly, in the range direction $\beta_y = 4\pi\beta/c$ rad/m and therefore $\beta = c/(2 \cdot \Delta R)$ as usual.

Figure 8.31a continues the example of Fig. 8.29b. The range-Doppler algorithm was unable to focus scatterers near the swath ends because their distance from the scene center exceeded the DOF. Applying the range migration algorithm to the same scenario fully focuses all of the scatterers. A rather demanding example is illustrated in Fig. 8.31b. This example simulates a UHF (600 MHz) radar system with a 750-MHz bandwidth and 51° of integration, giving range and cross-range resolutions of 0.2 m. The UHF frequency and very fine resolution might suggest a foliage penetration ("FOPEN") application, though the extremely short standoff and small scene are less realistic and more for simulation convenience. The five scatterers shown are in a 10 × 10-m region with the scene center only 30 m from the radar flight path. The DOF is a mere 0.64 m in this scenario and the RD algorithm fails badly (not shown). The RM algorithm successfully focuses all five scatterers to the expected resolution. Sidelobe levels are as expected, though the sidelobes are no longer aligned along the principal axes. This is expected in extreme scenarios.

Figure 8.32 shows the effect of the Stolt mapping for this example. The left-hand drawing is a set of sample points in 2D (K_u, K_R') space obtained by a 2D Fourier transform of the raw (u, t) fast-time/slow-time data collection. (The sampling grid is unrealistically sparse to make the individual sample locations easier to view.) The right-hand drawing is the nonuniform sample grid in the (K_x, K_y) space resulting from Stolt mapping.

Figure 8.33 shows the same (K_x, K_y) sample grid (gray circles), but still more sparse for clarity. Superimposed is the desired uniform (K_x, K_y) grid (black squares). The 5 × 5 cycles/m spatial bandwidth supports the 0.2 × 0.2-m resolution, and the grid is centered on the range two-way center spatial frequency of $(2/c)600$ MHz = 4 cycles/m. This figure emphasizes that interpolation will be necessary to move the (K_x, K_y) data from the nonuniform grid produced by the Stolt mapping to the new, uniform grid. The figure also illustrates that only a one-dimensional interpolation is required, in the K_y dimension; the K_x sample coordinates are unchanged. On the other hand, the sample spacing in the K_y dimension after the Stolt mapping is different for every value of K_x (column of samples), so the interpolation details have to be updated for each column. Another interpolation detail is highlighted by the two black sample locations in the upper corners. Since these are outside the ROS of the available (K_u, K_R') data, they cannot be predicted well. Instead, it is necessary to either use a target sampling grid that can be inscribed within the available data ROS, possibly giving up some range resolution, or to "over-collect" by gathering data over a range spatial bandwidth greater than indicated by the desired image range resolution. Specifically, suppose the maximum desired values of K_x and K_y are K_{xmax} and K_{ymax}. Then the maximum value of K_u required to perform the Stolt mapping is also K_{xmax} but the maximum value of K_R' is $\sqrt{K_{xmax}^2 + K_{ymax}^2} > K_{ymax}$. Finally, note that the interpolation is being applied to rapidly varying, complex frequency data. High-quality bandlimited interpolators are required to

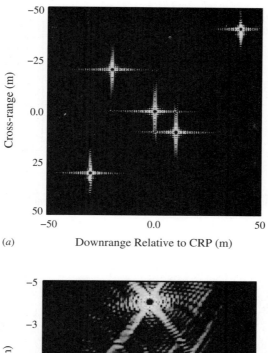

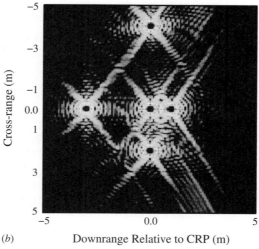

FIGURE 8.31 (a) Image of the same example scenario as Fig. 8.29b using the full range-migration algorithm. (b) RM algorithm image of a challenging UHF radar scenario. (Images courtesy of Dr. Gregory A. Showman.)

avoid introducing image artifacts (Carrara, 1995). As will be seen shortly, similar issues arise in implementing the polar format algorithm for spotlight SAR.

The need for the Stolt mapping in the RM algorithm was recognized by specifying the desired spectrum, deriving the actual spectrum after RD processing, and then identifying a frequency variable transformation that would convert the latter into the former. This same algorithm development strategy was also used in Chap. 5 to develop the keystone transformation for range migration compensation.

Similar to the RD algorithm, additional approximations can be made to the Stolt mapping that result in modified range migration algorithms that perform interpolation in the

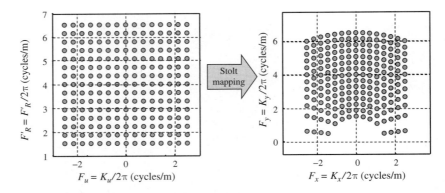

FIGURE 8.32 Illustration of the Stolt mapping's effect on sample locations in two-dimensional spatial frequency space.

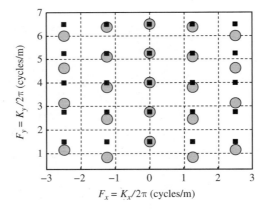

FIGURE 8.33 Overlay of desired uniform spatial frequency sample grid onto nonuniform Stolt mapping grid, illustrating the need for interpolation of the Stolt data in the K_y (range spatial frequency) dimension and issues with the region of support.

spatial domain instead of the spatial frequency domain. Extensive additional detail on the Stolt mapping, its rationale, and its effects on the phase structure and ROS of the various data sets and spectra and are given in Showman (2013) and Carrara et al. (1995). Both also discuss the *chirp scaling algorithm*, a more computationally efficient form of the approximate Stolt mapping version of the RM algorithm that uses properties of linear FM waveforms to avoid explicit interpolations.

8.4 Spotlight SAR Data Characteristics

The spotlight SAR scenario was illustrated in Fig. 8.11. Many spotlight SAR systems utilize a linear FM waveform with stretch processing. Assume an LFM pulse that sweeps from $F_t - \beta/2$ to $F_t + \beta/2$ Hz in the time interval $-\tau/2$ to $+\tau/2$ seconds. Adapting to the notation of this chapter, it was shown in Chap. 4 [Eq. (4.110)] that on a pulse taken from aperture position u, an ensemble of scatterers at ranges $\delta R_i = c \cdot \delta t_i/2$ relative to a CRP range R_0 results in the following signal at the stretch mixer output:

$$\begin{aligned} y(t) &= \sum_i \tilde{\rho}_i \exp\left(-j\frac{4\pi R_i}{\lambda_t}\right) \exp\left[-j2\pi\frac{\beta}{\tau}\delta t_i(t-t_0)\right] \exp\left[j\pi\frac{\beta}{\tau}(\delta t_i)^2\right] \\ &= \exp(-j4\pi R_0/\lambda_t) \sum_i \tilde{\rho}_i \exp\left[-j\left(\Omega_t + 2\pi\frac{\beta}{\tau}(t-t_0)\right)\delta t_i\right] \exp\left[j\pi\frac{\beta}{\tau}(\delta t_i)^2\right] \end{aligned} \quad (8.65)$$

where $t_0 = 2R_0/c$ and $\tilde{\rho}_i$ is the complex reflectivity of the echo from the ith scatterer[10] located at range $R_0 + \delta R_i$. Extending this formula to a continuum of scatterers gives

$$y(t) = w(t)\exp(-j4\pi R_0/\lambda) \times \cdots$$

$$\cdots \times \int_{-\infty}^{+\infty} \tilde{\rho}(\delta t) \exp\left[-j\left(\Omega_t + 2\pi\frac{\beta}{\tau}(t-t_0)\right)\delta t\right] \exp\left[j\pi\frac{\beta}{\tau}(\delta t)^2\right] d(\delta t) \quad (8.66)$$

In this equation, $w(t)$ is a rectangular window function that limits the data extent to that of the reference chirp in the stretch mixer. Specifically

$$w(t) = \begin{cases} 1, & t_0 - \dfrac{L_s}{c} - \dfrac{\tau}{2} \leq t \leq t_0 + \dfrac{L_s}{c} + \dfrac{\tau}{2} \\ 0, & \text{otherwise} \end{cases} \quad (8.67)$$

Now consider the Fourier transform of $\tilde{\rho}(\delta_t)$,

$$\tilde{P}(\Omega) = \int_{-\infty}^{-\infty} \tilde{\rho}(\delta t) \exp(-j\Omega \cdot \delta t) d(\delta t) \quad (8.68)$$

Comparing Eq. (8.68) to Eq. (8.66) and using the time limits established by Eq. (8.67) shows that

$$y(t) = w(t)\exp(-j4\pi R_0/\lambda_t)\tilde{P}\left[\Omega_t + \frac{2\pi\beta}{\tau}(t-t_0)\right], \quad t_0 - \frac{L_s}{c} - \frac{\tau}{2} \leq t \leq t_0 + \frac{L_s}{c} + \frac{\tau}{2} \quad (8.69)$$

provided that the residual video phase (RVP) term $\exp[j\pi\beta(\delta_t)^2/\tau]$ can be ignored; this condition will be considered later. Equation (8.69) shows that, like linear ramp FMCW, stretch processing of an LFM waveform has swapped the time and frequency domains in a sense: the mixer output is a time-domain voltage whose numerical values trace out the value of the Fourier transform $\tilde{P}(\Omega)$ of the scene reflectivity versus range, $\tilde{\rho}(\delta t)$. Put another way, the combination of the LFM waveform with a stretch processor acts as a spectrum analyzer for the reflectivity distribution.

The range of frequencies evaluated by this spectrum analyzer is determined by evaluating the argument of $\tilde{P}(\cdot)$ for the maximum and minimum values of t allowed by the window. Converting to cyclical frequency units, the result is

$$\begin{aligned} y(t) &= \tilde{P}(F), \quad F \in F_t \pm \left[1 + \frac{L_s}{(c\tau/2)}\right]\frac{\beta}{2} \\ &\approx \tilde{P}(F), \quad F \in \pm\frac{\beta}{2} \end{aligned} \quad (8.70)$$

---

[10] $\tilde{\rho}_i$ includes range weighting and other range equation factors; see Sec. 2.8.

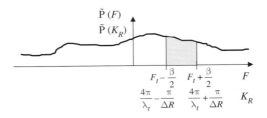

FIGURE 8.34 Portion of the range profile bandwidth measured by the stretch processor, in both temporal and spatial frequency units.

The second line assumes that $L_s \ll c\tau/2$, that is, the swath length is small compared to the pulse duration. Stretch processing systems are usually designed such that this is the case. Equation (8.70) states the remarkable result that the time domain output of the stretch mixer traces out the Fourier transform $\tilde{P}(F)$ of the range profile $\tilde{\rho}(\delta t)$ over the frequency interval $[F_t - \beta/2, F_t + \beta/2]$. This is intuitively satisfying since this is exactly the instantaneous frequency interval over which the LFM pulse sweeps. It is a specific example of the received signal "spectral model" for wideband signals developed in Sec. 2.9.

$\tilde{P}(F)$ can be rescaled into units of spatial frequency in radians per meter via the transformation $F \to cK'_R/4\pi$. Using $\Delta R = c/2\beta$, the mixer output range becomes $K'_R \in (4\pi/\lambda_t) \pm (\pi/\Delta R)$ radians per meter or, in cycles per meter, $\beta_R \in (2/\lambda_t) \pm (1/2\Delta R)$. Figure 8.34 is a notional illustration of the portion of the range profile spectrum $\tilde{P}(F)$ or $\tilde{P}(K'_R)$ measured on a single pulse by the stretch processor.

As discussed in Sec. 2.8.5, $\tilde{\rho}(\delta t)$ is a transformation of the scatterers in a two-dimensional scene into the one-dimensional range profile. This concept is illustrated in Fig. 8.35. The complex reflectivities of all of the scatterers on the isorange contour corresponding to a delay of $t_0 + \delta t_0$ are weighted by the antenna pattern and integrated to form the value of the range profile at time $t_0 + \delta t_0$, $\tilde{\rho}(\delta t_0)$. "CRP" again marks the central reference point at

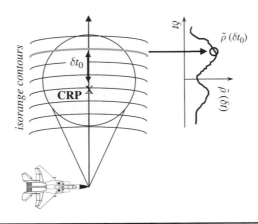

FIGURE 8.35 Projection of a two-dimensional scene into a one-dimensional angle-averaged range profile.

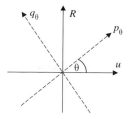

FIGURE 8.36 Coordinate systems for defining a two-dimensional projection.

range $R_0 = ct_0/2$. If the width of the illuminated area is much less than the nominal range, $R_0\theta_{az} \ll R_0$, the lines of integration become nearly straight. The range averaging then approximates a *projection* in the tomographic sense of the two-dimensional scene into a one-dimensional function.

Consider the two coordinate systems shown in Fig. 8.36. The (p_θ, q_θ) axes are rotated by θ radians with respect to the (u, R) axes. The coordinate systems are related by the equations

$$\begin{aligned} u &= p_\theta \cos\theta - q_\theta \sin\theta \text{ m} \\ R &= p_\theta \sin\theta - q_\theta \cos\theta \text{ m} \end{aligned} \qquad (8.71)$$

A projection of a two-dimensional effective reflectivity scene $\rho'(u, R)$ into a one-dimensional cross-range-averaged reflectivity range profile $\tilde{\rho}_\theta(p_\theta)$ is defined as

$$\tilde{\rho}_\theta(p_\theta) = \int_{-\infty}^{+\infty} \rho'(p_\theta \cos\theta - q_\theta \sin\theta, p_\theta \sin\theta + q_\theta \cos\theta) \, dq_\theta \qquad (8.72)$$

In essence, the integration over q_θ "collapses" the 2D function ρ' along the q_θ dimension and into a 1D function along the p_θ coordinate.

The *projection-slice theorem* of Fourier analysis then states that the one-dimensional Fourier transform of the projection is a slice of the two-dimensional Fourier transform of the original function $\rho'(u, R)$ (Dudgeon and Mersereau, 1984):

$$\int_{-\infty}^{+\infty} \tilde{\rho}_\theta(p_\theta) \exp(-jp_\theta U) \, dp_\theta = P'(Q\cos\theta, Q\sin\theta) \qquad (8.73)$$

where Q is the frequency variable corresponding to p_θ.

Figure 8.37 illustrates the consequences of the LFM/stretch data acquisition and the projection-slice theorem for spotlight SAR. The scene $\rho'(u, R)$ on the left represents a patch of terrain, perhaps containing a road, building, stand of trees, and small lake. The drawing on the right represents the Fourier transform $P'(K_u, K_R')$ of the scene in spatial frequency units. The radar views the scene from a particular angle. The radar and antenna effectively project the two-dimensional scene into a one-dimensional function. The LFM waveform and stretch processing measure a portion of the spectrum of that one-dimensional function. The projection-slice theorem allows that spectral segment to be interpreted as a measurement of the two-dimensional Fourier transform of the scene along the same angle, as shown on the right side of the sketch. Because of the limited bandwidth of the stretch measurement, only the segment of the total slice highlighted in a lighter shade of gray is actually measured (see Fig. 8.34).

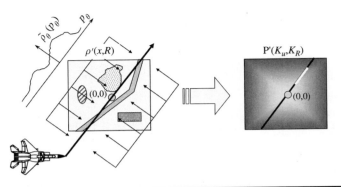

FIGURE 8.37 Spotlight SAR data acquisition model.

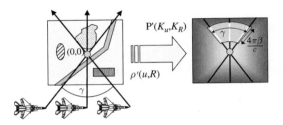

FIGURE 8.38 Spotlight SAR data acquisition collects an annular region in the spatial frequency domain.

As the aircraft flies along the synthetic aperture, it continues to transmit pulses while steering the antenna to remain aimed at the CRP of the region of interest. Each successive pulse measures a segment of the spectrum of a projection of $\rho'(u, R)$ at a new angle and therefore a segment of a slice of $P'(K_u, K'_R)$ at that same angle. Over a series of pulses the system therefore collects data over an annular region of the two-dimensional spectrum of the image. This model of the spotlight SAR data set is depicted in Fig. 8.38. The extent of the annulus is $4\pi\beta/c$ radians per meter $= 2\beta/c$ cycles per meter in the radial direction. The spatial range resolution is then $(2\beta/c)^{-1} = \Delta R$ meters as expected. The width of the annulus at its center $K'_R = 4\pi/\lambda_t$ is $(4\pi/\lambda_t)\gamma$ radians per meter, so the spatial cross-range resolution is $\lambda_t/2\gamma$, again as expected. Finally, note that the spectral data are on a polar, rather than rectangular, grid in (K_u, K'_R). For this reason, the data set is referred to as *polar format data*.

An interesting question is the number of projections required to reconstruct the image; the answer determines the required PRF. Suppose that the final image $\rho'(u, R)$ desired is $L_u \times L_s$ meters (cross-range × range). To avoid aliasing artifacts over a region of this size, Nyquist sampling theory requires that samples of $P'(K_u, K'_R)$ be no more than $1/L_u$ cycles per meter $= 2\pi/L_u$ radians per meter apart in K_u; similarly, samples should be no more than $1/L_s$ cycles per meter $= 2\pi/L_s$ radians per meter apart in K'_R. Consider the K_u dimension. At the center of the annulus ($K'_R = 4\pi/\lambda_t$), a sample spacing of $2\pi/L_u$ corresponds to an angular spacing between successive slices of $\lambda_t/2L_u$ radians. The number of slices required to span the total angular extent of the annulus is then

$$N_\gamma = \frac{\gamma}{(\lambda_t/2L_u)} = L_u\left(\frac{2\gamma}{\lambda_t}\right) = \frac{L_u}{\Delta CR} \qquad (8.74)$$

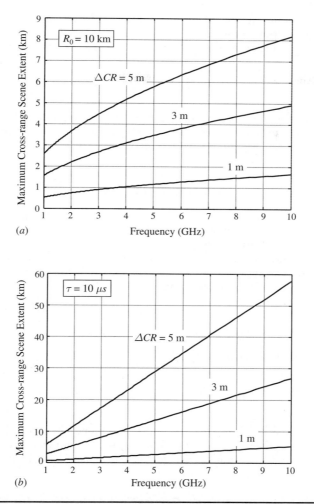

Figure 8.39 Scene size limitations in the polar format algorithm. (a) Limit due to isorange contour curvature. (b) Limit due to residual video phase.

which is simply the number of resolution cells in the cross-range extent. The linear distance traveled by the radar platform between slices and the resulting PRI are

$$\Delta u = R_0 \frac{\lambda_t}{2L_u} \text{ m} \quad \Rightarrow \quad PRI = \frac{\lambda_t R_0}{2vL_u} \text{ seconds} \tag{8.75}$$

If L_u is set equal to its practical maximum of the illuminated beamwidth $R_0\theta_{az} = R_0\lambda_t/D_{az}$, the platform travel between pulses is $D_{az}/2$, the same minimum value required in stripmap SAR to avoid Doppler ambiguities. As explained earlier, a higher cross-range sampling rate is often used in stripmap and also in spotlight SAR to minimize ambiguities.

The minimum number of range samples needed (equivalent to time samples of the stretch mixer output for each pulse) is the radial extent of the annulus divided by the sample spacing in K'_R which reduces to the number of range cells in the range extent,

$$N_R = \frac{(4\pi\beta/c)}{(2\pi/L_s)} = \frac{L_s}{\Delta R} \text{ bins} \tag{8.76}$$

Since the stretch mixer output traces out the two-dimensional Fourier transform slice, the interval $1/L_R$ between samples in K'_R specifies the time Δt between samples at the mixer output, using the mapping of time to spatial frequency $K_u = (4\pi\beta/c\tau)$,

$$\Delta t = \frac{c\tau}{4\pi\beta} \cdot \frac{2\pi}{L_R} = \frac{c\tau}{2\beta L_R} \text{ seconds} \tag{8.77}$$

It is important to remember that this spotlight data model assumes an LFM waveform and a stretch receiver. In addition, two assumptions were made in deriving the results above. The first is that the RVP term in Eq. (8.66) can be ignored. The second is that the curvature in the isorange contours can be ignored in modeling the mixer output $\tilde{\rho}(\delta t)$ as a projection $\tilde{\rho}(p_\theta)$. Both assumptions are valid provided the scene size is not too large. The limitation due to isorange curvature can be shown to be

$$L_u < 2 \cdot \Delta CR \sqrt{\frac{2R_0}{\lambda_t}} \text{ m} \tag{8.78}$$

while the limitation due to RVP is

$$L_u < \frac{2 \cdot \Delta CR \cdot F_t}{\sqrt{\beta/\tau}} \text{ m} \tag{8.79}$$

Both are derived in Jakowatz (1996, App. B). Figure 8.39 plots these two constraints for $R_0 = 10$ km, $\tau = 10$ μs, and $\beta/\tau = 0.75$ MHz/μs. The limit on scene size due to RVP is very loose; the limit due to isorange curvature is more constraining in this example.

8.5 The Polar Format Image Formation Algorithm for Spotlight SAR

Given the spotlight SAR data model, the basic image formation algorithm is fairly obvious: inverse Fourier transform the available two-dimensional spectral region of $P'(K_u, K'_R)$ to estimate the image $\rho'(u, R)$. However, a conventional inverse DFT algorithm assumes the spectral data are on a rectangular grid in (K_u, K'_R), while the spotlight data are on a polar grid. Consequently, an extra step is required to interpolate the data from a polar to a rectangular grid before the inverse discrete Fourier transform (IDFT) can be applied.[11] In practice, the data will often be windowed in both dimensions after interpolation in order to reduce sidelobes in the final image. The resulting *polar format* (PF) *algorithm* is diagrammed in Fig. 8.40.

The key step in the PF algorithm is the polar-to-rectangular interpolation of the Fourier data. While a two-dimensional interpolation would be ideal, separable schemes using successive one-dimensional interpolations in two different dimensions are generally

---

[11] In principle, the keystone grid could be directly measured by the SAR system by varying the mixer output sample rate on successive pulses in a carefully controlled fashion.

Introduction to Synthetic Aperture Imaging 551

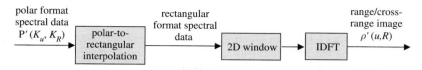

FIGURE 8.40 Polar format image formation algorithm for spotlight SAR.

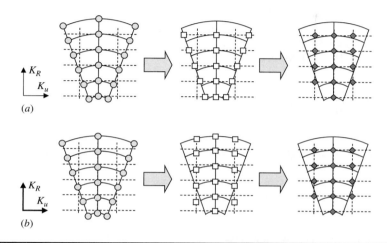

FIGURE 8.41 Two schemes for separable polar-to-rectangular interpolation. (a) Radial-keystone interpolation. (b) Angular-range interpolation.

preferred for simplicity. Two separable approaches are commonly used. In either approach, a rectangular grid of desired (K_u, K_R') sample locations is first established based on scene size and resolution considerations. The rectangular region covered by this grid must be contained within the annular region of available data. The first approach, illustrated in Fig. 8.41a, may be the more commonly used of the two. The complex polar format spectral data are interpolated along each radial to the *keystone* grid depicted in the center of the figure. After this step, the data are evenly spaced in K_R' but not in K_u. A second series of one-dimensional interpolations, now in the K_u dimension, align the samples to the desired grid in K_u. The second approach, depicted in Fig. 8.41b, interpolates first along constant-radius lines to align data on the desired K_u sample locations. The intermediate grid is then interpolated in the K_R' dimension to the final rectangular grid.

In the radial-keystone scheme, the interpolation is based on a separable two-dimensional sinc kernel that arises from the bandlimited nature of the data. Practical implementations window the sinc kernel to both limit sidelobes and provide a finite-length interpolating kernel in each dimension. Interpolating kernels typically range from 7th to 11th order, and the total computational load for polar-to-rectangular interpolation often rivals that of the two-dimensional inverse FFT used for the final image formation. The angular-range scheme uses a periodic asinc kernel in the angular interpolation and a sinc kernel in the range interpolation. A comparison of several of these schemes is given in Munson et al. (1985). A detailed discussion of the implementation of the radial-keystone approach is given in Jakowatz et al. (1996, Chap. 3).

At higher radar frequencies the annular region of polar format spectral data are further from the origin in (K_u, K_R') space. If the resolution is relatively coarse the required bandwidth is also small, so the extent of the annulus in range and angle is small. In this case the polar grid may be nearly rectangular, in which case the computationally expensive and quality-limiting polar-to-rectangular interpolation can be avoided and the image formed with a simple two-dimensional IDFT. This approach is called the *rectangular format algorithm* (RFA). It is essentially the DBS algorithm, but with an inverse DFT required in the range dimension to transform the stretch receiver output back to the fast-time domain. Consequently, the constraints to DBS processing apply to the RFA algorithm as well.

While the polar format algorithm is very commonly used in many operational spotlight SAR systems, other classes of algorithms can be applied to spotlight image formation. Application of the range migration and chirp scaling algorithms to spotlight SAR is described in Carrara et al. (1995). Backprojection can also be applied to spotlight SAR; see, for instance, Carrara et al. (1995), Munson et al. (1983), and Jakowatz et al. (2008).

8.6 Backprojection

In recent years interest has grown in *backprojection* (BP) methods for both stripmap and spotlight modes due to their capability in addressing extreme imaging parameters such as very low RFs in the UHF or even VHF regions, short standoff ranges resulting in large wavefront curvature and even near-field operation, nonrectangular data formats, very fine resolution requirements, avoidance of interpolations (a major source of image degradations), and ability to form images from nonrectilinear flight paths and on irregular sampling grids and non-flat surfaces such as existing digital elevation models. Backprojection methods are discussed in Desai and Jenkins (1992), Jakowatz et al. (2008), Frey et al. (2009), and Showman (2010).

The image formation algorithms discussed so far (DBS, RD, RM, and PF) are all Fourier-domain implementations of various approximations to a matched filter based on differing assumptions about the imaging scenario parameters such as resolution, RF, standoff range, degree and variability of range curvature, and so forth. Backprojection is a spatial domain implementation of the matched filter that does not make these approximations. Suppose the radar platform flies a linear collection path like that shown in Fig. 8.15b, again limiting the analysis to the two-dimensional case for simplicity. Consider an ideal scene $i(x, y)$ that is a superposition of P scatterers at different locations and having different complex amplitudes:

$$i(x,y) = \sum_{p=0}^{P-1} A_p \delta_D(x - x_p, y - y_p) \quad (8.80)$$

As the radar collects data in the aperture position and fast-time domain (u, t), each scatterer will contribute a PSR that can be modeled after pulse compression as in Eq. (8.36). The collected data will be

$$y(u,t) = \sum_{p=0}^{P-1} h_p(u,t) = \sum_{p=0}^{P-1} A_p \exp[-j(2\Omega_t/c)R_p(u)]\delta_D(t - 2R_p(u)/c)$$

$$R_p(u) = \delta_D\left[t - \frac{2}{c}\sqrt{(u - x_p)^2 + y_p^2}\right] \quad (8.81)$$

where $h_p(u, t)$ is the point spread response for scatterer p. Figure 8.42 illustrates a notional idealized data set for a scene containing three scatterers. Figure 8.42a shows the "ground truth," that is, the actual scatterer locations. The imaged area is 50 × 50 m at a standoff range of 100 m to the center of the scene from the 100 m long synthetic aperture flight path, as shown in the inset. Figure 8.42b shows the real parts of the PSRs of the three scatterers as observed from the synthetic aperture along the u axis; notice the differences in shape and

Introduction to Synthetic Aperture Imaging 553

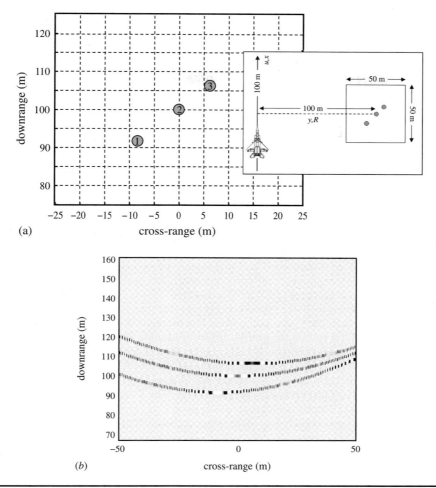

FIGURE 8.42 (a) Notional backprojection imaging scenario showing location of three scatterers. Inset shows geometry of image area and radar flight path. (b) Data collected by radar showing the real part of the post-pulse compressions point spread responses of the three scatterers. Grayscale variations reflect the signal amplitude.

orientation. (The PSR curves have been artificially increased in thickness for better visibility.) The oscillating shades of gray along each PSR represent the amplitude variations due to the $\exp[-j4\pi R_p(u)/\lambda_t]$ phase history on each echo.

The key realization is that if there is a scatterer in the scene at a particular coordinate (x_p, y_p) the baseband radar data matrix will contain a copy of the corresponding PSR $h_p(u,t)$ for that scatterer as it is observed from the radar moving along the synthetic aperture. Recalling that a correlator[12] maximizes signal-to noise ratio for a signal in WGN, this suggests that a way to test for the presence and generate an image of a scatterer at a specific image pixel

---

[12]The term "correlator" is used instead of "matched filter" because the latter implies shift-invariant processing, which is not the case here. Despite this quibble the term "matched filter" is widely used in the literature.

location (x_p, y_p) is to multiply the data (u, t) by the conjugate of the PSR for that specific location, $h_p^*(u, t)$, and integrate (sum) over u and t. If there was indeed a scatterer at that location, this operation will produce a peak at (x_p, y_p) with amplitude proportional to $|A_p|$. The PSRs of scatterers at other image locations will not align with $h_p^*(u, t)$ and will produce little or no contribution to the pixel value at (x_p, y_p). Repeating this operation for each pixel in the output image will build the full SAR scene,

$$d(x, y) = \int_{-\infty}^{\infty} \int_{-\infty}^{\infty} y(u,t) h_{x,y}^*(u,t)\, du\, dt \quad \forall (x, y)$$

$$h_{x,y}^*(u, t) = \exp[+j(2\Omega_t/c) R_{x,y}(u)] \delta_D\left[t - \frac{2}{c} R_{x,y}(u)\right]$$

(8.82)

where, in a change of notation, $h_{x,y}(u,t)$ is the PSR for a scatterer at an arbitrary (x, y) and $R_{x,y}(u)$ is the range from aperture position $(u, 0)$ to (x, y).

Continuing the three-scatterer example, Fig. 8.43 shows in gray just the region of support of the PSRs of the three scatterers in the collected data. Suppose the BP algorithm is computing the output image value at coordinates (6.25, 106.25), which is the location of scatterer #3. The matched filter response will be the conjugate of the scatterer #3 PSR. This is shown as the black dashed line marked "3 ref" and is slightly offset for visibility. Multiplying the complete data set (all three scatterer PSRs) by this matched filter will coherently integrate the scatterer #3 signature, and only that signature, to estimate the image value at coordinates (6.25, 106.25).

Also shown in the figure as the black line marked "4 ref" is the matched filter ROS that would be applied by the BP process when estimating the image at coordinate (−10, 100). There having been no scatterer at that location, there is no matching PSR in the data. The correlation operation will accumulate only minor contributions from the #2 and #3 scatterers where the matched filter crosses their PSRs. Figure 8.44 is a three-dimensional linear scale rendering of the finished backprojection image. The three scatterers are well-resolved and correctly located. Sidelobes are present and show a distinct "X"-shaped structure. That shape is due to the relatively high degree of range curvature in the PSRs for this particular example.

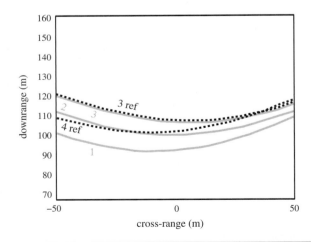

FIGURE 8.43 Backprojection of scatterer signatures. Solid gray lines are the regions of support of the contributions to the data of the three scatterers in Fig. 8.42a. The black dashed line "3 ref" is the BP PSR when forming the image pixel at the location of scatterer #3. The PSR "4 ref" is that of a potential scatterer at (−10,100). It does not coincide to any of the scatterer signatures in the data since there was no scatterer at that location.

Introduction to Synthetic Aperture Imaging

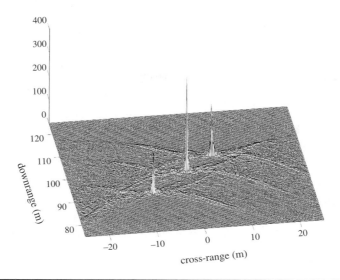

FIGURE 8.44 Image of the three scatterers formed by the backprojection algorithm.

As another example, BP image formation was applied to the same low frequency, short standoff, and fine resolution scenario used to illustrate the range migration algorithm in Sec. 8.3.5. The result fully focuses all five scatterers in the scene to the expected resolution. In fact, the backprojection image is visually indistinguishable from the RM result of Fig. 8.31b except for some fine differences in sidelobe amplitude (but not orientation) and so is not repeated here.

Why is this process called *backprojection*? A scatterer at a single pixel location in image space generates a signature spread across multiple samples in the radar data space, as shown in Fig. 8.42b. Put another way, the scatterer energy is "projected" onto the returns from a number of pulses, one at each synthetic aperture position, albeit in a different range bin for each. The correlation operation selects the range bin for each pulse that contains a contribution from that scatterer and recombines or "backprojects" them to re-form its impulse-like image.

Equation (8.82) suggests that a full two-dimensional correlation must be computed for each output pixel. This would be computationally forbidding. However, the post-pulse compression PSR, and therefore the matched filter ROS, is actually only one-dimensional, albeit on a curved ROS. Substituting the definition of $h_{x,y}(u,t)$ into the correlation double integral in Eq. (8.82) and carrying out the integration over t reduces the BP algorithm for one output pixel to

$$d(x,y) = \int_{-\infty}^{\infty} y\left(u, \frac{2}{c} R_{x,y}(u)\right) \exp[+j(2\Omega_t/c) R_{x,y}(u)] \, du \quad \forall (x,y) \tag{8.83}$$

In practice the data and image coordinates u, t, x, and y are all discretized. Formation of the complete image (all pixels) then becomes a triple summation:

$$d[n_x, n_y] = \sum_{n_x=0}^{N_x-1} \sum_{n_y=0}^{N_y-1} \sum_{n_u=0}^{N_u-1} y\left(n_u, \frac{2}{c} R_{n_x,n_y}(n_u)\right) \exp\left[+j(2\Omega_t/c) R_{n_x,n_y}(n_u)\right]$$

$$n_x \in [0, N_x - 1], n_y \in [0, N_y - 1], n_u \in [0, N_u - 1] \tag{8.84}$$

Equation (8.84) emphasizes the major drawback of backprojection: its very high $O(N^3)$ computational load. The Fourier-based algorithms (DBS, RD, RM, PF) all benefit from the applicability of the FFT, but the BP algorithm is a shift-variant time domain correlation so that Fourier methods are not applicable. Two-dimensional Fourier imaging algorithms using the FFT would be expected to have a computational load of $O(N^2 \log_2 N)$, a reduction compared to BP of a factor of $O(N/\log_2 N)$. For typical image and data dimensions ranging from perhaps 1000 to 8000 samples on a side, this factor varies from about 100 to 600. As an example, the difference in computer run time to generate the nearly indistinguishable BP and RM versions of Fig. 8.31*b* was a factor of 360. Considerable effort has been expended in developing "fast" backprojection algorithms. A good starting point in the literature is the $O(N^2 \log_2 N)$ algorithm described in Ulander et al. (2003).

8.7 Interferometric SAR

One of the newest developments in synthetic aperture radar is the ability to do fine-resolution imaging in three dimensions using interferometric techniques, commonly called *IFSAR* or *InSAR*. The basic approach employs two complex SAR images of a scene formed using two displaced receive apertures and either one or two transmit apertures. For example, the receive apertures may be physically separate antennas, each transmitting and receiving independently, or only one may be used as the transmit aperture for both. Another implementation uses a phased array divided into two subapertures. The full array is used for transmit, while each subaperture receives independently. Alternatively, IFSAR can be implemented using images collected from a conventional single aperture system on multiple passes. An introductory discussion is in Richards (2013).

8.7.1 The Effect of Height on a SAR Image

The output of a SAR image formation algorithm is a complex-valued two-dimensional image: both an amplitude and phase for each pixel. Conventional two-dimensional SAR imaging discards the phase of the final image, displaying only the magnitude information. In IFSAR, the pixel phase data are retained. Because SAR image formation is a nominally linear process, the complex amplitude $f(x, y_g)$ of a pixel at ground coordinates (x, y_g) and elevation of $z = 0$ can be viewed as the product of four factors,

$$f(x, y_g) = A \cdot G \cdot \exp(-j4\pi R_f(x, y_g)/\lambda_t) \rho(x, y_g)$$
$$= A \cdot |G| \cdot |\rho(x, y_g)| \cdot \exp\left\{ j\left[\phi_G + \phi_\rho(x, y_g) - \frac{4\pi}{\lambda_t} R_f(x, y_g) \right] \right\} \quad (8.85)$$

In this equation, ρ is the complex reflectivity of the pixel, the complex exponential term is the phase shift due to the range to the pixel, R_f is that range from the aperture phase center to the ground coordinates (x, y_g), G is the complex gain of the receiver and SAR image formation algorithm, and A contains all range equation factors. The phase of the pixel is therefore

$$\phi_f(x, y_g) = \phi_G + \phi_\rho(x, y_g) - \frac{4\pi}{\lambda_t} R_f(x, y_g) \text{ rad} \quad (8.86)$$

Now consider the two scatterers **P1** and **P2** at ground ranges y_{g1} and y_{g2}, as illustrated in Fig. 8.45. Both are at the same slant range R, but one is at elevation $z = h_0$ relative to some unknown reference plane while the other is at an elevation $z = h_0 + \Delta h$ ($h_0 = 0$ in the figure). They are observed from two distinct radar apertures at an altitude $z = h_0 + Z$ that are separated horizontally by a *baseline B*. Each aperture independently transmits a radar waveform,

Introduction to Synthetic Aperture Imaging

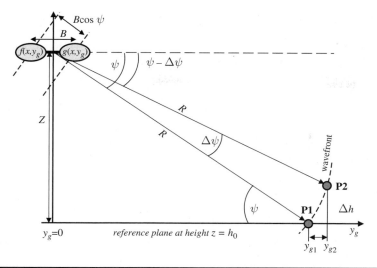

FIGURE 8.45 Geometry for interferometric height estimation.

receives the data, and forms a complex SAR image of the scene; these images are denoted $f(x, y_g)$ and $g(x, y_g)$.[13] The baseline should be orthogonal to the flight path; therefore, the direction of the aircraft motion is into the page.[14] The radar range is great enough that the incoming wavefront can be considered planar. If the depression angle from the middle of the baseline to **P1** is ψ, the difference in range to the two aperture phase centers is well approximated as $B\cos\psi$. The difference in received phase of the echo from **P1** at the two apertures then becomes, using Eq. (8.86),

$$\phi_{fg} \equiv \phi_f - \phi_g \approx -\frac{4\pi}{\lambda_t} B\cos\psi \quad \text{rad} \tag{8.87}$$

ϕ_{fg} is called the *interferometric phase difference* (IPD).

Now consider scatterer **P2**, which has the same slant range as **P1** but is elevated by Δh meters from the reference plane. The echo from **P2** arrives at the center of the radar baseline at the same time as that from **P1**. Consequently, conventional 2D SAR image formation algorithms will image both **P1** and **P2** as being at the same y_g coordinate, namely y_{g1}.[15] Assuming an approximately planar wavefront at range R, Fig. 8.45 shows that the difference in actual reference ground plane coordinates is approximately

$$y_{g2} - y_{g1} \approx \Delta h \tan\psi \quad \text{m} \tag{8.88}$$

---

[13]The results here are generalized to include use of a common transmit aperture and two independent receive apertures in Richards (2013). The effect of a common transmit aperture is to change the expression for $\Delta\phi_{fg}$ in Eq. (8.87) to $-(2\pi/\lambda_t)B\cos\psi$; this change then ripples through many subsequent formulas. Both separate and combined transmit aperture versions of IFSAR are of practical importance.

[14]The two apertures could also be displaced vertically or at an angle, with similar results. In general, the baseline direction must have a component orthogonal to the platform velocity vector.

[15]SAR images are naturally formed in the slant plane, but are usually projected into the ground plane for display.

This imaging of the elevated scatterer at a shifted ground range coordinate is termed *foreshortening* or *layover*[16] because the scatterer appears to have been shifted toward the radar. As illustrated, the foreshortening is only in the range coordinate. In squinted operation, foreshortening occurs in both range and cross-range; details are given in Sullivan (2000) and Jakowatz et al. (1996).

A conventional radar having a single transmit/receive phase center at the center of the baseline cannot distinguish the difference in height of **P1** and **P2** and so cannot "undo" the foreshortening distortions. However, the IPD between the two phase centers of an IFSAR system can be used to recognize the difference. In particular, the difference in the depression angle from the center of the IFSAR baseline to **P2** versus the angle to **P1** due to Δh can be found by differentiating Eq. (8.87) with respect to the grazing angle:

$$\frac{d\phi_{fg}}{d\psi} = \frac{4\pi}{\lambda_t} B \sin\psi \quad \Rightarrow \quad \Delta\psi = \frac{\lambda_t}{4\pi B \sin\psi} \Delta\phi_{fg} \quad \text{rad} \tag{8.89}$$

This equation states that a *change* in the IPD of $\Delta\phi_{fg}$ implies a change in depression angle to the scatterer of $\Delta\psi$ radians. To relate this depression angle change to an elevation change, consider Fig. 8.45 again. Examining two right triangles shows that $Z = R\sin\psi$ and $Z - \Delta h = R\sin(\psi - \Delta\psi)$. Eliminating R and applying a trigonometric identity gives

$$Z - \Delta h = \frac{Z\sin(\psi - \Delta\psi)}{\sin\psi} \quad \Rightarrow$$

$$\frac{\Delta h}{Z} = \frac{\sin\psi - \sin(\psi - \Delta\psi)}{\sin\psi} = \frac{2\sin\left(\frac{\Delta\psi}{2}\right)\cos\left(\frac{2\psi - \Delta\psi}{2}\right)}{\sin\psi} \tag{8.90}$$

$$\approx \frac{\Delta\psi}{\tan\psi}$$

The last step assumed $\Delta\psi$ small and applied a small angle approximation. Finally, using Eq. (8.90) to eliminate $\Delta\psi$ in Eq. (8.89) gives a measure of the relationship between the change in the IPD for a given pixel and a change in scatterer elevation above the reference plane (Carrara et al., 1995),

$$\Delta h = \frac{\lambda_t Z \cot\psi}{4\pi B \sin\psi} \Delta\phi_{fg} \quad \text{m} \tag{8.91}$$

Equation (8.91) is the basic result of IFSAR. It relates a *change* in the phase difference between echoes of a scatterer at two offset receive apertures to a change in height of that scatterer.

There are many ways to use these results to estimate scatterer height. One is described here, another in Richards (2013). Start by noting that even if the terrain is perfectly flat the IPD will still vary with range because ψ will vary. This variation of the IPD with range for flat terrain is called the *flat earth phase difference*, $\phi_{fg}^{FE}(x, y_g)$. It can be calculated from Eq (8.87) as

$$\phi_{fg}^{FE}(x, y_g) \approx -\frac{4\pi}{\lambda_t} B \cos[\psi(y_g)] = -\frac{4\pi B}{\lambda_t}\left(\frac{y_g}{R(y_g)}\right) = -\frac{4\pi B y_g}{\lambda_t \sqrt{Z^2 + y_g^2}}$$

$$= -\frac{4\pi B}{\lambda_t \sqrt{1 + (Z^2/y_g^2)}} \quad \text{rad} \tag{8.92}$$

---

[16] Layover is a special case of foreshortening in which scatterers are actually imaged out of order in the ground coordinate. However, the term layover is often used interchangeably with foreshortening.

A *flattened* or *leveled* IPD is then formed as

$$\phi'_{fg}(x,y_g) = \phi_{fg}(x,y_g) - \phi_{fg}^{FE}(x,y_g) \tag{8.93}$$

Any remaining variations in the IPD will be due to height variations in the scene relative to the reference plane. Consequently, the height difference between any two pixels will be proportional to the difference in the leveled phase at those two locations,

$$h(x_2,y_{g2}) - h(x_1,y_{g1}) = \frac{\lambda_t Z \cot\psi}{4\pi B \sin\psi}[\phi'_{fg}(x_2,y_{g2}) - \phi'_{fg}(x_1,y_{g1})] \tag{8.94}$$

A height map is formed by picking a reference point (x_0, y_{g0}) in the image and defining the height at that point, $h(x_0, y_{g0})$, as h_0. The remainder of the height map is then estimated according to

$$\hat{h}(x,y_g) = h_0 + \frac{\lambda Z \cot\psi}{4\pi B \sin\psi}[\phi'_{fg}(x,y_g) - \phi'_{fg}(x_0,y_{g0})] \tag{8.95}$$

The result is a *relative* height map expressing the height at each pixel as the change in height from h_0. Often this is adequate. Conversion to absolute height in some coordinate system, if needed, is discussed shortly.

The two SAR images required for IFSAR processing can be generated in one of two basic ways. In *one-pass IFSAR*, the approach described above, the radar platform has two displaced receive apertures so that the data for both images are collected on a single pass. This approach has the advantages of operational simplicity, much greater ease of aligning the trajectories of the two apertures, and no decorrelation of the scene between passes. *Two-pass IFSAR* uses a conventional single-receiver SAR system and flies two separate passes past the scene to be imaged. This approach requires careful offset alignment of the two flight paths to establish a constant baseline B, which can be difficult to achieve in airborne systems but is very feasible for orbiting radars. The chief advantage of two-pass IFSAR is that it can be implemented with existing conventional single-receiver SAR sensors. Note that two-pass IFSAR necessarily has separate transmit apertures, while one-pass IFSAR typically has a single transmit aperture (though a one-pass system with separate transmit apertures could be built).

8.7.2 IFSAR Processing Steps

Formation of an IFSAR image involves several major steps:

- Formation of the two individual SAR images, $f[l, m]$ and $g[l, m]$
- *Registration* of the two images
- Formation of the wrapped interferometric phase difference map $((\phi_{fg}[l, m]))_{2\pi}$
- Smoothing of $((\phi_{fg}[l, m]))_{2\pi}$ to reduce phase noise
- Two-dimensional *phase unwrapping* to obtain $\phi_{fg}[l, m]$ from $((\phi_{fg}[l, m]))_{2\pi}$
- Scaling of the unwrapped phase map $\phi_{fg}[l, m]$ to obtain the height map $\Delta h[l, m]$
- *Orthorectification* to correct foreshortening based on the relative height information

The images are formed using any SAR imaging algorithm appropriate to the collection scenario and the resolution and scene size goals. Because the height estimation depends on the

Chapter Eight

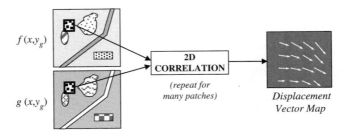

FIGURE 8.46 Generation of image pair warping function.

difference in phase of the echo from each pixel at the two apertures, it is important to ensure that like pixels are compared. The slightly different geometries of the two offset apertures will result in slight image distortions relative to one another, so an image registration procedure is used to warp one image to align well with the other. Many registration procedures have been developed in the image processing and radar imaging literature. One that is popular in IFSAR computes a series of correlations between small subimages of each SAR map to develop a warping function. This concept is illustrated in Fig. 8.46. The actual resampling is typically done with simple bilinear interpolators. The procedure is described in detail in Jakowatz et al. (1996).

Once the two images are registered, the next step is to compute the IPD. A major complication is that the radar signal processor can only measure the phase of a signal sample as the arctangent of the ratio of its imaginary and real parts. Therefore, only the wrapped phase for each receiver, $((\phi_f))_{2\pi}$ and $((\phi_g))_{2\pi}$, can be measured, where the notation $((\cdot))_{2\pi}$ indicates arithmetic modulo 2π. Consequently, the IPD ϕ_{fg} and its variation $\Delta\phi_{fg}$ are also measured modulo 2π.

The wrapped IPD map is easily computed as

$$((\phi_{fg}[l,m]))_{2\pi} = \arg\{f[l,m]g^*[l,m]\} \text{ rad} \tag{8.96}$$

Because any noise in the IPD map will become noise in the height map, local averaging is often applied to the IPD map to smooth phase noise. A simple 3×3, 5×5, or 7×7 local average is typical. The cost is a loss of spatial resolution in the IPD map and thus ultimately in the height map.

The two-dimensional phase unwrapping step to recover $\phi_{fg}[l,m]$ from $((\phi_{fg}[l,m]))_{2\pi}$ is the heart of IFSAR processing. Unlike many two-dimensional signal processing operations such as FFTs, it cannot be decomposed into one-dimensional unwrapping operations on the rows and columns. Two-dimensional phase unwrapping is an active research area. A brief introduction is given in Richards (2013) and a very thorough discussion is given in Ghiglia and Pritt (1998). Most unwrapping techniques can be classified as either *path following methods* or *minimum norm methods*; the latter are often based on fast transform techniques. An example of a minimum norm algorithm is the *discrete cosine transform* (DCT) method given in Ghiglia and Romero (1994). This technique finds the unwrapped phase function that, when rewrapped, minimizes the mean squared error between the gradients of the rewrapped phase function and the original measured wrapped phase.

The method begins by computing the wrapped gradients in the cross-range (*l*) and range (*m*) dimensions of the raw wrapped phase history data; these are then combined into a "driving function" *d*[*l*, *m*]:

$$\Delta x[l,m] = \begin{cases} ((((\phi_{fg}[l+1,m]))_{2\pi} - ((\phi_{fg}[l,m]))_{2\pi}))_{2\pi}, & 0 \le l \le L-2 \\ & 0 \le m \le M-1 \\ 0, & \text{otherwise} \end{cases}$$

$$\Delta y_g[l,m] = \begin{cases} ((((\phi_{fg}[l,m+1]))_{2\pi} - ((\phi_{fg}[l,m]))_{2\pi}))_{2\pi}, & 0 \le l \le L-1 \\ & 0 \le m \le M-2 \\ 0, & \text{otherwise} \end{cases} \quad (8.97)$$

$$d[l,m] = (\Delta_{yg}[l,m] - \Delta_{yg}[l-1,m]) + (\Delta_x[l,m] - \Delta_x[l,m-1]).$$

Let *D*[*k*, *p*] be the two-dimensional DCT of the driving function. The estimate of the unwrapped phase is then obtained as the inverse DCT of a filtered DCT spectrum,

$$\hat{\phi}_{fg}[l,m] = \text{DCT}^{-1}\left\{\frac{D[k,p]}{2\left[\cos\left(\frac{\pi k}{M}\right) + \cos\left(\frac{\pi p}{N}\right) - 2\right]}\right\} \quad (8.98)$$

This function is then used in Eq. (8.91) to estimate the terrain height map Δ*h*[*l m*].

As a simple idealized demonstration of this algorithm, consider the simulated terrain profile of a hill shown in Fig. 8.47. This function was created as the outer product of two one-dimensional Hann window functions. A simulation of one-pass IFSAR data collection produces the wrapped interferometric phase function of Fig. 8.48*a*. In addition, noise has been added to a rectangular patch of the phase data to simulate a low-reflectivity or degraded area. Straightforward application of Eqs. (8.97) and (8.98) produces the unwrapped

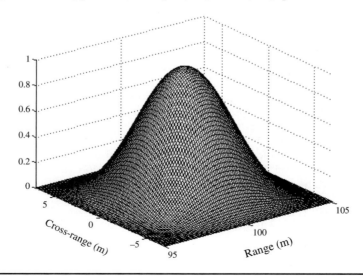

FIGURE 8.47 Height profile of simulated hill.

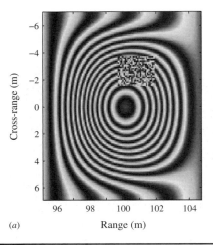 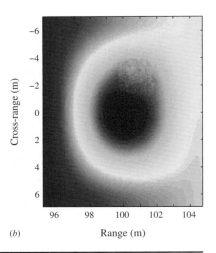

FIGURE 8.48 Example of DCT-based two-dimensional phase unwrapping on "hill" height map with noisy data patch. (a) Wrapped phase map $((\varphi_{fg}))_{2\pi}$. (b) Estimated unwrapped phase $\hat{\varphi}_{fg}$.

interferometric phase map estimate of Fig. 8.48b. The unwrapping is successful even in the noisy region. When converted back to terrain height, this phase map accurately reconstructs the hill height profile, although the height estimate is noisy in the degraded region.

IFSAR processing produces only relative height variations. Absolute height can be estimated by a variety of techniques. The most common is simply the use of surveyed reference points within an image; height relative to a reference point is then easily converted to absolute height. Another method uses two interferometric phase maps having different scale factors and therefore different ambiguity intervals in height. Multiple maps with different scale factors have been obtained using at least two different methods. In systems having a waveform bandwidth in excess of that required for the desired range resolution, the fast-time bandwidth can be split in half, and IFSAR processing completed separately for each half of the data. The effective center wavelength λ_t will be different for the two data sets, giving two different scale factors in Eq. (8.91). Another technique applicable to radar systems having three or more receive apertures forms one IFSAR height map using the first and second apertures, and another using the first and third apertures. The baselines for these two pairs are different, again giving rise to two different scaling factors in Eq. (8.91). These techniques are analogous to the use of multiple repetition rates to disambiguate aliased Doppler shifts or ranges in pulse Doppler processing.

The last step in IFSAR processing is orthorectification, which corrects the displacement of image pixels due to foreshortening. For each pixel in the SAR image $f[l, m]$, the corresponding height pixel $\Delta h[l, m]$ is used to estimate the foreshortening $-\Delta h \cdot \tan\psi$ present in that pixel. The image is then resampled in the range dimension to shift each pixel by $+\Delta h \cdot \tan\psi$ to its correct range position.[17]

Figure 8.49 is a *digital elevation model* (DEM), a three-dimensional visualization of an area generated from an IFSAR image, showing the San Andreas Fault, Transverse Ranges,

---

[17]If operated in squint mode, the image will also exhibit foreshortening in the cross-range dimension, so pixels must be shifted in both dimensions. See Jakowatz et al. (1996) for details.

Figure 8.49 Three-dimensional visualization of IFSAR data collected during the NASA Shuttle Radar Topography Mission (SRTM). (Image courtesy of NASA. SRTM image PIA03376.)

and Mojave Desert inland of the Los Angeles and Ventura basins in the United States. This image was generated from data collected by the Shuttle Radar Tomography Mission in 2000.

An important metric in quantifying IFSAR quality and describing error sources is the *coherence* $\gamma[l, m]$ of the two images $f[l, m]$ and $g[l, m]$. Coherence is a pixel-by-pixel correlation coefficient. A typical definition is

$$\gamma = \frac{E\{f \cdot g^*\}}{\sqrt{E\{|f|^2\} E\{|g|^2\}}} \qquad (8.99)$$

In the absence of errors, $|\gamma| = 1$ and $\arg\{\gamma\} = \phi_{fg}$ at each pixel. The actual value of coherence is a random variable due to additive noise, scene temporal decorrelation, baseline decorrelation, and other effects. Generally, coherence can be factored into a product of terms reflecting different error sources. The higher the total coherence, the better the IFSAR height estimate quality. Coherence magnitude values of 0.6 or larger are usually desirable.

The coherence due to thermal noise is given by Zebker and Villasenor (1992):

$$\gamma_{\text{noise}} = \frac{1}{1 + \chi^{-1}} \qquad (8.100)$$

This equation can be used for both thermal and quantization noise. An SNR of $\chi = 10$ dB gives a noise coherence of $\chi_{\text{noise}} = 0.91$. A second major contributor to degraded coherence is *baseline decorrelation*. This refers to the effect of the slightly different viewing angles of the two sensors on the radar cross section (RCS) of a given clutter cell. It is exactly the same RCS decorrelation effect discussed in connection with many-scatterer targets in Chap. 2. The angular decorrelation interval of Eq. (2.62) can be adapted to determine a *critical perpendicular baseline* $B_{c\perp}$ that would result in complete decorrelation of the two images (see Richards, 2013). "Perpendicular" refers to the baseline component perpendicular to the line of sight. The critical perpendicular baseline and the coherence factor that results from an actual perpendicular baseline $B_\perp$ are

$$\gamma_{\text{baseline}} = 1 - \frac{B_\perp}{B_{c\perp}}, \quad B_{c\perp} = \frac{\lambda_t R}{2\Delta R \cdot \tan\psi} \qquad (8.101)$$

where ΔR is the usual slant range resolution and separate transmitters are used. To keep coherence high it is desirable to design the system so that $B_\perp \ll B_{c\perp}$.

Another decorrelation source is *temporal decorrelation* in two-pass systems, which is primarily an issue for older single-receiver spaceborne IFSARs. Stable terrain regions can have decorrelation intervals as observed from satellites of several days, but water surfaces decorrelate almost completely in milliseconds. Forested areas also decorrelate fairly quickly due to leaf motion and may exhibit coherences on the order of 0.2. Atmospheric conditions can also affect IPD coherence. Changes in water vapor content and distribution can alter propagation delays in the atmosphere between passes in two-pass systems. Delay changes map directly into phase changes. Weather such as rain both decreases the echo strength and SNR and increases the weather clutter interference, affecting both one- and two-pass systems.

The effects of low coherence can be combatted by averaging multiple looks. Many of the same averaging techniques used to combat speckle can be employed (see Sec. 8.8.3), such as subbanding the data and averaging multiple lower-resolution IFSAR images, or spatial averaging of the IPD as described earlier.

The relative phase of two SAR images can be used in other ways (Richards, 2013). IFSAR presumes that there is no change in the imaged scene between passes, so phase difference variations are due only to height variations viewed from slightly different aspect angles. Another application of great interest is *coherent change detection* (CCD). CCD compares two images taken from the *same* trajectory at different times, from a few minutes apart to many hours or days apart. The pixel-by-pixel correlation coefficient will be close to one for pixels with the same complex reflectivity in both maps and close to zero for pixels whose complex reflectivity has changed between the two collections. An image formed from these correlation coefficients can provide a very sensitive indicator of activity in an observed area. Finally, *terrain motion mapping* also computes correlation coefficients between two images. However, it is assumed that the reflectivity is unchanged between the two collections, so any complex image changes are due to changes in the height of the scene. The time lag between collections may be days to years. Terrain motion mapping has been used to study such phenomena as glacier movement, land subsidence, volcanic activity, and seismic activity.

8.8 Other Considerations

8.8.1 Motion Compensation

By assuming the radar platform's position in the x coordinate is $u = vt$ and its z coordinate is a constant h, it has been implicitly assumed that the radar platform is flying a straight, level, and constant velocity path. In practice and despite best efforts, this is never quite true. For airborne systems especially, atmospheric turbulence, minor maneuvers and course corrections, crosswinds, vibration, antenna gimbal transient motions, and similar effects all cause deviations of the SAR antenna phase center from this ideal. However, this assumption is built into the design of the image formation processor, which is based on Eq. (8.27) or one of its various approximations. To the extent that the actual $R(u)$ does not follow the model of $R(u)$ used in the processor design, the actual SAR data will not match the expected PSR. The differences caused by deviations from the ideal flight path manifest themselves as phase errors in the processing.

Specifically, suppose the transmitted waveform is of the form $A\exp(j2\pi F_t t)$ and the range to some scatterer is R on a particular pulse. Then the received signal is

$$y(t) = A' \exp\left[j2\pi F_t \left(t - \frac{2R}{c}\right)\right] \qquad (8.102)$$

where A' absorbs all of the range equation factors. Now suppose that the phase center of the radar antenna is, for whatever reason, displaced from the intended path by a distance ε along the line of sight to the scatterer; the received signal will instead be

$$\tilde{y}(t) = \tilde{A}\exp\left[j2\pi F_t\left(t - \frac{2(R-\varepsilon)}{c}\right)\right] = \frac{\tilde{A}}{A'}\exp\left[-j4\pi\frac{\varepsilon}{\lambda_t}\right]y(t) \qquad (8.103)$$

Because ε is small, the amplitude ratio in Eq. (8.103) will be very nearly unity, so the primary effect of motion errors is a phase rotation to the data. All of the fast-time samples for a given pulse are rotated by the same phase factor.

The job of *motion compensation* is to estimate ε and correct the data by a simple counter phase rotation:

$$\hat{y}(t) = \exp\left[+j4\pi\frac{\varepsilon}{\lambda_t}\right]\tilde{y}(t) \approx y(t) \qquad (8.104)$$

The difficult part is estimating ε to the required accuracy. A displacement of only $\lambda_t/4$ meters (0.3 in at X band) corresponds to a two-way range change of $\lambda_t/2$ meters and therefore a 180° phase reversal. Thus, path deviations must be estimated to a small fraction of a wavelength to minimize phase error effects. Two factors ease this challenge somewhat. First, a constant displacement of the entire flight path has no effect; the same factor is added to the phase of all data samples, contributing a complex constant with no effect on focusing quality. Only variations in ε are significant. Second, any one pixel receives contributions from data only over the aperture time, on the order of a few tenths of a second to a few seconds in most SAR systems. Path drifts over longer time periods do not affect image focus.[18]

Phase errors are frequently categorized as either low- or high-frequency, based on the variation of the phase error in the slow-time dimension (Lacomme et al., 2001). Low-frequency errors are those that repeat on a time scale greater than the aperture time so that any periodicity of low-frequency errors is not apparent during formation of an image. High-frequency errors are subdivided into deterministic and noise-like errors. In the former, periodic structure is evident during the aperture time. Low-frequency errors produce net phase tapers across the aperture, affecting the resolution, gain, and cross-range accuracy of the PSR. For example, a linear phase taper is equivalent to an uncompensated Doppler shift and can produce a cross-range displacement of scatterers according to Eq. (8.40). Examples of the resolution and gain loss effects of quadratic phase errors were shown in Figs. 8.23 and 8.25a. High-frequency errors affect primarily the sidelobes of the PSR. Table 8.2 describes the major effect of the most important phase errors.

Phase errors are corrected by estimating the displacement $\varepsilon(u)$ at each aperture position u using a position measurement system, and in finer-resolution systems by also using *autofocus* algorithms. A typical way to develop position estimates is to use a motion compensation system with some or all of the elements in Fig. 8.50. Data from the aircraft or spacecraft's *inertial navigation system* (INS) provide the first level estimate of deviations from a straight line, constant-velocity flight path. The INS tracks the platform centroid, not the antenna phase center. Higher precision systems mount an additional strapdown *inertial measurement unit* (IMU) onto the antenna structure (Kennedy, 1988a). The IMU accounts more accurately for motions of the antenna relative to the airframe; these relative motions include not only

---

[18]Long-term drift may be quite important, however, to image interpretation. Targets or land features detected in a SAR image may need to be precisely geolocated for targeting or for surveying purposes. It is then essential to have accurate information on the absolute position of the radar platform.

Phase Error Class	Dominant Effect on PSR
Low Frequency	
Linear	Cross-range displacement
Quadratic	Mainlobe broadening, amplitude loss
Cubic	Mainlobe asymmetry and cross-range displacement
High Frequency	
Deterministic periodic	Discrete "paired echo" high sidelobes
Random	Increased sidelobe level

TABLE 8.2 Effects of Various Phase Errors on the SAR Point Spread Response

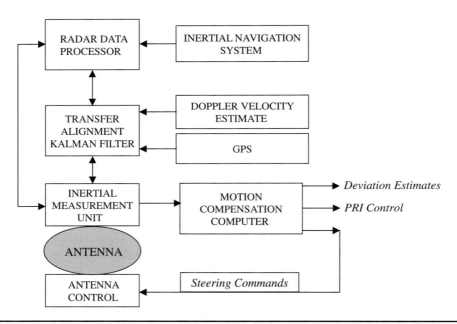

FIGURE 8.50 Generic motion compensation instrumentation package.

intentional scanning but also vibration and airframe flexure. Inertial antenna mounts can also be used to stabilize the antenna as much as possible. Additional independent kinematic estimates such as radar-derived Doppler velocity estimates and *global positioning system* (GPS) data may also be used. In this system, the INS provides an initial attitude reference to the IMU. IMU data are sent to the *radar data processor* (RDP), which corrects for the lever arm displacement between the INS and IMU. The difference in INS and corrected IMU position estimates is input to the Kalman *transfer alignment* filter, whose output is fed back to the IMU to update its state. The Kalman filter typically has on the order of 20 state variables. The updated IMU state is fed to the motion compensation computer, which uses these data to compute the position error ε and the associated phase correction. The motion compensation computer typically also computes antenna steering commands to stabilize the line of sight.

	1σ Value		Root Sum Square (RSS) Total	Quadratic Phase Error Formula, rad	Quadratic Phase Error, rad
Error Source	INS	IMU			
Cross-range velocity error, v_e (m/s)	0.24	0.06	0.247	$\dfrac{\pi v v_e T_a^2}{\lambda_t R_0}$	0.43
LOS accelerometer bias, A_b (μg)	70	80	106.3	$\dfrac{\pi A_b T_a^2}{2\lambda_t}$	0.22
Platform tilt, α (mrad)	0.2	0.1	0.22	$\dfrac{\pi \alpha a g T_a^2}{2\lambda_t}$	0.45
Boresight accelerometer scale factor error, s (PPM)	150	150	212	$\dfrac{\pi s A T_a^2}{2\lambda_t}$	0.13
Total RSS quadratic phase error, radians					0.67

Source: Kennedy (1988b).

TABLE 8.3 Motion Compensation Error Budget

A motion compensation error budget can be developed for the various INS/IMU errors. Table 8.3 is an example of such an analysis (Kennedy, 1988b). The parameter values used were an aperture time of 2 seconds, velocity of 152 m/s, standoff range of 37 km, gravitational acceleration $g = 9.8$ m/s^2, and a line-of-sight acceleration A of 3 m/s^2. These parameters imply a cross-range resolution of 1.8 m at 10 GHz. In this example, the total error is 0.67 rad, well under the $\pi/2$ standard suggested earlier. Since the phase error contributions all scale as T_a^2/λ_t, the motion error margins must be made tighter as radar frequency increases and especially as resolution is made finer. Because of its dependence on range, the velocity error tends to dominate at shorter ranges; at longer ranges the other error effects become more significant. Another analysis of error contributions, in terms of longitudinal and latitudinal motions, is given in Lacomme et al. (2001).

In addition to estimates of the path deviation ε, motion compensation systems often provide two additional types of control data. The first is fine pulse-to-pulse PRI adjustments to provide uniform sample spacing along the velocity vector, that is, in the x or u dimension. This technique eliminates the need to interpolate data in the slow-time dimension to obtain uniform sample spacing in finer-resolution systems. The second, applicable primarily to squinted SAR and spotlight SARs, computes the range walk relative to a reference point on the ground and varies the time delay from the time of pulse transmission to the time when the A/D converter at the receiver output is triggered to begin fast-time sampling. The delay is chosen such that the echo from some reference point, generally in the middle of the imaged scene, always occurs in the same range bin. This technique, called *motion compensation to a point*, reduces or eliminates the computation required for range walk correction. Both techniques require more complicated receiver control.

8.8.2 Autofocus

In fine-resolution SAR systems, the motion compensation systems discussed previously often may not provide sufficient phase correction to achieve good image quality. Autofocus algorithms attempt to estimate residual phase errors present in the data after motion compensation and compensate them to improve image quality. A number of techniques have been suggested; see Carrara et al. (1995) for an overview. Two common and representative methods are the map drift method and the phase gradient method. The most basic form of the map drift algorithm is designed specifically to correct quadratic phase errors. The algorithm divides the SAR data into two portions corresponding to the first and second halves of the synthetic aperture and forms a lower-resolution image from each. If quadratic phase error is present, peaks in the images will be shifted away from their correct locations but in opposite directions. Cross-correlating the two images produces a correlation peak whose offset from the origin is directly proportional to the amount of residual quadratic phase error. The full data set is then phase-corrected with the conjugate of the estimated quadratic error function and the image reformed from the modified data. The algorithm is able to correct quadratic phase errors up to a few tens of radians. Extensions to the map drift algorithm, which divide the data into more than two subapertures, are capable of estimating higher-order polynomial phase error terms.

A very basic analysis of a simple map drift algorithm can be outlined as follows. Suppose a slow-time phase history sequence $y[m]$ is contaminated with a quadratic phase error sequence,

$$y'[m] = y[m]\exp\{j4\alpha[m - (M-1)/2]^2/(M-1)^2\}, \quad 0 \le m \le M-1 \quad (8.105)$$

The parameter α is the maximum value of the quadratic phase error, which occurs at the two ends of $y'[m]$. Assume for convenience that M is even, and define the two half-sequences

$$\begin{aligned} y_1'[m] &= y'[m], & 0 \le m \le M/2 - 1, \\ y_2'[m] &= y'[m + M/2], & 0 \le m \le M/2 - 1 \end{aligned} \quad (8.106)$$

The phase error in $y_1'[m]$ can be decomposed into a quadratic component that is symmetric about the midpoint of $y_1'[m]$ and a linear component, which is just the straight line connecting the first and last sample of the phase curve in $y_1'[m]$. It is straightforward to show that this linear phase component is

$$\begin{aligned} \phi_{\text{lin}1}[m] &= \alpha - \left(\frac{2}{M-2}\right)\left(1 - \frac{1}{(M-1)^2}\right)\alpha m \\ &\equiv \alpha - \alpha' m \text{ rad} \end{aligned} \quad (8.107)$$

The linear phase term of $y_2'[m]$ has an identical quadratic component, but a linear component of the opposite sign (see Prob. 30). Because each sequence is half of the original sequence $y'[m]$, their DFTs will be similar. The linear phase terms, however, will shift the DFTs of $y_1'[m]$ and $y_2'[m]$, $Y_1'[k]$ and $Y_2'[k]$, by $-\alpha'K/2\pi$ and $+\alpha'K/2\pi$ samples, respectively, where K is the DFT size. A cross-correlation of $Y_1'[k]$ and $Y_2'[k]$ will therefore produce a peak at lag $k_0 = \alpha'K/\pi$ radians. An estimate $\hat{\alpha}$ of the quadratic phase error coefficient α can then be computed from the spectral correlation peak as

$$\begin{aligned} \alpha' &= \frac{\pi k_0}{K} \\ \hat{\alpha} &= \left(\frac{M-2}{2}\right)\left(1 - \frac{1}{(M-1)^2}\right)^{-1} \cdot \alpha' \end{aligned} \quad (8.108)$$

In practice, the estimate is not exact due to nonquadratic terms in the phase error and noise in the data. A corrected data sequence is then formed as

$$\hat{y}[m] = y'[m]\exp\{-j4\hat{\alpha}[m-(M-1)/2]^2/(M-1)^2\}, \quad 0 \le m \le M-1 \quad (8.109)$$

The map drift algorithm is applied iteratively, typically requiring on the order of three to six iterations. The iteration is stopped when the peak estimated phase error is less than $\pi/2$ (or some other appropriate quality threshold).

Figure 8.51 illustrates the map drift algorithm on data derived from a real scene. The original image in Fig. 8.51a was transformed to a simulated range-compressed phase history domain by assigning a random phase to each image pixel and then performing a Fourier transform in the cross-range dimension. A simulated phase error consisting of a quadratic term with a maximum of 5π radians across the aperture was then applied in the cross-range dimension and the image formed again by an inverse Fourier transform, giving the severely

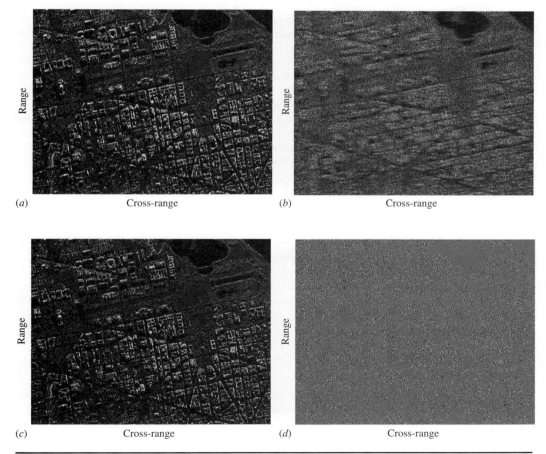

FIGURE 8.51 Illustration of map drift autofocus. (a) Original image. (b) Image formed from data with simulated noisy quadratic phase errors. (c) Image after map drift autofocus. (d) Error image. (Original image courtesy of Sandia National Laboratories.)

blurred image in Fig. 8.51*b*. Figure 8.51*c* is the result after applying map drift autofocus to the blurred image; it is visually indistinguishable from the original. The algorithm was halted when the maximum residual estimated phase error threshold was less than $\pi/2$ radians; this required three iterations. Finally, Fig. 8.51*d* is the difference image between the original and the autofocused image. While some correlation between the difference image and the original is visible on close inspection, the nearly uniform difference image confirms the quality of the phase error correction.

The *phase gradient algorithm* (PGA) does not assume a polynomial form for the phase error and appears to offer excellent, robust performance (Wahl et al., 1994; Carrara et al., 1995). PGA uses the blurred image of the strongest scatterers in the SAR scene to estimate the actual cross-range phase error function. More specifically, the algorithm estimates the first derivative of the cross-range phase error from a SAR image that is converted to (K_u, R) units by performing an inverse DFT in cross range. The phase error derivative is integrated to get the actual phase error function. Its negative is then applied to the data as a compensating phase function. As with map drift, the phase gradient algorithm is applied iteratively. The PGA also includes several preprocessing steps to improve the phase error estimation.

PGA is based on the following simple thought process. Regardless of the actual image formation algorithm used, model the imaging process as a simple one-dimensional DFT of the slow-time phase history data in each range bin, as in DBS. Now suppose the ideal complex image in a particular range bin consisted of a single bright point scatterer located at the origin in cross range. The corresponding slow-time data for that range bin, found by an IDFT of the image row, would be simply a complex constant. Any nonlinear phase component added to the slow-time data will result in a blurred image of that scatterer. The PGA tries to estimate the (presumably blurred) image of an isolated point scatterer and then use that estimate to derive and then compensate the slow-time phase error.

A simplified analysis of the PGA algorithm gives a more complete description of its operation. Consider an arbitrary range bin in an ideal, unblurred image and assume a dominant (bright) isolated point scatterer in that range bin is located at cross-range coordinate $k = k_0$. The model of the complex image domain data in this range bin is

$$Y[k] = A \cdot \delta[k - k_0] + W[k] \tag{8.110}$$

where $W[k]$ is a random process representing other clutter scatterers. The dominant scatterer complex amplitude A and location k_0 will vary in different range bins, but for notational simplicity the dependence on the range bin index l is not shown. Though not needed for the initial sketch of the algorithm, the phase of A will become important shortly. In addition, the low-level clutter image $W[k]$ will be ignored for now.

If the cross-range compression algorithm is assumed to be approximately the Fourier transform of the range-compressed phase history data implemented using a K-point DFT, the corresponding phase history data $y[m]$ will be the IDFT of the cross-range image:

$$y[m] = \text{IDFT}\{Y[k]\} = \frac{A}{K} \exp\left[j\frac{2\pi}{K}k_0 m\right] \tag{8.111}$$

Equation (8.111) represents the ideal phase history in the absence of phase errors. When phase errors $\phi_e[m]$ are present $y[m]$ is replaced by

$$\begin{aligned} y_e[m] &= \exp(j\phi_e[m])y[m] \\ &= \frac{A}{K}\exp\left(j\frac{2\pi}{K}k_0 m\right)\exp(j\phi_e[m]) \end{aligned} \tag{8.112}$$

Denote the DFT of the phase error modulation function $\exp(j\phi_e[m])$ as $E[k]$. The DFT of the phase history data including the phase errors is the actual observed complex cross-range image row due to the point scatterer. Applying the modulation property of the DFT gives

$$Y_e[k] = \text{DFT}\{y_e[m]\} = A \cdot E[k - k_0] \tag{8.113}$$

Thus, the point target image is degraded into a replica of the DFT of the phase error function, centered at the point target location. That is, $E[k]$ becomes the image-domain blurring PSR due to the phase errors. Equation (8.113) is also equivalent to saying that the blurred data is the convolution of the unblurred point scatterer and the blur PSR.

The blurred image of isolated point targets can be used to estimate $E[k]$ and therefore the phase error function $\phi_e[m]$. Begin by finding the peak amplitude of $Y_e[k]$; suppose this occurs at $k = k_p$. It is expected that the peak of $E[k]$ occurs at or near $k = 0$; therefore it should be the case that $k_p \approx k_0$. The cross-range slice is then circularly shifted left by k_p samples, giving a new image row that is an estimate of the blur PSR:

$$\tilde{Y}_e[k] = A \cdot E[((k - k_0 + k_p))_K] \approx A \cdot E[k] \tag{8.114}$$

with corresponding IDFT

$$\tilde{y}_e[m] = \frac{A}{K}\exp\{j\tilde{\phi}_e[m]\} = \frac{A}{K}\exp\{j\phi_e[m]\}\exp\{j2\pi(k_0 - k_p)m/K\}$$
$$\approx \frac{A}{K}\exp\{j\phi_e[m]\} \tag{8.115}$$

The phase $\tilde{\phi}_e[m]$ of the slow-time phase history corresponding to the shifted image row approximately equals the actual phase error function. The original slow-time phase history is then corrected by compensating it with the estimated phase:

$$\tilde{y}[m] = \tilde{y}_e[m]\exp\{-j\tilde{\phi}_e[m]\} \tag{8.116}$$

Finally, the deblurred image row is estimated by transforming back to the image domain,

$$\tilde{Y}[k] = \text{DFT}\{\tilde{y}[m]\} \tag{8.117}$$

If $\tilde{\phi}_e[m]$ is a good estimate of $\phi_e[m]$, the result will be

$$\tilde{Y}[k] \approx A \cdot \delta[((k - k_0 + k_p))]_K \approx A \cdot \delta[k - k_0] = Y[k] \tag{8.118}$$

Note that if the blur PSR $E[k]$ does not have its peak at the origin, $k_p \neq k_0$ and the corrected image slice will still be shifted by $(k_p - k_0)$ samples from the correct position. This does not hurt image quality but does impact geolocation accuracy.

The PGA adds a number of extensions to this basic idea to produce a robust algorithm. Because actual imagery will often not display sufficiently well-isolated and bright point scatterers to develop a good estimate of $E[k]$ in a given image row, it is necessary to repeat the process above on multiple cross-range rows. Independent estimates of the blur PSR in each row can then be coherently averaged to improve the estimate of $E[k]$. Further improvement can be obtained by limiting the integration to a subset of image rows that have the brightest peaks and averaging only those. Furthermore, the width of the blur PSR can be estimated from those relatively strong peaks and used to limit the averaging only to the portion of the image line believed to represent the blur PSR, plus a margin.

The algorithm is applied iteratively, improving the phase estimate and more effectively deblurring the image at each iteration until some stopping criterion is reached. At each iteration, the estimated blur PSR width is updated. Ideally it should get narrower, becoming more impulse-like, if the algorithm is converging well. Also, any linear trend and bias in the current iteration estimate of the phase error are removed before accumulating it into the total phase error estimate. The linear trend removal ensures that the phase corrections do not shift the image in the cross-range direction. The bias removal, while common, does not affect image focus.

But why is it called the phase *gradient* algorithm? The procedure outlined above directly estimates and compensates the motion-induced phase error $\phi_e[m]$. The phase gradient is not calculated and does not appear to be needed. The issue arises when coherently integrating across multiple range bins. Consider again the simple single-scatterer model resulting in the blurred complex image of Eq. (8.115). The motion error phase $\phi_e[m]$ is the same in each range bin but the complex amplitudes, and in particular their phases, are not. Coherently integrating across range bins is therefore as likely to destructively add the individual phase functions as to constructively add them. However, since the complex amplitude phase is a constant in slow time, the slow-time gradient of the phase $\Delta\phi_e[m]$ depends only on $\phi_e[m]$. Thus, $\phi_e[m]$ can be estimated by estimating $\Delta\phi_e[m]$ and then integrating that function to get the actual phase error.

There are several phase gradient estimators that can be used (Wahl et al., 1994). A simple maximum likelihood estimator is the first difference (Jakowatz and Wahl, 1993). Reintroducing the range bin index l and including integration over multiple range bins, this is

$$\Delta\tilde{\phi}_e[m] = -\arg\left\{\sum_l \tilde{y}_e^*[l, m-1]\tilde{y}_e[l, m]\right\}$$
$$\approx \sum_l (\tilde{\phi}_e[m] - \tilde{\phi}_e[m-1]) \tag{8.119}$$

Another common algorithm is the linear unbiased minimum variance estimator

$$\Delta\tilde{\phi}_e[m] = \frac{\sum_l \mathrm{Im}\{\tilde{y}_e^*[l, m] \cdot (\tilde{y}_e[l, m] - \tilde{y}_e[l, m-1])\}}{\sum_l |\tilde{y}_e[l, m]|^2} \tag{8.120}$$

The phase error itself is estimated by integrating the gradient:

$$\tilde{\phi}_e[m] = \begin{cases} 0, & m = 0 \\ \sum_{q=1}^{m} \Delta\tilde{\phi}_e[q], & \text{otherwise} \end{cases} \tag{8.121}$$

Figure 8.52 illustrates application of the PGA to a simple synthetic image. The original image in Fig. 8.52a is simply a collection of randomly located point scatterers of varying amplitude and phase in complex white uniform noise. A phase error consisting of quadratic, sinusoidal, and noise terms (Fig. 8.53) was applied to the cross-range DFT of these data, and the image re-formed via a cross-range IDFT. The result, seen in Fig. 8.52b, is severely blurred in cross range. Application of the PGA produces the corrected image of Fig. 8.52c after three iterations; additional iterations produce little further improvement in this example. The smearing has been greatly reduced.

The actual phase error function $\phi_e[m]$ applied in this example is shown in Fig. 8.53. It is the sum of quadratic, sinusoidal, and noise terms with about 18 rad (nearly 6π) of total phase variation across the aperture. The estimated phase error function after one and three iterations is also shown, as is the residual phase error $\tilde{\phi}_e[m] - \phi_e[m]$ after three iterations. The residual phase error is well within the $\pm\pi/2$ bounds denoted by the horizontal lines.

Introduction to Synthetic Aperture Imaging 573

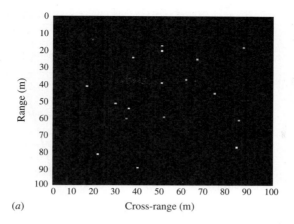

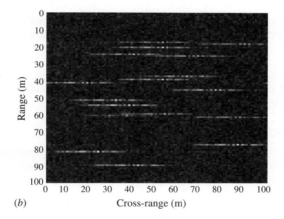

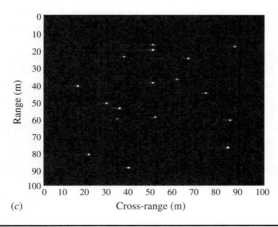

FIGURE 8.52 Illustration of phase gradient algorithm autofocus. (*a*) Synthetic image. (*b*) Image blurred by non-polynomial phase error. (*c*) Image after three iterations of PGA autofocus using LUMV gradient estimator.

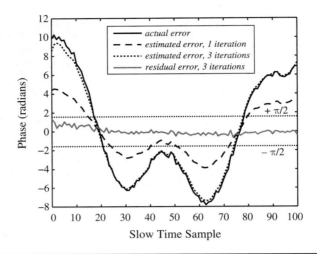

FIGURE 8.53 Actual phase error for example of Fig. 8.52; estimated error after one iteration of ML estimator; and estimated error and residual after three iterations.

Additional issues and details regarding PGA are addressed in Jakowatz et al. (1996) and Carrara et al. (1995).

8.8.3 Speckle Reduction

Like any coherent imaging system, SAR produces images contaminated by "speckle," an aptly-named multiplicative noise. Speckle is the natural result of the coherent combination of echoes from many different scatterers to form an image pixel. If the amplitude distribution of the real and imaginary parts (I and Q channels) of the received signal is Gaussian and the phase distribution is uniform, conditions ensured by the law of large numbers when many scattering centers are involved, then the pixel amplitude will be Rayleigh distributed, as shown in Chap. 2. Thus, pixels representing areas with the same average RCS (mean echo amplitude) can give rise to different pixel amplitudes. These variations are not due to thermal, quantization, or other noise sources, but are nonetheless considered "noise" because of their appearance and effect on image quality.

Speckle is reduced through various forms of filtering and averaging schemes (Oliver and Quegan, 2004). One of the most effective is noncoherently integrating multiple uncorrelated images, or *looks*, of the same scene. This process reduces the pixel variance, reducing the amplitude variations among pixels representing the same RCS. Uncorrelated looks can be obtained using multiple passes. Obtaining well-aligned passes is difficult for airborne systems due to atmospheric turbulence and limited navigation precision. For spaceborne systems, orbits are well known and well behaved, but considerable time can pass between repeat passes so that data acquisition time can be lengthy. In a single pass, transmitter frequency or multiple polarization channels can be used to obtain uncorrelated looks. Another method takes advantage of the fact that in many stripmap modes the maximum aperture time available exceeds that required to meet resolution goals. In this case, the slow-time data can be divided into multiple subapertures, each long enough to form an image of the proper resolution (Cumming and Wong, 2005). Images are calculated for each subaperture and combined. Typically, 4 to 10 looks might be used to obtain effective speckle reduction.

Figures 8.54a and b demonstrates the image enhancement obtained by integration of 9 simulated looks.

If multiple looks are not available, another method averages adjacent pixels of a fine-resolution image, typically using a 3 × 3 or 5 × 5 window, to form one lower resolution but reduced-speckle pixel, as illustrated in Fig. 8.54c. Figure 8.54d shows the result of a 3 × 3 median filter applied to a single look image. This approach sacrifices less resolution than spatial averaging. More sophisticated techniques include a variety of adaptive filters and statistical methods (Oliver and Quegan, 2004; Richards, 2009).

8.8.4 Moving Targets

SAR algorithms are designed to image stationary terrain and objects. They do so by implementing a receiver filter matched to the point spread response of a stationary target when viewed by a radar following a particular data collection protocol such as stripmap or spotlight SAR. Different imaging algorithms correspond to differing approximations to the scatterer PSR, as appropriate to differing scenarios. Moving objects within the scene will exhibit PSRs that are mismatched to the SAR matched filter and therefore will not focus properly.

Consider again the two-dimensional or slant plane sidelooking stripmap SAR scenario. A simple, approximate approach to understanding the effect of constant-velocity scatterer motion on its SAR image is to break that motion into two components: a velocity component v_{LOS} toward the radar along the line of sight and another component v_{at} in the along-track direction. These two components are not necessarily orthogonal. Consider the LOS component first. Convert the quadratic approximation to the stationary scatterer slow-time phase history $\phi(u)$ of Eq. (8.37) into instantaneous Doppler frequency in hertz $F_D(t)$ by using $u = vt$ and differentiating:

$$\phi(t) = \phi(u)|_{u \to vt} \approx -\left(\frac{4\pi}{\lambda_t}\right)\left(R_p + \frac{v^2}{2R_p}t^2 - \frac{x_p v}{R_p}t + \frac{x_p^2}{2R_p}\right) \text{ rads}$$

$$F_D(t) = \frac{1}{2\pi}\frac{d\phi(t)}{dt} \approx \left(\frac{2vx_p}{\lambda_t R_p} - \frac{2v^2}{\lambda_t R_p}t\right) \text{ Hz}$$

(8.122)

In this equation, recall that x_P is the cross-range coordinate of the scatterer at the middle of the synthetic aperture. In a sidelooking system, x_P is therefore the cross-range displacement from the boresight.

Now suppose the scatterer is moving toward the platform flight path with a velocity v_{LOS}. This will increase the Doppler shift at the middle of the synthetic aperture by the usual $2v_{LOS}/\lambda_t$ Hz. The instantaneous Doppler becomes

$$\begin{aligned}F_D(t) &\approx \left(\frac{2vx_p}{\lambda_t R_p} - \frac{2v^2}{\lambda_t R_p}t\right) + \frac{2v_{LOS}}{\lambda_t} = \frac{2}{\lambda_t}\left(\frac{x_p}{R_p}v + v_{LOS}\right) - \frac{2v^2}{\lambda_t R_p}t \\ &= \frac{2}{\lambda_t R_p}\left(x_p + \frac{v_{LOS}}{v}R_p\right)v - \frac{2v^2}{\lambda_t R_p}t \text{ Hz}\end{aligned}$$

(8.123)

The role of x_P has been replaced by the quantity $(x_p + (v_{LOS}/v)R_p)$, meaning that the scatterer cross-range position in the image will be shifted by $+(v_{LOS}/v)R_p$ meters. Recalling the relation between Doppler shift and cross-range position in DBS, this is as expected: the

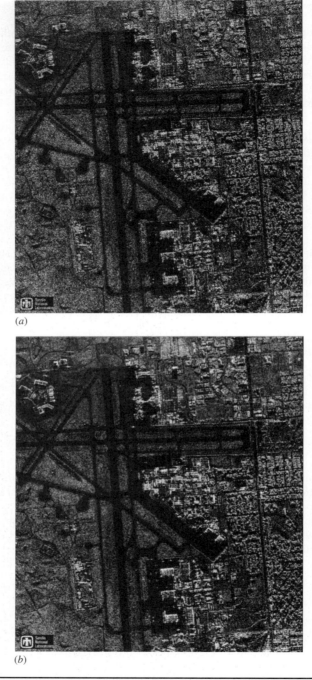

FIGURE 8.54 Speckle reduction. (a) Simulated single-look speckled image derived from Fig. 8.1a. (b) Full resolution image obtained by integration of nine looks. (c) Reduced-resolution image obtained with 3 × 3 spatial filtering. (d) Image obtained with 3 × 3 median filter. (Original image courtesy of Sandia National Laboratories.)

Introduction to Synthetic Aperture Imaging

(c)

(d)

FIGURE 8.54 (Continued)

increase in scatterer Doppler frequency is interpreted as a wider cross-range coordinate. This analysis does not predict any change in the quadratic phase component of the PSR, suggesting that radial velocity will not cause defocusing. A more careful analysis will show that there is indeed a change in the quadratic phase term as well, more significant for longer synthetic apertures than shorter ones.

Now consider scatterer motion in the along-track dimension only, in the same direction as the radar at velocity v_{at}. This will make the effective velocity of the radar relative to the moving scatterer less than its velocity relative to the stationary scatterers, which in turn will mean that the imaging algorithm will have the wrong quadratic phase rate of variation for the moving scatterer. To quantify this, let $x_P = 0$ for simplicity and modify the phase history in Eq. (8.122) to incorporate the scatterer motion:

$$\phi(t) \approx -\left(\frac{4\pi}{\lambda_t}\right)\left[R_p + \frac{(v - v_{at})^2}{2R_p}t^2\right] \text{ rad} \tag{8.124}$$

The difference between this modified phase history and the unmodified ($v_{at} = 0$) history is

$$\delta\phi(t) \approx -\left(\frac{4\pi}{\lambda_t}\right)\left\{\left[R_p + \frac{(v - v_{at})^2}{2R_p}t^2\right] - \left[R_p + \frac{v^2}{2R_p}t^2\right]\right\}$$

$$= \left(\frac{2\pi}{\lambda_t R_p}\right)(2v \cdot v_{at} - v^2)t^2 \text{ rad} \tag{8.125}$$

Equation (8.125) gives the change in quadratic phase shift due to along-track scatterer motion.[19] Quadratic phase error effects were discussed in Sec. 8.3.2, where it was shown that peak errors of more than $\pi/2$ rad can lead to significant degradation of resolution and sidelobes. The maximum value of $\delta\phi(t)$ occurs at the ends of the synthetic aperture, $t = \pm T_a/2$. A more complete description of target motion effects is given in Axelsson (2004).

Figure 8.55 is a simulation that demonstrates these two effects. A sidelooking stripmap system views a scene whose central reference point is at a range of 10 km. Three point scatterers are located at ranges of -300, 0, and $+200$ m relative to the CRP and cross ranges of 0 m. The scatterer at the CRP is stationary. The one at -300 m has an along-track velocity v_{at} of $+100$ m/s but no LOS velocity component. The scatterer at $+200$ m has no along-track motion but has a line of sight velocity of $+5$ m/s (toward the radar). The simulated data assumes a 10-GHz radar using an LFM pulse on a platform traveling 150 m/s. A simple DBS algorithm with azimuth dechirp was used to form the image. The waveform and SAR parameters were selected to give range and cross-range resolutions of 3 m. The required aperture time is 0.333 second.

Figure 8.55 shows that the stationary center scatterer is fully focused as expected. The anticipated cross-range shift of the scatterer with the LOS velocity is $(v_{LOS}/v)R_P = (5/150)(10,000) = 0.33$ km. The top scatterer of the figure clearly shows this shift while remaining fully focused. The bottom scatterer shows the cross-range smearing, but no cross-range positional shift, due to the along-track motion. The maximum quadratic phase shift for these parameters is about 1.4π rad, resulting in smearing more severe than the π radian worst case in Fig. 8.22a.

Figure 8.56a is a classic real example of the effect of a line-of-sight velocity component. The radar was an X-band airborne system. The bright diagonal streak in the image is a

---

[19]Notice the similarity of the term involving v_{at} to the "cross-range velocity error" formula in Table 8.3. When the v_{at} term is evaluated at the maximum relevant value of time, $t = T_a/2$, the two are identical.

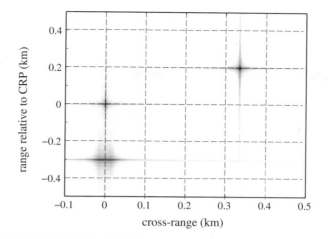

FIGURE 8.55 Simulation illustrating effect on SAR image of scatterer motion. The center scatterer is a fully focused stationary reference. The upper scatterer motion is along the LOS to the radar and results in a cross-range displacement while remaining focused. The lower scatterer motion is parallel to the radar flight path and results in cross-range smearing. See text for parameters.

railroad train. Because of the LOS component of its motion, it is imaged as being off the railroad tracks, which lie to its left. Figure 8.56b illustrates the effect of along-track motion. The circled area is a portion of a road near the Jefferson Memorial in Washington, D.C. Cars on this portion of the road are moving approximately parallel to the radar flight path, resulting in severe cross-range smearing of their signatures.

A number of techniques have been proposed to better detect and image moving targets within a scene. One attempts to identify "chips" within a SAR image that exhibit smearing. Candidate patches are extracted and then iteratively refocused using

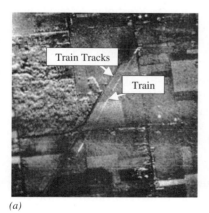

(a)   *(b)*

FIGURE 8.56 (a) X-band airborne SAR image of a moving train. The train image is displaced in cross-range, making it appear to be off the tracks. (b) Airborne SAR image of the Jefferson Memorial area. Cars traveling on the portion of the road parallel to the flight path are severely smeared in cross-range.

different parameters and a measure of image sharpness (Fienup, 2001). A similar approach refocuses the chips with a series of predefined matched filters and chooses the best output. If multiple receive apertures with an along-track baseline component are available, the ground moving target detection algorithms such as clutter suppression interferometry and along-track interferometry discussed in Chap. 5 can also be applied to SAR imagery.

8.9 Summary

Moving radars are capable of forming fine-resolution two-dimensional imagery using the synthetic aperture concept. The chief advantage of radar as compared to other imaging sensors is its 24/7, essentially all-weather operation.

The basic principle of SAR imaging was presented in terms of synthesizing a very large array antenna. This viewpoint is sufficient to develop a number of basic relationships and constraints for SAR systems, such as the basic cross-range resolution formula of $\lambda_t R/2D_{SAR}$, the minimum sidelooking stripmap SAR resolution of $D_{az}/2$, and the conflict between high area coverage rates and fine resolution.

SAR was initially developed as an extension of classical range-Doppler processing, leading to the Doppler beam sharpening algorithm as the first approach to actually forming an image. However, as finer resolution is sought, this viewpoint and the DBS algorithm both break down due to increases in range migration and in quadratic and higher-order terms in the phase history. In response, radar researchers and developers have evolved a hierarchy of image formation algorithms with increasing capabilities. The range-Doppler and range migration algorithms were presented for sidelooking stripmap imaging as examples of matched filtering approaches able to partially or fully handle range curvature effects and designed to take advantage of Fourier transform efficiencies in implementation. For spotlight SAR using LFM waveforms and stretch processing, the tomographic-style polar format algorithm was developed. More general are the emerging backprojection algorithms that take advantage of advances in computing power to implement explicit space-variant correlators that can handle a wide variety of scenarios. All of these algorithms were developed only for sidelooking SAR in this chapter; the references discuss extensions to squinted SAR geometries.

If the radar has two receive apertures offset in a direction orthogonal to the platform velocity vector, interferometric SAR can use pairs of SAR images to estimate terrain height variations, creating a three-dimensional image. Basic IFSAR processing involves a number of steps. Brief introductions to the key steps of image registration and two-dimensional phase unwrapping were given, as well as a discussion of image coherence and the factors affecting it.

Several issues affecting image quality for all SAR types were discussed. The most important may be motion compensation. All SAR imaging algorithms assume a particular platform trajectory relative to the scene and a resulting phase history or point spread response for each scatterer. Image formation algorithms are designed to correctly focus data exhibiting that PSR. Real platforms inevitably deviate from that ideal trajectory, resulting in phase errors relative to the expected phase history and, in turn, degradations in image quality. Typical motion compensation instrumentation used to estimate and compensate for these errors was described. However, in fine resolution systems autofocus techniques are necessary to achieve the intended image quality. Two workhorses in this area, the map drift and phase gradient methods, were described and demonstrated here.

SAR imaging also assumes the "target" (terrain) is comprised of stationary scatterers, an assumption that is clearly not true in many situations. It was seen that radial motion components lead to cross-range displacement of moving scatterer images, while tangential motion results in cross-range smearing. Finally, it was noted that the GMTI techniques described in Chap. 5 can be used with multichannel SAR imaging to detect moving targets within a scene and correct their cross-range positions.

References

Ausherman, D. A., "Digital vs. Optical Techniques in Synthetic Aperture Radar (SAR) Data Processing," *Optical Engineering*, vol. 19, no. 2, pp. 157–167, Mar./Apr. 1980.

Ausherman, D. A. et al., "Developments in Radar Imaging," *IEEE Transactions on Aerospace & Electronic Systems*, vol. AES-20, no. 4, pp. 363–400, Jul. 1984.

Axelsson, S., "Position correction of moving targets in SAR imagery," *Proceedings of SPIE*, vol. 5236, pp. 80–92, 2004.

Bamler, R., "A Comparison of Range-Doppler and Wavenumber Domain SAR Focusing Algorithms," *IEEE Transactions on Geoscience and Remote Sensing*, vol. 30, no. 4, pp. 706–713, Jul. 1992.

Brown, W. M., and L. J. Porcello, "An Introduction to Synthetic-Aperture Radar," *IEEE Spectrum*, vol. 6, no. 9, pp. 52–62, Sep. 1969.

Cafforio, C., C. Prati, and F. Rocca, "SAR Data Focusing Using Seismic Migration Techniques," *IEEE Transactions on Aerospace & Electronic Systems*, vol. AES-27, no. 2, pp. 194–207, Mar. 1991.

Carrara, W. G., R. S. Goodman, and R. M. Majewski, *Spotlight Synthetic Aperture Radar*. Artech House, Norwood, MA, 1995.

Charvat, G. L., *Small and Short-Range Radar Systems*. CRC Press, New York, 2014.

Cook, D. A., "Spotlight Synthetic Aperture Radar," Chap. 6 in W. L. Melvin and J. A. Scheer, (eds.), *Principles of Modern Radar: Advanced Techniques*. SciTech Publishing, 2013.

Cumming, I. G., and F. H. Wong, *Digital Processing of Synthetic Aperture Radar Data*. Artech House, Boston, MA, 2005.

Curlander, J. C., and R. N. McDonough, *Synthetic Aperture Radar: Systems and Signal Processing*. Wiley, New York, 1991.

Cutrona, L. J. et al., "On the Application of Coherent Optical Processing Techniques to Synthetic-Aperture Radar," *Proceedings of the IEEE*, vol. 54, no. 8, pp. 1026–1032, Aug. 1966.

Desai, M. D., and W. K. Jenkins, "Convolution Backprojection Image Reconstruction for Spotlight Mode Synthetic Aperture Radar," *IEEE Transactions on Image Processing*, vol. 1, no. 4, pp. 505–517, Oct. 1992.

Dudgeon, D. E., and R. M. Mersereau, *Multidimensional Digital Signal Processing*. Prentice Hall, Englewood Cliffs, NJ, 1984.

Elachi, C., *Spaceborne Radar Remote Sensing: Applications and Techniques*. IEEE Press, New York, 1988.

Fienup, J. R., "Detecting Moving Targets in SAR Imagery by Focusing," *IEEE Transactions on Aerospace and Electronic Systems*, vol. 37, no. 3, pp. 794–809, Jul. 2001.

Franceschetti, G., and R. Lanari, *Synthetic Aperture Radar Processing*. CRC Press, New York, 1999.

Freeman, A. et al., "The Myth of the Minimum SAR Antenna Area Constraint," *IEEE Transactions on Geoscience and Remote Sensing*, vol. 38, no. 1, pp. 320–324, Jan. 2000.

Frey, O., C. Magnard, and M. Rüegg, "Focusing of Airborne Synthetic Aperture Radar Data from Highly Nonlinear Flight Tracks," *IEEE Transactions on Aerospace & Electronic Systems*, vol. AES-47, no. 6, pp. 1844–1858, Jun. 2009.

Ghiglia, D. C., and M. D. Pritt, *Two-Dimensional Phase Unwrapping: Theory, Algorithms, and Software*. Wiley, New York, 1998.

Ghiglia, D. C., and L. A. Romero, "Robust Two-Dimensional Weighted and Unweighted Phase Unwrapping That Uses Fast Transforms and Iterative Methods," *Journal of Optical Society of America*, vol. 11, no. 1, pp. 107–117, Jan. 1994.

Gough, P. T., and D. W. Hawkins, "Unified Framework for Modern Synthetic Aperture Imaging Algorithms," *International Journal of Imaging Systems & Technology*, vol. 8, pp. 343–358, 1997.

Harger, R. O., *Synthetic Aperture Radar Systems: Theory and Design*. Academic Press, New York, 1970.

Henderson, F. M., and A. J. Lewis (eds.), *Manual of Remote Sensing: Principles and Applications of Imaging Radar*, 3d ed., Vol. 2. Wiley, New York, 1998.

Jakowatz, C. V., Jr., and D. E. Wahl, "An Eigenvector Method for Maximum Likelihood Estimation of Phase Errors in SAR Imagery," *Journal of Optical Society of America*, vol. 10, no. 12, pp. 2539–2546, Dec. 1993.

Jakowatz, C. V. et al., *Spotlight Mode Synthetic Aperture Radar*. Kluwer, Boston, 1996.

Jakowatz, C. V., Jr., D. E. Wahl, and D. A. Yocky, "Beamforming as a Foundation for Spotlight-Mode SAR Image Formation by Backprojection," *Proceedings SPIE vol. 6970, Algorithms for Synthetic Aperture Radar Imagery XV*, Mar. 2008.

Kennedy, T. A., "The Design of SAR Motion Compensation Systems Incorporating Strapdown Inertial Measurement Units," *Proceedings of the IEEE 1988 National Radar Conference*, pp. 74–78, Apr. 1988a.

Kennedy, T. A., "Strapdown Inertial Measurement Units for Motion Compensation for Synthetic Aperture Radars," *IEEE AESS Magazine*, vol. 3, no. 10, pp. 32–35, Oct. 1988b.

Kirk, J. C., Jr., "Discussion of Digital Processing in Synthetic Aperture Radar," *IEEE Transactions on Aerospace & Electronic Systems*, vol. AES-11, no. 3, pp. 326–337, May 1975.

Lacomme, P., J.-P. Hardange, J.-C. Marchais, and E. Normant, *Air and Spaceborne Radar Systems*. William Andrew Publishing, Norwich, NY, 2001.

Munson, D. C., Jr., J. D. O'Brien, and W. K. Jenkins, "A Tomographic Formulation of Spotlight-Mode Synthetic Aperture Radar," *Proceedings of the IEEE*, vol. 71, no. 8, pp. 917–925, Aug. 1983.

Munson, D. C., Jr. et al., "A Comparison of Algorithms for Polar-to-Cartesian Interpolation in Spotlight Mode SAR," *Proceedings of the IEEE International Conference on Acoustics, Speech, and Signal Processing*, vol. 10, pp. 1364–1367, 1985.

Munson, D. C., Jr., and R. L. Visentin, "A Signal Processing View of Strip-Mapping Synthetic Aperture Radar," *IEEE Transactions on Acoustics, Speech, and Signal Processing*, vol. 27, no. 12, pp. 2131–2147, Dec. 1989.

Oliver, C., and S., Quegan, *Understanding Synthetic Aperture Radar Images*. SciTech Publishing, Raleigh, NC, 2004.

Raney, R. K., "A New and Fundamental Fourier Transform Pair," *Proceedings of IEEE 12th International Geoscience & Remote Sensing Symposium* (IGARSS '92), pp. 106–107, May 26–29, 1992.

Richards, J. A., *Remote Sensing with Imaging Radar*. Springer, Heidelberg, 2009.

Richards, M. A., "Nonlinear Effects in Fourier Transform Processing," Chap. 6 in J. A. Scheer and J. L. Kurtz (eds.), *Coherent Radar Performance Estimation*. Artech House, Norwood, MA, 1993.

Richards, M. A., "Derivation of the Range-Doppler Algorithm Frequency Response," technical memorandum, Oct. 7, 2008. Available at http://radarsp.com.

Richards, M. A., "Interferometric SAR and Coherent Exploitation," Chap. 8 in W. L. Melvin and J. A. Scheer (eds.), *Principles of Modern Radar: Advanced Techniques*. SciTech Publishing, 2013.

Schleher, D. C., *MTI and Pulsed Doppler Radar*. Artech House, Norwood, MA, 1991.

Sherwin, C. W., J. P. Ruina, and R. D. Rawcliffe, "Some Early Developments in Synthetic Aperture Radar Systems," *IRE Transactions on Military Electronics*, vol. MIL-6, no. 2, pp. 111–115, Apr. 1962.

Showman, G. A., "An Overview of Radar Imaging," Chap. 21 in M. A. Richards, J. A. Scheer, and W. A. Holm (eds.), *Principles of Modern Radar: Basic Principles*. SciTech Publishing, Raleigh, NC, 2010.

Showman, G. A., "Stripmap SAR," Chap. 7 in W. L. Melvin and J. A. Scheer (eds.), *Principles of Modern Radar: Advanced Techniques*. SciTech Publishing, Edison, NJ, 2013.

Soumekh, M., *Synthetic Aperture Radar Signal Processing with MATLAB Algorithms*. Wiley, New York, 1999.

Stimson, G. W., *Introduction to Airborne Radar*, 2d ed. SciTech Publishing, Mendham, NJ, 1998.

Stolt, R. H., "Migration by Fourier Transform," *Geophysics*, vol. 43, no. 1, pp. 23–48, Feb. 1978.

Sullivan, R. J., *Microwave Radar: Imaging and Advanced Concepts*. Artech House, Norwood, MA, 2000.

Ulander, L. M. H., H. Hellsten, and G. Stenstrom, "Synthetic-Aperture Radar Processing Using Fast Factorized Backprojection," *IEEE Transactions on Aerospace & Electronics Systems*, vol. AES-39, no. 3, pp. 760–776, Jul. 2003.

Wahl, D. E., P. H. Eichel, D. C. Ghiglia, and C. V. Jakowatz, Jr., "Phase Gradient Autofocus—A Robust Tool for High Resolution SAR Phase Correction," *IEEE Transactions on Aerospace & Electronics Systems*, vol. AES-30, no. 5, pp. 827–835, Jul. 1994.

Wiley, C. A., "Pulsed Doppler Radar Methods and Apparatus," U.S. patent no. 3,196, 436, 1965 (originally filed 1954).

Wiley, C. A., "Synthetic Aperture Radars—A Paradigm for Technology Evolution," *IEEE Transactions on Aerospace & Electronic Systems*, vol. AES-21, pp. 440–443, 1985.

Wu, C., K. Y. Liu, and M. Jin, "Modeling and a Correlation Algorithm for Spaceborne SAR Signals," *IEEE Transactions Aerospace & Electronics Systems*, vol. AES-18, no. 5, pp. 563–574, Sep. 1982.

Zebker, H. A., and J. Villasenor, "Decorrelation in Interferometric Radar Echoes," *IEEE Transactions of Geoscience Remote Sensing*, vol. 30, no. 5, pp. 950–959, Sep. 1992.

Problems

1. Compute the sidelooking SAR synthetic aperture size D_{SAR} and aperture time T_a for cross-range resolutions of 50, 5, and 0.5 m for a 10-GHz airborne radar at a range of 10 km and a 5-GHz LEO satellite at a range of 770 km. Assume the aircraft velocity is 100 m/s and the satellite velocity is 7500 m/s.

2. Compute the integration angle γ in degrees needed to obtain cross-range resolutions of 100, 10, and 1 m at RFs of 1 and 10 GHz.

3. Compute the maximum swath length L_s for an elevation beamwidth of 10° at a grazing angle of $\delta = 30°$ and ranges of 10 and 50 km.

4. Compute the elevation beamwidth in degrees required to obtain a swath length of 100 km from an altitude of 770 km. Assume a grazing angle of $\delta = 30°$ and ignore earth curvature.

5. What is the area coverage rate of the radar in Prob. 4 in km²/h? Assume $v = 7500$ m/s.

6. For a spaceborne SAR having $v = 7500$ m/s and a grazing angle of $\delta = 30°$, what is the finest cross-range resolution possible if the swath length required is 100 km?

584 Chapter Eight

What is the longest swath length possible if the cross-range resolution is required to be 1 m? Assume sidelooking stripmap operation.

7. What is the upper bound on the number of images per hour obtainable by a spotlight SAR with a platform velocity of 50 m/s, an RF of 35 GHz, a standoff range of 10 km, and a cross-range resolution of 0.25 m? Assume that 10 percent of the time is used for re-steering the antenna between spots or other "overhead" operations.

8. The series expansion of the square root leading to Eq. (8.28) (and eventually to the DBS and range-Doppler algorithms) requires $|u - x_p| \ll R_p$. Interpret "much less than" to mean $|u - x_p| < R_p/100$ (a tight standard). Determine whether a radar on the space shuttle (250 km altitude) having a resolution of $\Delta CR = 10$ m at an RF of 5 GHz meets this requirement. Repeat for an airborne radar having a resolution of $\Delta CR = 0.25$ m at an RF of 16 GHz and a range of 5 km.

9. Develop formulas for range walk and range curvature normalized to range resolution, $\Delta R_w / \Delta R$ and $\Delta R_c / \Delta R$. Assume "square pixels," that is, $\Delta R = \Delta CR$. If the radar resolution is improved by a factor of 10× in both dimensions, by what factors do the normalized range walk and range curvature increase?

10. It is desired to use DBS to form an image 10 km on a side with resolution of 3 m. The RF is 10 GHz and the standoff range is 10 km. Considering the range walk and quadratic phase constraints, can these specifications be achieved with good focus quality? Assume "square pixels."

Problems 11 to 17 consider the same airborne sidelooking stripmap SAR system. For simplicity the problem is treated as two-dimensional rather than three-dimensional so that the slant plane and the ground plane are the same. The goal of the SAR is to image a swath extending from 10 to 20 km slant range from the aircraft; thus the swath length is $L_s = 10$ km. The resolution goal in both range and cross range is 1.0 m. The aircraft is flying straight and level at a constant velocity of 150 m/s. The radar operates at K_u band (16 GHz). The physical antenna azimuth beamwidth is $\theta_{az} = 2°$.

11. Estimate the azimuth size of the physical antenna aperture, D_{az}, in meters. *Hint:* It is not determined by the desired resolution, which may or may not meet the stripmap constraint that $\Delta CR \geq D_{az}/2$; use other information provided to determine D_{az}. What pulse bandwidth β is required to obtain the desired range resolution? (Do not allow for any weighting for range or cross-range sidelobe control.)

12. What is the maximum Doppler bandwidth of the slow-time data? What is the bound on the PRF due to the Doppler bandwidth? What is the bound on the PRF due to the swath length?

13. What is the aperture time T_a required to achieve the desired cross-range resolution ΔCR at the near edge of the swath? At the far edge? What are the corresponding synthetic aperture lengths D_{SAR}?

14. Assume the waveform is an LFM pulse with a time-bandwidth product $\beta \tau = 50$; its bandwidth β was found in Prob. 11. How many range (fast time) samples are required to process the full swath length? How many pulses will be transmitted in an aperture time at the near and far edges? Assume a PRF equal to the Nyquist rate for the Doppler bandwidth of the data.

15. What is the range walk for a scatterer within the swath? Does it vary over the swath? Express the answer in both meters and in range resolution cells. What is the maximum range curvature for a scatterer within the swath? Does it occur for scatterers at the near or far edge of the swath? Express the answer in both meters and in range resolution cells.

Introduction to Synthetic Aperture Imaging

16. What is the depth of focus for this system? Can a single point spread response (PSR) be used to focus well everywhere in the swath? If not, what is the minimum number of subswaths needed to achieve good focus throughout?

17. Consider Doppler beam sharpening processing (probably not a good choice for this fine a resolution, but relatively easy for hand calculations). Suppose that Fourier analysis (an FFT) of the slow-time data at range $R = 15$ km reveals a peak in the spectrum at $F_0 = 106.7$ Hz. What is the cross-range position x_0 relative to the antenna boresight of the scatterer that caused the peak?

18. A sidelooking imaging radar system achieves a cross-range resolution of $\Delta CR = 1$ m. The radar is an L-band (1-GHz) system with an azimuth antenna width $D_{az} = 3$ m. It is mounted on an aircraft that flies at 100 m/s. The center of the imaged area is at a range of 50 km. Must the radar be operated in stripmap mode, spotlight mode, or can either be used? Justify the answer.

19. Regardless of the answer to the previous problem, assume that spotlight mode is used. What is the line-of-sight rotation angle γ in degrees required to achieve the 1 m cross-range resolution? What are the resulting aperture size D_{SAR} in meters and aperture time T_a in seconds?

20. Starting with Eq. (8.33) for range curvature, derive Eq. (8.58). Note that R_P in Eq. (8.33) and R_0 in Eq. (8.58) represent the same quantity.

21. Consider an L band (1 GHz) designed to achieve range and cross-range resolutions of 0.5 m each. No weighting for sidelobe control is used. Sketch the region in the image (K_x, K_y) space over which data must be collected in order to support this resolution. Label all significant spatial frequencies.

22. Continuing the previous problem, sketch and label the region in the radar's (K_u, K'_R) data collection space which must be acquired in order to support the image space region of the previous problem when the Stolt mapping is applied.

23. In the example of Fig. 8.32, the spatial frequency data covers a region from about 1.5 to 6.5 cycles/m in two-way range spatial frequency F'_R and about ± 2.5 cycles/m in cross-range spatial frequency F_u. What image spatial frequency coordinates (F_x, F_y) will the corner sample at $(F_u, F'_R) = (2.5, 1.5)$ be mapped to by the Stolt transformation? What do you think should be done with this and similar collected data samples?

24. Equation (8.69) depends on the RVP term being negligible. Because this is a quadratic phase term in the integrand, it has negligible effect if $\pi(\beta/\tau)(\delta t_i)^2 < \pi/4$ for the maximum absolute value of δt_i. For a scene of depth L_u ($\pm L_u/2$ around the CRP), this is L_u/c. Derive an upper bound on L_u in terms of β and τ. Note: The result will not be the same as Eq. (8.79), which requires further assumptions but will have some similarities.

25. Consider a spotlight SAR using the polar format algorithm. The system operates at K_u band (16 GHz) from a standoff range $R_0 = 30$ km and is designed for a range and cross-range resolution of 0.2 m. The LFM pulse sweep rate is $\alpha = \pi(7.5 \times 10^{12})$ Hz/s. Compute the limits on scene size for using the PF algorithm due to isorange curvature and residual video phase.

26. Determine the ambiguous terrain height change Δh_{ua} for a K_u-band (16-GHz) IFSAR system at an altitude of $Z = 3$ km and a nominal grazing angle of $\psi = 20°$. The IFSAR baseline is $B = 1$ m. Δh_{ua} is the change in terrain height that produces a change of 2π in the IPD $\Delta \phi_{fg}$.

27. Determine the SNR needed to obtain coherence values $\gamma_{noise} = 0.5, 0.8,$ and 0.95.

28. Compute the critical baseline $B_{c\perp}$ for a C-band (5-GHz) LEO radar. The range is $R = 770$ km, the range resolution is 30 m, and the grazing angle is $\psi = 15°$. If the baseline coherence is to be no less than 0.9, what is the maximum spacing of two parallel orbits that can be used for IFSAR processing?

29. Table 8.2 states that a linear phase error term in slow-time results in a cross-range displacement of the scatterer image. Consider a sidelooking 10 GHz SAR and a scatterer directly abreast of the SAR at a range of 5 km. Assume the platform velocity is 100 m/s. Suppose that crosswinds cause the aircraft to drift such that there is a velocity component of +5 m/s along the LOS toward the imaged scatterer. The scatterer should be imaged as being at $x = 0$. Using Doppler beam sharpening as a simple model of the processing, at what coordinate x will the scatterer actually be imaged? Include the sign of x and state whether the shift is in the direction of the aircraft motion or its opposite.

30. Confirm that the slope of the linear phase component of $y_2'[m]$ of Eq. (8.106) is the negative of the slope of $y_1'[m]$, which can be obtained from Eq. (8.107).

31. Suppose a sidelooking SAR images a car approaching with a line-of-sight velocity of $v_{LOS} = 25$ m/s. The radar platform velocity is 100 m/s. By how much will the image of the car be displaced in the cross-range dimension? What is the maximum LOS velocity so that the displacement does not exceed one-half of a standard U.S. interstate highway lane width of 12 ft? Assume the nominal range R_P is 3 km and the cross-range resolution is 2 m.

32. Continuing the previous problem, suppose the radar frequency is 10 GHz and the target is at a range of 10 km. What is the maximum along-track velocity component of the car if the maximum quadratic phase error across the synthetic aperture is to be limited to π radians? Repeat for a range of 1 km.

CHAPTER 9

Introduction to Array Processing

In Chap. 3, the concept of the radar datacube was introduced to describe the data collected in a coherent processing interval. Figure 3.19, repeated in part here as Fig. 9.1, illustrates the datacube $y[l, m, n]$ with its independent axes of fast time, slow time, and antenna phase center. The radar signal processing described up to this point has dealt almost entirely with a fast-time/slow-time matrix for a single antenna phase center. When a multiple phase center antenna (a phased array) is used, the radar also collects spatially distributed samples of the echo waveform. Just as the temporal sampling of the slow-time axis enables analysis and processing of signals in a given range bin based on their temporal Doppler frequency content, the phase center axis enables analysis and processing of signals within a range bin based on their spatial frequency content, which is equivalent to the angle of arrival. Operating on the phase center axis provides a form of spatial filtering to complement Doppler filtering.

In this chapter, a basic introduction to *virtual arrays* (VAs), *beamforming,* and *space-time adaptive processing* (STAP) is presented. Virtual arrays provide a consistent way to describe the effect of a wide range of conventional and unconventional transmit and receive array structures. Beamforming refers to the coherent combination of data from multiple phase centers to provide selectivity in the angle of arrival (AOA), that is, to form and steer an antenna beam. STAP combines both spatial and temporal filtering on a moving radar platform to discriminate targets from both clutter and jamming. More detail on virtual elements and virtual arrays can be found in Richards (2018) and Richards (2019). Excellent introductions to adaptive beamforming and STAP are Aalfs (2013) and Melvin (2013), respectively.

9.1 Virtual Arrays

The ideas of a virtual element (VE) and a virtual array were introduced briefly in Chap. 1. To recap, consider a transmitting element or aperture (**T**) and a receive element or aperture (**R**). Presuming far-field operation, the VE associated with this pair is a monostatic **T-R** element located halfway between them. The VE has the property that the **T-R** path length, and therefore the received phase history, is to a very good approximation the same as that of the actual **T-R** pair. This means that the **T-R** pair can be replaced for analysis purposes by the VE. A distribution of multiple transmit and/or receive elements defines a collection of VEs that comprise the VA for that configuration.

Figure 9.2*a* shows an irregular transmit array comprising four numbered elements and a three-element receive array. Concentrating on the first transmit element only, the dashed lines connect it with each of the receive elements. The black squares are the midpoints between that transmit element and each receive element. These three points are the VEs

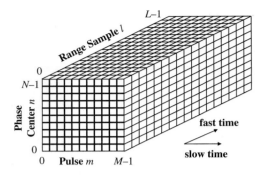

FIGURE 9.1 Radar datacube for a single CPI.

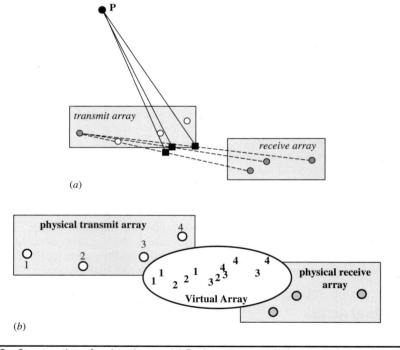

FIGURE 9.2 Construction of a virtual array. (a) Dark squares mark the three virtual elements created from the combination of the dark gray transmit element and the three receive elements. The two-way path length from the VE to the scatterer **P** and back is a good approximation to the path length from the transmit element to **P** and then to the corresponding receive element. (b) The full VA. VEs are indicated by numbers showing which transmit element generated them.

contributed to the VA by the first transmit element. The path length from one of these VEs to the scatterer **P** and back is a good approximation to the path length from the corresponding transmit element to **P** and back to the corresponding receive element. Figure 9.2b shows the complete VA, with the numbers representing the VEs contributed by the like-numbered transmit element.

Introduction to Array Processing

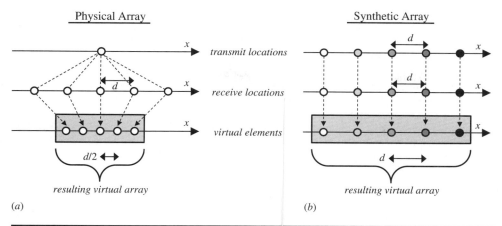

FIGURE 9.3 (a) Transmitting from a single phase center and receiving with a physical array generates a VA that is one-half the size of the physical array. (b) A SAR system generates and receives from one phase center at a time, moving it on each successive transmission. The resulting VA is the same size as the receive array, and twice the size of the VA associated with the physical array of the same size in (a).

Calculation of the VA element locations can be formalized as follows. Denote the set of N_T transmit element locations as $\{\mathbf{x}_{Tn}\}$, $n \in [0, N_T - 1]$, where $\mathbf{x}_{Tn}$ is the spatial coordinates of element n. A similar notation is used for the set of R receive elements. It is convenient to represent these sets of locations by spatial "locator functions" $L_T(\mathbf{x}_T)$ and $L_R(\mathbf{x}_R)$ that are simply sums of Dirac impulse functions, one at each transmit or receive element location. The locator function for the virtual array is then the sum of impulse functions at each midpoint (VE location),[1]

$$L_{TR}(\mathbf{x}_{TR}) = \sum_{n=0}^{N_T-1} \sum_{m=0}^{N_R-1} \delta\left(\mathbf{x} - \frac{(\mathbf{x}_{Tn} + \mathbf{x}_{Rm})}{2}\right) \equiv L_{mid}(\mathbf{x}_T, \mathbf{x}_R) \qquad (9.1)$$

Figure 1.11 showed a 1D example of constructing a uniform linear array (ULA) from a bistatic system with one transmit phase center and a separate, and separated, receive ULA. Recall that the resulting VA element spacing and total size were each one half that of the receive array.

In another example, it was noted in Chap. 8 that a synthetic aperture of size D_{SAR} provides cross-range resolution one-half that of a physical array of the same extent. Consideration of the virtual arrays shows why this is the case. First consider the physical array with spacing d between elements shown in Fig. 9.3a, which indicates the x-axis location of the transmit phase center, receive phase centers, and resulting VEs. (All of these elements are colinear but are shown displaced vertically for clarity.) Because the whole array is used for transmit, the transmit phase center is a single VE at the center of the array, which then pairs with each physical receive element to contribute a VE to the virtual array. The result is a VA one-half the extent of the physical array.

---

[1] It is often said that the VA element locations are the convolution of the transmit and receive locations. Equation (9.1) shows that this is not quite true. While reminiscent of a convolution of impulse functions, the division of the sum of transmit and receive locations by two makes this not a true convolution.

A standard SAR system transmits and receives a pulse from the same location along the flight path, then moves to the next location for the next pulse and so forth. Consequently, on each pulse, the transmit and receive phase centers are at the same x coordinate and therefore so is the resulting VE. This SAR data collection process constructs a VA identical to the synthetic array, as shown in Fig. 9.3b. Therefore, the virtual array corresponding to the SAR array is twice the size of the VA corresponding to a physical array. The larger SAR VA explains the reduced beamwidth and improved angular resolution of the synthetic array over the physical array.

Two more examples show how $N_T + N_R$ physical elements can be used to construct a VA having $N_T N_R$ virtual elements, suggesting a way to reduce array hardware costs. Figure 9.4 shows how a sparse physical array can generate a filled virtual array. The physical array has five physical elements: two transmit, two receive, and one capable of both. Effectively, however, there are $N_T + N_R = 3 + 3 = 6$ elements, since the shared one does double duty. The five physical elements form a sparse array because two elements needed to complete a filled ULA, indicated by the dashed circles, are missing. The corresponding VA is a fully filled array of $N_t N_R = 9$ elements.

The last example, shown in Fig. 9.5, constructs a 2D filled array from two orthogonal 1D arrays, with $N_T = N_R = 3$. The construction is straightforward. Again, $N_T + N_R$ physical elements generate $N_T N_R$ VEs forming a VA that is fully filled and has an element spacing half that of the generating arrays.

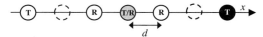

FIGURE 9.4 Creating a filled, non-redundant nine-element virtual array using a five-element sparse physical array. The shading of the VEs in the VA indicate the generating transmit element. The dashed circles highlight the missing elements in the physical array.

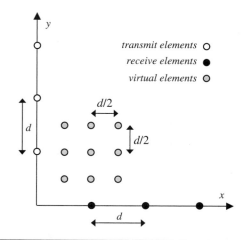

FIGURE 9.5 Creating a filled, non-redundant nine-element 2D virtual array from separate 1D three-element transmit and receive arrays.

9.2 Beamforming and Beam Steering

Consider a plane wave signal with temporal variation $x(t)\exp(j\Omega_t t)$ impinging on a uniform linear one-dimensional array (ULA) at an angle of arrival θ relative to the array normal, as shown in Fig. 9.6. The complex envelope $x(t)$ is the possibly wideband baseband waveform $a(t)\exp[j\varphi(t)]$. Because the lag in the time of arrival from one element to the next is $d\sin\theta/c$, the signal observed at the nth array element output is

$$\bar{y}_n(t) = x(t - nd\sin\theta/c)\exp[j\Omega_t(t - nd\sin\theta/c)] \qquad (9.2)$$

Thus, the set of array output signals are in general not in phase with one another. Adding them to form a composite output is as likely to result in destructive interference as constructive interference. Figure 9.7a shows the output signals for a three-element array when a pure sinusoidal pulse waveform impinges on the array. Each element output begins one-quarter cycle later than the previous, which would occur for instance if $d = \lambda_t/2$ and $\theta = 30°$. The angled dashed line marks the beginning of each output, emphasizing the time skew. The vertical dashed line emphasizes that at the same point in time the three outputs are not in phase.

Beamforming is the operation of modifying and combining the individual element outputs into a single composite output so as to provide a higher-amplitude output for signals arriving from a specified AOA while minimizing the amplitude at the beamformer output of signals from other directions. That is, the beamformer should create the effect of an antenna gain pattern with a high-gain mainlobe in a user-specified direction θ_0, low gain sidelobes in other directions, and possibly even nulls in the direction of interfering signals. This process is referred to as *steering* the array in the direction θ_0.

There are two fundamental ways to steer the array mainlobe: time delay steering and phase steering. Both have the goal of causing the outputs $\bar{y}_n(t)$ to align in phase with one another for signals coming from the design direction θ_0 so that a coherent integration gain can be achieved. Signals coming from other directions will not combine coherently. Time delay steering does this by time-shifting the $\bar{y}_n(t)$ to compensate for the differential propagation delays. Phase steering does not correct time alignment but does shift the phase of the signals to achieve coherency. Both are contemplated by Fig. 9.6. The boxes labeled "ϕ or t"

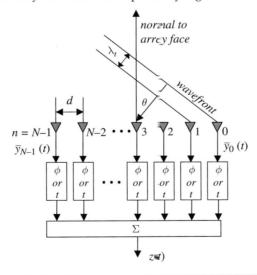

FIGURE 9.6 Wavefront impinging on a ULA. The ULA element signals are summed after being either time delayed or phase shifted to form the array output signal.

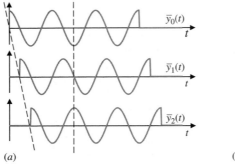
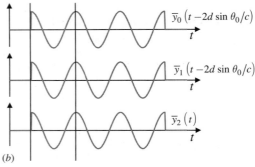

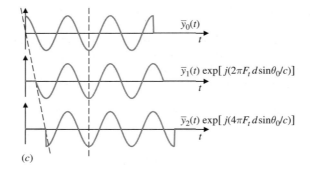

FIGURE 9.7 Effects of time delay steering versus phase steering. (a) Outputs from a three-phase center ULA similar to that of Fig. 9.6 showing the incremental time delay from one output to the next. (b) Time delay steering delays all signals to the latest start time, placing them in phase with one another for effective coherent integration. (c) Phase steering does not adjust the signal start times, but does shift their phases so that they are in phase with one another.

represent either a time delay operation or a phase shift. Whichever is used, if successful the outputs will be in phase and the summer will produce a single fast-time output signal $z(t)$ with N times the amplitude of the individual element outputs $\bar{y}_n(t)$.

9.2.1 Time Delay Steering

In time delay steering, the output of the nth array element is delayed by $(N - n - 1)d \sin\theta_0/c$ seconds and then combined with the other element outputs. This compensates the differential time of arrival at the array elements for a signal coming from AOA θ_0. The array output of the time delay beamformer when steered to AOA θ_0 for a signal from an arbitrary AOA θ is

$$\begin{aligned} z(t) &= \sum_{n=0}^{N-1} \bar{y}_n\left(t - (N-n-1)\frac{d\sin\theta_0}{c}\right) \\ &= \sum_{n=0}^{N-1} x\left(t - n\frac{d\sin\theta}{c} - (N-n-1)\frac{d\sin\theta_0}{c}\right) \times \cdots \\ &\quad \cdots \times \exp\left[j\Omega_t\left(t - n\frac{d\sin\theta}{c} - (N-n-1)\frac{d\sin\theta_0}{c}\right)\right] \end{aligned} \quad (9.3)$$

The beamformer output for signals whose actual AOA θ matches the selected AOA θ_0 is then

$$\begin{aligned}z(t) &= \sum_{n=0}^{N-1} x\!\left[t-(N-1)\frac{d\sin\theta_0}{c}\right]\exp\!\left[j\Omega_t\!\left(t-(N-1)\frac{d\sin\theta_0}{c}\right)\right] \\ &= N\cdot x\!\left[t-(N-1)\frac{d\sin\theta_0}{c}\right]\exp\!\left[j\Omega_t\!\left(t-(N-1)\frac{d\sin\theta_0}{c}\right)\right]\end{aligned} \qquad (9.4)$$

giving a coherent gain of N for the signal output. Figure 9.7b illustrates the effect of time delay steering on the element signals of Fig. 9.7a. The time skew has been eliminated, placing them all in phase again.

Time delay steering is generally not used in current radar systems because it is difficult to implement. The delays required are short, being multiples of $d\sin\theta_0/c$. With the element spacing $d = \lambda_t/2$ as usual and assuming a maximum θ_0 of 60°, this quantity is only 0.433 ns at 1 GHz and 43.3 ps at 10 GHz. The technology to implement variable time delays (so the steering angle can be varied) with such small granularity is too large and expensive to implement at the element level with current analog technology. Another possible implementation approach that may become feasible is to digitize the individual element signals. A sampling rate of 2.3 gigasamples per second would give a sample spacing of 0.433 ns. Sampling at this rate or higher combined with digital interpolation could be used to achieve fractional-sample time delays for other steering angles at the cost of high computation rates.

9.2.2 Phase Steering

The alternative to time delay steering is phase steering. Suppose $x(t)$ is simply a constant A. This is a narrowband (in fact, monochromatic) signal. Equation (9.2) becomes

$$\bar{y}_n(t) = A\exp[j\Omega_t(t-nd\sin\theta/c)] = A\exp[j(2\pi F_t t - 2\pi F_t nd\sin\theta/c)] \qquad (9.5)$$

which shows that the only difference between one element output and the next is an incremental phase shift of $-2\pi F_t nd\sin\theta/c$ rad (once the output has begun from each element). This in turn suggests that the array can be steered to achieve maximum coherent integration of signals from AOA θ_0 by phase shifting the element outputs, usually an easier operation to implement. In contrast to Eq. (9.3), the phase-steered array computes

$$z(t) = \sum_{n=0}^{N-1} \bar{y}_n(t)\exp(+j2\pi F_t nd\sin\theta_0/c) \qquad (9.6)$$

Continuing the example, this produces the three signals in Fig. 9.7c. While their time intervals of support still vary, during the time when all three are active they are in phase so that full coherent integration gain can be realized.

A limitation of phase steering arises when the incoming signal is not narrowband but has a significant instantaneous bandwidth, perhaps 10 percent or more of the RF. A simple model of $x(t)$ for the present purpose is simply a sum of some number of sinusoidal frequency components:

$$x(t) = A\sum_l \exp(j2\pi F_l t) \qquad (9.7)$$

where the frequencies F_l span some total bandwidth β. Assuming phase steering with this wideband waveform as input, the beamformer output will be

$$\begin{aligned}
z(t) &= \sum_{n=0}^{N-1} \bar{y}_n(t)\exp(+j2\pi F_t nd\sin\theta_0/c) \\
&= \sum_{n=0}^{N-1} \{x(t - nd\sin\theta/c)\exp[+j2\pi F_t(t - nd\sin\theta/c)]\}\exp(+j2\pi F_t nd\sin\theta_0/c) \\
&= A\sum_{n=0}^{N-1}\left\{\sum_l \exp[j2\pi F_l(t - nd\sin\theta/c)]\right\}\exp[+j2\pi F_t(t - nd\sin\theta/c)] \times \cdots \\
&\quad \cdots \times \exp(+j2\pi F_t nd\sin\theta_0/c) \\
&= A\sum_{n=0}^{N-1}\sum_l \exp[j2\pi(F_t + F_l)(t - nd\sin\theta/c)]\exp(+j2\pi F_t nd\sin\theta_0/c)
\end{aligned} \tag{9.8}$$

These phase-shifted element outputs will add in phase to maximize gain only if their phases are equal for each n, which occurs if $(F_t + F_l)\sin\theta = F_t\sin\theta_0$. When $F_l \neq 0$, this will occur for some value of θ other than θ_0; that is, the mainbeam of the array peaks at different angles for different frequency components of a wideband waveform. This means, for instance, that a wideband linear frequency-modulated waveform will cause the mainbeam to actually scan in angle during the pulse duration, creating an amplitude taper on the received signal and a modest signal-to-noise ratio (SNR) loss.

To quantify this effect, express the angle at which the maximum gain occurs for frequency $F_t + F_l$ as $\theta_0 + \delta\theta$. Then, using the small angle approximations $\sin(\delta\theta) \approx \delta\theta$ and $\cos(\delta\theta) \approx 1$,

$$\begin{aligned}
F_t\sin\theta_0 &= (F_t + F_l)\sin(\theta_0 + \delta\theta) \\
&= (F_t + F_l)[(\sin\theta_0\cos\delta\theta + \cos\theta_0\sin\delta\theta)] \\
&\approx (F_t + F_l)[(\sin\theta_0 + \delta\theta\cdot\cos\theta_0)] \Rightarrow \\
\delta\theta &= -\left(\frac{F_l}{F_t + F_l}\right)\tan\theta_0
\end{aligned} \tag{9.9}$$

The maximum absolute value of F_l will be $\beta/2$. The maximum shift in the steering angle is then

$$\delta\theta = -[1 + (2/\beta_{\text{rel}})]^{-1}\tan(\theta_0) \text{ rad} \tag{9.10}$$

where β_{rel} is the fractional bandwidth β/F_t. Wider waveform bandwidth, lower RF, and larger nominal steering angles all increase the maximum pointing error. Limiting it to one-half the 3-dB beamwidth requires

$$\begin{aligned}
[1 + (2/\beta_{\text{rel}})]^{-1}\tan(\theta_0) &\leq \frac{\lambda_t}{D\cos\theta_0} \Rightarrow \\
\beta_{\text{rel}} &\leq \frac{2}{(D\sin\theta_0/\lambda_t) - 1} \approx \frac{2\lambda_t}{D\sin\theta_0}
\end{aligned} \tag{9.11}$$

For example, a 10-GHz radar with a $D = 1$ m antenna at a nominal scan angle of 45° requires a fractional bandwidth of 8.49 percent or less to limit pointing angle variations to no more than one-half the 3-dB beamwidth.

Because phase shifters are relatively easy to implement at the element level and are effective for relatively narrowband signals, phase steering is the dominant beam steering method. If necessary, wideband beam steering effects can be countered by breaking the signal spectrum into multiple narrower subbands and adjusting the phase shifts for each to maintain a fixed steering angle. Subbanding can be accomplished with filterbanks or by using stepped frequency waveform techniques. There are many more elaborate array architectures that can also be used to control steering effects. One example uses element-level phase steering with subarrays, combined with time-delay steering of the subarray outputs. An introduction to array architectures is available in Bailey (2010).

9.2.3 Narrowband Phase Beamforming

Now consider the set of samples $y[n]$ formed from the individual array signals, sampled at a common time t_0, as shown in Fig. 9.8,

$$\begin{aligned} y[n] \equiv \bar{y}_n(t_0) &= A\exp\{j[\Omega_t(t_0 - nd\sin\theta/c) + \varphi_0]\} \\ &= \hat{A}\exp(-j\Omega_t nd\sin\theta/c) = \hat{A}\exp(-j2\pi nd\sin\theta/\lambda_t), \quad n = 0,\ldots,N-1 \end{aligned} \quad (9.12)$$

$\hat{A}$ incorporates the other constant phase terms. Assembling the N element samples into column vector form gives a *spatial snapshot* of the array at a fixed time:

$$\begin{aligned} \mathbf{y} &= \hat{A}[y[0] \quad y[1] \quad \cdots \quad y[N-1]]^T \\ &= \hat{A}[1 \quad \exp(-j2\pi d\sin\theta/\lambda_t) \quad \cdots \quad \exp(-j2\pi d(N-1)\sin\theta/\lambda_t)]^T \\ &= \hat{A}[1 \quad \exp(-jk_\theta) \quad \cdots \quad \exp[-j(N-1)k_\theta]]^T \\ &\equiv \hat{A}\mathbf{a}_s(\theta) \end{aligned} \quad (9.13)$$

where $k_\theta \equiv 2\pi d\sin\theta/\lambda_t$ is the normalized spatial frequency in radians per sample as projected into the plane of the array face and $\mathbf{a}_s(\theta)$ is the *spatial steering vector*. Thus, there is a one-to-one relationship between the AOA of a plane wave and the spatial frequency across the array face. The range of θ is $\pm\pi$, so the range of k_θ is $\pm 2\pi d/\lambda_t$. It is also useful to define the normalized spatial frequency in cycles per sample, $f_\theta = k_\theta/2\pi$. If the phase centers are spaced by $d = \lambda_t/2$, common for individual elements, the range of k_θ is $\pm\pi$ and of f_θ is ± 0.5. If phase centers correspond to subarrays their spacing will be larger and the range of k_θ and f_θ will be larger.

Conventional nonadaptive beamforming is implemented as a weighted sum of the element signals, $z = \mathbf{h}^T\mathbf{y}$,[2] where $\mathbf{h}$ is the vector of complex weights

$$\mathbf{h} = [h_0 \quad h_1 \quad \cdots \quad h_{N-1}]^T \quad (9.14)$$

A special case of interest occurs when $\mathbf{h}$ takes the form

$$\begin{aligned} \mathbf{h} &= [w_0 \quad w_1\exp[+jk_\theta] \quad \cdots \quad w_{N-1}\exp[+j(N-1)k_\theta]]^T \\ &= [w_0 \quad w_1 \quad \cdots \quad w_{N-1}]^T \odot \mathbf{a}_s^*(\theta) \\ &= \mathbf{w} \odot \mathbf{a}_s^*(\theta) \end{aligned} \quad (9.15)$$

---

[2] Most STAP literature defines the filter operation as $z = \mathbf{h}^H\mathbf{y}$. The convention $z = \mathbf{h}^T\mathbf{y}$ is retained here for consistency with the discussion of vector matched filtering in Chap. 5 and the usual definition of convolution in digital signal processing literature. A consequence is that the forms for $\mathbf{h}$ obtained here are the conjugate of the results in common STAP literature.

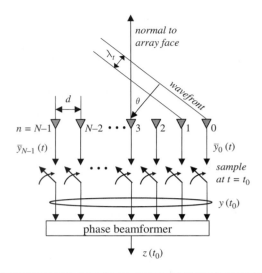

Figure 9.8 Digital beamformer. The array output signals are sampled at a series of time instants, forming a series of spatial snapshots that are operated on by the phase beamformer to form steered beams.

The symbol ⊙ represents the Hadamard (element-by-element) product of two vectors. If **a** and **b** are two N-element column vectors, then

$$\mathbf{a} \odot \mathbf{b} \equiv [a_0 b_0 \quad a_1 b_1 \quad \cdots \quad a_{N-1} b_{N-1}]^T \tag{9.16}$$

Equation (9.15) represents **h** as the Hadamard product of two factors: a shape vector **w** that provides for any weighting for sidelobe control and the conjugate of a steering vector $\mathbf{a}_s(\theta)$ that provides maximum coherent integration for signals arriving from angle θ.

Suppose the weights are matched to an AOA of θ_0; the array is said to be "steered" to θ_0. The response of a beamformer steered to θ_0 to an incoming wavefront at angle θ is

$$z(\theta) = \mathbf{h}^T \mathbf{y} = \hat{A} \sum_{n=0}^{N-1} w_n \exp[j(k_\theta - k_{\theta_0})n] \tag{9.17}$$

The scalar output $z(\theta)$ is just the discrete Fourier transform (DFT) of the weight sequence $\{h_n\}$, shifted to a center spatial frequency of k_{θ_0} and multiplied by $\hat{A}$. When all of the weight amplitudes $\{w_n\} = 1$, this is a standard asinc pattern in the variable $\sin\theta$,

$$z(\theta) = \exp\left[j(N-1)\frac{\pi d}{\lambda_t}(\sin\theta - \sin\theta_0)\right] \left\{ \frac{\sin\left[N\left(\frac{\pi d}{\lambda_t}\right)(\sin\theta - \sin\theta_0)\right]}{\sin\left[\left(\frac{\pi d}{\lambda_t}\right)(\sin\theta - \sin\theta_0)\right]} \right\} \tag{9.18}$$

More commonly, the weights are chosen to reduce antenna sidelobes at the cost of a modest SNR loss and degraded resolution and gain in the form of a wider mainbeam. Figure 9.9 illustrates the antenna pattern $|z(\theta)|$ for an array steered to $\theta_0 = 30°$ with $N = 11$ phase centers,

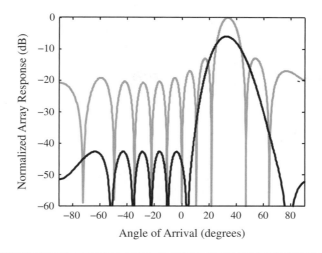

FIGURE 9.9 Antenna pattern with and without Hamming weighting. $N = 11$ phase centers, steering angle $\theta_0 = 30°$.

both with and without Hamming weighting. In the latter case, **w** is simply a vector of all ones; in the former case, it is a Hamming window function. The windowed case exhibits the same effects seen in windowing for sidelobe control in range and Doppler: much lower sidelobes at the cost of reduced peak gain and loss of resolution due to a wider mainlobe.

The beamformer output of Eq. (9.18) is a function of $\sin\theta$. This is a consequence of the spatial frequency k_θ being proportional to $\sin\theta$. A function of $u = \sin\theta$ is often said to be a function in *sine space*, a concept also mentioned briefly in Sec. 7.3.4.[3] In this chapter, most figures involving antenna patterns are plotted against θ instead of u. The result is that the patterns are "stretched" near $\theta = \pm\pi$ by the nonlinear transformation from u to θ, as seen for instance in Fig. 9.9.

The previous results can be obtained as the matched filter solution to a spatial filtering problem using the same results developed for Doppler filtering in Chap. 5. Suppose it is desired to maximize the output $z(\theta)$ when the input is a snapshot of a plane wave having AOA θ, plus white noise. The mathematics is exactly the same as used for vector matched filtering in Chap. 5 and the same results can be applied. Specifically,

$$\mathbf{h} = \kappa \mathbf{S}_I^{-1} \mathbf{t}^* \tag{9.19}$$

where
 $\mathbf{S}_I$ = covariance matrix of the interference at the N phase center outputs
 $\mathbf{t}$ = model of the desired target signal vector (i.e., steering vector)
 κ = arbitrary constant

Recall that $\mathbf{S}_I = \mathbf{S}_I^H$ for covariance matrices. Using this matched filter **h** in the filtering operation $z = \mathbf{h}^T \mathbf{y}$ shows that

$$z = \mathbf{h}^T \mathbf{y} = \left(\mathbf{S}_I^{-1} \mathbf{t}^*\right)^T \mathbf{y} = \mathbf{t}^H \left(\mathbf{S}_I^{-1}\right)^T \mathbf{y} \tag{9.20}$$

---

[3]Not to be confused with the variable u used for cross-range platform position in SAR in Chap. 8.

If the signal of interest is a monochromatic plane wave with AOA θ, the target model **t** is (letting $\hat{A} = 1$ without loss of generality) exactly the spatial steering vector $\mathbf{a}_s(\theta)$ defined in Eq. (9.13). When the interference is independent identically distributed (i.i.d.) white noise with variance σ_w^2 at each element, $\mathbf{S}_I = \sigma_w^2 \mathbf{I}$. Choosing $\kappa = \sigma_w^2$ gives

$$\mathbf{h} = [1 \quad \exp[+jk_\theta] \quad \cdots \quad \exp[+j(N-1)k_\theta]]^T = \mathbf{a}_s^*(\theta) \qquad (9.21)$$

The filter output is then

$$z(\theta) = \mathbf{h}^T \mathbf{y} = \sum_{n=0}^{N-1} y[n]\exp(+jk_\theta n) = \sum_{n=0}^{N-1} y[n]\exp(-j\tilde{k}_\theta n) \qquad (9.22)$$

which shows that the antenna pattern is the discrete-time Fourier transform (DTFT) of the data snapshot evaluated at $\tilde{k}_\theta \equiv -k_\theta$. If the observed signal $y[n]$ is in fact a snapshot of a plane wave at AOA θ_0, **y** is of the form of Eq. (9.13) with $\theta = \theta_0$ and $z(\theta)$ is again given by Eq. (9.17). The peak output occurs when the filter steering vector matches the actual steering vector of the input, that is, at $z(\theta_0)$, and is just N times the amplitude of $y[n]$. A plot of $z(\theta)$ as a function of AOA θ gives the (unweighted) normalized antenna gain pattern of Fig. 9.9. The beamformer signal processing gain G_{sp} is just this coherent gain N in the unwindowed case. It is reduced by the particular window's processing loss in the windowed case.

While the fast Fourier transform (FFT) can be used to efficiently compute the antenna pattern for a given steering vector **t**, the FFT output will be in terms of $\tilde{k}_\theta$ and must be flipped to obtain z as a function of k_θ. The flipped pattern will be sampled at constant increments of spatial frequency k_θ corresponding to constant increments in $\sin\theta$. Thus, the samples are not spaced uniformly in angle θ. If pattern samples at constant increments in θ are required, $z(\theta)$ must be computed explicitly using $z(\theta) = \mathbf{h}^T \mathbf{y}$ and varying **y** for each desired value of AOA.

Digital beamforming enables several advantageous capabilities for phased array antennas. One is the ability to form multiple simultaneous beams on receive. The same snapshot data **y** can be weighted with several different weight vectors **h**, effectively forming multiple simultaneous receive beams steered in different directions. Another major capability is the opportunity to compute the weights adaptively based on the interference environment characteristics. This makes possible adaptive nulling of jammers, discussed next. Other enhanced capabilities are mentioned in Aalfs (2013).

9.2.4 Adaptive Beamforming

The vector matched filter approach leads directly to a means for designing array weight vectors **h** that can steer zeros of the antenna pattern, often called *nulls*, in specific directions to cancel interference sources. This capability is useful in combating jammers, which are hostile interfering signal sources that seek to degrade radar performance by any of a number of mechanisms, such as degrading the signal-to-interference ratio (SIR) by increasing the noise level, or creating false detections to overwhelm the radar with false targets. One of the most common forms of jamming is a simple noise jammer. This device radiates a relatively high-power waveform at the victim radar from a specific air-, space-, or ground-based platform. The jammer waveform is a random noise process that is white over the receiver bandwidth of the victim radar.[4] From the radar's point of view, the jammer signal is therefore a white

---

[4]Effective jamming therefore requires knowledge of the victim radar frequency and bandwidth. Unless the jammer has a priori knowledge of the precise portion of the band used by the radar or can estimate it by detecting and analyzing the radar signal, it may have to spread its energy over a bandwidth wider than that actually used by the radar.

noise process arriving from some specific AOA. A major advantage enjoyed by the jammer in interfering with target detection is that it must only propagate in one direction, as opposed to the two-way propagation of the transmitted waveform and target echo. Consequently, the jammer signal suffers only a R^{-2} spreading loss instead of the R^{-4} loss of the target signal, making it imperative to try to attenuate the jammer signal on reception. Forming a beampattern with a null at the jammer's AOA accomplishes this.

The antenna pattern that maximizes the output SIR in the presence of both white noise and jamming is still given by Eq. (9.19); however, the model for the interference covariance matrix $\mathbf{S_I}$ must be extended to incorporate a model for the noise jammer. The temporal variation of the jammer signal $J_n(t)$ received from AOA θ at each array phase center can be expressed as

$$J_n(t) = \sigma_J w(t) \exp\{j[\Omega_t(t - nd\sin\theta/c) + \varphi_0]\} \tag{9.23}$$

where σ_J^2 is the jammer power and $w(t)$ is a unit variance zero mean white random process. A snapshot of the array response to the jammer gives

$$\begin{aligned}J[n] \equiv J_n &= \sigma_J w(t_0)\exp\{j[\Omega_t(t_0 - nd\sin\theta/c) + \varphi_0]\} \\ &= \sigma_J \hat{w}(t_0)\exp(-j\Omega_t nd\sin\theta/c) = \sigma_J \hat{w}(t_0)\exp(-j2\pi nd\sin\theta/\lambda_t) \\ &= \sigma_J \hat{w}(t_0)\exp(-jk_\theta n)\end{aligned} \tag{9.24}$$

where $\hat{w}(t_0)$ has absorbed the $\exp[j(\Omega_t t_0 + \varphi_0)]$ phase terms. In vector form, this is

$$\begin{aligned}\mathbf{J} &= \sigma_J \hat{w}(t_0)[1 \quad \exp(-jk_\theta) \quad \cdots \quad \exp[-j(N-1)k_\theta]]^T \\ &= \sigma_J \hat{w}(t_0)\mathbf{a}_s(\theta)\end{aligned} \tag{9.25}$$

The covariance matrix of the jammer signals is then

$$\begin{aligned}\mathbf{S_J} &= E\{\mathbf{J}^*\mathbf{J}^T\} = \sigma_J^2 E\{|\hat{w}(t_0)|^2\}\mathbf{a}_s^*(\theta)\mathbf{a}_s^T(\theta) \\ &= \sigma_J^2 \mathbf{a}_s^*(\theta)\mathbf{a}_s^T(\theta) \\ &= \sigma_J^2 \begin{bmatrix} 1 & \exp(-jk_\theta) & \cdots & \exp[-j(N-1)k_\theta] \\ \exp(+jk_\theta) & 1 & \ddots & \exp[-j(N-2)k_\theta] \\ \vdots & \vdots & \ddots & \vdots \\ \exp[+j(N-1)k_\theta] & \exp[+j(N-2)k_\theta] & \cdots & 1 \end{bmatrix}\end{aligned} \tag{9.26}$$

$\mathbf{S_J}$ is a Hermitian matrix, $\mathbf{S_J} = \mathbf{S_J}^H$. It is also positive semidefinite but has a rank of only one because the columns are all linearly dependent; in fact, they are all simple multiples of the first column.

The total interference covariance matrix for the sum of receiver noise and some number P of mutually uncorrelated jammers is

$$\mathbf{S_I} = \sigma_w^2 \mathbf{I} + \sum_{p=0}^{P-1} \mathbf{S}_{J_p} \tag{9.27}$$

The matched filter output can then be maximized by using the interference model of Eq. (9.27) in Eq. (9.19).

As an example, consider a case with $N = 16$ antenna phase centers. Assume that two jammers are present, one at an AOA of $+18°$ with a *jammer-to-signal ratio* (JSR) of $+50$ dB, and

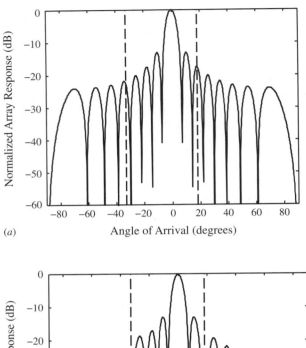

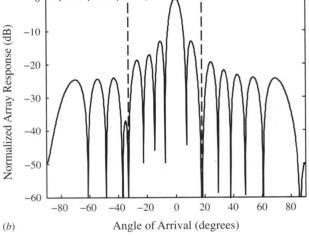

Figure 9.10 Antenna pattern for $N = 16$ phase centers. Two jammers present at angles indicated by the dashed lines. (a) Without adaptation. (b) With adaptation.

another at $-33°$ and a JSR of $+30$ dB. The SNR $= 0$ dB, resulting in a total SIR at the element level (before any beamforming) of -50.04 dB. Figure 9.10 shows the resulting antenna patterns when the antenna is steered to $0°$ (normal to the array face). In Fig. 9.10a, no adaptation was used; that is, the beamformer weights were designed using $\mathbf{S}_I = \sigma_w^2 \mathbf{I}$. A standard asinc pattern results. The two dotted vertical lines indicate the angles of arrival of the two jammers. Although in the sidelobes, both are on or near sidelobe peaks and, given their high power, will result in significant jamming energy in the beamformer output. In Fig. 9.10b, the beamformer weights were designed using a covariance matrix $\mathbf{S}_I$ computed via Eq. (9.27). The resulting pattern places deep nulls at the location of both jammers, effectively canceling them.

The SIR achieved by the optimum beamformer can be computed using the result of Eq. (5.8), namely

$$SIR = \frac{\mathbf{h}^H \mathbf{t}^* \mathbf{t}^T \mathbf{h}}{\mathbf{h}^H \mathbf{S}_I \mathbf{h}} \qquad (9.28)$$

For the case where the interference is noise only, $\mathbf{S}_I = \sigma_w^2 \mathbf{I}$, $\mathbf{h} = \mathbf{a}_s^*(\theta)$, and $\mathbf{t} = \hat{A} \mathbf{a}_s(\theta)$. Since $\mathbf{a}_s^T(\theta) \mathbf{a}_s^*(\theta) = \sum_n |\mathbf{a}_{sn}(\theta)|^2$, where $\mathbf{a}_{sn}(\theta)$ is the nth element of $\mathbf{a}_s(\theta)$, Eq. (9.17) reduces to

$$SIR = \frac{\left(|\hat{A}_s|^2 \sum_{n=0}^{N-1} |\mathbf{a}_{sn}(\theta)|^2\right)^2}{\sigma_w^2 |\hat{A}_s|^2 \sum_{n=0}^{N-1} |\mathbf{a}_{sn}(\theta)|^2} = \frac{|\hat{A}_s|^2 N^2}{\sigma_w^2 N} = N \frac{|\hat{A}_s|^2}{\sigma_w^2} \qquad (9.29)$$

Equation (9.29) shows again that a coherent integration gain of $G_{sp} = N$ is obtained against white noise using the N-element snapshot (with no window), while for a more general interference environment the gain is the SIR of Eq. (9.28) divided by the pre-beamforming SIR. For the example of Fig. 9.10 without jammers the interference is just the noise, so the pre-beamforming SIR is the single-element SNR $|\hat{A}_s|^2/\sigma_w^2 = 0$ dB. The post-beamforming SIR is 16 = 12.04 dB and the SIR gain is therefore also $G_{sp} = 12.04$ dB. With jammers present, the pre-beamforming SIR is -50.04 dB and the post-beamforming SIR is 15.64 (11.94 dB), so the gain is 61.98 dB. The slight reduction in post-beamforming SIR (0.1 dB) when the jammers are present occurs because some of the *degrees of freedom* (DOF) are consumed canceling the jammers instead of providing coherent integration gain for the signal.

This variation in peak gain in the direction of the desired target signal $\mathbf{t}$ can be compensated by requiring that $\mathbf{h}^T \mathbf{t} = 1$, resulting in what is often called a *distortionless* beamformer. Using this condition in Eq. (9.19) gives

$$1 = \mathbf{h}^T \mathbf{t} = \left(\kappa \mathbf{S}_I^{-1} \mathbf{t}^*\right)^T \mathbf{t} = \kappa \mathbf{t}^H \left(\mathbf{S}_I^{-1}\right)^T \mathbf{t} = \kappa \mathbf{t}^H \left(\mathbf{S}_I^{-1}\right)^* \mathbf{t} \Rightarrow$$

$$\kappa = \frac{1}{\mathbf{t}^H \left(\mathbf{S}_I^{-1}\right)^* \mathbf{t}}, \quad \mathbf{h} = \frac{\mathbf{S}_I^{-1} \mathbf{t}^*}{\mathbf{t}^H \left(\mathbf{S}_I^{-1}\right)^* \mathbf{t}} \qquad (9.30)$$

Since κ is simply a scalar, this choice does not affect the shape of the antenna pattern $z(\theta)$ but merely scales it up or down in amplitude so that $z(\theta_0) = 1$, where θ_0 is the AOA of the target signal $\mathbf{t}$. This design approach can be extended to constrain the pattern gain at multiple AOAs; details are given in Van Trees (2002).

If the interference is white, $\mathbf{S}_I = \sigma_w^2 \mathbf{I}$, the optimum weight vector is a scaled version of the steering vector, $\mathbf{h} = \kappa \mathbf{t}^*$. An interesting interpretation of the optimum adaptive beamformer, when the interference is *not* white, starts by noting that the inverse of any covariance matrix of interest can be factored in the form $\mathbf{S}_I^{-1} = \mathbf{V}^H \mathbf{V}$; $\mathbf{V}$ is upper triangular and is called the Cholesky matrix. That is, $\mathbf{V}$ is the "square root" of $\mathbf{S}_I^{-1}$ in a sense. It is also the case that $\mathbf{S}_I$, $\mathbf{V}$, and their inverses are Hermitian. Using this decomposition and Eqs. (9.19) and (9.20) with $\kappa = 1$ gives the beamformer output as

$$\begin{aligned} z(\theta) = \mathbf{h}^T \mathbf{y} &= \mathbf{t}^H \left(\mathbf{S}_I^{-1}\right)^T \mathbf{y} \\ &= \mathbf{t}^H [\mathbf{V}^H \mathbf{V}]^T \mathbf{y} = \mathbf{t}^H \mathbf{V}^T \mathbf{V}^* \mathbf{y} = (\mathbf{V}^* \mathbf{t})^H (\mathbf{V}^* \mathbf{y}) \\ &\equiv \tilde{\mathbf{t}}^H \tilde{\mathbf{y}} \end{aligned} \qquad (9.31)$$

The last line shows that the optimum filter output can be interpreted as a transformed steering vector $\tilde{\mathbf{t}} = \mathbf{V}^*\mathbf{t}$ applied to similarly transformed data $\tilde{\mathbf{y}} = \mathbf{V}^*\mathbf{y}$. Notice that the weight vector is just the transformed steering vector in this formulation; the interference covariance matrix does not appear explicitly. Put another way, the effective covariance matrix is apparently an identity matrix, suggesting that the transformed interference in $\tilde{\mathbf{y}}$ is white.

To see that this is indeed the case, consider the transformed data snapshot $\tilde{\mathbf{y}}$ when the data is interference only. Note that $\mathbf{S}_I^{-1} = \mathbf{V}^H\mathbf{V}$ implies also that $\mathbf{S}_I = \mathbf{V}^{-1}(\mathbf{V}^H)^{-1}$. The covariance of $\tilde{\mathbf{y}}$ is

$$\begin{aligned}\tilde{\mathbf{S}}_I &= E\{\tilde{\mathbf{y}}^*\tilde{\mathbf{y}}^T\} = E\{(\mathbf{V}^*\mathbf{y})^*(\mathbf{V}^*\mathbf{y})^T\} = E\{(\mathbf{V}\mathbf{y}^*)(\mathbf{y}^T\mathbf{V}^H)\} \\ &= \mathbf{V} \cdot E\{\tilde{\mathbf{y}}^*\tilde{\mathbf{y}}^T\} \cdot \mathbf{V}^H = \mathbf{V} \cdot \mathbf{S}_I \cdot \mathbf{V}^H \\ &= \mathbf{V} \cdot [\mathbf{V}^{-1}(\mathbf{V}^H)^{-1}] \cdot \mathbf{V}^H = \mathbf{I}\end{aligned} \quad (9.32)$$

Equation (9.32) states that transforming the data snapshot with the operator $\mathbf{V}^*$ *whitens* the interference. The appropriate weight vector is then just a scaled version of the similarly transformed steering vector as was discovered in Eq. (9.31). Applying the original optimal weight vector $\mathbf{h}$ to the original snapshot is therefore equivalent to solving a modified problem in which the data is transformed to whiten the interference and then filtered using an identically transformed steering vector.

Computing the adaptive weights requires finding $\mathbf{h} = \mathbf{S}_I^{-1}\mathbf{t}^*$, which is equivalent to solving the linear system of equations $\mathbf{t}^* = \mathbf{S}_I\mathbf{h}$. Various algorithms exist for doing this in a numerically stable and somewhat efficient matter. Generally, the computational load is $O((MN)^3)$, where the notation $O(\cdot)$ means "on the order of" and MN is the order of the system of equations (Arakawa and Bond, 2008).

How many jammers can be cancelled? For each distinct jammer signal $\mathbf{J}_p$, the vector matched filter chooses $\mathbf{h}$ such that $\mathbf{h}^T\mathbf{J}_p = 0$ (Guerci, 2014). If P jammers are present, this creates P such conditions. The distortionless constraint $\mathbf{h}^T\mathbf{t} = 1$ adds a $(P+1)$st condition. There are therefore $P+1$ equations in the N unknowns of the filter vector $\mathbf{h}$. A solution will exist so long as $P + 1 \leq N$. Thus, an N-phase center array can cancel up to $N-1$ jammers.

If a jammer is located in the mainbeam of the antenna pattern, the adaptive pattern is seriously degraded. In the previous example, the 3-dB beamwidth is approximately 6°. If the jammer at −33° is moved instead to −2° and the distortionless beamformer of Eq. (9.30) is applied the pattern of Fig. 9.11 results. The two jammers are nulled, but the antenna pattern peak has been shifted from the target AOA of 0°. The distortionless constraint still guarantees that the antenna pattern gain is unity at 0°, but the peak is now +5.15 dB and occurs at $\theta = 3.3°$. In this example, the post-beamforming SIR is reduced from 15.64 (11.94 dB) to only 3.64 (5.61 dB).

The elements of the steering vector and thus of the optimum adaptive weights depend on the RF through the wavelength λ_t or the spatial frequency k_θ. The weights that steer the beam to a particular AOA for one RF steer it to a different AOA for a different RF. (This is the same issue as the unintentional beam steering with wideband phase steered arrays discussed earlier.) Because of this dependence on λ_t, a single set of phase weights is incapable of cancelling a jammer over a wide bandwidth. A very common way to deal with this issue is to use *subbanding*. This technique, diagrammed in Fig. 9.12, divides the array spatial data as a function of fast time into K subbands, each relatively narrowband. A separate steering vector and adaptive beamformer is computed in each subband according to the effective value of λ_t for that band and applied to that data. After cancellation, the subband data is recombined to restore the full bandwidth with the jammer cancelled. Because the temporal bandwidth in each subband is $1/K$ times the full temporal bandwidth, the fast-time sampling

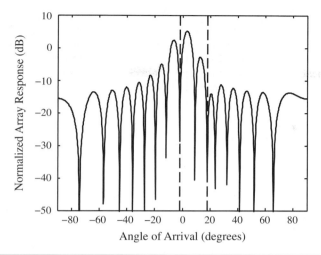

FIGURE 9.11 Adaptive pattern with one jammer in the mainlobe of the adapted pattern.

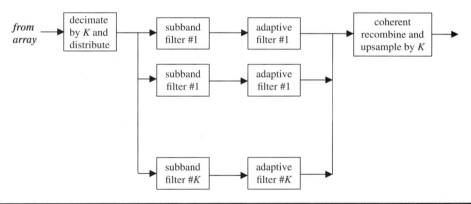

FIGURE 9.12 Subbanding approach for adaptive cancellation of interference in wideband signals.

rate can be reduced by the factor K in each subband. The full sampling rate is restored in the single output channel. This technique can be implemented quite efficiently using *polyphase filter banks* (Harris, 2004). Another approach to wideband cancellation uses fast-time filtering in each spatial channel (Aalfs, 2013).

9.2.5 Adaptive Beamforming with Preprocessing

The discussion so far has applied adaptive interference calculation directly to the individual array element signals, an approach called *element space* processing. Some systems perform fixed conventional beamforming on the element signals first and then apply adaptive processing to the individual beam outputs. This technique is called *beamspace processing*. Figure 9.9 was an example of forming a single fixed beam; a bank of such beams would provide coverage of the entire angle space. Beamspace processing tends to concentrate each interferer into the output of a small number of beams and can therefore reduce the dimensionality of the adaptive processing problem. Because of its cubic dependence on the DOF, this reduction can lead to major savings in computational load.

Beams are formed as linear combinations of the element signals as shown in the example of Eq. (9.17). A set of R beams formed from an N-element array can thus be represented as an $R \times N$ transformation matrix $\mathbf{T}$ acting on the $N \times 1$ spatial snapshot $\mathbf{y}$ to form a new $R \times 1$ vector of beam outputs $\tilde{\mathbf{y}}$ (Ward, 1994; Melvin, 2013):

$$\tilde{\mathbf{y}} = \mathbf{T}\mathbf{y} \tag{9.33}$$

For instance, a set of conventional "DTFT beams" at spatial frequencies $k_{\theta_0}, k_{\theta_1}, \ldots, k_{\theta_{P-1}}$ can be formed with the transformation matrix

$$\mathbf{T} = \begin{bmatrix} 1 & \exp[-jk_{\theta_0}] & \exp[-j2k_{\theta_0}] & \cdots & \exp[-j(N-1)k_{\theta_0}] \\ 1 & \exp[-jk_{\theta_1}] & \exp[-j2k_{\theta_1}] & \cdots & \exp[-j(N-1)k_{\theta_1}] \\ \vdots & \vdots & \vdots & \ddots & \vdots \\ 1 & \exp[-jk_{\theta_{(P-1)}}] & \exp[-j2k_{\theta_{(P-1)}}] & \cdots & \exp[-j(N-1)k_{\theta_{(P-1)}}] \end{bmatrix} \tag{9.34}$$

If the $\{k_{\theta_p}\}$ are evenly spaced over the interval $(-\pi, +\pi]$, $\mathbf{T}$ computes the P-point DFT of the element snapshot.

To see how to apply adaptive processing to the preprocessed beamspace data, note that the snapshot $\mathbf{y}$ is now replaced with $\tilde{\mathbf{y}}$. This applies to whatever signals are present at the array face, so the transformation is applied to the interference as well as the target signals. The new interference covariance matrix becomes

$$\begin{aligned} \tilde{\mathbf{S}}_I &= E\{\tilde{\mathbf{y}}^* \tilde{\mathbf{y}}^T\} \\ &= E\{(\mathbf{T}\mathbf{y})^*(\mathbf{T}\mathbf{y})^T\} = \mathbf{T}^* E\{\mathbf{y}^* \mathbf{y}^T\} \mathbf{T}^T \\ &= \mathbf{T}^* \mathbf{S}_I \mathbf{T}^T \end{aligned} \tag{9.35}$$

The adaptive weight vector and the filtered output are

$$\begin{aligned} \tilde{\mathbf{h}} &= \kappa \tilde{\mathbf{S}}_I^{-1} \tilde{\mathbf{t}}^* \\ \tilde{z}(\theta) &= \tilde{\mathbf{h}}^T \tilde{\mathbf{y}} \end{aligned} \tag{9.36}$$

where $\tilde{\mathbf{t}} = \mathbf{T}\mathbf{t}$ is the transformed target steering vector.

To illustrate this process, a DFT matrix $\mathbf{T}$ with $R = 16$ beams was applied to the same $N = 16$ element example used previously. Figure 9.13 shows the magnitudes of the adaptive weights with and without the beamspace transformation. No window for sidelobe control was included. Without the transformation, the weights are all similar in magnitude, suggesting that all of the data is important to obtaining good cancellation results. With the transformation, most of the weights are close to zero, suggesting that many channels could be discarded with little impact on the adapted pattern. Figure 9.14 shows the adapted pattern with sidelobe jammers when the beamspace weights are thresholded at 5 percent of their maximum value, corresponding to the dashed line in Fig. 9.13. This discards three-quarters of the elements in $\tilde{\mathbf{h}}$, keeping only elements 0, 4, 13, and 14. Both in the absence of jammers and when the jammers are located at $+18°$ and $-33°$, the resulting adapted antenna patterns are very close to those formed using the full 16 DOF in Fig. 9.10a and b.

The thresholded beamspace adaptive weight computation requires solving only a fourth-order system of equations, while the element space version requires the solution of a 16th-order system. The computational complexity of the weight vector computation will

Introduction to Array Processing

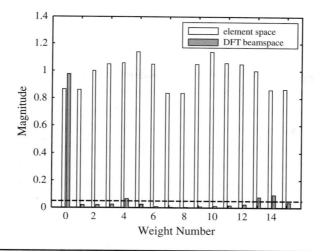

FIGURE 9.13 Magnitude of adapted weights with sidelobe jammers for element space and DFT beamspace processing.

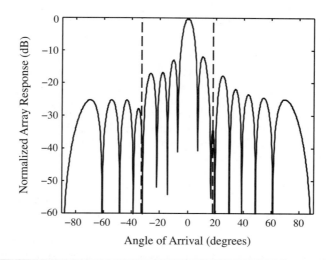

FIGURE 9.14 Adapted pattern using four beams obtained with a 16-beam DFT fixed beamformer preprocessor and a 5 percent magnitude threshold. Compare to Fig. 9.10b.

therefore decrease by a factor of about $(16/4)^3 = 64\times$. The beamspace approach has the extra step of applying the beam formation preprocessor $\mathbf{T}$, but in many cases the total computational load is still substantially reduced, especially if $\mathbf{T}$ has a structure that can be implemented efficiently, such as an FFT matrix.

The approach described here can be used with any linear transformation $\mathbf{T}$. Additional examples include beams uniformly spaced in θ instead of k_θ or approaches that form conventional beams and then combine sums of adjacent beams to achieve better cancellation. The technique can also be extended to include windowing for sidelobe control and gain constraints similar to the distortionless constraint discussed previously.

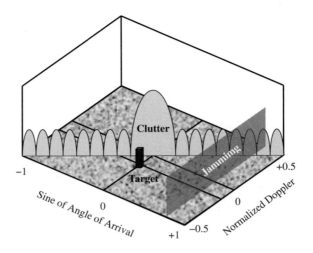

FIGURE 9.15 Space-time signal environment.

9.3 Space-Time Signal Environment

In a multiphase center radar, filtering is possible in both Doppler shift and angle of arrival. It is therefore important to characterize the data in a given range bin in terms of Doppler and AOA. Figure 9.15 is a notional sketch of the general behavior of noise, jamming, clutter, and moving targets in the space-time environment. Receiver noise has no structure in time or frequency, and therefore appears as a uniform noise floor throughout the angle-Doppler space. As described in the previous section, broadband noise jammers are localized in AOA but spread across the entire Doppler spectrum. This is reflected in Fig. 9.15 as a ridge of energy at a particular AOA spread across all Doppler frequencies. Because the jamming energy is present at all values of Doppler, discrimination against jamming must be based primarily on spatial filtering as described in the previous section.

Clutter is more complicated. Assume a platform with velocity v and a sidelooking radar,[5] as shown in Fig. 9.16. The height dimension is neglected for simplicity. In a given range bin, clutter scatterers anywhere on the isorange circle corresponding to the range bin of interest contribute to the total clutter return. (If the system is range ambiguous, there will be multiple isorange rings contributing to a given range bin.) A clutter scatterer directly on the radar boresight is at a squint angle of 90° with respect to the velocity vector; therefore, the Doppler shift for that scatterer is zero. More generally, scatterers at an angle of θ radians with respect to the antenna boresight will have a Doppler shift of $(2v/\lambda_t)\sin\theta$ Hz. Note that there are two such patches that will produce the same Doppler shift, one in the radar look direction, and one behind it in the *backlobes*. The backlobes are often ignored due to low antenna gain in that direction, but in some systems they must be considered. If the backlobe return is ignored, there is a one-to-one relationship between AOA and Doppler shift for clutter,

$$F_D = \frac{2v}{\lambda_t}\sin\theta \quad \text{Hz} \tag{9.37}$$

---

[5]These results generalize readily to non-sidelooking cases, but for simplicity only the sidelooking case is considered in this chapter.

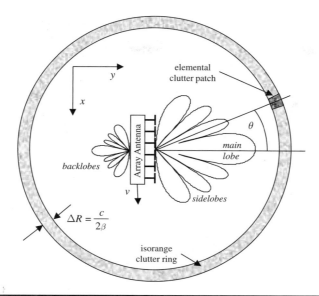

FIGURE 9.16 Clutter contributing to the angle-Doppler spectrum.

or in normalized units

$$f_D = \left(\left(\frac{2vT}{\lambda_t}\sin\theta\right)\right)_{1.0} \quad \text{cycles/sample} \tag{9.38}$$

where T is the PRI and the notation $((\cdot))_{1.0}$ indicates arithmetic modulo 1.0. Echo from stationary ground clutter that has a Doppler shift of F_D Hz can therefore be presumed to be arriving at an angle of θ radians with respect to the sidelooking antenna boresight. Consequently, clutter echoes tend to fall on a diagonal ridge in a Doppler-$\sin\theta$ space as was diagrammed in Fig. 9.15. The amplitude of ground clutter from different AOAs is determined by the clutter reflectivity and antenna gain in that direction. In homogenous clutter it is therefore largest near boresight and lower in the sidelobes and backlobes of the antenna.

Equation (9.38) can be rewritten as

$$\omega_D = 2\pi f_D = \left(\left(\frac{2vT}{d}k_\theta\right)\right)_{2\pi}$$
$$= ((\gamma_c k_\theta))_{2\pi} \quad \text{rad/sample} \tag{9.39}$$

where $\gamma_c \equiv 2vT/d$ is the slope of the clutter ridge when plotted in (k_θ, ω_D) coordinates. It also represents the number of times the clutter ridge spans the range of $-\pi$ to $+\pi$ radians per sample (-0.5 to $+0.5$ cycles per sample) in normalized Doppler as the AOA varies from $-\pi$ to $+\pi$.

Because F_D is proportional to $\sin\theta$ rather than θ itself, the clutter ridge is a straight line in Doppler-$\sin\theta$ space; however, if plotted as a function of θ instead of $\sin\theta$, the ridge curves visibly as the AOA approaches $\pm 180°$. This effect is visible in Fig. 9.17a, which is a simulation of the angle-Doppler spectrum for a medium pulse repetition frequency (PRF) radar. Two jammers are present, at approximately $-40°$ and $+60°$. The clutter ridge is diagonal through the center of the spectrum but curves noticeably at about $\pm 60°$. Also, the discrete time

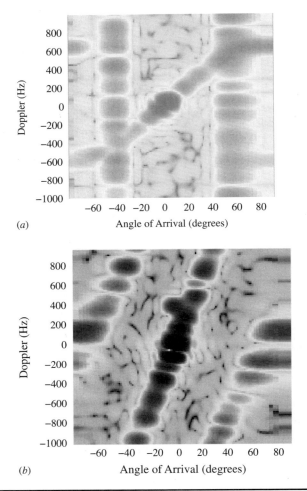

FIGURE 9.17 Simulation of angle-Doppler spectrum. Amplitude is in dB. (a) Illustration of clutter ridge curvature and two jammers. (b) Illustration of clutter ridge aliasing. (Figure courtesy of Dr. W. L. Melvin, GTRI.)

Doppler spectrum is periodic with period equal to the PRF. Thus, if the magnitude of F_D of Eq. (9.37) exceeds $PRF/2$, the clutter ridge will alias. This effect is seen in Fig. 9.17*b* that simulates a low PRF angle-Doppler spectrum. The clutter ridge is broken into three segments.

The angle-Doppler characteristics of the echo from a (possibly) moving point target depend on both the radar platform motion and the target motion. Assuming the target radar cross section is small enough that detection is likely only if it is in the radar mainbeam, the target can be presumed to be within a few degrees of the radar boresight. The Doppler shift will depend on the total radial velocity. If the target is stationary and directly on the boresight, the Doppler shift will be zero and it will fold in with the clutter. However, if the target is moving, it will separate from the clutter on the Doppler axis as shown in Fig. 9.15. This fact illustrates a key reason for the interest in space-time processing techniques, especially for the detection of relatively slow-moving ground targets ("slow movers"). As seen in Chap. 5, if only Doppler processing is used, the target Doppler shift must typically exceed the Doppler width of the clutter spectrum to achieve a signal-to-interference ratio adequate for detection.

If the platform velocity is high or the mainbeam relatively wide, ground clutter can fill most of the Doppler spectrum, making detection of slow movers very difficult. Figure 9.15 shows that introducing spatial processing gives a second dimension in which to separate the target from the clutter. Any Doppler shift of the target echo causes it to compete with clutter arriving from a different AOA; the added capability of filtering based on AOA then allows separation of target and clutter having the same Doppler shift.

9.4 Space-Time Signal Modeling

Space-time adaptive processing applies vector matched filtering to the combined slow-time/phase center data set in each range bin. It is usually assumed that pulse compression, if used, has been applied prior to STAP processing. The two-dimensional slice of the datacube in range bin l_0, $y[l_0, m, n]$ is called a *space-time snapshot* (or just *snapshot*) of the data. To proceed, the $M \times N$ two-dimensional snapshot is converted to an $MN \times 1$ column vector by stacking the columns:

$$\mathbf{y} = \begin{bmatrix} y[l_0,0,0] \\ y[l_0,0,1] \\ \vdots \\ y[l_0,0,N-1] \\ y[l_0,1,0] \\ y[l_0,1,1] \\ \vdots \\ y[l_0,1,N-1] \\ \vdots \\ y[l_0,M-1,0] \\ y[l_0,M-1,1] \\ \vdots \\ y[l_0,M-1,N-1] \end{bmatrix} \quad (9.40)$$

This process of converting the data from a given range bin to a one-dimensional vector is illustrated in Fig. 9.18.

Next, the filter weight vector $\mathbf{h}$ must be designed using Eq. (9.19) or (9.30). The target model vector $\mathbf{t} = \mathbf{t}(f_D, \theta)$ must represent the expected signal phase history from a target at some specific Doppler shift f_{Dt} and AOA θ_t of interest. Define a *temporal steering vector*

$$\mathbf{a}_t(f_{Dt}) = [1 \quad \exp[-j2\pi f_{Dt}] \quad \cdots \quad \exp[-j2\pi(M-1)f_{Dt}]]^T \quad (9.41)$$

This is simply the model for the slow-time data sequence corresponding to a target at normalized Doppler frequency f_{Dt}. The two-dimensional snapshot of the data from a target at Doppler f_{Dt} and AOA θ_t would have the temporal variation of $\mathbf{a}_t(f_{Dt})$ in each row and the spatial variation of $\mathbf{a}_s(\theta_t)$ in each column. The snapshot therefore has the form of the outer product of $\mathbf{a}_s$ with $\mathbf{a}_t$,

$$\begin{aligned} y[l_0,m,n] &= \mathbf{a}_s(\theta_t)\mathbf{a}_t^T(f_{Dt}) \\ &= [a_{t0}(f_{Dt})\mathbf{a}_s(\theta_t) \quad a_{t1}(f_{Dt})\mathbf{a}_s(\theta_t) \quad \cdots \quad a_{t(M-1)}(f_{Dt})\mathbf{a}_s(\theta_t)] \end{aligned} \quad (9.42)$$

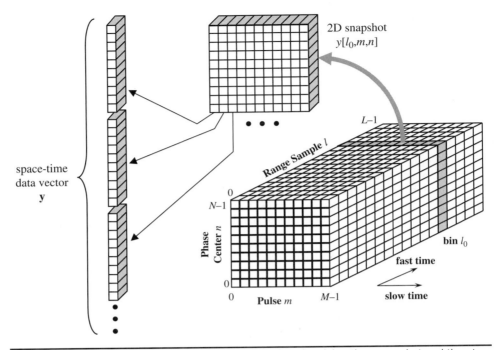

FIGURE 9.18 Mapping of the datacube range bin to a fast-time/slow-time snapshot and then to a one-dimensional vector.

where $a_{tm}(f_{Dt})$ is the mth element of $\mathbf{a}_t(f_{Dt})$. When this matrix is vectorized by stacking the columns, the result is the Kronecker product of the desired spatial and temporal steering vectors (Guerci, 2002; Melvin, 2013), denoted by the symbol $\otimes$ and given by

$$\mathbf{t} = \mathbf{a}_t(f_{Dt}) \otimes \mathbf{a}_s(\theta_t) = \begin{bmatrix} a_{t0}(f_{Dt})\mathbf{a}_s(\theta_t) \\ a_{t1}(f_{Dt})\mathbf{a}_s(\theta_t) \\ \vdots \\ a_{t(M-1)}(f_{Dt})\mathbf{a}_s(\theta_t) \end{bmatrix} \qquad (9.43)$$

Next, a model for the $MN \times MN$ covariance matrix $\mathbf{S_I}$ of the interference is needed. The interference is the sum of receiver noise ($\mathbf{n}$), jammer ($\mathbf{J}$), and clutter ($\mathbf{c}$) components. It is assumed that these are all uncorrelated with one another, with the result that the total interference covariance is the sum of the covariances of the three components, $\mathbf{S_I} = \mathbf{S_n} + \mathbf{S_J} + \mathbf{S_c}$.

Receiver noise is assumed as usual to be i.i.d. zero-mean complex Gaussian at each phase center and time sample. It follows that $\mathbf{S_n} = \sigma_w^2 \mathbf{I}_{MN}$, a scaled MNth-order identity matrix.

Now consider a single noise jammer signal. The spatial variation was given in Eq. (9.25). The temporal variation can be modeled as

$$\mathbf{a}_{tJ} = \begin{bmatrix} a_{tJ_0} & a_{tJ_1} & \cdots & a_{tJ(M-1)} \end{bmatrix}^T \qquad (9.44)$$

where the $\{a_{tJm}\}$ are uncorrelated i.i.d. random variables with equal power σ_J^2. The covariance of the jammer temporal variation is

$$\mathbf{E}\{\mathbf{a}_{tJ}^*\mathbf{a}_{tJ}^T\} = \sigma_J^2 \mathbf{I}_M \tag{9.45}$$

If the AOA of the jammer is θ_J, the space-time data vector for the jammer component is

$$\mathbf{J} = \sigma_J^2 \mathbf{a}_{tJ} \otimes \mathbf{a}_{sJ}(\theta_J) \tag{9.46}$$

and its covariance matrix can be shown to be (Ward, 1994)

$$\mathbf{S}_J = \mathbf{E}\{\mathbf{J}^*\mathbf{J}^T\} = \sigma_J^2 \mathbf{I}_M \otimes \mathbf{a}_{sJ}^*(\theta_J)\mathbf{a}_{sJ}^T(\theta_J) \tag{9.47}$$

which is a block-diagonal matrix. If P multiple uncorrelated jammers are present, $\mathbf{S}_J$ is the sum of P terms of the form in Eq. (9.47), each with its own AOA θ_{Jp}.

The clutter signal is the sum of contributions from all of the clutter scatterers within the isorange ring of interest (or multiple rings, if the system is ambiguous in range). Strictly speaking, this is an integral of the pertinent scatterers as with the angle-averaged reflectivity of Chap. 2. However, in STAP the integrated clutter is generally approximated as the sum of Q elemental clutter patches (see Fig. 9.16), each typically of an angular extent approximately equal to the radar beamwidth. For clutter patch q, the space-time data vector becomes

$$\mathbf{c}_q = \sigma_{cq}^2 \mathbf{a}_{tc}(f_{Dcq}) \otimes \mathbf{a}_{sc}(\theta_{cq}) \tag{9.48}$$

where σ_{cq}^2 represents the power of the qth clutter patch as determined by the radar range equation. σ_{cq}^2 is proportional to the antenna gain in the direction of that patch, $G(\theta_q)$. The normalized Doppler shift and AOA of the clutter patch are related through Eq. (9.38). The total clutter vector is

$$\mathbf{c} = \sum_{q=0}^{Q-1} \mathbf{c}_q = \sum_{q=0}^{Q-1} \sigma_{cq}^2 \mathbf{a}_{tc}(f_{Dcq}) \otimes \mathbf{a}_{sc}(\theta_{cq}) \tag{9.49}$$

The covariance matrix of the clutter is

$$\begin{aligned}\mathbf{S}_c &= \mathbf{E}\{\mathbf{c}^*\mathbf{c}^T\} = \sum_{q=0}^{Q-1} \sigma_{cq}^2 \mathbf{c}_q^* \mathbf{c}_q^T \\ &= \sum_{q=0}^{Q-1} \sigma_{cq}^2 \left[\mathbf{a}_{tc}^*(f_{Dcq})\mathbf{a}_{tc}^T(f_{Dcq})\right] \otimes \left[\mathbf{a}_{sc}^*(\theta_{cq})\mathbf{a}_{sc}^T(\theta_{cq})\right]\end{aligned} \tag{9.50}$$

This is an $M \times M$ block matrix, where each "element" of the block matrix is the $N \times N$ cross-covariance of the spatial snapshots from two different PRIs. $\mathbf{S}_c$ can be factored as (Ward, 1994)

$$\begin{aligned}\mathbf{S}_c &= \mathbf{C}\mathbf{\Sigma}_c\mathbf{C} \\ \mathbf{C} &= [\mathbf{c}_0 \quad \mathbf{c}_1 \quad \cdots \quad \mathbf{c}_{Q-1}] \\ \mathbf{\Sigma}_c &= \text{diag}\left(\left[\sigma_{c0}^2 \quad \sigma_{c1}^2 \quad \cdots \quad \sigma_{c(P-1)}^2\right]\right)\end{aligned} \tag{9.51}$$

The discussion of clutter so far assumes that it is uncorrelated in the spatial dimensions (range and cross-range) but perfectly correlated in slow time. It is indeed common to model the clutter as uncorrelated in space, assuming that clutter patches are separated by a distance on the order of a resolution cell. However, as discussed in Chap. 2 and again in the analysis of Doppler processing in Chap. 5, the clutter echo from any given ground patch cannot reasonably be modeled as constant in slow time over the time scale of a coherent processing interval. Natural clutter exhibits reflectivity fluctuations in time due to *internal clutter motion* (ICM, also called *intrinsic clutter motion*). This is simply the physical movement of scatterers due to wind or waves. The radar system itself contributes sources of temporal modulation such as antenna scanning modulation or pulse-to-pulse or sweep-to-sweep instabilities. All of these temporal reflectivity fluctuations cause a broadening of the clutter power spectrum and decorrelation of the temporal snapshots.

ICM is easily incorporated into the model of Eq. (9.39) by replacing the temporal snapshot for patch q, $\sigma_{cq}^2 \mathbf{a}_{tq}(f_{Dcq})$, with a modified temporal snapshot that replaces the constant amplitude σ_{cq}^2 with time-varying amplitudes $\sigma_{cq}^2 \boldsymbol{\alpha}_q$, where

$$\boldsymbol{\alpha}_q = \begin{bmatrix} \alpha_{0q} & \alpha_{1q} & \cdots & \alpha_{(M-1)q} \end{bmatrix}^T \tag{9.52}$$

Let $\mathbf{A}_q$ be the covariance matrix of the temporal fluctuations $\boldsymbol{\alpha}_q$; $\mathbf{A}_q$ is an $M \times M$ Toeplitz matrix. The modified space-time data vector for the qth clutter patch is then

$$\mathbf{c}_q = \sigma_{cq}^2 [\boldsymbol{\alpha}_q \odot \mathbf{a}_{tc}(f_{Dcq})] \otimes \mathbf{a}_{sc}(\theta_{cq}) \tag{9.53}$$

and the clutter covariance matrix becomes

$$\mathbf{S}_c = \sum_{q=0}^{Q-1} \sigma_{cq}^2 [\mathbf{A}_q \odot \mathbf{a}_{tc}^*(f_{Dcq}) \mathbf{a}_{tc}^T(f_{Dcq})] \otimes [\mathbf{a}_{sc}^*(\theta_{cq}) \mathbf{a}_{sc}^T(\theta_{cq})] \tag{9.54}$$

An alternative approach for modeling ICM using covariance matrix tapers is described in Sec. 9.7.

Various models for the temporal correlation of the data are available in the literature. One that is popular in STAP simulations is the Billingsley model (Billingsley, 2002). This model assumes that the clutter temporal power spectrum is the sum of a two-sided exponential function and an impulse at the origin in Doppler frequency space:

$$S_c(F) = \sigma_c^2 \left[\frac{a}{a+1} \delta_D(F) + \frac{1}{a+1} \left(\frac{b\lambda_t}{4} \right) \exp\left(-\frac{b\lambda_t}{2} |F| \right) \right] \tag{9.55}$$

where a is the ratio of the DC to AC components, and b is a parameter dependent primarily on wind conditions.

The corresponding autocorrelation function is

$$s_c(\tau) = \sigma_c^2 \left\{ \frac{a}{a+1} + \frac{1}{a+1} \left[\frac{(b\lambda_t)^2}{(b\lambda_t)^2 + (4\pi\tau)^2} \right] \right\} \tag{9.56}$$

Experimental data are available to choose the parameters a and b to fit various scenarios distinguished by the type of clutter, radar wavelength, weather conditions, and so forth. Simple autoregressive filters can be used to implement the model in simulations (Mountcastle, 2004).

9.5 Processing the Space-Time Signal

9.5.1 Optimum Matched Filtering

Optimal space-time processing of the two-dimensional data snapshot consists of the following steps. For a range bin of interest,

1. Form the interference covariance matrix $\mathbf{S_I}$. In practice, $\mathbf{S_I}$ must be estimated from the radar data; discussion of this process is deferred to Sec. 9.5.4.
2. Select a Doppler shift and angle of arrival at which to test for the presence of a target signal, and form the appropriate target space-time steering vector $\mathbf{t}$ using Eq. (9.43).
3. Compute the optimal filter weight $\mathbf{h}$ using either Eq. (9.19) or (9.30), depending on the normalization desired.
4. Form the space-time data vector $\mathbf{y}$, as illustrated in Fig. 9.18, and described in Eq. (9.40).
5. Apply the weight vector to the data to obtain the test statistic $z = z(f_D, \theta) = \mathbf{h}^T \mathbf{y}$.

The test statistic can then be used for detection or for other purposes such as angle estimation. If used for detection, typically $|z|$ or $|z|^2$ is computed and compared to an appropriate threshold computed using the techniques of Chap. 6.

The previous procedure computes the optimum (maximum SIR) test statistic for a single Doppler frequency and AOA and in a single range bin. It therefore must be repeated for each Doppler and AOA of interest, and then for each range bin. Within a given range bin the covariance matrix will remain constant and can be reused, but steps 2 through 5 must be repeated for each (f_D, θ) combination. $\mathbf{S_I}$ should ideally be recomputed for each range bin. Because the product MN of the number of pulses or sweeps in the CPI and the number of phase centers in the antenna can easily be in the hundreds, the procedure above implies solving a system of hundreds of linear equations for each Doppler-AOA point of interest and each range bin.

While the optimum weight vector is given by $\mathbf{S_I}^{-1}\mathbf{t}^*$, in practice it is often desirable to include windowing of the data to reduce sidelobes as was done in the spatial beamforming case in Fig. 9.9 and for Doppler processing (Chap. 5) and pulse compression (Chap. 4). Combined angle-Doppler weighting can be included by computing the weight vector as

$$\mathbf{h} = k\mathbf{S_I}^{-1}\mathbf{t}_w^* \qquad (9.57)$$

where the windowed target steering vector is

$$\mathbf{t}_w = (w_f \otimes w_\theta) \cdot \mathbf{t} \qquad (9.58)$$

and $\mathbf{w}_f$ and $\mathbf{w}_\theta$ are the temporal and spatial weight vectors. That is, the conventional steering vector $\mathbf{t}$ is multiplied by a space-time window vector that is the Kronecker product of the desired Doppler and angle weighting functions.

9.5.2 STAP Metrics

A common metric used to visualize STAP filtering performance is the *adapted pattern* (Ward, 1994). This is simply a two-dimensional plot of $|z(f_D, \theta)|^2 = |\mathbf{h}^T \mathbf{t}|^2 = \mathbf{h}^H \mathbf{t}^* \mathbf{t}^T \mathbf{h}$ as f_D and θ are stepped over a regular grid. If the array has uniformly spaced phase centers and a constant PRI is used, the adapted pattern can be computed as the two-dimensional DFT of the weight vector $\mathbf{h}$ after it is remapped to a two-dimensional snapshot form, allowing the use of the

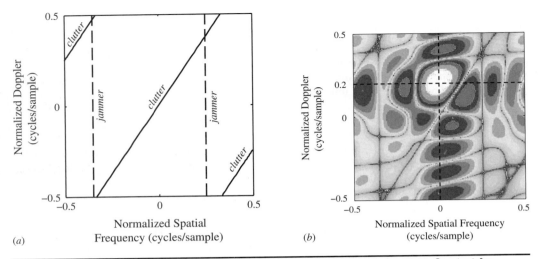

FIGURE 9.19 (a) Clutter and jamming loci for idealized example. (b) Adapted pattern. See text for details.

FFT to efficiently compute the pattern.[6] However, the angle samples will be evenly spaced in $\tilde{k}_\theta$ rather than in θ itself. If it is preferred to display the pattern in terms of f_D and θ, one-dimensional FFTs can be applied to the temporal dimension of **h** but the angle dimension samples must be computed individually.

Figure 9.19 illustrates the adapted pattern for an idealized example.[7] The case considered corresponds to a sidelooking radar at an RF of 675 MHz. The platform velocity is 50 m/s and the PRF is 200 pulses per second. There are eight phase centers spaced by $\lambda_t/2$ meters and eight pulses in the CPI so that $M = N = 8$. The clutter ridge slope β equals 1.5 in this example. Jammers are present at $+30°$ and $-44.43°$, corresponding to normalized spatial frequencies of $+0.25$ and -0.35 cycles per sample. The *clutter-to-noise ratio* (CNR) is $+40$ dB, as is the *jammer-to-noise ratio* (JNR) for both jammers. Part *a* of the figure shows the loci of the jammer and clutter energy in angle-Doppler space. Figure 9.19*b* shows the resulting adapted pattern when a target is present at an AOA of $0°$ and a normalized Doppler shift of 0.2 cycles per sample. The adapted pattern clearly shows vertical nulls at the two jammer AOAs and a three-part diagonal null corresponding to the clutter ridge. The adapted pattern has a strong peak at the target location of $0°$ and 0.2 cycles per sample. Because no windowing was used in this example, high Doppler and spatial sidelobes are evident as horizontal and vertical ridges extending from the peak. Figure 9.20 plots Doppler and spatial cuts through the adapted pattern taken at the target coordinates, showing the nulls in the spatial pattern at the jammer and clutter locations and in the Doppler pattern at the clutter ridge location.

Another important class of metrics involves SIR when the interference is noise only and when it is noise plus clutter and/or jamming. Metrics in this class include SIR and *SIR loss*, which is the reduction in SIR for the clutter-plus-jamming case compared to the noise-only case. Each of these is a function of target angle and Doppler; typically they are plotted as a function of Doppler for a fixed AOA.

---

[6]Note that the DFT will compute the pattern in terms of $\tilde{k}_\theta$, not k_θ.
[7]Figures 9.19 through 9.21 were generated using LL_STAP©, a MATLAB® program for demonstrating basic STAP patterns developed by Massachusetts Institute of Technology Lincoln Laboratory (MIT/LL).

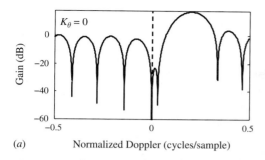 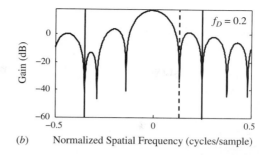

Figure 9.20 Pattern cuts from the adapted pattern of Fig. 9.19. (a) Doppler pattern. (b) Spatial pattern. Both cuts pass through the actual target location; cut lines are shown in Fig. 9.19b.

Consider a target that produces an SNR of χ_t for a single pulse and phase center, that is, for a single sample of the two-dimensional snapshot. The optimum SNR is then

$$\chi_0 = MN\chi_t \tag{9.59}$$

The factor MN is the coherent integration gain achievable by combining the MN samples of the snapshot. The SIR is, from Eq. (9.28),

$$SIR = \frac{\mathbf{h}^H \mathbf{t}^* \mathbf{t}^T \mathbf{h}}{\mathbf{h}^H \mathbf{S}_I \mathbf{h}} = \frac{|z|^2}{\mathbf{h}^H \mathbf{S}_I \mathbf{h}} \tag{9.60}$$

From Chap. 5, when using the optimum weight vector this becomes just

$$SIR_{max} = \mathbf{t}^T \mathbf{S}_I^{-1} \mathbf{t}^* \tag{9.61}$$

The SIR loss is then defined as

$$L_{SIR} = \frac{SIR}{\chi_0} \tag{9.62}$$

If the steering vector used in the STAP filter design exactly matches the target Doppler and angle, SIR_{max} is used in Eq. (9.62); however, this definition of L_{SIR} can be used with suboptimal filter designs as well, in which case Eq. (9.60) is used as the numerator of L_{SIR}. Note that L_{SIR} is defined here as a number less than one (negative dB), in keeping with common practice in the STAP literature.

L_{SIR} is a function of Doppler and angle of arrival. It is typically plotted as a function of Doppler with the array steered in the correct target direction. Figure 9.21 is a plot of L_{SIR} versus Doppler for the interference scenario of Fig. 9.19. Each point on this curve is obtained by positing a target at the corresponding Doppler shift and then evaluating Eq. (9.62) using the matched steering vector for that target Doppler. When the target velocity is far from the clutter ridge, for example $|f_D| > 0.3$, the loss is minimal, about 0.7 dB at the plot edges. As the target velocity nears the clutter at zero Doppler the losses increase, reaching over 50 dB at $f_D = 0$.

This plot serves as the basis for two additional metrics (Ward, 1994). *Minimum detectable velocity* (MDV) is the velocity closest to the clutter notch at which some acceptable SIR loss is achieved. *Minimum detectable Doppler* (MDD) is the corresponding Doppler shift in either

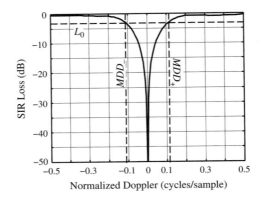

FIGURE 9.21 SIR loss versus Doppler for the scenario of Fig. 9.13.

absolute or normalized units. Clearly, MDV is $\lambda_t/2$ times the MDD in hertz or $\lambda_t/2T$ times the MDD in normalized frequency units of cycles per sample. The frequency at which the SIR loss becomes acceptable is not necessarily symmetric about the clutter notch. Using normalized frequency units, define the upper and lower MDD as

$$MDD_+ = \min_{f_D}\{f_D \text{ such that } L_{SIR} \geq L_0, f_D > 0\}$$
$$MDD_- = \max_{f_D}\{f_D \text{ such that } L_{SIR} \geq L_0, f_D < 0\} \quad (9.63)$$

where L_0 is the maximum acceptable loss threshold. This definition assumes that the clutter notch is at $f_D = 0$; it is straightforward to generalize it to clutter notches at other frequencies. The MDD is then defined as the average of these two offsets, taking into account the fact that MDD_- is negative:

$$MDD = \frac{1}{2}(MDD_+ - MDD_-) \quad (9.64)$$

The choice of L_0 is a system design decision. Based on radar range equation considerations, $L_0 = -12$ dB would correspond to a 50 percent reduction in detection range, while $L_0 = -5$ dB would correspond to a 25 percent reduction. In Fig. 9.21, a value of $L_0 = -3$ dB has been selected. In this example, this choice results in $MDD_+ = 0.115$ cycles per sample and $MDD_- = -0.112$ cycles per sample, giving $MDD = 0.1135$.

Another metric related to MDD is the *usable Doppler space fraction* (UDSF). This is the fraction of the Doppler space over which the SIR loss is acceptable, that is, $L_{SIR} > L_0$. UDSF is simply expressed in terms of the MDD in normalized frequency units:

$$\begin{aligned} UDSF &= 1 - (MDD_+ - MDD_-) \\ &= 1 - 2MDD \end{aligned} \quad (9.65)$$

In the example of Fig. 9.21, $UDSF = 0.773$, that is, the SIR loss is considered acceptable over 77.3 percent of the Doppler spectrum.

As a final example, Fig. 9.22 shows the adapted patterns obtained with the optimum filter for the two interference environments of Fig. 9.17. In the medium PRF case, there are now two vertical nulls in the Doppler dimension where the jammer energy was located. In

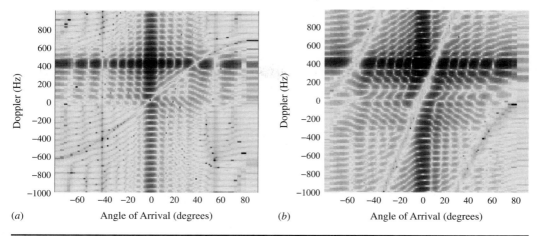

FIGURE 9.22 Result of optimal STAP processing of data in Fig. 9.17. (a) Medium PRF case with clutter and two jammers. (b) Low PRF case with aliased clutter. (Figure courtesy of Dr. W. L. Melvin, GTRI.)

addition, the STAP processing has implemented an S-shaped null[8] to follow the clutter ridge and attenuate the clutter energy. The large response at approximately $F_D = 400$ Hz and $\theta = 0°$ is a target that was not visible in the original data. The two vertical ridges of energy in Doppler and angle are the sidelobes of the target response. Similarly, the low PRF case shows the ability of STAP to implement a three-part null to remove aliased clutter, again revealing a hidden target.

9.5.3 Relation to Displaced Phase Center Antenna Processing

Section 5.9 introduced displaced phased center antenna processing in the context of slow-moving target indication from a moving radar platform. Displaced phase center antenna (DPCA) processing combines data from multiple pulses and multiple antenna phase centers to form a clutter-cancelled output; as such, it appears to be related to STAP processing. To identify this connection, consider nonadaptive DPCA processing using two phase centers separated by a distance d_{pc} so that $N = 2$. Assume that the DPCA condition is met with a time slip of M_s pulses; that is,

$$M_s = \frac{d_{pc}}{2vT} \quad \Rightarrow \quad \gamma_c = \frac{2vT}{d_{pc}} = \frac{1}{M_s} \tag{9.66}$$

The space-time snapshot in two-dimensional form is a $2 \times M$ array, with spatial index $n = 0$ corresponding to the fore phase center and $n = 1$ corresponding to the aft phase center as shown in the upper portion of Fig. 9.23.

A nonadaptive DPCA processor is constrained to be of the form

$$z[m] = y_f[m] - y_a[m + M_s] \tag{9.67}$$

where $y_f[m]$ and $y_a[m]$ are the fore and aft phase center outputs, respectively. It is sufficient to work with a single "sub-CPI," that is, an interval of M_s pulses, that spans the two pulses

---

[8]The null would follow a straight line in $(f_D, \sin\theta)$ space.

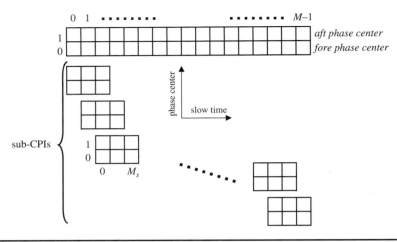

FIGURE 9.23 Schematic diagram of two-dimensional space-time snapshot for a DPCA processor, and its decomposition into sub-CPIs.

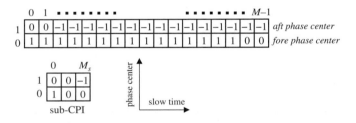

FIGURE 9.24 Full-CPI weight vector **h** and sub-CPI weight vector **h**₁.

combined in the DPCA processor. The complete CPI snapshot is then constructed of a series of $M - M_s$ overlapping sub-CPIs as shown in the bottom portion of Fig. 9.23 for the case $M_s = 2$. The weight vector $\mathbf{h}_1$ that implements Eq. (9.67) for a single sub-CPI and the corresponding weight vector $\mathbf{h}$ for the entire CPI are shown in Fig. 9.24 in two-dimensional form.

Viewed as a two-dimensional discrete function, the sub-CPI weight vector $\mathbf{h}_1$ can be written as

$$\mathbf{h}_1 \Rightarrow h_1[m,n] = \delta[m] - \delta[m + M_s, n-1] \tag{9.68}$$

The adapted pattern is the two-dimensional DTFT of this function,

$$\begin{aligned} H_1(\omega_D, \tilde{k}_\theta) &= \sum_{m=0}^{M_s-1} \sum_{n=0}^{1} h_1[m,n] \exp[-j(m\omega_D + n\tilde{k}_\theta)] \\ &= 1 - \exp[-j(-M_s\omega_D - \tilde{k}_\theta)] = 1 - \exp[j(M_s\omega_D + \tilde{k}_\theta)] \\ &= 1 - \exp[j(M_s\omega_D - k_\theta)] \end{aligned} \tag{9.69}$$

Note that this function has a zero at $\omega_D = \gamma_c k_\theta$, which is exactly the expression for the clutter ridge in Eq. (9.39). Figure 9.25 shows $|H_1(f_D, k_\theta)|$ for the case $M_s = 2$. Clearly, the DPCA

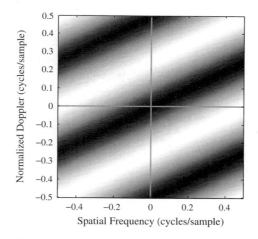

FIGURE 9.25 Adapted pattern for nonadaptive DPCA processor with $M_s = 2$.

processor implements a null along the clutter ridge. Repeating the single sub-CPI DPCA process for the additional sub-CPIs within the total CPI simply provides coherent integration by a factor of $M - M_s$. On the other hand, the constrained form of the weight vector [Eq. (9.68)] prevents any combining of samples from different phase centers on the same pulse so that the DPCA processor cannot provide any spatial beamforming capability. As a result, DPCA processing can cancel the clutter but not jammers.

The model of Eq. (9.69) implicitly assumes an omnidirectional antenna pattern for each of the antennas or subarrays that form the DPCA phase centers. A more realistic model that accounts for the subarray patterns can be obtained by repeating the analysis with the output from each phase center filtered in angle by the antenna pattern of the antenna or subarray that forms that phase center. Denoting the fore and aft subarray antenna patterns as $E_f(k_\theta)$ and $E_a(k_\theta)$, respectively, Eq. (9.69) can be generalized to

$$H_1(\omega_D, k_\theta) = E_f(k_\theta) - E_a(k_\theta) \exp[-j(M_s \omega_D - k_\theta)]$$
$$= E_f(k_\theta)\{1 - \exp[-j(M_s \omega_D - k_\theta)]\} \quad (9.70)$$

where the last step holds only if $E_f(k_\theta) = E_a(k_\theta)$. If the two patterns differ, the first line of Eq. (9.70) shows that there will not be a null at the clutter ridge location as desired. This emphasizes the need for carefully matched subarray patterns in DPCA processing.

Adaptive DPCA improves on these results by applying the vector matched filtering framework, as described in Sec. 5.9.4. The adaptive DPCA processor is still constrained to a weight vector structure that prohibits combining of phase center outputs from the same pulse and therefore, like nonadaptive DPCA, provides no spatial beamforming. In addition, the use of the target model $\mathbf{t} = [1 \ 0]^T$ provides a result that is optimized "on average" over all target Doppler frequencies, rather than for any specific target Doppler.

9.5.4 Adaptive Matched Filtering

In Chap. 6, it was seen that the knowledge of the interference power was required to set the detection threshold. Constant-false-alarm-rate (CFAR) techniques were introduced to estimate the noise level from the radar data. The same issue exists with STAP processing, which requires knowledge of the interference covariance matrix $\mathbf{S_I}$ to compute the optimal weight

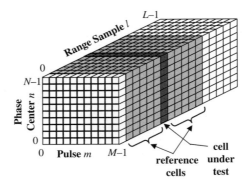

FIGURE 9.26 Datacube showing selection of reference cells for estimating the interference covariance matrix.

vector **h** and perform the matched filtering. Again, it is not realistic to assume that $\mathbf{S}_I$ can be known a priori, so it must be estimated from the radar data.

The most common approach to STAP when $\mathbf{S}_I$ is unknown is called the *sample matrix inverse* (SMI) method (Ward, 1994). This technique is directly analogous to cell-averaging CFAR. Figure 9.26 illustrates a datacube with the cell under test (CUT) indicated, as well as a number of adjacent reference range bins. The data in these reference cells are assumed to consist of i.i.d. interference having the same statistics as the interference in the CUT. It is also assumed that the reference cells do not contain any target signals.[9] Consequently, they can be used to compute a sample mean estimate of $\mathbf{S}_I$, denoted $\hat{\mathbf{S}}_I$. Specifically, the sample covariance of the data from a single reference cell is

$$\hat{\mathbf{S}}_l = \mathbf{y}_l^* \mathbf{y}_l^T \qquad (9.71)$$

If L_S reference cells are used, the estimate of $\mathbf{S}_I$ is simply

$$\hat{\mathbf{S}}_I = \frac{1}{L_s} \sum_{l=1}^{L_s} \hat{\mathbf{S}}_l \qquad (9.72)$$

Filter weight vectors computed using $\hat{\mathbf{S}}_I$ instead of $\mathbf{S}_I$ will be suboptimum because of the imperfect interference estimate.

The reduction in SIR due to the use of $\hat{\mathbf{S}}_I$ instead of $\mathbf{S}_I$ is denoted L_{CFAR}. It has been shown to be a beta-distributed random variable with expected value (Reed et al., 1974; Nitzberg, 1984)

$$\mathrm{E}(L_{\text{CFAR}}) = \frac{L_s + 2 - P}{L_s + 1} \qquad (9.73)$$

Note that, as defined, $\mathrm{E}(L_{\text{CFAR}})$ will be less than 1 (negative dB). It is plotted as a function of L_s/P, the size of the reference window relative to the number of DOF (snapshot size), in Fig. 9.27. This figure shows that the number of reference cells must be twice the

---

[9]If desired, guard cells can be included immediately adjacent to the CUT to prevent target contamination of the interference estimate.

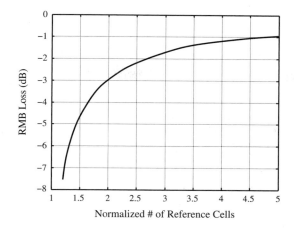

FIGURE 9.27 RMB estimate of SIR loss due to estimation of covariance matrix $\mathbf{S}_I$ for $P = 256$.

number of DOF in the processor to limit the loss due to covariance estimation to 3 dB. To limit the loss to 1 dB requires $L_s > 5P$. The conclusion that the reference window size should usually be two to five times the snapshot size is known as the "Reed-Mallet-Brennan" or *RMB rule*. The shape of the curve is a very weak function of P; the example shown is for $P = 256$, corresponding for example to a system with $N = 8$ phase centers and $M = 32$ pulses in the CPI. These results apply only when there is no mismatch between the actual target steering vector and the model vector **t** used in the filter design; this also implies no use of windowing for reduced sidelobes. Generalizations of Eq. (9.73) for mismatched and windowed target model vectors are discussed in Boroson (1980) and Kelly (1989).

Once the covariance matrix has been estimated, the weight vector is computed in the usual fashion. One particular choice of the scale factor κ of interest when using a square-law detector is the *adaptive matched filter* (AMF) (Kelly, 1986; Chen and Reed, 1991; Robey et al., 1992)

$$\kappa = \frac{1}{\sqrt{\mathbf{t}^H \left(\hat{\mathbf{S}}_I^{-1}\right)^* \mathbf{t}}} \tag{9.74}$$

Notice the similarity to Eq. (9.30). With this choice, the filter output becomes

$$|z|^2 = \frac{\left|\mathbf{t}^H \left(\hat{\mathbf{S}}_I^{-1}\right)^* \mathbf{y}\right|^2}{\mathbf{t}^H \left(\hat{\mathbf{S}}_I^{-1}\right)^* \mathbf{t}} \tag{9.75}$$

It can be shown that a threshold test applied to this test statistic exhibits CFAR behavior. It is also claimed that the AMF is more robust to targets in the training data than alternative tests such as the generalized likelihood ratio test (Steinhardt and Guerci, 2004).

The adapted pattern when using the SMI technique is subject to pattern degradations such as elevated sidelobes and distorted mainbeams, especially when L_s is relatively small. In addition, the pattern may vary from update to update, a condition called *weight jitter*. If $L_s < P$, then $\hat{\mathbf{S}}_I$ may also become nonsingular. A common extension of the basic SMI technique that addresses these issues is *diagonal loading*, in which a bias term is added to the diagonal elements of the estimated covariance matrix (Carlson, 1988):

$$\hat{\mathbf{S}}_I \Rightarrow \hat{\mathbf{S}}_I + \varepsilon \mathbf{I} \qquad (9.76)$$

Since diagonal loading adds a factor that has the same structure as the covariance matrix of white noise, its effect is to increase the apparent noise floor of the data. The loading factor ε is typically set 10 to 30 dB above the actual noise level σ_w^2 (Kim et al., 1998). Diagonal loading tends to ensure nonsingularity of $\hat{\mathbf{S}}_I$ and reduce distortion of the adapted pattern but also reduces the depth of the nulls (Guerci, 2014).

9.6 Reduced-Dimension STAP

It was shown in Sec. 9.2.5 that preprocessing of the element data could be used to reduce the dimensionality of the adaptive processing equations. The same technique can be applied to STAP processing. However, reduced-dimension processing is especially important in STAP because the snapshot dimensionality can be quite large, possibly in the hundreds. This aggravates two problems. First, the RMB rule states that the number of reference range bins required to maintain acceptable losses due to covariance estimation is on the order of $2P$ to $5P$. Consequently, complex systems (many phase centers, long CPIs) can have long reference windows, with the result that the training data are very unlikely to be statistically homogeneous as assumed in the SMI algorithm due to terrain variations. Second, the computational load is of order P^3. Reducing the dimensionality by only a factor of two produces nearly an order of magnitude reduction in computational load in solving the SMI equations. Increasing it by a factor of two for better performance (for instance, by reducing straddle losses or covariance matrix estimation losses) would similarly increase the load by nearly an order of magnitude.

Because STAP operates on two data dimensions, preprocessing can be applied in the slow-time dimension, phase center dimension, or both. Figure 9.28 illustrates a taxonomy of reduced dimension STAP techniques based on the choice of preprocessing options (Ward, 1994). There are many variants of each general class. As an example, Fig. 9.29

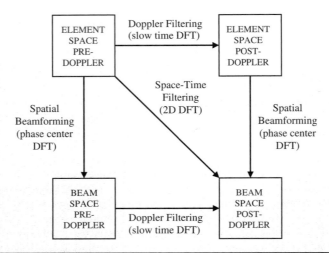

Figure 9.28 Taxonomy of reduced-dimension STAP algorithms. (After Ward, 1994.)

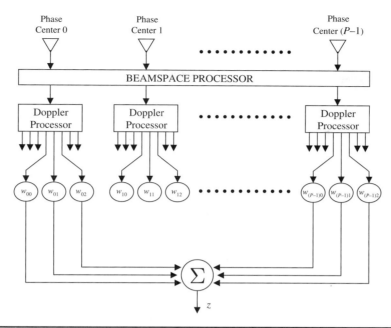

FIGURE 9.29 Structure of a particular beamspace post-Doppler STAP processor.

illustrates a particular variant of the beamspace post-Doppler class of STAP processors. DFTs in both dimensions are used to form a grid of fixed angle-Doppler bins. For a given target model vector **t**, adaptation is performed using only a small number of angle and Doppler bins around the target AOA and Doppler frequency, analogous to the earlier example of adaptive beamforming using DFT beamspace preprocessing. Beamspace post-Doppler architectures can isolate the interference in both angle and Doppler and substantially reduce the order of the SMI equations. Typical implementations use three to five bins in each of the Doppler and angle dimensions, giving 9 to 25 total degrees of freedom.

9.7 Advanced STAP Algorithms and Analysis

Only the most basic STAP algorithm, the SMI approach based on the vector matched filter, has been introduced in this chapter, and only metrics related to the adapted pattern and SIR have been used to evaluate its performance. A deeper understanding of the character of the interference and the behavior of STAP algorithms requires analysis of the eigenstructure of the signal environment, a topic beyond the scope of this book but thoroughly covered in texts devoted to STAP such as those of Klemm (2006), Van Trees (2002), and Guerci (2014). The latter includes a taxonomy of STAP algorithms proposed for radar data.

A number of ideas in advanced STAP are based on the realization that the covariance matrix of neither the clutter nor the jammers is full rank. For instance, the rank of an ideal covariance matrix for J independent jammers is MJ (Guerci, 2014). The rank of the clutter-only covariance matrix under ideal conditions (no crab, constant velocity) can be estimated as (Brennan and Staudaher, 1992)

$$\text{rank}\{\mathbf{S}_c\} \approx N + (M-1)\gamma_c \qquad (9.77)$$

where γ_c is the clutter ridge slope of Eq. (9.39). Because γ_c is usually in the range of zero to three, this is well below the full rank $P = NM$. Thus, both the clutter and jamming signals can be represented with a relatively few basis vectors in the P-dimensional signal space. Eigenanalysis provides a convenient means for decomposing the signal components, defining a reduced-dimension representation, and analyzing the performance of the resulting algorithms.

This approach leads to algorithms that are fundamentally different from the SMI approach. Suppose that the combined clutter and jamming covariance matrix has Q dominant eigenvalues, with the remainder at or near the noise floor eigenvalue level. The *principal components* (PC) method forms the adaptive filter weight vector as a linear combination of the Q eigenvectors corresponding to the dominant eigenvalues, with weights related to the eigenvalue associated with each eigenvector used. The PC method and similar *subspace projection* techniques can construct high-quality adapted patterns without the degradation of sidelobes that often occurs with SMI techniques when estimated covariance matrices are used. Furthermore, because the filter vector is derived from only Q eigenvectors, these algorithms provide another means of dimensionality reduction. Algorithms based on projecting data into lower-dimensional spaces are generally called *reduced rank* STAP techniques (as opposed to the reduced dimension techniques of Sec. 9.6).

Even with effective rank reduction, the nonstationarity and heterogeneity of the data due to error effects to be discussed in Sec. 9.8, and more fundamentally to variations in the physical clutter scene over the reference windows, remain major limiting factors in STAP performance. Significant research effort has been focused on *knowledge-aided* (KA) STAP (Weiner et al., 1998; Guerci, 2002; Guerci, 2020). KA STAP attempts to use auxiliary sources of information to improve the interference covariance estimate for the CUT. For example, map data can be used to identify changes in terrain type, roadways, and other variations in the characteristics of the clutter in the reference cells. Preprocessing algorithms can then excise some cells from the covariance matrix estimation process in an attempt to provide a more homogeneous set of training data and an estimate of $\hat{\mathbf{S}}_I$ more consistent with the actual covariance matrix $\mathbf{S}_I$ in the CUT. The high level architecture of one KA STAP system is shown in Fig. 9.30. The knowledge-aided preprocessing edits the datacube to provide a modified datacube with more homogeneous statistics. Any of the conventional STAP algorithms can

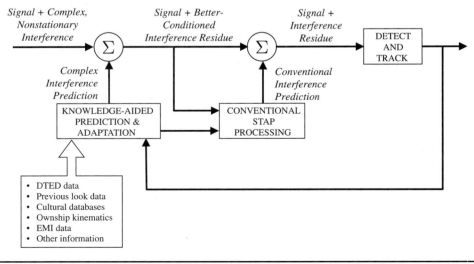

Figure 9.30 Basic architecture of a knowledge-aided STAP processor. (After Guerci, 2002.)

then be applied to cancel the now better-behaved clutter. In another example, *digital terrain elevation data* (DTED) maps can be used along with terrain type maps to predict clutter characteristics. In-band *electromagnetic interference* (EMI) can be identified using data on known emitters such as television stations and wireless services that may corrupt portions of the radar frequency band. The EMI can then be removed by prefiltering of the data in the receiver.

9.8 Limitations to STAP

In Chap. 5, it was seen that the performance of MTI is degraded by several factors, some internal to the radar system and some external to it. Internal factors include pulse-to-pulse variability in pulse amplitude, oscillator phase drifts, I/Q channel errors, and antenna scanning modulation. External factors are principally internal clutter motion and clutter heterogeneity.

All of these factors also degrade STAP performance, but additional complications exist due to the multiphase center radar and the moving platform. For instance, mismatches will exist among the N channels of the receive array. These can be classified further as angle-independent mismatches, which are differences in the frequency responses $H_n(\Omega)$ of the channels, and angle-dependent mismatches. The latter arise from a variety of sources, including element placement errors, wideband dispersion, and mutual coupling of elements. Platform motion effects such as misalignment of the array face with the platform velocity vector due to platform crab angle create additional degradations. Because adaptive weights are not generally updated every CPI, changing geometry between the moving platform and stationary or moving jammers can cause a jammer to move out of the adapted filter null until the weights are updated again, an effort referred to as the "stale weights" problem.

All of these effects tend to increase the rank of the interference covariance matrix, a phenomenon called *interference subspace leakage* (ISL) because the expanded rank implies that the size of the subspace required to represent the interference is increased. The rank increase implies an increase in reference window size requirements. Many ISL effects can be modeled as *covariance matrix tapers* (CMTs) applied to the true covariance matrix. A CMT is a $P \times P$ matrix **G** that combines with the ideal interference covariance matrix in a Hadamard product to give a modified and higher-rank covariance matrix:

$$\tilde{\mathbf{S}}_I = \mathbf{S}_I \odot \mathbf{G} \tag{9.78}$$

G can often be modeled in turn as the Hadamard product of several component CMTs representing different ISL effects (Guerci, 2014):

$$\mathbf{G} = \mathbf{G}_0 \odot \mathbf{G}_1 \odot \cdots \odot \mathbf{G}_{G-1} \tag{9.79}$$

for some number G of component effects. A one-dimensional example of this in adaptive beamforming defines an N-element CMT vector as samples of a sinc function (Mailloux, 1995; Zatman, 1995). Applying this CMT to the estimated covariance matrix widens the interference notch in the adapted pattern, providing increased immunity to the stale weight problems caused by changing jammer AOAs. As another example, in the space-time case it can be shown that an appropriate CMT for angle-independent channel mismatch is the weighted sum of a matrix of all ones and an identity matrix (Guerci, 2014). CMTs can also be used to represent the effects of internal clutter motion. A CMT based on the Billingsley model, discussed earlier, is obtained by sampling Eq. (9.56),

$$\mathbf{G}(m,n) = s_c(|m-n|T) \tag{9.80}$$

where T is the repetition interval.

However defined, CMTs can be used in at least two ways. The first uses the CMT in Eq. (9.78) to improve the model of the interference covariance in standard STAP algorithms. The second develops new algorithms that take account of the covariance structure, including the CMT component, to provide improved performance. Examples are given in Guerci (2014).

As noted earlier, clutter heterogeneity is an especially significant concern in STAP because of the large reference window size often required. Losses due to clutter heterogeneity can range from an insignificant 0.1 dB to as much as 16 dB for realistic scenarios (Melvin, 2000). Approaches to combating these losses include reduced dimension and reduced rank techniques described previously to limit reference window size, and knowledge-assisted algorithms to improve the homogeneity of the data used to estimate the clutter covariance in the CUT.

9.9 Summary

Array antennas sample received signals from a given range bin across a spatial aperture. Just as slow-time temporal sampling makes possible signal filtering and discrimination based on Doppler frequencies, this spatial sampling enables filtering based on AOA.

Virtual arrays provide a consistent way to describe and analyze both conventional physical arrays and more elaborate structures such as synthetic array or sparse arrays. As an example, the VA idea was used to explain the factor of two difference in the beamwidth of synthetic and physical arrays of the same extent.

The remainder of the chapter concentrated on uniform linear arrays. It was shown that the antenna mainbeam can be steered using either time delays or phase shifters, though phase steering presents some difficulties with wideband waveforms. Antenna beams can be formed and steered, and their sidelobes controlled through weighting, in exactly the same manner used for Doppler filtering. The vector matched filter mathematics used for Doppler processing in Chap. 5 is equally applicable to both fixed and adaptive AOA filtering.

Adaptive filtering based on Doppler differences can be combined with adaptive filtering based on AOA, leading to space-time adaptive processing. STAP filtering is also based on the same vector matched filter mathematics. It can be very effective in enabling detection of relatively slow moving targets in complex interference environments comprising noise, clutter, and even jammers. The computational cost of basic STAP is very high but can be reduced through the use of various preprocessing methods to reduce the dimensionality of the STAP equations.

Adaptive beamforming and STAP are rich areas of radar signal processing, as well as the subject of much continuing research. This chapter has provided only a brief introduction to the basic ideas. Additional information is available in the references.

References

Aalfs, D., "Adaptive Digital Beamforming," Chap. 9 in W. L. Melvin and J. A. Scheer (eds.), *Principles of Modern Radar: Advanced Techniques*. SciTech Publishing, Edison, NJ, 2013.

Arakawa, M., and R. A. Bond, "Computational Characteristics of High Performance Embedded Algorithms," Chap. 5 in D. R. Martinez, R. A. Bond, and M. M. Vai (eds.), *High Performance Embedded Computing Handbook: A Systems Perspective*. CRC Press, Boca Raton, FL, 2008.

Bailey, C. D., "Radar Antennas," Chap. 9 in M. A. Richards, J. A. Scheer, and W. A. Holm (eds.), *Principles of Modern Radar: Basic Principles*. SciTech Publishing, Edison, NJ, 2010.

Billingsley, J. B., *Low Angle Radar Land Clutter*. William Andrew Publishing, Norwich, NY, 2002.

Boroson, D. M., "Sample Size Consideration in Adaptive Arrays," *IEEE Transactions on Aerospace & Electronic Systems*, vol. AES-16, no. 4, pp. 446–451, Jul. 1980.

Brennan, L. E., and F. M. Staudaher, "Subclutter Visibility Demonstration," technical report RL-TR-92-21, Adaptive Sensors, 1992.

Carlson, B. D., "Covariance Matrix Estimation Errors and Diagonal Loading in Adaptive Arrays," *IEEE Transactions on Aerospace & Electronic Systems*, vol. AES-24, no. 3, pp. 397–401, Jul. 1988.

Chen, W. S., and I. S. Reed, "A New CFAR Detection Test for Radar," *Digital Signal Processing*, vol. 1, pp. 198–214. Academic Press, New York, 1991.

Guerci, J. R., "Knowledge Aided Sensor Signal Processing and Expert Reasoning," *Proceedings of Knowledge Aided Sensor Signal Processing and Expert Reasoning* (KASSPER) Workshop, Apr. 2002.

Guerci, J. R., *Space-Time Adaptive Processing for Radar*, 2d ed. Artech House, Norwood, MA, 2014.

Guerci, J. R., *Cognitive Radar: The Knowledge-Aided Fully Adaptive Approach*, 2d ed. Artech House, Boston, MA, 2020.

Harris, F. J., *Multirate Signal Processing for Communication Systems*. Prentice-Hall, New York, 2004.

Kelly, E. J., "An Adaptive Detection Algorithm," *IEEE Aerospace & Electronic Systems Magazine*, vol. 28, no. 1, pp. 115–127, Mar. 1986.

Kelly, E. J., "Performance of an Adaptive Detection Algorithm: Rejection of Unwanted Signals," *IEEE Transactions on Aerospace & Electronic Systems*, vol. AES-25, no. 2, pp. 122–133, Mar. 1989.

Kim, Y. L., S. U. Pillai, and J. R. Guerci, "Optimal Loading Factor for Minimal Sample Support Space-Time Radar," *Proceedings* of 1998 *IEEE International Conference on Acoustics, Speech, and Signal Processing* (ICASSP), vol. 4 , pp. 2505–2508, May 1998.

Klemm, R., *Principles of Space-Time Adaptive Processing*, 3d ed. Institution of Electrical Engineers (IEE), London, 2006.

Mailloux, R. J., "Covariance Matrix Augmentation to Produce Adaptive Array Pattern Troughs," *Electronics Letters*, vol. 31, no. 10, pp. 771–772, 1995.

Melvin, W. L., "Space-Time Adaptive Radar Performance in Heterogeneous Clutter," *IEEE Transactions on Aerospace & Electronic Systems*, vol. AES-36, no. 2, pp. 621–633, Apr. 2000.

Melvin, W. L., "Clutter Suppression Using Space-Time Adaptive Processing," Chap. 10 in W. L. Melvin and J. A. Scheer (eds.), *Principles of Modern Radar: Advanced Techniques*. SciTech Publishing, Edison, NJ, 2013.

Mountcastle, P. D., "New Implementation of the Billingsley Clutter Model for GMTI Data Cube Generation," *Proceedings of IEEE 2004 Radar Conference*, pp. 398–401, Apr. 2004.

Nitzberg, R., "Detection Loss of the Sample Matrix Inversion Technique," *IEEE Transactions on Aerospace & Electronic Systems*, vol. AES-26, no. 6, pp. 824–827, Nov. 1984.

Reed, I. S., J. D. Mallet, and L. E. Brennan, "Rapid Convergence Rate in Adaptive Arrays," *IEEE Transactions on Aerospace & Electronic Systems*, vol. AES-10, no. 16, pp. 853–863, Nov. 1974.

Richards, M. A., "Virtual Arrays, Part 1: Phase Centers and Virtual Elements," technical memorandum, Mar. 2018. Available at http://radarsp.com.

Richards, M. A., "Virtual Arrays, Part 2: Virtual Arrays and Coarrays," technical memorandum, Feb. 2019. Available at http://radarsp.com.

Robey, F. C. et al., "A CFAR Adaptive Matched Filter Detector," *IEEE Transactions on Aerospace & Electronic Systems*, vol. AES-28, no. 1, pp. 208–216, Jan. 1992.

Steinhardt, A., and J. Guerci, "STAP for RADAR: What Works, What Doesn't, and What's in Store," *Proceedings of IEEE Radar Conference*, pp. 469–473, Apr. 2004.

Van Trees, H. L., *Optimum Array Processing: Part IV of Detection, Estimation, and Modulation Theory*. Wiley, New York, 2002.

Ward, J., "Space-Time Adaptive Processing for Airborne Radar," Technical Report 1015, Massachusetts Institute of Technology, Lincoln Laboratory, Dec. 13, 1994.

Weiner, D. D., G. T. Capraro, C. T. Capraro, G. B. Berdan, and M. C. Wicks, "An Approach for Using Known Terrain and Land Feature Data in Estimation of the Clutter Covariance Matrix," *Proceedings of IEEE 1998 National Radar Conference*, pp. 381–386, Dallas, TX, May 1998.

Zatman, M., "Production of Adaptive Array Troughs through Dispersion Synthesis," *Electronics Letters*, vol. 31, no. 25, p. 2141, Dec. 1995.

Problems

1. Section 9.1 compared the size of the virtual arrays for synthetic and physical ULAs of N elements spanning an array extend D. The one-way antenna voltage pattern of such an array when it is phase-steered to an angle θ_0 as a function of angle θ is given by Eq. (9.18) (with appropriate adjustment for the differences in the two VAs).

    a. Show that the peak receive voltage gain of the array, whether synthetic or physical, is N. At what AOA does this gain occur?

    b. Either analytically or numerically, show that the gain of the peak receive sidelobe is approximately 13.2 dB below the mainbeam peak for both the synthetic and physical arrays for $N \geq 20$.

    c. For the physical array case, consider the two-way voltage pattern for the echo from a scatterer on the mainbeam peak (i.e., at angle θ_0). What will be the peak gain? What will be the peak sidelobe level relative to that gain, in dB? What is the Rayleigh beamwidth? Note: It is not necessary to work out the detailed math of the two-way pattern. It should be possible to deduce the result based on the one-way pattern and mode of operation.

    d. Repeat part (c) for the synthetic array case.

    e. Summarize the pros and cons of the synthetic and physical arrays in terms of angular resolution, sidelobe levels, and gain.

2. Equation (9.18) gave the beamformer response as a function of AOA θ when the array is steered to angle θ_0. Let $d = \alpha \lambda_t / 2$ for some integer α. Show that the magnitude of the pattern $z(\theta)$ will have peaks at α different AOAs θ_0 in the interval $[-\pi/2, \pi/2]$.

3. Verify that the pre-beamforming SIR in the example of Fig. 9.10 is -50.04 dB.

4. Consider an adaptive beamformer with $N = 2$ phase centers in an interference environment consisting of noise and one jammer. The noise and jammer powers are σ_n^2 and σ_j^2. The jammer AOA is θ_j radians. Define the noise-to-jammer ratio as $NJR = \sigma_n^2 / \sigma_j^2$. By factoring out the jammer power, write the total interference covariance matrix in terms of σ_j^2, NJR, and θ_j.

5. For the beamformer and interference environment in the previous problem, compute the optimum matched filter vector **h** as a function of the target angle θ_t. All overall scale factors, for instance those resulting from matrix inversion, can be ignored. What is the form of **h** as $NJR \to 0$ (i.e., the jammer dominates the interference)? Now let the target angle $\theta_t \to \theta_j$. What is the form of **h** in this case? Interpret the result.

6. Now assume that $NJR \to \infty$ (noise limited environment) in Prob. 4. Determine the limiting form of **h** for this case. Interpret the result.

7. Suppose the interference covariance matrix for a second-order ($N = 2$) adaptive beamformer is such that its inverse is

$$S_I = \begin{bmatrix} 17 & j8 \\ -j8 & 17 \end{bmatrix}$$

Find the matrix square root V such that $S_I = V \cdot V$. *Hint*: At least one valid V is of the form

$$V = \begin{bmatrix} a & \exp(j\theta) \\ \exp(-j\theta) & a \end{bmatrix}$$

Use this information to find the values of a and θ. (Systematic ways to find V without assuming a particular structure for V use the eigenvalues and eigenvectors of S_I or its Cholesky decomposition and are not covered here.)

8. Using the result for V from the previous problem, calculate $\tilde{S}_I$ using the second expression on the second line of Eq. (9.32) and show that it is an identity matrix to verify that applying the transformation V whitens the interference covariance matrix S_I of that problem. MATLAB® or a similar computational tool can be used for the calculations.

9. Sketch the STAP clutter ridge in (k_θ, ω_D) coordinates assuming the following system parameters: $v = 150$ m/s, $T = 1$ ms, $d = 0.3$ m, and $F_t = 1$ GHz. What is the value of the clutter ridge slope γ_c in this case? Repeat for $d = 0.15$ m and $T = 0.25$ ms. Plot the ridge for the range of k_θ corresponding to AOAs of $-\pi$ to $+\pi$.

10. Write out explicitly the target steering vector t of Eq. (9.43) for a small case with $M = 3$ pulses and $N = 2$ phase centers. The target vector should be steered to an angle of $\theta_t = 30°$ and normalized Doppler shift $f_D = 0.25$ cycles per sample. Assume $d = \lambda_t/2$.

11. Consider a small STAP radar system operating at X band (10 GHz) with $M = 4$ slow-time samples in the CPI and $N = 2$ phase centers. The goal is to compute the optimum matched filter coefficients for detecting a target directly on boresight of the antenna ($\theta_t = 0°$) and having a normalized Doppler shift of $f_{Dt} = 0.25$ cycles per sample.

    a. What is the temporal steering vector for the target, a_t? Simplify the resulting expression (e.g., $\exp(j\pi) = -1$).

    b. What is the spatial steering vector for the target, a_s? Simplify.

    c. What is the complete target model vector t?

12. For the STAP system in the previous problem, assume the interference is white noise with power σ_w^2 only (no clutter, no jammers). What will be the optimum beamforming filter coefficient vector h? Do not simplify or combine constants.

13. According to the RMB rule, how many range bins should be used in the STAP system of Prob. 11 to estimate the interference covariance matrix S_I^{-1} so that the mismatch loss is less than 1 dB? Use the exact formula of Eq. (9.73). How does the exact result compare to the rule-of-thumb estimate?

14. Verify that for a ULA and constant PRI, computing the adapted pattern of a STAP processor over a uniform grid in f_D and k_θ is equivalent to computing the magnitude squared of the two-dimensional DFT of the outer product of the weight vector of the spatial and temporal steering vectors when it is remapped to a two-dimensional function.

15. Using MATLAB® or another computational tool, compute the adapted pattern for the case of noise interference only and the target parameters from Prob. 11. Let $\kappa = 1$ and assume no windows are used for sidelobe control in either Doppler or angle. Use a two-dimensional DFT of size 30×30 or larger to get reasonable detail in the pattern.

16. Consider the optimum weight vector for the jammer cancellation example of Prob. 5 for the general case (*NJR* neither zero nor infinite) as a function of *NJR* when $\theta_t = \theta_j$, that is, at the null. What is the expected value of the output power $|z|^2$ when the input is a jammer signal from angle θ_j? The jammer signal can be modeled using Eq. (9.24) or (9.25).

17. Recompute $E\{|z|^2\}$ for the conditions of Prob. 16, but with a diagonal loading matrix of $\varepsilon \cdot \sigma_n^2 \mathbf{I}$ added to the covariance matrix, where $\varepsilon > 0$. Is the output power from the jammer input increased, unchanged, or decreased?

APPENDIX A
Selected Topics in Probability and Random Processes

A.1 Probability Density Functions and Likelihood Functions

The reader is assumed to be familiar with the concept of the probability density function (PDF) $p_x(x)$ for a continuous random variable x. In brief, the PDF describes the allowable range of values of x and the probability that x falls in a particular interval. Specifically, the probability that x falls in the interval x_1 to x_2 is computed as

$$\Pr\{x_1 \leq x \leq x_2\} = \int_{x_1}^{x_2} p_x(x)\,dx \qquad (A.1)$$

Setting $x_1 = x_0 - \delta x/2$ and $x_2 = x_0 + \delta x/2$ and taking the limit as $\delta x \to 0$ shows that the probability that x falls in a narrow interval around the value x_0 is

$$\Pr_{\delta x \to 0}\left\{x_0 - \frac{\delta x}{2} \leq x \leq x_0 + \frac{\delta x}{2}\right\} \approx p_x(x_0) \cdot \delta x \qquad (A.2)$$

This equation shows that the probability that $x \approx x_0$ is proportional to the PDF evaluated at x_0. Thus the PDF tells us not only the range of x but also the relative probability of observing a measurement of the random variable near a given value of x. For example, a uniform PDF over the interval [0, 1] dictates that the random variable described by that PDF will never take on a value less than zero or greater than 1, but that a value in any small interval in between those limits is equally probable. On the other hand, a zero-mean, unit-variance Gaussian PDF indicates that the random variable so described could take on any real value whatsoever. However, values near zero are much more probable than values greater than, say, $+3$ or less than -3. The symmetry of the PDF also implies that positive and negative values are equally probable.

Another specific example of Eq. (A.1) of importance in radar is the probability that x exceeds some threshold value T, a calculation that arises frequently in detection theory. This is the probability that x lies between T and $+\infty$:

$$\Pr(x > T) = \int_{T}^{\infty} p_x(x)\,dx \qquad (A.3)$$

and is often described as a "right-tail probability" because it is the area under the right-hand tail of the PDF from T to $+\infty$.

A possibly less familiar use of the PDF is as a *likelihood function*, an interpretation important in estimation theory. Consider a Gaussian random variable x with mean A and variance σ^2, $x \sim N(x; A, \sigma^2)$.[1] The PDF is

$$p_x(x) = \frac{1}{\sqrt{2\pi\sigma^2}} \exp[-(x-A)^2/2\sigma^2] \quad (A.4)$$

x might model a noisy measurement of a fixed value A. Suppose that Eq. (A.4) is known to be an appropriate model for the PDF of x, and that the parameter σ^2 is known but A is not. Also suppose it is desired to estimate A based on a measurement of x. Consider a plot of the PDF *as a function of A* for a given known value of σ^2 and an observed value of x, for example, $x = 3$. Denoting this "likelihood function" as $\ell(A|\sigma^2, x = 3)$ gives

$$\ell(A|\sigma^2, x = 3) = p_x(3) = \frac{1}{\sqrt{2\pi\sigma^2}} \exp[-(3-A)^2/2\sigma^2] \quad (A.5)$$

Figure A.1 plots $\ell(A|\sigma^2, x = 3)$ for $\sigma^2 = 1$. Again, note this is a function of A, not of x; the value of x is fixed by the measurement that has been made. This plot indicates that the most likely value of A, given that $x = 3$, is $A = 3$. Specifically, the probability that the measured value of x will be in a small neighborhood of width δx around the value 3 is, from Eq. (A.5), $\ell(A = 3|\sigma^2, x = 3) \cdot \delta x = 0.399 \cdot \delta x$. For any other value of A that probability is less. For instance, given the measurement $x = 3$, Fig. A.1 shows the probability that A is close to 6 will be $\ell(A = 6|\sigma^2, x = 3) \cdot \delta x = 0.004432 \cdot \delta x$, nearly two orders of magnitude lower. Clearly, the estimate of A that will produce the highest probability of the observation $x = 3$ is, for this simple problem, $A = 3$, the peak of the likelihood function. While it is possible that the measurement $x = 3$ would be observed if the actual value of A were 6, that is a much less likely event and therefore not a very reasonable estimate of A. The estimate of A obtained by choosing the value that maximizes the likelihood function for a given observed x is called the *maximum likelihood estimate*.

Any monotonically increasing transformation can be applied to the likelihood function without changing the value of A which maximizes it. In many problems of interest, the log-likelihood function $\ln[\ell(\cdot|\cdot)]$ is convenient to work with. Continuing the above example, the log-likelihood function is

$$\ln[\ell(A|\sigma^2, x = 3)] = \frac{-(3-A)^2}{2\sigma^2} - \frac{1}{2}\ln(2\pi\sigma^2) \quad (A.6)$$

Differentiating Eq. (A.6) with respect to A and setting the result equal to zero quickly gives $A = 3$ as the maximum likelihood estimate again.

A.2 Common Probability Distributions in Radar

Many probability distributions have been used in modeling radar target and interference signals, and their number and sophistication is constantly increasing in attempts to better model observed phenomena. This section provides basic information on some of the most frequently

---

[1] The symbol $\sim$ means "is distributed as" and the notation $N(x; a, b)$ denotes a normal or Gaussian distribution with mean a and variance b.

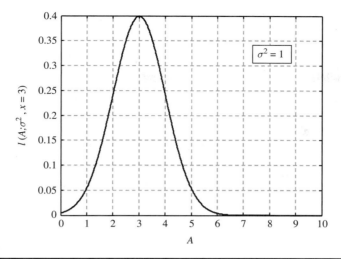

Figure A.1 The likelihood function of Eq. (A.5) when the observed $x = 3$ and $\sigma^2 = 1$.

encountered distributions in radar signal processing. The focus is PDFs most commonly used to represent the signal power (magnitude-squared), voltage or amplitude (magnitude), and phase of complex-valued radar receiver output signals. The corresponding characteristic functions, useful in calculating the distributions of sums of random variables, are given as well in several cases. The use of these distributions in radar signal modeling and the many extensions and modifications made for that purpose are discussed in the appropriate chapters. Most of the results in this section can be found in Papoulis and Pillai (2001) and Omura and Kailath (1965).

A.2.1 Power Distributions

Chi-Squared, Exponential, Erlang, and Gamma

Chi-squared distributions have a long history in modeling the power of radar target and interference signals. For example, versions of the chi-squared PDF are used for the Swerling target models and the standard model of white noise. The Erlang PDF is a generalization of the chi-squared PDF, and the gamma PDF is a generalization of both. The exponential PDF is a special case of the chi-squared and thus of the gamma and Erlang as well. This section describes the relationship between these four PDFs.

Consider a random variable x with mean $\bar{x}$ and variance σ_x^2. The common definition of the chi-squared PDF with N degrees of freedom (DOF) (also called the chi-squared of duo-degree N) is

$$p_x(x) = \chi^2(x; N) = \begin{cases} \dfrac{x^{N/2-1}}{2^{N/2}\Gamma\left(\dfrac{N}{2}\right)} \exp(-x/2), & x \geq 0 \\ 0, & x < 0 \end{cases} \quad (A.7)$$

where the notation $\chi^2(x; N)$ has been defined to indicate that the random variable x has a χ^2 PDF with parameter N. The mean and variance of x are

$$\bar{x} = N, \quad \sigma_x^2 = 2N \quad (A.8)$$

The N-DOF chi-squared PDF arises in radar as the PDF of a random variable x that is the sum of the squares of N independent zero-mean, unit-variance Gaussian random variables,

$$x = \sum_{i=1}^{N} x_i^2 \qquad (A.9)$$

where the x_i are independent identically distributed (i.i.d.) random variables (RVs) and $x_i \sim N(x; 0, 1)$. Of particular interest are the cases $N = 2$ and $N = 4$ because of their use in the Swerling target radar cross section (RCS) fluctuation models (see Chap. 2).

It is more useful to consider a slight generalization that allows for an arbitrary mean. For a general random variable x the PDF, mean, and variance of the new random variable y obtained by the linear transformation $y = ax + b$ are related to those of x according to Papoulis and Pillai (2001):

$$\begin{aligned} p_y(y) &= \frac{1}{|a|} p_x\left(\frac{y-b}{a}\right) \\ \bar{y} &= a\bar{x} + b \\ \sigma_y^2 &= |a|^2 \sigma_x^2 \end{aligned} \qquad (A.10)$$

If all the x_i in Eq. (A.9) have variance σ^2 instead of 1, the new sum-of-squares variable $x' = \sigma^2 x$ so that the PDF of x' is (dropping the "prime" in the notation)

$$p_x(x) = \chi^2(x; N, \sigma^2) = \begin{cases} \dfrac{x^{N/2-1}}{(2\sigma^2)^{N/2} \Gamma\left(\dfrac{N}{2}\right)} \exp(-x/2\sigma^2), & x \geq 0 \\ 0 & x < 0 \end{cases} \qquad (A.11)$$

with mean and variance

$$\bar{x} = N\sigma^2, \quad \sigma_x^2 = 2N\sigma^4 \qquad (A.12)$$

This generalized chi-squared PDF of Eq. (A.11) is a function of two parameters, N and σ^2, rather than just one, N.

Figure A.2a illustrates the chi-squared PDF for various values of N. In this plot, the variance of the underlying Gaussian random variables has been set to $\sigma^2 = 1/N$ so that all the PDFs have a common mean $\bar{x} = 1$. Notice that as N increases, the PDF becomes more Gaussian in shape, as would be expected for the sum of a large number of independent random variables of any distribution.

The *characteristic function* (CF) of a PDF $p_x(x)$ is defined as

$$C(q) = \int_{-\infty}^{\infty} p_x(x) \exp(jqx) dx \qquad (A.13)$$

The CF is essentially the Fourier transform of the PDF, though with the sign of the complex exponential chosen opposite from the definition commonly used in electrical engineering texts. The CF corresponding to the generalized chi-squared PDF is

$$C_{\chi^2}(q) = \frac{1}{(1 - j2\sigma^2 q)^{N/2}} \qquad (A.14)$$

Selected Topics in Probability and Random Processes

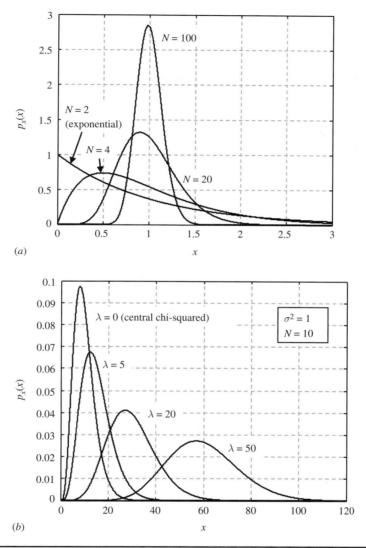

FIGURE A.2 The central and noncentral chi-squared families of PDFs. (a) Central chi-squared PDFs with unit mean and varying N. (b) Noncentral chi-squared PDFs with $\sigma^2 = 1$, $N = 10$, and varying noncentrality parameters λ.

The exponential PDF with mean γ, $\text{Exp}(x; \gamma)$, is

$$p_x(x) = \text{Exp}(x; \gamma) = \begin{cases} \dfrac{1}{\gamma} \exp(-x/\gamma), & x \geq 0 \\ 0, & x < 0 \end{cases} \tag{A.15}$$

The mean and variance of x are

$$\bar{x} = \gamma, \quad \sigma_x^2 = \gamma^2 \tag{A.16}$$

Its characteristic function is

$$C_{\text{Exp}}(q) = \frac{1}{1 - j\gamma q} \quad (A.17)$$

The exponential PDF is a special case of the generalized chi-squared with $N = 2$. This choice reduces it to a single-parameter distribution. In order that the mean of the exponential variate be γ when $N = 2$, it is necessary that $\gamma = 2\sigma^2$. Thus, $\text{Exp}(x; \gamma) = \chi^2(x; 2, \gamma/2)$.

In radar, the exponential PDF most often arises in connection with the standard model of noise in a coherent radar receiver. If the I and Q channel noise processes are assumed to be i.i.d. $N(x; 0, \sigma^2)$, the magnitude-squared of the complex noise (noise power) is the sum of the I and Q voltages squared and so will be distributed as $\chi^2(x; 2, \sigma^2)$. The Swerling 1 and 2 target fluctuation models also use the exponential PDF to describe the RCS fluctuations of a target having many scatterers with none dominant, as discussed in Chap. 2.

The gamma PDF $\Gamma(x; \alpha, \beta)$ is a more general distribution that includes the chi-squared and exponential as special cases, but can fit a wider range of experimental data. It is

$$p_x(x) = \Gamma(x; \alpha, \beta) = \begin{cases} \dfrac{x^{\alpha-1}}{\beta^\alpha \Gamma(\alpha)} \exp(-x/\beta), & x \geq 0 \\ 0, & x < 0 \end{cases} \quad (A.18)$$

where $\Gamma(\cdot)$ is the usual gamma function. The mean and variance of x are

$$\bar{x} = \alpha\beta, \quad \sigma_x^2 = \alpha\beta^2 \quad (A.19)$$

The CF corresponding to the gamma PDF is

$$C_\Gamma(q) = \frac{1}{(1 - j\beta q)^\alpha} \quad (A.20)$$

The generalized chi-squared PDF is related to the gamma PDF as $\chi^2(x; N, \sigma^2) = \Gamma(x; N/2, 2\sigma^2)$. Since $\text{Exp}(x; \gamma) = \chi^2(x; 2, \gamma/2)$, it follows also that $\text{Exp}(x; \gamma) = \Gamma(x; 1, \gamma)$.

The Erlang or Erlang-k PDF is sometimes mentioned in radar signal modeling. It lies between the gamma and generalized chi-squared models in that it restricts the parameter α of $\Gamma(x; \alpha, \beta)$ to an integer value k. The parameter β is arbitrary but is usually expressed in the form $1/\lambda k$ for integer k and some λ. With this choice, the Erlang PDF $E(x; k, \lambda) = \Gamma(x; k, 1/\lambda k)$ is

$$p_x(x) = E(x; k, \lambda) = \begin{cases} \dfrac{(\lambda k)^k x^{k-1}}{\Gamma(k)} \exp(-\lambda k x), & x \geq 0 \\ 0, & x < 0 \end{cases} \quad (A.21)$$

Recall that $\Gamma(k) = (k-1)!$ for integer k. The mean and variance of x are

$$\bar{x} = \frac{1}{\lambda}, \quad \sigma_x^2 = \frac{1}{k\lambda^2} \quad (A.22)$$

and the characteristic function is

$$C_E(q) = \frac{1}{(1 - jq/\lambda k)^k} \quad (A.23)$$

Noncentral Chi-Squared

The $\chi^2(x; N, \sigma^2)$ distribution is also called the central chi-squared with N degrees of freedom. The traditional *noncentral* chi-squared PDF with N degrees of freedom, $\chi^2_{nc}(x; N, \lambda)$, describes the sum of squares of normal random variables with different, nonzero means but identical unit variances:

$$x = \sum_{i=1}^{N} x_i^2 \tag{A.24}$$

with the $x_i \sim N(x; \mu_i, 1)$. The resulting PDF is

$$p_x(x) = \chi^2_{nc}(x; N, \lambda) = \begin{cases} \frac{1}{2}\left(\frac{x}{\lambda}\right)^{\frac{N-2}{4}} \exp[-(x+\lambda)/2] \cdot I_{\frac{N}{2}-1}\left(\sqrt{\lambda x}\right), & x \geq 0 \\ 0, & x < 0 \end{cases} \tag{A.25}$$

where the *noncentrality parameter* λ is defined as

$$\lambda = \sum_{i=1}^{N} \mu_i^2 \tag{A.26}$$

and is the magnitude-squared of the vector of means $\boldsymbol{\mu} = [\mu_1 \cdots \mu_N]^T$. $I_\alpha(\cdot)$ is the modified Bessel function of the first kind and order α.

Again, it is useful to slightly generalize this to the case where the x_i all have the same non-unit variance σ^2. This is done by scaling the x_i by the factor σ so that now $x_i \sim N(x; \sigma\mu_i, \sigma^2)$. The new sum variable and noncentrality parameter are

$$x' = \sum_{i=1}^{N} x_i'^2 = \sigma^2 \sum_{i=1}^{N} x_i^2 = \sigma^2 x, \quad \lambda' = \sum_{i=1}^{N} (\sigma\mu_i)^2 = \sigma^2 \sum_{i=1}^{N} \mu_i^2 = \sigma^2 \lambda \tag{A.27}$$

Applying these scalings to Eq. (A.25) and again dropping the primes from the notation, the generalized noncentral chi-squared PDF is

$$p_x(x) = \chi^2_{nc}(x; N, \lambda, \sigma^2) = \begin{cases} \frac{1}{2\sigma^2}\left(\frac{x}{\lambda}\right)^{\frac{N-2}{4}} \exp[-(x+\lambda)/2\sigma^2] I_{\frac{N}{2}-1}\left(\frac{\sqrt{\lambda x}}{\sigma^2}\right), & x \geq 0 \\ 0, & x < 0 \end{cases} \tag{A.28}$$

with mean and variance

$$\bar{x} = \sigma^2 N + \lambda, \quad \sigma_x^2 = 2N\sigma^4 + 4\sigma^2\lambda \tag{A.29}$$

and characteristic function

$$C_{\chi^2_{nc}}(q) = \frac{1}{(1 - j2\sigma^2 q)^{N/2}} \exp\left(\frac{j2\lambda\sigma^2 q}{1 - j2\sigma^2 q}\right) \tag{A.30}$$

This PDF reduces to the (central) chi-squared when $\lambda = 0$ and approaches a Gaussian as $N \to \infty$. Figure A.2b illustrates the noncentral chi-squared PDF for fixed values $N = 10$ and $\sigma^2 = 1$ and various values of λ. The nonzero mean of the noncentral chi-squared PDF allows modeling of signal statistics when the underlying complex data includes a persistent component.

Weibull, Log-Normal, and K Power

The Weibull, log-normal, and K densities are used when longer-tailed PDFs are needed, implying that the phenomenon being modeled has a higher rate of occurrence of large values than predicted using the distributions above. Such data is often referred to as being "spiky." These PDFs have become popular for modeling land and sea clutter returns at shallow grazing angles, especially at finer resolutions and higher radar frequencies. They are sometimes used to model fine resolution, high frequency target echoes as well.

One common form of the Weibull PDF is

$$p_x(x) = W(x;\alpha,\beta) = \begin{cases} \dfrac{\alpha}{\beta}\left(\dfrac{x}{\beta}\right)^{\alpha-1} \exp(-x/\beta^\alpha), & x \geq 0 \\ 0, & x < 0 \end{cases} \quad (A.31)$$

The mean, median x_m, and variance of x are

$$\bar{x} = \beta \cdot \Gamma\left(1+\dfrac{1}{\alpha}\right), \quad x_m = \beta(\ln 2)^{1/\alpha}, \quad \sigma_x^2 = \beta^2\left\{\Gamma\left(1+\dfrac{2}{\alpha}\right) - \Gamma^2\left(1+\dfrac{1}{\alpha}\right)\right\} \quad (A.32)$$

The Weibull characteristic function is

$$C_W(q) = \sum_{n=0}^{\infty} \dfrac{(j\beta q)^n}{n!} \Gamma\left(1+\dfrac{n}{\alpha}\right) \quad (A.33)$$

The parameter α is called the "shape parameter," while β is the "scale parameter." As α varies from 1 to 2, the Weibull varies from an exponential PDF to a Rayleigh PDF. However, α is not restricted to that range; it can take on any nonnegative value. Figure A.3a illustrates the Weibull PDF with $x_m = 1$ and several values of the shape parameter.

The log-normal PDF describes a random variable whose logarithm (in any base) is normally distributed. Using the natural logarithm, the PDF is

$$p_x(x) = LN(x;\alpha,\beta) = \dfrac{1}{\sqrt{2\pi} \cdot \beta x} \exp\left[-\dfrac{(\ln x - \alpha)^2}{2\beta^2}\right] \quad (A.34)$$

Its mean, median, and variance are

$$\bar{x} = \exp(\alpha + \beta^2/2), \quad x_m = \exp(\alpha), \quad \sigma_x^2 = [\exp(\beta^2) - 1]\exp(2\alpha + \beta^2) \quad (A.35)$$

Figure A.3b illustrates the log-normal PDF with $x_m = 1$ and several values of β^2.

The K distribution is generally applied to amplitude y rather than power $x = y^2$ in radar literature, and so its definition is deferred to the next section. The PDF of the power variable x associated with an amplitude variable y can be obtained from the PDF of y by the transformation (remembering that y is also nonnegative) $p_x(x) = p_y(\sqrt{x})/2\sqrt{x}$ (Papoulis and Pillai, 2001). Applying this to Eq. (A.50) gives the following PDF for the power of a random variable with a K-distributed amplitude:

$$p_x(x) = \begin{cases} \dfrac{c}{\sqrt{x} \cdot \Gamma(a)}\left(\dfrac{c\sqrt{x}}{2}\right)^a K_{a-1}(c\sqrt{x}), & x \geq 0 \\ 0 & x < 0 \end{cases} \quad (A.36)$$

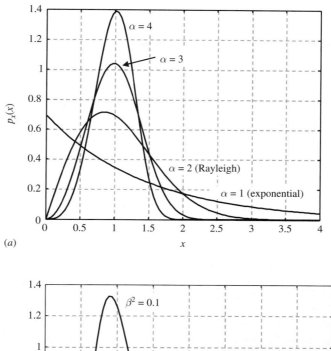

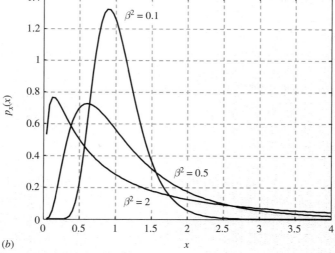

FIGURE A.3 The "long-tailed" Weibull and log-normal families of PDFs. (*a*) Weibull PDFs with a unit median and varying values of the shape parameter α. (*b*) Log-normal PDFs with unit median and varying values of β^2.

$K_{a-1}(\cdot)$ is the modified Bessel function of the second kind and order $a - 1$. Equation (A.36) is not quite in the form of a K distribution; here it will be called the "K power distribution." The shape and scale parameters a and c are those described with the K distribution for amplitude in the next section. The mean, mean-square, and variance of x are (Jakeman and Pusey, 1976)

$$\bar{x} = \left(\frac{2}{c}\right)^2 a, \quad \overline{x^2} = \frac{8a(a+1)}{c^2}, \quad \text{var}(x) = \overline{x^2} - \bar{x}^2 = \frac{16a(a+2)}{c^4} \quad \text{(A.37)}$$

Appendix A

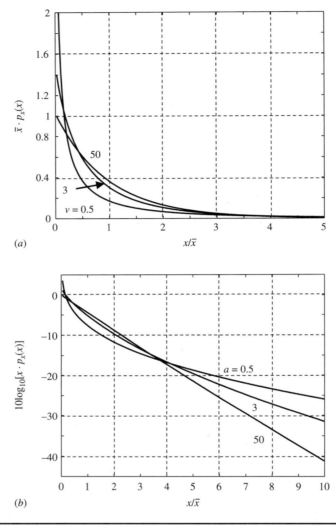

FIGURE A.4 The K-power PDF with scale parameter $c = 2$ and three values of the shape parameter a. (a) Linear scale plot with both graph coordinates normalized by the mean of the distribution as indicated. For large a the PDF converges to an exponential PDF. (b) The same data on a decibel scale. Smaller values of a result in a longer high-reflectivity "tail" on the PDF.

Figure A.4a illustrates the K-power distribution with the scale parameter $c = 2$ and three values of the shape parameter a. Both graph coordinates are normalized by the mean of the distribution as indicated. Notice that unlike the log-normal and Weibull PDFs, the K-power PDF does not have a distinct peak. As $a \to \infty$ the K-power PDF converges to an exponential PDF. For smaller a ($a > 0$), the PDF initially decays quickly as x increases, but then the rate of decay slows and the PDF exhibits a long tail. This behavior is more easily seen in Fig. A.4b, which shows the same data on a decibel scale. In this plot, a straight line is an exponential function; the case $a = 50$ produces a PDF that is very nearly straight. As a gets smaller, the PDF still decays monotonically as seen in Fig. A.4a, but the decay rate deviates increasingly from an exponential.

A.2.2 Amplitude Distributions

Given a random variable x representing the magnitude-squared and thus the power of a complex-valued voltage, the quantity $y = \sqrt{x}$ represents the amplitude (magnitude) of the complex voltage. The actual complex voltage cannot be obtained from the power without also knowing the signal phase. Nonetheless, the amplitude alone is often of interest, for instance, when using a linear detector instead of a square-law detector in a radar receiver. Using standard results from random variable theory and noting that the power x is always nonnegative, the PDF of y can be obtained from the PDF of x with the transformation

$$p_y(y) = 2y \cdot p_x(y^2) \tag{A.38}$$

Rayleigh

Applying this result, a signal whose power is described by the exponential PDF with mean γ has an amplitude described by the Rayleigh PDF:

$$p_y(y) = R(y;\gamma) = \begin{cases} \dfrac{2y}{\gamma} \exp(-y^2/\gamma), & y \geq 0 \\ 0, & y < 0 \end{cases} \tag{A.39}$$

The mean and variance of y are

$$\bar{y} = \frac{1}{2}\sqrt{\pi\gamma}, \quad \sigma_y^2 = \left(1 - \frac{\pi}{4}\right)\gamma \tag{A.40}$$

Its characteristic function is

$$C_R(q) = \left(1 + j\frac{\sqrt{\pi\gamma}}{2} q\right) \exp(-\gamma q^2/4) \tag{A.41}$$

Central Chi

As with the exponential and chi-squared distributions, the generalizations of the Rayleigh amplitude distribution are the central and noncentral chi distributions. Using the transformation of Eq. (A.38) in Eq. (A.11) gives the central chi distribution:

$$p_y(y) = \chi(y;N,\sigma^2) = \begin{cases} \dfrac{2y^{N-1}}{(2\sigma^2)^{N/2}\Gamma\left(\dfrac{N}{2}\right)} \exp(-y^2/2\sigma^2), & y \geq 0 \\ 0, & y < 0 \end{cases} \tag{A.42}$$

The mean, mean-square, and variance are

$$\bar{y} = \sqrt{2\sigma^2} \cdot \frac{\Gamma[(N+1)/2]}{\Gamma(N/2)}, \quad \overline{y^2} = 2\sigma^2 \frac{\Gamma(N/2+1)}{\Gamma(N/2)} = N\sigma^2, \quad \sigma_y^2 = \overline{y^2} - \bar{y}^2 \tag{A.43}$$

Equation (A.42) retains the parameters N and σ^2 that emphasize the connection to the amplitude of the sum of N squared $N(x; 0, \sigma^2)$ variates. As will be seen in the discussion of the K distribution, the chi PDF is also used to empirically model variations of the local mean of a nonstationary Rayleigh process. In this case, there is no particular connection to the sum of squared Gaussians so the PDF is better expressed in the more neutral form

$$p_y(y) = \chi(y;a,b) = \begin{cases} \dfrac{2b^{2a}}{\Gamma(a)} y^{2a-1} \exp(-b^2 y^2), & y \geq 0 \\ 0, & y < 0 \end{cases} \quad \text{(A.44)}$$

where the substitutions $b = \sqrt{1/2\sigma^2}$ and $a = N/2$ have been made. In this form, a is the shape parameter and b is the scale parameter. The first and second moments are

$$\bar{y} = \frac{1}{b} \cdot \frac{\Gamma(a + 1/2)}{\Gamma(a)}, \quad \overline{y^2} = \frac{1}{b^2} \frac{\Gamma(a + 1)}{\Gamma(a)} = \frac{a}{b^2}, \quad \sigma_y^2 = \overline{y^2} - \bar{y}^2 \quad \text{(A.45)}$$

Noncentral Chi

The noncentral chi distribution is the result of applying Eq. (A.38) to the noncentral chi-squared PDF, and thus represents the PDF of the magnitude of the sum of N random variables distributed as $N(x; \sigma \cdot \mu_i, \sigma^2)$. Defining the magnitude of the vector of means as $v = \sqrt{\lambda}$, the PDF is

$$p_y(y) = \chi_{nc}(y; N, v, \sigma^2)$$

$$= \begin{cases} \dfrac{v}{\sigma^2} \left(\dfrac{y}{v}\right)^{\frac{N}{2}} \exp[-(y^2 + v^2)/2\sigma^2] I_{\frac{N}{2}-1}\left(\dfrac{vy}{\sigma^2}\right), & y \geq 0 \\ 0, & y < 0 \end{cases} \quad \text{(A.46)}$$

The moments of the noncentral chi PDF do not take a simple form. The mean, mean-square, and variance are

$$\bar{y} = \sqrt{2\sigma^2} \exp(-v^2/2\sigma^2) \frac{\Gamma[(N+1)/2]}{\Gamma(N/2)} {}_1F_1\left(\frac{N+1}{2}, \frac{N}{2}, \frac{v^2}{2\sigma^2}\right)$$

$$\overline{y^2} = 2\sigma^2 \exp(-v^2/2\sigma^2) \frac{\Gamma[N/2+1]}{\Gamma(N/2)} {}_1F_1\left(\frac{N}{2}+1, \frac{N}{2}, \frac{v^2}{2\sigma^2}\right) \quad \text{(A.47)}$$

$$= N\sigma^2 \exp(-v^2/2\sigma^2) {}_1F_1\left(\frac{N}{2}+1, \frac{N}{2}, \frac{v^2}{2\sigma^2}\right)$$

$$\sigma_y^2 = \overline{y^2} - \bar{y}^2$$

where ${}_1F_1(\cdot, \cdot, \cdot)$ is the confluent hypergeometric function[2] and the gamma function recurrence relation $\Gamma(x+1) = x \cdot \Gamma(x)$ was used to get the second form of $\overline{y^2}$. The noncentral chi PDF reduces to the Rayleigh PDF when $v = 0$, $\gamma = 2\sigma^2$, and $N = 2$.

---

[2]The confluent hypergeometric function ${}_1F_1(\alpha, \beta, x)$ is also called Kummer's function and denoted $M(\alpha, \beta, x)$. See Olver et al. (2010) for more information.

Rice

The Rice (also called Rician) distribution is the amplitude distribution that corresponds to the generalized noncentral chi-squared power distribution with $N = 2$, that is, it is the noncentral chi with $N = 2$. The PDF becomes

$$p_y(y) = \text{Ri}(y; N, v, \sigma^2) = \chi_{nc}(y; 2, v, \sigma^2)$$
$$= \begin{cases} \dfrac{y}{\sigma^2} \exp[-(y^2 + v^2)/2\sigma^2] I_0\left(\dfrac{vy}{\sigma^2}\right), & y \geq 0 \\ 0, & y < 0 \end{cases} \quad \text{(A.48)}$$

The mean, mean-square, and variance are given by Eq. (A.47) with $N = 2$,

$$\bar{y} = \sqrt{2\sigma^2} \exp(-v^2/2\sigma^2) \Gamma\left(\frac{3}{2}\right) \cdot {}_1F_1\left(\frac{3}{2}, 1, \frac{v^2}{2\sigma^2}\right)$$
$$= \sqrt{\frac{\pi \sigma^2}{2}} \exp(-v^2/2\sigma^2) {}_1F_1\left(\frac{3}{2}, 1, \frac{v^2}{2\sigma^2}\right) \quad \text{(A.49)}$$
$$\overline{y^2} = 2\sigma^2 \exp(-v^2/2\sigma^2) {}_1F_1\left(2, 1, \frac{v^2}{2\sigma^2}\right)$$
$$\sigma_y^2 = \overline{y^2} - \bar{y}^2$$

The relations $\Gamma(1/2) = \sqrt{\pi}$ and $\Gamma(x + 1) = x \cdot \Gamma(x)$ have been used to get the second form of $\bar{y}$ in Eq. (A.49). The Rice PDF reduces to the Rayleigh PDF when $v = 0$.

Weibull, Log-Normal, and K

Applying the transformation of Eq. (A.38) to the Weibull PDF of Eq. (A.31) shows that the amplitude PDF that corresponds to a Weibull power PDF is also Weibull but with the parameters modified to $\alpha' = 2\alpha$ and $\beta' = \sqrt{\beta}$. The earlier formulas for the mean, median, variance, and characteristic function can be used with these new parameters.

Similarly, applying Eq. (A.38) to Eq. (A.34) yields a log-normal distribution with new parameters $\alpha' = \alpha/2$ and $\beta' = \beta/2$ for the amplitude corresponding to a log-normal power variate. The mean, median, and variance of Eq. (A.35) may be used for the amplitude variate with these new values.

The K distribution is a "compound PDF" composed from two more basic PDFs. Specifically, assume the signal amplitude follows a Rayleigh PDF with mean z, appropriate to a "many-scatterer" model. Suppose also that z is itself modeled as a random variable described by a central chi distribution with shape and scale parameters a and c. The resulting PDF for the amplitude y is the K distribution (Watts, 1985):

$$p_y(y) = \int_0^\infty p_y(y|z) p_z(z) dz = \begin{cases} \dfrac{2c}{\Gamma(a)} \left(\dfrac{cy}{2}\right)^a K_{a-1}(cy), & y \geq 0 \\ 0, & y < 0 \end{cases} \quad \text{(A.50)}$$

where $K_{a-1}(\cdot)$ is the modified Bessel function of the second kind and a and c are the shape and scale parameters of the distribution, respectively. The mean, mean-square, and variance of the K distribution are

$$\bar{y} = \frac{\sqrt{\pi}}{c} \cdot \frac{\Gamma(a + 1/2)}{\Gamma(a)}, \quad \overline{y^2} = \frac{4a}{c^2}, \quad \sigma_y^2 = \overline{y^2} - \bar{y}^2 \quad \text{(A.51)}$$

The rationale for the compound PDF approach is discussed in Sec. 2.3.4.

A.2.3 The Unfortunate Tendency in Radar to Call Power Distributions by the Name of the Amplitude Distribution

It is an unfortunate fact that it is common in the radar community to apply the correct name for a amplitude PDF to the PDF for power that results when the amplitude variable is squared. For example, a random variable representing RCS and described by the exponential PDF may be referred to as a Rayleigh RCS. The reader must be careful to realize that such a reference probably means Rayleigh amplitude, but exponential power. Similarly, a variable described by a noncentral chi-squared PDF with $N = 2$ may be referred to as a Rice variate.

Although not an error, another source of confusion arises when the PDF of both an amplitude variable and its corresponding power variable have the same general form but different parameters. As seen above, this occurs with Weibull and log-normal variates. When discussing these PDFs, extra caution is needed to determine whether the parameters used are modeling amplitude or power.

A.2.4 Phase Distributions

Uniform

The uniform PDF $U(x; x_1, x_2)$ describes a random variable that takes on any value in an ordered interval (x_1, x_2) with equal likelihood. It is

$$p_x(x) = U(x; x_1, x_2) = \begin{cases} \dfrac{1}{x_2 - x_1}, & x_1 \leq x \leq x_2 \\ 0, & x < 0 \end{cases} \quad (A.52)$$

The mean and variance of x are

$$\bar{x} = \frac{1}{2}(x_2 - x_1), \quad \sigma_x^2 = \frac{(x_2 - x_1)^2}{12} \quad (A.53)$$

Its characteristic function is

$$C_U(q) = \frac{\exp(jx_1 q) - \exp(-jx_2 q)}{j(x_2 - x_1)q} = \exp\{j[(x_2 + x_1)/2]q\}\operatorname{sinc}\left[\left(\frac{x_2 - x_1}{2}\right)q\right] \quad (A.54)$$

The most common use of the uniform PDF in radar is to describe a completely random phase that is uniformly distributed over $[0, 2\pi)$ radians (equivalently, $(-\pi, \pi]$ radians). Another common usage is to describe quantization error (see App. B).

Tikhonov

The Tikhonov PDF, also called the Von Mises PDF, is given in Van Trees (1968) as

$$p_x(x; \alpha) = T(x; \alpha) = \begin{cases} \dfrac{1}{2\pi I_0(\alpha)} \exp[\alpha \cos(x - \mu)], & -\pi < x \leq \pi \\ 0, & \text{otherwise} \end{cases} \quad (A.55)$$

Most often, the moments of the complex variable $z = \exp(jx)$ are considered rather than the moments of x itself. The moments of z are called the *circular moments* of x. They are

$$\bar{z} = \frac{I_1(\alpha)}{I_0(\alpha)} \exp(j\mu), \quad \sigma_z^2 = 1 - \frac{I_1(\alpha)}{I_0(\alpha)} \quad (A.56)$$

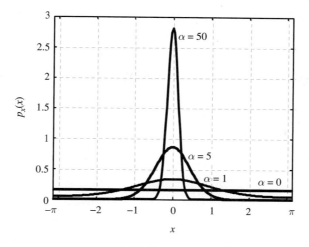

FIGURE A.5 The Tikhonov PDF for phase with $\mu = 0$.

$I_0(\cdot)$ and $I_1(\cdot)$ are the modified Bessel functions of the first kind and orders zero and 1, respectively. The mean of x itself is the argument of $\bar{z}$:

$$\bar{x} = \mu \qquad (A.57)$$

The Tikhonov distribution provides a family of PDFs suitable for modeling phase and phase errors that vary from the uniform PDF when $\alpha = 0$ to a fixed nonrandom phase of μ as $\alpha \to \infty$. The PDF is defined over the interval $(-\pi, \pi]$ radians and its integral over that interval is unity so that it is a strictly valid PDF for phase. However, the same formula can be used over any interval of length 2π (e.g., 0 to 2π) because of the periodicity of the cosine function. Figure A.5 illustrates the Tikhonov PDF for $\mu = 0$ and several different values of the parameter α.

A.3 Estimators and the Cramèr-Rao Lower Bound

Consider an observed signal $x(t)$ that is the sum of a target component $s(t)$ and a noise component $w(t)$,

$$x(t) = s(t) + w(t) \qquad (A.58)$$

$s(t)$ is a function of one or more parameters Θ_i. These might be, for example, the time delay, amplitude, Doppler shift, or angle of arrival of the target component. Once a target is detected, estimation of these parameters will usually be the next goal of the radar. In many cases, the estimated values may serve as inputs to tracking algorithms. Thus, the quality of these estimates is of great interest.

Suppose $x(t)$ is sampled multiple times (intrapulse or intrasweep, and/or over multiple pulses or sweeps) to give a column vector of N observations,

$$\mathbf{x} = [x_0 \quad x_1 \quad \cdots \quad x_{N-1}]^T \qquad (A.59)$$

Now define an *estimator f* of a parameter Θ_i as some function or procedure that produces an estimated value $\hat{\Theta}_i$ from the data $\mathbf{x}$,

$$\hat{\Theta}_i = f(\mathbf{x}) \tag{A.60}$$

Because $\mathbf{x}$ is random, the estimate $\hat{\Theta}_i$ is itself a random variable and therefore has a probability density function with a mean and variance. The *accuracy* of the estimator is the average error in the mean, $E\{\hat{\Theta}_i - \Theta_i\}$, also known as the *bias*. The *precision* of the estimator is the standard deviation $\sigma_{\hat{\Theta}_i}$.

Two desirable properties of an estimator are that it be *unbiased* and *consistent*. These mean that

$$\begin{aligned} E\{\hat{\Theta}_i\} &= \Theta_i \quad \text{(unbiased)} \\ \lim_{N \to \infty} \{\sigma_{\hat{\Theta}_i}^2\} &\to 0 \quad \text{(consistent)} \end{aligned} \tag{A.61}$$

In this text, only unbiased estimators will be considered, and so the estimate quality issue becomes how large or small is the variance of the estimate and how it behaves versus the amount of data N. The derivation of the real, single-parameter CRLB in this section closely follows Peebles (1998). Most of the other results follow Kay (1993).

A.3.1 The Cramèr-Rao Lower Bound on Estimator Variance

The *Cramèr-Rao Lower Bound* (CRLB) is a famous and very important result that establishes the minimum variance of an unbiased estimator. Any particular unbiased estimator must then have a variance (square of precision) equal to or greater than the CRLB, and the quality of an unbiased estimator can be judged by how close its actual variance comes to achieving the CRLB. An estimator that is unbiased and achieves the CRLB is said to be *efficient*.

Consider a signal dependent on a single scalar parameter Θ. Denote the N-dimensional joint PDF of the observations $\mathbf{x}$, given the actual value of Θ, as $p_\mathbf{x}(\mathbf{x}|\Theta)$. The assumption of an unbiased estimator requires

$$E\{\hat{\Theta} - \Theta\} = \int \cdots \int (\hat{\Theta} - \Theta) p_\mathbf{x}(\mathbf{x}|\Theta) d\mathbf{x} = 0 \tag{A.62}$$

Using Leibniz' integral rule to differentiate this equation with respect to Θ gives

$$\frac{\partial E(\hat{\Theta} - \Theta)}{\partial \Theta} = \int \cdots \int (\hat{\Theta} - \Theta) \frac{\partial p_\mathbf{x}(\mathbf{x}|\Theta)}{\partial \Theta} d\mathbf{x} - \underbrace{\int \cdots \int p_\mathbf{x}(\mathbf{x}|\Theta) d\mathbf{x}}_{=1} = 0 \tag{A.63}$$

The second multiple integral equals 1 because it has just the PDF $p_\mathbf{x}(\mathbf{x}|\Theta)$ as the integrand, and the integral of any valid PDF is 1. Note also that for any function $g(\mathbf{x}, \Theta)$,

$$\begin{aligned} \frac{\partial}{\partial \Theta} \ln[g(\mathbf{x}, \Theta)] &= \frac{1}{g(\mathbf{x}, \Theta)} \frac{\partial g(\mathbf{x}, \Theta)}{\partial \Theta} \Rightarrow \\ \frac{\partial g(\mathbf{x}, \Theta)}{\partial \Theta} &= \left\{ \frac{\partial}{\partial \Theta} \ln[g(\mathbf{x}, \Theta)] \right\} g(\mathbf{x}, \Theta) \end{aligned} \tag{A.64}$$

Using Eq. (A.64) in Eq. (A.63) gives

$$\int \cdots \int (\hat{\Theta} - \Theta) \frac{\partial \ln[p_\mathbf{x}(\mathbf{x}, \Theta)]}{\partial \Theta} p_\mathbf{x}(\mathbf{x}, \Theta) d\mathbf{x} = 1 \tag{A.65}$$

This relationship is the consequence of the unbiased estimator assumption.

The Schwarz inequality in integral form can be applied to Eq. (A.65). The scalar version was used in discussing the matched filter in Chap. 4. The multivariable version is

$$\left\{ \int \cdots \int A(\mathbf{x}) B(\mathbf{x}) \, d\mathbf{x} \right\}^2 \leq \int \cdots \int A^2(\mathbf{x}) \, d\mathbf{y} \cdot \int \cdots \int B^2(\mathbf{x}) \, d\mathbf{x} \tag{A.66}$$

with equality if and only if $A(\mathbf{x}) = \alpha B(\mathbf{x})$ for some scalar α. Choose

$$A(\mathbf{x}) = (\hat{\Theta} - \Theta)\sqrt{p_\mathbf{x}(\mathbf{x}|\Theta)}, \quad B(\mathbf{x}) = \frac{\partial \ln[p_\mathbf{x}(\mathbf{x}|\Theta)]}{\partial \theta} \sqrt{p_\mathbf{x}(\mathbf{x}|\Theta)} \tag{A.67}$$

to get

$$\left\{ \int \cdots \int (\hat{\Theta} - \Theta) \frac{\partial[\ln p_\mathbf{x}(\mathbf{x}|\Theta)]}{\partial \Theta} p_\mathbf{x}(\mathbf{x}|\Theta) d\mathbf{x} \right\}^2 \leq$$
$$\int \cdots \int (\hat{\Theta} - \Theta)^2 p_\mathbf{x}(\mathbf{x}|\Theta) d\mathbf{x} \cdot \int \cdots \int \left\{ \frac{\partial[\ln p_\mathbf{x}(\mathbf{x}|\Theta)]}{\partial \Theta} \right\}^2 p_\mathbf{x}(\mathbf{x}|\Theta) d\mathbf{x} \tag{A.68}$$

From Eq. (A.65), the left-hand side of Eq. (A.68) equals 1. The first multiple integral on the right hand side is the estimator variance $\sigma_{\hat{\Theta}}^2$, while the second multiple integral is, by definition, $\mathbf{E}\{\partial[\ln p_\mathbf{x}(\mathbf{x}|\Theta)]/\partial\Theta\}^2$. Using these relationships and rearranging Eq. (A.68) then gives the CRLB,

$$\sigma_{\hat{\Theta}}^2 \geq \frac{1}{\mathbf{E}\{\partial[\ln p_\mathbf{y}(\mathbf{y}|\Theta)]/\partial\Theta\}^2} \tag{A.69}$$

An alternate form of the CRLB is also common. If $p_\mathbf{x}(\mathbf{x}|\Theta)$ is twice differentiable and obeys some other mild regularity conditions, it can be shown that

$$\mathbf{E}\left\{ \left[\frac{\partial \ln[p_x(x|\Theta)]}{\partial \Theta}\right]^2 \right\} = -\mathbf{E}\left\{ \frac{\partial^2 \ln[p_x(x|\Theta)]}{\partial \Theta^2} \right\} \tag{A.70}$$

which gives another version of the CRLB,

$$\sigma_{\hat{\Theta}}^2 \geq \frac{1}{-\mathbf{E}\{\partial^2 \ln[p_x(x|\Theta)]/\partial\Theta^2\}} \tag{A.71}$$

The choice between Eq. (A.69) and Eq. (A.71) is a matter of convenience, depending on the functional form of $\ln[p_x(x|\Theta)]$.

The denominator of Eq. (A.71) is called the *Fisher information* $I(\Theta)$ for the data $\mathbf{x}$. The CRLB is therefore the inverse of $I(\Theta)$. Equation (A.71) shows that signals whose log-likelihood function has a high curvature with respect to the parameter Θ will have a large Fisher information and a small value of $\sigma_{\hat{\Theta}}^2$.

Consideration of $I(\Theta)$ leads to an interesting observation about i.i.d. samples. Suppose N i.i.d. observations x_i of an RV dependent on a parameter Θ are available. For a single observation $I(\Theta) = -\mathbf{E}\{\partial^2 \ln[p_x(x|\Theta)]/\partial\Theta^2\}$. Because the N observations are independent and identically distributed, $p_x(\mathbf{x}|\Theta) = [p_x(x|\Theta)]^N$ so that $I(\Theta)$ is N times larger, $-E\{\partial^2 \ln[p_x(x|\Theta)]^N/\partial\Theta^2\} = -N \cdot E\{\partial^2 \ln[p_x(x|\Theta)]/\partial\Theta^2\}$. Consequently, the CRLB for N i.i.d. observations is $1/N$ times the CRLB for a single observation.

A.3.2 The CRLB for Transformed Parameters

If the CRLB for a particular parameter Θ in some estimation problem is known, it is easy to find the CRLB for a related parameter $\Phi = g(\Theta)$ for some function g. The result is

$$\sigma_\Phi^2 = \left[\frac{\partial g(\Theta)}{\partial \Theta}\right]^2 \sigma_\Theta^2 \tag{A.72}$$

As a trivial example, if the CRLB for estimating time delay t_0 is known, the CRLB for the corresponding range $R_0 = c \cdot t_0/2$ is $\sigma_{R_0}^2 = (c/2)^2 \sigma_{t_0}^2$.

It is easily shown that if the estimator for Θ is efficient, it is not the case in general that the estimator for Φ is also efficient. However, if the transformation g is affine (a linear transformation followed by a translation), efficiency of the estimator is maintained. Even in the case of a nonlinear transformation, the transformed estimator will be *asymptotically efficient*; that is, it will become efficient as $N \to \infty$.

A.3.3 Signals in Additive White Gaussian Noise

The CRLB takes on a special form for the very important case of a signal in additive white Gaussian noise (AWGN). Assume the measurement vector $\mathbf{x}$ is composed of the N real-valued signal + noise samples

$$x[n] = s[n;\Theta] + w[n], \quad n = 0, \ldots, N-1 \tag{A.73}$$

where Θ is the real-valued parameter to be estimated and the variance of $w[n]$ is σ_w^2. Because the noise is white and Gaussian, the PDF $p_x(\mathbf{x}|\Theta)$ is a multidimensional real-valued Gaussian function

$$p_x(\mathbf{x}|\Theta) = \frac{1}{\left(2\pi\sigma_w^2\right)^{N/2}} \exp\left\{\frac{-1}{2\sigma_w^2} \sum_{n=0}^{N-1} (x[n] - s[n;\Theta])^2\right\} \tag{A.74}$$

so that

$$\ln[p_x(\mathbf{x}|\Theta)] = -\frac{N}{2}\left(2\pi\sigma_w^2\right) - \frac{1}{2\sigma_w^2} \sum_{n=0}^{N-1} (y[n] - s[n;\Theta])^2 \tag{A.75}$$

The first and second partial derivatives of $\ln[p_x(\mathbf{x}|\Theta)]$ with respect to the unknown parameter Θ are

$$\frac{\partial \ln[p_x(\mathbf{x}|\Theta)]}{\partial \Theta} = \frac{1}{\sigma_w^2} \sum_{n=0}^{N-1} (x[n] - s[n;\Theta]) \frac{\partial s[n;\Theta]}{\partial \Theta}$$

$$\frac{\partial^2 \{\ln[p_x(\mathbf{x}|\Theta)]\}}{\partial \Theta^2} = \frac{1}{\sigma_w^2} \sum_{n=0}^{N-1} \left\{(x[n] - s[n;\Theta]) \frac{\partial^2 s[n;\Theta]}{\partial \Theta^2} - \left(\frac{\partial s[n;\Theta]}{\partial \Theta}\right)^2\right\} \tag{A.76}$$

The expected value of the second partial derivative is

$$\mathrm{E}\left\{\frac{\partial^2 \{\ln[p_\mathbf{x}(\mathbf{x}|\Theta)]\}}{\partial \Theta^2}\right\} = \frac{-1}{\sigma_w^2}\sum_{n=0}^{N-1}\left(\frac{\partial s[n;\Theta]}{\partial \Theta}\right)^2 \quad (A.77)$$

This result follows because the term $(x[n] - s[n;\Theta])$ is just the noise $w[n]$, which is zero mean. Finally, using Eq. (A.77) in Eq. (A.71) gives the CRLB for a real-valued signal in real white Gaussian noise:

$$\sigma_{\hat{\Theta}}^2 \geq \frac{\sigma_w^2}{\sum_{n=0}^{N-1}\left(\frac{\partial s[n;\Theta]}{\partial \Theta}\right)^2} \quad \text{(real-valued signal in real AWGN)} \quad (A.78)$$

The denominator of Eq. (A.78) shows that signals that are more sensitive to changes in the value of the parameter Θ (large magnitudes of the derivative) will have smaller values of $\sigma_{\hat{\Theta}}^2$, suggesting that systems having finer resolution in a parameter of interest will also exhibit finer precision than coarser-resolution systems. This point is illustrated more clearly in Chap. 7.

A.3.4 Signals with Multiple Parameters in AWGN

The CRLB can be generalized to describe estimators of multiple simultaneous scalar parameters by defining a vector parameter Θ. For example, it might be desirable to simultaneously estimate the amplitude, frequency, and initial phase of a sinusoid in additive WGN. The Fisher information for a signal $\mathbf{x}$ dependent on a real-valued parameter vector Θ is now the $N \times N$ matrix:

$$[\mathbf{I}(\Theta)]_{ij} = -\mathrm{E}\left[\frac{\partial^2 \{\ln[p_\mathbf{x}(\mathbf{x}|\Theta)]\}}{\partial \Theta_i \, \partial \Theta_j}\right] \quad (A.79)$$

The vector parameter CRLB then states that the covariance matrix of any estimate $\hat{\Theta}$ of Θ must satisfy

$$\mathbf{C}_{\hat{\Theta}} - \mathbf{I}^{-1}(\Theta) \geq 0 \quad (A.80)$$

where $\mathbf{C}_{\hat{\Theta}}$ is the covariance matrix of $\hat{\Theta}$ and the notation "≥ 0" means that the left-hand side is positive semidefinite. In particular, consideration of the diagonal elements gives

$$\sigma_{\hat{\Theta}_i}^2 \geq [\mathbf{I}^{-1}(\Theta)]_{ii} \quad (A.81)$$

If the signal $\mathbf{x}$ has a Gaussian PDF with mean vector $\boldsymbol{\mu}$ and general covariance matrix $\mathbf{C}_\mathbf{x}$ (not necessarily white noise), both possibly dependent on a real-valued parameter vector Θ, the Fisher information matrix becomes (Kay, 1993, Chap. 3)

$$[\mathbf{I}(\Theta)]_{ij} = \frac{1}{2}\mathrm{tr}\left[\mathbf{C}_\mathbf{x}^{-1}(\Theta)\frac{\partial \mathbf{C}_\mathbf{x}(\Theta)}{\partial \Theta_i}\mathbf{C}_\mathbf{x}^{-1}(\Theta)\frac{\partial \mathbf{C}_\mathbf{x}(\Theta)}{\partial \Theta_j}\right] + \left[\frac{\partial \boldsymbol{\mu}(\Theta)}{\partial \Theta_i}\right]^T \mathbf{C}_\mathbf{x}^{-1}(\Theta)\left[\frac{\partial \boldsymbol{\mu}(\Theta)}{\partial \Theta_j}\right] \quad (A.82)$$

where

$$[\mathbf{C}_x(\Theta)]_{ij} = E[(x_i - \bar{x}_i)(x_j - \bar{x}_j)]$$

$$\frac{\partial \boldsymbol{\mu}(\Theta)}{\partial \Theta_i} = \left[\frac{\partial [\boldsymbol{\mu}(\Theta)]_0}{\partial \Theta_i} \quad \frac{\partial [\boldsymbol{\mu}(\Theta)]_1}{\partial \Theta_i} \quad \cdots \quad \frac{\partial [\boldsymbol{\mu}(\Theta)]_{N-1}}{\partial \Theta_i} \right]^T$$

$$\frac{\partial \mathbf{C}_x(\Theta)}{\partial \Theta_i} = \begin{bmatrix} \frac{\partial [\mathbf{C}_x(\Theta)]_{0,0}}{\partial \Theta_i} & \frac{\partial [\mathbf{C}_x(\Theta)]_{0,1}}{\partial \Theta_i} & \cdots & \frac{\partial [\mathbf{C}_x(\Theta)]_{0,N-1}}{\partial \Theta_i} \\ \frac{\partial [\mathbf{C}_x(\Theta)]_{2,0}}{\partial \Theta_i} & \frac{\partial [\mathbf{C}_x(\Theta)]_{2,1}}{\partial \Theta_i} & \cdots & \frac{\partial [\mathbf{C}_x(\Theta)]_{2,N-1}}{\partial \Theta_i} \\ \vdots & \vdots & \ddots & \vdots \\ \frac{\partial [\mathbf{C}_x(\Theta)]_{N-1,0}}{\partial \Theta_i} & \frac{\partial [\mathbf{C}_x(\Theta)]_{N-1,1}}{\partial \Theta_i} & \cdots & \frac{\partial [\mathbf{C}_x(\Theta)]_{N-1,N-1}}{\partial \Theta_i} \end{bmatrix} \quad (A.83)$$

Now consider the common additional restriction that the signal is of the form $\mathbf{x} = \mathbf{s}(\Theta) + \mathbf{w}$ with $\mathbf{s} = [s[0;\Theta] \ s[1;\Theta] \ \cdots \ s[N-1;\Theta]]^T$ being real and deterministic and $\mathbf{w}$ being i.i.d. real-valued Gaussian noise with zero mean and covariance matrix $\mathbf{C}_w = \sigma_w^2 \mathbf{I}$. Then $\mathbf{x}$ is Gaussian with covariance $\mathbf{C}_x = \mathbf{C}_w$ and mean $\boldsymbol{\mu} = \mathbf{s}$. Furthermore, for most problems of interest $\mathbf{C}_w$ and therefore $\mathbf{C}_x$ does not depend on Θ, so that Eq. (A.82) reduces to[3]

$$[\mathbf{I}(\Theta)]_{ij} = \frac{1}{\sigma_w^2} \left[\frac{\partial \mathbf{s}(\Theta)}{\partial \Theta_i} \right]^T \cdot \left[\frac{\partial \mathbf{s}(\Theta)}{\partial \Theta_j} \right]$$

$$= \frac{1}{\sigma_w^2} \sum_{n=0}^{N-1} \frac{\partial s[n;\Theta]}{\partial \Theta_i} \frac{\partial s[n;\Theta]}{\partial \Theta_j} \quad \begin{array}{l} \text{(real-valued signal in real} \\ \text{AWGN, multiple parameters)} \end{array} \quad (A.84)$$

A.3.5 Complex Signals and Parameters in AWGN

Similar results can be derived for the case where the signal and noise are complex-valued and the parameters of interest are real or complex. One way to derive the needed results defines the derivative of a complex function with respect to a complex variable. This can be done in more than one way, but all produce some surprising results. For example, if Θ is a complex variable, the definition used in Kay (1993) results in $\partial(|\Theta|^2)/\partial \Theta = \Theta^*$ and $\partial \Theta^*/\partial \Theta = 0$ Another method is to treat complex parameters as two real parameters, $\Theta_R = \text{Re}\{\Theta\}$ and $\Theta_I = \text{Im}\{\Theta\}$. This allows the mixture of both real and complex parameters in Θ and the use of conventional calculus. Either method, used correctly, produces the same results. The latter approach is used here.

The general CRLB in terms of PDFs given in Eq. (A.69) or Eq. (A.71) for the scalar case and Eqs. (A.79) and (A.81) in the vector case still apply. For a complex signal dependent on a real parameter vector Θ in complex AWGN the Fisher information matrix can be shown to be (Kay, 1993, Chap. 15)

$$[\mathbf{I}(\Theta)]_{ij} = \frac{2}{\sigma_w^2} \text{Re} \left\{ \left[\frac{\partial \mathbf{s}(\Theta)}{\partial \Theta_i} \right]^H \cdot \left[\frac{\partial \mathbf{s}(\Theta)}{\partial \Theta_j} \right] \right\} = \frac{2}{\sigma_w^2} \text{Re} \left\{ \sum_{n=0}^{N-1} \left[\frac{\partial s[n;\Theta]}{\partial \Theta_i} \right]^* \left[\frac{\partial s[n;\Theta]}{\partial \Theta_j} \right] \right\} \quad (A.85)$$

---

[3]An exception would be if the noise variance σ_w^2 was the parameter to be estimated.

Because Θ is real, $\partial s^*[n;\Theta]/\partial \Theta_i = \{\partial s[n;\Theta]/\partial \Theta_i\}^*$. The diagonal elements, which can be compared to the single parameter real case of Eq. (A.78), are

$$[\mathbf{I}(\Theta)]_{ii} = \frac{2}{\sigma_w^2} \sum_{n=0}^{N-1} \left|\frac{\partial s[n;\Theta]}{\partial \Theta_i}\right|^2 \quad \text{(complex signals in complex AWGN, real parameters)} \quad \text{(A.86)}$$

The CRLBs for the individual parameters are still given by Eq. (A.81).

Comparing Eqs. (A.85) and (A.86) to Eq. (A.84) shows that the CRLBs for signals in AWGN have the same general form in the complex case as in the real case but are smaller by a factor of 2. Thus, the real CRLB is *not* a special case of the complex CRLB.

Because a complex parameter Θ is treated as two real parameters Θ_R and Θ_I, the results above produce the CRLBs for Θ_R and Θ_I. However, the CRLB for Θ is usually the quantity of interest. It is easy to show that $\sigma_\Theta^2 = \sigma_{\Theta_R}^2 + \sigma_{\Theta_I}^2$, so the CRLB for Θ is simply the sum of the CRLBs for Θ_R and Θ_I.

A.3.6 Finding Minimum Variance Estimators

Equations (A.69) and (A.71) give the minimum variance of an unbiased estimator. Clearly, it would be good to know how to construct an estimator that achieves this minimum. While this is not always possible, it is possible to see what form it must take if it exists. The minimum variance will be achieved when the condition for equality in the Schwarz inequality is met. From Eqs. (A.66) and (A.67), this will occur when

$$A(\mathbf{x}) = (\hat{\Theta} - \Theta)\sqrt{p_{\mathbf{x}}(\mathbf{x}|\Theta)} = \alpha \frac{\partial \ln[p_{\mathbf{x}}(\mathbf{x}|\Theta)]}{\partial \Theta} \sqrt{p_{\mathbf{x}}(\mathbf{x}|\Theta)} = B(\mathbf{x}) \quad \text{(A.87)}$$

so that the estimator takes the form

$$\hat{\Theta} = \Theta + \alpha \frac{\partial \ln[p_{\mathbf{x}}(\mathbf{x}|\Theta)]}{\partial \Theta} \quad \text{(A.88)}$$

for some α.

This estimator has a serious problem: the estimate $\hat{\Theta}$ depends on knowing the actual value Θ of the parameter being estimated! Obviously, if Θ is known, there is no need to estimate it. In many cases, but not all, this problem can be solved by judicious choice of the scalar α.

To clarify, consider a minimum variance estimate $\hat{A}$ of the mean A in the constant-plus AWGN example of Sec. 7.1. Using the intermediate result in Eq. (7.13) so that the dependence on the data x_i is explicit in Eq. (A.88) gives

$$\hat{A} = A + \alpha\left\{\frac{1}{\sigma_w^2}\sum_{i=0}^{N-1}(x_i - A)\right\} = \left(1 - \frac{\alpha N}{\sigma_w^2}\right)A + \frac{\alpha}{\sigma_w^2}\sum_{i=0}^{N-1} x_i \quad \text{(A.89)}$$

This estimate is indeed unbiased for any choice of α:

$$\mathrm{E}\{\hat{A}\} = A + \frac{\alpha}{\sigma_w^2}\left[\sum_{i=0}^{N-1} \mathrm{E}\{x_i - A\}\right] = A \quad \text{(A.90)}$$

The dependence of $\hat{A}$ on A can be removed by choosing $\alpha = \sigma_w^2/N$, giving the unbiased minimum variance estimator

$$\hat{A} = \frac{1}{N}\sum_{i=0}^{N-1} x_i \qquad (A.91)$$

Note that the required choice of α in fact was the CRLB for this problem. It is true in general that $\alpha = I^{-1}(\Theta)$, the CRLB, so that Eq. (A.88) is more specifically

$$\hat{\Theta} = \Theta + \frac{1}{I(\Theta)}\frac{\partial \ln[p_x(\mathbf{x}|\Theta)]}{\partial \Theta} \qquad (A.92)$$

Equation (A.92) can be generalized for the vector parameter case. The minimum variance estimator must be of the form

$$\hat{\Theta} = \Theta + \mathbf{I}^{-1}(\Theta)\frac{\partial \ln[p_x(\mathbf{x}|\Theta)]}{\partial \Theta} \qquad (A.93)$$

Equations (A.92) and (A.93) are both necessary and sufficient for the existence of a minimum variance estimator. Thus, an estimator that satisfies these equations, however it was found, is efficient.

A.4 Random Signals in Linear Systems

A rigorous definition of a *random signal* is subtle; see Papoulis and Pillai (2001) for a discussion. For this text, it is adequate to define a random signal as a continuous or discrete signal $x(t)$ or $x[n]$ whose value at each instant of t or n is a random variable. If the PDF of that random variable is the same for each value of t or n, the random signal is *stationary*. The principal concern here is the effect of linear shift-invariant (LSI) systems such as filters on the properties of a random signal. Discrete signal notation will be used, but all the results translate readily to the continuous case.

A.4.1 Correlation Functions

The mean of a stationary random process x is denoted $\mathbf{E}\{x\}$, m_x, or $\bar{x}$, as convenient. The (stochastic) *cross-correlation function* of two possibly complex random signals $x[n]$ and $y[n]$ is defined as

$$s_{xy}[k] = \mathbf{E}\{x[n]y^*[n+k]\} \qquad (A.94)$$

The variable k is called the correlation *lag*. This definition is obviously very similar to the deterministic autocorrelation function described in App. B, with a stochastic expectation instead of a deterministic averaging.

The *autocorrelation function* (ACF) results when $y = x$, that is, x is correlated with itself:

$$s_x[k] = \mathbf{E}\{x[n]x^*[n+k]\} \qquad (A.95)$$

The *normalized ACF* is defined as

$$\rho_{xx}[k] = \frac{1}{s_x[0]}s_x[k] \qquad (A.96)$$

The ACF has several useful properties. The average power of x is given by the zero lag:

$$s_x[0] = \mathbf{E}\{x[n]x^*[n]\} = \mathbf{E}\{|x[n]|^2\} \tag{A.97}$$

s_x has a symmetry property

$$s_x[-k] = s_x^*[k] \tag{A.98}$$

a "shape property"

$$|s_x[k]| \leq s_x[0] \tag{A.99}$$

and in many cases of interest an asymptotic limit property,

$$\lim_{k \to \infty} \{s_x[k]\} = |m_x|^2 \tag{A.100}$$

The shape property states the ACF always has a maximum at the zero lag; it can reach the same amplitude at other lags but can never exceed it. It also implies that the normalized ACF has a maximum value of one at zero lag and is bounded between ±1 at other lags. An infinite-duration sinusoid is an example of a signal that matches the maximum autocorrelation amplitude at lags other than zero, specifically at lags equal to integer multiples of the signal period. The asymptotic limit property applies only to signals whose samples tend to become uncorrelated as the time between them (the lag) becomes large. This is true for most signals of interest, but the infinite-duration sinusoid is an example of a signal, which does not have this behavior. If a signal does exhibit this decorrelation for large lags, the limit property states that the ACF approaches the square of the process mean in the limit. For zero mean random processes, this means that the ACF must decay to zero.

A.4.2 Correlation and Linear Estimation

A frequent problem is estimating a random variable y based on measurements of another random variable x. An example would be estimation of a random process at sample n_2 based on its value at n_1. A common approach is to find the minimum mean-squared error (MMSE) estimator, which is the estimate $\hat{y}$ that minimizes the mean-squared error (MSE):

$$\varepsilon = \mathbf{E}\{(y - \hat{y})^2\} \tag{A.101}$$

Only real-valued variables are considered for now. It is frequently convenient to restrict the estimator to be linear so that $\hat{y}$ is of the form

$$\hat{y} = ax + b \tag{A.102}$$

Specifying the estimator is then a matter of specifying the constants a and b. Linear estimators are convenient because a and b can be found by solving linear equations and because they will turn out to depend only on the second-order moments of x and y, whereas more general estimators require knowledge of the joint PDFs of x and y, information that is less likely to be known.

The MMSE linear estimator can be found by using Eq. (A.102) in Eq. (A.101) and then setting the derivatives of ε with respect to a and b equal to zero so as to find its minimum value. The expectation and partial derivative operators commute, so

$$\begin{aligned} \frac{\partial \varepsilon}{\partial a} &= -2\mathbf{E}\{xy\} + 2a\mathbf{E}\{x^2\} + 2b\mathbf{E}\{x\} = 0 \\ \frac{\partial \varepsilon}{\partial b} &= -2\mathbf{E}\{y\} + 2a\mathbf{E}\{x\} + 2b = 0 \end{aligned} \tag{A.103}$$

It is convenient to first solve the second equation for b and then use that result in the first equation to get a. The result is (using the overbar notation for expected values)

$$a = \frac{\overline{xy} - \overline{x} \cdot \overline{y}}{\sigma_x^2}, \quad b = \overline{y} - a\overline{x} \tag{A.104}$$

This solution for a is reminiscent of the definition of the normalized *correlation coefficient* between two random variables x and y, which is

$$\rho_{xy} = \frac{E\{(x - \overline{x})(y - \overline{y})\}}{\sigma_x \sigma_y} = \frac{\overline{xy} - \overline{x} \cdot \overline{y}}{\sigma_x \sigma_y} \tag{A.105}$$

Consequently, $a = (\sigma_y/\sigma_x)\rho_{xy}$. Combining these results gives the MMSE linear estimator as

$$\hat{y} = \rho_{xy} \frac{\sigma_y}{\sigma_x}(x - \overline{x}) + \overline{y} \tag{A.106}$$

It is instructive to consider two limiting cases. It can be shown that $|\rho_{xy}|^2 \leq 1$ (Hayes, 1996). If x and y are uncorrelated, then $\rho_{xy} = 0$ and $\hat{y} = \overline{y}$. The lack of correlation implies that knowing x gives no information regarding the value of y, so the best linear estimate is just to guess the mean value of y. At the opposite extreme, suppose $\rho_{xy} = 1$, meaning that x is highly correlated with y. In this event, the best linear estimate of y is obtained by adding the deviation of x from its mean, scaled by the ratio of standard deviations of x and y, to the mean of y.

An important example occurs when the variables x and y are samples of the same random process at different times, that is, $x = w[n]$ and $y = w[n + k]$. If the random process is stationary, then $\sigma_x = \sigma_y$ and ρ_{xy} is the normalized autocorrelation function evaluated at lag k, $\rho_w[k]$. Equation (A.106) then becomes

$$\widehat{w[n + k]} = \rho_w[k](w[n] - \overline{w}) + \overline{w} \\ = \rho_w[k]w[n] + (1 - \rho_w[k])\overline{w} \tag{A.107}$$

If ρ_w evaluated at a lag of m samples is zero, the MMSE estimate of $w[n + k]$ is just the mean of w. If $\rho_w[k] = 1$, the MMSE estimate of $w[n + k]$ is $w[n]$. Thus, a high value of the normalized ACF predicts that the value of the future sample will be close to that of the current sample. A small value of ρ does not indicate that the future value will necessarily be different from the current value, but rather that it cannot be predicted based on the current value.

A.4.3 Power Spectrum

The *power spectral density* (PSD) or *power spectrum* of a random process is the Fourier transform (DTFT in the discrete case) of the autocorrelation function:

$$S_x(\omega) = \mathbf{F}\{s_x[k]\} \tag{A.108}$$

Its properties include

$$E\{|x|^2\} = s_x[0] = \frac{1}{2\pi} \int_{-\pi}^{\pi} S_x(\omega) d\omega$$

$$S_x^*(\omega) = S_x(\omega) \\ S_x(-\omega) = S_x(\omega) \text{ if } x(n) \text{ is real} \\ S_x(\omega) \geq 0 \tag{A.109}$$

The first property follows from Parseval's theorem and states that the average power of a random process equals the integral of the PSD, which tends to explain the term PSD. The second property states that the PSD is real-valued, even if x is complex. The third states that when x is real, the PSD is symmetric about the origin. The fourth states that it is nonnegative, again consistent with the idea of a measure of power versus frequency. These last three properties are also seen in the magnitude-squared of the Fourier transform of any deterministic signal.

A.4.4 White Noise

White noise is a random signal having the specific ACF (discrete case)

$$s_x[k] = \sigma_x^2 \delta[k] \tag{A.110}$$

where $\delta[k]$ is the discrete impulse function. It follows from Eq. (A.108) that the PSD of white noise is

$$S_x(\omega) = \sigma_x^2 \tag{A.111}$$

Thus the PSD of white noise is a constant for all frequencies. This fact accounts for the term "white noise," in analogy to the idea that white light contains all visible colors (wavelengths or frequencies) in equal proportion.

White noise is necessarily zero mean. To see this, suppose x has a nonzero mean m_x. Then x could be written as the sum of a zero-mean term $\tilde{x}$ and m_x, $x = \tilde{x} + m_x$. The autocorrelation would take the form

$$s_x[k] = s_{\tilde{x}}[k] + |m_x|^2 \tag{A.112}$$

The constant term in s_x will result in an impulse $2\pi |m_x|^2 \delta_D(\omega)$ in $S_x(\omega)$, where $\delta_D(\omega)$ is the continuous-variable Dirac impulse function. The resulting PSD is not a constant for all ω, so x cannot be white noise.

A.4.5 The Effect of LSI Systems on Random Signals

Suppose a random signal $x[n]$ is passed through an LSI system characterized by its impulse response $h[n]$ and frequency response $H(\omega)$. How are characteristics such as the mean or the power spectrum altered?

The output $y[n]$ is given by the convolution sum:

$$y[n] = \sum_{l=-\infty}^{\infty} h[l] x[n-l] \tag{A.113}$$

The mean of y, m_y, is

$$\begin{aligned} m_y &= \mathrm{E}\left\{ \sum_{l=-\infty}^{\infty} h[l] x[n-l] \right\} = \sum_{l=-\infty}^{\infty} h[l] \, \mathrm{E}\{x[n-l]\} \\ &= m_x \sum_{l=-\infty}^{\infty} h[l] = m_x H(0) \end{aligned} \tag{A.114}$$

If m_x is thought of as the "DC value" ("direct current," the zero frequency component) of x, then the DC value of y is just m_x multiplied by the frequency response at DC.

The ACF of y is

$$\begin{aligned}
s_y[k] &= \mathrm{E}\{y[n]y^*[n+k]\} \\
&= \mathrm{E}\left\{\left(\sum_{l=-\infty}^{\infty} h[l]\,x[n-l]\right)\left(\sum_{m=-\infty}^{\infty} h^*[m]\,x^*[n+k-m]\right)\right\} \\
&= \sum_{l=-\infty}^{\infty} h[l] \sum_{m=-\infty}^{\infty} h^*[m]\,\mathrm{E}\{x[n-l]x^*[n+k-m]\} \\
&= \sum_{l=-\infty}^{\infty} h[l] \sum_{m=-\infty}^{\infty} h^*[m]\,s_x[k-m+l]
\end{aligned} \tag{A.115}$$

Making the change of variables $p = m - l$ gives

$$\begin{aligned}
s_y[k] &= \sum_{l=-\infty}^{\infty} h[l] \sum_{p=-\infty}^{\infty} h^*[l+p] s_x[k-p] = \sum_{p=-\infty}^{\infty}\left(\sum_{l=-\infty}^{\infty} h[l]h^*[l+p]\right) s_x[k-p] \\
&= \sum_{p=-\infty}^{\infty} s_h[p] s_x[k-p] = s_h[k] * s_x[k]
\end{aligned} \tag{A.116}$$

The output autocorrelation is the stochastic autocorrelation of the input random signal convolved with the deterministic autocorrelation of the LSI system impulse response. The relationship between power spectra immediately follows

$$S_y(\omega) = S_h(\omega) \cdot S_x(\omega) = |H(\omega)|^2 S_x(\omega) \tag{A.117}$$

The power in the output random signal can be found from Eqs. (A.97), (A.109), and (A.117):

$$\mathrm{E}\{|y[n]|^2\} = s_y[0] = \frac{1}{2\pi}\int_{-\pi}^{\pi} S_y(\omega)\,d\omega = \frac{1}{2\pi}\int_{-\pi}^{\pi} |H(\omega)|^2 S_x(\omega)\,d\omega \tag{A.118}$$

The white noise input case is of special interest. If $x[n]$ is a white noise signal with power σ_x^2 and ACF $\sigma_x^2\,\delta[k]$, the autocorrelation of $y[n]$ will be

$$s_y[k] = s_h[k] * s_x[k] = s_h[k] * \sigma_x^2 \delta[k] = \sigma_x^2 s_h[k] \tag{A.119}$$

and the output power will simplify to

$$\sigma_y^2 = s_y[0] = \sigma_x^2 s_h[0] = \sigma_x^2\left(\sum_{n=-\infty}^{\infty} |h[n]|^2\right) \tag{A.120}$$

An important consequence is that if the input to an LSI system is a white random signal, the output will no longer be white (assuming the system is nontrivial, $h[n] \neq \delta[n]$). This is illustrated in the following example.

Suppose $h[n]$ is the impulse response of a four-point averager so that $h[n] = 1$ for $n = 0$, 1, 2, 3 and zero otherwise. Let $x[n]$ be 100 samples of WGN with $\sigma_x^2 = 1$ and pass it through the averaging filter to get $y[n]$. The normalized autocorrelation and PSD of the random data are computed at the input and output of the filter. Since these functions are themselves random because of the random input, the experiment is repeated 100 times and the various sample ACFs and PSDs are averaged to get the results shown in Fig. A.6. Figure A.6a shows the average normalized ACF of the inputs, which well-approximates the ideal unit-amplitude impulse. Shown in Fig. A.6b is the average sample PSD, which well-approximates the

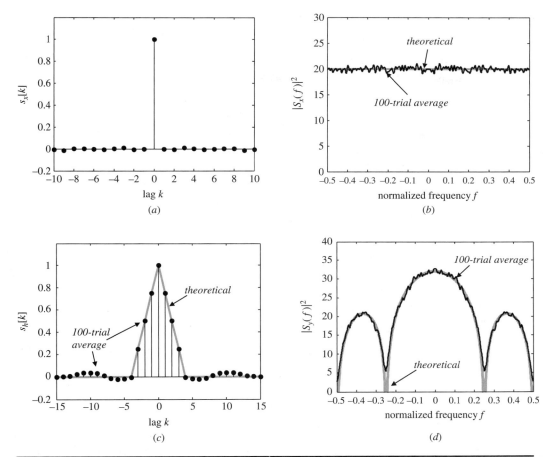

Figure A.6 Effect of four-point averaging filter on 100 samples of white noise. Averages of 100 sample realizations shown. (a) Normalized input autocorrelation $s_x[k]$. (b) Input PSD $S_x(f)$ and ideal white-noise PSD. (c) Normalized output autocorrelation $s_y[k]$ and theoretical result $s_h[k]$. (d) Output PSD $S_y(f)$ and theoretical result $S_h(f) = |H(f)|^2$.

theoretical white noise spectrum also shown. Figure A.6c shows that the average output normalized ACF closely matches the theoretical result, which is just $s_h[k]$. Finally, Fig. A.6d shows that the average output PSD closely matches the expected result $S_h(\omega) = |H(\omega)|^2$.

As another example, continuous-time white noise with a power spectral density of σ_x^2 passed through a unity gain ideal lowpass filter with a total bandwidth of β Hz (cutoff frequency of $\pm \beta/2$ Hz) will have a PSD that is white only over $\pm \beta/2$ Hz. The corresponding ACF is

$$s_y(t) = \sigma_x^2 \frac{\sin(\pi \beta t)}{\pi t} = \sigma_x^2 \beta \cdot \mathrm{sinc}(\pi \beta t) \tag{A.121}$$

where the sinc function is defined as $\mathrm{sinc}(z) = \sin(\pi z)/\pi z$. The filtered white noise is still zero mean but is no longer white. However, because it is bandlimited to β Hz the ACF of the output noise is zero at lags of integer multiples of $1/\beta$ seconds. Consequently, the samples of a discrete-time sequence obtained by sampling the filtered noise at the Nyquist rate of β samples per second will be uncorrelated with one another, so that the discrete-time ACF will be $\sigma_y^2 = \sigma_x^2 \beta \delta[k]$. The sampled noise is therefore white. Oversampling will result in noise having a sampled sinc function ACF and therefore a PSD that is bandlimited to some degree on the normalized frequency scale.

References

Hayes, M. H., *Statistical Digital Signal Processing and Modeling*. Wiley, New York, 1996.

Jakeman, E., and P. Pusey, "A Model for Non-Rayleigh Sea Echo," *IEEE Transactions Antennas and Propagation*, vol. 24, no. 6, pp. 806–814, Nov. 1976.

Kay, S. M., *Fundamentals of Statistical Signal Processing, Vol. 1: Estimation Theory*. Prentice-Hall, Upper Saddle River, New Jersey, NJ, 1993.

Olver, F. W. J., D. W. Lozier, R. F. Boisvert, and C. W. Clard (eds.), *NIST Handbook of Mathematical Functions*. Cambridge University Press, New York, 2010.

Omura, J., and T. Kailath, "Some Useful Probability Distributions," Technical Report 7050-6, Stanford Electronics Laboratories, Sep. 1965. Available as free downloadable PDF from the U.S. Defense Technical Information Center at www.dtic.mil as report AD747128.

Papoulis, A., and S. U. Pillai, *Probability, Random Variables and Stochastic Processes*, 4th ed. McGraw-Hill, New York, 2001.

Peebles, P. Z., Jr., *Radar Principles*. Wiley, New York, 1998.

Van Trees, H. L., *Detection, Estimation, and Modulation Theory, Part 1*. Wiley, New York, 1968, p. 338.

Watts, S., "Radar Detection Prediction in Sea Clutter Using the Compound K-Distribution Model," *IEE Proceedings*, vol. 132, pt. F, no. 7, pp. 613–620, Dec. 1985.

APPENDIX B

Selected Topics in Digital Signal Processing

A few basic concepts and signal processing operations appear over and over again, even if sometimes in disguise, in radar signal processing. In this section, several are reviewed with special emphasis on a few aspects not often emphasized in some other application fields.

B.1 Fourier Transforms

The Fourier transform is as ubiquitous in radar signal processing as in most other signal processing fields. Frequency domain representations are often used to separate desired signals from interference; the Doppler shift is a frequency domain phenomenon of critical importance; and it will be seen that in some radar systems, especially imaging systems, the collected data is related to the desired end product by a Fourier transform.

Both continuous and discrete signals are of interest and therefore Fourier transforms are required for both. Consider a signal $x(u)$ that is a function of a continuous variable in one dimension called the *signal domain*.[1] Its Fourier transform, denoted $X(\Omega)$, is given by

$$X(\Omega) = \int_{-\infty}^{\infty} x(u) \exp(-j\Omega u)\, du, \quad \Omega \in (-\infty, \infty) \tag{B.1}$$

and is said to be a function in the *Fourier domain*. The inverse transform is

$$x(u) = \frac{1}{2\pi} \int_{-\infty}^{\infty} X(\Omega) \exp(+j\Omega u)\, d\Omega, \quad u \in (-\infty, \infty) \tag{B.2}$$

In Eqs. (B.1) and (B.2) the frequency variable Ω is in radians per unit of u. For example, if u is time then Ω is the usual radian frequency in units of radians per second; if u is a spatial variable in meters, then Ω is spatial frequency in units of radians per meter. While it is most common for the signal and Fourier domains to correspond to time and frequency respectively, that is not always the case. For instance, in radar $x(u)$ can be an antenna aperture current distribution, with the signal domain variable u being the aperture position in wavelengths; then the Fourier domain variable Ω is related to the signal angle of arrival relative to the antenna boresight direction (see Sec. 1.3.3).

---

[1] To unify discussion of sampling in time, frequency, and space and to maintain generality, in this discussion the signal to be sampled will be referred to as a function in the *signal domain*, and its Fourier transform as a function in the *Fourier domain*.

659

An equivalent transform pair using a cyclical frequency variable $F = \Omega/2\pi$ is

$$X(F) = \int_{-\infty}^{\infty} x(u)\exp(-j2\pi Fu)\, du, \quad F \in (-\infty, \infty) \tag{B.3}$$

$$x(u) = \int_{-\infty}^{\infty} X(F)\exp(+j2\pi Fu)\, dF, \quad u \in (-\infty, \infty) \tag{B.4}$$

If the signal domain is time, F is in cycles per second, or hertz. Ω and F will sometimes be referred to as *analog frequencies*, representing their correspondence to continuous-variable signals.

There are many excellent textbooks on Fourier transforms and their properties. Two classics are Papoulis (1987) and Bracewell (1999). Papoulis uses primarily the radian frequency notation, while Bracewell uses cyclical frequency.

A useful example in radar is the Fourier transform of a constant-frequency complex exponential pulse. Define

$$x(t) = \begin{cases} A\exp(j\Omega_0 t), & -\tau/2 \le t \le \tau/2 \\ 0, & \text{otherwise} \end{cases} \tag{B.5}$$

Inserting Eq. (B.5) into the definition Eq. (B.1) gives

$$\begin{aligned} X(\Omega) &= \int_{-\tau/2}^{+\tau/2} A\exp(+j\Omega_0 t)\exp(-j\Omega t)\, dt = A\int_{-\tau/2}^{+\tau/2} \exp[-j(\Omega - \Omega_0)t]\, dt \\ &= \frac{A}{j(\Omega - \Omega_0)}\left\{\exp\left[+j(\Omega - \Omega_0)\frac{\tau}{2}\right] - \exp\left[-j(\Omega - \Omega_0)\frac{\tau}{2}\right]\right\} \end{aligned} \tag{B.6}$$

Applying Euler's formula gives the result

$$\begin{aligned} X(\Omega) &= \frac{2A}{(\Omega - \Omega_0)}\sin\left[(\Omega - \Omega_0)\frac{\tau}{2}\right] \\ &= A\tau\frac{\sin[\pi\tau(F - F_0)]}{\pi\tau(F - F_0)} \equiv A\tau\,\text{sinc}\,[\tau(F - F_0)] \end{aligned} \tag{B.7}$$

where $\text{sinc}(z) \equiv \sin(\pi z)/\pi z$ is consistent with the MATLAB® definition of the sinc function. The main portion of this function is shown in Fig. B.1 for $F_0 = 5$ MHz and $\tau = 1$ μs. The peak occurs at $F = F_0$ and has a value of $A\tau$. It is easy to see from Eq. (B.7) that the Rayleigh width (peak to first null) is $1/\tau$ Hz. It can be shown numerically that the 3 dB width is $0.89/\tau$ Hz and that the magnitude of the largest sidelobe (first negative peak) relative to the mainlobe magnitude is -13.26 dB.

Conditions on a signal for the existence of its Fourier transform are discussed in Bracewell (1999). For the purposes of this text, it can be assumed that the Fourier transform of any signal of interest exists, and that the signal and its Fourier transform form a unique one-to-one pair.

The Fourier transform for continuous signals is important for some analyses, particularly those relating to establishing sampling requirements, but most actual processing will be performed with discrete-variable signals. There are two classes of Fourier transforms for

Selected Topics in Digital Signal Processing

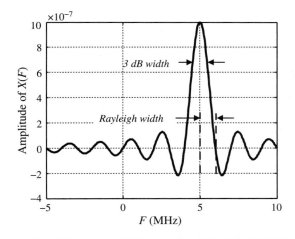

FIGURE B.1 The Fourier transform of a 1-μs long, 5-MHz complex exponential pulse.

discrete-variable signals. Directly analogous to the continuous-variable case is the following transform pair for a discrete-variable signal $x[m]$:[2]

$$X(\omega) = \sum_{m=-\infty}^{\infty} x[m]\exp(-j\omega m), \quad \omega \in (-\infty, \infty) \tag{B.8}$$

$$x[m] = \frac{1}{2\pi} \int_{-\pi}^{\pi} X(\omega)\exp(+j\omega m)\,d\omega, \quad m \in [-\infty, \infty] \tag{B.9}$$

In this pair, m is the signal index in units of samples and ω is a *continuous* frequency variable with units of radians per sample (not radians per second or radians per meter).[3] The frequencies ω and f will sometimes be referred to casually as *digital* frequencies because the corresponding signals are discrete. They are also called *normalized* frequencies for reasons to be explained in Sec. B.3.

A related Fourier transform pair using a normalized cyclical frequency $f = \omega/2\pi$ in cycles per sample is given by

$$X(f) = \sum_{m=-\infty}^{\infty} x[m]\exp(-j2\pi f m), \quad f \in (-\infty, \infty) \tag{B.10}$$

---

[2] The fairly common convention in digital signal processing texts of enclosing the independent variable in square brackets when it is discrete and parentheses when it is continuous will be followed. Thus $x(u)$ is a function of a continuous variable, while $x[m]$ is a function of a discrete variable.

[3] Another fairly common, though definitely not universal, DSP convention observed here is to use upper case variables to denote analog frequencies (F or Ω, in hertz or radians per second if the signal domain variable represents time) and lower case variables to denote "digital" frequency variables (f or ω). Many DSP texts consider the signal domain index m to be unitless and therefore the units of f or ω to be simply cycles or radians. The preference here is to consider m to have units of samples so that f and ω have units of cycles per sample or radians per sample.

$$x[m] = \int_{-0.5}^{0.5} X(f)\exp(+j2\pi fm)\,df, \quad m \in [-\infty, \infty] \tag{B.11}$$

The function $X(\omega)$ or $X(f)$ is called the *discrete-time Fourier transform* (DTFT) of $x[m]$. It is readily seen from Eq. (B.8) that $X(\omega)$ is continuous in the frequency variable ω with a period of 2π radians per sample; that is, the DTFT repeats itself every 2π radians per sample. Consequently, though it is defined for all ω, normally only the principal period $-\pi \le \omega \le \pi$ is discussed and illustrated. Similarly, $X(f)$ has a period of one cycle per sample and the principal period is $-0.5 \le f < 0.5$. The properties of DTFTs are described in modern digital signal processing textbooks, for example Oppenheim and Schafer (2010).

Similar to the continuous-time case, the DTFT of a sampled constant-frequency complex exponential is of interest. Defining the waveform as

$$x[m] = A\exp(j\omega_0 m), \quad 0 \le m \le M-1 \tag{B.12}$$

a derivation very similar to that leading to Eq. (B.7) gives

$$\begin{aligned}X(\omega) &= A\frac{\sin[(\omega-\omega_0)M/2]}{\sin[(\omega-\omega_0)/2]}\exp[-j(\omega-\omega_0)(M-1)/2] \\ X(f) &= A\frac{\sin[\pi(f-f_0)M]}{\sin[\pi(f-f_0)]}\exp[-j\pi(f-f_0)(M-1)]\end{aligned} \tag{B.13}$$

The result is shown for $f_0 = 0.285$, $M = 20$, and $A = 1$ as the solid curve in Fig. B.2a or b. Examination of Eq. (B.13) and Fig. B.2 shows that this DTFT is very similar to the sinc function of Eq. (B.7) in appearance, especially around the mainlobe. Unlike the sinc function, however, the sidelobes do not decay indefinitely because of the requirement that the DTFT be periodic in frequency. The phase term in Eq. (B.13) arises because the pulse was not centered on the origin; if M were odd and it was centered, the phase term would not be present.

While the DTFT is well-suited for mathematical analysis of discrete-time signals, $X(\omega)$ and $X(f)$ are not suitable for computation and manipulation in a digital processor because they cannot be computed for all of the uncountably infinite values of the continuous frequency variable ω or f. A finite, discrete set of frequency values is needed. The *discrete Fourier transform* (DFT, not to be confused with the DTFT) is a computable Fourier transform defined for *finite-length* discrete-variable signals. The K-point DFT and its inverse for an N-point signal $x[n]$ are given by[4]

$$X[k] = \sum_{n=0}^{N-1} x[n]\exp(-j2\pi nk/K), \quad k \in [0, K-1] \tag{B.14}$$

$$x[n] = \frac{1}{N}\sum_{k=0}^{K-1} X[k]\exp(+j2\pi nk/K), \quad n \in [0, N-1] \tag{B.15}$$

---

[4]The definitions of Eqs. (B.14) and (B.15) are unusual in that most DSP texts define the DFT with $K = N$ and introduce the idea of "zero padding" to account for the case where there are more DFT frequency samples than there are signal samples. The preference here is to distinguish these two numbers in the definition.

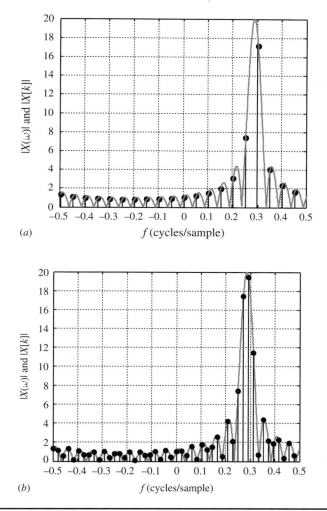

FIGURE B.2 Sampling of the DTFT of a 20-sample complex sinusoid by the DFT. (a) $K = 20$-point DFT. (b) $K = 48$-point DFT.

Inspection shows that for $x[n]$ of finite duration N samples, $X[k]$ is simply a sampled version of $X(\omega)$ or $X(f)$, with K frequency samples distributed uniformly across the period $[0, 2\pi)$ in ω or $[0, 1)$ in f,

$$X[k] = X(\omega)\big|_{\omega=\frac{2\pi}{K}k} = X(f)\big|_{f=\frac{1}{K}k}, \quad k \in [0, K-1] \tag{B.16}$$

Because the DTFT is periodic, so is the DFT, with period K samples.

It is usually preferable to plot the spectrum of a signal on a symmetric scale that places zero frequency in the center. Because of the periodicity of the DTFT, the DFT samples corresponding to frequencies in the interval $f \in [0.5, 1)$ cycles per sample are identical to those

Appendix B

Signal Variable	Transform Variable	Transform
Continuous	Continuous	Fourier Transform (FT)
Sample in Time		
Discrete	Continuous	Discrete-Time Fourier Transform (DTFT)
Sample in Frequency		
Discrete	Discrete	Discrete Fourier Transform (DFT)
Fast Algorithm for Computing the DFT		
Discrete	Discrete	Fast Fourier Transform (FFT)

TABLE B.1 Relationship between Fourier Transform Versions

in the interval $f \in [-0.5, 0)$. The upper half of the DFT samples ($k > \lceil K/2 \rceil$) are therefore often plotted first, followed by the lower half of the samples.[5]

Figure B.2 illustrates the difference between the DTFT and the DFT for a signal consisting of an $M = 20$ sample complex exponential pulse with a frequency $f = 0.285$ cycles per sample. The solid curves are the DTFT of the data over the interval $f \in [-0.5, 0.5]$. The small circles represent the DFT values, which have been rearranged as discussed above. In Fig. B.2a, a $K = 20$-point DFT was used, resulting in a sparsely sampled representation of the DTFT that misses the true peak frequency and amplitude by a significant amount and provides only a crude representation of the sidelobe structure. In Fig. B.2b, a $K = 48$-point DFT was used. The underlying DTFT is unchanged; it is determined only by the data $x[m]$. The larger DFT simply samples it on a denser frequency grid. A still larger DFT would trace out the details of the DTFT quite accurately.

Only the DFT frequency samples are actually available for digital processing and computation. Because the frequency of the DTFT peak does not generally coincide with one of those frequency samples, there is an apparent reduction in the peak value of the Fourier transform. This reduction is called a *straddle loss* (because the DFT samples "straddle" the true peak location). In Fig. B.2a the peak DTFT value is 20, but the peak 20-point DFT sample value is only 17.17, a reduction of 1.32 dB. Because the more densely sampled 48-point DFT in Fig. B.2b cannot miss the DTFT peak by as much, it exhibits a straddle loss of only 0.26 dB. The worst case occurs when $K = M$ and the DTFT peak is halfway between two DFT samples. In this case, the straddle loss is 3.9 dB.

So far three versions of the Fourier transform have been discussed: the original "analog" Fourier transform, the DTFT, and the DFT. Most readers will be aware of the *fast Fourier transform* (FFT). This is not a fourth variation, but an efficient algorithm for computing the DFT (Oppenheim and Schafer, 2010). Table B.1 summarizes the relationship between the various Fourier transforms.

B.2 Windowing

Equations (B.12) and (B.13) defined a complex sinusoidal pulse and its DTFT. Figure B.3a plots the magnitude of the DTFT for the case $\omega_0 = \pi/4$, $M = 20$ samples, and $A = 1$. So long as $M \geq 4$, the Rayleigh (peak-to-null) mainlobe bandwidth is $1/M$ cycles per sample; this is also the 4-dB bandwidth. The two-sided width of the mainlobe at the -3 dB points is $0.89/M$ cycles per sample.

---

[5]The MATLAB® function fftshift exists precisely to perform this rearrangement of the output of the fft function for plotting convenience.

Selected Topics in Digital Signal Processing

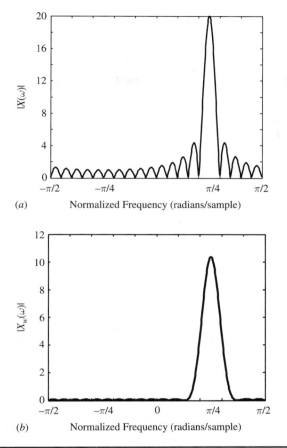

FIGURE B.3 Magnitude of the discrete-time Fourier transform of an ideal moving target slow-time data sequence with $\omega_0 = \pi/4$, $M = 20$ samples, and $A = 1$. (a) No window. (b) Hamming window.

These mainlobe width measures are the principal determinant of the resolution in the Fourier domain variable. Both are inversely proportional to the signal domain data extent M. Thus, resolution in the Fourier domain is determined by the extent of the signal domain data, as first discussed in Chap. 1. Longer time observations, larger antenna apertures, or longer synthetic apertures (see Chap. 8) result in finer resolution in frequency, angle, or spatial coordinates.

The first sidelobe of the DTFT is 13.2 dB below the response peak, often considered to be too high. It is common to use a data window to weight the slow-time data samples $x[m]$ prior to computing the DTFT or DFT.[6] To analyze the window's effect, replace $x[m]$ by $w[m]x[m]$ in Eq. (B.8). $X(\omega)$ becomes

$$X_w(\omega) = A \sum_{m=0}^{M-1} w[m] \exp[-j(\omega - \omega_0)m] = W(\omega - \omega_0) \qquad (\text{B.17})$$

where the notation $X_w(\omega)$ is used to emphasize that the spectrum is computed with a window applied to the data. This is simply the Fourier transform of the window function itself,

---

[6]Windowing of the data is also called *weighting, tapering, shading,* or *apodization* in various contexts.

shifted to be centered on the sinusoid frequency ω_0 rather than at zero. In fact, the asinc function of Fig. B.3a is also just the Fourier transform of the rectangular window function (equivalent to no window).

Figure B.3b illustrates the effect on the DTFT of applying a Hamming window to the same data used in Fig. B.3a. The sidelobe levels are greatly reduced, but at the cost of an increase in mainlobe width (degraded resolution) and a decrease in peak amplitude and SNR. It is straightforward to compute the reduction in peak amplitude and the SNR loss given the window function $w[n]$. Consider the peak gain first. From Eq. (B.13), the peak value of $|X(\omega)|^2$ when no window is used is $A^2 M^2$. Evaluating Eq. (B.17) at $\omega = \omega_0$ gives the peak power when a window is used:

$$|X_w(\omega_0)|^2 = \left| A \sum_{m=0}^{M-1} w[m] \exp[-j(0)m] \right|^2 = A^2 \left| \sum_{m=0}^{M-1} w[m] \right|^2 \tag{B.18}$$

The ratio $|X(\omega_0)|^2 / |X_w(\omega_0)|^2$, called the *loss in processing gain* (LPG), is

$$LPG = \frac{M^2}{\left| \sum_{m=0}^{M-1} w[m] \right|^2} \tag{B.19}$$

With this definition $LPG \geq 1$, so the loss in dB is a positive number. Using Eq. (B.19), the LPG can be computed for any window. Values of 3 to 7 dB are typical. Details depend on the specific window function, but the LPG is typically a weak function of the window length M, higher for small M and rapidly approaching an asymptotic value for large M (on the order of 100 or more).

While the window reduces the peak amplitude of the DTFT substantially, it also reduces noise power if the signal includes additive noise. A more significant metric than LPG is *processing loss* (PL), the reduction in SNR at the peak of the DTFT for a pure sinusoidal input in additive white noise. Denoting the SNR with and without the window as χ and χ_w, respectively, it is possible to separate the effects of the window on the target and noise components of the signal:

$$\frac{\chi}{\chi_w} = \frac{(S/N)}{(S_w/N_w)} = \left(\frac{S}{S_w}\right)\left(\frac{N_w}{N}\right) = LPG\left(\frac{N_w}{N}\right) \tag{B.20}$$

To determine the window's effect on the noise power, suppose $x[m]$ is a zero mean stationary white noise with variance σ_x^2. Then the windowed noise power is

$$\begin{aligned}
N_w &= E\left\{ \left(\sum_{m=0}^{M-1} w[m]x[m]\right)\left(\sum_{l=0}^{M-1} w^*[l]x^*[l]\right) \right\} \\
&= E\left\{ \left(\sum_{m=0}^{M-1} |w[m]x[m]|^2\right) + \left(\sum_{m=0}^{M-1}\sum_{\substack{l=0 \\ m \neq l}}^{M-1} w[m]w^*[l]x[m]x^*[l]\right) \right\} \\
&= \left(\sum_{m=0}^{M-1} |w[m]|^2 E\{|x[m]|^2\}\right) + \left(\sum_{m=0}^{M-1}\sum_{\substack{l=0 \\ m \neq l}}^{M-1} w[m]w^*[l]E\{x[m]x^*[l]\}\right) \\
&= \sigma_x^2 \left(\sum_{m=0}^{M-1} |w[m]|^2\right)
\end{aligned} \tag{B.21}$$

Window	Window Length M	3-dB Mainlobe Width (Relative to Rectangular Window)	LPG (dB Relative to Rectangular Window)	Peak Sidelobe, dB	PL, dB	Worst-Case Straddle Loss, dB
Rectangular	20	1.0	0.0	13.2	0.0	3.9
	100	1.0	0.0	13.3	0.0	3.9
Hann	20	1.71	6.5	31.5	2.0	1.3
	100	1.65	6.1	31.5	1.8	1.4
Hamming	20	1.52	5.7	40.4	1.5	1.6
	100	1.49	5.4	42.6	1.4	1.7
Kaiser, $\beta = 3$	20	1.26	3.6	25.3	0.7	2.4
	100	1.25	3.4	24.0	0.6	2.5
Dolph-Chebyshev 50-dB equiripple	20	1.55	5.8	50	1.6	1.6
	100	1.53	5.7	50	1.5	1.7

TABLE B.2 Properties of Some Common Data Windows

The last line results because $\mathbf{E}\{x[m]x^*[l]\} = 0$ when $m \neq l$ and σ_w^2 when $m = l$. The unwindowed noise power N can be obtained from Eq. (B.21) by setting $w[m] = 1$ for all m, giving $N = M\sigma_w^2$. Using Eq. (B.19) and these values for N and N_w in Eq. (B.20) gives the processing loss as

$$PL = \frac{M \sum_{m=0}^{M-1} |w[m]|^2}{\left| \sum_{m=0}^{M-1} w[m] \right|^2} \quad (B.22)$$

Like the loss in peak gain, the processing loss is a weak function of M that is higher for small M but quickly approaches an asymptotic value. As an example, for the Hamming window the loss in SNR is 1.75 dB for a very short ($M = 8$) window, decreasing asymptotically to about 1.35 dB for very long windows.

Table B.2 summarizes the four key properties of 3-dB mainlobe width, peak sidelobe level, loss in processing gain, and processing loss for several common windows. Also shown is worst-case straddle loss, discussed in the previous section.

Windows and their properties are used throughout radar signal processing. In this text, they will be seen in Chap. 4 in the context of linear frequency-modulated (LFM) signal range sidelobe control, in Chap. 5 in the context of the Doppler spectrum, in Chap. 8 in the context of radar imaging, and in Chap. 9 in the context of digital beamforming. An extensive description of common window functions and their characteristics is in Harris (1978).[7] That paper includes a table similar to Table B.2, but with many more types of windows and several additional metrics for each.

---

[7]Harris uses a slightly different definition of the windows than is now conventional in data analysis and simulation packages for reasons having to do with DFT symmetry properties. Effectively, his version of an M-point shaped window (e.g., Hamming) is the first M points of the $(M + 1)$-point symmetric window commonly used. The difference is of little consequence, especially as M gets large.

B.3 Sampling, Quantization, and A/D Converters

The subject of this text is *digital* processing of radar signals. A digital signal is one that is discretized in two independent ways. Both are necessary to represent the signal in a digital processor. The first discretization represents the signal as a function of a discrete, rather than continuous, variable: one discrete variable for one-dimensional signals, two variables for two-dimensional signals, and so forth. Most discrete-variable signals of interest here will be obtained by *sampling* a continuous-variable physical quantity at selected, usually equispaced, values of the independent variable. An example is a discrete-time sampled version of the continuous-time output voltage of an antenna and receiver, but sampled functions of frequency and spatial position are of concern as well. A discrete representation is necessary to allow a finite-length signal to be represented in a digital processor by a finite number of values.

The second discretization is *quantization* of the signal's values. Each sample of a continuously varying signal can take on an infinity of possible values. Quantization maps the continuous amplitude of a signal to one of a finite set of values so that each sample can be represented in the digital processor with a finite number of bits. The number of permissible values is determined by the number of bits available in the quantized signal representation and the encoding used. Figure B.4 illustrates the distinction between continuous, sampled, quantized, and digital (sampled *and* quantized) signals.

B.3.1 Sampling

The most fundamental question in developing a sampled-variable representation is "How many samples are enough?" That is, how should the sampling interval be chosen? The Nyquist sampling theorem provides the answer. It states that if the Fourier transform $X(F)$

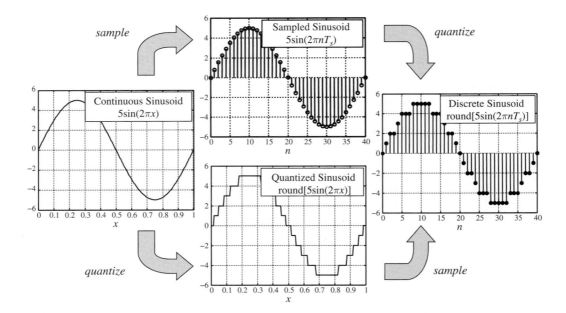

FIGURE B.4 Relationship between continuous signals, sampling, quantization, and discrete signals.

of a signal $x(u)$ is bandlimited to an interval in the Fourier domain of total width β_F cyclical frequency units (equivalently, β_Ω radian frequency units), then the signal can be recovered from a set of samples taken at a sampling interval

$$T_s < \frac{1}{\beta_F} = \frac{2\pi}{\beta_\Omega} \text{ seconds} \tag{B.23}$$

by an appropriate interpolation operation. That is, the sampling rate $F_s = 1/T_s$ must satisfy the simple relation

$$F_s > \beta_F \text{ samples/second} \tag{B.24}$$

This formula will be used in this text to establish sampling rates for pulse echoes, Doppler processing, and phased array antenna element spacing, among other things.

The Nyquist theorem is easy to derive. Details supporting the following outline are provided in Oppenheim and Schafer (2010). Start by modeling the process of sampling $x(u)$ as multiplication by an infinite Dirac impulse train:

$$x_s(u) = x(u) \left[\sum_{n=-\infty}^{+\infty} \delta_D(u - nT_s) \right] \tag{B.25}$$

T_s is the sampling interval in units of u. For instance, if the independent variable is time, $u = t$, T_s is in seconds. The sampled signal $x_s(u)$ then has the Fourier transform in cyclical frequency units given by

$$X_s(F) = \frac{1}{T_s} \sum_{k=-\infty}^{+\infty} X\left(F - \frac{k}{T_s}\right) = \frac{1}{T_s} \sum_{n=-\infty}^{+\infty} X(F - kF_s) \tag{B.26}$$

where $F_s = 1/T_s$ is the sampling rate. Equation (B.26) illustrates a very important effect of sampling: infinite *replication* of the original spectrum at an interval of F_s units in cyclical frequency ($2\pi F_s$ in radian units). This Fourier domain replication occurs any time a signal is sampled at regular intervals. Because the forward and inverse Fourier transforms are dual operations, this also occurs in reverse: if one samples the Fourier domain signal, then the signal is replicated in the original signal domain.

The fact that sampling in one domain results in replication in the complementary domain is an important property of Fourier analysis in its own right, independent of the Nyquist theorem that will follow. It provides the basis for an alternative to the usual modulation technique of time-domain multiplication by sinusoids for shifting a signal in the frequency domain and is used, for instance, for one method of efficient formation of digital in-phase and quadrature signals described in Chap. 3.

Replication also occurs when "sampling" a discrete-variable signal by decimation. Consider a signal $x[n]$ with DTFT $X(f)$. Define a *decimated* signal by subsampling $x[n]$:

$$y[n] = x[nM], \quad -\infty < n < +\infty, \quad M \text{ integer} \tag{B.27}$$

The signal $x[n]$ is said to be decimated by the factor M to produce $y[n]$. The DTFT of $y[n]$ is related to that of $x[n]$ according to (Oppenheim and Schafer, 2010):

$$Y(f) = \frac{1}{M} \sum_{k=0}^{M-1} X\left(\frac{f-k}{M}\right) \tag{B.28}$$

The summation is finite because the DTFT already repeats periodically. This is exactly the same type of scaling and replication of the spectrum seen also in the sampling of a continuous-variable signal.

Resuming the derivation of the Nyquist theorem, the next step is to relate the analog and digital signals and their spectra to one another. A discrete-time signal is formed from the sampled data by the simple assignment

$$x[n] = x(nT_s), \quad n \in [-\infty, +\infty] \tag{B.29}$$

The DTFT of $x[n]$, computed using Eq. (B.8) or Eq. (B.10), will be simply $X_s(F)$ expressed on a "digital" frequency scale,

$$X(f) = \frac{1}{T_s} \sum_{k=-\infty}^{+\infty} X((f+k)F_s) \tag{B.30}$$

in cyclical units. Comparing Eq. (B.26) with Eq. (B.30) shows that the analog and digital frequency scales are related according to $F = fF_s$, so that

$$\begin{aligned} f &= \frac{F}{F_s} = FT_s \\ \omega &= 2\pi \frac{F}{F_s} = \Omega T_s \end{aligned} \tag{B.31}$$

Equation (B.31) provides the basis for converting units between analog and digital frequencies. The first form, $f = F/F_s$, is the reason that the digital frequencies ω and f are often called *normalized* frequencies: f is the frequency in hertz normalized to the sampling frequency, while ω is its radian counterpart. Also, combining Eqs. (B.16) and (B.31) shows that in continuous-variable units the K-point DFT frequency sample locations are equivalent to frequencies of $k/KT_s = (k/K)F_s$ cycles per unit (hertz if T_s is in seconds) or $2\pi k/KT_s = 2\pi(k/K)F_s$ radians per unit, where k is the DFT sample index number and ranges from zero to $K-1$.

Figure B.5 illustrates the spectrum replication effect of sampling and the relation of the resulting DTFT $X(f)$ to the original spectrum $X(F)$. A notional Fourier transform $X(F)$ is shown in Fig. B.5a. $X(F)$ is not assumed to be bandlimited or to be centered at $F = 0$. Because $x(u)$ may be complex-valued, $X(F)$ also is not assumed to exhibit Hermitian symmetry. Figure B.5b shows the replicated analog spectrum $X_s(F)$, while Fig. B.5c is the corresponding DTFT $X(f)$ or $X(\omega)$. Note that $X(f)$ is periodic with period 1 and $X(\omega)$ with period 2π as required for spectra of discrete-variable signals.

The Nyquist theorem relies on the fact that a signal and its Fourier transform form a unique one-to-one pair. Consequently, reconstruction of the original spectrum $X(F)$ is equivalent to reconstructing $x(u)$. Consideration of Fig. B.5 makes clear the condition needed to enable recovery of the original spectrum: the replicas must not overlap. If each spectrum replica is distinct, then the sampled signal can be passed through a linear shift-invariant system (a filter) with a frequency response that selects only the original copy of the spectrum. The output of that system must then be the original signal.

This strategy leads to two conditions, one on the signal and one on the sampling rate. First, $x(u)$ must be strictly bandlimited to some finite bandwidth β_F. In cyclical frequency units,

$$X(F) \equiv 0, \quad F < F_c - \frac{\beta_F}{2} \text{ and } F > F_c + \frac{\beta_F}{2} \tag{B.32}$$

where F_c is the center of the region of support of $X(F)$. If this is not the case, no sampling rate will separate the spectrum replicas; they will always overlap. Assuming $x(u)$ is bandlimited,

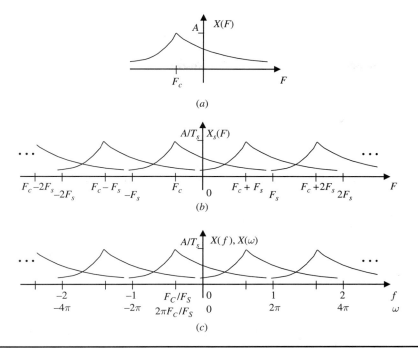

Figure B.5 Spectrum replication effect of sampling and relation between spectra. (a) Original non-bandlimited spectrum $X(F)$. (b) Spectrum of sampled signal, $X_s(F)$. (c) DTFT of sampled data, $X(F)$ or $X(\omega)$.

then the sampling rate must be chosen large enough to ensure that the spectrum replicas do not overlap. Figure B.6 shows a notional bandlimited spectrum and the corresponding sampled spectrum $X_s(F)$. By inspection, the condition needed to avoid overlap is $F_c + \beta_F/2 < F_c + F_s - \beta_F/2$, which gives the simple result $F_s > \beta_F$ already quoted in Eq. (B.24).

If the original $x(u)$ is a *baseband* signal, meaning its spectrum is centered at $F_c = 0$, then the region of support of $X(F)$ is $(-\beta_F/2, \beta_F/2)$. In this event, Eq. (B.24) expresses the conventional wisdom that the sampling frequency should be at least twice the highest frequency component in the signal. However, a more direct and versatile interpretation of Eq. (B.24) is that the sampling frequency should be greater than the total spectral width of the signal. This form of the rule is more easily applied to non-baseband signals.

In outlining the derivation of the Nyquist theorem, no assumption has been made that $x(u)$ is real-valued. The theorem applies equally well to real or complex signals. In the case of a complex signal of total spectral width β_F Hz, the Nyquist criterion implies collecting at least β_F *complex* samples per second, equivalent to $2\beta_F$ real samples per second.

It was also not assumed that the original signal spectrum was baseband, that is, centered at $F_c = 0$. The spectrum can be offset to any location on the frequency axis without changing the required sampling rate; only the bandwidth matters. However, if the original spectrum is not centered at zero, none of the replicas will necessarily be centered at zero in the discrete-variable signal's spectrum unless appropriate relationships between the sampling frequency, spectrum bandwidth, and spectrum offset are maintained. In the end, it is usually desirable to have the information-bearing portion of the spectrum (the gray shaded part of Fig. B.6a) end up centered at the origin for ease of processing.

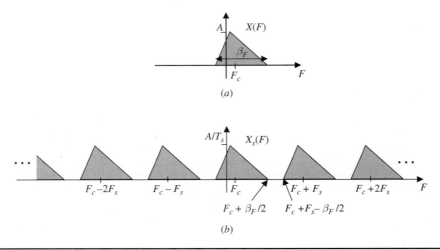

FIGURE B.6 Illustration of Nyquist sampling requirement for bandlimited signals. (a) Original bandlimited spectrum $X(F)$. (b) Spectrum of sampled signal, $X_s(F)$.

The Nyquist theorem is not peculiar to sampling of time domain signals. It is a consequence of the replication effect of sampling on the Fourier transform of the sampled signal, and can equally well be applied to sampling of frequency spectra or spatial signals. For example, when one samples a frequency spectrum, the corresponding time domain signal is replicated. So long as those time domain replicas are of finite duration and do not overlap, the original spectrum can be recovered, meaning that the frequency spectrum is adequately sampled. This fact was used in Sec. 3.6.1 to establish the minimum DFT size for computing the Doppler spectrum.

B.3.2 Quantization

Binary representations of numeric data can be categorized as either fixed point or floating point. In fixed-point representations, the b bits of a binary word are used to represent 2^b distinct and evenly spaced numerical values.[8] In a floating-point representation, some number e of the b total bits is used to represent an exponent, and the remaining $m = b - e$ bits are used to represent the mantissa. Here, fixed-point representations are of primary interest, as this is the representation output by most analog-to-digital (A/D) converters.

The numeric value corresponding to each of the 2^b possible binary words in a fixed-point encoding is determined by the arithmetic coding scheme that is used and the quantization step size Δ. The step size is the change in input value in volts for each increment or decrement of the binary word value. Thus, the value of the quantized data sample is simply Δ times the binary number output by the quantizer.

The two most common encodings are called *sign-magnitude* encoding and *two's complement* encoding. In sign-magnitude encoding the most significant bit of the binary word represents the sign of the data sample; usually a value of zero represents a positive number, while a value of one represents a negative number. The remaining $b - 1$ bits encode the magnitude of the sample. Thus, sign-magnitude encoding can represent numbers from

---

[8]"Nonlinear quantization" using unevenly spaced values of output voltage is common in some applications. Linear quantization (constant level spacing) is the norm in radar.

$(-2^{b-1} - 1)\Delta$ to $(+2^{b-1} - 1)\Delta$. Note that there are two codes for the value zero, corresponding to $+0$ and -0.

Two's complement is a somewhat more complex encoding that has advantages for the design of digital arithmetic logic. In two's complement encoding, there is only one code for zero. The extra code value allows the representation of one more negative number, so the range of values becomes $-2^{b-1}\Delta$ to $(+2^{b-1} - 1)\Delta$. For any reasonable number of bits b, the difference in range is not significant. Details are given in Ercegovac and Lang (2003).

The choice of Δ and b is determined by the desired values of the *signal-to-quantization noise ratio* (SQNR) and dynamic range, and is also constrained by A/D converter technology. Consider the step size Δ first. Quantization entails rounding or truncating the analog sample value to one of the allowed quantized values. The difference between the unquantized and quantized samples is the *quantization error*. Although it is a deterministic function of the input data and A/D converter parameters, the behavior of the quantization error signal is usually complex enough that it is treated as a random variable uncorrelated from one sample to the next and independent of the original unquantized signal. Assuming rounding, the quantization error for each sample can vary between $\pm\Delta/2$, and it is commonly assumed that errors anywhere in that range are equally likely. Thus, the quantization noise process is modeled as a uniform random process over this range so that the quantization noise power is $\sigma_q^2 = \Delta^2/12$.

To choose Δ, consider an A/D converter with a white noise input of power σ_w^2. Ideally the noise power at the output of the A/D would be nearly the same as at the input so that quantization noise could be considered negligible. Let the step size $\Delta = \alpha \cdot \sigma_w$ for some α. The increase in noise power at the A/D output due to quantization noise will be a factor of

$$\frac{\sigma_w^2 + \sigma_q^2}{\sigma_w^2} = 1 + \frac{\Delta^2/12}{\sigma_w^2} = 1 + \frac{\alpha^2}{12} \tag{B.33}$$

Figure B.7 plots the increase in decibels versus α. Values of α less than 1.76 produce a noise power increase of 1 dB or less. In practice, α is often chosen to be in the range of 0.25 to 0.5

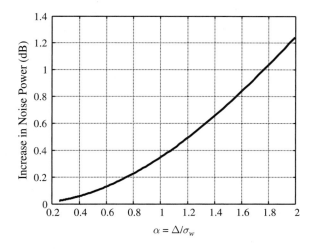

FIGURE B.7 Effect of the quantization step size on the ratio of output power to input power for an A/D converter with a noise signal input.

to ensure that the receiver noise toggles one or two least significant bits. This in turn ensures that very small signals at or below the noise level that should be reinforced by coherent integration are not suppressed instead.

This style of analysis is extended considerably in McClellan and Purdy (1978) to consider the saturation effects that result when Δ is made too small and the underflow that results when Δ is too large. It remains the case that Δ should normally be chosen to be around one-quarter to one-half the input noise standard deviation for good results.

Now consider the choice of b. This is a question of the *dynamic range* of the A/D converter. One definition of dynamic range is the ratio of the largest representable magnitude to the smallest nonzero magnitude. For the two's complement case this is, in both linear and decibel units,

$$DR = \frac{2^{b-1}\Delta}{\Delta} = 2^{b-1}$$
$$DR \text{ (dB)} = 20\log_{10}(2^{b-1}) = (b-1)20\log_{10}(2) = 6.02b - 6.02$$
(B.34)

For the sign-magnitude case $DR = 2^{b-1} - 1 \approx 2^{b-1}$ for b more than just a few bits so that Eq. (B.34) still applies. This result shows that the dynamic range of the numbers that can be represented at the A/D converter output increases by about 6 dB per bit. A 60-dB dynamic range would require 11 bits.

An alternative view of dynamic range comes from treating the input signal as a random process with variance σ_s^2. Express the maximum input signal required to be represented without saturation (overflow) as $k \cdot \sigma_s$. For a Gaussian-distributed input signal, k would normally be chosen in the range of 3 to 4 so that there is only a small probability of saturation. Equating this to the maximum representable number and recalling that $\Delta = \alpha \cdot \sigma_w$ gives

$$k \cdot \sigma_s = 2^{b-1}\Delta = 2^{b-1}\alpha \cdot \sigma_w$$
(B.35)

Defining the input signal-to-noise ratio $SNR_{in} = \sigma_s^2/\sigma_w^2$, substituting in Eq. (B.35), and converting to decibels gives

$$SNR_{in} = \frac{2^{2b-2}\alpha^2}{k^2} \Rightarrow SNR_{in} \text{ (dB)} = 6.02b - 20\log_{10}\left(\frac{2k}{\alpha}\right)$$
(B.36)

Solving for the number of bits gives

$$b = \left\lceil \frac{SNR_{in} \text{ (dB)} + 20\log_{10}(2k/\alpha)}{6.02} \right\rceil$$
(B.37)

Equation (B.37) expresses the number of bits needed to quantize a signal with a given SNR at the quantizer input without significant saturation. For a typical choice of $\alpha = 0.5$ and $k = 4$, the second term in the numerator is 24.08 dB. For $\alpha = 1$ and $k = 3$, it is 15.56 dB. For the first case, representing signals with an SNR at the input of zero dB requires 4 bits. If the input SNR can be as high as 10 dB, this rises to 6 bits.

It is important to realize that this analysis gives the number of bits required to represent the signal present at the A/D converter input. It does not take into account increases in dynamic range due to processing gains achieved by coherent integration, matched filtering, and DFTs. If a post-A/D conversion operation achieves a coherent integration gain factor of N, an additional $\log_2 N$ bits will be required in fixed-point digital arithmetic to represent the operation result without overflow.

Another metric frequently cited is the SQNR, mentioned previously. For the Gaussian input signal, the signal power is σ_s^2. Using the first relation in Eq. (B.35) and $\sigma_q^2 = \Delta^2/12$ and converting to decibels gives

$$\frac{\sigma_s^2}{\sigma_q^2} = \frac{(2^{b-1}\Delta/k)^2}{\Delta^2/12} = \frac{12 \cdot 2^{2b-2}}{k^2} \Rightarrow$$
$$SQNR \text{ (dB)} = 6.02b - 20\log_{10}(k) - 4.77 \quad (B.38)$$

Like the dynamic range, the SQNR also increases by 6 dB per bit.

Once the data are quantized, a choice must be made between implementing the subsequent processing in fixed- or floating-point arithmetic. Generally, floating-point arithmetic requires more digital logic to implement and is therefore slower and consumes more power. However, robust mathematical algorithms are easier to develop for floating-point arithmetic because numerical overflow and underflow are much less likely. Early radar digital processors relied mostly on fixed-point arithmetic, at least in the early processing stages that tend to be more computationally intensive, because it is faster. Increasing numbers of modern systems are using floating-point processing because the dramatic increases in processor power have made it possible to implement the desired algorithms in real time. When portions of the signal processing require greater speed or power efficiency than is available in floating point processors, field programmable gate array (FPGA) or graphical processing unit (GPU) technology may sometimes provide a good alternative.

B.3.3 A/D Conversion Technology

As has been seen, the required sampling rate is determined through the Nyquist criterion by the instantaneous bandwidth β_F of the receiver output, perhaps increased by a safety margin of 10 to 20 percent to allow for the imperfect bandlimiting of real waveforms and antialiasing filters, while the number of bits b in the digital sample is determined by the maximum dynamic range and quantization noise requirements of the system. Except in a few specialized situations, the minimum number of bits required for a signal sampled at its Nyquist rate is at least 6 and preferably 8, while 12 bits or more are desired in many applications (Merkel and Wilson, 2003).

In high-resolution radars the resulting sampling rates can be very high, often tens to hundreds of megasamples per second and in some cases even reaching rates of gigasamples per second. The use of digital I/Q or digital IF techniques for non-baseband sampling can increase the required rate relative to the waveform bandwidth by a factor of 2.5× to 4× (see Chap. 3), further exacerbating the problem. Many very wideband systems use a specialized linear FM waveform processing method called *stretch processing* (see Chap. 4) to reduce the analog signal bandwidth prior to A/D conversion, often by an order of magnitude or more.

In general, the higher the required sampling rates, the fewer the number of bits available in current A/D converter technology. Figure B.8 summarizes the state of the art in A/D converters in 2006 (Walden, 2008). The quantity plotted is the effective number of bits (ENOB) versus sampling rate. ENOB is inferred from measured SQNRs as described in Walden (1999), which also discusses many other A/D metrics and limiting factors. Another source of A/D technology state of the art is Murmann (2020). The Walden data suggests that A/D converter wordlengths tend to drop about 2 to 3 bits per decade of sampling rate. However, in 2006 an ENOB of 8 bits was achievable at rates up to about 1 gigasample per second, and an ENOB of 12 bits was achievable at rates up to approximately 100 megasamples per second. These are quite adequate for most radar applications.

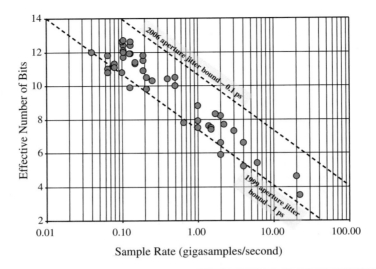

Figure B.8 Analog-to-digital converter performance. (Data courtesy of Dr. Robert Walden and The Aerospace Corporation. Used with permission.)

B.4 Spatial Frequency

The idea of spatial frequency is an important concept in any study involving propagating waves, and it will be needed to analyze spatial sampling and space-time adaptive processing. A simplified intuitive introduction to the concept is given here and in Richards (2018). For a more complete discussion, see Johnson and Dudgeon (1993).

Consider a sinusoidal electromagnetic wave propagating in the $+x$ direction with wavelength λ_t and velocity c, as shown in Fig. B.9. An observer at a fixed spatial position x_0 will see successive crests of the electric field at a time interval (period) of $T = \lambda_t/c$ seconds; thus the temporal frequency of the wave is the usual $F_t = 1/T = c/\lambda_t$ Hz or $2\pi c/\lambda_t$ radians per second.

A spatial period can also be defined; it is simply the interval between successive crests in space for a fixed observation time. Put another way, it is the change in range between the source and the observer that produces one full cycle (2π radians) of phase change in the wave. The spatial period of the pulse is obviously λ_t meters. The spatial frequency is therefore $1/\lambda_t$ cycles per meter or $2\pi/\lambda_t$ radians per meter. It is common to call the latter quantity the *wavenumber* of the pulse and to denote it with the symbol K.[9] In this text, the wavenumber in the direction of propagation, that is, in the range direction, is denoted K_R specifically.

Because position in space and velocity are three-dimensional vector quantities in general, so is the wavenumber. For simplicity of illustration, consider the two-dimensional version of Fig. B.9 shown in Fig. B.10. The pulse, now propagating at an angle in an x-y plane, still has a range wavenumber $K_R = 2\pi/\lambda_t$ in the direction of propagation. However, measured in the $+x$ direction, the wavenumber is $K_x = (2\pi/\lambda_t)\sin\theta$, where θ is the angle of

---

[9] Most radar and array processing literature uses a lower case k for spatial frequency. Here upper case K is used in keeping with the convention to use uppercase letters for analog quantities and lower case for "digital" quantities, that is, those normalized by a sampling interval.

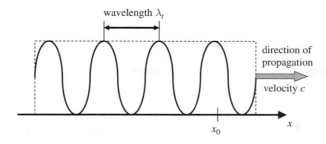

FIGURE B.9 Propagating electromagnetic wave in one dimension.

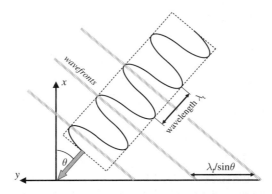

FIGURE B.10 Propagating electromagnetic wave in two dimensions.

incidence of the pulse measured relative to the $+y$ axis. Similarly, the same signal has a wavenumber in the y direction of $K_y = (2\pi/\lambda_t)\cos\theta$.[10] Note that as $\theta \to 0$, the wavelength in the x dimension tends to ∞ so that $K_x \to 0$.

The extension to three dimensions of space is straightforward. The total wavenumber is related to the components in the obvious way,

$$K = \sqrt{K_x^2 + K_y^2 + K_z^2} \tag{B.39}$$

and always equals $2\pi/\lambda_t$. Note that the temporal frequency remains c/λ_t Hz regardless of the direction of propagation.

In the common case of a monostatic radar, the change in radar-scatterer range required to produce a 2π change in phase at the receiver is only $\lambda_t/2$ m, implying a spatial frequency of $1/(\lambda_t/2) = 2/\lambda_t$ cycles/m or $4\pi/\lambda_t$ rad/m. This *two-way range spatial frequency* is denoted here as K_R' to distinguish it from the one-way range spatial frequency K_R. Two-way spatial frequency is discussed further in Richards (2018).

---

[10]Incidence angle will often be measured with respect to the normal to the y axis, that is, the x axis, because this is convenient and conventional in analyzing antenna patterns. If the antenna aperture lies in the y dimension, then an incidence angle of $\theta = 0$ indicates a wave propagating normal to the aperture, that is, in the boresight (x) direction.

B.5 Correlation

Correlation is an operation that compares one signal against a reference signal to determine their degree of similarity. *Cross-correlation* is the correlation of two different signals; *autocorrelation* is the correlation of a signal with itself. Correlation is frequently defined in both a probabilistic sense, as a descriptive property of random signals, and as a deterministic operation performed on actual digital signals. If a random process is *ergodic*, the two interpretations are closely linked; see Oppenheim and Schafer (2010) or any text on random signals and processes for an introduction to these concepts. The probabilistic meaning is discussed in App. A. The deterministic processing operation is of concern here.

Consider two signals $x[n]$ and $y[n]$ with DTFTs $X(f)$ and $Y(f)$. Their deterministic cross-correlation is defined as

$$s_{xy}[k] = \sum_{n=-\infty}^{+\infty} x[n]\, y^*[n+k], \quad -\infty < k < +\infty \tag{B.40}$$

If $x[n] = y[n]$, this is the autocorrelation of $x[n]$, denoted $s_x[k]$. The particular value $s_x[k_0]$ is called the k_0-th correlation *lag*. It is straightforward to show that the deterministic cross-correlation of $x[n]$ and $y[n]$ is identical to the convolution of $x[n]$ and $y^*[-n]$.

The Fourier transform of the cross-correlation function is called the *cross-power spectrum*, and is just the product

$$S_{xy}(f) = \mathbf{F}\{s_{xy}[k]\} = X(f)Y^*(f) \tag{B.41}$$

The Fourier transform of the autocorrelation function (ACF) is usually called simply the *power spectrum* or the *power spectral density* (PSD). Notice that the power spectrum S_x is the squared-magnitude of the Fourier transform of the underlying signal x,

$$S_x(f) = \mathbf{F}\{s_x[m]\} = X(f)X^*(f) = |X(f)|^2 \tag{B.42}$$

Thus, the power spectrum is not dependent on the phase of the signal spectrum.

The extension to two-dimensional signals is obvious:

$$s_{xy}[l,k] = \sum_{m=-\infty}^{+\infty} \sum_{n=-\infty}^{+\infty} x[m,n]y^*[m+l,n+k], \quad -\infty < l,k < +\infty \tag{B.43}$$

$$S_{xy}(f_l, f_m) = \mathbf{F}\{s_{xy}[l,m]\} = X(f_l, f_m)Y^*(f_l, f_m) \tag{B.44}$$

Graphically, correlation corresponds to overlaying the two constituent signals; multiplying them sample-by-sample; and adding the results to get a single value. One of the two signals is then shifted and the process repeated, creating a series of output values which form the correlation sequence $s_{xy}[k]$. This is illustrated notionally in Fig. B.11, which shows the cross-correlation of two functions: $x[n]$, which is nonzero for $0 \le n \le M-1$, and $y[n]$, nonzero for $0 \le n \le N-1$. Note that $s_{xy}[k]$ will be nonzero only for $1 - N \le m \le M - 1$.

It is sometimes convenient to define the *normalized correlation* function:

$$\rho_x[k] \equiv \frac{s_x[k]}{s_x[0]} \tag{B.45}$$

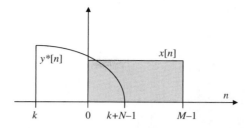

FIGURE B.11 Computation of the deterministic cross-correlation function $s_{xy}[k]$.

Normalized versions of the cross-correlation function and of the two-dimensional auto- and cross-correlation functions are defined similarly.

The properties of correlation functions are described in many standard texts. Here only two properties of particular importance are presented, without proof:

$$s_x[k] \leq s_x[0] \tag{B.46}$$

$$s_{xy}[k] = s_{yx}^*[-k] \tag{B.47}$$

The first states that the zero lag ($k = 0$) of an ACF is always the peak value. In terms of Fig. B.11, this corresponds to the case where the signal is being correlated with itself and the two replicas are completely overlapping. In this case Eq. (B.40) specializes to

$$s_x[0] = \sum_{n=-\infty}^{+\infty} x[n]x^*[n] = \sum_{n=-\infty}^{+\infty} |x[n]|^2 = E_x \tag{B.48}$$

Note that $s_x[0]$ is necessarily real, even if $x[n]$ is complex. Furthermore, $s_x[0]$ is the total energy in $x[n]$, E_x.

The second property above establishes the Hermitian symmetry of any auto- or cross-correlation function. From properties of the Fourier transform, it follows that the power spectrum will have even symmetry if the correlation function is also real-valued.

Computing the zero autocorrelation lag is a weighted integration of the signal samples. In coherent integration it is necessary to align the phases of the signal samples to maximize the coherent sum. This is exactly what happens in Eq. (B.48). If $x[n]$ is represented in the magnitude-phase form $A\exp(j\varphi)$, then multiplication by $x^*[n] = A\exp(-j\varphi)$ cancels the phase component so that the products are all real positive values and therefore add in phase. Thus when performing an autocorrelation, the zero lag term is the weighted coherent integration of the signal samples. This observation will be useful in understanding matched filtering and synthetic aperture imaging.

Figure B.12 illustrates one of the important uses of correlation in radar signal processing: detecting and locating desired signals in the presence of noise. The sequence in Fig. B.12a is the sum of a zero mean, unit variance Gaussian random noise signal with the finite duration pulse $y[n] = 2$, $9 \leq n \leq 38$ and zero otherwise. The signal-to-noise ratio is 6 dB. While there is some evidence of the presence of $y[n]$ in the interval $9 \leq n \leq 38$ in Fig. B.12a, it does not stand out well above the noise and cannot be located precisely, nor can its amplitude be accurately estimated. However, the cross-correlation of the signal with $y[n]$ shown in Fig. B.12b displays a clear peak at lag $k = 9$, indicating the presence of the signal $y[n]$ in the noisy signal of Fig. B.12a and correctly locating it nine samples from the origin. Thus, correlation can be used to aid in identification and location of known signals in noise, that is, as a way to "look for" a signal of interest in the presence of interference.

680 Appendix B

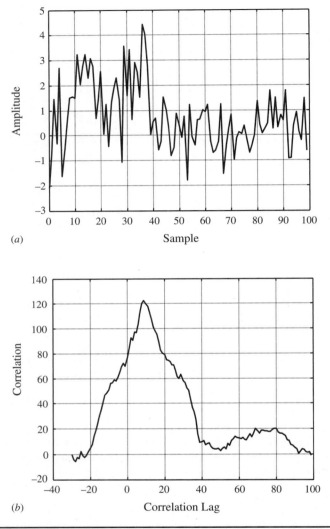

FIGURE B.12 Identifying signals in noise using cross-correlation. (*a*) Signal consisting of rectangular pulse in noise. (*b*) Result of correlating with reference rectangular pulse. See text for details.

B.6 Vector-Matrix Representations and Eigenanalysis

B.6.1 Basic Definitions and Operations

It will sometimes be convenient to represent a finite length signal as a vector rather than in indexed sequence notation. That is, if the signal $x[n]$ is defined for $0 \leq n \leq N-1$, it can be denoted in column vector form as

$$\mathbf{x} = [x[0] \ x[1] \ \ldots \ x[N-1]]^T \tag{B.49}$$

where the superscript T denotes the matrix transpose. Signal vectors will usually be defined as column vectors, and boldface notation will be used to indicate vectors (lower case) and matrices (upper case).

Many important signal processing operations can be expressed in matrix-vector form. One of particular importance is calculation of a single output sample of a *finite impulse response* (FIR) filter, also called a *tapped delay line* or *nonrecursive* filter. Suppose the filter impulse response is denoted $h[n]$, $0 \leq n \leq L - 1$. The filter output is given by the convolution sum

$$y[n] = \sum_{l=0}^{L-1} h[l] x[n - l] \tag{B.50}$$

Vector notation can be used to represent the coefficients $h[l]$ by the L-element column vector $\mathbf{h}$:

$$\mathbf{h} = [h[0] \; h[1] \; \ldots \; h[L-1]]^T \tag{B.51}$$

Now define the L-element vector $\mathbf{x}_n$ of the L most recent signal samples as

$$\mathbf{x}_n = [x[n] \; x[n-1] \; \ldots \; x[n-L+1]]^T \tag{B.52}$$

Equation (B.50) can then be written as the vector inner product

$$y[n] = \mathbf{h}^T \mathbf{x}_n \quad \text{or} \quad y_n = \mathbf{h}^T \mathbf{x}_n \tag{B.53}$$

This notation will be convenient in discussing matched filters and array processing in Chaps. 4 and 9.

Two special product operations on vectors are common in some radar signal processing. The symbol $\odot$ represents the *Hadamard product*, which is the element-by-element product of two vectors or matrices. Specifically, if $\mathbf{a}$ and $\mathbf{b}$ are two N-element column vectors, then $\mathbf{a} \odot \mathbf{b}$ is the N-element column vector

$$\mathbf{a} \odot \mathbf{b} \equiv [a_0 b_0 \; a_1 b_1 \; \ldots \; a_{N-1} b_{N-1}]^T \tag{B.54}$$

The *Kronecker product*, denoted by the symbol $\otimes$, is the replication of one vector weighted by the elements of the other. If $\mathbf{a}$ is an M-element column vector and $\mathbf{b}$ is an N-element column vector, then $\mathbf{a} \otimes \mathbf{b}$ is the MN-element column vector

$$\mathbf{a} \otimes \mathbf{b} \equiv \begin{bmatrix} a_0 \cdot \mathbf{b} \\ a_1 \cdot \mathbf{b} \\ \vdots \\ a_{M-1} \cdot \mathbf{b} \end{bmatrix} \tag{B.55}$$

It is frequently necessary to compute the power or energy of various signals in vector notation. The energy $E_\mathbf{x}$ and power $P_\mathbf{x}$ in a vector $\mathbf{x}$ are given by

$$E_\mathbf{x} = \mathbf{x}^H \mathbf{x} = \sum_{n=0}^{N-1} |x_n|^2, \quad P_\mathbf{x} = \frac{1}{N} E_\mathbf{x} = \frac{1}{N} \mathbf{x}^H \mathbf{x} = \frac{1}{N} \sum_{n=0}^{N-1} |x_n|^2 \tag{B.56}$$

where the superscript H denotes the Hermitian (conjugate) transpose. The energy or power in the result of an inner product (filtering) operation $y = \mathbf{h}^T \mathbf{x}$ is

$$E_y = P_y = |y|^2 = y^* \cdot y^T = \mathbf{h}^T \mathbf{x}^* \mathbf{x}^T \mathbf{h} \tag{B.57}$$

E_y and P_y are the same because y is scalar ($N = 1$).

For a random vector **w** (noise samples, for example), the expected value is required to get meaningful expressions for energy or power:

$$E_\mathbf{w} = \mathrm{E}\{\mathbf{w}^H \mathbf{w}\}, \quad P_\mathbf{x} = \frac{1}{N} E_\mathbf{x} = \frac{1}{N} \mathrm{E}\{\mathbf{w}^H \mathbf{w}\} \tag{B.58}$$

The power or energy in a filtered noise vector $y = \mathbf{h}^T \mathbf{w}$ then becomes

$$\begin{aligned} E_y = P_y &= \mathrm{E}\{\mathbf{h}^T \mathbf{w}^* \mathbf{w}^T \mathbf{h}\} \\ &= \mathbf{h}^T \mathrm{E}\{\mathbf{w}^* \mathbf{w}^T\} \mathbf{h} = \mathbf{h}^T \mathbf{S}_\mathbf{w} \mathbf{h} \end{aligned} \tag{B.59}$$

where the *covariance matrix* $\mathbf{S}_\mathbf{w}$ has been defined as the outer product[11]

$$\mathbf{S}_\mathbf{w} \equiv \mathrm{E}\{\mathbf{w}^* \mathbf{w}^T\} \tag{B.60}$$

Covariance matrices have numerous useful properties. For example, $\mathbf{S}_\mathbf{w}$ is Hermitian ($\mathbf{S}_\mathbf{w} = \mathbf{S}_\mathbf{w}^H$) and Toeplitz (all the elements on a diagonal are identical). The inverse is also Hermitian, $\mathbf{S}_\mathbf{w}^{-1} = (\mathbf{S}_\mathbf{w}^{-1})^H$. Some more important properties are deferred until after a brief discussion of eigenanalysis.

The last form of Eq. (B.59) is an example of a *quadratic form* of a matrix. The quadratic form of a Hermitian matrix **A** with respect to a vector **x** is the real-valued scalar quantity

$$Q_\mathbf{A}(\mathbf{x}) = \mathbf{x}^H \mathbf{A} \mathbf{x} = \sum_{i=0}^{N-1} \sum_{j=0}^{N-1} x_i^* a_{ij} x_j \tag{B.61}$$

The name arises because $Q_\mathbf{A}(\mathbf{x})$ is a second-order polynomial in the elements of **x**.[12] If $Q_\mathbf{A}(\mathbf{x}) > 0$ for any **x**, **A** is said to be *positive definite*. If $Q_\mathbf{A}(\mathbf{x}) \geq 0$, **A** is said to be *positive semi-definite*. Matrices can also be negative definite or negative semidefinite. A matrix that is none of these is called indefinite.

The signal-to-interference (SIR) ratio for a signal **x** and interference **w** is just $P_\mathbf{x}/P_\mathbf{w} = E_\mathbf{x}/E_\mathbf{w}$. If the input to a filter **h** is the sum of **x** and **w**, the SIR of the output y will be the ratio of the powers in the filtered signal and noise from Eqs. (B.57) and (B.59), which takes on the often-seen form

$$\mathrm{SIR} = \frac{\mathbf{h}^T \mathbf{x}^* \mathbf{x}^T \mathbf{h}}{\mathbf{h}^T \mathbf{S}_\mathbf{w} \mathbf{h}} \tag{B.62}$$

B.6.2 Basic Eigenanalysis

Consider the system of linear equations described by an $N \times N$ matrix **A**,

$$\mathbf{A}\mathbf{e} = \lambda \mathbf{e} \tag{B.63}$$

---

[11] Strictly speaking, this is the *autocorrelation matrix*. The covariance matrix is obtained as the autocorrelation of the interference after the mean is subtracted out; see Hayes (1996) for details. The interference may almost always be assumed to have zero mean, in which case they are the same. The covariance terminology seems to be more prevalent in the radar community.

[12] Since Q must be real, the second-order terms in the polynomial all appear in magnitude-squared form, $|x_i|^2$.

Any vector **e** which satisfies this equation is called an *eigenvector* of **A** and the scalar λ is called the corresponding *eigenvalue* (not to be confused with the wavelength). An eigenvector is thus a vector that is unmodified (to within a scale factor) when operated on by **A**. It is frequently the case that eigenvectors are normalized to unity norm, $\|\mathbf{e}\| = 1$.

Rearranging Eq. (B.63) gives

$$(\mathbf{A} - \lambda \mathbf{I})\mathbf{e} = 0 \tag{B.64}$$

For Eq. (B.64) to hold for a nonzero vector **e**, it must be the case that $\mathbf{A} - \lambda\mathbf{I} = 0$ so that in turn

$$\det(\mathbf{A} - \lambda\mathbf{I}) \equiv p(\lambda) = 0 \tag{B.65}$$

The *characteristic polynomial* $p(\lambda)$ will be an Nth-order polynomial in λ. Its roots are the eigenvalues of **A**.

Following are some useful properties of eigenvalues and eigenvectors, including a number which apply to Hermitian matrixes and therefore to covariance matrices. Proofs of these properties can be found in Hayes (1996).

For any matrix **A**:

- Nonzero eigenvectors $\mathbf{e}_i$ corresponding to distinct eigenvalues λ_i are linearly independent.
- If the rank of **A** is M, there will be M nonzero eigenvalues and $N - M$ eigenvalues with a value of zero.

For a Hermitian matrix **A** (includes covariance matrices):

- The eigenvalues of **A** are real.
- If **A** is the covariance matrix of a wide-sense stationary process (normally a reasonable assumption), the eigenvalues are also nonnegative. Consequently, **A** is a nonnegative definite matrix.
- Eigenvectors corresponding to distinct eigenvalues are orthogonal, $\mathbf{e}_i^T\mathbf{e}_j = 0$.
- (Spectral theorem) **A** can be decomposed in terms of its eigenvectors and eigenvalues as

$$\mathbf{A} = \mathbf{V}\mathbf{\Lambda}\mathbf{V}^H = \sum_{i=0}^{N-1} \lambda_i \mathbf{e}_i \mathbf{e}_i^H \tag{B.66}$$

where **V** is a matrix whose columns are the normalized eigenvectors and $\mathbf{\Lambda}$ is matrix having the eigenvalues on the diagonal and zero elsewhere.

- If $\mathbf{B} = \mathbf{A} + \alpha\mathbf{I}$, then **A** and **B** have the same eigenvectors. The eigenvalues of **B** are $\lambda_i + \alpha$, where the λ_i are the eigenvalues of **A**. This property is useful when considering covariances matrices of the sum of a signal with white noise.

B.6.3 Eigenstructure of Sinusoids in White Noise

Many important problems in radar model the signal as a sum of sinusoids in random interference. Examples include Doppler processing, adaptive beamforming, and space-time adaptive processing. Certain important classes of processing algorithms rely on the structure of the covariance matrix of such signals. A brief summary of the main results is given here. A much more complete but still concise discussion is given in Hayes (1996).

Assume a length-N signal $x[n]$ is the sum of a signal component $s[n]$ consisting of K complex sinusoids and white noise $w[n]$ of variance σ_w^2:

$$x[n] = s[n] + w[n] = \sum_{i=0}^{K-1} A_i \exp(j\omega_i n) + w[n] \quad (B.67)$$

The amplitudes A_i are complex with uncorrelated phases randomly distributed on $(-\pi, \pi]$. The frequencies ω_i and magnitudes $|A_i|$ are nonrandom but unknown. Given a vector $\mathbf{x}$ consisting of N consecutive samples of $x[n]$, the covariance matrix of $\mathbf{x}$ will be the sum of the covariance matrices of $\mathbf{s}$ and $\mathbf{w}$,

$$\mathbf{S_x} = \mathbf{S_s} + \mathbf{S_w} = \sum_{i=0}^{K-1} P_i \boldsymbol{\omega}_i \boldsymbol{\omega}_i^H + \sigma_w^2 \mathbf{I} = \boldsymbol{\Omega P \Omega} + \sigma_w^2 \mathbf{I} \quad (B.68)$$

In this equation, $\mathbf{I}$ is the $N \times N$ identity matrix, $\boldsymbol{\omega}_i = [1 \; \exp(j\omega_i) \; \ldots \; \exp[j(N-1)\omega_i]]^T$, $\boldsymbol{\Omega} = [\boldsymbol{\omega}_0 \ldots \boldsymbol{\omega}_{K-1}]$ is a rank K matrix, and $\mathbf{P} = \text{diag}\{P_0, P_1, \ldots, P_{K-1}\}$ is a diagonal matrix of the sinusoid powers $P_i = |A_i|^2$. While the $\boldsymbol{\omega}_i$ will not generally equal the eigenvectors of $\mathbf{S_x}$, they will lie in the space spanned by those eigenvectors, which is called the *signal subspace*. The subspace spanned by the remaining eigenvectors is called the *noise subspace*. Because of the last property of covariance matrices enumerated above, the $N - K$ noise subspace eigenvalues will equal σ_w^2 while the K signal subspace eigenvalues will equal the eigenvalues of $\boldsymbol{\Omega P \Omega}$ plus σ_w^2.

Figure B.13 illustrates these properties for a signal that is the sum of 15 samples of each of five unit-amplitude complex sinusoids with randomly chosen normalized frequencies and phases and unit-variance, zero mean complex Gaussian noise. Thus $N = 15$ and $K = 5$. The solid line with round markers indicates the eigenvalues of the theoretical covariance matrix of Eq. (B.68). The square markers are the eigenvalues of a covariance matrix estimated by averaging 1000 sample covariance matrices formed from simulated signals in noise. The eigenvalues of the simulated data match the theoretical eigenvalues well. Both show that there are 10 noise eigenvalues with values near 1. The remaining five signal eigenvalues are larger as expected. The associated eigenvectors define the signal and noise subspaces.

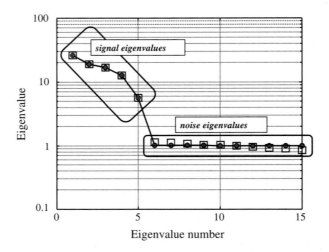

FIGURE B.13 Example of theoretical (round markers) and simulated (square markers) signal and noise eigenvalues for five sinusoids plus white noise.

B.7 Instantaneous Frequency

Consider a real or complex sinusoidal function with an arbitrary phase function $\psi(t)$, $x(t) = A\cos\psi(t)$ or $x(t) = A\exp[j\psi(t)]$. The *instantaneous frequency* of this waveform is defined as

$$\Omega_i(t) = \frac{d\psi(t)}{dt} \text{ rad/s}, \quad F_i(t) = \frac{1}{2\pi}\frac{d\psi(t)}{dt} \text{ Hz} \tag{B.69}$$

As an example, a conventional constant-frequency sinusoid $A\cos(2\pi F_0 t)$ has $\psi(t) = 2\pi F_0 t$ so that $F_i(t) = F_0$ Hz, as should be expected. Note that the phase function of a constant-frequency sinusoid is linear in time.

Instantaneous frequency is sometimes useful in understanding the behavior of more complex signals. Consider a signal having a quadratic phase function $\psi(t) = \alpha t^2 + \beta t + \gamma$. The instantaneous frequency in radians per second is $\Omega_i(t) = \alpha t + \beta$. Thus, a quadratic phase sinusoid has a frequency that varies linearly with time as in the linear FM "chirp" pulses or FMCW sweeps of Chap. 4. As a final example, consider the following sinusoid and its instantaneous frequency:

$$\begin{array}{l} x(t) = A\sin[2\pi F_0 t + \alpha\sin(2\pi\beta t)] \quad \Rightarrow \quad \psi(t) = 2\pi F_0 t + \alpha\sin(2\pi\beta t) \\ F_i(t) = F_0 + \alpha\beta\cos(2\pi\beta t) \end{array} \tag{B.70}$$

This waveform is a sinusoid whose frequency varies sinusoidally by $\pm\alpha\beta$ Hz around a nominal frequency of F_0 Hz, with that variation occurring at a rate of β cycles per second.

B.8 Decibels

Many radar parameters are commonly expressed in decibel (dB) units. Examples include antenna gains, radar cross sections, signal-to-interference ratios, and sidelobe levels. The advantage of using decibel units is that it reduces the numerical range of parameters that have very large variations. Radar cross section is a good example. Average RCS values of interest in radar can range from 10^{-5} m^2 for insects to over 10^6 m^2 for a large ship such as an aircraft carrier viewed at certain aspect angles. This range of 11 orders of magnitude becomes a range of only 110 dB, from -50 dB to $+60$ dB.

Decibels in radar are a measure of *relative* power of an electrical signal. Consider a signal having a power of P watts. The power in decibels is defined as

$$P \text{ (dB)} = 10\log_{10}\left(\frac{P}{P_{\text{ref}}}\right) \tag{B.71}$$

where P_{ref} is the reference power level. A value P_{dB} in decibels is easily converted back to a value P in non-decibel units (often called *linear units*) by inverting Eq. (B.71),

$$P = P_{\text{ref}} \cdot 10^{P_{\text{dB}}/10} \tag{B.72}$$

The units of dB are often modified to reflect the reference value used. For instance, if $P_{\text{ref}} = 1$ mW and $P = 100$ mW, then P might be expressed as 20 dBm, with the abbreviation dBm interpreted as "decibels relative to 1 milliwatt." Another example is dBW for decibels relative to 1 W. It is often the case, especially in signal processing, that the absolute units of a signal are not important, just its numerical value. For this reason, the reference value P_{ref} is often taken equal to 1 and the units of the decibel value are left as just dB with no modifier.

In electrical engineering, signals of interest are often voltages. In accordance with Ohm's law, the power of a quantity V in volts is proportional to V^2. A quantity expressed in volts can be converted between decibel and linear scales using the relations:

$$P \text{ (dB)} = 10\log_{10}\left(\frac{|V|^2}{|V_{\text{ref}}|^2}\right) = 20\log_{10}\left(\frac{|V|}{|V_{\text{ref}}|}\right) \quad \text{(B.73)}$$

$$|V| = |V_{\text{ref}}| \cdot 10^{P_{\text{dB}}/20}$$

The use of $|V|^2$ instead of just V^2 allows for the possibility that the voltage V is complex. Note that converting from decibels back to the linear scale can recover only the magnitude of V. The phase of a complex voltage is lost in the conversion to decibels, as is the sign of a real-valued voltage. An arbitrary signal $x(t)$ or $x[n]$, for instance the output of a radar receiver, is usually considered to be a voltage unless otherwise specified. There are exceptions. The output of a square-law detector (see Chap. 6) would be considered to be in power units, while the units of the output of a logarithmic detector would be the logarithm of power.

In addition to signal powers, decibel scales are commonly used in radar to describe antenna pattern features, time- and frequency-domain filter response features, and radar cross sections. For example, the power gain at the peak of the first sidelobe, relative to the peak mainlobe power gain, of an antenna with an ideal uniform illumination is about 0.047, corresponding to -13.26 dB. This same value surfaces frequently, for instance as the peak sidelobe level of the slow-time DTFT of a constant-Doppler target or the peak range sidelobe level of a matched filter for a linear FM waveform with a high time-bandwidth product. RCS is defined in terms of the ratio of two values of power density (see Chap. 2), and so becomes a factor in power calculations in radar range equations. Because RCS has units of m^2, on a decibel scale RCS is often given units of decibels relative to 1 m^2, written dBm. The distinction between the usage of dBm for dB relative to 1 square meter and dB relative to 1 milliwatt is usually clear from context.

Expressing parameters in decibels simplifies multiplication, division, and exponentiation calculations, as shown here:

$$\begin{aligned} z = x \cdot y &\Leftrightarrow z_{\text{dB}} = x_{\text{dB}} + y_{\text{dB}} \\ z = x/y &\Leftrightarrow z_{\text{dB}} = x_{\text{dB}} - y_{\text{dB}} \\ z = x^\alpha &\Leftrightarrow z_{\text{dB}} = \alpha \cdot x_{\text{dB}} \end{aligned} \quad \text{(B.74)}$$

These transformations were the basis for slide rules. They were of great utility before the age of handheld scientific calculators and high-speed computers but are of less importance today. Nonetheless, it is useful to be familiar with a few key correspondences between linear and decibel values given in Table B.3.

With this table and the arithmetic properties of decibels, one can approximate the linear value of any parameter given in dB without resorting to a scientific calculator. For example, the linear equivalent of 7 dB can be determined as

$$\alpha_{\text{dB}} = 7\,\text{dB} = 3\,\text{dB} + 3\,\text{dB} + 1\,\text{dB} \Rightarrow \alpha = 2 \cdot 2 \cdot 1.25 = 5$$

or

$$\alpha_{\text{dB}} = 7\,\text{dB} = 10\,\text{dB} - 3\,\text{dB} \Rightarrow \alpha = 10/2 = 5 \quad \text{(B.75)}$$

which is quite close to the actual linear value corresponding to 7 dB of 5.0119.

Decibel Value	Linear Factor
0	1
1	≈1.25 (1.2589)
3	≈2 (1.9953)
10	10
10α (e.g., 30 if $\alpha = 3$)	10^{α} (e.g., 1000 if $\alpha = 3$)
-10α (-30 if $\alpha = 3$)	$10^{-\alpha}$ (e.g., 0.001 if $\alpha = 3$)

TABLE B.3 Linear Equivalences to Selected Decibel Values

References

Bracewell, R. N., *The Fourier Transform and Its Applications*, 3d ed. McGraw-Hill, New York, 1999.

Ercegovac, M., and T. Lang, *Digital Arithmetic*. Morgan-Kauffman, San Francisco, CA, 2003.

Harris, F. J., "On the Use of Windows for Harmonic Analysis with the Discrete Fourier Transform," *Proceedings of the IEEE*, vol. 68, no. 1, pp. 51–83, Jan. 1978.

Hayes, M. H., *Statistical Digital Signal Processing and Modeling*. Wiley, New York, 1996.

Johnson, D. H., and D. E. Dudgeon, *Array Signal Processing: Concepts and Techniques*. Prentice-Hall, Englewood Cliffs, NJ, 1993.

McClellan, J. H., and R. J., Purdy, "Applications of Digital Signal Processing to Radar," Chap. 5 in A. V. Oppenheim (ed.), *Applications of Digital Signal Processing*. Prentice-Hall, Englewood Cliffs, NJ, 1978.

Merkel, K. G., and A. L., Wilson, "A Survey of High Performance Analog-to-Digital Converters for Defense Space Applications," *Proceedings 2003 IEEE Aerospace Conference*, vol. 5, pp. 2415–2427, Mar. 8–15, 2003.

Murmann, B., "ADC Performance Survey 1997–2020," [Online]. Available at http://web.stanford.edu/~murmann/adcsurvey.html.

Oppenheim, A. V., and R. W. Schafer, *Discrete-Time Signal Processing*, 3d ed. Pearson, Englewood Cliffs, NJ, 2010.

Papoulis, A., *The Fourier Integral and Its Applications*, 2d ed. McGraw-Hill, New York, 1987.

Richards, M. A., "Spatial and Temporal Frequency," technical memorandum, Jan. 2018. Available at http://radarsp.com.

Walden, R. H., "Analog-to-Digital Converter Survey and Analysis," *IEEE Journal on Selected Areas in Communications*, vol. 17, no. 4, pp. 539–550, Apr. 1999.

Walden, R. H., "Analog-to-Digital Conversion in the Early 21st Century," in B. Wah (ed.), *Encyclopedia of Computer Science and Engineering*. Wiley, New York, 2008.

Problems

1. What is the Nyquist sampling rate (minimum rate to avoid aliasing) for the signal $x(t)$ having the spectrum $X(F)$ shown in the figure below? Sketch a block diagram of a system for recovering a new signal $\hat{x}(t)$ from samples of $x(t)$ taken at the Nyquist rate, such that the spectrum $\hat{X}(F)$ has the same shape as $X(F)$ but is centered at $F = 0$.

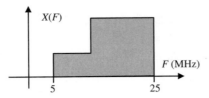

2. In some cases, the spectrum replication property of sampling can be used as a substitute for demodulation. Given a signal $x_a(t)$ with the spectrum shown, what is the lowest sampling rate that will ensure both no aliasing of the spectrum and that one of the spectrum replicas is centered at $F = 0$?

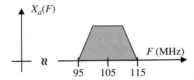

3. How many bits are required in an A/D converter to provide a dynamic range of at least 40 dB? What is the expected SQNR with this number of bits, assuming $k = 3$?

4. Numerical values of spatial frequency differ greatly from the usual temporal frequency values. What is the spatial frequency in cycles per meter for a 1-GHz electromagnetic wave?

5. What is the instantaneous frequency in hertz of the waveform $x(t) = \exp[j \cdot \exp(-\alpha t)]$?

6. Determine a phase function $\psi(t)$ such that the instantaneous frequency in hertz of the waveform $x(t) = \cos[\psi(t)]$, $-\tau/2 \leq t \leq \tau/2$ seconds, sweeps linearly from $-\beta/2$ Hz to $+\beta/2$ Hz.

7. Suppose signal x_1 is 30 dB greater in power than signal x_2. What is the ratio of their power in linear units (i.e., not in dB)? What is the ratio of the corresponding voltages?

Index

Page numbers in *italics* indicate figures; page numbers in **bold** indicate tables; page numbers followed by *n* indicate footnotes.

A

α-β filter, 37–38, 475–480
  correction stage, 480
  prediction stage, 480
  track filtering of noisy measurements, 37
  tracking, 470
  velocity estimates, 482
Accuracy, 4, 65
  amplitude, 291
  angle, 47
A/D conversion. *See* Analog-to-digital conversion
Adaptive displacement phase center antenna. *See* Displace phase center antenna
Additive white Gaussian noise (AWGN), 430, 492, 648
Albersheim's equation, 373-375, 381, 383, 419
Ambiguity
  Doppler, 32, 243, *268*, **316**. *See also* Doppler spectrum
  function. *See* Ambiguity function
  range, 138
  resolution. *See* Ambiguity resolution
  velocity, 162, 241, 320. *See also* Ambiguity, Doppler
Ambiguity function
  complex ambiguity function, 177, 179, 191–193, 195–196, 202
  of Costas waveform, 221–223
  definition, 177–181
  ideal, 180
  of linear FM pulse, 200, 216
  properties, 177–181
  of pulse burst, 184–186
  of simple pulse, 181–184

Ambiguity function (*Cont.*)
  of stepped chirp waveform, 220–221
  thumbtack, 180, 222
Ambiguity resolution
  Chinese remainder theorem, 318, 341
  coincidence algorithm, 319–320, 342
  congruences, 318
  ghosts, 320
Analog-to-digital conversion, 5, 672, *676*
AN/FPS-108 radar, 10
Angle estimation, 174
  2D angle, 468
  lobing bound, 464–466
  lobing-based, 458–469, 498
  maximum likelihood estimation, 22, 52, 320, 359, 393, 412, 414, 588, 632
  phase-based, *329*
Angle of arrival (AOA), 330–*331*, 333, 455–458, 467, 469, 537, 587, 591–593, 596–603, 605–609, 613–615, *617*, 623, 625–627
Antenna
  array, 149
  array factor, 16, 216, 508
  beamwidth, 8, 13, 25–26, 39, 101–102, 112, 115, 133, 164, 510
  boresight, 10, 51–52, 72, 131, 149, 337, 503*n*4, 585, 606–607
  effective aperture, 14, 49, 51, 509
  fan beam, 13–14
  gain, 14, 32, 48, 50–51, 82, 84–85, 104, 109, 338, 364, 418, 591, 598, 606–607, 611, 685
  grating lobes, 149*n*14, 220
  monopulse, 139
  one-way pattern, 16, 23, 149, 628
  pencil beam, 13–14, 26, 50, 53

Index

Antenna (*Cont.*)
  phase center, 3, 16, 33, *139–140*, 161, 256–257, *324–325*, *328*, 330–332, 334, 507, 519, 556–558, 564–565, 587–*588*
  phase front, 16
  phased array, 16
  power gain, 6, 13, 51, 90–91, 256, 271, 686
  power pattern, 10, 51, 53, 101, 104, 106, 109
  spatial spectra, *104*
  two-way pattern, 104, 151, 554
Asinc function, 15, 81, 144–145, 147, 190–191, 194, 217–218, 282, 285–288, 301, 326, 529, 666
ASR-9 radar, 323
ASR-12 radar, 323
Atmospheric attenuation, 8, *8–9*, 49
Auction methods, for cost measurement, 491
Autocorrelation matrix, 682
Autofocus, *31*, 565, 567–574, 580
  map drift algorithm, 568–569
  phase gradient algorithm, 570, 572–573
Automatic detection. *See* Constant false alarm rate detection
Autoregressive (AR) model, 86–87, 294–*295*, 612
Azimuth angles, 3, 12, 25, 51, 102, 117, 149, 337, 455, 485, 503

B

Bandwidth
  3-dB, 12–13, 16, 26, 34, 44–45, 53, 56, 113, 122–124, 132, 134, 144, 147, 149–151, 163–164, 250, 280, 283, 337, 504, 518, 530, 594, 602, **667**
  Doppler. *See* Doppler bandwidth
  null-to-null, 122, 124, 144, 308, 504, 518
  Rayleigh, *12–13*, 23–27, 34, 66, *71*, **79**, 87–89, 122–124, 144, 163–175, 177, 190, 197–203
  root-mean-square (RMS), 435
  of a simple pulse, 45, 120, 197, 204, 216, 223
Bandwidth-time product. *See* Time-bandwidth product
Barker coded waveform. *See* Biphase-coded waveform
Barker codes, 225–227, 229–231, 234, *234*, 242, 248
Baseband signal, 6, 108, 121, 128, 167, 237, 671
Baseline decorrelation, 563
Batch processing, 472, 474
Bayesian estimation, 476–477, 487
Beam limited resolution, 515

Beamforming, *31*, 33, 64, 587
  adaptive, *31*, 33, 64, 587, 596, 603–606, 624–626, 683–684
  beamspace processing, 603, *605*
  degrees of freedom, 243, 601, 623, 633, 637
  distortionless, 601–602, 605
  element space, 603–605, 622
  jammers, 598–608, 611, 614, 617, 619, 623, 625–626
  matched filter, 30, 33
  nonadaptive, 595
  nulls, 34, 267, 298, 591, 598, 600, 614, 616, 622
  preprocessing, 243, 308, 333, 540, 570, 603, 622–624, 626
  signal-to-interference ratio (SIR), 4, 171, 367, 568
Billingsley model, 86, 114, 612, 625
Biphase-coded waveform, 223, 248
  ambiguity function of Barker code, *227*, 231–*232*
  nested Barker code, 227, 229
  pseudorandom sequences, 230
Bistatic radar, 2
Blind Doppler shift, 134, 267, 270–271, 273
Blind velocity, 134
Blind zone maps, 295, 297–300, 317, 342

C

Cell-averaging CFAR, 392–394
  adaptive, 412–413
  analysis of, 394–398
  CFAR loss, 398–399, 406, 408, 411–416, 418, 424
  concept, 398–399, 404
  distribution-free, 417–418
  effect of false alarms, 418–419
  extensions of, 404–408
  greatest-of cell-averaging. *See* Greatest-of cell-averaging CFAR
  least-of cell-averaging. *See* Smallest-of cell-averaging CFAR
  limitations, 398–404
  masking, 406–408, 411, 413, 415, 424
  masking loss, *401*, 424
  order statistic, 408–412
  performance, 406–408, 411–412, 414
  self-masking, *402*, 415
  smallest-of cell-averaging. *See* Smallest-of cell-averaging CFAR
  target masking, 168, 180, *401*, 405–407, 424
  temporal, 414–417

Cell-averaging CFAR (*Cont.*)
  threshold multiplier, linear detector, *367*, 394–398, 400, 404, 406–407, 417, 419, 424
  two-parameters PDFs, 413–414
Central chi distribution, 88, 641, 643
Central reference point, 206, 209, 526, 546, 578
Centroid interpolation, *442–443*, 453
Chaff, 48, 81
Channel mismatch. *See* I/Q errors
Characteristic function (CF), 67, 370–372, 376, 633–634, 636–638, 641, 643–644
Chinese remainder theorem, 318, 341
Chip, 10, 223–226, 230, *233–234*, 248, 579–580
Chirp. *See* Linear FM waveform
Circulator, 6, 118
Clutter, 31
  altitude line, *254*, 256, 296
  Billingsley model, 86, 612, 625
  compound RCS models, 87–88
  constant gamma model, 83, 114
  cubic power spectrum, *85*
  defined, 5n3
  Doppler spectrum, 80–81
  GTRI model, 84, 114
  internal motion, 72, 85
  K distribution, 88
  log-normal, 67–71, 87–88
  main lobe, *607*
  modeling, 81–82
  power spectrum, 82, 85–86, 89–90, 92, 612
  side lobe, 254
  signal-to-clutter ratio, 84–85
  temporal and spatial corrections, 85–87
  volume, 52, 58–60
  Weibull, 67–71, 82, 87–88
Clutter map, *31*, 321–323, 335
  CFAR, 414–417
  in moving target detector, 322–323, 416
  threshold, 322
Clutter mapping, *31*, 281, 321–323, 335, 414, 416, 419
Coherent change detection, 564
Coherent processing interval (CPI), 32, 77, 96, 117, 131, *131*, 161, 184, 256, 295, 389, 425, 587, 612
Coherent scattering, 99–101
Coherent signals, 20–*21*, 186
Conical scan, 468, *468*, 498
Constant false alarm rate (CFAR) detection, 334, 390–418
  adaptive CFAR, 412–413

Constant false alarm rate (CFAR) detection (*Cont.*)
  adaptive matched filter, 619–622
  cell under test (CUT), 392, 620
  cell-averaging. *See* Cell-averaging CFAR
  CFAR windows, *394*
  for clutter map, 414
  distribution-free, 417–418
  guard cells, 394, 396n18, *402*, 404, 408, 411, 417, *620*
  lagging window, 396–397, 403–404
  log CFAR, 407–408
  order statistic. *See* Order statistic CFAR
  power on false alarm probability, 390–392
  rank-based. *See* Order statistic CFAR
  reference cells, *394*, 396–398, 404, 406–408, 412–413, 417–418, 424
  reference window, 394–398, 404, 406–408, 414, 418
  switching, 407, 455
  trimmed mean, 406
  two-parameter, 413–414
Continuous wave, 1–2, 8–*9*, 47, 117–118, 125, 167–168, 206, 235–236, 250, 257
Correlation, 678–680
  autocorrelation, 62n11, 678–679
  cross-correlation, 678
  cross-power spectrum, 678
  and linear estimation, 653–654
  normalized, 653–654
  power spectral density (PSD), 678
Correlator performance estimate, 440
Cosine function, 16, 19, 23, 360, 645
Covariance matrix
  of clutter, 263, 277
  of jammers, 599
  of preprocessed data, 343
  rank, 623
Cramèr-Rao Lower Bound (CRLB), 429–431
  complex signals in white Gaussian noise, 368, 497
  definition, 429–430
  on estimator variance, 646–648
  signals in white Gaussian noise, 430, 492
  signal-to-noise ratio, 431–432
  for sinusoid parameters, 446–448
  time delay, 453
  for transformed parameters, 495, 648
Cross-power spectrum, 678
Cross-range resolution, 12, 26, 44–45, 503–505
Cumulative distribution function, 409, 417, 452
Cumulative probability, 387

D

Data association, 38
  auction methods, 491
  gating, 489–493
  global nearest neighbor, 491
  nearest neighbor, 490–491
  probabilistic data association (PDA), 491
  strongest neighbor, 491
Data processing, 37, *118*, 345–346, 348
Data turning, 130, 143, *143*, 165, 284, 288
Datacube, 2, 32, 117, 138–140, 161, 236, 244, 256, 331–332, 587–*588*, 609–*610*, *620*, 624
Dechirp processing. *See* Stretch processing
Decibels, 685–687
Decimated signal, 669
Decimation, *156*, 159–160, 162, 165, *603*, 669
Deramp processing. *See* Stretch processing
Detection
  Bayesian approach, 359
  in coherent receivers, 177, 355–358
  coincidence detection, *319*, 385
  constant false alarm rate (CFAR). *See* Constant false alarm rate detection
  detector loss, 364
  fluctuating targets, 375–378
  likelihood ratio test. *See* Likelihood ratio test
  linear detector, 364–367, 374–375, 397–398, 406, 408
  Marcum's Q function, 361, 419
  M-of-N detection, 315
  Neyman-Pearson criterion, 345–346, 348
  nonfluctuating targets, 368–373
  null hypothesis, 344
  SNR loss, 174, 213, 217, 235, 363, 594, 596, 666
  square law detector, 75, 364–365, *367*
  sufficient statistic. *See* Sufficient statistic
  Swerling 1 example, 379–380
  Swerling 2 example, 381
  Swerling cases, 380–381
  threshold, 36–37, 330
  unknown parameters, 355, 358–364
  unknown phase example, 364–365, 384, 423
Difference signal in angle tracking, 460
Digital IF. *See* Digital I/Q
Digital I/Q, *31*, 151–161
  Lincoln Laboratory method, *159*
  Rader's method, *157*, 159–161, 165
Digital signal processing, 2, 40, 267, 343, 595n2, 659–688, 661n2
Digital sinc function. *See* Asinc function

Digital Terrain Elevation Data (DTED), 625
Dirichlet function. *See* Asinc function
Discrete cosine transform, 560
Displaced phase center antenna processing, 617–619
  adaptive, *619*–622
  relation to STAP, *617*, 622–626
  time slip, 325
Doppler
  ambiguity, 32, 243–244, *316*
  blind, 297–298, 317, 389
  minimum detectable, 277
  mismatch, 176–177, 180–184
  spatial, 97–99, 132, 141, 153, 240, 414, 416
  tolerance, 184, 230, 242–243
  unambiguous, 134, **162**, 194–195, 241, 246, 267–268, **313**, **316**
Doppler bandwidth, 132–134, 161, 164, 241, 516–518, 584
  due to intrinsic clutter motion, 612
  due to platform motion, 133, 256, 520
Doppler beam sharpening (DBS), 501, 514
  azimuth dechirp, 531–533, 578
  beam sharpening ratio, 527
  geometric distortion, 527–528
  quadratic phase errors, 514, 565, 568–*569*
  range curvature, 533–539
  range migration correction, 532
  secondary range compression, 532, 539
Doppler bins, *76*–78, *281*, 284, 288, 320–321, 331–334, 342, 623
Doppler estimation, 288–294
  accuracy, 335
  CRLB, 442
  discrete Fourier transform, 141, 288–294
  interpolation, 288–294
  maximum likelihood, 22, 393
Doppler processing, *31*–35, 99, 115
Doppler resolution, 37, 139, *162*, 168–169, 177, 184, 190, 194–196, 204, 246, 282, 307, 316, 341, 367, 503, 513
Doppler shift, 3, 16, 23, 28
  cone angle, 95, 253, 256, 280
  estimation. *See* Doppler estimation
  spatial Doppler, 97–99
  stop-and-hop approximation, 95–97
Doppler spectrum, 80–81
  Billingsley model, 86, 612, 625
  clear region, 251–*252*
  clutter region, 251–*252*

Index

Doppler spectrum (*Cont.*)
  Gaussian model, 85–86
  micro, 132*n*8
  moving platform effects, 253–256
  sampling, 239–240
  skirt region, 251–*252*
  spectral width, 280, 300
Downchirp, 198, 205, 239–240, 243, 250
Duplexer, 6, *6*, 49
Dwell, *31*–32, 77–78, 140–141, 161–162, 268–269
  time, 140–141
Dynamic range, 4–5, 23, 50, 64, 124, 205, 309, 407, 673–675, 688

E

Edge threshold detector, 443
Efficient estimator, 430, 433–434, 494–495
Eigenanalysis, 624
  basic, 682–683
  of sinusoids in white noise, 683–684
Eigenvector, 489, 624, 629, 683–684
Electromagnetic interference (EMI), 35, 47, 92, 242, 366, 390, 625
Elevation angle, 3, 3*n*2, 25, 51, 102, 117, 337, 485, 515
Error function, 349
  complementary, 349
Estimator, 426–429
  accuracy, 427
  angle. *See* Angle estimation
  consistent, 428
  Cramèr-Rao Lower Bound (CRLB), 429–431
  Doppler signal. *See* Doppler estimation
  efficient, 430
  frequency. *See* Doppler estimation
  maximum likelihood, 432–434
  minimum variance, 428
  precision, 427
  properties, 427–429
  range. *See* Range estimation
  sequential least squares, 470–475
  time delay. *See* Range estimation
  unbiased, 428
Extended Kalman filter, 475, 486, 498

F

False alarm, 4, 29–30, 110, 132, 390–392, 470, 487, 493, 619
Fast Fourier transform (FFT), 4, 130, 142, 191, 280, 340, 534, 598, 664, *664*
Fast time, 30–32, 36, 40, 119–131
Feature-aided tracking, 492
FFT. *See* Fast Fourier transform
Fine Doppler estimation, 288–294
Finite impulse response, 160, 210, 259, 262, 273, 416, 681
FIR filter. *See* Finite impulse response
Fisher information matrix, 446, 448, 496–497, 649–650
Flat earth phase difference, 558
Fourier transform, 4, 435
  continuous, **664**
  in digital signal processing, 659–664
  discrete, 526, 529, 550
  discrete time, 281–285, 529, 598
  fast. *See* Fast Fourier transform
  reciprocal spreading, 598, 662
Frequency agility, 4, 66, 74, 76–77, 378
Frequency estimation error, 292, *450*, 453
Frequency models, 93–99
Frequency-modulated continuous wave (FMCW), *9*, 206, 257, 434

G

Gain
  antenna, 14, 32, 48, 50–51, 82, 84–85, 104*n*14, 109, 338, 364, 418, *459*, 591
  coherent integration, 28–30, 219, 591, 593, 601, 615, 674
  noncoherent integration, 29–30, 372, 374–375, 388, 424
  pulse Doppler processing, 286
  signal processing. *See* Signal processing gain
Gain scheduling, 485
Gating, 314, **316**, 481, 489–491, 493
Gaussian model, 85
Ghosts, 315, 320, *320*
Glint, 466, 468
Global positioning system, 566
Greatest-of cell-averaging CFAR, *405*, 417
Ground plane, 54–*55*, 116, 503, 514, 518–519, 557, 584

H

Hadamard product, 596, 625, 681
Hamming window, *210*–213, 211*n*10, *211*, 282–283, 285, 291–293, 340–341, *529*–530, 597
Hermitian, 157, 161, 262, 275, 355, 599, 601, 670, 679, 681–683
High-resolution range profile, 219

History of radar. *See* Radar history
Hypothesis testing, 343–354

I

I channel. *See* In-phase channel
IF sampling. *See* Digital I/Q
IIR filter. *See* Infinite impulse response
Imaging. *See* Synthetic aperture radar
Improvement factor, 273
    vector method, 273, 339
Inertial measurement unit (IMU), 565
Inertial navigation system (INS), 565
Infinite impulse response, 262, 273, 416
Instantaneous frequency, 34–35, 97, 125–127,
    197, 202, 207–*208*, 211, 214–216, *238*, *240*,
    520, 528, 546, 685
Integration, 12, 28–30
    binary, 366–368, *388*, 479
    coherent, 22, 366–368
    noncoherent, 22, 366–368
Integration gain, 28–30, 45, 189, 196, 218
    coherent, 28–30, 219, 372
    noncoherent, 28–30, 372
Interacting multiple models (IMM), 484
Interferometric phase difference (IPD),
    557, 559
Interferometric SAR, 503, 556–564, 580
    coherent change detection, 564
    concept, 556–564
    effect of height, 556–559
    layover, 558, 558n16
    one-pass, 559, 561
    orthorectification, 559
    phase unwrapping, 559
    pixel phase, 556
    processing steps, 559–564
    registration, 559, 570
    terrain motion mapping, 564
    two-pass, 564
Internal clutter motion. *See* Clutter
Interpolation, 4, *31*, *442*–443, 453, 669
    angular-range interpretation, *551*
    asinc interpolation kernel, 289
    bandlimited, 331
    digital, 593
    Doppler. *See* Doppler interpolation
    linear, 231n13, 249
    polar-to-rectangular interpolation, *551*–552
    quadratic, *290*, 340
    range, 341, 551
    spatial frequency, 551

Interpolation (*Cont.*)
    Stolt, 541–542
    straddle loss, 174
    two-dimensional, 550–551
I/Q errors
    correcting, 154–157
    image component, 156, 165
I/Q imbalance. *See* I/Q errors

J

Jamming, 92

K

K distribution, 88, 384, 638–639, 641, 643
Kalman filter (KF), 470, 475, 480–487, 493, 498,
    566
    α-β filter as, 470, 475, 480–487
    extended, 475
    properties, 470
Kronecker product, 227, 610, 613, 681

L

Lagrange multipliers, 346
Least-of cell-averaging CFAR. *See* Smallest-of
    cell-averaging CFAR
Least-squares estimate (LSE), 472
Likelihood ratio test (LRT), 345–353
    generalized likelihood ratio test, 359n10, 366
    log likelihood ratio test, 347
Linear detector, 76, 364–367, 374–375, 397–398,
    406, 408, 423, 425, 641
Linear FM waveform, *35*, 39, 242
    ambiguity function, *203*–205
    chirp waveform, 34, *124*, 153, 156, 202
    Rayleigh resolution, 199–200
Linear time-invariant system, 270
Lobing, 64–65
    conical scan, 468, 498
    lobe switching, 455
    monopulse, 459–460
    sequential, *459*–461
Lobing bound, 464–466
Loss
    amplitude, 184, *566*
    antenna, 91
    atmospheric. *See* Atmospheric attenuation
    brightness, 529
    CFAR. *See* Constant false alarm rate detection
    conversion, 21

Loss (*Cont.*)
  detector, 364
  mismatch, 13. *See also* Doppler, mismatch processing. *See* Processing loss
  in processing gain. *See* Loss in processing gain
  propagation. *See* Atmospheric attenuation
  SNR, 174, 214, 217, 235, 363–364
  straddle, 143–147, 664
Loss in processing gain (LPG), 212, 235, 247, 282–283, 335, 666–667
Low Earth orbit, 39, 94, 504, *511*

M

Mahalanobis distance, 489–490
Major-minor method, 317
Maneuverability index, 482
Marcum's Q function. *See* Detection
Masking, 4–5, 180, *210*, 243, *401*, *403*, 411, 413
  self-masking, 415
  target, 210, 399, 401–407
Matched filter, 4
  all-range, 173–174
  clutter suppression, 263–265
  impulse response, 34, 170–173, 186, 191, 210–212
  impulse response shaping, 213
  for moving targets, 175–177
  output, 171–180
  peak SNR, 438
  for pulse burst, 177, 184–186
  pulse-by-pulse processing. *See* Pulse-by-pulse processing
  range resolution, 174–175
  received pulse, 436
  for simple pulse, 171–173
  waveform, 169–171
  whitening, 171
Maximum likelihood estimate (MLE), 359*n*10, 393, 412, 414, 437, *446*, 632
  of interference power, 398, 406
  of phase gradient, 572
  of sinusoid parameters, 448–451
Mean-square error (MSE) bound, 445
Measurement precision, 186, 424, 451, 455, 466, 497
Minimum detectable Doppler. *See* Space-time adaptive processing
Minimum detectable velocity. *See* Space-time adaptive processing
Minimum norm methods, phase unwrapping, 560

Minimum variance unbiased (MVU) estimator, 428, 432
M-of-N method, 315, 367, 487
Monopulse tracking, 139
Monostatic radar, 3, 5–6, 16, 26, 95, 118, 296–297, 677
Monte Carlo simulation techniques, 408
Moore's law, 5, 5*n*5, 143, 502
Motion compensation, 564–567
  autofocus. *See* Autofocus
  global positioning system (GPS), 566
  inertial measurement unit (IMU), 565
  inertial navigation system (INS), 565
  Kalman filter, *566*
  phase errors, 565–566
  to a point, 567
  PRI control, *566*
Moving target detector (MTD), 322–323, 416
Moving target indication (MTI), 34, 253, 334
  blind speeds, 262, 266–273
  clutter attenuation, 273–275
  combined with pulse Doppler, 280–281
  concept, 277
  figures of merit, 273–277
  improvement factor. *See* Improvement factor
  limitations. *See* Moving target indication limitations
  limitations to, 278–280
  in moving target detector, 322–323
  optimum filter, 266
  pulse cancellers, 259–262
  signal model, 286–288
  subclutter visibility, 273
  transient effects, 311
  visibility factor, 277
MTI. *See* Moving target indication
Multipath, 84, 106–108, 116
Multiple hypothesis tracking, 492
Munson, David, 105, 502, 551–552

N

Nearest neighbor, cost measurement, 490–491
NEXRAD, 60, 112, 305
Neyman-Pearson criterion. *See* Detection
Noise
  DFT of, 285–286
  effective temperature, 91
  figure, 91
  power spectrum of, 89–92
  signal-to-noise ratio (SNR), 84, 89–92

Noise equivalent bandwidth, 90–91, 112
Noise power, 22, 28–29, 89–92, 334
Noise subspace, 684
Noisy measurement, 37, 470–471, 473, 478, 485, 632
Noncoherent scattering, 105–106
Null hypothesis. *See* Detection
Number fluctuations model, 88
Nyquist rate, 122–123, 130, 134, 138, 151, 157, 161, 290–*291*, 441, 658, 675, 687
   in 3-dB beamwidths, 150
   in Doppler, 141–143, 147
   in fast time, 122

O

Order statistic, 408–411, 413–414, 417–418, 424
Order statistic CFAR, 407–412, 414, 420, 424
   CFAR loss, 411
   performance, 411
   threshold multiplier, 411
Oscillator, 18
   local, 5–*6*, 18, 20, 22, 278
   stable local, 20

P

Path following method, phase unwrapping, 560
Phase history, 22, 28–29, 95–97, 99, 109, 117, 121, 240, 264–265, 286, 553, 561, 578, 580, 587
   signal, 609
   slow-time, 132, 271, 568–571, 575
   spatial, 456
   two-dimensional SAR, 521
Phase unwrapping, 559–560, 562, 580
Phenomenology, 30–32, 38, 40, 82, 274, 355, 499
Pilot signal, 155–156
Polar format algorithm, 210, 543, *549*, 552, 580, 585
   keystone grid, 551
   polar-to-rectangular interpolation, 550–552
   radial-keystone interpolation, *551*
Polarization scattering matrix, 57
Power spectral density (PSD), 90–91, 169, 434, 654, 657, 678
Power spectrum, 85*n*8, 275*n*7, 294, 300, 332, 654–655, 678
   clutter, 279, 338–339, 612
   cubic model, I85
   Doppler, 302–304, 340

Power spectrum (*Cont.*)
   Gaussian, 261
   Gaussian clutter, 276
   interference, 171
   range-Doppler, 251
   sampled clutter, 274
   slow-time, *301*
   temporal, 612
Precision, 4, 23, 425, 427–*428*, 436, 439*n*2, 439*n*3, *439*–445, *465*–466
   phase, 452
Principle of stationary phase, 200, 214, 535
Probabilistic data association, 491
Probability density functions, 30, 32, 110, 345, 348, *351*, 631–645
   central chi, 641
   chi-squared, 633–636
   Erlang, 633–636
   exponential, 633–636
   gamma, 633–636
   Gaussian, 631–634, 637, 642
   K distribution, 638–639, 641–643
   log-normal, 638–640, 643
   noncentral chi, 642–643
   noncentral chi-squared, 637
   for phase, 279–280, 645
   for power, 88, 644
   Rayleigh, 641
   Rice, 643
   Rician, 643
   Tikhonov, 644–645
   uniform, 644
   voltage, **71**, 633
   Weibull, 638–640, 643
Probability of detection P_D, 4, 29, 321, 324, 344–345, *352*, 357–358, 361–363, 371, 373, 376–380, *383*, 385–386, 390–391, 398–399, *401*, 411, 413, 418–419, 423
Probability of false alarm P_{FA}, 4, 29, 45, 344–345, *352*, 358, 361–363, 366, 371, 376, *380*, 385–386, 390–*391*, 394, 409–*410*, 418–419
Probability of miss, 137, 344, 385
Process noise, 477–478, 481–482, 484–486, 488
Processing gain G_p, 92, 186, 199, 209, 212, 217, 224, 227, 235, 274, 282–283, 335, 341, 368, 374, 520, 598, 674
   loss in, 666–667
   pulse Doppler, 286

Index

Processing loss (PL), 212, 247, 282–283, 286, 335, 341, 598, 666–667
Projection-slice theorem, 547
Pulse burst waveform, 184–196
   ambiguity function, 191–195
   Doppler response, 190–191
   pulse-by-pulse processing, 186–187
   range ambiguity, 187–190
   relation of ambiguity function to slow time spectrum, 195–196
Pulse cancellers. *See* Moving target indication
Pulse compression, *31*, 34, 97, 140, *140*, 196–210, 213, 221, 242–244, *306*–307, **316**, 418, 505, 520, 525, 534, 537
   phase-modulated pulse compression waveforms, 223–235
   post-pulse compression, 555
Pulse Doppler processing, 280–300
   combined with MTI, 286–288
   CPI-to-CPI stagger, 295–300
   DFT of noise, 285–286
   DTFT of moving target, 281–283
   fine Doppler estimation. *See* Doppler interpolation
   as matched filter, 286–288
   modern spectal estimation in, 294–295
   in moving target detector, 322–323
   PRF regimes, 311–314
   sampling the DTFT, 283–285
   signal processing gain, 286
   steady-state operation, 311
   transient effects, 310–311
Pulse pair processing, 294, 300–305
   spectral width, 300–304, 342
Pulse repetition frequency (PRF), 8, 76, 131, 251, 296, 340, 516, 607
Pulse repetition interval, 8, 98, 131, 184, 193, 195, 269, 434, 531
Pulse-by-pulse processing, 186–187, 189–190, 196, 218

Q

Q channel. *See* Quadrature channel
Q function. *See* Detection
Quadratic interpolation, 290–291, 340, 453–*454*
Quadrature channel, 154–155
Quantization, 672–675
Quantization error, 492, 644, 673

R

Radar
   AN/FPS-108, 10
   antenna. *See* Antenna
   applications, 2, 40–42
   ASR-9, 323
   ASR-12, 323
   basic functions, 40–41
   bistatic, 2
   Chain Home, 1
   continuous wave, 1–2, 8–9, 47, 117–118
   detection, 36–37
   frequency bands, 7
   imaging. *See* Synthetic aperture radar
   literature, 40–42
   measurement and track filtering, 37–38
   monostatic, 2–3, 5–6, 16, 26, 95, 118, 120, 296–297, 677
   NEXRAD, 60, 112, 305
   phenomenology, 32
   range equation. *See* Radar range equation
   receiver. *See* Receiver
   SCR-270. *See* Receiver
   signal conditioning and interference suppression, 32–35
   time scales, 30–32
   transmitter, 5–10
Radar cross section (RCS), 49, 321, 563, 634
   of clutter. *See* Clutter
   common PDFs, for RCS, 67
   compound models, 87–88
   correlation, **79**
      pulse-to-pulse, 77–78
      scan-to-scan, 78
   definition, 49
   meteorological, 58–60
   spectral model, 108–109
   stastical description, 60–76
   statistical models. *See* Target models
   Swerling models. *See* Swerling models
   target fluctuation models, 76–79
      effect of, 80–81
   typical values, 58
Radar history
   Chain Home, 1
   Hülsmeyer, Christian, 1
   Page, Robert, 1
   Radiation Laboratory, 1
   SCR-270, 1

Radar history (*Cont.*)
  Taylor, Albert, 1
  Watson-Watt, Sir Robert, 1
  Young, Leo, 1
Radar range equation, simple point target, 48–51
Radial velocity *v*, 3, 37, 48, 95, 117, 127*n*5, 134, 175, *240*, 264–265, 267, 281, 286, 305–306, 321, 334, 340–341, 446, 492, 578, 608
Range
  ambiguity. *See* Ambiguity
  bins, *39*, 93, 98*n*11, 121–122, 121*n*2, 129–130, 136, 248, 251–253
  cells, 121, 313, 341, 392–394, 416, 550
  gates, 121
  swath, 120–121, 135, 138–139, 514, 523, 535–536
  unambiguous, 113, 120, 135–138, 161–**162**, 164, 188
  window, 120–121, 129, 206–209, 238–241, 247
Range estimation, 294, 434–446
  CRLB, 434–436
  leading edge, 25, 135
  MLE, 433
  MSE bound, 436–446
  practical, 436–446
  a priori bound, 445
  sampling bound, 442–443
  split-gate, 445
Range migration, *31*, 98*n*11, 115, *310*, 520–521
  range curvature, 520–521
  range walk, 520–521, 527, 531–532
Range profile (RP), 104–109, 155, 161, 209, 219, 238, 242, 256, 259, 308, 546–547
  FMCW radar, 125–130
  measuring a, 119–125
  multiple, 131–138
Range resolution, 8, 10, *25–27*, 34–36, 39, 48, 51, 53–56
Range skew, 207–208, 242, 247
Range swaths, 120–121, 135, 138–139, 514, 516, 523, 535–536
Range window, 120–121, 124, 129, 206–209, 214, 220, 238–239, 241, 247, 332
Range-Doppler algorithm, 502, 514, 533–542
Range-Doppler coupling. *See* Linear FM waveform
Rank sum double quantizer, 417

Real aperture radar, 39
Real-beam imaging, 503
Receiver, 2–6, 17–22, *31*
  effective temperature, 91
  superheterodyne, 6, 21–22
Receiver operating characteristic (ROC), 350, 352, 354, 357, 372, 419
Receiver output, 3, 28, 47, 51, 54, 90–92, 102–*103*, 127, 149, 161, 168–169, 207–208, 330, 425, 434, 443, 503, 552, 567, 633, 675
Reed-Mallett-Brennan (RMB) rule, 621
Reflectivity, **60**
  area clutter, 82, 84, 114
  baseband complex, 99–100
  effective, 100–102, 104, 106, 174*n*2, 547
  factor, 58–60, 113
  meteorological reflectivity factor, 60
  projections, 106
  volume clutter, 52, 58, 84, 105, 317
Reflectivity factor, 58, 60, **60**, 113
Region of interest (ROI), 511, 532, 548
Replication, 142, 159, 162, 193, 319, 669–672, 681, 688
Residual video phase. *See* Spotlight SAR
Resolution, 4–5, 5*n*4, 7–8, 10
  3-dB, 283, *530*
  cell, 23–27, 51–56, 58
  cross-range, 12, *26*, 39–40, 44–45, 503–505
  Doppler, 37, 139, **162**, 168–169, 177, 184, 190, 194–196, 204, 246, 282
  range, 8, 10, 25–27, 34–36, 39–40, 45, 48, 54–56, 101
  Rayleigh, 19, 164, 175, 177, 190, 199–205, 209–210, 214, 220, 224, 226, 246
  velocity, 139, 165, 168, 177, 249–250, 320
Rice distribution, 78
Root-mean-square (RMS) bandwidth, 435

S

Sampling, 668–672
  in angle, 149–151
  in Doppler frequency, 141–148
  the DTFT, 283–285
  in fast time, 294, 567
  Nyquist rate, 148
  SAR coverage, 514–518, *663*
  in slow time, 164, 282, *309*
  spatial array, 148–149
Sampling bound (SB), 441–443, *450*–455, 497

Index

ScanSAR, 509, 515
Schwarz inequality, 170, 262–263, 647, 651
Sequential least squares, 470–472, 475–476, 480, 493
Sequential lobing, 459–461, 468
Shnidman's equation, 367, 381–383, 419, 423
Shuttle imaging radar–C. See SIR-C radar
Signal subspace, 684
Signal-to-clutter ratio (SCR), 84–85, 114, 251, 262, 273–274, 277, 407
Signal-to-interference ratio (SIR), 4, 22, 47, 171, 367, 418, 598, 608, 682, 685
Signal-to-quantization noise ratio (SQNR), 673
Sign-magnitude encoding, 672
Sinc-squared two-way voltage, 425
SIR-C radar, 501, 522–523
Slant plane, 503, 518–519, 557, 575, 584
Slant range, 54–56, 112, 114, 116, 338, 518, 556–557, 564, 584
Smallest-of cell-averaging CFAR, 405, 417
Snapshot, 456
    space-time, 609, 617–618
    spatial, 595–596
    temporal, 612
Space-time adaptive processing (STAP), 31, 33, 140, 326
    adapted pattern, 613–617
    adaptive matched filter, 619–622
    Billingsley model, 612
    clutter ridge slope, 624
    covariance matrix. See Covariance matrix
    covariance matrix taper (CMT), 612, 625
    displaced phase center antenna processing, 617–619
    knowledge-aided, 624
    limitations, 625–626
    metrics, 613–617
    minimum detectable Doppler (MDD), 615
    minimum detectable velocity (MDV), 615–616
    optimum matched filtering, 613
    reduced dimension, 622–623
    reduced rank, 624
    relation to DPCA, 617
    sample matrix inverse, 620
    signal model, 609–612
    SIR loss, 615–616
    space-time signal modeling, 639
    spatial frequency, 676
    subspace projection, 624

Space-time adaptive processing (STAP) (Cont.)
    temporal correlation, 612
    test statistic, 613
    usable Doppler space fraction (UDSF), 277
Spatial Doppler, 97–99, 132, 141, 153, 240, 414, 416
Spatial frequency, 619, 676–677, 676n9, 688
Spatial snapshot, 456, 595–596, 604, 611
Spotlight SAR, 548
    cross-range resolution, 503–505
    data characteristics, 518–523
    mapping rate, 517
    polar format algorithm, 550–552
    residual video phase, 585
    sampling requirements, 503
Squint, 132–133, 164, 459–460, 462–466, 468–469, 513
Staggered PRF
    blind speeds, 266–273
    CPI-to-CPI, 78, 268, 295n8
    pulse-to-pulse, 295n8
    stagger ratio, 269–270
    staggers, 269–271
Stationary point, 201
Steady-state operation, 311
Steering vector, 597–598, 601–604
    space-time, 613
    spatial, 595, 598, 629
    target, 604, 613, 621, 629
    temporal, 609–610
    transformed, 601–602
Stepped chirp waveform, ambiguity function, 220–221
Stepped frequency waveform, 216–220
    ambiguity function, 219–221
    definition, 217
    linear, 217–221
    pulse-by-pulse processing, 218
Stop-and-hop assumption. See Doppler shift
Straddle loss, 143–148
    definition, 146
    in Doppler, 284
Stretch processing, 31, 205–206, 209–210, 216, 220, 236, 238, 242
Stripmap SAR
    chirp scaling algorithm, 552
    cross-range chirp, 520
    data characteristics, 518–523
    data set size, 514
    definition, 502

Stripmap SAR (*Cont.*)
 geometry, 518–523
 image formation algorithms, 524–525
 mapping rate, 517
 point scatterer response (PSR), 518
 range migration algorithm, 540–544
 space-bandwidth product, 520
 subswaths, 530
Strongest neighbor, cost measurement, 491
Sufficient statistic, 348–351, 354–357, 360, 366, 370
Sum signal, in angle tracking, 460
Surface moving target indication (SMTI), 323
Swath length, 139, 509, **509**, 514–518, 522, 539, 546, 583–584
Swerling models, 79–80, 366–367, 378–**379**, 381, 388–**389**
Synthetic aperture radar, 32, 38–40
 aperture time, 508–514
 autofocus. *See* Autofocus
 coverage and sampling, 514–518
 cross-range resolution, 509–518
 Doppler beam sharpening. *See* Doppler beam sharpening
 Doppler viewpoint, 512–514
 interferometric SAR. *See* Interferometric SAR
 looks, 501, 564
 mapping rate, 517
 motion compensation. *See* Motion compensation
 point scatterer response (PSR). *See* Stripmap SAR
 quadratic phase errors, 568–*569*
 range curvature. *See* Range migration
 range migration. *See* Range migration
 range walk. *See* Range migration
 range-Doppler algorithm, 502, 514
 scanSAR, **509**, 515
 sidelooking, 506–507, 509, 513
 speckle, 499, 574–575
 speckle reduction, 574–575
 spotlight. *See* Spotlight SAR
 squinted SAR, 507, 515, 567, 580
 stripmap. *See* Stripmap SAR
 swath constraint, 517
 swath length, 515–518
Synthetic array radar. *See* Synthetic aperture radar
System loss factor, 49

T

Target masking, 168, 180, 210, *210*, 243, 399, *399*, 401, *401*, 406–407, 424
Target models, Swerling. *See* Swerling models
Target visibility. *See* Moving target indication, visibility factor
Terrain motion mapping, 564
Threshold detection. *See* Detection
Time delay estimation, 435–437, 440–442, 445–*446*, 453, 497
Time delay measurement, 442
Time-bandwidth product, 124, 197, 199, 242, 246, 439, 520, 522, 584, 686
Tracking
 α-β filter, 37, 475
 association. *See* Data association
 cycle, 487–492
 extended Kalman filter, 475, 486
 gain scheduling, 485
 initiation, 485
 initiation process, 487
 Kalman filter, 475
 least-squares estimate, 472
 monopulse tracking antenna, 139
 in spherical coordinates, 472
 state, 473–474
Tracking index, 482, 498
Two's complement encoding, 672–674
Two-sample estimator, 443

U

Unambiguous range. *See* Range
Upchirp, 198, *198*, 205, 239–240, 243, 250, 522

V

Vector representation, 265

W

Walker, Jack, 502
Waveform
 biphase-coded. *See* Biphase-coded waveform
 Costas frequency codes, 223
 linear FM (LFM). *See* Linear FM waveform
 nonlinear FM, 197, 214–216, 236, 242
 phase coded, 168, 200, 223–224, 233–234, 242, 248
 polyphase coded. *See* Polyphase coded waveform

Waveform (*Cont.*)
    pulse burst. *See* Pulse burst waveform
    simple pulse, 45, 177, 197
    stepped frequency. *See* Stepped frequency waveform
Wavefront. *See* Antenna, phase front
Weiss-Weinstein analysis, 440
White noise, 90, 92, 286, 655
Whitening, 81, 171, 481
Wiley, Carl, 501, 525

Z

Zero padding, 142–143, 165, 284, 289, 662*n*4
Zero velocity filter, 321, *323*
Ziv-Zakai bound (ZZB), *439*–440, 453